Semiconductors

Martin I. Pech-Canul
Nuggehalli M. Ravindra
Editors

Semiconductors

Synthesis, Properties and Applications

Springer

Editors
Martin I. Pech-Canul
Centro de Investigación y de Estudios Avanzados del Instituto Politécnico Nacional—Unidad Saltillo—Ingeniería Cerámica
Ramos Arizpe, Coahuila, Mexico

Nuggehalli M. Ravindra
Department of Physics
New Jersey Institute of Technology
Newark, NJ, USA

ISBN 978-3-030-02169-6 ISBN 978-3-030-02171-9 (eBook)
https://doi.org/10.1007/978-3-030-02171-9

Library of Congress Control Number: 2018957645

This Springer imprint is published by the registered company Springer Nature Switzerland AG
The registered company address is: Gewerbestrasse 11, 6330 Cham, Switzerland

Preface

It has been over two centuries since the term "semiconducting" was used by Alessandro Volta in 1782. In 1833, Michael Faraday reported the decrease in resistance with increase in temperature in silver sulfide. Similar results of copper sulfide and a detailed analysis of the temperature dependence of the electrical conductivity in Ag_2S and Cu_2S were published by Johann Wilhelm Hittorf in 1851. This behavior of Ag_2S and Cu_2S was in contrast to metals in which resistance increased with increase in temperature. Around the same period, in 1839, Alexander Edmund Becquerel discovered the photovoltaic effect in an experiment in which silver chloride was placed in an acidic solution and illuminated while connected to platinum electrodes. The first results of photoconductivity were reported by Willoughby Smith in selenium in 1873. In 1874, Ferdinand Braun discovered the semiconductor point-contact rectifier using a metal-galena junction.

The year 2018 is the celebration of the 160th birthday of Jagadish Chandra Bose. His patent on PbS point-contact rectifiers, in 1904, marked the beginning of lead chalcogenide-based infrared detectors. It has been 140 years since Edwin Herbert Hall discovered the deflection of charge carriers in solids under the influence of magnetic fields in 1878. In 1926, Julius Edgar Lilienfeld patented the field effect in semiconductors in a three-electrode amplifying device based on copper sulfide. Quantum mechanics and its role in the theory of electronic semiconductors were published by Alan Wilson in 1931. The discovery of the p–n junction and the solar cell by Russel Ohl, in 1941, was the beginning of a transformation in scientific research in semiconductors. This led to the invention of the germanium bipolar point-contact transistor by William Shockley, John Bardeen, and Walter Brattain in 1947 and the first integrated circuit by Jack Kilby in 1958. Today, semiconductors have revolutionized the everyday lifestyle of people from across the world.

Semiconductors are omnipresent and contribute to the global economy on a very large scale. Aeronautics, communications, computers, defense, energy, health, instrumentation, and transportation are just a few of the examples of applications of semiconductors.

It must be emphasized that semiconductors represent a sector of the global economy in which the time lag from basic research to industry practice is minimal. This is due to the significant number of applications and the resulting market economy. Along with the materials related issues, challenges such as wafer/device scaling, wafer/device/circuit/systems integration, high speed/frequency, low power requirements, packaging, Joule heating/thermal management, improvement in yield, and cost to manufacture are being constantly addressed. The recent developments in 0D, 1D, and 2D materials, devices, and structures have contributed to significant creativity in semiconductors.

In putting together this book, all aspects of semiconductors have been considered—from fundamentals to applications. It is the result of the unified efforts of an inter-institutional and interdisciplinary group of authors, whose research interests coalesce in the realm of semiconductors. The book comprises ten chapters, covering the fundamentals and synthesis/processing, the properties and current or potential applications, emphasizing the processing–structure–property correlations, in consonance with the "Central paradigm of materials science and engineering." The essence of the central paradigm is revisited, but within the perspective of sustainability, in which the recycling/reusing aspects, incorporated within the materials processing cycle, are suggested to apply to semiconductors. Although specialized terminologies are used when appropriate and necessary, efforts have been made to present the chapters in a scientific, technical, and comprehensible language to non-specialized readers.

The coordinating authors are earnestly thankful to the editorial team, Ms. Anita Lekhwani, Senior Editor, Ms. Faith Pilacik, Editorial Assistant, Mr. Brian Halm, Project Coordinator, and Ms. Ania Levinson, formerly of Springer for their wholehearted support and patience throughout the publication process. Likewise, all contributing authors are gratefully acknowledged for their steadfast dedication and perseverance, and to their corresponding universities or research institutions, for their valuable support. Finally, our heartfelt gratitude to those individuals who in one way or the other helped to finalize the book. We thank our colleagues who provided valuable insights or suggested pertinent literature.

We hope that this book will help and serve the general readers, students, academicians, technologists, practitioners, and entrepreneurs in their work to address the ever-growing evolution in semiconductors and their applications.

Ramos Arizpe, Coahuila, Mexico — Martin I. Pech-Canul
Newark, NJ, USA — Nuggehalli M. Ravindra
July 2018/August 2018

Contents

Introduction

Standing at the cutting edge in the science and technology of semiconductors is a challenging task, not only by the plethora of topics to cover within the subject matter at once but also by the fast-moving pace at which the advancements are unfolded. The near ubiquity of semiconductors is perceived in all walks of life, from mobile phones, home appliances, personal computers, to automotive, navigation, aviation, and military equipment, as well as in medical diagnostic instruments. The semiconductor industry is perhaps the most influential driver of today's society. The fast-growing amount of information on the wide variety of topics is due to the in-depth current investigations and to past publications in scientific journals and specialized books or more general textbooks. Their invaluable usefulness is acknowledged.

In addition to covering the fundamentals in Chap. 1, this book addresses the synthesis/processing aspects, the properties, and current or potential applications of a variety of existing and emerging semiconductor materials. And albeit one of the purposes of this text is to provide the readers with most of the semiconductor materials and their processing/fabrication routes, there might be some materials or processes unintentionally omitted and others extensively addressed. An effort has been made, nonetheless, to include the most recent references on the subject matter.

Chapter 1 reviews the core theoretical concepts of Solid State Physics, beginning with a quantum mechanical description of the fundamental constituents of solid state matter, namely, electrons and atoms and serves to lay the foundation to adequately assimilate the essentials of semiconductor materials.

Chapter 2 is devoted to the processing techniques for bulk crystal growth and thin films, including epitaxial growth as well as polycrystalline and amorphous thin films. A wide range of processing techniques and their variants are discussed, including MBE, ALE, ELO, PECVD, LPCVD, CVD, and laser ablation. Specific sections are devoted to self-assembly, wafer preparation methods, ion implantation, and vacuum deposition techniques.

Chapter 3, Characterization Techniques, focuses on presenting the most common techniques that are used in semiconductor materials, namely, compositional characteristics, structural aspects, and optical properties.

Chapter 4 addresses the synthesis, properties, and applications of vanadium oxides. These compounds undergo insulator-to-metal phase transition (IMT) accompanied by structural changes, in response to their extreme sensitivity to external stimuli such as pressure, temperature, or doping. Unlike vanadium pentoxide (V_2O_5), vanadium dioxide (VO_2) and vanadium sesquioxide (V_2O_3) exhibit IMT in their bulk phases. In order to illustrate their attractive applications, VO_x with x equal to 1.8 has been chosen for the sensing element of the Honeywell microbolometer structure.

In Chap. 5, Graphene—Properties, Synthesis, and Applications, the electronic, optical, and thermoelectric properties of graphene and graphene nanoribbons, as a function of the number of layers, doping, chirality, temperature, and lattice defects, are described. Some aspects related to synthesis and applications are presented.

Chapter 6 is devoted to transition metal dichalcogenides (TMDCs)—their properties and applications. Since the discovery of graphene, 2D materials including TMDCs represent, perhaps, the fastest growing area of research in materials science and engineering and condensed matter physics. In particular, this chapter focuses on the properties and applications of sulfides and selenides of tungsten and molybdenum.

Chapter 7 deals comprehensively with group II–V semiconductors, involving compounds formed by group IIB metallic elements (Cd, Zn, and Hg) with group VI non-metallic elements (O, S, Se, and Te). Due to its extraordinary optoelectronic properties which have led to unique device applications in thin-film photovoltaics, nanophotodetectors, and lasers, cadmium sulfide (CdS) has set the standard requirements for material properties in its applications to solar cells. Zinc Oxide (ZnO), another important II–VI semiconductor, has exceptionally good properties which has led to wide applications in varactors, phosphors, sensors, and optoelectronic devices. Zinc oxide is treated in more detail in Sect. 7.2, recognizing it as a biosafe key technological material. This has become the focal point in nanoscience and nanotechnology in the fabrication of sensors, transducers and catalysts, transparent electronics, ultraviolet (UV) light emitters, piezoelectric devices, chemical sensors, and spin electronics (or spintronics). Semiconductor devices made of HgTe and related binary, ternary, and quaternary compounds are discussed in Sect. 7.3.

Chapter 8, Other Miscellaneous Semiconductors and Related Binary, Ternary, and Quaternary Compounds, discusses the electronic, optical, and elastic/mechanical properties of various semiconductor alloys. These materials have contributed significantly to the development of Bandgap Engineering. The applications of these semiconductor alloys include solar cells, solid-state lasers, detectors, Light Emitting Diodes (LEDs), and Optoelectronic Integrated Circuits (OEICs).

Chapter 9, Organic Semiconductors, addresses one of the most promising material classes for applications in electronics, optoelectronics, and flexible electronics. The processing encompasses vacuum deposition and solution deposition techniques, with their respective variety of routes, including vacuum thermal evaporation and laser deposition, for the former, and dip coating, spin coating, printing, and spray coating for the latter.

In Chap. 10, the highlights of the research opportunities that have been reported in the current literature and the summary of some of the new topics, mentioned in the previous chapters, are presented.

Perhaps, the salient feature of this book is the stimulus of the cross-disciplinary/collaborative work, together with the sensitivity to the EHS (Environment, health, and safety) aspects, and inclusiveness of the recycling/reuse of components within the central paradigm of materials science and engineering. The confluence of these three vectors may help to respond more efficiently to the current technological and social demands in the constantly evolving field of semiconductors and their applications.

Chapter 1
Semiconductor Fundamentals

P. Horley, P. J. Gonçalves Ribeiro, J. A. Aguilar Martínez and V. J. Rocha Vieira

1.1 Introduction

The remarkable progress in semiconductor microelectronics that essentially shaped technological developments of the twentieth century was made possible due to a better understanding of the physical phenomena on an atomic scale, and in particular, due to the development of quantum mechanics. The more precise experimental techniques and new characterization equipment allowed considerable gain in knowledge, with theoretical and experimental physicists working together to explain and understand a range of physical phenomena that are taking place in semiconductors. The results of those studies were explained in numerous research papers, and by now, they form a part of excellent textbooks that have played an important role in the education of a new generation of physicists [1–14]. This chapter provides a discussion of the main quantum mechanical concepts and physical phenomena that are important

P. Horley (✉)
Centro de Investigación en Materiales Avanzados CIMAV S.C.,
Unidad Monterrey, Parque de Investigación e Innovación Tecnológica,
Alianza Norte 202, 66-628 Apodaca, Nuevo León, Mexico
e-mail: paul.horley@cimav.edu.mx

P. J. Gonçalves Ribeiro · V. J. Rocha Vieira
Centro de Física e Engenharia de Materiais Avançados (CeFEMA),
Departamento de Física, Instituto Superior Técnico,
Universidade de Lisboa, Av. Rovisco Pais, 1049-001 Lisboa, Portugal
e-mail: vrv@cefema.tecnico.ulisboa.pt

P. J. Gonçalves Ribeiro
Russian Quantum Center, Novaya street 100 A,
Skolkovo, Moscow area 143-025, Russia
e-mail: ribeiro.pedro@gmail.com

J. A. Aguilar Martínez
Universidad Autónoma de Nuevo León, Facultad de Ingeniería
Mecánica y Eléctrica (FIME), Centro de Investigación e Innovación
en Ingeniería Aeronáutica (CIIIA), Aeropuerto Internacional del Norte,
Carretera a Salinas Victoria Km. 2.3, C.P. 66-600 Apodaca, Nuevo León, Mexico
e-mail: josue.aguilarm@uanl.mx

M. I. Pech-Canul and N. M. Ravindra (eds.), *Semiconductors*,
https://doi.org/10.1007/978-3-030-02171-9_1

for understanding solid-state physics of semiconductors. Although the bibliography is by no means exhaustive, we refer to it for further reading.

1.2 Principles of Quantum Mechanics

1.2.1 Energy Quantization

By the end of the nineteenth century, there were a set of problems that the existing theory—now known as classical physics—seemed to address wrongly. Part of them could have been resolved if energy was postulated to exist in discrete quantities—or quanta. During the twentieth century, further investigation exposed the limitations of the classical theory and eventually led to the development of Quantum Mechanics.

One of these problems was the black body radiation. It was described by Lord Kelvin as one of the clouds over classical physics, together with the negative result of the Michelson-Morley experiment, which led to the development of special relativity. The explanation of the discrete nature of atomic spectra and the low-temperature behavior of specific heat in solids, in violation of the Dulong–Petit law, were also problematic. These factors led to the introduction of the so-called quantization rules, setting restrictions to the allowed values of some quantities, such as energies or angular momentum. Quantized variables were associated to conserved quantities and, in the old quantum theory, the quantization was done in the action variables of analytical mechanics.

The quantum theory was introduced in 1925 by Heisenberg, with its matrix mechanics, and by Schrödinger, in 1926, with its wave mechanics. These two formulations are equivalent, as was shown by Schrödinger already in 1926. Later on, in 1948, Feynman introduced a third formulation of quantum mechanics, the path integral formulation. The quantization rules could now be understood as resulting from boundary conditions whose solutions must be satisfied by quantum wave equations.

1.2.2 Wave Functions and Quantum Equations in a Nutshell

In quantum mechanics, particle and wave properties coexist and are manifested by the type of measurement being performed. This duality is illustrated by Planck's $E = \hbar\omega$ and de Broglie's $\mathbf{p} = \hbar\mathbf{k}$ relations, relating the two quadrivectors $(\frac{E}{c}, \mathbf{p})$ and $(\frac{\omega}{c}, \mathbf{k})$ that characterize particles and waves respectively.

Quantum mechanics is a probabilistic theory whose central object is the wave function $\psi(\mathbf{r}, t)$ describing the physical state of a particle. The wave equation gives a deterministic evolution for $\psi(\mathbf{r}, t)$ from some initial state $\psi(\mathbf{r}, t = 0)$. If the position of the particle is measured at time t, $|\psi(\mathbf{r}, t)|^2\, d^3r$ gives the probability of finding a particle in a box of volume d^3r around position $\mathbf{r}$. Just after the measurement, if the particle is found at position $\mathbf{r}'$ at time t', the wave function becomes delta-peaked at that point, i.e., $\lim_{t\to t'^+} \psi(\mathbf{r}, t) = \delta(\mathbf{r} - \mathbf{r}')$, a phenomenon denominated as wave function collapse. If instead, the linear momentum of a particle is measured, and $\mathbf{p}'$ is found at time t', the wave function becomes the corresponding momentum state $\lim_{t\to t'^+} \psi(\mathbf{r}, t) \propto e^{i\mathbf{p}'\cdot\mathbf{r}/\hbar}$.

As a result of the particle-wave duality, momentum and position cannot be found simultaneously in a deterministic way. This generalizes to other conjugate variables for which the quantum theory does not allow a precise joint determination. The accuracy bounds are given by the celebrated Heisenberg uncertainty relations $\Delta x \Delta p \geq \frac{\hbar}{2}$ and $\Delta t \Delta E \geq \frac{\hbar}{2}$. Experiments further confirmed that, at the quantum level, the measurement process disturbs the system in a nonnegligible manner and, as a result, no measurement can lead to a better resolution.

The equation determining the evolution of the wave function respects the usual properties of particles in classical mechanics. The relations $E = \frac{\mathbf{p}^2}{2m}$ and $\left(\frac{E}{c}\right)^2 = \mathbf{p}^2 + (mc)^2$ of nonrelativistic and relativistic classical mechanics become dispersion relations for wave functions. Following this principle, the Schrödinger and Klein–Gordon equation are obtained by replacing $E \to i\hbar\frac{\partial}{\partial t}$ and $\mathbf{p} \to \frac{\hbar}{i}\frac{\partial}{\partial \mathbf{r}}$ in the nonrelativistic and relativistic relations, respectively. The Schrödinger equation reads as

$$i\hbar\frac{\partial \psi(\mathbf{r}, t)}{\partial t} = H\psi(\mathbf{r}, t), \tag{1.1}$$

with $H = p^2/(2m)$ being the Hamiltonian of the system. The Klein–Gordon equation is obtained similarly considering the relativistic relation instead. In addition, the relativistic Dirac equation can be obtained by searching for a Hamiltonian of the form $H = c\boldsymbol{\alpha} \cdot \mathbf{p} + mc^2\beta$, where spatial derivatives appear in first order that respects the relativistic spectral relation $\left(\frac{E}{c}\right)^2 = \mathbf{p}^2 + (mc)^2$. This can only be accomplished if the coefficients $\boldsymbol{\alpha}$ and β are anti-commuting matrices composed minimally of four components that square to the identity. In the so-called Dirac representation, the Dirac equation is $\left(i\hbar\gamma^\mu \frac{\partial}{\partial r^\mu} - mc\right)\psi = 0$, where $(\gamma^0 = \beta, \boldsymbol{\gamma} = \beta\boldsymbol{\alpha})$, $\gamma^0 = \begin{pmatrix} \sigma_0 & 0 \\ 0 & -\sigma_0 \end{pmatrix}$, $\boldsymbol{\gamma} = \begin{pmatrix} 0 & \boldsymbol{\sigma} \\ -\boldsymbol{\sigma} & 0 \end{pmatrix}$, with $\sigma^{x,y,z}$ being the Pauli matrices and σ_0 the identity. The presence of at least four components of the Dirac equation describing relativistic fermionic fields relates with the existence of antiparticles with opposite quantum numbers and spin.

In the presence of electric $\mathbf{E} = -\nabla V - \frac{1}{c}\frac{\partial \mathbf{A}}{\partial t}$ and magnetic $\mathbf{B} = \nabla \times \mathbf{A}$ fields, derived from the scalar and vector potentials $(V, \mathbf{A})$, E and $\mathbf{p}$ are replaced by $E - eV$ and $\mathbf{p} - \frac{e}{c}\mathbf{A}$. The wave equations are obtained by making the replacement $E - eV \to i\hbar\frac{\partial}{\partial t} - eV$ and $\mathbf{p} - \frac{e}{c}\mathbf{A} \to \frac{\hbar}{i}\frac{\partial}{\partial \mathbf{r}} - \frac{e}{c}\mathbf{A}$. This procedure is called minimal coupling.

In the old quantum mechanics, in order to describe the spectrum of an atom in the presence of magnetic fields, besides the three quantum numbers associated with the angular momentum, to its third component and to the radial distance, it was necessary to introduce a fourth quantum number, the spin, in an *adhoc* manner. One of the triumphs of the Dirac equation was to explain the spin, and, in particular, the gyromagnetic factor, giving the coupling of an electron to the magnetic field.

The Pauli–Schrödinger Hamiltonian

$$H = \frac{1}{2m}\left(\mathbf{p} - \frac{e}{c}\mathbf{A}\right)^2 + eV - \frac{e\hbar}{2mc}2\frac{\boldsymbol{\sigma}}{2}\cdot\mathbf{B}, \quad (1.2)$$

having the correct, $g = 2$, gyromagnetic factor, can be obtained taking the nonrelativistic limit of the Dirac equation or more simply, rewriting $\mathbf{p}^2 = (\boldsymbol{\sigma}\cdot\mathbf{p})^2$, in the kinetic term of the Schrödinger equation, and making the minimal coupling replacement, since $\boldsymbol{\sigma}\cdot\mathbf{p}\,\boldsymbol{\sigma}\cdot\mathbf{A} + \boldsymbol{\sigma}\cdot\mathbf{A}\,\boldsymbol{\sigma}\cdot\mathbf{p} = 2\mathbf{A}\cdot\mathbf{p} + \hbar\boldsymbol{\sigma}\cdot\mathbf{B} - i\hbar(\nabla\cdot\mathbf{A}) = \mathbf{B}\cdot(\mathbf{L} + 2\frac{\hbar\boldsymbol{\sigma}}{2})$, with $\mathbf{L} = \mathbf{r}\times\mathbf{p}$, the angular momentum, and $\mathbf{A} = \frac{1}{2}\mathbf{B}\times\mathbf{r}$. This, in fact, follows the derivation of the nonrelativistic limit of the Dirac equation.

Coming back to the probabilistic interpretation of the wave function, note that the total probability is conserved under evolution. This can be requested under the so-called continuity equation $\frac{\partial\rho}{\partial t} + \nabla\cdot\mathbf{j} = 0$, relating the probability density $\rho = \psi^\dagger\psi$ and the probability current $\mathbf{j} = \frac{1}{2m}\left\{\psi^\dagger\left(\mathbf{p} - \frac{e}{c}\mathbf{A}\right)\psi + \left[\psi\left(\mathbf{p} - \frac{e}{c}\mathbf{A}\right)\right]^\dagger\psi\right\}$. Multiplying this equation by e, the electric charge of the particle, these densities can also be interpreted as the charge and current densities as well as the equation of continuity of charge. However, it should be stressed that when, in an actual experiment, the particle is detected, one always finds its total charge and not a fraction of it.

1.2.3 Electronic Wave Functions in Atoms, Molecules, and Bulk Materials

The simplest model of an atom can be obtained by considering the nonrelativistic system of a single proton–electron pair interacting via a Coulomb potential. In this case, the solutions of the Schrödinger equation can be obtained analytically. The quantized energies obtained in this way present deviations due to relativistic effects when compared with the spectrum of the hydrogen atom. In the so-called real hydrogen atom, besides the Coulomb potential, it is necessary to take into account additional interaction terms, namely, the first relativistic correction to the kinetic energy $H_{Krel} = -\frac{1}{8}\frac{((\mathbf{p})^2)^2}{m^3c^2}$ (since $E = mc^2\sqrt{1 + \left(\frac{\mathbf{p}}{mc}\right)^2} \approx mc^2 + \frac{(\mathbf{p})^2}{2m} - \frac{1}{8}\frac{((\mathbf{p})^2)^2}{m^3c^2}$), the spin-orbit interaction $H_{SO} = \frac{1}{2m^2c^2}\frac{1}{r}\frac{dV}{dr}\mathbf{S}\cdot\mathbf{L}$ (with the relativistic Thomas factor $\frac{1}{2}$), the Darwin term $H_D = \frac{\hbar^2e^2}{8m^2c^2}\delta^3(\mathbf{x})$, with a delta function at the origin, affecting the s wave functions only, and the Zeeman interaction $H_Z = \frac{e}{2mc}(\mathbf{L} + 2\mathbf{S})$, for the interaction with a magnetic field, also with the spin degree of freedom, with the correct gyromagnetic factor, if the net spin is not zero (as in the so-called anomalous effect). Besides this, one can also consider the hyperfine splitting due to the atomic nucleus. These interactions are normally taken as perturbation terms and, as referred above, their derivation can be done by taking the classical limit of the Dirac equation.

The consideration of heavier atoms and molecules becomes more complicated and requires more sophisticated methods, such as the variational self-consistent Hartree and Hartree–Fock approximations. Moreover, the quantum statistics of the particles, namely, electrons, must be taken into account and requires the use of Slater determinants. The Hund's rules stipulate how to determine the ground state of a multi-electron atom.

For electrons in a crystal, electric fields created by the ions can be considered to be static giving rise to a background electric potential. In this case, the free space Hamiltonian appearing in Schrödinger's equation (1.1) gets modified by $H = p^2/(2m) + eV(x)$, where V is the ionic electric potential. Atomic moments can also be responsible for local variations in the magnetic field in terms of the Hamiltonian that couple spin and momentum operators—the so-called spin-orbit coupling.

1.2.4 Scattering

As an illustration of quantum mechanics in action, we consider the scattering of a particle by a potential barrier in one dimension. This example allows to understand the concept of quantum tunneling—a process that is strictly impossible in classical mechanics. The formalism developed here will be generalized in further sections to periodic potentials to illustrate the emergence of the band structure of metals.

Consider the Hamiltonian $H = p^2/(2m) + V(x)$ in 1d, in a region where the potential is, at least approximately, constant $V(x) = \hbar u/2m$. In this region, the eigenstates of H have to be of the form: $\psi(x) = Ae^{ikx} + Be^{-ikx}$ with an energy given by $(k^2 + u)/(2m)$. The flux of probability of this solution is $\mathbf{j} = \frac{\hbar k}{m}(|A|^2 - |B|^2)$.

To the left of the scattering region (Fig. 1.1a), we denote the wave function as $\psi_<(x) = Ae^{ikx} + Be^{-ikx}$ and to the right as $\psi_>(x) = Ce^{ikx} + De^{-ikx}$. Due to the conservation of probability, the fluxes at the boundaries of a region under consideration must be equal. This determines important relations for the so-called

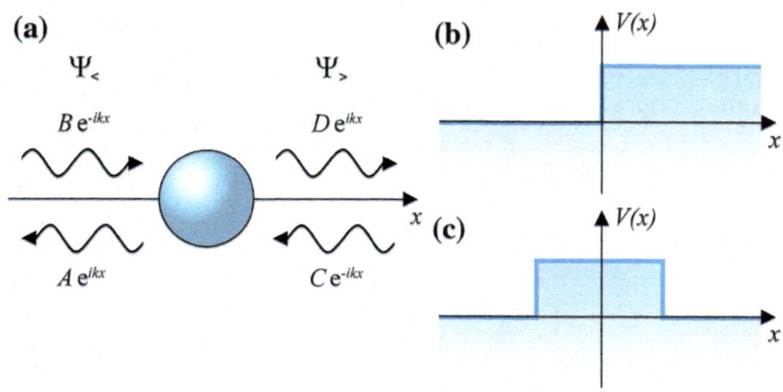

Fig. 1.1 Scattering: **a** scattering amplitudes of the wave function; **b** step potential; **c** potential barrier

scattering matrix S that relates the outgoing amplitudes (A, D) with the incoming amplitudes (B, C). Equivalently, the transfer matrix $\bar{M}$ gives the amplitudes in one of the boundaries (C, D) in terms of the amplitudes at the other (A, B). Since the potential in the different regions, in general, is different, it is preferable to use normalized amplitudes $\bar{A}$, $\bar{B}$, rewriting the wave function as $\psi_<(x) = \frac{\bar{A}}{\sqrt{k}}e^{ikx} + \frac{\bar{B}}{\sqrt{k}}e^{-ikx}$, and $\psi_>(x) = \frac{\bar{C}}{\sqrt{k}}e^{ikx} + \frac{\bar{D}}{\sqrt{k}}e^{-ikx}$, in the left and right regions, respectively. We denote the coefficients associated with these normalized amplitudes by the use of a bar on top, when different from the non-normalized coefficients. In this notation, the scattering matrix $\bar{S}$ of normalized coefficients is defined by

$$\begin{pmatrix} \overline{C} \\ \overline{B} \end{pmatrix} = \bar{S} \begin{pmatrix} \overline{A} \\ \overline{D} \end{pmatrix}, \quad \bar{S} = \begin{pmatrix} \bar{t} & r' \\ r & \bar{t}' \end{pmatrix}, \tag{1.3}$$

and the transfer matrix $\bar{M}$ is defined and is given by

$$\begin{pmatrix} \overline{C} \\ \overline{D} \end{pmatrix} = \bar{M} \begin{pmatrix} \overline{A} \\ \overline{B} \end{pmatrix}, \quad \bar{M} = \frac{1}{\bar{t}'} \begin{pmatrix} \bar{t}\bar{t}' - rr' & r' \\ -r & 1 \end{pmatrix}, \tag{1.4}$$

with r and $\bar{t}$ as the normalized reflection and transmission coefficients. Flux conservation implies that the matrix $\bar{S}$ is unitary and time inversion symmetry implies then that $\bar{t}' = \bar{t}$ yielding $\det(\bar{M}) = 1$. As a result, the $\bar{S}$ and $\bar{M}$ can be parameterized by

$$\bar{S} = e^{i\theta} \begin{pmatrix} \mu & -\nu^* \\ \nu & \mu \end{pmatrix}, \quad \bar{M} = \begin{pmatrix} \alpha & \beta^* \\ \beta & \alpha^* \end{pmatrix} \tag{1.5}$$

with real μ, $\mu^2 + |\nu|^2 = 1$, $\alpha = \frac{e^{i\theta}}{\mu}$ and $\beta = -\frac{\nu}{\mu}$ satisfying $|\alpha|^2 - |\beta|^2 = 1$.

When there are successive scattering processes, for example, one from a region 1 to a region 2 and another from region 2 to region 3, the overall transfer matrix is given by the product of the different transfer matrices and the overall scattering matrix can be obtained from it. The transfer matrix $\bar{M}_{31} = \bar{M}_{32}\bar{M}_{21}$ is parameterized by $\alpha_{31} = \alpha_{32}\alpha_{21} + \beta_{32}^*\beta_{21}$ and $\beta_{31} = \beta_{32}\alpha_{21} + \alpha_{32}^*\beta_{21}$, as a function of the parameters of the two elementary transfer matrices.

In solution to certain problems, it is convenient to refer the wave functions to specific points, characterizing the scattering regions. In this case, we define $\psi_<(x) = \frac{\bar{A}'}{\sqrt{k_1}}e^{ik_1(x-a)} + \frac{\bar{B}'}{\sqrt{k_1}}e^{-ik_1(x-a)}$, and $\psi_>(x) = \frac{\bar{C}'}{\sqrt{k_2}}e^{ik_2(x-b)} + \frac{\bar{D}'}{\sqrt{k_2}}e^{-ik_2(x-b)}$, with $\bar{A}' = e^{ik_1 a}\bar{A}$, $\bar{B}' = e^{-ik_1 a}\bar{B}$, $\bar{C}' = e^{ik_2 b}\bar{C}$, $\bar{D}' = e^{-ik_2 b}\bar{D}$. The new transfer matrix for these modified normalized amplitudes is then given by

$$\bar{M}' = \begin{pmatrix} e^{i(k_2 b - k_1 a)}\alpha & e^{i(k_2 b + k_1 a)}\beta^* \\ e^{-i(k_2 b + k_1 a)}\beta & e^{-i(k_2 b - k_1 a)}\alpha^* \end{pmatrix}. \tag{1.6}$$

As one may expect, when $k_2 = k_1$, α is affected by the distance $b - a$ of the two points, and β by their position, given by their center of gravity $\frac{b+a}{2}$.

The simplest potential we can consider is the step function potential, at point a, given by $V(x) = \frac{\hbar u_1}{2m}\theta(a-x) + \frac{\hbar u_2}{2m}\theta(x-a)$ (see Fig. 1.1b), where $\theta(x)$ is the Heaviside step function. The superposition of the two plane wave functions is a solution. It is continuous, with a continuous first derivative, and we find $\alpha = \frac{k_2+k_1}{2\sqrt{k_2 k_1}}e^{-i(k_2-k_1)a}$, and $\beta = \frac{k_2-k_1}{2\sqrt{k_2 k_1}}e^{i(k_2+k_1)a}$, with $k_1^2 + u_1 = k_2^2 + u_2$.

Another example is the barrier potential shown in Fig. 1.1c. It can be seen as a superposition of two-step potentials. For the symmetric barrier, of width l and centered at point a, given by $V(x) = \frac{\hbar u}{2m} + \frac{\hbar \Delta u}{2m}\theta(x-(a-\frac{l}{2}))\theta((a+\frac{l}{2})-x)$, we have

$$\alpha_{31} = \left(2kk'\cos k'l + i(k^2+k'^2)\sin k'l\right)\frac{e^{-ikl}}{2kk'} \tag{1.7}$$

$$\beta_{31} = -i(k-k')^2 \sin k'l \frac{e^{i2ka}}{2kk'}. \tag{1.8}$$

where $k_1 = k_3 = k$ and $k_2 = k'$, with $k_1^2 = k_2^2 + \Delta u = k_3^2$.

In the limit $l \to 0$ with $\Delta u = \hbar\lambda/l$, the potential converges to a delta function centered at point a: $V(x) = \frac{\hbar u}{2m} + \frac{\hbar^2}{m}\lambda\delta(x-a)$. In this case, the previous parametrizations allow simplification $\alpha = 1 - i\frac{\lambda}{k}$ and $\beta = i\frac{\lambda}{k}e^{i2ka}$, with $k_1 = k_2 = k$.

1.3 Fermi–Dirac Statistics

1.3.1 Fermions and Bosons

For a system containing many identical particles, the wave function must have some symmetry properties while considering exchange interaction between two or more particles. The Klein–Gordon and Dirac equations derived in the previous sections refer to single particles. To ensure causality, their generalization to many identical particles implies specific relations of the many body wave function under permutation of particles. More generally, this relation is established by the so-called spin-statistics theorem stating that the wave function of a system of identical integer (half-integer) spin particles is symmetric (antisymmetric) when the position of any two particles is swapped. Particles with wave functions which are symmetric (antisymmetric) under exchange are called bosons (fermions). As a consequence, half-integer spin particles, such as electrons, are subject to the Pauli exclusion principle and cannot occupy the same quantum state. This phenomenon has remarkable consequences in the physics of electrons in materials and specifically in metals, as we will see further. On the other hand, the number of bosons that can occupy a quantum state is unrestricted. This can give rise to the so-called Bose–Einstein condensation with a "macroscopic" fraction of bosons occupying the same quantum state.

1.3.2 *The Free Electron Gas*

Most of the properties of metals can be understood qualitatively by a simple model called the free electron gas. The valence electrons of the atoms in a metal are considered here to move freely within the system. As the Coulomb potential of the ions is completely disregarded, some of the aspects of crystalline materials such as the existence of energy bands cannot be understood within this model. Nonetheless, it elucidates the interplay between the kinetics of conduction electrons and Pauli's exclusion principle and provides a clear explanation for the low-temperature properties of metals.

For a free-particle in three dimensions, the Schrödinger equation has the form

$$-\frac{\hbar^2}{2m}\left(\partial_x^2+\partial_y^2+\partial_z^2\right)\psi(r)=E\psi(r) \tag{1.9}$$

with E the energy of the eigenstate $\psi(r)$. As in the one-dimensional case of the previous sections, $\psi(r)$ can be written as a combination of plane waves. For particles confined in a cubic box of linear size L, i.e., $x_i \in [0, L]$, we further have to impose that at the boundaries of the box $x_i = 0, L$, the wave function vanishes. This implies that the eigenfunctions will be of the form $\psi_k(r) \propto \sin(k_x x)\sin(k_y y)\sin(k_z z)$ where $k = \pi/L\left\{n_x, n_y, n_z\right\}$ with $n_i \in \mathbb{Z}^+$. For periodic boundary conditions, eigenfunctions can be chosen to be the plane waves $\psi_k(r) \propto e^{ik\cdot r}$ with $k = 2\pi/L\left\{n_x, n_y, n_z\right\}$ with $n_i \in \mathbb{Z}$. In either case, the energy is given by $E_k = \frac{\hbar^2}{2m}k^2$.

Due to the Pauli's exclusion principle, the ground state of a system with N electrons is the state where the lowest $N/2$ energy states are occupied by spin up and spin down electrons. For convenience, we will consider the case of periodic boundary conditions for which the ground state can be visualized as the interior of a sphere of radius k_F defined by $N = \sum_{\sigma=\pm 1}\sum_{|k|<k_F} 1$ (Fig. 1.2a). This sphere is called the Fermi surface. The states on this surface have $E = E_F = \frac{\hbar^2}{2m}k_F^2$, which is called Fermi

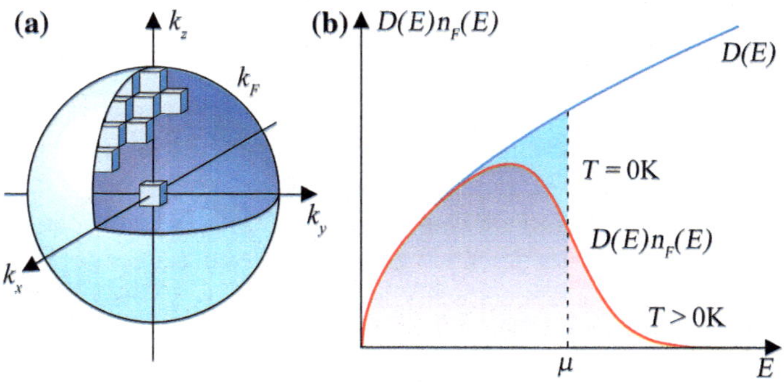

Fig. 1.2 Free electron gas approximation: **a** continuous approximation of the Fermi surface for system size going to infinity; **b** density of states for Fermi gas (blue line) and occupied Fermionic states at zero and finite temperature (filled areas)

energy. Inside the sphere, the states are occupied and outside the sphere, the states are empty. For a cube-shaped object with side L, we can compute the number of states inside the sphere by approximating the summation over momentum by an integral $\frac{1}{L^3}\sum_k \ldots \simeq \int \frac{d^3k}{(2\pi)^3}\ldots$, therefore we obtain

$$\frac{N}{L^3} \simeq \sum_{\sigma=\pm 1} \int \frac{d^3k}{(2\pi)^3}\theta(E_F - E_k) = 2\int_{|k|<k_F} \frac{d^3k}{(2\pi)^3} = \frac{L^3}{3\pi^2}k_F^3 \tag{1.10}$$

yielding $k_F = \left(3\pi^2 N/L^3\right)^{1/3}$ and $E_F = \hbar/(2m)\left(3\pi^2 N/L^3\right)^{2/3}$. The density of states is the number of states per unit energy

$$D(E) = \sum_{\sigma=\pm 1}\sum_k \delta(E - E_k) \simeq 2\int \frac{d^3k}{(2\pi)^3}\delta\left(E - \frac{\hbar^2}{2m}k^2\right) = \frac{3N}{2E_F}\left(\frac{E}{E_F}\right)^{1/2} \tag{1.11}$$

and thus $N = \int_{-\infty}^{E_F} D(E)$.

1.3.3 Heat Capacity

At finite temperature, the probability of a state of energy E to be occupied is given by the Fermi–Dirac Distribution

$$n_F(E) = \frac{1}{exp\left(\frac{E-\mu}{k_B T}\right) + 1} \tag{1.12}$$

where μ is the chemical potential as follows from the Boltzmann factor $exp\left(\frac{-(E-\mu)n}{k_B T}\right)$ for the relative occupation probability of a state with energy E by a number of particles, n and the Fermi–Dirac statistics of the electrons allows $n = 0, 1$ only.

The number of electrons and total energy E_T of a metal are given by

$$N = \int_{-\infty}^{\infty} D(E)n_F(E)dE, \tag{1.13}$$

$$E_T = \int_{-\infty}^{\infty} En_F(E)D(E)dE. \tag{1.14}$$

Equation (1.13) defines the chemical potential. Equation (1.14) is obtained by summing the energy times the probability of that energy state to be occupied times the number of states within a width dE around E. At low temperature, n_F becomes a step function and at finite temperatures, the sharp step becomes smeared out on an energy scale of the order of temperature (Fig. 1.2b). From the definition of μ, the Fermi energy E_F coincides with the chemical potential at zero temperature.

The heat capacity measures the rate of energy increase in an object as a function of temperature change and is obtained as

$$C = \partial_T E_{\mathrm{T}}. \tag{1.15}$$

In the framework of free electron gas model, we can compute E_{T} with $D(E)$ given by Eq. (1.11). In order to obtain explicit results at low temperatures, we can use the Sommerfeld expansion to evaluate the integrals

$$\int_{-\infty}^{\infty} dE\, n_f(E) f(E) = \int_{-\infty}^{\mu} dE f(E) + k_B T \sum_{n=0}^{\infty} c_n\, \partial_E^{2n+1} f(E)\big|_{E=\mu} (k_B T)^{2n+1} \tag{1.16}$$

where f is a regular function and $c_n = \frac{1}{2(2n+2)!} \int_{-\infty}^{\infty} dx \frac{x^{2n+2}}{1+\cosh(x)}$. Using this expression in Eqs. (1.13–1.14) and keeping the terms with the order up to T^2, we obtain $N = N\left(\frac{\mu}{E_F}\right)^{3/2}\left[1 + \frac{\pi^2 k_B^2 T^2}{8\mu^2}\right]$ and $E_T = N\mu\left(\frac{\mu}{E_F}\right)^{3/2}\left[\frac{3}{5} + \frac{3\pi^2 k_B^2 T^2}{8\mu^2}\right]$. Solving these for chemical potential, again considering only the terms up to T^2, we will get

$$\mu = E_F\left[1 - \frac{\pi^2 k_B^2 T^2}{12 E_F^2}\right], \quad E_T = N E_F\left[\frac{3}{5} - \frac{\pi^2 k_B^2 T^2}{4 E_F^2}\right]. \tag{1.17}$$

This yields heat capacity in the form

$$C/T = \gamma = N\frac{k_B^2 \pi^2}{2 E_F}. \tag{1.18}$$

At low temperatures, the linear variation of specific heat with temperature is a hallmark of metals. At high temperatures, the heat capacity may acquire corrections due to contributions of other degrees of freedom such as phonons. However, the experimentally measured values of the coefficient γ can deviate considerably from the predictions of Eq. (1.18) obtained in the framework of free electron gas model. This might be due to the specific band structure of the material that we will be addressing in the next chapters, or due to interactions, not taken into account in this simple model. As $E_F \propto 1/m$, the deviations of γ are often converted in an electronic effective mass through the relation $m^*/m = \gamma_{\mathrm{measured}}/\gamma$.

1.4 Concept of Band Structure

1.4.1 Solid Matter

In liquids and gases, the separation between molecules is large compared with typical molecular sizes. Therefore, for most practical purposes, molecules can be considered as isolated from each other. On the contrary, in solids, the atomic distance is of the order of atomic size and the notion of individual molecules is lost as the forces binding electrons within a molecule are comparable with intra-molecular interactions. In such conditions, the emergent collective behavior of the resulting material can be very different from a collection of individual molecules.

The classification of solids can rely on physical and chemical properties, atomic arrangements or the nature of inter-atomic binding energies. A common classification divides solids into crystalline and noncrystalline or amorphous in accordance with their atomic arrangements. In a crystalline solid, atoms are arranged in regular patterns. On the other hand, for noncrystalline materials, atoms are placed in an overall irregular manner, the crystalline order being lost over long distances even if locally atoms seem to be arranged in an ordered fashion. Among crystalline solids, a further classification separates them into single-crystalline and the polycrystalline materials. In the first, the crystalline order extends over the entire crystal, while in the latter, order exists over small regions of the crystal with small single-crystalline regions surrounded by grain boundaries.

Classifications of solids can also be based on their capacity to conduct electrical currents. Insulators have low electrical conductivity while in metals, this quantity is high. Strictly speaking, for an ideal insulator, the electrical conductivity vanishes at zero temperature while it is finite for metals. For finite temperatures, some corrections are required but the electrical conductivity remains nearly independent of temperature over a wide range of temperatures around room temperature (in metals, however, corrections due to phonons and impurities arise at very low temperatures). Semiconductors, while exhibiting ideal insulating behavior near zero temperature, manifest electrical conductivity with strong dependence on temperature.

In the following sections, we will discuss the properties of crystalline solids as those are most relevant to describe the physics of semiconductors.

1.4.2 Direct and Reciprocal Lattices and the Concept of Brillouin Zone

An n-dimensional lattice is defined by n noncoplanar basis vectors $a_1, a_2, \ldots, a_n$ such that two points in the lattice r and r' are related by $r' = r + \sum_i u_i a_i$ with $u_i \in \mathbb{Z}$. The collection of points r' obtained in this way by varying the integer coefficients u_i constitutes a lattice. The vectors $a_1, a_2, \ldots, a_n$ are called primitive translation vectors if they form a lattice cell with minimum volume. In this case, the n-dimensional parallelogram formed by them defines a primitive cell. The crystal structure can be described by a lattice with a group of atoms attached to every lattice point. The group of atoms is known as the basis. A crystal structure is a lattice plus the motif (Fig. 1.3). A basis of atoms is defined by the atomic positions $r_i = \sum_i^n x_i a_i$ with $x_i \in \mathbb{R}$.

The primitive cell contains the pattern that is repeated. A canonical way of choosing a primitive cell consists of taking for its boundaries, the points on the perpen-

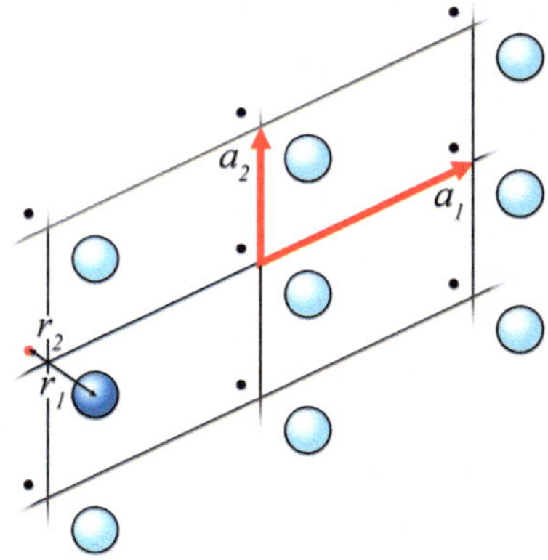

Fig. 1.3 Primitive cell translation vectors a_1, a_2 and atom position vectors r_1, r_2

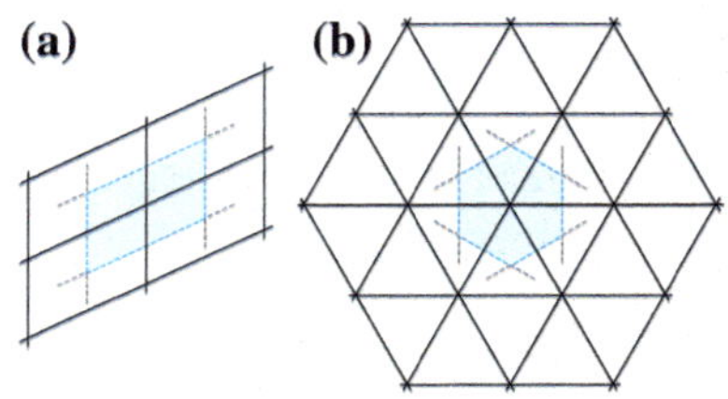

Fig. 1.4 Primitive cell **a** for an oblique lattice and Wigner-Seitz cell **b** for a triangular lattice

dicular space to the vector linking two neighboring lattice sites that passes by the midpoint between them—this is called the Wigner–Seitz construction (Fig. 1.4).

Lattices can be classified according to their point symmetries, i.e., if they remain invariant under spatial rotations around a lattice site. A generic lattice may present no symmetries; however, special lattices can be invariant under a group of rotations and or inversion around a lattice site. In a 2D case, there are four such special lattices, and thus one can classify a lattice as belonging to five different symmetry classes also known as Bravais lattices. In 3D, there are 14 different Bravais lattices. Additional symmetry elements, such as screw axes, glide planes, or time and space inversion, can also be useful to further classify the systems and to explain the properties of materials.

In order to describe wave propagation in solids, we will study the structure of the reciprocal space. A lattice in real (or position) space is referred to as the direct lattice. The reciprocal lattice is the Fourier Transform of the Bravais lattice. The reciprocal lattice is defined as the set of points G that, for all points $R = \sum_i n_i a_i$ with integer n_i of the direct lattice, obey $e^{iG \cdot R} = 1$. A basis for the reciprocal lattice can be given by a set of b_j vectors, with $j = 1, 2, \ldots, n$, such that $a_i \cdot b_j = 2\pi \delta_{i,j}$. A vector G can thus be written as $G = \sum_i m_i b_i$ with integer m_i's.

A Brillouin zone is any primitive unit cell of the reciprocal lattice. However, as in the direct lattice case, there is a canonical way of choosing a Brillouin zone using the Wigner–Seitz construction on the reciprocal lattice. The Brillouin zone obtained in this way is referred to as the first Brillouin zone.

1.4.3 Bloch Functions

Ions on a lattice form a periodically arranged potential for the conduction electrons. We neglect possible electron–electron interaction and assume the positions of the ions to be fixed, which is a good approximation as ionic masses are much larger than those of electrons. Under these assumptions, electrons in crystalline solids obey the Schrödinger equation with a periodic electrostatic potential. Consequently, translations by an integer number of lattice vectors commute with the Hamiltonian and the wave functions can be chosen to be eigenfunctions of the operators T_R, the generators of lattice translations. Different translations T_R and $T_{R'}$ commute with each other and can be simultaneously diagonalized. Since they are unitary, their eigenvalues lie on a unit circle. Altogether, this implies that we can parametrize their eigenvalues by a phase

$$(T_R\psi)(x) = \psi(r+R) = e^{ik\cdot R}\psi(r). \tag{1.19}$$

In this context, Bloch showed that for a Schrödinger equation with a periodic potential, one can use eigenstates which are also eigenstates of the translation operator, i.e., of the form

$$\Psi_k^\alpha(r) = e^{ik\cdot r}u_k^\alpha(r) \tag{1.20}$$

with $u_k^\alpha(r+R) = u_k^\alpha(r)$, a periodic function, and the crystalline momentum k chosen within the first Brillouin zone. This statement, known as Bloch's theorem, is a restatement of Eq. (1.19). The superscript α distinguishes different solutions called "energy bands" with the same k but different energies. Since u is periodic, it can be written as $u_k^\alpha(r) = \sum_G \tilde{u}_{G,k}^\alpha e^{iG\cdot r}$ and thus

$$\Psi_k^\alpha = \sum_G \tilde{u}_{G,k}^\alpha e^{i(G+k)\cdot r}. \tag{1.21}$$

This formula can be reinterpreted in the following way: momentum is a good quantum number up to the lattice momenta G—as momenta that differ by G interfere among themselves only when the crystalline momentum k, chosen within the first Brillouin zone, is conserved.

1.4.4 Kronig–Penney Model

In order to illustrate the ideas of the previous section, we consider a simple model that permits an explicit analytic solution. The Kronig–Penney model consists of a 1D system in which the atomic potentials are approximated by a square well as in Fig. 1.5:

$$\left[-\frac{\hbar}{2m}\partial_x^2 + U(x)\right]\psi(x) = E\psi(x). \tag{1.22}$$

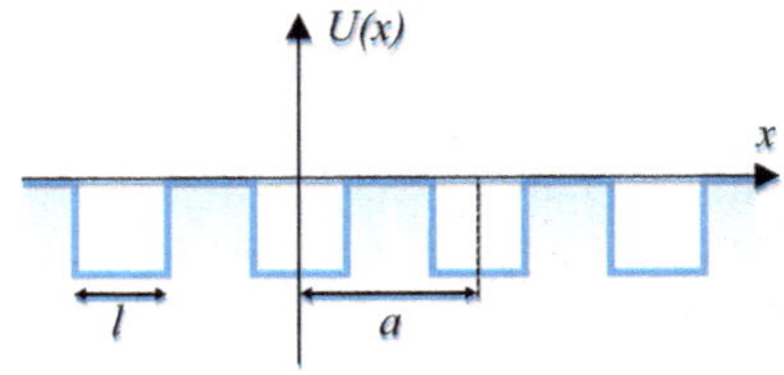

Fig. 1.5 Potential of the Kronig–Penney model

The wave function is $\psi(x) = \frac{\bar{A}'_n}{\sqrt{k}} e^{ik(x-na)} + \frac{\bar{B}'_n}{\sqrt{k}} e^{-ik(x-na)}$, in the region outside the wells (where the potential is zero), in the interval $na \leq x < (n+1)a$, and $\bar{A}'_n = e^{ikna}\bar{A}_n$, $\bar{B}'_n = e^{-ikna}\bar{B}_n$, in terms of the normalized amplitudes. The energy is $E(k) = \frac{(\hbar k)^2}{2m}$. The transfer matrix $\bar{M}'$ is then given by

$$\begin{pmatrix} \bar{A}'_{n+1} \\ \bar{B}'_{n+1} \end{pmatrix} = \begin{pmatrix} \alpha e^{ika} & \beta'^{*} e^{ika} \\ \beta' e^{-ika} & \alpha^{*} e^{-ika} \end{pmatrix} \begin{pmatrix} \bar{A}'_n \\ \bar{B}'_n \end{pmatrix}$$

with $\beta' = \beta e^{-i2kna}$, in terms of the parameters α and β. This new matrix does not depend on n, as can be seen from the transformation properties of α and β under translations.

The condition $\psi(x+a) = e^{iqa}\psi(x)$, of Bloch's theorem, where e^{iqa} is the translation operator eigenvalue, becomes

$$\begin{pmatrix} \bar{A}'_{n+1} \\ \bar{B}'_{n+1} \end{pmatrix} = \bar{M}' \begin{pmatrix} \bar{A}'_n \\ \bar{B}'_n \end{pmatrix} = e^{iqa} \begin{pmatrix} \bar{A}'_n \\ \bar{B}'_n \end{pmatrix}$$

i.e., the eigenvector problem of the transfer matrix $\bar{M}'$. The solution can then be written as $\psi(x) = e^{iqx}\phi(x)$, the product of the wave function e^{iqx} of momentum q, and a function $\phi(x)$ with the periodicity of the lattice, as stated in Bloch's theorem.

We must have $\det\left(\bar{M}' - e^{iqa} I_2\right) = e^{i2qa} - 2we^{iqa} + 1 = 0$, where $2w = \alpha e^{ika} + \alpha^{*} e^{-ika} = \alpha_r \cos(ka) - \alpha_i \sin(ka)$, with α_r and α_i being the real and imaginary parts of α, and using $|\alpha|^2 - |\beta'|^2$. Taking the real and imaginary parts, the eigenvalue equation reduces to $\cos(qa) = w$. It gives the condition to be satisfied by the crystalline momentum q in terms of the energy $E(k) = \frac{(\hbar k)^2}{2m}$.

In the case of the simplified Kronig–Penney model, placing a periodic array of delta function potentials at points $x = na$, we have $\alpha = 1 - i\frac{\lambda}{k}$ and $\beta' = i\frac{\lambda}{k}$ leading to the familiar dispersion relation $\cos(qa) = \cos(ka) + \frac{\lambda}{k}\sin(ka)$.

For each k, there is a set of energies (Fig. 1.6) that satisfy the quantization condition; those are different energy bands labeled by α in the previous section. For each band, the energy $E = E_\alpha(k)$ is a continuous function of k within the first Brillouin zone.

If the amplitude of the periodic potential is small, as compared to the kinetic energy, momenta differing by lattice momenta G almost do not interfere with each other. The dispersion relation resembles that of a free electron gas, folded within the first Brillouin zone. If the temperature is larger than the energy difference (or

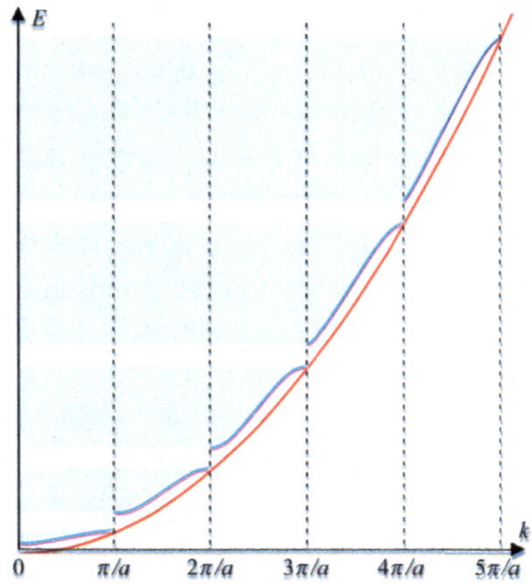

Fig. 1.6 Schematic band structure for the Kronig–Penney model

energy gap) between adjacent bands at the Brillouin zone edges, a free electron gas is a very good approximation and the properties derived in the previous sections are appropriate.

The contrast to the free election picture is the so-called tight binding regime. Here, a band, or a small set of bands, is sufficiently separated in energy from the rest of the spectrum that allows to discard all other bands. This is formally equivalent to having a finite number of sites per unit cell—the same number as the number of relevant bands—that electrons can occupy.

1.4.5 Metals, Insulators, and Semiconductors

As has been seen, the dispersion relation of electrons with respect to periodic potentials may contain energy gaps. Energy values within a gap correspond to forbidden energy regions that discretize the spectrum into bands.

At zero temperature, valence electrons fill the lowest available energy states. The number of crystalline momentum states in the Brillouin zone k is equal to the total number of unit cells in the system. Taking into account that there are two electronic spin states, if each unit cell has an even number of valence electrons, it is possible to completely fill an energy band. The energy of the last filled states is called Fermi energy E_F. If the electrons completely fill a set of bands and there is a finite energy gap with respect to the next unfilled band, the crystal is an insulator. This is because an infinitesimally small external electric field cannot cause current to flow, as an additional energy of the order of the gap has to be supplied to introduce or remove electrons from the system. On the contrary, if there are partially filled bands, electrons can populate nearby momentum states if a small electric field is applied, causing current to flow. A partially filled band in a crystal, where E_F lies within an energy band, is a metal.

In metals, the Fermi surface is represented by electrons with crystalline momenta k whose energies equal E_F. The set of filled states with energies below E_F is called the Fermi sea. The name Fermi surface is adequate for 3D systems as the manifold defined by $E(k) = E_F$ is two-dimensional. However, the term is also used for one and two dimensions where the dimensionality of the Fermi surface is 0 (i.e. a set of points) and 1, respectively. Electrons with momenta near the Fermi surface can be easily excited to fill empty states above E_F leaving a void in the Fermi sea. Electrons can also be injected or removed with infinitesimally small energies above or below E_F. Energies corresponding to temperatures much smaller than the bandwidth of electronic excitations, near the Fermi surface, account for most of the physical properties of metals.

Interactions, may induce other ways to inhibit electrons from moving freely and induce insulating behavior. To distinguish this mechanism leading to an insulating state, these are called band insulators. Having an even number of valence electrons per unit cell is a necessary but insufficient condition to obtain an insulator as energy bands may overlap.

A sharp distinction between insulators and metals is only possible at zero temperature. For band insulators at finite temperatures, there is always a finite probability for an electron to be excited to the conduction band. This probability tends to unity if the thermal excitation energy becomes comparable to the energy gap. Thermally excited electrons can move around freely and carry current in the presence of an applied electric field. For bandgap below $\sim$4 eV, thermal activation produces conduction electrons at room temperature. Therefore, around and above room temperature, the insulating property is lost. Such materials are therefore called semiconductors.

1.4.6 Beyond the Band-Structure Paradigm

Based on the band-structure picture discussed in the previous section, it can be concluded that only materials with an even number of valence electrons per unit cell could be insulators. This is, however, not the case. The discrepancy between a simplified model and real materials arises because of neglecting the Coulomb repulsion between electrons and the interaction of electrons with lattice vibrations—the phonons—of the underlying ionic lattice. Interactions are responsible for a set of states of matter that cannot be described within the band-structure picture.

A good example of this is a magnetically ordered state such as a ferromagnet. Due to interactions in ferromagnetic materials, it is energetically favorable for spins to align rather than to minimize the kinetic energy by filling low-energy bands. Therefore, the band-filling rules mentioned above no longer apply. As the band-structure picture breaks, ferromagnetic materials with an odd number of electrons per site can be insulators. Mott insulators are another example. Here, due to strong electron repulsion, there is a large energy penalty for electrons to inhabit the same atom. The ground state is a collection of electrons occupying one atom each and the energy gap to accommodate an extra electron is of the order of the intra-atomic electron repulsion. In contrast are superconductors in which electron–electron interactions mediated by phonons become attractive.

1.5 Intrinsic and Extrinsic Semiconductors

Solid state materials can be classified as metals (with carrier concentrations as high as 10^{22} cm^{-3}), dielectrics (with carrier concentration as low as 10^3 cm^{-3}); the intermediate concentration values are characteristic of semimetals and semiconductors. In semiconductors, the band gap is of a moderate size (0.2–2 eV) allowing to excite carriers into a conduction band with a number of external influences, such as illumination, temperature, etc. At $T = 0$ K, pure semiconductor will behave as a dielectric.

A semiconductor with a negligibly low level of impurities is called an *intrinsic semiconductor*. Addition of impurities can significantly increase the conductivity of the semiconductor, producing *extrinsic* material. In both cases, an electron transition from the valence band generates a vacancy called a hole. This vacancy can be occupied by another electron, allowing transfer of positive charge without moving the nuclei of the atoms. As the number of holes generated in the valence band is less than the number of electrons originally occupying the valence band, it is natural to expect that the holes will have lower drift mobility; that is, they will move less efficiently by an applied electric field $\mathscr{E}$. Due to the difference in charge sign, the force experienced by an electron and a hole in a semiconductor, subjected to an electric field, will have different directions $F = \pm e\mathscr{E}$.

The movement of the carriers in semiconductors will be influenced by the periodic potential of the crystalline lattice, which may facilitate (or complicate) acceleration of the particle along certain directions. Using the relation between the acceleration and energy of the particle

$$\frac{dv}{dt} = \frac{1}{\hbar}\frac{d^2E}{dk^2}\frac{dk}{dt} \tag{1.23}$$

and expressing the force acting on the particle as $F = d(\hbar k)/dt$, one can see that inverse of the term containing the second derivative of energy has the units of a mass. This allows to introduce the parameter called carrier effective mass m^* in the following way

$$\frac{1}{m^*} = \frac{1}{\hbar^2}\frac{d^2E}{dk^2} \tag{1.24}$$

As the second derivative is related to the curvature of the energy band $E(k)$, the carriers populating the bands with wide extremes will have large effective mass values, while the carriers in bands characterized by narrow extremes will have lower effective mass values. The effective mass $m^* \neq m_0$ does not mean that the carrier's mass actually changes; it rather illustrates that the carrier will require a larger (or smaller) force to be accelerated in a periodic potential in comparison to a free carrier. The experimental definition of carrier effective mass can be performed by subjecting a semiconductor sample to a magnetic field B, causing carrier precession with a cyclotron frequency $\omega_c = eB/m^*$; detection of the resonance absorption allows to measure ω_c and calculate m^*.

1.5.1 Intrinsic Carrier Concentration

For the case of an intrinsic semiconductor, every electron promoted to the conduction band will produce a hole in the valence band, so that the concentration of both types of carries will be equal. To calculate the carrier concentration in thermodynamic equilibrium, it is necessary to integrate the product of the density of states $N_{C,V}$, and the Fermi–Dirac distribution function $n_F(E)$ over the corresponding band (as both electrons and holes are Fermions). The integration can be simplified taking into account that $n_F(E)$ tends to zero for high energies. This allows to change the upper integration limit to infinity, simplifying the integration

$$n_0 = \int_{E_C}^{\infty} N_C n_{Fn}(E)dE = 2\left(\frac{m_e^* k_B T}{2\pi\hbar^2}\right)^{3/2} \exp\left(\frac{\mu - E_C}{k_B T}\right), \tag{1.25}$$

$$p_0 = \int_{-\infty}^{E_V} N_V n_{Fp}(E)dE = 2\left(\frac{m_h^* k_B T}{2\pi\hbar^2}\right)^{3/2} \exp\left(\frac{E_V - \mu}{k_B T}\right) \tag{1.26}$$

where the density of states in the corresponding bands are given by

$$N_{C,V} = 2\left[\frac{m_{e,h}^* k_B T}{2\pi\hbar^2}\right]^{3/2} \tag{1.27}$$

and $m_{e,h}^*$ represent the effective mass of electron and hole, respectively.

For an intrinsic semiconductor, both concentrations coincide so that intrinsic concentrations $n_i = p_i$ and are related to equilibrium concentrations by the law of mass action

$$n_i^2 = n_0 p_0 = 4\left(\frac{k_B T}{2\pi\hbar^2}\sqrt{m_e^* m_h^*}\right)^3 \exp\left(\frac{-E_g}{k_B T}\right). \tag{1.28}$$

The values n_i and p_i can thus be found as

$$n_i = p_i = \sqrt{np} = 2\left(\frac{k_B T}{2\pi\hbar^2}\sqrt{m_e^* m_h^*}\right)^{3/2} \exp\left(\frac{-E_g}{2k_B T}\right). \tag{1.29}$$

The chemical potential for an intrinsic semiconductor can be expressed as

$$\mu = \frac{1}{2}E_g + \frac{3}{4}k_B T \ln\frac{m_h^*}{m_e^*}, \tag{1.30}$$

meaning that, at zero temperature, μ is located in the middle of the band gap. If $m_h^* = m_e^*$, the position of the chemical potential does not depend on temperature. When the aforementioned effective mass ratio is different from unity, the chemical potential will drift in a linear fashion toward the corresponding band with increase in temperature.

1.5.2 Donors and Acceptors

The conductivity of a semiconductor can be significantly increased by doping it with elements belonging to another group of the periodic table. Even small concentrations of impurities (fractions of percent) may change conductivity by several orders of magnitude. If the impurity atom has more valence electrons than those of the semiconductor matrix, it is called a "donor" because it will "donate" that extra electron to the conductivity band. If the impurity atom has fewer valence electrons in comparison to the base semiconductor material, it will "accept" an electron from the valence band and hence it is called "acceptor". Semiconductor doped with a donor impurity will have a larger number of conduction electrons (those provided by the impurity in addition to those generated by electron transitions from valence into conduction band) and thus it is referred as n-type material. Semiconductor doped with an acceptor impurity is a p-type material.

Taking into account that the motion of this electron takes place in the medium with dielectric constant ε, one can obtain the following equation for the radius of the first orbital as

$$a = \frac{\varepsilon\varepsilon_0 h^2}{\pi e^2 m_e^*} n^2. \quad (1.31)$$

Numerical estimation of a results in values exceeding the lattice constant of the base semiconductor by hundreds of times, meaning that even for a small impurity atom concentration, their external orbitals may overlap, enabling free electron to "jump" from one impurity atom to another. These overlapping orbitals produce a level in the band gap of the semiconductor, the energy of which (counted from the corresponding band for donor and acceptor impurities) can be found as

$$E_{D,A} = \frac{13.6\,\mathrm{eV}\, Z^2}{\varepsilon^2} \frac{m_{e,h}^*}{m_0}. \quad (1.32)$$

The donor levels are usually (but not universally) located closer to the conduction band, and acceptor levels—closer to the valence band. The energy difference between the impurity level and the corresponding band is also called ionization energy. It defines the amount of energy required to promote an electron/hole from the impurity level into the conduction/valence band. The levels are called shallow when the energy of thermal excitation $k_B T$ at the device operation temperature is sufficient to excite the carriers. If every impurity atom donates (or accepts) an electron, the impurity will become fully ionized. The concentration of the carriers, in this case, can be found as

$$N_D = N_C e^{-\frac{E_C - E_F}{k_B T}} \text{ and } N_A = N_V e^{-\frac{E_F - E_V}{k_B T}}. \quad (1.33)$$

In the case of mixed conductivity, when the semiconductor is doped with both donor and acceptor impurities, the charge balance equation will be as follows

$$n - p = N_D - N_A. \tag{1.34}$$

Taking into account that the carrier concentrations obey the mass action law $np = n_i^2$, one obtains two quadratic equations for n and p with solutions

$$n = \frac{N_D - N_A}{2} + \sqrt{\left(\frac{N_D - N_A}{2}\right)^2 + n_i^2}, \tag{1.35}$$

$$p = \frac{N_A - N_D}{2} + \sqrt{\left(\frac{N_A - N_D}{2}\right)^2 + n_i^2}. \tag{1.36}$$

1.5.3 Recombination Mechanisms

The excitation of carriers represents a nonequilibrium state, which will tend to thermodynamic equilibrium via recombination, the process inverse to carrier generation. For the case of direct recombination, an electron from the conduction band will return to an unoccupied position in the valence band, releasing the transition energy. This can be either radiated in the form of a light quantum or transferred to lattice vibrations. The intensity of recombination process is related to the intrinsic carrier concentration

$$R_D = \gamma_D(np - n_i^2). \tag{1.37}$$

The quantity γ_D is the direct recombination coefficient. The lifetime of nonequilibrium carriers under radiative recombination in an intrinsic semiconductor is

$$\tau_0 = \frac{1}{\gamma_D(n_0 + p_0 + \Delta n)}. \tag{1.38}$$

If the concentration of nonequilibrium carriers is negligible in comparison with those in equilibrium, the formula can be simplified as

$$\tau_0 = \frac{1}{\gamma_D(n_0 + p_0)}. \tag{1.39}$$

For a donor/acceptor semiconductor with totally ionized impurities, the carrier lifetimes under direct recombination will be

$$\tau_{0n} = \frac{1}{\gamma_D N_D}, \qquad \tau_{0p} = \frac{1}{\gamma_D N_A} \tag{1.40}$$

When recombining carriers possess an energy larger than E_g, the energy will be transmitted to a third particle in a process called Auger recombination. The particle that receives the excess energy will become "hot" and upon returning to thermal equilibrium, it will pass this energy to the crystal lattice. For an n-type semiconductor, Auger recombination will most possibly transfer the energy to another electron, resulting in recombination rate

$$R_{An} = A_n n(np - n_i^2). \tag{1.41}$$

For a p-type semiconductor, the most likely recipient of the excess energy of the carrier will be a hole, leading to

$$R_{Ap} = A_p p(np - n_i^2). \tag{1.42}$$

Here, A_n and A_p are the Auger recombination coefficients for electrons and holes. Average carrier lifetimes, in this case, can be expressed in the following way

$$\tau_{0n} = \frac{1}{A_n(n_i^2 + n_0^2 + p_0\Delta n + 2n_0\Delta n + (\Delta n)^2)}, \tag{1.43}$$

$$\tau_{0p} = \frac{1}{A_p(n_i^2 + p_0^2 + n_0\Delta p + 2p_0\Delta p + (\Delta p)^2)}. \tag{1.44}$$

For n- and p-type semiconductors, the equilibrium and nonequilibrium concentration of minority carriers are negligibly small in comparison with that of majority carriers, allowing for the following simplification

$$\tau_{0n} = \frac{1}{A_n N_A^2}, \qquad \tau_{0p} = \frac{1}{A_p N_D^2}. \tag{1.45}$$

For intrinsic semiconductor with equal concentration of equilibrium carriers, the Auger recombination is equal to

$$R_A = (A_n n + A_p p)(np - n_i^2). \tag{1.46}$$

If the semiconductor has impurities that are forming energy levels within its band gap, the carriers may use such levels for multistage process called Shockley-Read-Hall recombination. The levels can serve as "traps" for the carriers, reducing the energy threshold required for recombination. The most intense recombination occurs when the trap level is located close to the middle of the band gap because such position offers most efficient reduction in recombination energy. The general formula for the Shockley-Read-Hall recombination rate is

$$R_{HSR} = \frac{np - n_i^2}{\tau_{HSR}^p(n + n_1) + \tau_{HSR}^n(p + p_1)}, \tag{1.47}$$

which depends on carrier concentration in the corresponding band when its quasi-Fermi level matches the trap energy E_r

$$n_1 = N_C \exp\frac{E_C - E_r}{k_B T}, \qquad p_1 = N_V \exp\frac{E_r - E_V}{k_B T}. \tag{1.48}$$

The carrier lifetimes are defined with the concentration of traps $N_r = n_r + p_r$ as

$$\tau^n_{HSR} = \frac{1}{\gamma_n N_r}, \qquad \tau^p_{HSR} = \frac{1}{\gamma_p N_r}, \tag{1.49}$$

with recombination constants $\gamma_{n,p} = v_{n,p}\sigma_{n,p}$ including thermal velocity v and cross section of the trapping center σ. When the concentration of recombination centers is low, one can neglect Δn_r in favor of Δn and Δp, simplifying the formula

$$\tau_{HSR} = \tau^n_{HSR}\frac{p_0 + p_1 + \Delta p}{p_0 + n_0 + \Delta p} + \tau^p_{HSR}\frac{n_0 + n_1 + \Delta n}{p_0 + n_0 + \Delta n}. \tag{1.50}$$

If all three recombination processes take place in the material, the mean carrier lifetime can be calculated as

$$\frac{1}{\tau_i} = \frac{1}{\tau^i_0} + \frac{1}{\tau^i_R} + \frac{1}{\tau^i_{HSR}} + \frac{1}{\tau^i_A}, \qquad i = \{p, n\}. \tag{1.51}$$

Here, the quantities τ^n_0 and τ^p_0 correspond to intrinsic lifetimes of the carriers.

1.6 Transport Properties

Under thermal equilibrium, when a semiconductor is not subjected to external influences, the motion of carriers is chaotic. On the time scale characteristic for electronic device applications, these thermal fluctuations will average themselves out. In order to obtain an ordered flux of carriers trough the device, it is necessary to subject the sample to electric or magnetic fields, which will exert the Lorentz force on the carriers and cause them to move in a corresponding direction. Alternatively, it is possible to create temperature gradient or illuminate the sample with light with the frequency sufficient for photo-excitation of carriers. Such external influence will establish a gradient of carrier concentration through the sample, favoring diffusion of carriers. By combining different external influences, the drift/diffusion transport phenomena can be fine-tuned depending on the desired device application. In this section, we discuss the transport properties including electrical conductivity, galvano-magnetic effects (Hall), thermoelectric effects (Seebeck, Peltier, Thomson) and thermomagnetic (Nernst, Ettingshausen, Righi-Leduc) effects.

1.6.1 Electrical Conductivity

Under an applied electric field, the conduction electrons experience the force

$$F = -e\mathscr{E}. \tag{1.52}$$

The holes have a positive charge so that they will experience a force in the opposite direction. The ordered motion of electrons and holes produces the corresponding currents

$$J_n = \sigma_n \mathscr{E} = e\mu_n n \mathscr{E} \tag{1.53}$$
$$J_p = \sigma_p \mathscr{E} = e\mu_p p \mathscr{E}, \tag{1.54}$$

where the quantities $\sigma_{n,p}$ and $\mu_{n,p}$ correspond to conductivity and drift carrier mobility, respectively. The drift mobility is defined with carrier's mean relaxation time and effective mass

$$\mu_n = e < \tau > /m_e^*, \qquad \mu_p = e < \tau > /m_h^* \tag{1.55}$$

The carrier mobility depends on the scattering mechanisms taking place in a semiconductor. Thus, for phonon scattering, the carrier mobility will be

$$\mu = \frac{4e}{3\sqrt{\pi}} \frac{\tau_0}{\sqrt{k_B m^{*5}}} T^{-3/2}. \tag{1.56}$$

Under impurity scattering, the carrier mobility can be written as

$$\mu = \frac{8\sqrt{2}\varepsilon^2}{\pi^{3/2} Z^2 e^3 N_I m^{*1/2}} \frac{(k_B T)^{3/2}}{\ln\left[1 + \left(3\varepsilon k_B T / e^2 Z N_I^{1/3}\right)^2\right]}. \tag{1.57}$$

For the case of scattering over neutral impurities, the relaxation time is independent of temperature so that the mobility value can be found as

$$\mu = \frac{e^2 m^*}{20\varepsilon \hbar^3 N_A}. \tag{1.58}$$

In the event of dislocation scattering, the carrier mobility is

$$\mu = \frac{eT^{-1/2}}{\sqrt{8\pi k_B m^* R N_D}}, \tag{1.59}$$

with N_D representing the density of dislocation lines of radius R.

In the case of real crystals for which several scattering mechanisms manifest themselves at the same time, it is important to account for the temperature dependence of the scattering phenomena. In the low-temperature regime, the dominating scattering mechanism involves impurity ions, while at high temperature, the phonon scattering will dominate.

The conductivity of a compensated semiconductor includes contributions from both types of carriers

$$\sigma = e(n\mu_n + p\mu_p). \tag{1.60}$$

The mean drift velocity of the carriers under an applied electric field $\mathscr{E}$ can be expressed as

$$v_{d_{n,p}} = \mu_{n,p}\mathscr{E}. \tag{1.61}$$

1.6.2 Hall Effect

The combined action of an electric current and a magnetic field applied perpendicular to each other produces a transverse potential difference known as the Hall effect (Fig. 1.7). A charged particle moving in the magnetic field B precesses with a cyclotron frequency ω_c. Introducing the carrier mean path l as a distance covered along carrier's trajectory with radius r, one obtains

$$l/r = \tau\omega_c. \tag{1.62}$$

In the case of weak magnetic fields, the carrier with a small mean lifetime τ will be able to recur only a small part of the cyclotron orbit, which means that its motion will be more or less straightforward. For the case of strong fields and large τ, the curvature of carrier's track will be more pronounced. For the current J_x flowing along a sample and with a magnetic field B_z applied perpendicular to it, the carriers will accumulate at the side surfaces due to the action of the Lorentz force. As one can see from Fig. 1.7, both electrons and holes will experience the force moving them in the same direction, creating a potential difference U_H with the sign depending on conductivity type of the material. Partial redistribution of the charge towards the sides of the sample will create an electric field $\mathscr{E}_y$ compensating the action of the magnetic field $e\mathscr{E}_y = evB_z$, enabling the carriers to move undisturbed between the

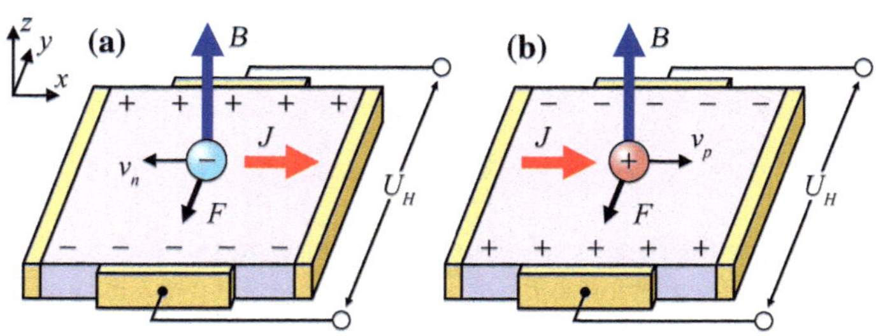

Fig. 1.7 Hall effect in **a** n-type and **b** p-type semiconductors. Note that due to the opposite direction of charge and velocity, both electrons and holes move toward the same side of the sample, leading to Hall potential difference U_H of different sign

terminals of the device. The Hall coefficient is defined as a ratio of perpendicular electric field to the product of electric current and magnetic field

$$R_H = \frac{\mathscr{E}_y}{J_x B_z}. \tag{1.63}$$

For a semiconductor with one type of carrier, the value of Hall coefficient is related to their concentration

$$R_{Hn} = -1/en \qquad \text{and} \qquad R_{Hp} = 1/ep, \tag{1.64}$$

providing a useful tool for direct measurement of carrier concentrations. The sign of the Hall coefficient indicates the conductivity of the sample ($n-$ or $p-$type). For the mixed conduction case, the expression for Hall coefficient R_H will be

$$R_H = \frac{R_{Hn}\sigma_n^2 + R_{Hp}\sigma_p^2}{(\sigma_n + \sigma_p)^2}. \tag{1.65}$$

As conductivity σ depends on carrier mobility μ, which in turn is governed by different scattering mechanisms at high and low temperatures, the Hall coefficient will also change nonlinearly with temperature. The temperature at which R_H changes its sign is called the inversion temperature.

The classical setup for measuring Hall effect requires a sample to be shaped as a rectangular slab. However, in real life many of the samples are grown as thin films over substrates, so that they may have rather complicated geometry. To conduct the measurements in such cases, the four-probe method proposed by van der Pauw is used. The contacts 1 and 2 serve for measurement of voltage difference; the current flowing through the sample is measured with the contacts 3 and 4. The magnetic field B is applied perpendicular to the sample plane. The formula for Hall coefficient, in this case, will be

$$R_H = \frac{V_{12}(B) - V_{12}(-B)}{2J_{34}B} d, \tag{1.66}$$

where d is the thickness of the sample. The van der Pauw method has an additional advantage, because the same four-probe arrangement can be used for measurement of resistance and resistivity of a thin-film sample.

1.6.3 Thermoelectric and Thermomagnetic Effects

The studies of temperature dependence of transport phenomena started with the discovery of the Seebeck effect. It consists of the development of a potential difference in a system of two conductors, the contacts between which are kept at different temperatures. The variation of potential difference ΔV with temperature difference ΔT is characterized by the Seebeck coefficient

$$S = \Delta V / \Delta T. \tag{1.67}$$

To increase the efficiency of a device operating under Seebeck effect, it is preferable to select materials with high values of S and high conductivity σ, while the thermal conductivity coefficient of the material should be low. The Seebeck coefficient for metals is about 10 μV/K, whereas for semiconductors, it may reach values of 100–1000 μV/K. Additionally, doping the material with a proper impurity may help to fine-tune the effect. For n-type material, the Seebeck coefficient is negative, while for p-type materials, it is positive. This means that in an n-type semiconductor, the electrons will move away from the hot contact, participating in heat transfer and creating a positive charge at the side of the hot contact. In a p-type semiconductor, to the contrary, a positive charge will be created at the cold contact side.

In the case of mixed conduction, the partial Seebeck coefficients S_n and S_p contribute to the total coefficient S as

$$S = \frac{S_n \sigma_n + S_p \sigma_p}{\sigma_n + \sigma_p}, \tag{1.68}$$

so that the sign of S will provide useful information concerning the type of majority carriers. In an intrinsic semiconductor, the Seebeck coefficient depends only on the ratio of carrier mobilities and the band gap E_g

$$S = -\frac{k_B}{e} \frac{\mu_n/\mu_p - 1}{\mu_n/\mu_p + 1} \left(2 + \frac{E_g}{2k_B T} \right). \tag{1.69}$$

An inverse effect can be observed by passing a current through two conducting materials. Depending on the choice of materials and current polarity, the heat can be generated or absorbed at the contacts. This effect is characterized by Peltier coefficient

$$\Pi = \Delta Q / J, \tag{1.70}$$

where ΔQ corresponds to the temperature variation per unit time achieved under passing current J. The Peltier effect is used for thermoelectric cooling. Despite its low efficiency, it offers a plausible solution in a form of full solid-state device without any moving parts. Additionally, it allows for providing targeted micro-cooling by creating Peltier contacts on top of electric circuitry parts that heat in excess. The Seebeck and Peltier coefficients are interdependent, as was shown by Thomson

$$\Pi = ST. \tag{1.71}$$

During his studies of the thermoelectric phenomena, Thomson predicted yet another effect that bears his name. Due to the variation of the Seebeck coefficient with temperature gradient in a solid-state object, with different temperatures at its extremities, the carriers passing through a temperature gradient will gain (or lose) energy depending on the direction of the current. This gives rise to "continuous Peltier effect" that can be described by Thomson coefficient

$$\tau = J^{-1}\frac{\partial Q/\partial x}{\partial T/\partial x}. \tag{1.72}$$

The Thomson coefficient depends on Peltier and Seebeck coefficients as

$$\tau = \frac{d\Pi}{dT} - S = T\frac{dS}{dT}. \tag{1.73}$$

The application of a magnetic field to a semiconductor sample, subjected to a temperature gradient, produces Nernst-Ettingshausen effect: when the temperature gradient is applied along the x-axis and the magnetic field is directed perpendicular to it along z-axis, the motion of the carriers due to the temperature gradient will be influenced by Lorentz force, resulting in accumulation of charge at the edges of the sample. This will produce a transverse electric field $\mathscr{E}_y$ canceling the action of the magnetic field. The qualitative measure of this effect is introduced by Nernst coefficient

$$Q_n = \frac{\mathscr{E}_y}{B_z(\partial T/\partial x)}. \tag{1.74}$$

In a semiconductor with two types of carriers, the total Nernst coefficient of the system will include contributions from electrons and holes

$$Q = \frac{\left[(Q_n\sigma_n + Q_p\sigma_p)(\sigma_n + \sigma_p) + \sigma_n\sigma_p(S_n - S_p)(R_{Hn}\sigma_n - R_{Hp}\sigma_p)\right]}{(\sigma_n + \sigma_p)^2}. \tag{1.75}$$

If an electric current J_x is running through the sample in the presence of a transverse magnetic field B_z, the carriers arriving at the sides of the sample will have different temperatures so that a temperature gradient $\partial T/\partial y$ will be generated across the sample. The intensity of this effect is characterized by the Ettingshausen coefficient

$$P_E = (J_x B_z)^{-1}\frac{\partial T}{\partial y}. \tag{1.76}$$

The relation between Nernst and Ettingshausen coefficients is the following

$$P_E K_n = Q_n T, \tag{1.77}$$

where the value of K_n corresponds to the thermal conductivity of the material. K_n is proportional to the temperature gradient and depends on concentration and average velocity of carriers

$$K_n = -\frac{nv\mathscr{E}}{\mathrm{grad}T}. \tag{1.78}$$

The thermally induced variation of Hall effect involves carrier motion assisted by longitudinal temperature gradient $\partial T/\partial x$ under the action of a perpendicular magnetic field B_z. In place of a transverse electric field, the system produces a transverse temperature gradient $\partial T/\partial y$ to compensate for the action of the magnetic field and achieve steady carrier transport along the sample. This phenomenon is known as Righi-Leduc effect. The strength of temperature gradient is characterized by the coefficient

$$R_L = \frac{\partial T/\partial y}{B_z \partial T/\partial x}. \tag{1.79}$$

Due to the relatively low values of the electron thermal conductivity in lightly doped semiconductors compared to their lattice thermal conductivity, the magnitude of Righi-Leduc effect is much smaller than that of the Hall effect. In highly-doped samples, it is possible to increase the electron thermal conductivity, but, as carrier mobility decreases, the magnitude of Righi-Leduc effect will nevertheless remain quite small.

1.7 Optical Properties

The light wave incident on a semiconductor material can be reflected, transmitted, absorbed, and scattered inside the crystal. In the majority of the cases, the wavelength of light used for studies of optical properties considerably exceeds the lattice constant of the material. The light propagating inside a semiconductor medium can cause photoluminescence—emission of radiation with a frequency different from that of the incident source. Light scattering may proceed with participation of acoustic phonons (Brillouin scattering) and optical phonons (Raman scattering). The scattering phenomena are usually weak and harder to measure in comparison with the basic optical effects such as reflection, absorption and transmission.

If the energy $\hbar\omega$ provided by the light is equal to or greater than the semiconductor band gap, it will excite carriers from the valence band into the conduction band, creating nonequilibrium electron-hole pairs in a direct-band gap semiconductor. For semiconductors with indirect band gap, the absorption process will be mediated by phonon interaction. The experimental measurement of absorption coefficient, as a function of $\hbar\omega$, provides a useful tool for calculation of the semiconductor band gap and the energy of phonons participating in indirect transitions. The photogeneration of nonequilibrium carriers leads to diverse phenomena including photoconductivity and photovoltaic effects.

1.7.1 Optical Transmission and Absorption

In order to describe the propagation of light through a solid sample, in addition to the simple solution to the wave equation, it is necessary to account for the term including the electrical conductivity σ

$$\nabla^2 E = \mu_0 \varepsilon \varepsilon_0 \frac{\partial^2 E}{\partial t^2} + \mu_0 \sigma \frac{\partial E}{\partial t}. \tag{1.80}$$

The wave in the medium will be characterized by a complex wave vector k^*, which is related to a complex refractive index

$$n^* = \sqrt{\varepsilon + i\frac{\sigma}{\omega \varepsilon_0}} = \sqrt{\varepsilon^*}, \tag{1.81}$$

and the complex dielectric constant ε^*. The refractive index can be expressed by introducing the extinction coefficient k_e

$$n^* = n + ik_e. \tag{1.82}$$

In experiments, it is much easier to measure the absorption coefficient α corresponding to a rate at which the light intensity diminishes with the distance x measured from the frontal illuminated surface of the sample

$$I(x) = I_0 e^{-\alpha x}. \tag{1.83}$$

The relationship between the absorption and extinction coefficients is as follows

$$\alpha = \frac{4\pi k_e}{\lambda_0}, \tag{1.84}$$

where the parameter λ_0 represents the wavelength of light in vacuum. For energies exceeding the semiconductor band gap, the absorption coefficient value is on the order of thousands per inverse millimeter, meaning that only thin samples will have a noticeable transparency.

If the electron-hole pair, produced under photo-excitation, maintains its interaction and moves in a coordinated way, it can be considered as a dual particle called exciton. The intensity and range of interaction differ between weakly bound Wannier–Mott excitons and strongly bound Frenkel excitons. The generation of excitons is detectable via characteristic peaks that are formed at the absorption edge in semiconductors with indirect band gap; this effect is more pronounced at low temperatures.

Depending on the angle of incidence, a fraction of light beam will be reflected from the sample surface. The intensity of reflection will depend on the quality of the surface preparation. In studies of transmission phenomena, Snell's law

$$\frac{\sin\theta_1}{\sin\theta_2} = \frac{n_2}{n_1} = \frac{\lambda_1}{\lambda_2} \tag{1.85}$$

defines the relation between the angles of incident θ_1 and transmitted beams θ_2, as well as wavelengths $\lambda_{1,2}$ and refraction constants $n_{1,2}$ of both media. For an ideal material surface under normal light incidence, the transmission and reflection coefficients are defined as

$$T = \frac{4n}{(n+1)^2 + k_e^2} \tag{1.86}$$

and

$$R = \left| \frac{(n-1)^2 + k_e^2}{(n+1)^2 + k_e^2} \right|. \tag{1.87}$$

Knowing the reflectance spectra over a wide range of wavelengths (measured under normal light incidence) allows the calculation of absorption spectra using Kramers–Kronig relations.

The most pronounced absorption takes place under illumination with the light quanta of energy that is sufficient for performing electron transition from the valence band into conduction band. If the maximum of the valence band is aligned with the minimum of the conduction band, carriers can be excited via direct transition with a constant wave vector (Fig. 1.8a), resulting in the absorption coefficient

$$\alpha_{DT} = K_{DT}\sqrt{\hbar\omega - E_g}. \tag{1.88}$$

By plotting the absorption coefficient squared versus light wave frequency and extrapolating the resulting dependence till its intersection with the abscissa axis, one can find the optical band gap of semiconductor E_g.

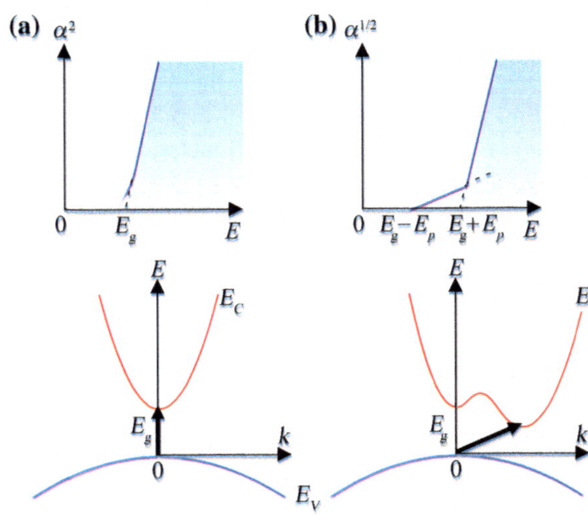

Fig. 1.8 Direct **a** and indirect **b** optical absorption

Under indirect transitions, in addition to exciting a carrier from valence to conduction band, it is also necessary to change its wave vector with the participation of a third particle (Fig. 1.8b). Usually, it is a phonon with momentum κ, so that

$$\mathbf{k}' = \mathbf{k} \pm \kappa. \tag{1.89}$$

The signs of the phonon wave vector correspond to emission or absorption of the phonon during the process of interaction. Both the phenomena will lead to different formulations of the absorption coefficient

$$\alpha_{IPA} = K_{IPA} \frac{(\hbar\omega - E_g + E_p)^2}{exp(E_p/k_BT) - 1}, \tag{1.90}$$

$$\alpha_{IPE} = K_{IPE} \frac{(\hbar\omega - E_g - E_p)^2}{1 - exp(-E_p/k_BT)}, \tag{1.91}$$

with α_{IPA} providing the contribution to the indirect absorption coefficient for the case of phonon absorption and α_{IPE} corresponding to the process when a phonon is emitted. The photon/phonon energy is denoted as $\hbar\omega$ and E_p, respectively. As phonon energy and bang gap of the semiconductor are squared in these formulae, they will be responsible for straight-line segments in a $\sqrt{\alpha}$ plot. The extrapolation of these segments till intersection with the abscissa axis provides the way to find the values of band gap and the energy of the phonon that mediates the indirect transition (Fig. 1.8b).

1.7.2 Photoconductivity and Photovoltaic Effects

In a non illuminated semiconductor, the conductivity depends on carrier concentration and mobility

$$\sigma = e(n_0\mu_n + p_0\mu_p). \tag{1.92}$$

The photo-generated carriers, Δn and Δp, contribute to nonequilibrium carrier concentrations $n = n_0 + \Delta n$ and $p = p_0 + \Delta p$, which modify the conductivity by

$$\Delta\sigma = e(\Delta n\mu_n + \Delta p\mu_p). \tag{1.93}$$

The increase in carrier concentration is related to the photogeneration rate and average carrier lifetime. For electrons, it is

$$\Delta n = g(x)\tau_n = \alpha I_0 \tau_n e^{-\alpha x}. \tag{1.94}$$

As $\Delta n \ll n_0$ and $\Delta p \ll p_0$, it means that the increase in carrier concentration is more pronounced for the minority carriers. Depending on the energy of light quanta, it is possible to achieve either complete transition of electrons from the valence to conduction band (resulting in intrinsic photoconductivity) or from impurity levels to the conduction band by absorption of low-energy light quanta (extrinsic photoconductivity). As the energy difference between the impurity levels and the corresponding band is considerably smaller than the band gap, the observation of the extrinsic photoconductivity may be complicated due to the influence of thermal excitation as a competitive carrier generation mechanism. Thus, the clearest observation of extrinsic photoconductivity can be made at very low temperatures. The increment in material conductivity for extrinsic photoconductivity is achieved by carriers that are excited from donor and acceptor levels, Δn_D and Δp_A

$$\Delta\sigma_n = e\Delta n_D \mu_n \tag{1.95}$$

$$\Delta\sigma_p = e\Delta p_A \mu_p \tag{1.96}$$

The excess carriers will take part in current transport phenomena upon application of a potential difference U between the terminals attached to the opposite ends of a semiconductor sample. In the case of a thin sample with thickness $d \ll 1/\alpha$, it is possible to assume a constant generation rate $g(x) \equiv g_0$. Thus, we can estimate the total concentration of generated carriers as the product of the generation rate and the volume of the sample $G_E = g_0 V_0$. If the distance between the contacts applied to the sides of the sample is L, the total photocurrent through the device is

$$I_{ph} = \frac{eUG_E}{L^2}(\mu_n\tau_n + \mu_p\tau_p) = eUG_E S, \tag{1.97}$$

with the photosensitivity factor S depending on mobility of the carriers and their lifetime, $S = \mu\tau/L^2$. As photocurrent is proportional to S, more intense photoconductivity effect will be observed for materials that are characterized with both large carrier mobility and average lifetime. As was mentioned in the section describing recombination mechanisms, inverse carrier lifetime can be calculated as the sum of inverse lifetimes characteristic to direct, Auger and Shockley-Read-Hall recombination mechanisms. For thin samples, surface effects may play a considerable role, so that surface recombination should be accounted for in the calculation of carrier lifetime. To diminish the carrier loss due to recombination process, it is preferable to reduce the distance between the terminals L.

For thick samples, the coordinate dependence of the generation rate $g(x)$ cannot be neglected, leading to a more complicated formula for I_{ph}. One of the principal uses of photoconductivity effect includes optical sensors, for which the band gap energy and position of impurity levels define the desired window of sensitivity.

1.7.3 Photovoltaic Effect

In the case of a thick sample illuminated from one of its sides, it is necessary to account for the variation of nonequilibrium carrier generation $g(x)$ with coordinate

$$g(x) = -\frac{dI}{dx} = \alpha I_0 e^{-\alpha x}, \tag{1.98}$$

Here, I_0 corresponds to the intensity of the incident light and α is the absorption coefficient of the semiconductor. Due to different generation rate at different depth, the gradient of carrier concentration will be in the sample, which will lead to diffusion currents aiming to restore equilibrium. The diffusion coefficients for electrons D_n and holes D_p are defined with Einstein relations

$$D_{n,p} = \frac{k_B T}{e} \mu_{n,p}. \tag{1.99}$$

The electron and hole currents including drift mobility and diffusion terms are

$$J_n = en\mu_n \mathcal{E} + eD_n \frac{dn}{dx}, \tag{1.100}$$

$$J_p = ep\mu_p \mathcal{E} - eD_p \frac{dp}{dx}. \tag{1.101}$$

The concentration gradient will vanish for equal carrier mobilities. However, as electron mobility is usually higher than that of the holes, it will favor the presence of a concentration gradient. Introducing the ratio of mobilities $b = \mu_n/\mu_p$, one can obtain the expression for the total current density as

$$J = e(bn + p)\mu_p \mathcal{E}_D + (b-1)eD_p \frac{d\Delta n}{dx}, \tag{1.102}$$

with the first term representing carrier drift under the action of Dember field $\mathcal{E}_D$ established due to a carrier concentration gradient. The second term describes the diffusion under the nonequilibrium carrier concentration gradient $\Delta n = n - n_0$. For open circuit conditions ($J = 0$), the Dember field can be expressed as

$$\mathcal{E}_D = -\frac{b-1}{bn+p} \frac{k_B T}{e} \frac{d\Delta n}{dx}. \tag{1.103}$$

In the case when the carrier mobilities are equal, $\mu_p = \mu_n \rightarrow b = 1$, Dember field vanishes. The potential difference (Dember voltage) associated with $\mathcal{E}_D$ can be obtained by integrating the anterior formula

$$V_D = \int_0^L -\mathcal{E}_D dx = \frac{b-1}{bn_0 + p_0} \frac{k_B T}{e} \left[\Delta n(0) - \Delta n(L)\right], \tag{1.104}$$

where $\Delta n(0) - \Delta n(L)$ represents the difference in nonequilibrium carrier concentration at the frontal ($x = 0$) and rear ($x = L$) surfaces of the device. For a thick sample, these quantities can be expressed by light absorption law, diffusion lengths of the carriers, surface recombination rates and average carrier lifetime.

The resulting formula

$$V_D = \frac{b-1}{bn_0 + p_0} \frac{k_B T}{e} \frac{\alpha L_n I_0 \tau_n}{\hbar\omega(\alpha L_n + 1)(L_n + s\tau_n)} \tag{1.105}$$

shows that Dember voltage does not depend on any applied field (in contrast to photoconductivity effect). This means that the illumination of a thick sample with a suitable light can be used for the generation of electric current. The photovoltaic effect is the principle of operation of solar cells.

To enhance the lifetime of photo-generated electron-hole pairs, it is important to reduce the recombination probability by separating the carriers, which can be accomplished by an electric field. One of the possible ways to achieve this is by creating an electric field inside the semiconductor, by making a junction between p- and n-type materials. Upon contact, the electrons will drift from the n-type semiconductor into the p-type semiconductor that has lower concentration of electrons; in the vicinity of the boundary, the holes will move in the opposite direction. Such redistribution of carriers, caused by diffusion process, will cause depletion of the regions adjacent to the contact, forming a space-charge region characterized by diffusion potential difference

$$U_d = e(2\varepsilon\varepsilon_0)^{-1}(N_D w_n^2 + N_A w_p^2). \tag{1.106}$$

The efficiency of carrier separation can be fine-tuned by changing the depletion region parameters, such as penetration depth of space-charge region into p- and n-type material, w_p and w_n.

$$w_n = \sqrt{\frac{2\varepsilon\varepsilon_0}{eN_D} \frac{N_A U_d}{N_A + N_D}} \tag{1.107}$$

$$w_p = \sqrt{\frac{2\varepsilon\varepsilon_0}{eN_A} \frac{N_D U_d}{N_A + N_D}}. \tag{1.108}$$

As one can see, these quantities are related to each other by

$$w_p N_A = w_n N_D. \tag{1.109}$$

The thickness of p- and n- layers can be adjusted to position the metallurgical boundary at the most appropriate depth. As photo-generation rate decays exponentially from the illuminated surface, it is reasonable to move the junction closer to the front side so that a larger number of photo-generated carriers will be influenced by the electric field of the junction reducing their recombination.

The carrier concentration gradient established through the device will produce diffusion currents transporting the carriers towards the contacts at which they can be collected. The carrier diffusion equations can be written as:

$$\frac{d^2n}{dx^2} = \frac{n - n_0}{L_n^2} - \frac{I_0\alpha}{D_n}e^{-\alpha x} \tag{1.110}$$

$$\frac{d^2p}{dx^2} = \frac{p - p_0}{L_p^2} - \frac{I_0\alpha}{D_p}e^{-\alpha x}, \tag{1.111}$$

The maximum conversion efficiency can be calculated depending on the band gap of the semiconductor. The total generation current produced by a device with thickness L can be found by integrating the generation rate $g(x)$ as

$$e\int_0^L g(x)dx = e\alpha I_0 \int_0^L e^{-\alpha x}dx = eI_0(1 - e^{-\alpha L}). \tag{1.112}$$

However, these idealized characteristics are hard to achieve; the current loss may occur due to the bulk and surface recombination, non-ideality of the junction, dislocations, and non-ohmicity of the contacts.

References

1. Ashcroft NW, Mermin ND (1976) Solid state physics. Holt, Rinehart and Winston, Brooks Cole, New York
2. Gasiorowicz S (2003) Quantum physics. Wiley, Hoboken
3. Eisberg R, Resnick R (1985) Quantum physics of atoms, molecules, solids, nuclei, and particles. Wiley, Hoboken
4. Fox M (2010) Optical properties of solids. Oxford University Press, Oxford
5. Kittel C (2005) Introduction to solid state physics. Wiley, Hoboken
6. Li S (2006) Semiconductor physical electronics. Springer, New York
7. Miller D (2008) Quantum mechanics for scientists and engineers. Cambridge University Press, Cambridge
8. Schäfer W, Wegener M (2002) Semiconductor optics and transport phenomena. Springer, Berlin
9. Shalimova KV (1985) Physics of semiconductors. Energoatomizdat, Moscow
10. Singh J (2007) Electronic and optoelectronic properties of semiconductor structures. Cambridge University Press, Cambridge
11. Sze SM, Ng KK (2007) Physics of semiconductor devices. Wiley, Hoboken
12. Yu P, Cardona M (2010) Fundamentals of semiconductors: physics and materials properties. Springer, Berlin
13. Ziman JM (1979) Principles of the theory of solids. Cambridge University Press, Cambridge
14. Ziman JM (1995) Elements of advanced quantum theory. Cambridge University Press, Cambridge

Chapter 2
Processing Techniques

Barbara Cortese, Luciano Velardi, Ilaria Elena Palamà, Stefania D'Amone, Eliana D'Amone, Gianvito de Iaco, Diego Mangiullo and Giuseppe Gigli

The semiconductor industry is nowadays under constant pressure to produce devices which are cheaper, smaller, more powerful, and efficient. Moreover, current advances in the production of thin layers have made a whole new range of devices manufacturable. Thin films on substrates are generally prepared using bulk growth methods or physical vapor deposition (PVD) and chemical vapor deposition (CVD). Growth and processing techniques of materials including semiconductors are reviewed in this chapter.

2.1 Bulk—Crystal Growth

Bulk crystal growth techniques are highly attractive for producing semiconductors with the highest purity as well with tunable electrical properties. Bulk crystal growth

B. Cortese (✉)
Nanotechnology Institute, CNR-NANOTEC, University La Sapienza, P. zle Aldo Moro, Rome, Italy
e-mail: barbara.cortese@nanotec.cnr.it

L. Velardi
P.LAS.M.I Lab@Bari, Nanotechnology Institute, CNR-NANOTEC, Via Amendola 122/D 70,126, Bari, Italy

I. E. Palamà · S. D'Amone · E. D'Amone · G. de Iaco · D. Mangiullo · G. Gigli
Nanotechnology Institute, CNR-NANOTEC, Via per Monteroni, 73100 Lecce, Italy

G. Gigli
Department Matematica e Fisica 'Ennio De Giorgi', University of Salento, Via per Monteroni, 73100 Lecce, Italy

G. Gigli
Italian Institute of Technology (IIT)—Center for Biomolecular Nanotechnologies, via Barsanti, Arnesano, Italy

M. I. Pech-Canul and N. M. Ravindra (eds.), *Semiconductors*,
https://doi.org/10.1007/978-3-030-02171-9_2

Table 2.1 Classification of bulk crystal growth techniques

Phase transformation	Bulk crystal growth	Growth method	Materials
Growth from liquid Liquid→Solid phase transformation	Melt Growth	Czochralski	Silicon, III–V Compounds, Germanium, Garnets, Lithium niobate
		Float-zone	Silicon
		Bridgman	Metals, III–V Compounds, II-VI Compounds, Alkali halides, Germanium
		Verneuil	Ruby
	Solution Growth	Low-temperature solution	Phosphates, Triglycine sulfate (TGS)
		High-temperature solution	Diamond
		Hydrothermal	Quartz, Ruby, ZnS, Calcite crystals

is defined as the process of "arranging atoms, ions, molecules, or molecular assemblies into regular three-dimensional periodic arrays" [1, 2]. Defects (i.e., vacancies, interstitial impurity atoms, etc.), which can generate during crystal growth, can greatly affect the structural, chemical, electronic, and scattering properties of single crystals. This chapter primarily focuses on bulk growth methods which can be roughly divided into two major groups: melt growth and growth from solution, as summarized in Table 2.1.

2.1.1 Melt Growth

Melt growth involves a controlled phase change from liquid to solid at the melting point or solidus of the material and is used to obtain elemental semiconductors and metals, oxides, halides, chalcogenides, etc., at relatively high growth rates [2]. Alloy semiconductors are often more difficult to grow from a melt due to segregation and buoyancy, that are thermally or compositionally driven. Various methods have been developed and each method of growth from the melt has its advantages and disadvantages.

2.1.1.1 The Czochralski Technique

Approximately, 95% of the monocrystalline silicon manufactured today is obtained using the Czochralski (CZ) growth technique. This method was named after its

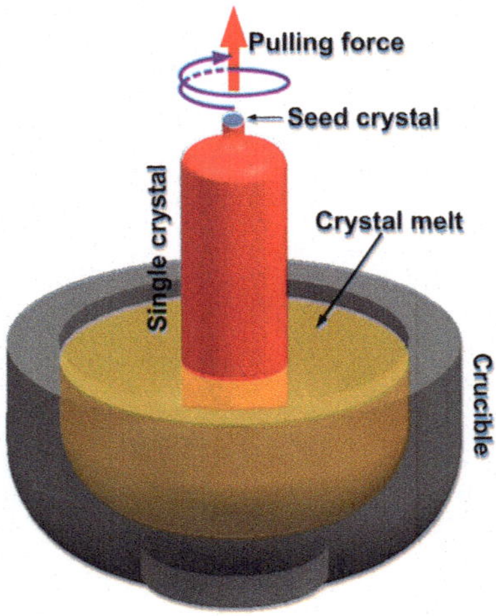

Fig. 2.1 Schematic of a conventional Czochralski growth process of a single crystal. Adapted from [4]

inventor, J. Czochralski in 1917 [3]. The conventional growth technique and apparatus is shown in Fig. 2.1.

The CZ growth apparatus consists of three main components: a crucible, a heater, and a pulling rod positioned axially above the crucible. The growth process begins by melting the required material in a crucible. A seed crystal attached to the end of the pulling rod is dipped into the melt, and then slowly lifted from the melt-free surface while simultaneously being rotated. The melt temperature and pulling speed are adjusted to form a small meniscus at the end of the seed in order to eliminate dislocations that are generated by the seed/melt contact shock and to reduce the effects of thermal asymmetry on the crystal. The lifting rates (from a few tenths of a millimeter per hour to tens of centimeters per hour) differ among materials and the grown crystal size. In some cases, the crucible is also rotated to allow for a better melt homogeneity. The main advantage of the Czochralski technique is the absence of a crystal container. Also, the lack of contact between the grown crystal rod and the crucible during the cooling process, after the growth, reduces the thermal stresses (due to different thermal expansion rates of the crystal and the crucible). This is significant in the production of dislocation-free crystals because of the lack of stresses that are normally generated by the mismatch in the thermal expansion coefficients of the container and the crystal.

2.1.1.2 Float-Zone Crystal Growth

Float-zone crystal growth (also referred to as floating zone) has been employed to obtain crystals with the highest purity and minimum cracks. Cracks are caused by thermal strain induced during cooling which is strongly decreased due to the fact that the molten material does not touch any container surface. During this process, only a small fraction of the feed rod, molten Al, is heated above its melting point, forming

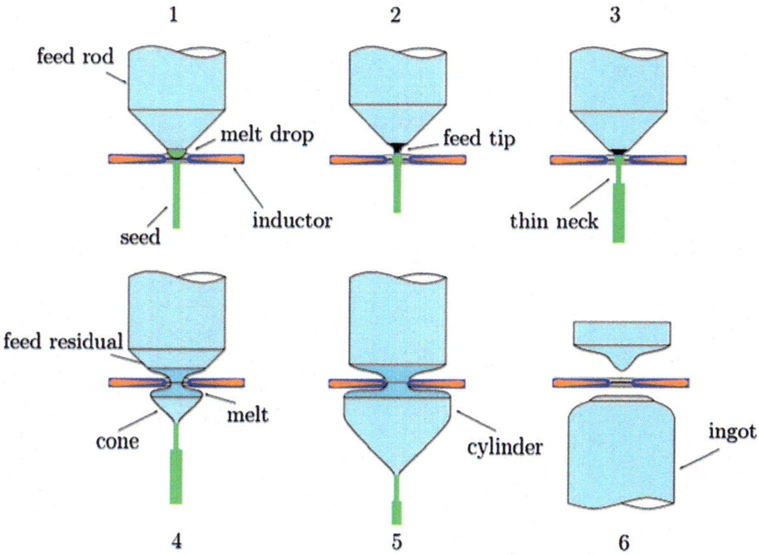

Fig. 2.2 Stages of the float zone process. Generation of the melt drop (1), formation of the feed tip (2), formation of the neck (3), development of the cone (4), growth of the cylinder (5) and closing the crystal (6). Solid material in white, melt in gray

a free molten zone (the float-zone or the liquid bridge). The melt is resolidified into a single crystal by pulling at the liquid–solid interface (the crystallization front) of the grown crystal at a very slow rate. As the melting point of the material typically ranges from 500 °C to over 1000 °C, this process is commonly conducted in an inert gas environment (i.e., argon) to avoid unwanted oxidation from the atmosphere.

A variety of zone melting techniques have been developed over the years. Four basic designs are used: horizontal boat, enclosed vertical, temperature gradient zone melting, and vertical floating zone [5, 6]. However, methods of heating also differ: arc image halogen lamp, arc discharge, electron beam, and laser heating [7]. The heating method most frequently used is radio frequency (RF) induced heating with specially shaped external RF coils.

The float zone process is conducted in six steps: (1) generation of the melt drop, (2) formation of the feed tip, (3) creating the neck, (4) forming the cone, (5) growth of the cylinder, and (6) closing the crystal, as shown in Fig. 2.2.

The seed and the feed rod are attached to vertical spindles, which allow vertical and rotary movement and horizontal displacement. The monocrystalline seed is slowly dipped into the melt drop. After melting together, the feed rod and the seed are pulled upwards. Crystal growth starts with a thin neck which occurs as the seed and feed rod are moved downwards through the inductor slit hole with increased pull rates. When the thin neck has a sufficient length of 30–50 mm, the cone starts to form. The feed rod and crystal are rotated in opposition in order to maintain a steady melt flow and reduce thermal inhomogeneities, resulting in a circular crystal. The feed rod descends and the molten phase flows through the inductor hole down along the solid feed residual into the melt. The diameter of the cone is enlarged until the desired final dimension of the crystal is reached. During the growth of the crystal in the shape of a cylinder, the crystal diameter is held constant by maintaining constant power, rotation and pull rates of feed rod and crystal. The completion of the crystal phase is determined by moving the feed rod upwards until it separates from the molten zone while the heater power is slowly decreased to control the crystallization of the melt. This step is done meticulously to avoid melt volume expansion while crystallizing and the consequent breakage of the crystal and introduction of dislocations by plastic deformation. The cylindrical crystal with constant diameter is called ingot which is sliced into wafers to be used as substrates for high-power electronic components.

This crystal growth technique is simple and suitable for uniform doping. Although it appears as a superior growth technique, disadvantages are the difficulties in maintaining a stable molten zone, limited dimension of crystal diameter, expensive, and mechanically sophisticated machine which requires skill and experience to be used. Another drawback of this technique is the inadequateness to grow alloys or compound semiconductors such as GaAs and GaSb due to volatilization of the higher vapor pressure elements.

2.1.1.3 The Bridgman–Stockbarger Technique

The Bridgman–Stockbarger method growth technique also referred to as the unidirectional solidification method or vertical gradient freezing (VGF) is used for single element growth and purification and for binary (or ternary) semiconductor crystal growth of two (or three) group III/V elements. The charge materials are loaded into a crucible, melted, and resolidified by the translation of a positive temperature gradient across the sample. The gradient translation is obtained either by moving the crucible, as shown in Fig. 2.3, from the hot zone of the furnace towards the cold zone or by moving the furnace across the immobile crucible.

Another alternative to obtain the gradient translation is by reducing the furnace temperature keeping both crucible and furnace stationary, referred to as the gradient freeze method. This method requires a multi-zone furnace in order to translate the temperature profile in a controlled and steady manner. The orientation of the crystal depends on the orientation of the seed crystal at the cold end of the crucible. At the beginning of the melting of the charge material in the crucible, a small part of the seed is also melted to improve adhesion between the seed and the melt. Alternatively, the growth of the crystal is initiated in a narrowly confined region at the cold end of the crucible. If the confined region is sufficiently long, a single orientation is obtained due to the preferential growth in certain directions. This technique is used for low melting temperature materials (CaF_2, NaCl, AgCl, $CaWO_4$), and high melting temperature ferroelectrics, such as tungsten bronze and perovskite, and can be promptly scaled up in size.

This melt growth process is carried out in either vertical or horizontal configuration [9]. The main advantage of using the vertical Bridgman technique with respect to

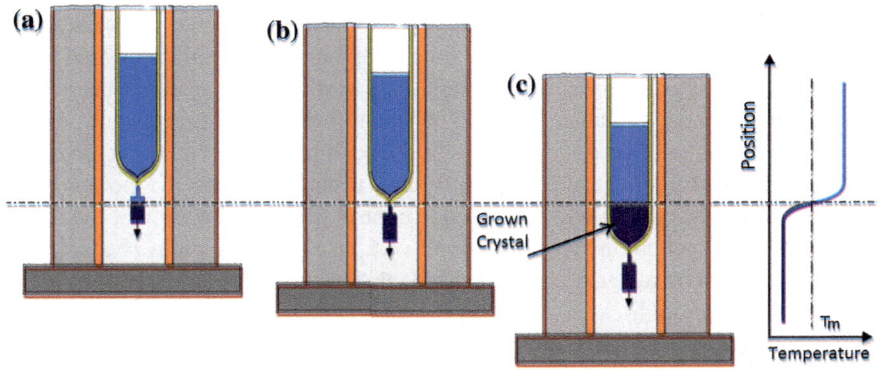

Fig. 2.3 Phases of the bridgman–stockbarger technique. Adapted from [8]

the horizontal Bridgman growth is the cross-sectional shape of the grown crystal, which is circular while the horizontal generates "D" shaped cross-sections. Circular wafers are preferred for large-scale epitaxial growth and device fabrication. However, the advantage of the horizontal Bridgman technique is the high crystalline quality obtained due to reduced thermal stresses. Disadvantages of the Bridgman growth are that the element composition along the length of a Bridgman-grown ternary semiconductor is often inconsistent due to the inherent segregation of many ternary alloys.

2.1.1.4 The Verneuil Technique

This technique also called flame fusion is preferentially used for high melting temperature materials, and was first used to grow crystals [10]. It is commonly used to grow all types of sapphire with different dopants and qualities. If the material is pure, it is called white sapphire; when doped with chromium oxide, it assumes a red color and is called ruby; if it presents other colors, it is labeled sapphire. Basically, the technique consists of melting grains or powder through injection into a flame burning hydrogen and oxygen and pulling the crystal downwards out of a melt drop. The flame is the heat source for the process. The main advantage of this process is that it requires no crucible, and hence can be used for the growth of high melting point crystals. However, it presents many disadvantages as it does not produce crystals of high quality as other methods of growth from the melt, as well as the difficulty of controlling the growth parameters, internal stresses, subgrain boundaries, and high dislocation densities that are related to high thermal gradients and small liquid volume of this technique.

2.1.2 *Solution Growth*

Growth from solution is mainly a diffusion-controlled process that is used to grow materials at temperatures below their melting points and crystals that present polymorphous structure as well as compounds which are not thermally stable [11]. This method of growth requires the dissolution of chemical components that form the crystal in a liquid medium, or flux, and subsequent slow crystallization with decrease in temperature. Solution growth techniques differ according to the solubility of the crystal material: water solution growth at room temperature, flux growth, and hydrothermal approaches (where high temperature and pressure are required) and are classified according to the solvent type (i.e., water, multicomponent water solutions, melts of some chemical substances). Moreover, the solubility of the solute in the chosen solvent must be established before beginning the growth process. In fact, the solubility data at various temperatures are crucial to determine the level of supersaturation, the driving force of the crystallization process. If the solubility is too high, the growth of bulk single crystals is complex while on the other hand, a smaller solubility constraints the size and growth rate of the crystals. Therefore, crystallization depends on the temperature of the solution, the composition of the solution, or the chemical reaction. Among the many advantages, growth from solution is used for materials that do not melt congruently, which decompose before melting or undergo a solid-state phase transformation before melting or which have a very high melting point.

The conventional basic apparatus for solution growth involves the use of a large thermostat tank heated at the base by an infrared lamp and controlled in temperature. Large quantities of solution in large container are normally used and only a small fraction of the solute is converted into a bulk single crystal.

Depending on the process temperature, growth methods can be classified mainly into low-temperature solution (80–90 °C), and high-temperature solution (flux growth, hydrothermal method, in which temperatures are over 800 °C).

2.1.2.1 High-Temperature Solution Growth

In high-temperature solutions, the material components are dissolved in a suitable solvent, and crystallization occurs as the solution becomes critically supersaturated. This occurs via solvent evaporation, by cooling the solution or by a transport process in which the solute flows from the hottest to the coldest section.

Flux Growth Technique

Flux growth is commonly used for the growth of oxide crystals. The process consists of heating the container at a temperature of the flux and the solute so that all the solute materials will dissolve for several hours after which it is lowered very slowly.

Hydrothermal Growth

The hydrothermal technique is commonly used for the growth of quartz, ruby, calcite crystals, and sphalerite (cubic zinc-blende) ZnS. It is characterized by high pressure (hundreds or thousands of atmospheres) and temperature. (400–600 °C). Growth is typically carried out in steel autoclaves with gold or silver linings that are classified by the pressure used. Growth depends on the concentration gradient given by the temperature difference between the nutrient and growth areas. The high-pressure condition limits the quality and the dimensions of the crystals grown by this technique. A critical difficulty of this technique is the recurrent inclusion of OH^- ions into the crystal, which makes them inappropriate for many applications. Advantages of this process are that the final crystals show almost no stress, or plastic deformations, and other structural defects due to the relatively low temperatures.

2.1.2.2 *Crystallization from Low-Temperature Solutions*

Low-temperature solution method is based on lowering the temperature of the solution in the crystallization zone in order to achieve supersaturation. This technique is commonly used to grow materials such as potassium-sodium tartrate (seignette salt), tryglicine sulfate, potassium aluminum sulfate, and potassium dihydrogen phosphate (KDP) crystals.

Temperature gradient is obtained by creating two zones with different temperatures: one for dissolving the material, the other for crystallization. The conventional apparatus is a tall reservoir with a feed liquid in its lower part (solution zone) and a seed crystal in its upper part (crystallization zone). The exchange between the solution zone and the crystallization zone is driven by the difference in density values of the saturated and non-saturated solutions.

The main drawbacks of this method are the slow growth rate and the ease of solvent inclusion into the growing crystal. The latter can be minimized by controlling growth conditions. On the other hand, the process of solution growth generates crystals of good quality for a variety of applications. Moreover, since growth is carried out at RT (room temperature), structural imperfections are relatively minimized [11].

2.2 Thin Films—Epitaxial Growth—MBE, ALE, ELO

The demand for high-performance semiconductor devices that can be used for advanced technology applications depends highly on the epitaxial growth techniques. In recent years, there has been considerable activity on the deposition of thin films on a variety of substrate materials. Epitaxial growth of thin films is, in fact, a key process for a wide range of applications in electronics, optoelectronics and several industries.

Epitaxy is the growth of a crystalline material on a substrate, such that the new material is ordered in the same way as the substrate. Epitaxial growth of thin films occurs when a metastable material (in a solid, liquid, or vapor form) is nucleated onto a crystalline substrate. The newly formed crystalline layer is referred to as the epilayer. The past decades have seen a convergence of various vapor-phase epitaxy techniques, such as molecular-beam epitaxy, atomic layer epitaxy and epitaxial lift-off (MBE, ALE, ELO) respectively. In this chapter, three major epitaxial growth processing techniques are discussed.

2.2.1 Molecular-Beam Epitaxy (MBE)

Molecular-beam epitaxy (MBE) is a nonequilibrium technique in which a monolayer-scale control is obtained [12]. The advantages of this technique is that precisely controlled molecular beams are deposited onto a heated substrate at temperatures much lower than that required for equilibrium growth techniques. The possibility of a precise level of control over the growth process of high-quality crystalline semiconductor offers many opportunities to implement device structures which have not been practical or realizable in the past. This technique is widely used for processing semiconductors based on III-V materials. MBE requires ultrahigh vacuum conditions and works in association with techniques which can be operated in situ as growth proceeds—principally, reflection high-energy electron diffraction (RHEED) and reflection anisotropy spectroscopy (RAS). Other techniques can be employed (e.g., ion scattering, EDX, and STM) and practically any surface science tool can be used ex-situ by a process of sample transfer under vacuum from the growth chamber to the characterization chamber. The core of MBE is the vacuum chamber where the materials are grown. Ultrahigh vacuum chambers, with background pressures in the range of 10^8–10^{10} mbar, and low partial pressures of impurity gases such as DI (deionized) water, CO and CO_2 are typically required. This is because growth in MBE is achieved by the use of molecular beams and is entirely dependent on the arrival of atoms at the surface. As such, the quality of the growth material is highly sensitive to the arrival of impurity atoms. Unwanted background gases, such as nitrogen or oxygen, can have undesirable effects on the properties of the semiconductors and in order to overcome this problem, the use of an ultrahigh vacuum system is necessary. Generally, MBE systems have additional chambers in between the load lock and the growth chamber, which allows further isolation and preparation and storage of substrates prior to growth.

The molecular beams required for the growth are generated by heating a crucible of effusion cells containing a single source material, see Fig. 2.4. The material purity is critical to the final quality of the epitaxial layer. The material is sublimated (or evaporated in the case of a liquid source) from effusion cells, thus forming molecular beams that are incident on a heated sample. A typical MBE vacuum chamber is shown in Fig. 2.4.

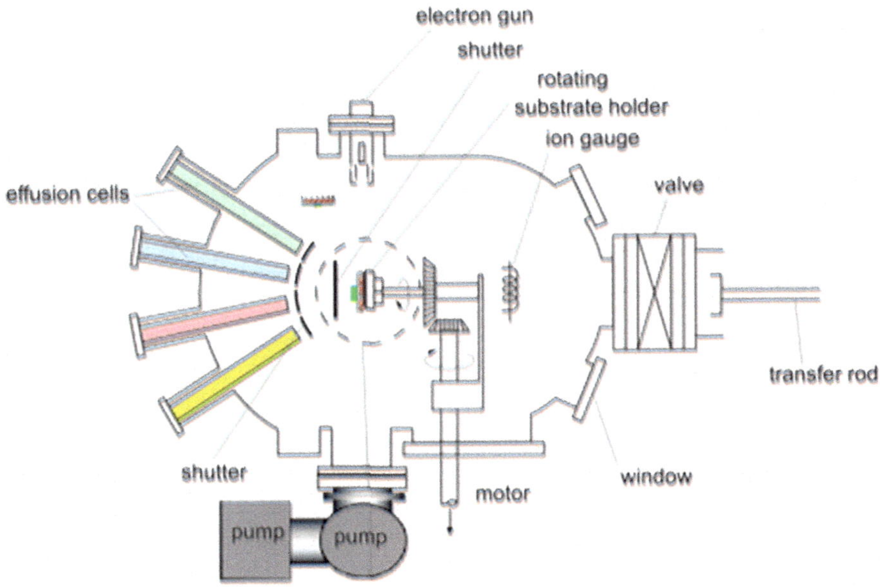

Fig. 2.4 Basic schematic of a MBE system showing how the various constituents are co-evaporated onto the substrate surface

The effusion cells consist of a crucible containing the source material, typically constructed from pyrolytic boron nitride and a heating source with shutters to enable control over the atomic species being deposited. Heating can be typically performed by either ohmic heating from a filament wound around the crucible or by electron beam heating. The latter consists of thermionic electrons which are accelerated toward the source material through a potential of a few kV and deposit tens of watts of power, inducing heating in the source material [13]. In general, the type of heating required is imposed by the necessary flux of material and the melting point of the materials (i.e., indium has a low melting point (156 °C) but requires temperatures in excess of 700 °C, therefore, an ohmic resistive filament heating is used. On the other hand, tungsten requires temperatures over 2000 °C, therefore, electron beam heating will be applied [14].

Epitaxy in MBE is driven primarily by surface kinetics. Once the atom or a molecule reaches the surface, the molecule may be physisorbed onto the surface through weak, physical bonds such as van der Waals forces or chemisorbed on the surface if an electron exchange process occurs. If the molecule does not desorb, then the energy available from the heated substrate will cause it to diffuse on the surface, promoting growth. In order to form a crystallographically oriented material, the incoming molecules must arrive or diffuse to epitaxial sites and become chemisorbed at that site.

The formation of new material by chemisorption and incorporation may yield different growth modes, depending on a number of growth parameters such as substrate

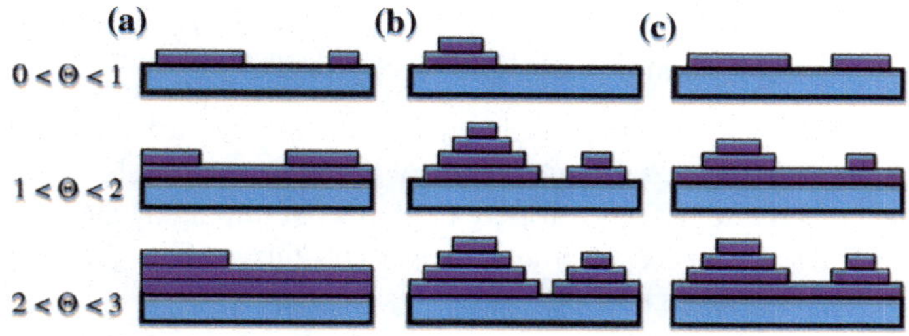

Fig. 2.5 The three idealized growth modes: **a** Frank–van der Merwe (layer-by-layer growth), **b** Volmer–Weber (3D island growth) and **c** Stranski–Krastanov (3D islands with a wetting layer). θ denotes the degree of coverage

temperature, deposition rate, growth temperature, strain, and surface energies of the different materials. Ideally, one can generally distinguish between three different growth modes:

(1) Layer or Franck–van der Merwe growth mode: ideal for high-quality thin films, in which each growing atomic layer is completed before the next begins to grow (Fig. 2.5a).
(2) Island or Volmer–Weber growth mode: involves the formation of 3D islands on the surface. The deposited atoms cannot diffuse past the island boundaries (Fig 2.5b).
(3) Layer plus island or Stranski–Krastanov growth mode: This type is a combination of layer and island growth mode (Fig. 2.5c).

The Franck–van der Merwe (FM) growth mode is a layer-by-layer growth mode, mostly found in lattice-matched combinations of material systems. The interatomic interactions between substrate and epilayer materials are stronger and more attractive than those in between the different atomic species within the epilayer material; thus atoms attach preferentially to surface sites resulting in atomically smooth, fully formed layers. This layer-by-layer growth is two dimensional, indicating that complete films form prior to the growth of subsequent layers [15, 16]. The Volmer–Weber growth mode leads to the formation of three-dimensional clusters or islands due to stronger atom–atom interactions with respect to the surface [15]. Growth of these clusters, along with coarsening, will cause rough multilayer films to grow on the substrate surface. This growth mode is characteristic of highly mismatched combinations of semiconductors. The Stranski–Krastanov growth mode is a two-step process categorized by both 2D layer and 3D island growth renowned as "layer-plus-island growth". At first, several monolayers of films of adsorbates grow in a layer-by-layer trend on a substrate. Transition from the layer-by-layer to island-based growth occurs at a critical layer thickness which is highly dependent on the chemical and physical properties, such as surface energies and lattice parameters, of the substrate and epilayer [17]. Many variations of the idealized models exist, such as coherent or incoherent 3D island growth, or layer-by-layer growth with strain relief by dislocations or large-scale rippling.

There are several parameters within MBE, which can be independently adjusted and monitored in order to improve the quality of growth:

- Flux rate (the number of atoms arriving at the substrate surface) controlled by adjusting the evaporation/sublimation rate of the source materials
- Substrate temperature (affects diffusive properties of impinging atoms)
- Source temperature (speed of atomic arrival and flux rate) is directly adjusted depending on the heating method used.

The temperature of the substrate is crucial to the effectiveness of the chemical reactions taking place on its surface. The atomic processes effective during the growth, including adsorption, desorption, migration, and reaction, are sensitive to temperature fluctuations. Higher temperatures result in surface atoms that are highly motile, and thus lead to more highly ordered material. Therefore, it is very important to achieve accurate temperature control of the substrate to maintain a constant growth rate and ensure superior crystalline quality of the material produced. Some disadvantages are that major diffusion of atoms smoothens the interfaces and substrates require particularly high temperatures during preparation and the following growth. One such process is called annealing, in which the sample is again heated in order to recrystallize and repair damage to the crystal structure. The flux rate and the source temperature also affect the growth rate as they influence the arrival rate of molecules at the surface. In order to reduce the diffusion of atoms on the surface, lower temperatures are used for the growth of strained layers. This does reduce the probability of the layer relaxing and eliminates the advantages of introducing strain into the system. The drawback is that a major number of defects (i.e., point defects) are introduced during growth due to the decreased mobility of deposited atoms.

Typical MBE growth rates are 1 Å per second; therefore, using the correct growth rates and having control of the molecular beams using the shutter systems, multilayer structures can be grown reliably and repeatedly.

2.2.2 Atomic Layer Epitaxy (ALE)

Atomic layer epitaxy (ALE) is a chemical vapor-phase thin film deposition method that was developed around the 1980s [18]. Some confusion enfolds the term atomic layer epitaxy, as the word epitaxy originates from Greek and means "on-arrangement". In this denotation, epitaxy describes the growth of single crystalline layers on a single crystalline surface. ALE, instead, covers the deposition of amorphous, polycrystalline, and single crystalline thin films. Therefore, over the years, the term ALE deposition of amorphous films has been adapted to atomic layer deposition (ALD) as a result of the fact that most films grown were not epitaxial to their underlying substrates [19–21].

As a general introduction to the ALE process, Fig. 2.6 summarizes the basic ALE reaction cycle leading to the formation of a monolayer. This processing method is based on alternate exposure of the substrate surface to different precursors. In

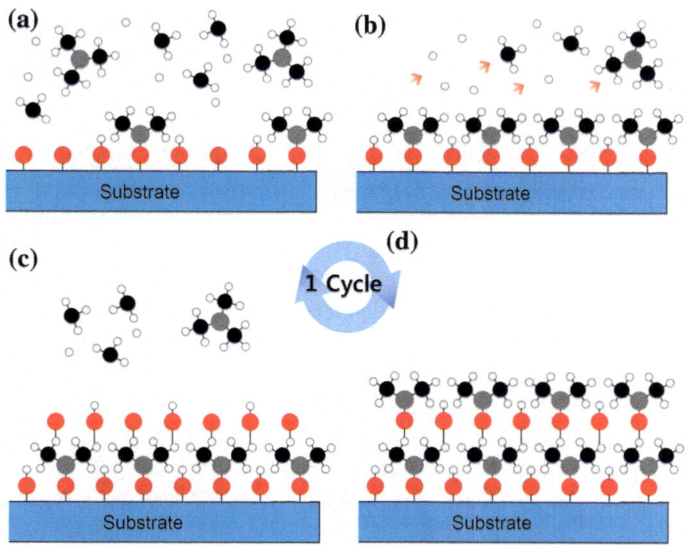

Fig. 2.6 Schematic of ALE process. **a** The first precursor is pulsed on the substrate and reacts with the functionalized surface. **b** Excess of precursor and reaction by-products are purged with inert gas. **c** The second precursor is pulsed and reacts with the surface and excess of precursor and reaction by-products are purged with inert carrier gas. **d** Steps a–c are repeated until the desired material thickness is achieved

the ideal case, the reactant vapors are pulsed onto the substrate alternately one at a time and purged with an inert gas between the reactant pulses. As a result, the surface exposed to the precursor is saturated via chemisorption or by reaction with the functional surface groups. The excess of precursor molecules and the released ligands as well as the volatile byproduct molecules are removed from the reactor by a purging step, which leaves behind only the precursor monolayer that is adsorbed on the substrate surface. The second precursor then reacts with the earlier deposited monolayer, liberating ligands, and producing the desired solid layer. The deposition cycle is completed with a second purge step in which the excess precursor and volatile byproduct molecules are removed from the reactor. Repetition of this reaction cycle leads to deposition of a solid film layer-by-layer.

Formation of a complete monolayer during one cycle is often prevented by the bulkiness of precursor molecules, and as a consequence, the growth rate is less than one monolayer per cycle [22]. Nevertheless, the film thickness is proportional to the number of deposition cycles used and an accurate control of the thickness can be achieved, i.e., the film growth is self-controlled. This represents an important advantage of ALE processing technique, i.e., the self-limiting adsorption of a given precursor onto the substrate surface of which ALE gives a precise thickness control at the angstrom or monolayer level. In addition, complex, large-area substrates can be uniformly coated with excellent uniformity and reproducibility. Another benefit is the possibility to obtain high-quality material at relatively low growth temperatures.

Disadvantages of this method are the low deposition rate, although this can be overridden as the required thickness of the films is of the order of a few nanometers. Low precursor utilization efficiency is also an issue, but it may be compensated by reactor and precursor design.

Another important aspect of any ALE process, in particular with respect to the morphology of the growing film, is that it can be distinguished in layer-by-layer growth, also referred to as Frank–van der Merwe growth, island growth or Volmer–Weber growth mode, and random deposition. The majority of ALE processes proceed either by island growth or random deposition. In this case, new material is deposited statistically both in empty places between islands growing on the substrate as well as at sites on the evolving film, due to equal adsorption coefficients on all those sites. The random deposition mode is, therefore, able to generate smoother films in most cases, as compared to films deposited in an island growth fashion, which normally exhibit greater roughness [23, 24].

As previously outlined, precursor chemistry holds a significant function as they must chemisorb on the surface or react rapidly with the surface groups and react aggressively with each other. Therefore, they are volatile and thermally stable but they can be gases, liquids, or solids. This allows a practical deposition rate as saturation can be reached relatively quickly (less than 1 s). Moreover, not only the properties of a single precursor molecule are important but also the combination of the precursors plays a key role. Precursors can be metal, nonmetal, and metal alkoxides. Metal precursors (i.e., halides, chlorides, and metals alkyls) require reactivity at temperatures <500 °C and the presence of surface—OH groups [25, 26]. In many cases, the alkyl compounds (Al, Zn) behave almost ideally for the growth of oxide and sulfide films, but in some cases (Ga, In), the results have not been very promising due to the complex and unfavorable surface chemistry. Nonmetal precursors are usually hydrides (H_2O, H_2S, NH_3, and AsH_3) without problems in volatility and thermal stability. Metal alkoxides have been successfully employed in several oxide processes. Electropositive metals (alkaline earth metals, rare earth metals) pose a challenge for chemical thin film depositions because they do not have many volatile compounds.

Different reactor types have been employed for ALE growth. Low-pressure MBE reactors were used to grow epitaxial III–V or II–VI semiconductors on single crystalline substrates [27]. Several modifications, including MOVPE type sources for metal organics and hydrides in a MBE reactor [28], small MOVPE type reactor [29] or moving and rotating substrates for the pulsing of the reactants [30–32] have been developed.

ALE is a coating method especially suited to form ultra-thin films of several nanometers (2–10 nm) up to several ten nanometers (10–50 nm). Although in principle, thicker films are also possible, this is mostly impractical due to prohibitively long processes.

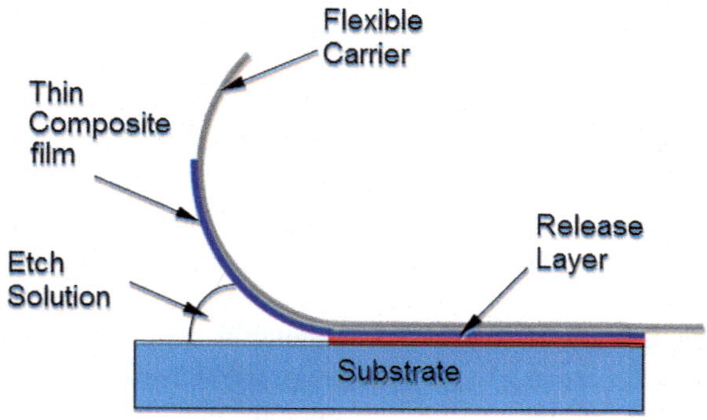

Fig. 2.7 Schematic of Epitaxial Lift-off (ELO) technique

2.2.3 Epitaxial Lift-off (ELO)

Epitaxial lift-off (ELO) is a processing technique that allows the integration of III-V films onto arbitrary material substrates. First reported in the 1970's [33], this technique enables thin epitaxial layers grown on substrates to be "peeled off" from the host substrate and deposited onto a new substrate, Fig. 2.7.

Separation from the substrate may be achieved chemically (i.e., using hydrofluoric acid (HF) to selectively etch the sacrificial layer inserted between the device layer and the substrate) or mechanically by etching away the whole substrate. In this way, the active part of a semiconductor heterostructure can be etched from its original substrate and deposited onto a host substrate. By stressing the epitaxial layer, this can bow up slightly maintaining an open channel for the etch to proceed [34]. The simplest and most common way to apply the stress is by applying a flexible carrier (i.e., a layer of wax) to the top of the structure, which also helps to support it after the etch. The flexible carrier, when cooled to room temperature, is under tension and the substrate is under compression; thus, as the release layer is etched away, the thin film curls up slightly, thus aiding the lift-off process. Once separated, the film is transferred to the host substrate to which it adheres through van der Waals bonding. This force is very short ranged, typically just a few nanometers, varying with the layer separation cubed.

Such a technique offers several important advantages, such as performance enhancement and cost reduction of III–V electronic and optoelectronic devices. Moreover, flexibility of fabrication owes to the fact that the epitaxial film can be transferred to new support substrates which can be thin, flexible, lightweight, and with higher thermal conductivity than the original growth substrate. This also allows to reprocess substrates and permit an etch selectivity and lattice-matched growth. The main disadvantages of this method are associated with the high surface roughness of the parent substrate when exposed to aggressive chemical etchants, to residues on the substrate after the ELO process, additional fabrication costs, and also a reduction of material from the top surface of the substrate which does not allow the substrates to be reused and the etching may damage the epilayers [35–38]. Besides, aggressive chemical etchant such as HF is a lethal and highly corrosive acid requiring special handling and extra protection. This process is also characterized by slow lateral etch rate of the sacrificial layer, taking hours to complete, which potentially increases the exposure risk for operators and the operating cost. In order to obviate these issues, a high-throughput technique known as "surface tension-assisted (STA) ELO" has been investigated. During this process, the film is drawn away from the substrate and flattened on the surface by surface tension. While the film is being pulled off from the surface, the space between the lifted-off film and the substrate is replenished via capillary force, which causes the process to continue to be reaction limited [39]. This process allows to reduce the amount of post-etching residues and to keep the surface smooth, leading to direct reuse of the substrate.

Typical applications are optical interconnects, displays, lasers, and LEDs. The possibility to separate the growth substrate from the active thin film plays a key role in the optimization of materials engineering. For example, the supporting substrate can be optimized separately from the epitaxially grown film. Although a rapid progress of this technique has been made over the years, a better and further understanding of the van der Waals bonding process and the resultant interface is still required.

2.3 Thin Films—Polycrystalline Films—PECVD, LPCVD, APCVD

Thin solid film fabrication process is one of the most important technical advances contributing to materials-processing technology. The attention for this technology has arisen as the result of the need for high-quality thin film deposition, high-throughput wafer processing, and low operating costs. The chemical vapor deposition (CVD) approach is a process by which a solid film on a substrate is formed by the reaction of vapor-phase chemicals (reactants) which contain the required constituents [40]. This method is suitable for high purity, large scale, uniform layers, produced at a relatively high growth rate, and does not react with or consume the substrate. Generally speaking, multispecies of chemically reacting gases flow continuously through a controlled environment of a reactor chamber, in which the substrate is placed. The growth rate of the thin film is determined by the competition between the diffusive and convective transport of species in the gas phase, the homogeneous and heterogeneous chemical kinetics, and the morphology of the gas–film interface. These processes generally differ in the means by which chemical reactions are initiated. Three of the most common and widely used CVD processes, presented in further detail in the subsections of this chapter, are as follows:

- Plasma-enhanced CVD (PECVD)—in which plasma is used to enhance the chemical reaction rates of the precursors. PECVD processing allows deposition at lower temperatures, which is often critical in the manufacture of semiconductors. The low temperatures also allow for the deposition of organic coatings, such as plasma polymers, that have been used for nanoparticle surface functionalization.

- Low-pressure CVD (LPCVD)—is CVD at sub-atmospheric pressures. Reduced pressures tend to reduce the unwanted gas-phase reactions and improve film uniformity across the wafer.
- Atmospheric pressure CVD (APCVD)—is CVD at atmospheric pressure.

When making the choice between these different deposition methods, several aspects must be taken into consideration: substrate temperature, deposition rate and film uniformity, morphology, the electrical and mechanical properties, and the chemical composition of the films.

2.3.1 Plasma-Enhanced Chemical Vapor Deposition (PECVD)

Plasma Enhanced Chemical Vapor Deposition (PECVD) is a deposition method used to produce thin films of various materials. It is, nowadays, used in a wide variety of industrial applications for the synthesis of nanostructured coatings to obtain functional surfaces such as, for example, anticorrosion, anti-adhesion, wear-resistant, scratch-resistant, self-lubricating, and antireflective, hydrophobic, hydrophilic or oleophilic [41, 42] coatings. PECVD has also been used to synthesize transition metals, transition metal silicides, and transition metal nitrides for the formation of conducting layers for barrier, contact, and interconnect applications. In particular, silicon dioxide and silicon nitride are insulating materials that are commonly used in the fabrication of electronic devices for separating one conductive layer from another. These films are also used as coatings to protect devices from atmospheric corrosion [43].

PECVD reactors generally use a radio frequency (RF) power (13.56 MHz or 380 kHz) source to create a plasma field in the deposition gas. The use of plasma frequently allows deposition at a much lower temperature than could otherwise be achieved, and/or it permits the use of source gases that would ordinarily be considered nonreactive. This is the major advantage of PECVD. The deposition is achieved by introducing reactant gases between two parallel electrodes, the first one placed at the ground and the second one to RF, as in Fig. 2.8. The capacitive coupling between the electrodes excites the reactant gases (into a plasma) and a chemical reaction induces the formation of a thin solid film on the substrate surface [44]. The substrate is placed on the grounded electrode and it is typically heated to 250–350 °C temperatures that are very low when compared with those required in CVD processes, 600–800 °C [45]. The pressure in the chamber varies in the range of 450–1400 mTorr. Typical values of the deposition rate are of the order of few hundred nm/min.

Further configurations to generate plasmas consider DC discharges between two electrodes (the space is filled with the reacting gases) or microwaves (MW). A microwave source typically operates at 2.45 GHz [46] and is used when highest deposition rates and/or high fragmentation of precursor material is desirable. Another configuration of PECVD is the so-called Inductively Coupled Plasma (ICP PECVD).

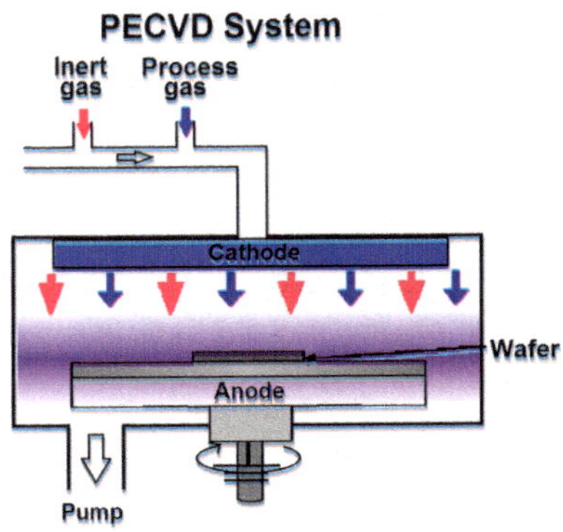

Fig. 2.8 Typical PECVD apparatus

Table 2.2 Films and precursors typically deposited using PECVD technique

Precursors	Film deposited
SiH_4 (pure)/NH_3/N_2	SiN_x
SiH_4 (pure)/CH_4	SiC
SiH_4 (pure)/Ar	a-Si
SiH_4 (pure)/N_2O/Ar	SiO_2
SiH_4 (pure)/NH_3/N_2O	SiN_xO_y

In this type of plasma source, the energy is supplied by electrical currents which are produced by electromagnetic induction. Compared with a conventional PECVD process, ICP allows control of more process parameters.

The films typically deposited using PECVD technique are silicon nitride (Si_xN_y), silicon dioxide (SiO_2), silicon oxynitride (SiO_xN_y), silicon carbide (SiC), and amorphous silicon (a-Si). The silicon source gas is silane (SiH_4), and it is combined with an oxygen source gas to form silicon dioxide or a nitrogen gas source to produce silicon nitride [45, 47]; see Table 2.2 for details.

The major advantage of PECVD is the ability to produce deposition processes that would normally occur only at much higher temperatures with CVD. The combination of vacuum pressure and low temperature produces better uniformity in the deposited layer. Disadvantages are that low temperatures during deposition in many cases result in incomplete desorption of by-products and other gases, particularly hydrogen, resulting in unwanted inclusions and outgassing causing thermal instability. Therefore, deposits of pure materials are often difficult to achieve. Batch size is also small: PECVD can only deposit the film on one side of 1–4 wafers. Further drawbacks include the formation of undesirable compressive stresses mostly at lower frequencies and the commonly more expensive and complicated nature of the equipment.

2.3.2 Low-Pressure Chemical Vapor Deposition (LPCVD)

Low-Pressure Chemical Vapor Deposition (LPCVD) is a technique used for the deposition of a wide range of possible thin film compositions on semiconductors ranging from a few nanometers to many micrometers. LPCVD systems are similar to other CVD systems as gaseous species react on a solid surface or wafer and the reaction produces a solid phase material. The reduced gas pressure enhances the mass transfer rate relative to the surface reaction rate, thus allowing the deposition of a uniform film in a relatively highly economical close spaced positioning of the substrate wafers in a standup fashion [48].

There are two types of reactions that take place in an LPCVD reactor: homogeneous and heterogeneous reactions. In a homogeneous reaction, the reactants react in the atmosphere of the reactor and not on the surfaces within the reactor causing nucleation in the gas phase in the plasma or particulates to precipitate on the substrate and therefore lead to a high defect count. This phenomenon is more frequent in silane-based reactions. In a heterogeneous reaction, the reactants diffuse to the substrate and are adsorbed on the substrate where adatom migration takes place. The by-products, therefore, desorb from the surface and diffuse into the main gas flow, thus reducing the presence of defects.

LPCVD systems can be divided into two different groups: hot and cold wall systems. Usually, LPCVD systems are hot-wall reactor systems. The advantages of hot-wall systems are a uniform temperature distribution and reduced convection effects. However, a downside of these systems is the formation of deposits on the wall which can lead to depletion of the deposition species and the formation of particles, which may contaminate the wafers. Deposits on the walls also lead to memory effects, i.e., material previously deposited on the walls may be deposited on the wafer. Hence hot-wall reactors must be dedicated systems and the tube must be periodically cleaned or replaced to minimize hitches with particulate matter. On the other hand, this issue may be overcome in cold wall systems as deposition on the walls are reduced, but due to poor temperature control, gas convection would be another hitch.

A hot-wall LPCVD system is shown in Fig. 2.9 [49]. A quartz tube is placed in a spiral heater that starts at very low pressure (around 0.1 Pa). The tube is then heated to the desired temperature and the reacting gaseous species are permitted to flow into one end of the tube at the pressure predetermined between 10 and 1000 Pa, and pumped out from the other end, spreading out around the hot substrates in the tube. The substrate temperature is extremely important and influences the reactions that take place. Temperature range is from 300 to 1150 °C [50].

The gas reacts with the substrates and forms a solid phase material and the excess material is pumped out of the tube.

LPCVD technique has a considerably superior thickness uniformity of the films and better step coverage compared to conventional atmospheric pressure CVD reactors. In fact, because of the intrinsically cleaner hot-wall operations and due to the wafers being positioned vertically, the formation and co-deposition of homo-

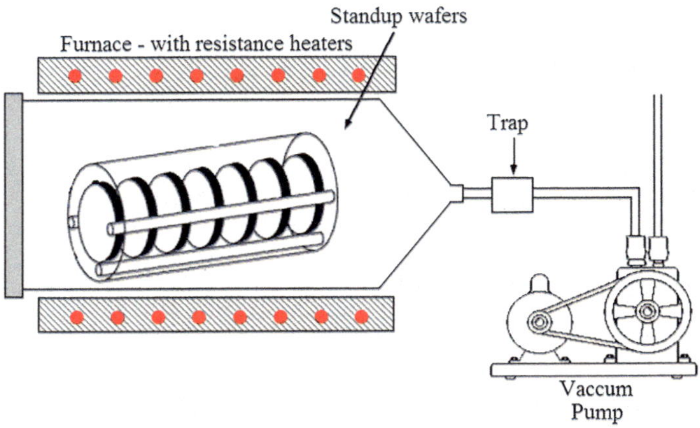

Fig. 2.9 Schematization of an LPCVD system

geneously gas-phase nucleated particles are minimized. Therefore, the films have fewer defects, such as particulate contaminants and pinholes [51]. The simplicity of the process permits the processing of large wafer batch sizes. Some downsides of this technique are that it can only be performed at high temperatures (550–850 °C) and it has a relatively low deposition rate.

When the pressure is lowered during LPCVD, the diffusion of the gas decreases proportionately to the reciprocal of the pressure. This means that the velocity of mass transport will decrease, meaning that the substrates can approach more closely and the deposited films show better uniformity and homogeneity [52]. Thus there is less dependence of the resulting layer on mass transport velocity.

Improved LPCVD systems are vertical furnaces in which the wafers are held by gravity [53]. The advantages are a more uniform wafer-to-wafer spacing in the reactor, uniform convective effects across the wafer, and most of all, reduced particle counts. Moreover, vertical CVD systems can be more easily integrated into automated factories and robotic handling since the wafers do not have to be tipped to the vertical. One disadvantage is the higher cost of these systems.

2.3.3 Atmospheric Pressure CVD (APCVD)

Atmospheric Pressure CVD is a CVD method at normal pressure known for its high reaction rates and simplicity as a CVD system [54, 55]. A schematization of this type of reactor is shown in Fig. 2.10. The substrates are provided continuously through the system on a heated chain track, and large-diameter wafers are easily handled. The reactant gases flow through the center section of the reactor and are confined by nitrogen curtains at the ends. Temperature range varies from 240 to 450 °C.

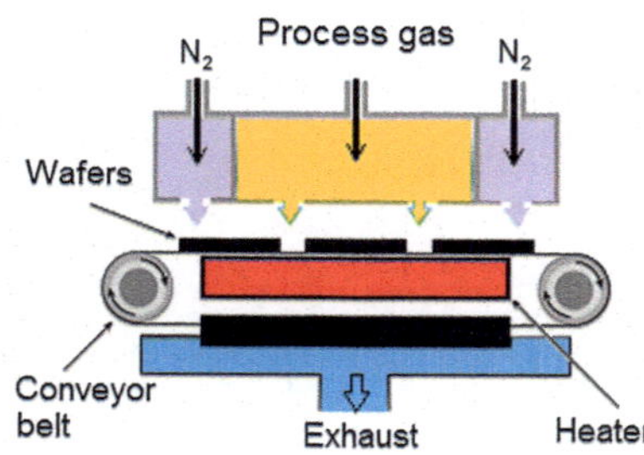

Fig. 2.10 Schematization of an APCVD reactor

The advantages offered by this method are low equipment cost, high dilution of toxic or flammable gases, large-area uniformity realized through control of temperature and gas flow and the potential for operation as continuous-feed belt systems. Disadvantages of APCVD operation comprise the need for high gas-flow rates, large size reactors, low uniformity and high levels of particulate contamination. Moreover, large number of pinhole defects can occur. APCVD is also extremely susceptible to oxidation due to the greater gas density and residence times. As such, vacuum-caliber reactor construction is required, eliminating many of the purported advantages of APCVD. In addition, tremendous quantities of liquid nitrogen would be required to control particle formation as well as segmenting the showerhead in order to keep the reactant gases separated until they are injected into the chamber, representing a higher utility cost.

As such, the maximum obtainable growth rates with APCVD are only ~ 0.1 μm/min; therefore, APCVD has faded from popularity, with the most significant developments occurring in the remaining two categories.

2.3.4 Conclusion

In conclusion, high quality of thin films pertains to the structure and stoichiometry, the desired physical properties, the step coverage, the pinhole density, and so on. PECVD process has more parameters to vary for process control; therefore, the physical properties of the films may be tailored to a specific application. However, improved purity and internal structure of the films can be obtained through LPCVD, although it requires high-temperature performance. A comparison between APCVD and LPCVD shows that the benefits of the low-pressure deposition processes are uniform step coverage, precise control of composition and structure, low-temperature processing, high enough deposition rates and throughput, and low processing costs. The main reason for using LPCVD instead of APCVD is the ratio of the mass transport velocity and the velocity of reaction on the surface. Disadvantage of using LPCVD and APCVD, however, is that processing temperature is high, which can be worked out using PECVD. A comparison of the characteristics and applications of the three CVD processes is summarized in Table 2.3.

Table 2.3 A comparison of the PECVD, LPCVD, APCVD methods for thin film deposition

Process	Advantages	Disadvantages	Pressure/temperature	Usage
PECVD	Fast, low temperature	Risk of chemical and particle contamination	200–600 Pa/300–400 °C	Low-temperature oxides, passivation nitrides
LPCVD	Excellent cleanliness, uniformity, and conformity, large wafer capacity	Low deposition rate, high temperature, requires vacuum system	100 Pa/550–600 °C	Silicon nitride, Doped/undoped high-temperature oxides, Polysilicon nitride, and oxide, tungsten
APCVD	Simple, fast, low temperature	Particle contamination, poor step coverage	10–100 kPa/350–1200 °C	Doped/undoped low-temperature oxides, Epi films

2.4 Amorphous Films—CVD, Laser Ablation

The word amorphous is derived from Greek, meaning without shape or form; therefore, an amorphous solid has no defined structural meaning, but is defined as "non-crystalline on any significant scale" [56].

Various amorphous oxide, carbide, and nitride thin films have attracted significant interest in the scientific community, and have been investigated for a number of applications. In the following section, techniques of deposition of these films such as chemical vapor deposition (CVD) and laser ablation will be discussed.

2.4.1 Chemical Vapor Deposition (CVD)

Chemical vapor deposition (CVD) is a process in which thin solid films are deposited on solid substrates by a heterogeneous chemical reaction of vapor-phase precursors [57, 58]. The chemical reactions occur homogeneously in the gas phase and/or heterogeneously on the substrate. Basically speaking, the key steps of CVD are schematized in Fig. 2.11: multispecies of the precursors are transported in the gas phase either by sublimation or evaporation through heating and introduced into the reaction chamber by vacuum or bubbling (i.e., an inert carrier gas is bubbled through the precursor mixture such as oxygen or nitrogen to assist in forming oxides or nitrides). As the gas flows over the heated substrate, surface reactions may occur during which further decomposition of the precursors takes place to form the material of interest in

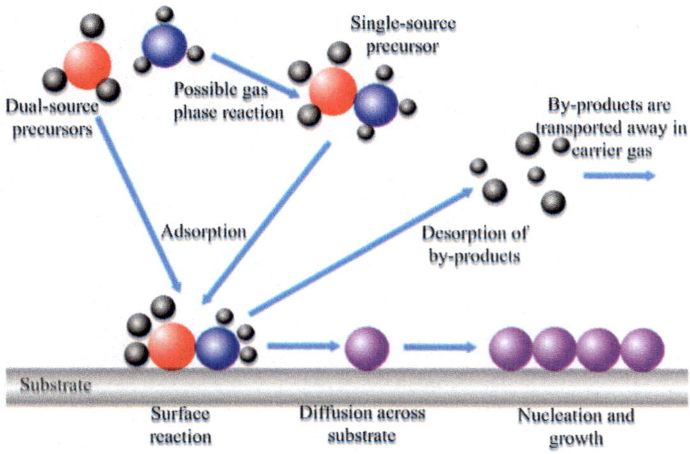

Fig. 2.11 Schematization of CVD process: which include (1) Evaporation and transport of reagents in the bulk gas flow region into the reactor; (2) Mass transport of the reactants to the substrate surface; (3) Adsorption of the reactants on the substrate surface; (4) Surface diffusion to growth sites; (5) Nucleation and surface chemical reactions leading to film growth; (6) Desorption and mass transport of remaining fragments of the decomposition away from the reaction zone

the form of a thin film. During this stage, volatile by-products of the precursors can also form, desorbing along with any unnecessary ligands, from the substrate and are transported out of the chamber via the inert carrier gas, or vacuum, as function of the type of CVD used. Additionally, atoms bind initially to the film only through weak van der Waals interactions and, therefore, may diffuse across the surface of the film. Heating the substrate (~200 °C) weakens the van der Waals bonds which are replaced as nucleation of the film occurs with several atoms bonding together. Moreover, substrate defects can act as nucleation points to which further atoms migrate, and grow in clusters which form the film over time.

Film growth rate is determined by the substrate temperature, pressure of the reactor and gas-phase composition [59]. The substrate temperature influences growth rate by the kinetics of the chemical reactions occurring either in the gas phase or on the substrate surface, known as surface reaction limited film growth. If the process is controlled by the surface reaction, the deposition mechanism of the solid film is described by the Arrhenius expression [60, 61]

$$GR = A\ \exp(E_A/RT) \tag{2.4.1}$$

where GR is the film growth rate, E_A is the activation energy, R is the gas constant and T is the temperature. According to this equation, there is a linear plot of ln(GR) versus 1/T, which denotes that the reaction rate increases with the increasing temperature, as shown in Fig. 2.12.

With increasing temperature, film growth rate is mostly determined by the mass transport of the precursors through the boundary layer to the growth surface, denoted

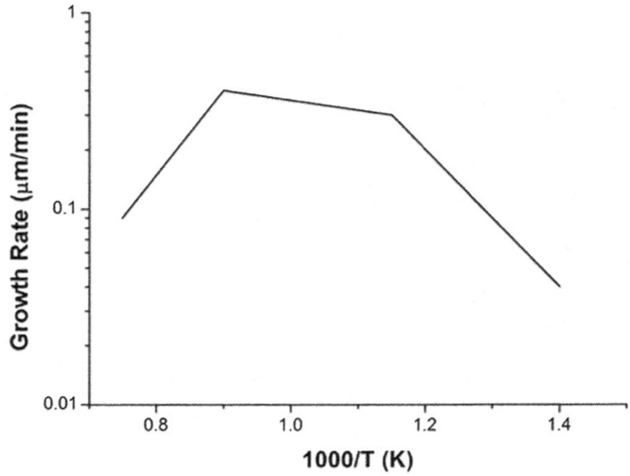

Fig. 2.12 Temperature dependence of growth rate

as mass transport limited film growth. This type of growth is advantageous in cold wall CVD systems. The reaction rate is expressed as follows

$$GR = A\,k_D/RT(P - Po) \qquad (2.4.2)$$

where A is the surface area of the substrate, k_D is the mass transport coefficient, P the partial pressure of the reactant and Po is equilibrium partial pressure of the reactant at a certain temperature.

With further increase in temperature, the growth rate decreases as the film precursors are desorbed from the film surface and/or depletion of reagents on the reactor walls due to gas-phase side reactions. These gas-phase side reactions become more significant at high temperatures and higher partial pressures of the reactants. If a CVD process is mass transport limited, then a typical rate-limiting step can be the diffusion of the precursor molecules through the boundary layer, which will be discussed later in this chapter.

Surface diffusion and nucleation affect the layer morphology of the thin films formed by CVD which can be epitaxial, amorphous, or polycrystalline. Growth of amorphous layers requires low temperatures, high growth rates, and a slow surface diffusion related to the incoming precursor flux. Epitaxial layers necessitate the growth of defect-free single crystal films on a substrate of the same material (homoepitaxy) or on a different material (heteroepitaxy). Epitaxial growth is more difficult as a process, as the initial substrate must be free from defects and surface contamination to avoid lattice mismatch between the crystalline spacings of the film and substrate. Growth of epitaxial layers requires high temperatures and low growth rates to quicken surface diffusion. Polycrystalline layers are grown at intermediate temperatures, which results in nucleation at many different sites resulting in islands which coalesce to form a polycrystalline layer.

Adhesion of CVD layers depends on various factors. In order to obtain high-quality films, precursors (or their reactive intermediates) must not react or decompose in the gas phase in order not to produce unreactive species or particulate material which may lead to poor film coverage and adhesion. A school of thought is a careful cleaning of the substrates before deposition as impurities on the surface can physically lead to a discontinuous loose film with poor adhesion. However, even particles formed in a CVD reaction, via homogeneous gas-phase reactions, result in loosely adhered films. In fact, particles consume precursors, depleting the quantity for thin film formation. Particles are also potential sources of defects resulting in pinhole centers. Growth rate and risk of contamination depend on the precursor used in the process. In general, precursors are gases, liquids or solids. Gas precursors are the easiest to handle since they can be simply delivered via mass flow controllers to the reactor. Liquids, with a high vapor pressure, are also favored as the amount can be controlled by the temperature of the reservoir or by reducing the flow. Solid precursors are sublimated by heating the solid. At low temperatures, steric bulk and low molecular weight precursors are ideal due to their high volatility, as the energy demands of the system remain low. Halides are precluded as they are prone to be incorporated into films as impurities [62].

A number of advantages are presented by this method. These advantages include the following: affordable equipment cost and operating expenses, its suitability to be used for both batch and semicontinuous operation, and compatibility with other processing steps along with quickly and easily changeable parameters. Additionally, films have excellent adhesion, are uniform and easily reproducible. Moreover, it enables the possibility to achieve surface-selective deposition and to coat three-dimensional features with high aspect ratio.

CVD processing presents several reactor designs as illustrated in Fig. 2.13. Operating pressures vary from tens of mTorr to 1 atm and reactor temperatures vary from room temperature to 1500 K [63]. Variants include atmospheric pressure (APCVD), low-pressure (LPCVD), metal-organic CVD (MOCVD) and plasma-enhanced (PECVD), which will be further described in this chapter. Variants are differentiated physically on the basis of pressure or use of plasmas, while MOCVD is distinguished by the chemical nature of the precursor gases. One advantage of metal organics is their generally high volatility at moderately low temperatures. Carbon contamination of film, however, is the uppermost drawback.

A key parameter that is used to describe the CVD process is the Knudsen number (Kn), which determines the mass transport properties of the reactor. Kn is expressed as the ratio of the mean free path, λ, and the reactor dimension D (the diameter of the tube) perpendicular to the flow direction. The mean free path is defined as the average distance traveled by a vapor molecule, between collisions with another molecule. $Kn < 0.01$ denotes a viscous flow; $0.01 < Kn < 1$ is indication of a transition flow and $Kn > 1$ indicates molecular flow.

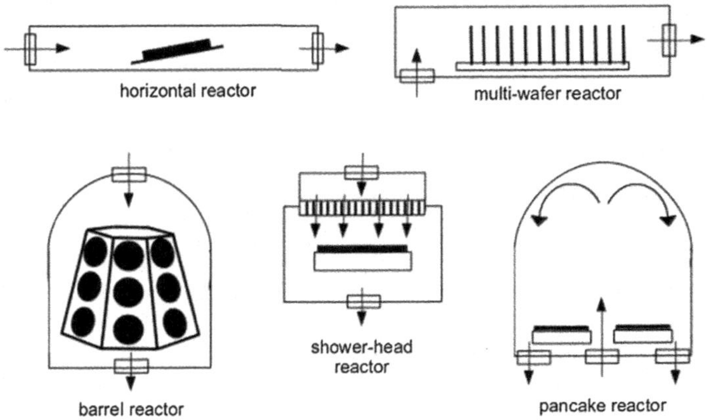

Fig. 2.13 Different CVD reactors

2.4.2 *Laser Ablation Method*

Laser ablation is one of the most versatile deposition techniques. This process uses laser beams to remove material from a solid surface by vaporization of condensed matter. A short pulsed high-power laser beam is focused onto the substrate of choice converting its constituents instantaneously into a vapor phase. As the vapor moves rapidly away from the target, it can be used either to grow a thin film or it can be analyzed by various spectroscopic techniques.

2.4.2.1 Pulsed Laser Deposition (PLD)

Pulsed laser deposition (PLD) is a growth method in which the photon energy of a laser is characterized by pulse duration and laser frequency. Fundamentally, a laser pulse is focused onto the target substrate surface (i.e., solid or liquid) in a vacuum chamber and thereby vaporizes the material from the surface. At a determined threshold power density, a considerable portion of the material is removed and partially ionized forming the ablation plume. The threshold power density that is necessary to create the plasma depends on the absorption properties of the substrate and the laser characteristics (laser wavelength and pulse duration). Typical values are a pulse length of 10 ns for excimer laser and around 500 fs for femtosecond excimer lasers. The material removed is subsequently focused towards a substrate in order to recondensate forming a thin film.

It is known that the interaction of plasma and ion beams with surfaces can change their chemical and physical properties in an innovative way which is not otherwise possible. For both economic and scientific reasons, the technique of pulsed laser deposition (PLD) is ideally suited for the preparation of thin films, cathodes for photoelectron, planar electronic devices in the form of integrated circuitry, etc., while

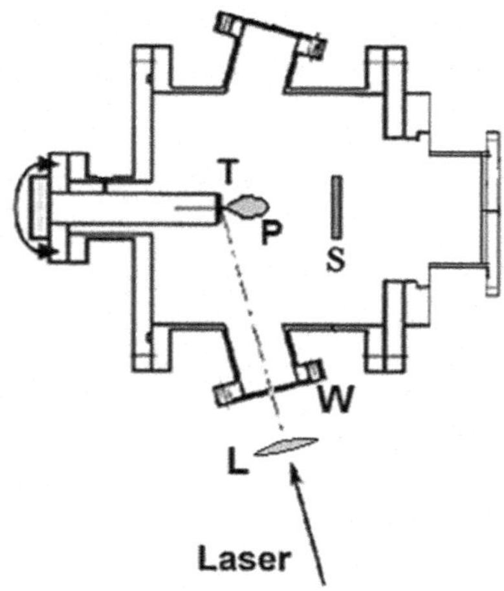

Fig. 2.14 Typical experimental setup for PLD apparatus: W: UV transparent window, T: rotating target, P: plasma plume. S: substrate, L: lens

ion implantation is an easy and good method for the modification, by doping process, of solid materials such as metals, semiconductors, polymers, etc.

In the last 20 years, PLD has aroused significant interest in scientific research, as one of the most flexible and promising film growth techniques. It is, in fact, a good alternative method for the production of composite materials and structures for photonic applications; and it is a versatile technique for the deposition and growth of high optical quality thin films, functional materials with tunable properties, and nanostructured materials.

2.4.2.2 Characteristics of the PLD

PLD is a technique for the deposition of thin films by pulsed plasma laser. The optimal range for laser wavelengths for the growth of thin films by PLD is in the UV range, from 200 nm to 400 nm because many materials that are used for the deposition show a strong absorption in this UV region, while naturally, the target is constituted by the material that is desired for preparing the thin film [64].

A typical experimental setup for PLD is shown in Fig. 2.14 and it consists of a vacuum chamber with quartz windows (transparent to the UV radiation), the target to be ablated and the substrate support. Externally to the system, a lens, also transparent to the UV, is placed to focus the laser beam on the target surface, in order to induce the laser ablation.

The ablated chemical species have energy ranging from 10 eV to more than 100 eV, depending on the constituent of the target material. The bombardment of the substrate by these energetic particles has the effect to modify its properties, and it can have consequences on its morphology, stoichiometry, and microstructure [65]. One of the unwanted effects of the PLD, is the formation of microparticles on the surface of the deposited film; such particles can be clusters formed during the vapor phase, solidified droplets (of the dimension of 1 μm) or objects of greater dimensions (about 10 μm) expelled by the target surface because of the thermal stress [66, 67].

2.4.2.3 Deposition Parameters

In a typical PLD apparatus, the fundamental parameters are the energy and the frequency of the laser pulses, the target–substrate distance, the temperature of the substrate, the pressure of the gas in the chamber (or background gas) [68]. The control of these parameters influences the quality of the deposited film. The pulse frequency determines the deposition rate: a high deposition rate is preferable but it can provoke the formation of a film that is rich in structural defects. Such defects can be eliminated, and therefore the film crystallinity can be improved if the time among two consecutive laser pulses is greater or equal to the time due to achieve the crystallization. The kinetic energy of the particles depends on the target–substrate distance, and the pressure of the background gas. If the distance of flight is too high, the chemical uniformity of the film can worsen. The temperature of the substrate is also very important, which must be controlled because it influences the film crystallization (amorphous, polycrystalline, epitaxial, etc.) and ensures a sufficient surface mobility of the arriving species. Control of the background pressure and gas allows to obtain a correct film chemistry.

The growth of thin films requires an appropriate choice of the substrate material. In fact, the substrate material should be chemically compatible with the film in order to obtain a good crystallographic lattice match between the film and substrate. Also, the thermal expansion coefficients must be similar with a thermodynamically and chemically stable surface.

2.4.2.4 Lasers Employed

Various pulsed lasers have been applied among which the first used were ruby lasers. Nowadays, lasers predominantly used are either Nd:YAG or excimer lasers. Solid-state Nd:YAG systems are chosen due to their many advantages such as being inexpensive, minimal maintenance requirements, minor divergence of the laser beam, and easy to incorporate in small commercial ablation systems. Excimer lasers require careful handling, more care and are rather large. However, the advantages are bigger power output and better-defined beam profile.

2.5 Self-Assembly—Langmuir–Blodgett

Self-assembly is a broad term that is used to describe a variety of processes in which an ordered arrangement of molecules and small components such as small particles occur spontaneously under the influence of forces such as chemical reactions, electrostatic attraction, and capillary forces. This technique is commonly used to form monolayers on substrates by surface modification and to alter the physicochemical property of the outermost surface of a material. The two most common methods for depositing monolayers are, namely, self-assembled monolayer (SAM) and Langmuir–Blodgett deposition. These methods will be discussed in the following section.

2.5.1 Self-assembled Monolayers (SAMs)

An organized self-assembled monolayer (SAM) is a single layer of molecules formed spontaneously on a substrate by the adsorption of molecular constituents from solution or from the gas phase onto solid surfaces [69] in which the molecules exhibit a high degree of orientation, molecular order, and packing. The understanding of the self-assembly process is important in order to construct a well-designed SAM and to obtain well-defined molecular layers with tailored properties. A schematization of the preparation and the constituents of a SAM is shown in Fig. 2.15. SAMs are prepared by immersion of a substrate into a solution of organic molecules. Typically the molecules that form SAMs, also called surfactants, consist of three parts: a headgroup, an endgroup (or tail-group), and a backbone or bridge [70], as shown in Fig. 2.15. The endgroup provides the functionality to the SAM as it constitutes the outer surface of the film. SAMs with different surface properties can be achieved by combining the spacer with a functional tail-group such as carboxyl, hydroxyl, porphyrin, or ferrocene, etc. The backbone or bridge connects headgroup and endgroup and affects the intermolecular separation and molecular orientation. The headgroup binds the molecule to the substrate surface forming a covalent bond. The headgroup-substrate is the primary driving force for the self-assembly process interaction and the strongest of all interactions [71]. In fact, a strong headgroup-substrate interaction results in an immobilization of the molecules on specific sites on the surface. The type of headgroup depends on the chemical composition of the substrate. There is a number of headgroups that bind to specific metals, metal oxides, and semiconductors. For example, organosilanes bind to hydroxylated surfaces via a Si-O bond, thiols bind to gold via a S-Au bond, and carboxylic acids bind to silver via an ionic $COO^{-}Ag^{+}$ bond. The most extensively studied class of SAMs is derived from the adsorption of thiol (-SH) molecules on noble metals and semiconductor surfaces. The thiol headgroup is one of the rather rare functionalities which forms a strong interaction with noble metals [72, 73]. Therefore, it is possible to utilize the thiol molecules to generate well-defined organic surfaces with alterable chemical functionalities displayed at the outer surface. An alternative to thiol is selenol (-SeH). Another class of elements used in SAMs is tellurium although SAMs formed from tellurium-containing compounds are not stable under ambient conditions as they tend to oxidize easily after the film formation [74, 75].

The self-assembly process can be divided into two stages. In the first stage which takes generally a few minutes, the molecules adsorb and form bonds between the headgroup and the substrate.

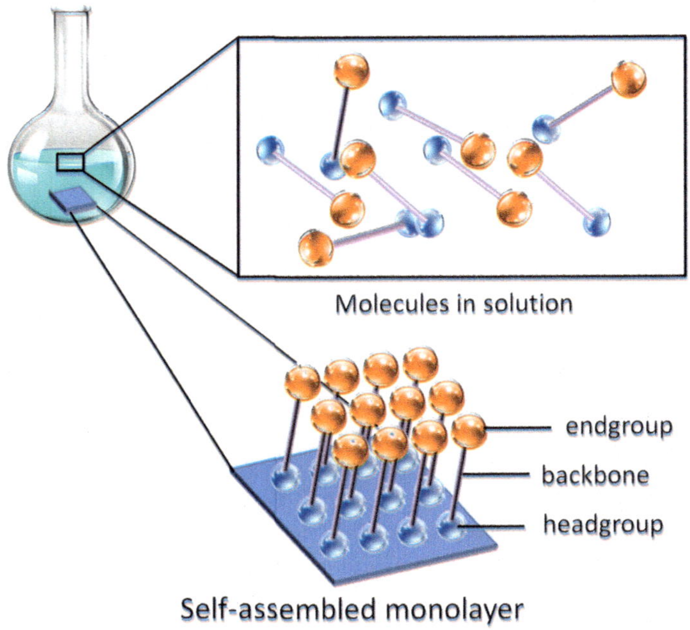

Fig. 2.15 Schematic of SAM preparation and the constituents of a SAM-molecule (headgroup, spacer, and tail-group)

The second stage lasts for several hours, wherein the molecules pinned to the surface, start organizing themselves into an ordered, densely packed layer [69] until the surface reaches saturation. At this stage, the molecules organize themselves to reach equilibrium and form an ordered film. Organization is driven by van der Waals interactions between the chains, as the system aims to optimize lateral interactions and reach potential energy minimum. The energy associated with the intermolecular interaction is in the order of tens of kJ/mol (but lower than 42 kJ/mol) [71]. This interaction plays a crucial role in the molecular packing on the surface.

Therefore, the structure of the monolayer is determined by the interplay of the headgroup–substrate and intermolecular interactions. Finally, mixed monolayers can be created. Mixed SAMs are obtained by coadsorption of molecules with different functionalities. In this way, the surface properties can be designed via the control of surface chemical functionalities. Mixed SAMs provide the opportunity to improve the quality of the monolayer that comprises of molecules whose physical dimension precludes a well-organized assembly. For example, a mixed SAM was obtained by diluting ferrocene alkanethiol in unfunctionalized alkanethiols [76]. Another advantage of using mixed SAMs is that it can be associated with microcontact printing to form patterns on the surfaces, that are useful for microfabrication and bioapplications [77–79].

Many variants of the self-assembly technique have emerged over the past decade. Bottom-up methods such as lithographically controlled wetting (LCW) or “confined dewetting lithography" (CDL) assemble the atomic or molecular components into controlled patterns on various chemically modified substrates as a result of dewetting or other instabilities and by confining surface organization to a defined geometry [80, 81]. An important surface tension driven process (STD) is the dewetting of thin polymer films in which a macromolecular layer retracts from a surface that does not favor spreading as a function of the polarity of solutions used [82]. The ability to control surface tension provides ample opportunity to provide a low-cost and convenient approach to generate organic/inorganic patterning over a large area.

2.5.2 *Langmuir–Blodgett (LB) Technique*

Langmuir–Blodgett (LB) deposition technique is a versatile and useful tool to create self-assembled systems with applications in electronics and optoelectronics. It consists of the deposition of preformed monolayers from a gas–liquid interface to a solid planar substrate [83, 69]. Pioneering work of Irving Langmuir first developed the theoretical and experimental concepts of the behavior of the molecules in insoluble monolayer floating on the interface, referred to as Langmuir monolayer or Langmuir film [84]. In collaboration with Kathleen Blodgett, Langmuir carried out a study on the transfer of fatty acids onto a solid substrate from a film on the air–water interface [85]. Transfer of the monolayer film (also referred to as a "Langmuir" monolayer) is achieved by passing the substrate vertically through the monolayer that is supported on the water subphase. However, complex self-assembled systems were obtained in a controlled manner only in the sixties, mostly owing to Kuhn's work [86].

The key feature of this technique is the ability of molecules to be trapped in the air–water interface. Amphiphilic molecules, typically comprising of a long hydrocarbon "tail" and a polar "headgroup", tend to be used in LB processes due to their ability to form insoluble, stable monolayer films at an air–water interface. The balance between the opposing solubilities of the polar headgroup (pulls the molecule into the subphase, i.e., the substance on which the monolayer is going to be formed) and hydrocarbon chain (remain orientated toward the air) act to maintain the monolayer stable at the interface. Changes in the polarity of the headgroup will influence the monolayer stability: weakly polar headgroups result in aggregation of the molecules into drops on the water surface while highly polar headgroups can cause the molecules to be too soluble in the subphase for stable monolayer formation. Most monolayer forming surfactants are brought onto a subphase surface by first dissolving them in a proper solvent (often chloroform). Mainly the subphase is demineralized–deionized water with 18.5 MΩ resistivity. This low ion-content ensures that the surfactant polar head will not be hybridized with minerals that are contained in natural water leaving, therefore, the properties of the monolayer constant.

The monolayers feature properties that can be obtained by measuring the surface pressure. Surface pressure is defined as the decrease in the surface tension of the liquid owing to the presence of the monolayer. It is normally measured using a Wilhelmy plate/electrobalance arrangement that monitors the force required for the

Fig. 2.16 Pressure–area isotherm and molecular configuration

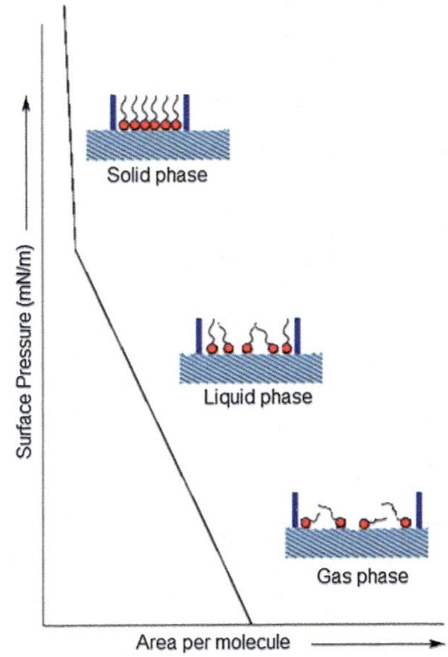

sensing plate to be kept constant versus changes in surface tension [87]. The pressure–area ($\pi - A$) isotherm, therefore, provides the "fingerprint" of the monolayer. A typical ($\pi - A$) is shown in Fig. 2.16 and usually consists of three distinct regions. In the low-pressure region (gas phase), the molecules behave as a two-dimensional gas; they are far enough apart on the water surface to undergo only weak interactions with their neighbors. In the gas phase, the molecules are not strongly interacting with each other. In the liquid phase region, generally called expanded monolayer phase, the hydrocarbon chains assume a more regular orientation with their polar groups that are in contact with the subphase. When the molecular area is decreased, the intermolecular distance between the molecules decreases, the molecules become more closely packed and start to interact with each other. At the solid phase, the molecules are completely organized, oriented with the hydrocarbon chains pointing away from the water surface and the surface pressure increases dramatically. Increasing the surface pressure, a maximum surface pressure (the collapse point) is reached after which the monolayer packing is no longer controlled.

The structure of the monolayer that is deposited on the substrate depends on the film transfer technique employed. The method most commonly used is the upstroke transfer of the monolayer to a hydrophilic substrate, by pulling the substrate upwards through a subphase-supported Langmuir monolayer. In the resulting monolayer, the polar headgroups adhere to the hydrophilic surface, as shown in Fig. 2.17a. Conversely, monolayers can also be deposited using the downstroke transfer in a "tail-down" fashion on a hydrophobic surface through vertical passing of the substrate through the monolayer, into the subphase, as shown in Fig. 2.17b.

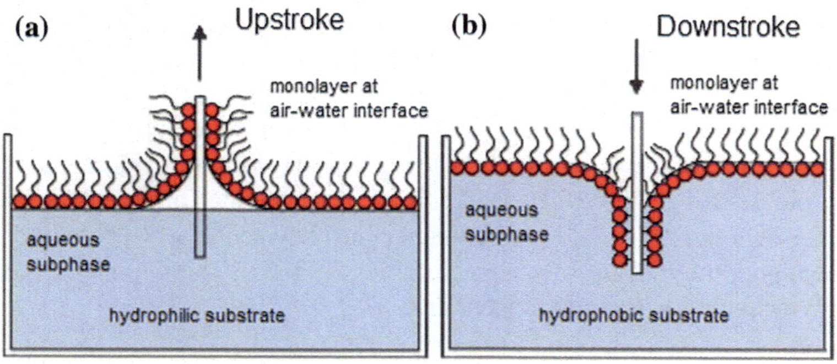

Fig. 2.17 Schematic showing mechanisms of transfer of LB monolayers from an aqueous subphase surface to a **a** Hydrophilic and **b** Hydrophobic substrate

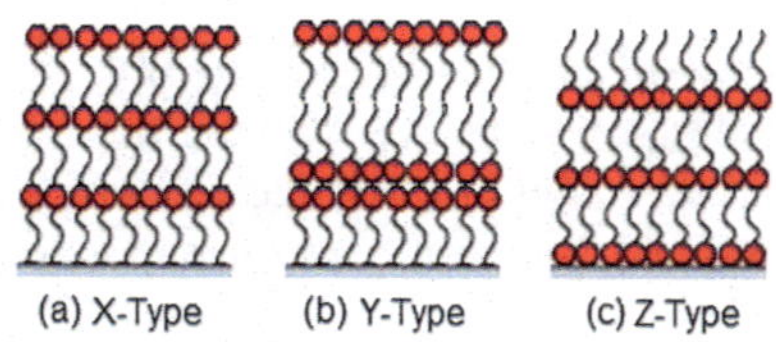

Fig. 2.18 Schematic representation of possible multilayered LB structures

The wettability of the substrate (i.e., hydrophilic/hydrophobic) is also important for processing film transfer. If a substrate is hydrophilic, transfer will only take place on the upstroke, referred to as Y-type films in which the molecules in successive layers adopt a head-to-head and tail-to-tail arrangement as shown in Fig. 2.18b. On hydrophobic substrates, transfer will take place on downstroke of the substrate. The resulting structure will stack head-to-tail arrangements of their constituent monolayers, known as X-type (downstroke deposition only) and Z-type (upstroke deposition only). Repeated upstroke and downstroke depositions of the Langmuir monolayer on the substrate leads to multilayer depositions of the monolayers.

Efficiency of transfer is quantitatively evaluated using a "transfer ratio" defined as the observed decrease in the Langmuir monolayer area (whilst held at a constant pressure), divided by the surface area of the substrate:

$$\text{Transfer ratio (TR)} = \frac{\text{decrease Langmuir monolayer surface area}}{\text{total surface area of substrate}}$$

However, this equation assumes that the molecular packing and order of the deposited film is the same as that in the initial subphase-supported monolayer without considering changes which may take place within the film structure during deposition. Poor stability of the Langmuir monolayer upon the subphase can also introduce an error into the transfer ratio. For example, poorly adhered monolayers deposited upon the upstroke may "peel off" from the substrate surface upon the subsequent downstroke and the transfer ratio does not account for "peeling" off of such monolayers.

Advantages presented by this technique are the precise control of the monolayer thickness, the homogeneous deposition of the monolayer over large areas, the possibility to obtain a multilayered structure by varying the layer composition and that the deposition can be obtained on almost any kind of solid substrate. However, there are several disadvantages, such as it can only be used for molecules that are water-insoluble and have surfactant-like properties. Another drawback is that the surface pressure has to be held constant and controlled accurately. Moreover, films can present poor thermal and mechanical stability [83, 69]. These limitations can be overcome by using variants of the technique.

2.5.3 Variants of the Langmuir–Blodgett Technique

A modification of the Langmuir–Blodgett technique is the horizontal deposition technique, named as Langmuir–Schaefer (LS) technique [88–90]. Conversely to the classic Langmuir–Blodgett technique, the solid substrate is placed parallel to the air–water interface, and the deposition is done by dipping the substrate horizontally through a floating monolayer from the gas phase (air) toward the liquid phase (monolayer). Thus, the monolayer is transferred with the polar head of the molecules in contact with the air and the hydrophobic part in contact with the solid. The Langmuir–Schaefer method does not produce a Y-type film or an expected X-type multilayer. Another variant is presented by the Kossi-Leblanc technique, in which the substrate is inclined to the air–water interface, usually with an angle of 40°, and then submerged on the subphase. The transfer occurs by dipping the substrate as in the LB deposition (Kossi et al. 1981).

2.6 Wafer Preparation Methods—RCA, Modified RCA

This section outlines the preparation methods that are used for wafer cleaning for electronic, optical, and optoelectronic applications. These are RCA standard cleaning and modified RCA technique. All the techniques have advantages and disadvantages and these will be highlighted where appropriate in the following sections.

2.6.1 Introduction and Background

The preparation of ultraclean silicon wafers is one of the key technologies in the fabrication of advanced semiconductor manufacturing [91]. Effective techniques for cleaning silicon wafers ensure minimal contamination prior to processing and are critically important because of the extreme sensitivity of the semiconductor surface and the increased complexity of the device features. Contaminants present on the surface of silicon wafers at the start of processing, or accumulated during processing, have to be removed at specific processing steps in order to obtain high performance and high-reliability semiconductor devices, and to prevent contamination of process equipment. Contaminants and impurities on the silicon wafer surfaces are present as molecular, ionic contamination, or particles. Molecular compounds are mostly particles or films of condensed organic vapors from lubricants, greases, photoresists, solvent residues, organic compounds from DI (deionized) water, fingerprints or plastic storage containers and inorganic compounds. Ionic contamination involves cations and anions caused mostly by inorganic chemicals that may be physically adsorbed or chemically bonded (chemisorbed), such as ions of sodium, fluorine, and chlorine. Particles may consist of silicon particles, dust, fibers, or metal debris from equipment, processing chemicals, factory operators, wafer handling, and film deposition systems. The main objective of wafer cleaning is, therefore, the removal of particles and chemical impurities from the semiconductor surface without inducing damage, degradation or deleteriously altering the substrate surface. Several techniques of wafer cleaning and surface conditioning have been used such as plasma, dry-physical, wet chemical, vapor phase, and supercritical fluid methods [92, 93]. Dry cleaning processes use gas-phase chemistry and rely on chemical reactions that are required for wafer cleaning, as well as other techniques such as laser, aerosols and ozonated chemistries. However, the most widely used method for cleaning silicon wafers in the initial stages of processing is based on aqueous–chemical processes that typically use hydrogen peroxide mixtures renowned as the "RCA cleaning process". Generally, dry cleaning technologies use less chemicals and are less hazardous for the environment but usually do not perform as well as wet methods, especially for particle removal. The discussion of the RCA cleaning process is as follows.

2.6.2 RCA Cleaning

RCA cleaning was developed by Werner Kern at RCA (Radio Corporation of America) laboratories in the late 1960s—hence the name [91]. The process consists of two consecutively applied solutions known as "RCA Standard Clean", SC-1 and SC-2.

Standard Clean-1

RCA-1 clean (or "standard clean-1", SC-1) is a procedure for removing organic residue and films from silicon wafers. The SC-1 solution for the first processing step consists of a mixture of water (H_2O), hydrogen peroxide (H_2O_2), and ammonium hydroxide (NH_4OH) usually in the range from 5:1:1 to 7:2:1 parts by volume. The ratio usually used is 5:1;1, see Box 1. The decontamination works based on sequential oxidative desorption and complexing with H_2O_2-NH_2OH-H_2O (RCA1). The ammonium hydroxide also serves to remove impurities by complexing some periodic group IB and IIB metals such as Cu, Au, Ag, Zn, and Cd, as well as some elements from other groups such as Ni, Co, and Cr. Actually, Cu, Ni, Co, and Zn are known to form amine complexes. In the process, it oxidizes the silicon at a very low rate and forms a new thin oxide layer on the silicon surface by oxidation at approximately the same rate. A second RCA-2 clean (SC-2) comprising of H_2O_2-HCl-H_2O is often used to further clean the surface.

Standard Clean-2

The SC-2 clean solution was designed to dissolve and remove alkali residue and any residual trace metal as well as metal hydroxides from the silicon surface. Decontamination is based on sequential oxidative desorption and complexing with H_2O_2-NCl-H_2O (RCA-2). Typically this is preceded by an RCA-1 clean to remove organic residues. The SC-2 solution consists of a mixture of water, hydrogen peroxide, and hydrochloric acid (HCl) usually in the range of 6:1:1 to 8:2:1 parts by volume. The ratio normally used is 6:1:1, for simplicity, see Box 2. Displacement resulting from solution is prevented by the formation of soluble metal complexes with the dissolved ions. The solution does not etch silicon or oxides and does not have the beneficial surfactant activity of SC-1 for removing particles. SC-2 has better thermal stability than SC-1 so that the treatment temperature and bath life do not need to be as closely controlled.

Box 1: STANDARD CLEAN No.1 (SC-1)

Materials needed

- Ammonium hydroxide
- Hydrogen peroxide
- Pyrex bath containers
- Hot plate

Preparation: Setup time for this process is about 5 minutes. This process takes about 20 minutes to complete.

- 5 parts DI water
- 1 part Ammonium Hydroxide (NH_4OH – 27-28%)
- 1 part Hydrogen Peroxide (H_2O_2 30%)

Process

- Add 500ml of DI water in a Pyrex beaker
- Add 100ml of ammonium hydroxide, and then heat to 75 to 80°C on hot plate
- Remove from hot plate and add 100ml of hydrogen peroxide.
- Solution will bubble vigorously after 1–2 minutes, indicating that it is ready for use.
- Place wafer(s) into the solution and hold for 15 When finished, transfer the wafer to a container with overflowing DI water from a tap to rinse and remove the solution
- After several water changes, remove the wafer under flowing water.

Note: RCA cleaning solution loses its effectiveness in 24 hours at room temperature (30 minutes at 70°C).To dispose of the RCA-1 solution, dilute with cold water, let cool and sit for 10 minutes, then pour down the drain with plenty of cold water to flush. Rinse all labware three times in clean water.

Box 2: STANDARD CLEAN No. 2(SC-2)

Materials needed

- Hydrogen chloride
- Hydrogen peroxide
- Pyrex bath containers
- Hot plate

Preparation: Setup time for this process is about 5 minutes. This process takes about 20 minutes to complete.

- 6 parts DI Water
- 1 part Hydrochloric Acid (HCl).
- 1 part Hydrogen Peroxide (H_2O_2 30%)

Process

- Add 600ml of DI water in a Pyrex beaker
- carefully add 100ml of HCl, and then heat to 75 to 85°C on hot plate
- Remove from hot plate and add 100ml of hydrogen peroxide.
- Solution will bubble vigorously after 1–2 minutes, indicating that it is ready for use.
- Place wafer(s) into the solution and hold for 10 minutes.
- When finished, transfer the wafer to a container with cleanDI water.

Note: To dispose of the RCA-2 solution, let cool to room temperature. Then pour in a labeled waste container. Close, but do not tighten, waste lid to allow escape of any additional gases that might be generated.

⚠ Hydrogen peroxide is an explosive chemical. Never leave the RCA process unattended. Do not store the hydrogen peroxide near the hotplate or any other source of heat. Any small spills should be wiped up immediately with wipes. Dispose the wipes in the corrosive waste container. In case of exposure to skin or eyes, flush immediately with water for 15 minutes. Remove all clothing that are exposed and flush with water. Report to staff immediately. Seek medical attention to ensure that the burns are minimal.

2.6.3 Modified RCA

The properties of RCA wet chemical processing is based on processing time, temperature, concentration, and process parameter. Process improvements to the original RCA cleaning procedure have been introduced over the years to achieve more efficient results [94–97]. One important development was the integration with megasonic cleaning system for the cleaning and rinsing of wafers [97]. The advantage of using megasonic agitation is related to the physical dislodgment of surface particles from the wafer in SC-1 cleaning. It also allows a considerable reduction in solution temperature and offers a much more effective rinsing of the surface with respect to simple immersion tank processing. Another modification consisted of reducing the NH_4OH concentration (e.g., 5:1:0.25 H_2O:H_2O_2:NH_4OH) in SC-1 by at least four-fold to allow to prevent micro-roughening of the silicon surface and to enhance particle removal [98, 99].

Additional dilution of SC-1 and SC-2 with DI water to various concentrations (typically tenfold) helps to attain effective cleaning [100]. Moreover, substitution of SC-2 with very dilute HCl at room temperature could still facilitate to remove residual metals and their hydroxides [98].

Additionally, an optional process step has been introduced by stripping the hydrous oxide film formed after SC-1 with high-purity particle-free, 1:50 HF for 10 s so as to reexpose the silicon surface for the subsequent SC-2 treatment.

The number of cleaning and rinsing steps has also been decreased by using the so-called Marangoni technique. The Marangoni principle involves the slow withdrawal of wafers from a DI water bath to an environment of isopropyl alcohol (IPA) and nitrogen such that only the portion of the surface that is at the interface of the liquid and vapor phase is "drying" at any one time. In this way, uncontrolled evaporative drying on the wafer is prevented. IPA drying provides a significant advantage in hydrophobic cleaning steps such as pre-gate, pre-silicide, and precontact cleans. Many advances are based on the use of the diluted chemistries and ozonated UPW (Ultra Pure Water) as a replacement of hydrogen peroxide or even sulphuric acid-based mixtures.

2.6.4 HF Step or Diluted Hydrofluoric Acid (HF or DHF @20–25 °C)

HF processing is often applied as the last step of the RCA cleaning sequence in order to create an oxide-free, hydrogen-passivated, hydrophobic silicon surface. It removes oxides from areas of interest, etches silicon oxides and dioxides, and reduces metals contamination of the surface. The method incorporates brief immersion of the SC-1/SC-2 cleaned wafers in a very dilute (1:100) ultrahigh-purity HF, followed by final rinsing and drying [101, 102]. Alternatively, wafers can be exposed to HF-IPA (isopropyl alcohol) vapor [103]. In either case, both methods lead to a very clean hydrogen-passivated, hydrophobic silicon surface which is suitable for the epitaxial growth of silicon layers in which no oxide traces can be tolerated. Sometimes buffered oxide etch, (BOE or BHF,/NH_4/HF/H_2O @60–80 °C) is used in place of DHF in some processes, but exposure to it can lead to NH_4F precipitation and contamination.

2.7 Diffusion

Diffusion is a well-known natural phenomenon. It delineates a thermodynamically driven process, defined as the mass transport occurring in response to a nonhomogeneous distribution of atoms in a solid (i.e., due to a gradient in the chemical potential of a component in the system), whereas atomic movements occur to lead the system back toward thermal equilibrium. Diffusion can be a slow process: diffusion coefficients in gases are around 10^{-5} m^2 s^{-1}, in liquids, about 10^{-9} m^2 s^{-1}, slower still in solids [104–106].

Comprehension of diffusion mechanisms is of fundamental importance and interest in the semiconductor technology field since, for example, the electrical properties of semiconductor devices strongly depend on the thermal stability of a p-n junction or an Ohmic contact.

2.7.1 Basic Models for Diffusion

Mass transport in solids can be discussed on the basis of Fick's first law of diffusion

$$J(x,t) = -D(x,t)\frac{\partial C(x,t)}{\partial x} \tag{2.7.1}$$

where $J(x, t)$ is the rate of transfer of species per unit area of the section, $C(x, t)$ is the concentration of diffusing species [atoms/m^3], $D(x)$ [m^2/s] is the diffusion coefficient and x is the direction perpendicular to the substrate surface.

Applying the continuity equation for mass conservation which states that the ratio of increase in concentration with time corresponds to the negative of the divergence of the particle flux, it results in

$$\frac{\partial C}{\partial t} = -\nabla J \tag{2.7.2}$$

In a one-dimensional case, the divergence is equivalent to the gradient, therefore, Fick's second law of diffusion may be derived as

$$\frac{\partial C}{\partial t} = D\frac{\partial^2 C}{\partial x^2} \tag{2.7.3}$$

where the diffusion coefficient is position independent but dependent on the concentration C and the temperature T. Equation 2.7.3 is defined as the concentration independent diffusion equation. Analytically, the mathematical solutions of the diffusion equation depend on the boundary conditions, which are controlled by the physical conditions of the experiment. Therefore, experimental profiles allow the determination of values of D considering only well-established initial conditions for diffusion.

In the eventuality that D is concentration dependent (i.e., in doped semiconductors), more sophisticated numerical methods and the Boltzmann–Matano analyses are required to extract the effective diffusivity D_{eff} from the experimental profiles [107].

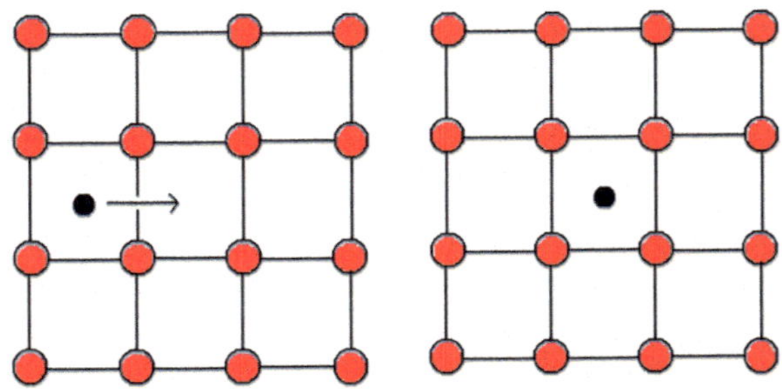

Fig. 2.19 Interstitial Mechanism

By definition, diffusion is a result of the random motion of the diffused particles which are always thermally activated. Thus, the diffusivity, D, is strongly temperature dependent and can be expressed by the Arrhenius expression

$$D(T) = D_0 \times \exp(-E_A/kT) \tag{2.7.4}$$

where the factor D_0 is constant, E_A is the activation energy, and k is the Boltzmann constant. This equation is valid for most diffusing species although the concentration dependence relies on the specific diffusing species [108, 109].

2.7.2 *Diffusion Mechanisms in Semiconductors*

In semiconductors, a considerable number of impurities, labelled A, are present as interstitial–substitutional, i-s, species.

Atoms in crystals oscillate around their equilibrium positions but under determinate conditions, an atom may change its site in the crystal when oscillations become large enough. These jumps cause diffusion in solids. Several atomic mechanisms of diffusion in crystals are described in the following sections.

2.7.2.1 Interstitial Mechanism

The most simple diffusion mechanism in semiconductors is the interstitial mechanism (Fig. 2.19). Such substitutional/interstitial interchange is often denoted as kick-out reactions and non-native point defects are involved in this mechanism.

This mechanism assumes that a diffusing atom jumps from an interstitial site to another (i.e., in the space between lattice atoms). This kind of diffusion is typically very fast and it is characteristic of very small atoms compared to host lattice atoms.

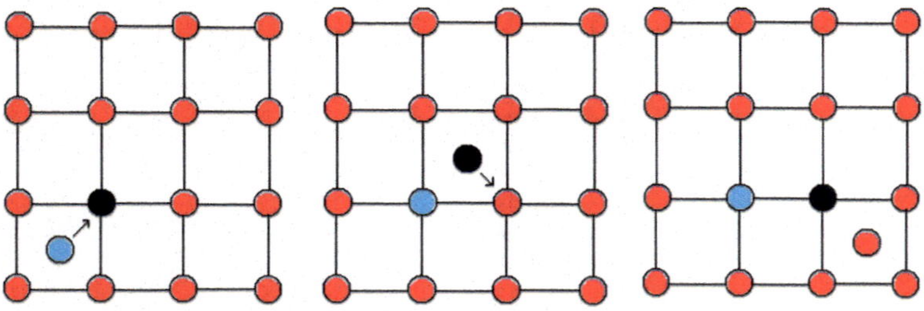

Fig. 2.20 Interstitialcy mechanism

2.7.2.2 Interstitialcy Mechanism

The interstitialcy mechanism or interstitial-mediated mechanism assumes that diffusion can be expressed as the following reaction:

$$A_s + I \leftrightarrow A_I$$

where A_s is an (self- or impurity) atom at a substitutional site and I is the self-interstitial. The substitutionally dissolved atom is exchanged by a self-interstitial and pushed into an interstitial site from which it changes over to a neigboring lattice site by pushing out that lattice atom, as shown in Fig. 2.20.

Within this mechanism, atoms dissolved at substitutional sites diffuse by interacting with self-interstitials displacing each other from substitutional sites to interstitial sites.

Diffusion through interstitial and interstitialcy mechanisms are very similar in nature, and in general are not distinguishable experimentally. Therefore, these diffusion mechanisms are often collectively referred to as I-type diffusion, or simply just interstitial diffusion.

2.7.2.3 Vacancy Mechanism

The most common diffusion mechanism is the vacancy mechanism (Fig. 2.21). Vacancy is a void in the ordinary lattice site. If a lattice vacancy is present adjacent to a substitutional atom, it can move to the vacant lattice site. This kind of mechanism can be described by the reaction

$$A_s + V \leftrightarrow AV$$

where V is the vacancy defect and AV is the vacancy-atom pair.

Similar to the interstitialcy vacancy mechanism, atoms diffuse in the lattice by exchanging places with vacancies and substitutionally dissolved atoms require the

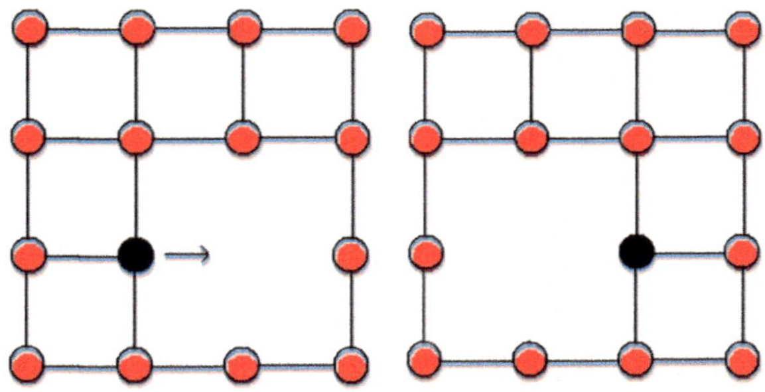

Fig. 2.21 Vacancy mechanism

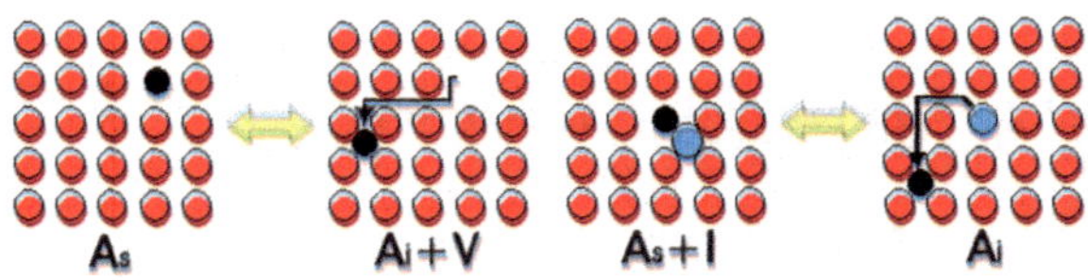

Fig. 2.22 Kick-out and dissociative (Frank-Turnbull) mechanisms

incorporation of defects in order to diffuse. However, diffusivity is enhanced by increasing the vacancy concentration.

2.7.2.4 Kick-Out and Dissociative Mechanisms

Diffusion may occur via kick-out or dissociative mechanisms. In this set of circumstances, the atom can relocate both to the substitutional and interstitial sites, as shown in Fig. 2.22.

The kick-out mechanism is the same as the interstitialcy mechanism but differs as an atom at an interstitial (A_I) state can diffuse long distances via direct interstitial mechanism before it is kicked back to a substitutional site. The reaction can be expressed as

$$A_s + I \leftrightarrow A_I$$

The dissociative mechanism is also denoted as Frank-Turnbull mechanism [110]; the change in the lattice site type occurs via interaction with vacancies, expressed as

$$A_s \leftrightarrow A_I + V$$

or rather the atom at a substitutional site can dissociate to an atom an interstitial site and a vacancy (formation of a Frenkel pair). Backward reaction occurs by recombination of the interstitial atom and the vacancy to A_s. The Frank-Turnbull mechanism is also denoted as a dissociative mechanism and even in this case, A_I can diffuse long distances via direct interstitial mechanism before recombining with a vacancy, and converting to a substitutional atom.

These two mechanisms have common large effective diffusion coefficients, as direct interstitial diffusion is very fast compared to point defect mediated diffusion (like interstitialcy and vacancy mechanisms). Such mechanisms describe the diffusion behavior of hybrid elements, such as gold, sulfur, zinc, and platinum that are mainly dissolved on substitutional sites but move as interstitial defects.

The main diffusion mechanism, being interstitial or vacancy mediated, depends on the dopant chemical species. The solution to these equations can be obtained by numerical methods.

Advantages of using this method are that no damage is created to the lattice and batch fabrication is possible. Drawbacks of diffusion are related to the solid solubility limit and it is a high-temperature process.

2.8 Ion Implantation

Ion implantation is the preferred technique that is used to introduce impurities into solids, i.e., dopants into semiconductor crystals, due to its many advantages: reproducibility, precise dose control, and tailored doping profiles in a uniform and reliable manner [111, 112]. It is, therefore, also one of the key processing tools in the semiconductor industry. Ion implantation involves the forced entry of ions of the desired doping element into a substrate through its surface by virtue of the kinetic energy of the ions (or momentum).

The ions which bombard the target have typically energies that range from keV to a few MeV. As the ions travel through the lattice, they interact with the lattice atoms through elastic collisions with electron clouds and inelastic electronic or nuclear collisions, causing disorder in the lattice structure, till the ions stop at a depth normally referred to as the range. The concentration and distance of penetration of the implanted ions can be controlled by varying ion current, time duration of implantation, and beam energy. After the energetic ions come to rest and equilibration has occurred, the implanted atoms can be in a position in which they serve to change the electronic properties of the substrate lattice (interstitial or preferably substitutional sites), i.e., doping occurs. In the early 1970s, it was found that ion implantation of metal surfaces could improve their wear, hardness, friction, and corrosion properties [111, 113–118]. Ion implantation presents no solubility constraints due to being a nonequilibrium process; therefore, any atomic species can be implanted into any target.

2.8.1 Energy Loss

As an ion penetrates a solid, it undergoes a series of inelastic electronic excitations as well as elastic collisions and interactions with the nuclei and electrons present in the target. The former process is known as the electronic energy loss and the latter is referred to as nuclear stopping. Typically, electronic energy loss is dominant in high energy implants whereas nuclear stopping dominates as the ions slow down [119]. The total energy loss is a combination of both processes. The nuclear stopping power is modeled as

$$S_{\mathrm{n}}(E) = 2.8 * 10^5 \frac{Z_1 Z_2}{\left(Z_1^{2/3} + Z_2^{2/3}\right)^{1/2}} \frac{m_1}{m_1 + m_2} \tag{2.8.1}$$

where $S_{\mathrm{n}}(E)$ is the nuclear stopping power as a function of energy, Z_1 and Z_2 are the atomic number of the ion and substrate, respectively, m_1 and m_2 the mass of the ion and the substrate, respectively. As seen from Eq. 2.8.1, nuclear stopping increases with increasing mass and decreasing energy. Nuclear stopping causes most of the damage due to ion implantation. The electronic stopping mechanism is expressed as

$$S_{\mathrm{e}}(E) = c v_{ion} = k E^{1/2} \tag{2.8.2}$$

where $S_{\mathrm{e}}(E)$ is the electronic stopping power as a function of energy, E is the energy of the ion, v_{ion} is the velocity of the ion, and c and k depend on the ion, substrate and the stopping mechanism. From this equation, it is clear that the electronic stopping increases with increasing implant energy.

The total energy loss is given by the combination of both processes as

$$\frac{\mathrm{d}E}{\mathrm{d}x} = \left(\frac{\mathrm{d}E}{\mathrm{d}x}\right)_e + \left(\frac{\mathrm{d}E}{\mathrm{d}x}\right)_n \tag{2.8.3}$$

which represents the rate at which an ion loses its energy.

Loss mechanisms are calculated with simulation programs, such as the Monte Carlo simulation program SRIM (The Stopping and Range of Ions in Matter).

2.8.2 *Range of Incident Ions*

The range of an incident ion depends on the rate at which it loses energy. The energy loss process is typically statistical. In noncrystalline or amorphous materials, implanted ions exhibit a Gaussian distribution; on the other hand, single crystals display channeling effects and are, therefore, more complicated. Thus, several computer codes (i.e., TRIM, or Transport of ions in matter) are used for calculating the range and range straggling of ions and estimation of damage created during ion implantation [111].

2.8.3 *Ion Implantation Damage*

Ion implantation produces damage through lattice displacement due to energy transfer to the lattice atoms from the primary ion or from recoiled ions during the implantation. Generally, this occurs when the energy transferred to a lattice atom through nuclear collisions is higher than 15 eV. The progressive phenomenon of generation of lattice displacements along the path of the ion through the crystal is denoted as the collision cascade. Lattice atom displacement generates both an interstitial and a vacancy (Frenkel pair) which may recombine during relaxation. The probability of recombination is proportional to several factors such as the separation distance between the interstitial and vacancy, temperature, and the presence of point defect traps. Subsequent recombination will depend on the implant conditions (i.e., ion mass, ion dose, wafer temperature, and ion dose rate), number of interstitials and vacancies. Therefore, upon annealing, various defects can form. These include the following: damaged crystalline lattice; a continuous buried amorphous layer centered around the peak of the damage profile or from the surface down to a depth determined by the implant conditions. An amorphous layer can be recrystallized upon annealing by solid phase epitaxial regrowth, (typically between 550 and 650 °C).

Damage/dopant distribution in the substrate can be also modeled through Monte Carlo simulation and SRIM (Stopping and Range of Ions in Matter). The latter software calculates the distribution of the dopant atoms into the matter and the damage cascade they produce, using quantum–mechanical treatment of the interaction between the ion and the atom of the substrate [120].

2.8.4 *Conventional Ion Implantation Techniques*

Ion implantation has been traditionally developed using ion gun facilities, in which the setup extracts ions from the source, accelerates and focuses them into a beam, which is directed on the target. Often the sample must be rotated and manipulated in the vacuum to obtain a homogeneity of the implantation process. A schematic drawing of a typical ion implanter is shown in Fig. 2.23.

A recent advancement in ion implantation technology is represented by Plasma Immersion Ion Implantation (PIII) [121–124]. In a conventional immersion type of PIII system, also called as the diode-type configuration [125], the sample is kept at a negative pulse potential since the positively charged ions of the electropositive plasma are the ones that get extracted and implanted at normal incidence. The sample holder, placed in a vacuum chamber, is electrically insulated from the chamber wall and it is connected to a high voltage (HV) power supply. The details of a typical PIII setup is shown in Fig. 2.24.

The PIII technique has been shown to be very efficient for the superficial modification of several materials that can be metal, semiconductors or insulating substrates, such as, for example, in the hardening of tools, in the ion implantation of semicon-

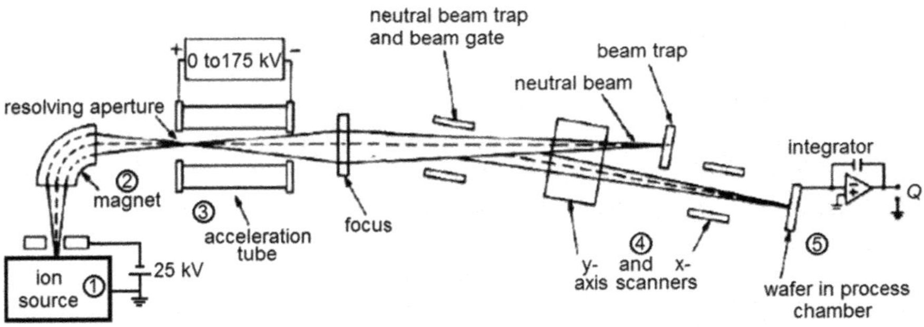

Fig. 2.23 Schematic drawing of a typical ion implanter, showing the ion source (1), the mass spectrometer (2), the high voltage accelerator column (3), the x- and y-axis deflection system (4), and the target chamber (5)

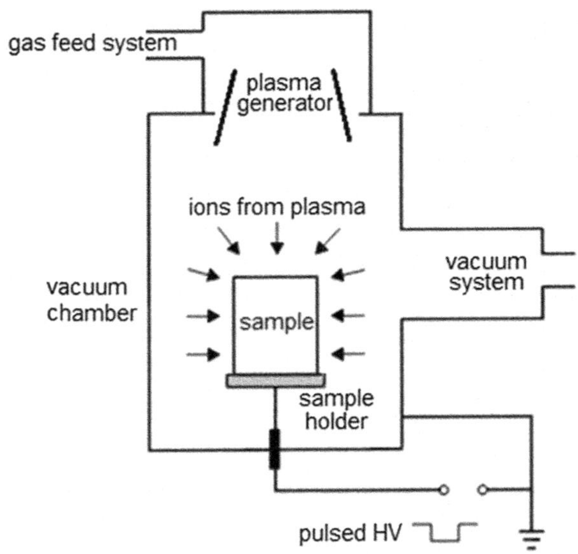

Fig. 2.24 Schematic illustration of the plasma immersion ion implantation technique

ductor devices, in improving the properties of resistance to corrosion and wear of metals and alloys, and most recently, in improving the biocompatibility of biomedical materials. Plasma is generally produced by thermionic cathodes, radiofrequency, microwaves, magnetron sputtering or vacuum arc discharges and recently by laser ablation process. From the 80s, many developments have been made in this field and also variations of such techniques have been devised [126, 127].

The following are the main advantages of using ion implantation: it is a low-temperature process with short process times; it allows precise dose and depth control and it allows implantations through thin layers of oxide/nitride. Disadvantages are the additional cost of annealing, implant damage enhances diffusion, and channeling.

2.9 Vacuum Deposition Techniques

Vacuum deposition techniques have become increasingly important for the production of coatings due to their high versatility and exceptional environmental friendliness. These techniques are all commonly characterized by the following principles of operation: samples are coated in chambers with pressures of the order of 10^{-5} mbar, where vapors of the materials which will form the coating, are generated and condense on the substrates. The condensation can occur in the presence of a process gas introduced into the chamber in a controlled mode. Thus, the formation of a compound different from the initial one (reactive deposition) can occur.

Advantages of using the vacuum deposition techniques are the high quality and reproducibility of the coatings, cleanliness in the chamber, and being ecologically clean. Vacuum deposition techniques can be subdivided into two main categories, Physical Vapor Deposition (PVD) and Chemical Vapor Deposition (CVD). PVD is fundamentally a family of techniques in which the vapors form the film by physical means (heating, sputtering), while in CVD techniques, vapors are obtained by means of dissociation of suitable gaseous species. Each of these groups comprise of many different techniques, i.e., PVD includes evaporation, sputtering, and molecular-beam epitaxy; amongst the CVD techniques to be mentioned are the thermal (conventional) CVD and the plasma activated one (PECVD) which have been reviewed earlier in this chapter.

2.9.1 Conventional Evaporation

The most basic form of vacuum evaporation is conventional evaporation. A schematic of a conventional vacuum evaporation system is shown in Fig. 2.25. With conventional evaporation process, the deposition material (source) is heated to its evaporation point in a high vacuum chamber and the vapor is exposed to the substrate in order to grow or deposit a thin film. Evaporation is caused by thermal or electron beam heating of a source material.

The heating element is designed to be made of refractory metals such as tungsten, molybdenum, and tantalum, or alloys in order to avoid contamination of the thin film. Deposition is controlled by regulating the current driven through the boat and a crystal thickness monitor (i.e., quartz crystal microbalance, QCM), which measures the thickness and rate of deposition of a thin film while it is being deposited. Since deposition takes place in a high vacuum, the source vapor travels a straight line, allowing the rate of deposition to have a simple geometric relation to the location of the source of $1/r^2$, where r denotes the distance between source and substrate. Before evaporation begins, the substrate is shielded from the source by a shutter, till an appropriate deposition rate (usually in the angstroms per second scale) is achieved via the control sequence. The shutter is subsequently opened to expose the substrate

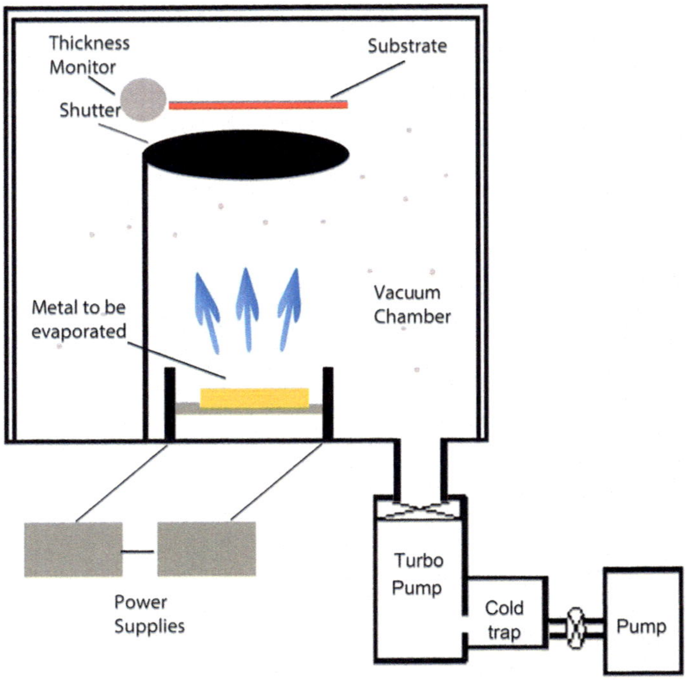

Fig. 2.25 Schematic of an evaporation system

to the source. When the desired thickness of the thin film is reached, the shutter is closed and the evaporation is stopped.

Despite the advantages of this technique, thermal and reactivity restrictions limit its implementation. In fact, with resistively heated sources, constraints are given by power limitations, reactivity with coated filaments, and exceedingly high melting points [128]. To overcome some of these difficulties, the implementation of electron beam evaporation has gained more and more importance [129, 130]. Molecular Beam Epitaxy (MBE) also represents a specific implementation of conventional evaporation or electron beam evaporation.

Classification of vacuum evaporation processes comprises also reactive evaporation and metalizing. Reactive evaporation utilizes a chemical reaction during the evaporation process wherein, as the source material is vaporized, it is intermixed with a reacting gas (e.g., oxygen or hydrogen) to produce the desired stoichiometry of coating on the substrate. This kind of process is delineated as a combination of both chemical and physical vapor depositions [131].

Metallizing is a common form of vacuum evaporation. This process involves the deposition of various metals, from aluminum and copper to gold and platinum [130].

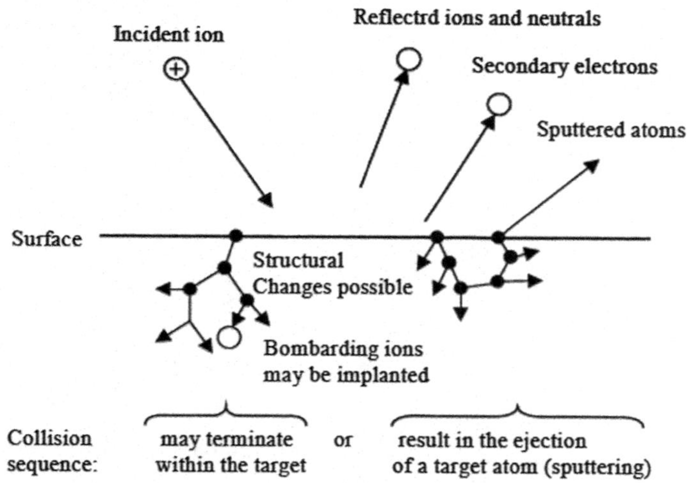

Fig. 2.26 Possible processes occurring at the target surface during sputtering

2.9.2 Sputtering

Sputter deposition is a process in which a solid target material is deposited as a solid onto a substrate under high vacuum. It is classified as a PVD technique as the material does not typically undergo any chemical composition changes during deposition.

The basis of this process relies on the collision between incident atoms, usually ions, with the surface of the target material and subsequent ejection of target atoms after striking the surface, leaving the target surface with relatively high energies (~10 eV). The sputter system, most generally used, supplies the incident ions in the form of plasma and attracts the ions with voltage applied to the target which is located at the cathode. Atoms are ejected from the surface of the target and deposited to condense on a substrate to form a thin film.

The efficiency of this process depends on the sputtering yield (Y) of the target material defined as the average number of atoms ejected from the surface per incident ion [132]. The sputtering yield is influenced by various factors, such as the mass of the sputtering ion, the kinetic energy of the incident ions, the binding energy of the target atoms, the angle of incidence, and the composition of the target, including crystallinity [133,134].

Other processes may also occur, as shown in Fig. 2.26: ions may be reflected, probably being neutralized in the process; the impact due to the ion may cause the target to eject an electron, (or a secondary electron); the ion may become buried in the target, (i.e., ion implantation); the impact of the ion may cause structural rearrangements in the target material.

Sputtering requires ultrahigh vacuum conditions to ensure that the mean free path of sputtered atoms is sufficient to travel from the target to the substrate [130].

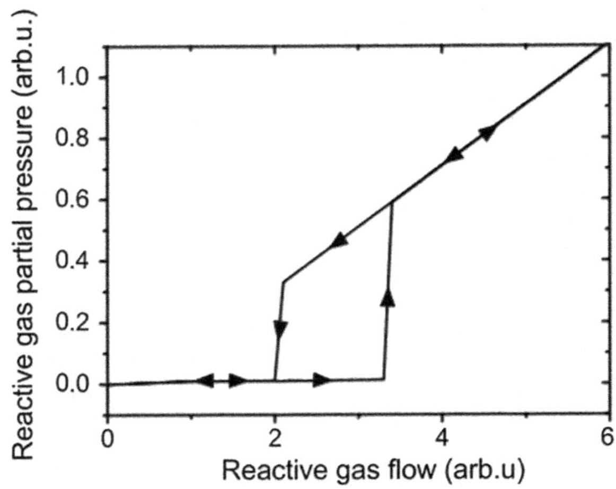

Fig. 2.27 Characteristic gas partial pressure plot versus reactive gas flow during sputtering

Sputtering process can employ DC, RF power, magnetron and reactive sputtering. Typically, DC sputtering uses DC volts (in the KV range) and is used to deposit metals and RF sputtering can be used for metals, semiconductors, and insulators, and utilizes RF power at a specific frequency of 13.56 MHz.

Reactive Sputtering

Reactive sputtering allows sputter deposition of a compound, such as a nitride or oxide, using a metallic target together with a reactive gas such as oxygen or nitrogen. Occasionally, a combination of an inert sputtering gas, such as argon, and a reactive gas is used. This method allows for the use of DC, or pulsed DC power and most often results in higher deposition rates.

The reactive sputtering process is still a rather complicated process in which efficacy is difficult to control [135]. The complexity of the process is caused by the reactions which may occur between the reactive gas and the chamber walls, and the demand for depositing stoichiometric films, while avoiding the formation of a compound phase on the target (so-called target poisoning). A reactive gas flow will lead to the formation of non-stoichiometric films, whereas an excessively high gas flow will lead to the growth of a compound also on the target surface, causing arcing on the target surface or a reduction in the deposition rate. This is for the same reasons as when depositing from a compound target directly. A characteristic reactive gas partial pressure plot versus reactive gas flow is reported in Fig. 2.27.

As the reactive gas is initially introduced, a slight increase in reactive gas partial pressure occurs due to gas consumption that is caused by the compound formation between the sputtered metal and the reactive gas at the chamber walls and the substrate. The threshold value is reached when the chamber walls are saturated and the pressure increases rapidly. Above this limit, the partial pressure varies linearly with the reactive gas flow. Decrease in the reactive gas flow will lead to a decrease in the partial pressure, forming a hysteresis curve.

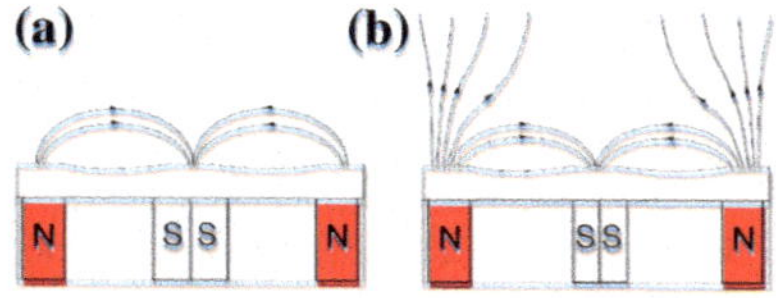

Fig. 2.28 Models of **a** balanced and **b** unbalanced magnetron systems

Magnetron sputtering

Magnetron sputtering represents a flexible deposition process with high throughput and can be used for high-purity thin film fabrication in semiconductors or for large-area coatings of structural materials.

Magnetron sputter deposition systems utilize *magnetrons*, with magnets placed directly behind the target as shown in Fig. 2.28. In most cases, the magnetron is run in "balanced" mode, in which the magnetic field strength of all the magnets is set such that all the field lines remain in closed loops, running from one magnet to another, as shown in Fig. 2.28a. The advantage of using this configuration is that it allows a better confinement of electrons yielding a high sputtering rate by strongly ionizing the feed gas. However, this arrangement may prevent the plasma from spreading. With the "unbalanced" mode, the field lines tend to end on the surfaces rather than on the permanent magnets.

Moreover, a portion of electrons may escape from the magnetron reducing the sputtering rate; on the other hand, plasma extends farther away from the magnetron surface. Further configurations involve the use of two or more magnetrons used in conjunction, i.e., a dual magnetron sputtering system can be arranged in a balanced "mirrored" configuration in which two balanced magnetrons of both magnetrons are mounted in the same way facing each other or in an unbalanced "closed-field" configuration, in which the magnets of both magnetrons are mounted in the opposite way (unbalanced), but the dual magnetron pair is balanced as a group [136, 137].

The mirrored arrangement is used when each magnetron is sputtering a different material, and the region between the magnetrons is used for deposition/reaction. In the closed-field configuration, the plasma is closer to the substrate and a lesser amount of target material is lost on the chamber walls, thus resulting in a cleaner chamber. The result is a dense sputtered coating.

One main advantage of magnetron sputtering is the additional kinetic energy of the sputtered atoms which leads to a more adherent coating and improved conformal coverage. Another advantage is the possibility to obtain compound thin films at much lower temperature [138, 139].

References

1. Bhat HL (2014) Introduction to crystal growth: principles and practice. CRC Press, Taylor and Francis

2. Dhanaraj G, Byrappa K, Prasad V, Dudley M (2010) Springer handbook of crystal growth. Springer, Springer Handbook of Crystal Growth
3. Brice JC (1965) The growth of crystals from the melt. Vol. V selected topics in solid state phys. In: Wohlfarth EP (ed) North Holland, Amsterdam
4. Lu J, Miao J (2012) Growth mechanism of carbon nanotubes: a nano Czochralski model. Nanoscale Res Lett 7:356
5. Brice JC (1986) Crystal growth processes. Blackie&Son Ltd
6. Gilman JJ (1963) The art and science of growing crystals. Wiley, New York
7. Feigelson RS (1983) In: Kaldis E (ed) Crystal growth of electronic materials. North-Holland, New York, pp 127–145
8. Miyagawa C, Kobayashi T, Taishi T et al (2013) Demonstration of crack-free c-axis sapphire crystal growth using the vertical Bridgman method. J Cryst Growth 372:95–99
9. Hoshikawa K, Taishi T, Ohba E et al (2014) Vertical Bridgman growth of sapphire crystals, with thin-neck formation process. J Cryst Growth 401:146–149
10. Wilke K (1988) Kristallzüchtung. In: J. Bohm (ed) VEB Deutscher Verlag der Wissenschaften, Berlin
11. Brice JC (1973) The growth of crystals frorn liquids. North-Holland, Amsterdam
12. Arthur JR (2002) Molecular beam epitaxy. Surf Sci 500:189–217
13. Chambers A (2005) Modern vacuum physics. Chapman & Hall (CRC), Foundations of Vacuum Science and Technology, Wiley-Interscience
14. Knodle WS, Chow R (1988) Molecular beam epitaxy: equipment and practice in handbook of thin film deposition processes and techniques: principles, methods, equipment and applications, 2nd edn. Noyes Publications, Norwich, New York, USA. Ch. 10 Ed Krishna Seshan
15. Oura K, Lifshits VG, Saranin AA et al (2003) Surface science: an introduction. Springer, Advanced Texts in Physics
16. Pimpinelli A, Villain J (1998) Physics of crystal growth. Cambridge University Press, Collection Alea-Saclay
17. Venables JA (2000) Introduction to surface and thin film processes. Cambridge University Press
18. Suntola T, Antson J, Pakkala A et al (1980) Atomic layer epitaxy for producing EL-Thin films SID Intern Symposium. Digest Techn Papers 11:108–109
19. George SM (2010) Atomic layer deposition: an overview. Chem Rev 110:111–131
20. Puurunen RL (2014) A short history of atomic layer deposition: tuomo suntola's atomic layer epitaxy. Chem Vap Deposition 20:332–344
21. Ritala M, Leskelä M (2001) Deposition and processing of thin films in handbook of thin film materials. In: Nalwa HS (ed) Vol 1. Academic Press, San Diego
22. Leskela M, Niinisto L (1990) In Atomic Layer Epitaxy. In: Suntola, T, Simpson M (ed) Blackie and Son Ltd., Glasgow
23. Nilsen O, Mohn CE, Kjekshus A et al (2007) Analytical model for island growth in atomic layer deposition using geometrical principles. J Appl Phys 102:024906
24. Puurunen RL (2004) Random deposition as a growth mode in atomic layer deposition. Chem Vap Deposition 10:159–170
25. Haukka S, Lakomaa EL, Root A (1993) An Ir and NMR study of the chemisorption of TiCl4 on silica. J Phys Chem 97:5085–5094
26. Matero R, Rahtu A, Ritala M et al (2000) Effect of water dose on the atomic layer deposition rate of oxide thin films. ThinSolid Films 368:1–7
27. Goodman CHL, Pessa MJ (1986) Atomic layer epitaxy. Appl Phys 60:R65–R81
28. Nishizawa J, Abe H, Kurabayashi T (1985) Molecular Layer Epitaxy. J Electrochem Soc 132:1197–1200
29. DenBaars SP, Beyler CA, Hariz A et al (1987) GaAs/AlGaAs quantum well lasers with active regions grown by atomic layer epitaxy. Appl Phys Lett 51:1530–1532
30. Bedair SM, Tischler MA, Katsuyama T et al (1985) Atomic layer epitaxy of III-V binary compounds. Appl Phys Lett 47:51–53

31. Tischler MA, Bedair SM (1990) Atomic layer epitay. Blackie and Son Ltd, Glasgow, pp 110–154
32. Usui A, Sunakawa H (1986) GaAS Atomic layer epitaxy by hydride VPE. Jpn J Appl Phys 25:L212–L214
33. Konagai M, Sugimoto M, Takahashi K (1978) High efficiency GaAs thin film solar cells by peeled film technology. J Cryst Growth 45:277–280
34. Yablonovitch E, Gmitter T, Harbison JP et al (1987) Extreme selectivity in the lift-off of epitaxial GaAs films. Appl Phys Lett 51:2222–2224
35. Bauhuis GJ, Mulder P, Haverkamp EJ et al (2010) Wafer reuse for repeated growth of III-V solar cells. Prog Photovolt Res Appl 18:155–159
36. Liu LM, Lindauer G, Alexander WB et al (1995) Surface preparation of ZnSe by chemical methods. J Vac Sci Technol B: Microelectron Nanometer Struct 13:2238–2244
37. Pinel S, Tasselli J, Bailbé JP et al (1998) Mechanical lapping, handling and transfer of ultra-thin wafers. J Micromech Microeng 8:338–342
38. Rei Vilar M, El Beghdadi J, Debontridder F et al (2005) Characterization of wet-etched GaAs (100) surfaces. Surf Interface Anal 37:673–682
39. Cheng CW, Shiu KT, Li N et al (2013) Epitaxial lift-off process for gallium arsenide substrate reuse and flexible electronics. Nature Commun 4:1577
40. Wolf S, Tauber RN (2000) Silicon processing for the VLSI Era. In: Process technology 2nd edn. vol 1. Lattice Press, Sunset Beach, CA
41. Caschera D, Cortese B, Mezzi A et al (2013) Ultra hydrophobic/superhydrophilic modified cotton textiles through functionalized Diamond-Like Carbon coatings for self-cleaning applications. Langmuir 29:2775–2783
42. Cortese B, Caschera D, Federici F et al (2014) Superhydrophobic fabrics For Oil/Water separation through a diamond like carbon (DLC) coating. J Mater. Chem. A 2:6781–6789
43. Moretti G, Guidi F, Canton R et al (2005) Corrosion protection and mechanical performance of SiO_2 films deposited via PECVD on OT59 brass. Anti-Corrosion Methods and Materials 52:266–275
44. Chapman BN (1980) Glow discharge processes: sputtering and plasma etching. Wiley, New York
45. Melliar-Smith CM, Mogab CJ (1978) Thin film processes. In: Vossen, JL, Kern W (ed). Academic Press, New York, pp 497–552
46. Acquafredda P, Bisceglie E, Bottalico D et al (2010) Characterization of polycrystalline diamond films grown by Microwave Plasma Enhanced Chemical Vapor Deposition (MWPECVD) for UV radiation detection. Nucl Instrum Methods Phy Res A 617:405–406
47. Kern W (1986) In microelectronic material and processes. In: Levy RA (eds). Kluwer Academic, New Jersey
48. Kern W, Schnable G (1979) Low-pressure chemical vapor deposition for very large-scale integration processing-A review. L IEEE Trans Electron Dev 26:647–657
49. Stoffel A, Kovács A, Kronast W et al (1996) LPCVD against PECVD for micromechanical applications. J Micromech Microeng 6:20–33
50. Jaeger RC (1993) Introduction to microelectronic fabrication. Addison-Wesley Publishing Company, Inc
51. Pierson HO (1992) Handbook of chemical vapor deposition. Noyes Publications, New Jersey
52. Fu XA, Dunning J, Zorman CA et al (2004) Development of a high-throughput LPCVD process for depositing low stress poly-SiC. Mater Sci Forum 457–460:305–308
53. Campbell SA (2001) The science and engineering of microelectronic fabrication. Oxford University Press, New York
54. Foggiato J (2001) Chemical vapor deposition of silicon dioxide films in handbook of thin film deposition processes and techniques, 2nd edn. Noyes Publications, Norwich, New York, USA, pp 111–150
55. Gesheva KA, Ivanova TM, Bodurov GK (2014) APCVD Transition metal oxides—functional layers in: smart windows. J Phys Conf Ser 559:012002
56. Kittel C (2005) Introduction to solid state physics. Wiley, Inc

57. Ohring M (1992) The materials science of thin films. Academic Press, London
58. Rees WS (1996) Introduction, in CVD of nonmetals. In: Rees WS (ed) Wiley-VCH Verlag GmbH, Weinheim, Germany
59. Krumdieck S (2008) Chemical vapor deposition: precursors and processes. In: Jones A, Hitchman ML (ed) RSC Publishing, Cambridge, UK
60. Blocher JM, Vuilland GE, Wahl G (1981) The electrochemical society: pennington. New Jersey, USA
61. O'Mara WC, Herring RB, Hunt LP (1990) Handbook of semiconductor silicon technology. Noyes Publications, New Jersey, USA
62. Niinisto L, Nieminen M, Päiväsaari J et al (2004) Advanced electronic and optoelectronic materials by atomic layer deposition: an overview with special emphasis on recent progress in processing of high-k dielectrics and other oxide materials. Phys stat sol (a) 201:1443–1452
63. Kleijn CR (2002) Chemical physics of thin film deposition processes for micro- and nanotechnologies. In: Pauleau Y (eds) Springer Netherlands, vol 55, pp 119–144
64. Bäuerle D (2000) Laser processing and chemistry. 3rd edn. Springer, Berlin
65. Scharf T, Krebs HU (2002) Influence of inert gas pressure on deposition rate during pulsed laser deposition. Appl Phys A 75:551–554
66. Proyer S, Stangl E, Borz M, Hellebrand B, Bauerle D (1996) Particulates on pulsed-laser deposited YBaCuO films. Physica C 257:1–15
67. Bierleutgeb K, Proyer S (1997) Pulsed-laser deposition of Y–Ba–Cu–O films: the influence of fluence and oxygen pressure. Appl Surf Sci 110:331–334
68. Rani JR, Mahadevan Pillai VP, Ajimsha RS, Jarajaj MK, Jayasree RS (2006) Effect of substrate roughness on photoluminescence spectra of silicon nanocrystals grown by off axis pulsed laser deposition. J Appl Phys 100:014302
69. Ulman A (1991) An introduction to ultrathin organic films from langmuir-blodgett to self-assembly. Acadamic Press, California
70. Zharnikov M, Grunze M (2001) Spectroscopic characterization of thiol-derived self-assembled monolayers. J Phys: Condens Matter 13:11333–11365
71. Ulman A (1996) Formation and structure of self-assembled monolayers. Chem. Rev. 96:1533–1554
72. Bain CD, Evall J, Whitesides GM (1989) Formation of monolayers by the coadsorption of thiols on gold: variation in the head group, tail group, and solvent. J Am Chem Soc 111:7155–7164
73. Ulman A (1989) Ultrathin organic films: from langmuir-blodgett to self-assembly. J Mat Ed 11:205–207
74. Nakamura T, Miyamae T, Nakai I et al (2005) Adsorption states of dialkyl ditelluride autooxidized monolayers on Au(111). Langmuir 21:3344–3353
75. Weidner T, Shaporenko A, Müller J et al (2007) Self-assembled monolayers of aromatic tellurides on (111)-oriented gold and silver substrates. J Phys Chem C 111:11627–11635
76. Watcharinyanon S, Moons E, Johansson LSO (2009) Mixed self-assembled monolayers of ferrocene-terminated and unsubstituted alkanethiols on gold: surface structure and work function. J Phys Chem C 113:1972–1979
77. Kumar A, Biebuyck HA, Whitesides GM (1994) Patterning self-assembled monolayers: applications in materials science. Langmuir 10:1498–1511
78. Raynor JE, Capadona JR, Collard DM et al (2009) Polymer brushes and self-assembled monolayers: versatile platforms to control cell adhesion to biomaterials. Biointerphases 4:FA3–16
79. Tan JL, Tien J, Chen CS (2002) Microcontact printing of proteins on mixed self-assembled monolayers. Langmuir 18:519–523
80. Cavallini M, Gentili D, Greco P et al (2012) Micro- and nanopatterning by lithographically controlled wetting. Nat Protoc 7:1668–1676
81. Celio H, Barton E, Stevenson KJ (2006) Patterned assembly of colloidal particles by confined dewetting lithography. Langmuir 22:11426–11435
82. Toro RG, Caschera D, Palamà IE et al (2015) Unconventional patterning by solvent-assisted surface-tension-driven lithography. J Colloid Interface Sci 446:44–52

83. Gaines GL Jr (1966) Insoluble monolayers at liquid-gas interfaces. Interscience, New York
84. Langmuir I (1917) The constitution and fundamental properties of solids and liquids. II Liquids J Am Chem Soc 39:1848–1906
85. Blodgett K (1935) Films built by depositing successive monomolecular layers on a solid surface. J Am Chem Soc 57:1007–1022
86. Kuhn H, Möbius D, Bücher H (1972) Spectroscopy of monolayer assemblies. In: Weissberger A, Rossiter B (eds) Physical methods of chemistry, Part III B, vol 1. Wiley, New York
87. Martin R, Szablewski M (1998) Tensiometers and langmuir-blodgett troughs operating manual, 4 edn. Nima Technology Ltd, The Science Park, Coventry, England
88. Lambert K, Capek RK, Bodnarchuk MI et al (2010) Langmuir-schaefer deposition of quantum dot multilayers. Langmuir 26:7732–7736
89. Langmuir I, Schaefer VJ (1938) Activities of urease and pepsin monolayers. J Am Chem Soc 60:1351–1360
90. Lee YH, Lee CK, Tan B et al (2013) Using the langmuir-schaefer technique to fabricate large-area dense SERS-active Au nanoprism monolayer films. Nanoscale 5:6404–6412
91. Kern W (2007) Overview and evolution of silicon wafer cleaning in handbook of silicon wafer cleaning technology, 2nd edn. William Andrew Publishing, Norwich, NY
92. Hess DW, Reinhardt KA (2008) Plasma stripping, cleaning and surface conditioning in handbook of silicon wafer cleaning technology, 2nd edn. William Andrew Publishing, Norwich, NY, pp 355–427
93. Verhaverbeke S, Messoussi R, Morinaga H et al (1995) Recent advances in wet processing technology and science. In: Proceedings Ultra Clean semicond. Processing technology materials research society symposium proceedings, vol. 386, Material research society
94. Kern W (1983) Hydrogen peroxide solutions for silicon wafer cleaning. RCA Eng 28:99–105
95. Kern W (1984) Purifying Si and SiO/sub 2/ surfaces with hydrogen peroxide. Semicond Int 7:94–99
96. Kern W (1990) The evolution of silicon wafer cleaning technology. J Electrochem Soc 137:1887–1892
97. Shwartzman S, Mayer A, Kern W (1985) Megasonic particle removal from solid state wafers. RCA Review 46:81–105
98. Anttila OJ, Tilli MV, Schaekers M et al (1992) Metal contamination removal on silicon wafers using dilute acidic solutions. J Electrochem Soc 139:1180
99. Meuris M, Heyns M, Kuper W et al (1991) Correlation of metal impurity content of ulsi chemicals and defect-related breakdown of gate oxides. ULSI Science and Technology, Pennington NJ, Electrochemical Society, pp 141–161
100. Smith SM, Varadarajan M, Christenson K (1996) The effects of dilute SC-1 and SC-2 chemistries on dielectric breakdown dor pre gate cleans in Proceedings fourth international symposium on cleaning technology in semiconductor device manufacturing. The Electrochem. Soc. Pennington, NJ
101. Ohmi TJ (1996) Total room temperature wet cleaning for si substrate surface. J Electrochem Soc 143:2957–2964
102. Verhaverbeke S, Alay J, Mertens P et al (1992) Surface characterisation of Si after HF treatments and its influence on the dielectric breakdown of thermal oxides in proceedings on chemical surface preparation, passivation, and cleaning, growth and processing. In: Symposium B., Spring Mtg. of MRS, San Francisco
103. Verhaverbeke HS, Schmidt HF, Meuris M et al (1993) Technology Conference Semicon/Europe '93, Geneva, Switzerland
104. Cussler EL (2005) Diffusion, mass transfer in fluid systems, 3rd edn. Cambridge University Press, New York
105. Poling B, Prausnitz J, O'Connell J (2004) The properties of gases and liquids, 5th edn. McGraw-Hill, New York
106. Taylor R, Krishna R (1993) Multicomponent mass transfer, 1st edn. Wiley, New York, NY
107. Crank J (1975) The mathematics of diffusion. Clarendon Press, Oxford

108. Bentzen A, Holt A, Christensen JS et al (2006) High concentration in diffusion of phosphorus in Si from a spray-on source. J Appl Phys 99:064502
109. Vick G, Whittle K (1969) Solid solubility and diffusion coefficients of boron in silicon. J Electrochem Soc 116:1142–1144
110. Frank FC, Turnbull D (1956) Mechanism of diffusion of copper in germanium. Phys Rev 104:617
111. Chason E, Picraux ST, Poate JM et al (1997) Ion beams in silicon processing and characterization. J Appl Phys 81:6513–6561
112. Parikh NR, Thompson DA, Carpenter GJC (1986) Ion implantation damage in CdS. Radiation Effects 98:289–300
113. Current MI (1996) Ion implantation for silicon device manufacturing: a vacuum perspective. J Vac Sci Tech A 14:1115–1123
114. Dearnaley G, Freeman JH, Nelson RS et al (1973) Implantation. American Elsevier Publishing Co., New York
115. Rimini E (1995) Ion implantation: basics to device fabrication. Kluwer Academic Publishers, Boston
116. Rubin L, Morris W (1997) High-energy ion implanters and applications take off. Semicond Internat 20:77–85
117. Ryssel H, Ruge I (1986) Ion implantation. Wiley, New York
118. Ziegler JF (2000) Ion implantation: science and technology. Ion Implant Technology Co., Edgewater
119. Dresselhaus MS, Kalish R (1992) Ion implantation in diamond, Graphite and Related Materials. Springer-Verlag, Berlin Heidelberg New York
120. Ziegler JF (1985) SRIM-stopping and range of ions in matter. Pergamon Press, New York
121. Chu PK, Qin S, Chan C et al (1996) Plasma immersion ion implantation—a fledgling technique for semiconductor processing. Mater Sci and Eng R 17:207–280
122. Conrad JR, Radtke JL, Dodd RA, Worzala J, Tran NC (1987) Plasma source ion-implantation technique for surface modification. J Appl Phys 62:4591–4596
123. Rej DJ, Faehl RJ, Matossian JN (1997) Key issues in plasma-source ion implantation. Surf Coat Technol 96:45–51
124. Sheuer JT, Shamim M, Conrad JR (1990) Model of plasma source ion implantation in planar, cylindrical, and spherical geometries. J Appl Phys 67:1241–1245
125. Liberman MA, Lichtenberg AJ (1994) Principles of plasma discharges and material processing. Wiley, New York
126. Anders A (2000) Handbook of plasma immersion ion implantation. Wiley, New York
127. Qi B, Lau YY, Gilgenbach RM (2001) Extraction of ions from the matrix sheath in ablation-plasma ion implantation. Appl Phys Lett 78:706–708
128. Powell CF (1966) Physical vapor deposition. Vapor deposition. In: Powell CF, Oxley JH, Blocher JM Jr(Ed) Wiley, New York, pp 221–248
129. Konuma M (1992) Film deposition by plasma techniques. Springer series on atoms and plasmas. Springer-Verlag, New York, vol 10
130. Stuart RV (1983) Vacuum technology, thin films, and sputtering. Academic Press, New York
131. Kern W, Schuegraf KK (1988) Deposition technologies and applications: introduction and overview. handbook of thin-film deposition processes and techniques. In: Schuegraf KK (ed) Park ridge. Noyes, New Jersey, pp 1–25
132. Behrisch R (1983) Topics in applied physics: sputtering by particle bombardment II, vol 47. Springer-Verlag, Berlin, Heidelberg, New York
133. Cuomo JJ, Rossnagel SM, Kaufman HR (1989) Handbook of ion beam processing technology. Noyes Publications
134. Lieberman MA, Lichtenberg AJ (2005) Principles of plasma discharges and materials processing, 2nd edn. Wiley, Hoboken, pp 308–310
135. Sproul WD, Christie DJ, Carter DC (2005) Control of reactive sputtering processes. Thin Solid Films 491:1–17

136. Gibson DR, Brinkley I, Wadell EM et al (2008) Closed field magnetron sputtering: new generation sputtering process for optical coatings. Proc SPIE Int Soc Opt Eng 7101:710108
137. Matthews A (2003) Plasma-based physical vapor deposition surface engineering processes. J Vac Sci Technol, A 21(5):S224–S231
138. Berg S, Nyberg T (2005) Fundamental understanding and modeling of reactive sputtering processes. Thin Solid Films 476:215–230
139. Li N, Allain JP, Ruzic DN (2002) Enhancement of aluminum oxide physical vapor deposition with a secondary plasma. Surf Coat Technol 149:161–170

Chapter 3
Characterization Techniques

Marla Berenice Hernández Hernández,
Mario Alberto García-Ramírez, Yaping Dan,
Josué A. Aguilar-Martínez, Bindu Krishnan and Sadasivan Shaji

3.1 Electrical Properties

3.1.1 Introduction

Nowadays, our society lives immersed in an oversaturated information bubble that is interconnected the whole time. This interconnection activity requires devices with key characteristics such as high switching speed, low power consumption, and large processing capabilities, among others. These devices are manufactured by the semiconductor industry. A few of the semiconductor devices that are providing such capabilities are transistors, diodes, varistors, piezoelectric materials, etc. By following the trends as stipulated by Moore's law, it is required that the semiconductor devices fulfill a few key characteristics and for that to happen, they need to be characterized

M. B. Hernández Hernández · J. A. Aguilar-Martínez (✉) · B. Krishnan · S. Shaji
Universidad Autónoma de Nuevo León, San Nicolás de los Garza, Mexico
e-mail: josue.aguilar74@gmail.com; josue.aguilarmrt@unal.edu.mx

M. B. Hernández Hernández
e-mail: marla.hernandezhr@uanl.edu.mx

B. Krishnan
e-mail: kbindu_k@yahoo.com

S. Shaji
e-mail: sshajis@yahoo.com

M. A. García-Ramírez
Universidad de Guadalajara, Guadalajara, Mexico
e-mail: seario@gmail.com

Y. Dan
University of Michigan–Shanghai Jiao Tong University, Shanghai, China
e-mail: yapping.dan@sjtu.edu.cn

M. I. Pech-Canul and N. M. Ravindra (eds.), *Semiconductors*,
https://doi.org/10.1007/978-3-030-02171-9_3

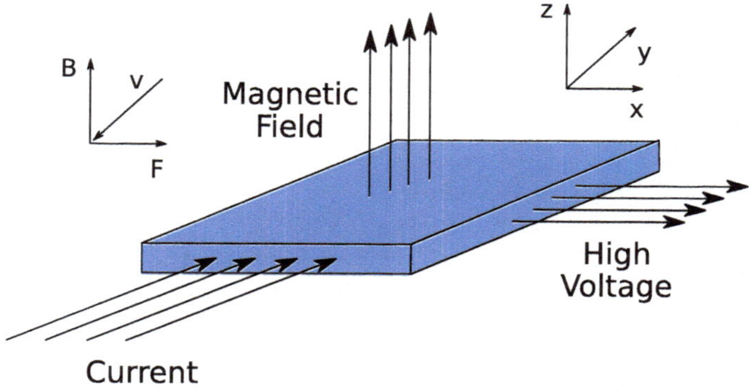

Fig. 3.1 Schematic diagram of the parameters and forces that interact to generate the Hall Effect through the Lorentz force

from raw material to the final product. A few key characterization techniques used in the semiconductor industry are described in this chapter.

Electrical characterization is an important tool for the semiconductor industry. It is possible to analyze and measure in situ a broad variety of electrical properties in order to get a deep understanding and knowledge of any material or device that will be put under test. The equipment considered to perform such analysis must be absolutely reliable and repeatable in performance and accuracy. The electrical conductivity is a primordial characteristic for any material. It is a function of material thickness, doping and temperature. The device/material characterization is performed through a set of tools based on several techniques such as direct current (dc), alternating current (ac), optical and magnetic methods. This chapter is divided into four main sections; we start with the electric characterization techniques, followed by the optical tools that are involved to gather key information from the material, structural techniques and properties obtained through magnetic field related measurement techniques.

3.1.2 Hall Effect

In semiconductors, a wide variety of parameters such as carrier mobility, carrier concentration, resistivity, conductivity either p- or n-type as well as the Hall coefficient, can be characterized.

In general, it is well known that an electron or hole can move in the presence of an electric field, as they do under the influence of a magnetic field through Lorentz force $\left(f_L = qv_p x \overrightarrow{B}\right)$. This shows that a charge is moving with a velocity. As a result of this vector product, we obtain a perpendicular vector parallel to $\overrightarrow{B}$ and v_p. The direction of the force is defined by the left-hand rule (see Fig. 3.1). For a negative charge, f_L will show an opposite direction.

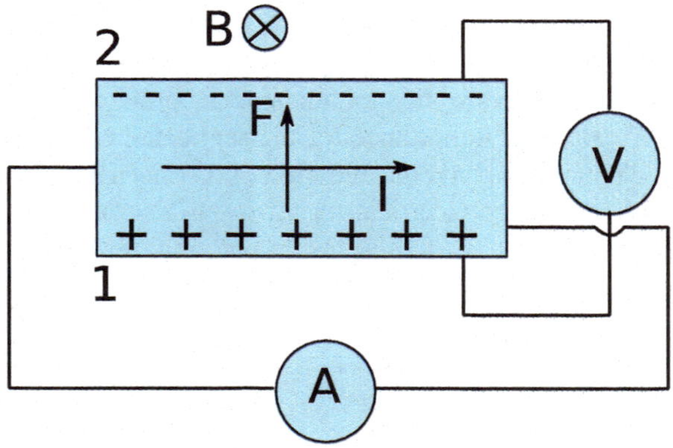

Fig. 3.2 Schematic diagram of a basic Hall Effect circuit

Therefore, through this technique, it is possible to find the carrier that is in place according to the applied $\overrightarrow{B}$. By combining the magnitude of the Lorentz force with the electric current density that is proportional to the carrier concentration and carrier velocity, the carrier concentration can be found.

Figure 3.2 shows a simple setup to measure the Hall Effect. A positive voltage is applied to the sample between electrodes 1 and 2. Through drift velocity, holes move from electrode 2 to electrode 1. The magnetic field will act on the holes due to the Lorentz force by deflecting them towards electrode 3. This will leave a net positive charge. In contrast, electrode 4 is negatively charged with ionized acceptors. As a result, electrodes 3 and 4 will display an electric field f_{H}. The generated field acts on the holes with a force in opposite direction as the Lorentz force.

Once the charge separation process is equilibrated and the drift velocity is zero, the force that produces this effect is called as the Hall electric field.

$$F_H = v_p B \tag{3.1}$$

The voltage difference or Hall voltage is defined as

$$V_H = F_H W = v_p B W \tag{3.2}$$

The sample under test should comply with the following requirements; it should be thin, as thick samples might affect the measurement. The dimensions of the sample must be larger than the electrode separation and it should be circular when possible. Also, the equipment to measure the Hall Effect requires a constant current power supply, high input impedance voltmeter, either a permanent magnet or an electromagnet, and a set of probes. One of the main issues that appears during the measurements is related to the large offset voltage caused by the experimental setup due to its nonsymmetric shape and unstable temperature.

3.1.3 Electrical Resistivity

The electrical resistivity is a fundamental property of the material. This characteristic is of particular importance for the semiconductor industry since it is a function of the impurity/dopant concentration of the material. The electrical conductivity is the reciprocal of the electrical resistivity. It is strongly related to the material doping, fabrication process, atmospheric condition, temperature, among others. To analyze such characteristic, there are a few reliable methods such as the two-point and four-point probes [1, 2].

3.1.3.1 Two-Point Probe

Two-point probe analysis is one of the methods that is used to measure the resistivity on a wire-based geometry or for uniform cross-section geometry. For this method, the voltage that is passing through the sample, while injecting a constant known current, is measured. Although this method is straightforward to implement, it has a few important drawbacks. The main one is that this method cannot be used for random shape materials. Also, semiconductors with soldering on the surface will affect the intrinsic electrical resistivity. This is also the case for metallic contacts that form Schottky barriers. This technique can be used for simple and swift analysis except for scientific or industry needs.

3.1.3.2 Four-Point Probe

As resistivity can be directly associated with the impurity content of the samples, a few methods can be used such as four-point probe or Hall effect measurements. By using Hall Effect, it is possible to measure the dopant concentration, carrier mobility, among others but the analysis usually destroys the sample. The four-point probe technique helps to perform the resistivity/conductivity analysis by controlling a few variables to obtain accurate measurements without destroying the sample. The sample array features four probes that will be in contact with the sample. Current will be injected through the external probes while the internal probes will measure the voltage as shown in Fig. 3.3.

The setup uses an equidistant four-point collinear probe that is in contact with an arbitrary sample shape. There are two main ways to measure the resistivity either on the surface or bulk. For bulk, the resistivity is calculated as a relationship between s and w, where s is the separation between tips and w is the sample thickness.

$$G(w/s) = \frac{2s}{w} In2 \tag{3.3}$$

$$\rho = \frac{\pi w}{In2}\left(\frac{V}{I}\right) = 4.5324w\left(\frac{V}{I}\right) \tag{3.4}$$

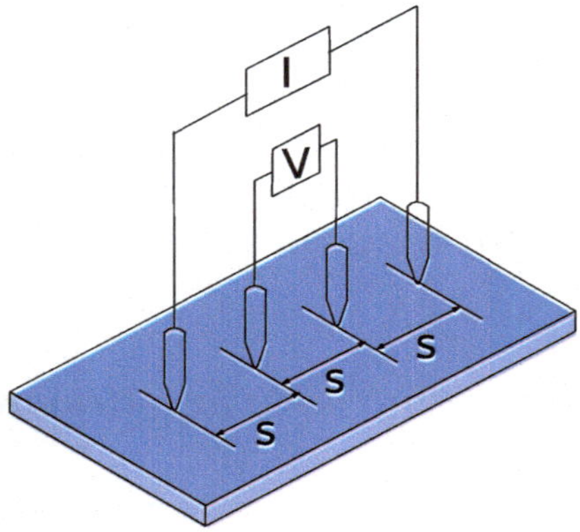

Fig. 3.3 Schematic diagram of a four probe setup that features an equidistant separation between tips as well as the internal voltage source and external current injection

where ρ is the bulk resistivity, V is the voltage measured, I is the current, and w is the sample thickness.

When the sample is a thin film or a few nm of coating, the surface resistivity is measured. As this measurement does not take into account the thickness of the material, the surface resistivity is analyzed through

$$\rho = \left(\frac{V}{I}\right) 2\pi s \tag{3.5}$$

where V is the floating voltage, I is the current at the outer probes, s is the separation between probes, and ρ is the resistivity.

As both the equations seem to be similar, the units of ρ are defined in Ω/square. A key characteristic of the four-point probe technique relies on the elimination of error measurements due to the probe resistance, spreading resistance under each probe, and contact resistance between each metal probe and the material under test.

3.1.4 Capacitance–Voltage Measurements

Capacitance–voltage or C–V characteristics can provide key information regarding the device that is being tested. There are three different main approaches to perform such measurements such as capacitance–voltage provided by multifrequency capacitance, capacitance–frequency, and capacitance–time [3]. To get a proper knowledge of the capacitance–voltage measurements, it is required to have experience with the frequency dependence of the measurements. Normal C–V measurements drive frequencies from tens of hertz up to a few Megahertz. In recent years, a novel set of equipment has been designed and tested. As a result, there are commercially available instruments that can perform very low-frequency measurements that are key to characterize the slow trapping and de-trapping phenomenon in some materials.

C–V measurement setup is straightforward. However, special attention is required to avoid issues concerning low capacitance measurements, instrument connections as well as instrument limitations. Commonly, capacitances for novel devices based on semiconductors or exotic materials such as graphene, BN-h, or MoS_2 are in the order of femto Farads (fF). Thus, very sensitive equipment is required to perform accurate measurements regardless of the material, size, or behavior.

3.1.5 Current–Voltage Measurements

This technique has been the main analysis tool that has been utilized in a variety of semiconductor material/device measurements. The materials/devices include intrinsic or n- or p-doped semiconductors, diodes, BJT (Bipolar Junction Transistors), high-speed memories, sensors or novel structures based on exotic materials such as graphene or MoS_2 [3].

I–V characteristic curves are commonly used to understand the fundamental electrical properties such as current-conduction mechanism or breakdown in a material or device. The results of measurements are also used to model mathematically the complete behavior of the material or device or circuit. This technique requires finding the relationship between the current that is passing through a device under test or a material and the voltage across the terminals. Simple devices such as solar cells, diodes, varistors, or resistors have been successfully characterized through this technique. It is important to clarify that the I–V characteristics for a device or material are not limited to two terminal devices but also to arrays, circuits and systems.

Novel devices such as those based on graphene or MoS_2 require a set of electrodes to properly characterize their full behavior under various conditions such as temperature, pressure, pH, current, voltage, among others [4]. By performing such measurements, several characteristics such as leakage, breakdown, nature of contact etc., can be obtained. In recent years, a number of techniques, based on current-voltage measurements, have been developed. These techniques include temperature and time dependent current-voltage (dynamic and static) measurements.

3.1.6 Deep Level Transient Spectroscopy

Deep level transient spectroscopy (DLTS) is a technique that has been developed to quantitatively evaluate the density as well as the energy trap levels in a semiconductor. The analysis allows to accurately differentiate between the traps that are associated with majority or minority carriers, various defects etc., as well as their concentrations, energy levels, and capture rates. The deep level refers to the energy states in the bandgap of a semiconductor.

The principle of DLTS is based on the dependence of capacitance, usually at 1 MHz, on the occupancy of traps within the space charge region in the semiconductor.

Through this technique, it is possible to obtain the trap concentration using the intensity of the capacitance peak. Moreover, the type of carriers, either holes or electrons, can be inferred from the changes in the sign of the capacitance. Moreover, the activation energy of the trap and the capture cross-section, due to the emission rate and temperature dependence, can be determined.

The measurements described above are performed by an automatic computer controlled spectrometer that features a transient lock-in amplifier measurement station. In a simultaneous way, it is also possible to perform both analyses such as capacitance–voltage and current–voltage measurements as well as obtain Arrhenius plots within the same equipment, besides other characteristics.

3.1.7 AC Impedance Spectroscopy

In order to obtain the critical electric properties of the material/device and the characteristics of the material itself, a few important characterization techniques have been implemented. The AC impedance method has been successfully used to characterize the electrical properties of polycrystalline, crystalline, and single crystal materials as well as electro-aqueous systems. Since polycrystalline materials show random grain sizes and a broad variety of grain boundary related properties, this technique allows separating the contributions of each one of the material constituents.

The main characteristic of this method is to analyze the AC response in a test system by measuring the impedance in the frequency domain. Impedance is a composition of real and imaginary parts for passive devices such as resistor Z, capacitors Z_C, and inductors Z_L. It is plotted as a vector in a complex plane. The imaginary part of the impedance is known as reactance, and the real part is the resistance. Reactance (X) has two forms, capacitive and inductive.

$$X_L = 2\pi f L \tag{3.6}$$

and

$$X_C = \frac{1}{2\pi f C} \tag{3.7}$$

where f is the applied frequency, L is the inductance, and C is the capacitance. On the other hand, a complementary method is defined by the complex admittance (Y) from which the complex electric modulus (M) and the complex permittivity can be obtained.

This technique is so versatile that several instruments can measure both real and imaginary parts of the impedance vector and through complex algorithms convert them into the desired parameters. Although this technique can yield a broad set of parameters, the permittivity as well as the impedance plots are preferred. A special importance must be taken into account for the test fixture due to the fact that its quality defines the total measurement accuracy limit. Also, it is possible to obtain quite a few parameters as a function of temperature such as electric impedance, phase angle, loss tangent, and capacitance.

3.2 Optical Properties

The optical properties of semiconductors can provide a large set of important characteristics that cannot be obtained from other characterization techniques. These properties of semiconductors lead to a wide range of applications as well as device implementation and fabrication. In semiconductor physics, there are two main properties that are defined as optical absorption coefficient (α) and optical bandgap (E_g) [5]. In general, the optical characterization is performed by using the UV-Visible absorption spectroscopy technique. In this technique, the beam attenuation of light, after it passes through a semiconductor sample or after reflection from a sample surface, is measured. The measurements of absorptance, transmittance, and reflectance of samples in the vacuum ultraviolet (VUV), ultraviolet (UV), visible and near-infrared (NIR) spectral regions are obtained.

Typically, such measurements are made in terms of percentage transmittance ($T\%$):

$$T = (I/I_O)100\% \tag{3.8}$$

where I is the transmitted light intensity and I_O is the initial incident light intensity.

The absorption of light by an optical medium with thickness t, is quantified by its absorption coefficient, α, which can be obtained using Beer's law:

$$I = I_O \mathrm{e}^{-\alpha t} \tag{3.9}$$

In general, the surface of the semiconductor can reflect a fraction of the incident light beam. The measurements of optical properties lead to transmittance and reflectance spectra. By combining these experimental measurements, the absorption coefficient can be obtained

$$\alpha = \frac{1}{t}\mathrm{In}\left[\frac{(1-R)^2}{T}\right] \tag{3.10}$$

By using Eq. (3.10), the absorption coefficient is determined. The optical bandgap E_g corresponding to the fundamental absorption is defined as the minimum energy required to promote an electron from the valence band into the conduction band [5]. For a given semiconductor, α and E_g are related as

$$(\alpha h\upsilon)^n = A\left(h\nu - E_g\right) \tag{3.11}$$

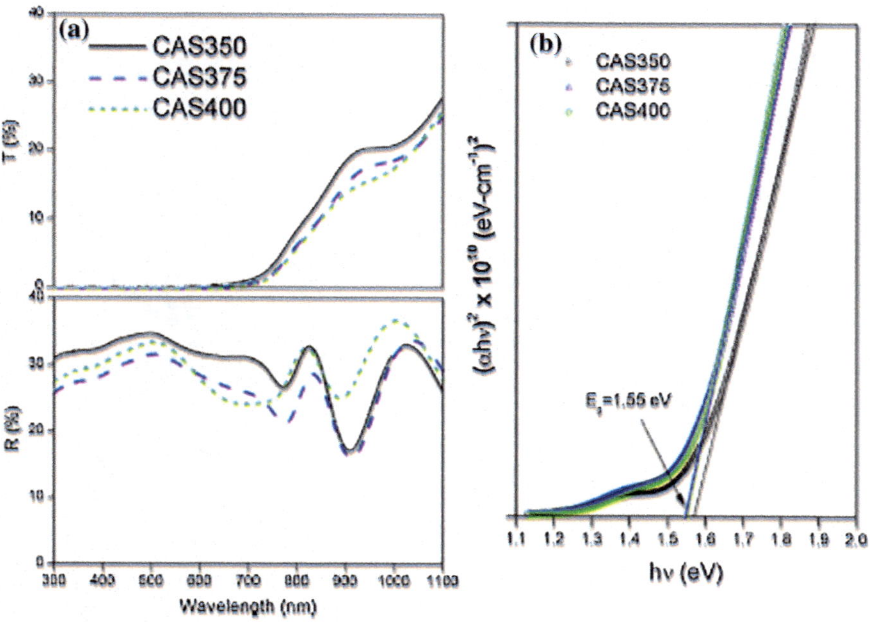

Fig. 3.4 **a** Transmittance (T) and reflectance (R) spectra, **b** evaluation of optical bandgap for $CuSbS_2$ thin films formed by heating glass/Sb_2S_3/Cu stack layers at 350, 375, and 400 °C. This illustrates the plot of $(\alpha h\upsilon)^2$ versus $h\upsilon$ (Tauc plot) for the samples giving good linear fit for $n = 2$. This implied that the fundamental optical absorption in the $CuSbS_2$ thin films was dominated by direct allowed transition [6] (With permission from Elsevier)

where $n = 2$, 1/2, 2/3 for direct allowed, indirect allowed, and direct forbidden transitions, respectively; h is the Planck's constant, ν is the frequency, and A is a constant. Thus, the optical bandgap is evaluated by extrapolating the straight-line portion of the Tauc plot (graph of $(\alpha h\nu)^n$ vs. $h\nu$) on the $h\nu$ axis. Figure 3.4 shows the bandgap evaluation of $CuSbS_2$ semiconductor thin films formed under different conditions [6].

Optical measurements of the absorption spectra of the CIGS nanoparticle colloids [7] are shown in Fig. 3.5. The optical bandgap values of these nanocolloids were obtained from their corresponding Tauc plots. These nanocolloids showed an increment in their bandgap values compared to bulk CIGS due to the quantum size effects. As an effect of the liquid media and laser ablation fluence (energy/cm^2), the bandgaps were different for each of these CIGS nanocolloids.

When semiconductors are modeled as thin films and as nanomaterials, due to their technological applications, the structural characterization, morphology, composition, and optical properties are vital for the analysis and implementation. Some of these experimental techniques are briefly described.

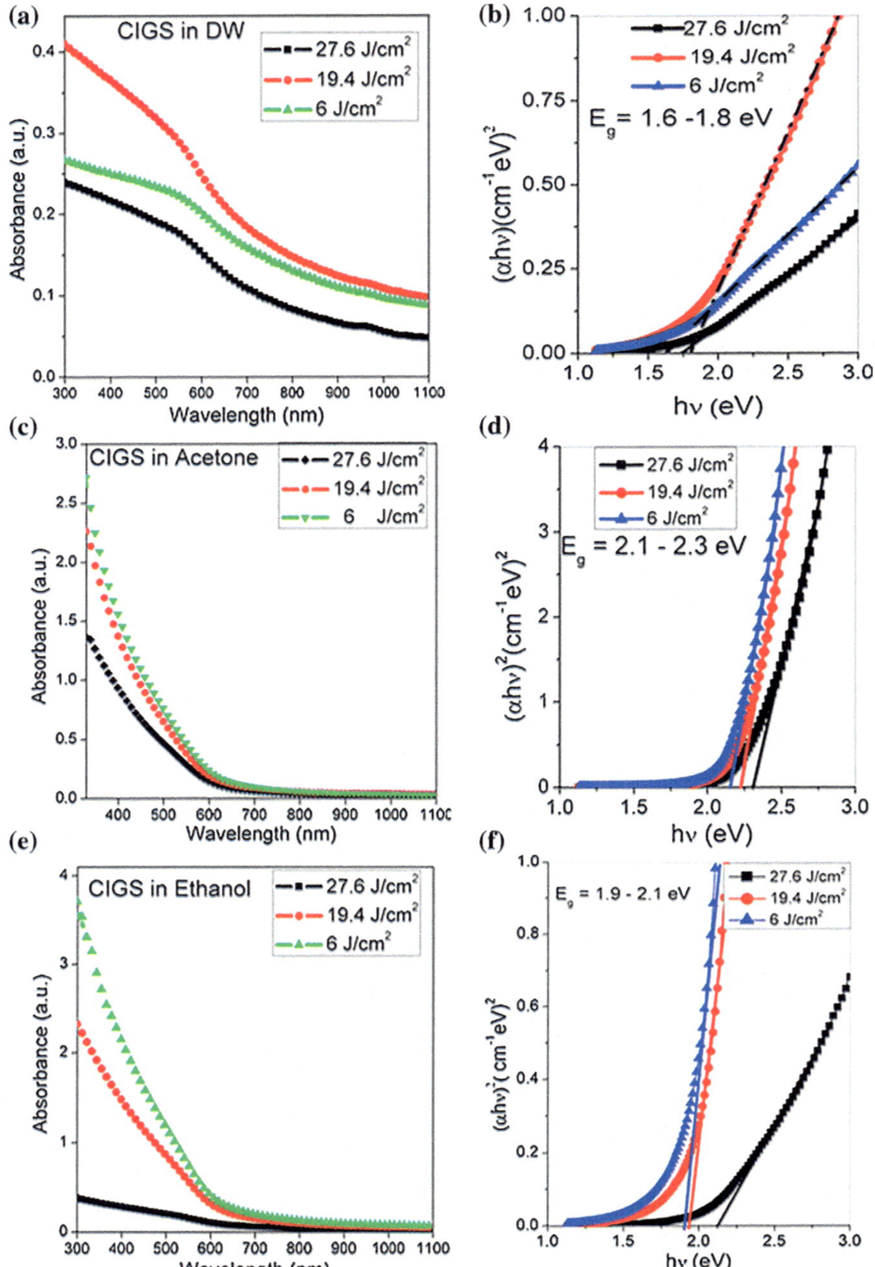

Fig. 3.5 UV–vis absorption spectra and the corresponding Tauc plots with bandgap evaluation for CIGS nanocolloids synthesized by 532 nm ablation using energy fluence of 27.6, 19.4 and 6 J/cm^2 in **a**, **b** distilled water, **c**, **d** acetone and **e**, **f** ethanol, respectively [7] (With permission from Elsevier)

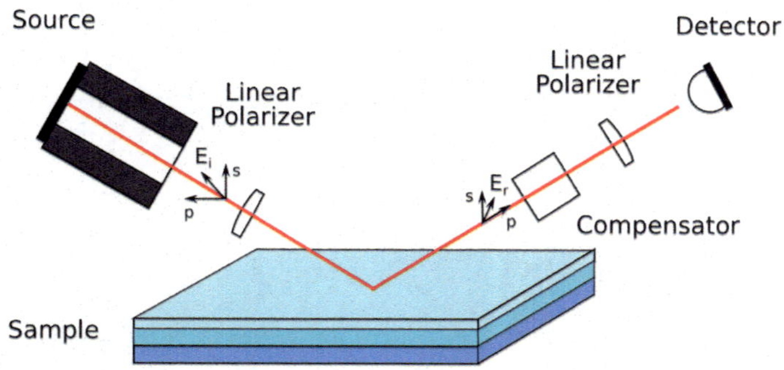

Fig. 3.6 Schematic diagram of a basic ellipsometer setup featuring a source (laser) a polarizer lens, the sample under analysis, the compensator and a detector

3.2.1 Ellipsometry

Ellipsometry is a nondestructive method that is commonly used to characterize a broad range of materials either in the form of thin films or bulk [8–10]. The basic setup for ellipsometry features a light source, a set of linear polarizers, and a compensator. Also, a quarter wave plate is used as shown in Fig. 3.6.

A plane-polarized beam is directed to the sample; an elliptically polarized beam is reflected from the sample surface. In this technique, the polarization changes (intensity and phase) of the reflected beam from the surface are measured. As a result of the measurement, two parameters, Psi (ψ) and Delta (Δ), known as ellipsometric parameters, are obtained. The obtained parameters are real values. These parameters are related to the ratio of the complex Fresnel reflection coefficients R_p and R_s for p- and s- polarized beam, respectively according to

$$\tan\psi e^{(i\Delta)} = \frac{R_p}{R_s} = \rho \tag{3.12}$$

$$\psi \tan^{-1}\left(\frac{R_p}{R_s}\right) \tag{3.13}$$

$$\Delta = \delta_p - \delta_s = \left(\arg r_p - \arg r_s\right) \tag{3.14}$$

where ψ, geometrically can be interpreted as the angle between the reflected polarization ellipse and the linear polarization direction of the inner beam. On the other hand, delta has a strong relationship with the main axes and the ratio of the polarization ellipse. Moreover, it represents the phase shift measurement between s- and p-beam components due to scattering from the sample. The ellipsometry technique allows us to know important information of the optical properties of multi-layers by varying the angle of incidence. The information encompasses a few parameters such as refractive index, permittivity for a single or a set of wavelengths [10–15] and optical constants of the components of the multi-layers. Through a computational set of algorithms, it is possible to determine the thickness of a single as well as multiple layers. By performing ellipsometry, it is possible to generate a spatial map of the sample point by point regarding the number of layers, the thickness of each layer, the permittivity as well as the refractive index, extinction coefficient and doping concentration.

3.3 Structural Properties

3.3.1 Introduction

The material structure relates to the arrangement of its constituent atoms or molecules. It is well known that the physical and electronic structure is intimately related to its properties which in turn determine the material performance in a specific application. In general, the material structure can be viewed at the nano, micro, and macro scale. This section mainly focuses on the structural characterization techniques of semiconductors: both in thin film and powder forms. There are various techniques to determine the structure of semiconductors directly or indirectly. The commonly used techniques are X-ray diffractometry, electron diffractometry, scanning electron microscopy, scanning tunneling microscopy, transmission electron microscopy, and atomic force microscopy. In the following subsections, we will explain the application of these techniques to analyze the structure of various semiconductors in the form of thin films and nanomaterials.

3.3.2 X-Ray Diffractometry

This is an indirect method that is used to analyze the structure of crystalline semiconductors. The suitability of this technique for probing structure is because of the interatomic distances (0.15–0.4 nm) in semiconducting materials or condensed matter, in general, which are in the same order of the wavelength of X-rays having energy in the range of 3–8 keV. Further, the technique is nondestructive and leaves the sample or device intact after the analysis.

The basic principle of the diffraction technique is based on the Bragg equation (3.15) given by

$$2d_{hkl} \sin\theta = \lambda \tag{3.15}$$

where d_{hkl} is the interplanar spacing with Miller indices (hkl), θ is the angle of diffraction (also known as Bragg angle), and λ is the X-ray wavelength. The equation was used by W. H Bragg and W. L Bragg in 1913 to describe the position of X-ray scattered peaks with respect to angular space. Since d_{hkl} for a given structure is

Table 3.1 The relation between d_{hkl} and the unit cell parameters (a, b, c and α, β, γ) for all the seven crystal structures

Crystal structure	Relation
Orthorhombic	$\frac{1}{d_{hkl}^2} = \frac{h^2}{a^2} + \frac{k^2}{b^2} + \frac{l^2}{c^2}$
Tetragonal	$\frac{1}{d_{hkl}^2} = \left[h^2 + k^2 + l^2\left(\frac{a}{c}\right)^2\right]\frac{1}{a^2}$
Cubic	$\frac{1}{d_{hkl}^2} = (h^2 + k^2 + l^2)\frac{1}{a^2}$
Monoclinic	$\frac{1}{d_{hkl}^2} = \frac{h^2}{a^2 \sin^2\gamma} + \frac{k^2}{b^2 \sin^2\gamma} - \frac{2hk\cos\gamma}{ab\sin^2\gamma} + \frac{l^2}{c^2}$
Hexagonal	$\frac{1}{d_{hkl}^2} = \left[\frac{4}{3}(h^2 + k^2 + hk) + l^2\left(\frac{a}{c}\right)^2\right]\frac{1}{a^2}$
Triclinic	$\frac{1}{d_{hkl}^2} = \left[\frac{h}{a}\begin{vmatrix} \frac{h}{a} & \cos\gamma & \cos\beta \\ \frac{k}{b} & 1 & \cos\alpha \\ \frac{l}{c} & \cos\alpha & 1 \end{vmatrix} + \frac{k}{b}\begin{vmatrix} 1 & \frac{h}{a} & \cos\alpha \\ \cos\gamma & \frac{k}{b} & \cos\alpha \\ \cos\beta & \frac{l}{c} & 1 \end{vmatrix} + \frac{l}{c}\begin{vmatrix} 1 & \cos\gamma & \frac{h}{a} \\ \cos\gamma & 1 & \frac{k}{b} \\ \cos\beta & \cos\alpha & \frac{l}{c} \end{vmatrix}\right] \cdot \begin{vmatrix} 1 & \cos\gamma & \cos\beta \\ \cos\gamma & 1 & \cos\alpha \\ \cos\beta & \cos\alpha & 1 \end{vmatrix}^{-1}$
Trigonal (rhombohedral)	$\frac{1}{d_{hkl}^2} = [(h^2 + k^2 + l^2)\sin^2\alpha + 2(hk + kl + lh)(\cos^2\alpha - \cos\alpha)]\frac{1}{A^2(1 + 2\cos^3\alpha - 3\cos^2\alpha)}$

related to its unit cell parameters (a, b, c and α, β, γ), as given in Table 3.1, from the diffraction pattern, knowing d_{hkl}, the unit cell type can be determined.

The identification of phases for most of the semiconductors is achieved by comparing the X-ray diffraction experimental patterns with patterns of a reference database. The most comprehensive database of powder diffraction patterns, the powder diffraction file (PDF), is maintained by the International Centre for Diffraction Data (ICDD) [16].

Figure 3.7 shows the determination of the crystalline nature of lead iodide thin films prepared by spin coating of the solution of lead iodide (PbI_2) powder in dimethyl formamide [17]. The diffraction patterns recorded for the PbI_2 thin films, formed at different spin speeds, are compared with that of standard pattern containing hexagonal PbI_2 (PDF # 070235).

Also, from the diffraction line broadening width (FWHM—Full Width at Half Maximum—β) of a given peak, the crystallite size of the material can be estimated using the Scherrer formula, given by,

$$D = K\lambda/\beta\cos\theta \tag{3.16}$$

where D is the average crystallite size, which may be smaller or equal to the grain size; K is a dimensionless shape factor, with a value close to unity. The shape factor has a typical value of about 0.9 but varies with the actual shape of the crystallite; λ is the X-ray wavelength, β, sometimes denoted as $\Delta(2\theta)$, is expressed in radians and is obtained by subtracting the instrumental line broadening from the measured peak width.

In Fig. 3.7, the values of average crystallite size (16–13 nm) are calculated using the Scherrer equation.

The XRD technique is applicable to probe the crystalline structure of nanoparticles. Figure 3.8 shows the structure of antimony sulfide nanoparticles, prepared by

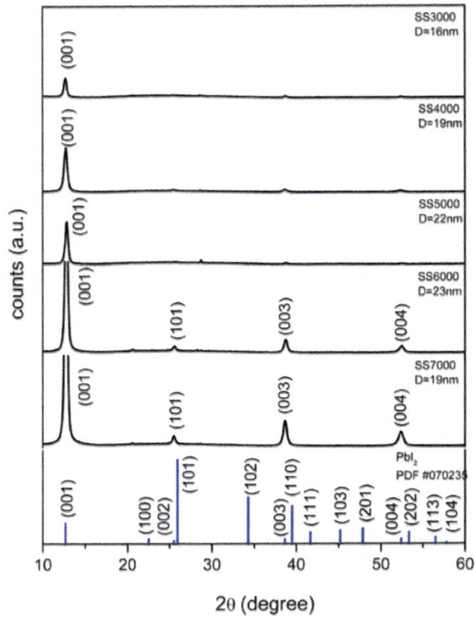

Fig. 3.7 X-ray diffraction patterns of PbI_2 thin films deposited by varying the spin speed (SS) 3000, 4000, 5000, 6000, and 7000 rpm [17]. The standard pattern of hexagonal PbI_2 (PDF 07-0235) is also included. Crystallite size calculated using the Scherrer equation for each case is noted in the respective pattern (With permission from Springer)

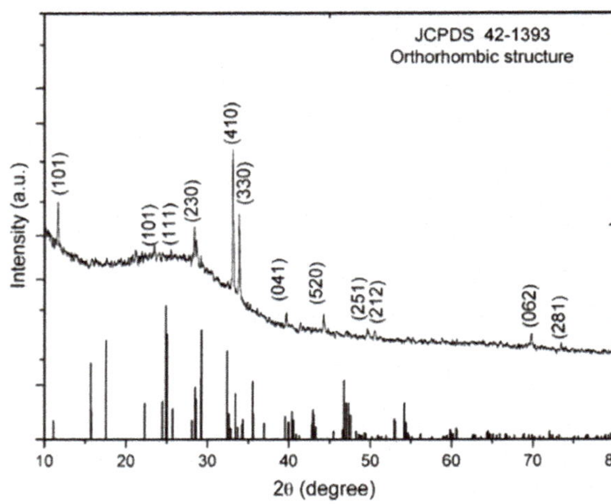

Fig. 3.8 XRD pattern of Sb_2S_3 nanoparticles produced by pulsed laser ablation (532 nm) of antimony sulfide pellets in isopropyl alcohol. The standard JCPDS data for orthorhombic Sb_2S_3 (PDF-42-1393) is included [18] (With permission from Springer)

pulsed laser ablation in liquid medium (PLALM), using 532 nm output laser beam in isopropyl alcohol [18]. The average crystallite size is calculated using the Scherrer equation.

For samples with nonuniform lattice strain (microstrain), the particle size and the lattice strain (ε) can be separately evaluated using the detailed diffraction peak broadening analysis. It is known that the line broadening occurs due to small grain size as well as due to microstrains that are associated with the sample, especially in thin films [19]. Hence, the FWHM of the given peak contains the effects of both particle

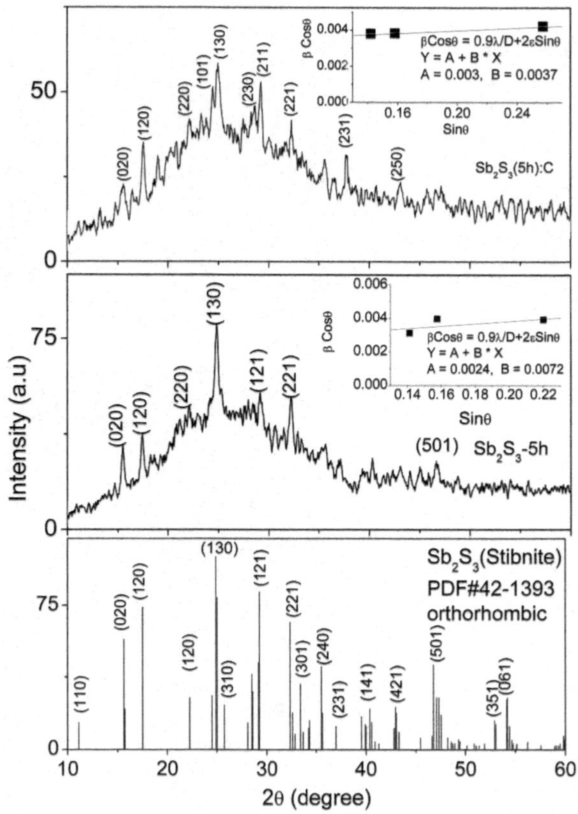

Fig. 3.9 XRD analysis of Sb_2S_3 thin films annealed in nitrogen for 30 min **a** Sb_2S_3-5 h **b** Sb_2S_3-½ h. The standard PDF data and the Williams-Hall plot are given in the inset of each pattern. The average crystallite size (D) and the microstrain (ε) values are given in each case [20] (With permission from Elsevier)

size and microstrain, provided the instrumental contribution in the broadening is determined using a standard sample. The size and strain effects can be separated by the Williamson–Hall plot [19] in which values of β cosθ versus sinθ are plotted. This plot can be fitted to a straight line. The inverse of the intercept of the line on β cosθ axis gives D/λ (the ratio of the average crystalline diameter (D) to the wavelength) from which the value of D can be extracted. The slope of the line indicates the microstrain (2ε). Figure 3.9 illustrates this procedure to determine the crystalline structure, particle size, and lattice strain from the XRD pattern. The figure shows the XRD analysis of antimony sulfide thin films prepared by chemical bath deposition at different conditions [20] (Fig. 3.9).

In the next section, the use of electron diffraction for structural characterization will be briefly discussed.

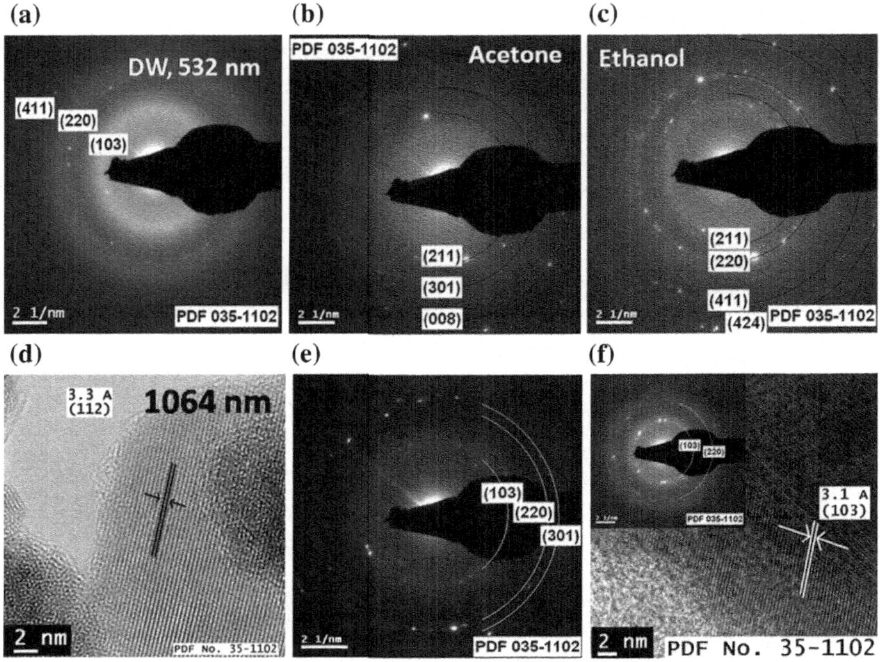

Fig. 3.10 Selected area electron diffraction (SAED) patterns and high-resolution TEM (HRTEM) micrographs of CIGS nanoparticles obtained by pulsed laser ablation (top) using 532 nm wavelength in **a** distilled water (DW, 19.4 J/cm^2), **b** acetone (27.6 J/cm^2) and **c** ethanol (6 J/cm^2), (bottom) using 1064 nm wavelength in **d** distilled water (DW, 20 J/cm^2), **e** acetone (6 J/cm^2) and **f** ethanol (J/cm^2) respectively. The crystalline planes were identified using JCPDS No. 35-1102 for chalcopyrite and is highlighted in the figures [7] (With permission from Elsevier)

3.3.3 Electron Diffraction and High-Resolution Transmission Electron Microscopy (HRTEM)

If a beam of accelerated electrons passes through a thin crystal, the crystal planes can act as a diffraction grating. This technique is a useful tool to probe the crystal structure of nanostructured materials. The use of this technique to analyze the structure of various nanoparticles or nanostructured materials is given in Fig. 3.10. Normally, this technique is coupled with a high-resolution transmission electron microscope and used as selected area electron diffraction (SAED) in which the diffraction pattern corresponding to a selected area of the sample is recorded. The use of SAED from a TEM micrograph is demonstrated in Fig. 3.10, for copper indium gallium selenide nanoparticles obtained by PLALM [7]. HRTEM image can give a direct view of crystalline planes, and the analysis can give the material structure. Figure 3.10d, f shows the HRTEM analysis of CIGS nanoparticles in which the lattice planes and interplanar distances obtained are in agreement with the JCPDS data.

3.3.4 Atomic Force Microscopy

The atomic force microscopy (AFM) technique is designed to measure local properties, such as height, surface RMS (root mean square) roughness, etc., by raster scanning of an area with a probe at the nanoscale. AFM images can reveal the topography of a sample surface in 2D and 3D, at atomic resolution.

In AFM, the deflection of a sharp tip that is attached to a microcantilever, which bends under the influence of a force, is measured by the reflection of a laser beam from the cantilever onto a position sensitive photodiode (PSPD). Thus, as an AFM tip passes over a sample surface, the resulting cantilever deflections are recorded by the PSPD. An AFM images the topography of a sample surface by scanning the cantilever over a region of interest. The raised and lowered features on the sample surface influence the deflection of the cantilever, which is monitored by the PSPD [21].

In AFM, there are two types of operations, viz., contact mode and tapping mode or dynamic mode. Most of the semiconductors are analyzed in the contact mode in which the images of the height are presented in 2D and 3D as shown in Fig. 3.11 [22].

Tapping or dynamic mode is used when the sample surface is soft. In this case, the AFM tip oscillates while interacting with the sample surface. Both height and phase images can be acquired in such cases as seen in Fig. 3.12. The 1 μm × 1 μm AFM images show the morphological features of $CuSbS_2$ semiconducting thin films formed by annealing chemical bath deposited Sb_2S_3 and thermally evaporated Cu stacked layers (Sb_2S_3/Cu) at different temperatures [6].

Another technique to probe the surface morphology is scanning electron microscopy. The following section describes the use of this technique to analyze semiconductor samples.

3.3.5 Scanning Electron Microscopy (SEM)

This technique is commonly known as SEM and is widely used for direct view of a sample surface. In this case, a finely focused electron beam scans a sample surface and produces images of its topography. Due to electron–sample interaction, various types of signals, originating from the specific emission volume of the sample, can be of secondary electrons, backscattered electrons, characteristic X-rays, and other photons of different energies, and can be used to examine many characteristics of the sample such as surface topography, crystallography, composition, etc. The most interesting imaging signals are the secondary electrons since they vary primarily because of the difference in surface topography [23].

Due to the surface charging effects, an ultrathin coating of electrically conducting metal—such as gold (Au), gold/palladium (Au/Pd), platinum (Pt), silver (Ag), chromium (Cr), or iridium (Ir), is deposited on the sample; a metallic coating is rec-

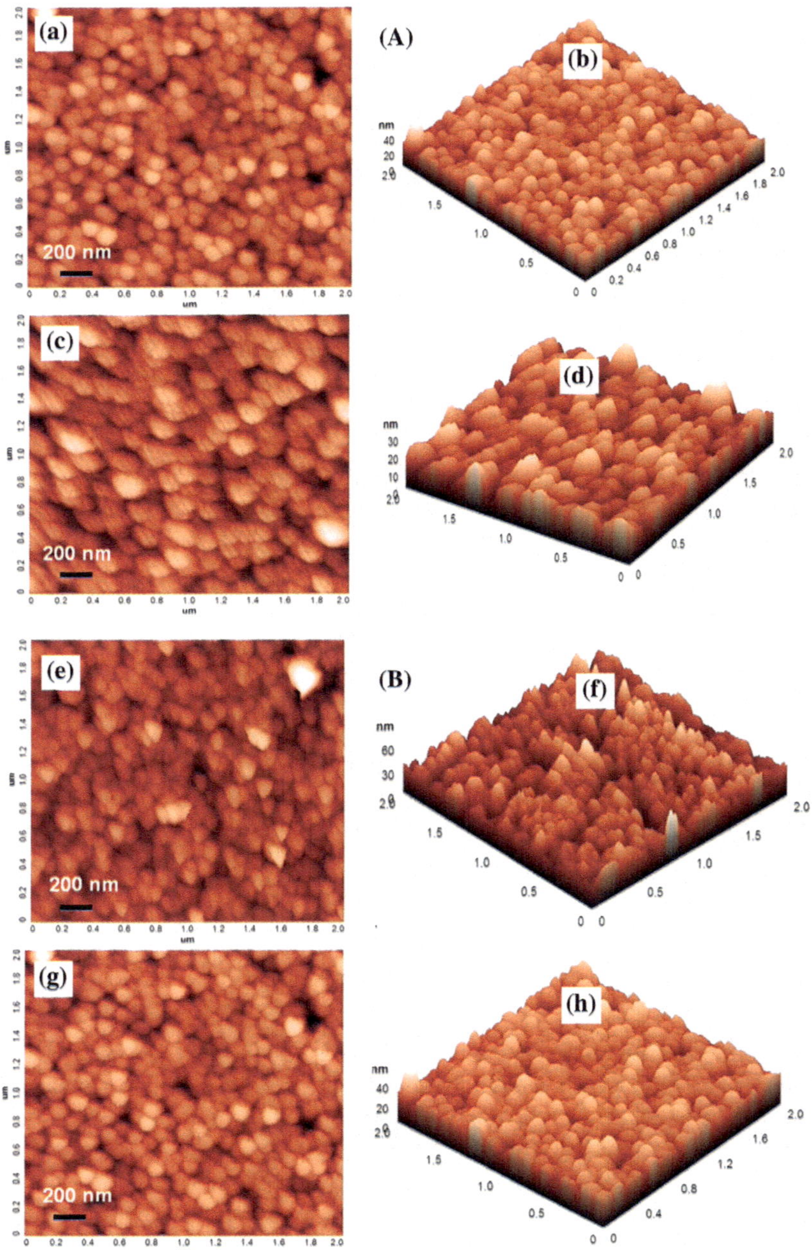

Fig. 3.11 **A** Atomic force micrographs (2D and 3D) of the CdS thin films grown by normal CBD and pulsed laser-assisted CBD (PLACBD), **a**, **b** normal CBD for 10 min, **c**, **d** PLACBD for 10 min under laser energy fluence of 12 mJ/cm^2 (532 nm). **B** Atomic force micrographs (2D and 3D) for the CdS thin films grown by pulsed laser-assisted CBD (PLACBD), **e**, **f** for 10 min under laser fluence of 27 mJ/cm^2 (532 nm), **g**, **h** for 10 min under laser fluence of 50 mJ/cm^2 (1064 nm) (All the images were recorded in the contact mode) [22] (With permission from Elsevier)

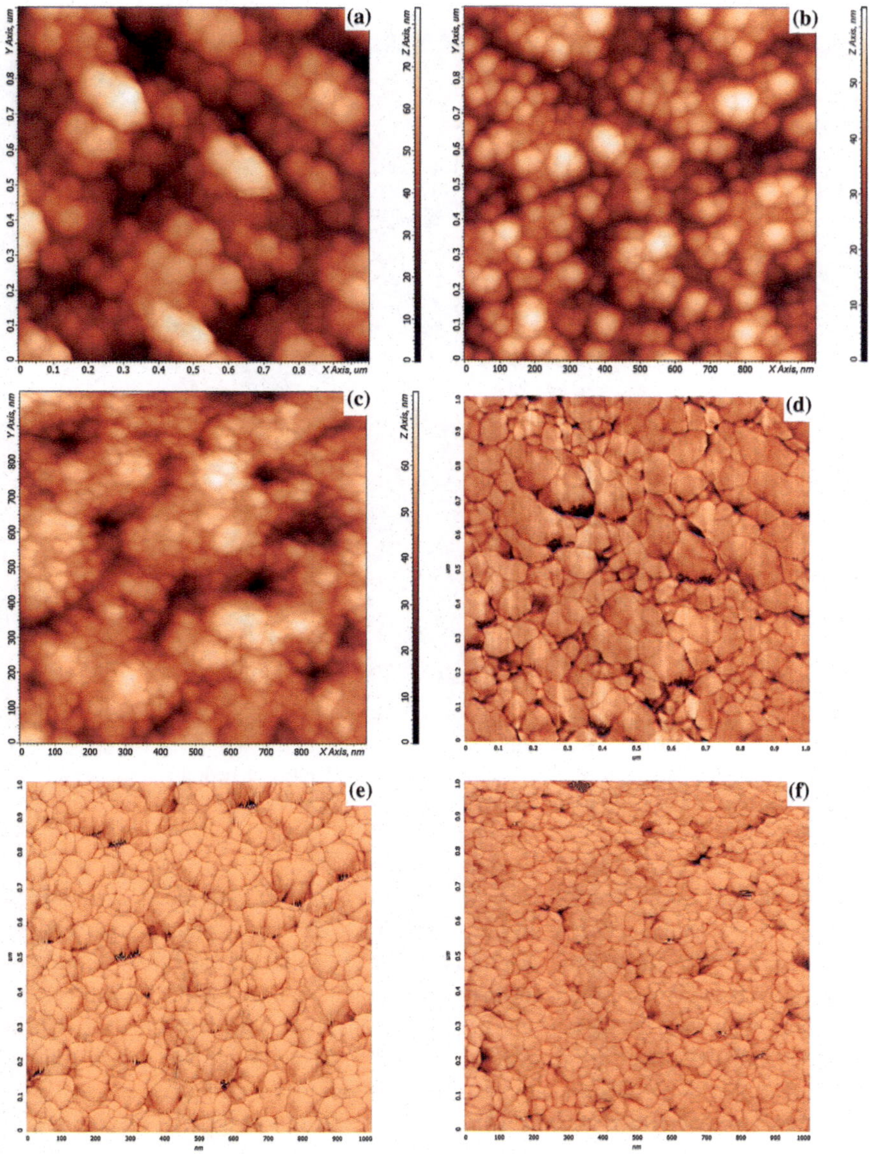

Fig. 3.12 Atomic force micrographs of glass/Sb_2S_3/Cu thin films annealed in low vacuum at different temperatures for 1 h (10^{-3} Torr) at **a** 350 °C, **b** 375 °C and **c** 400 °C in tapping mode, **d** 350 °C, **e** 375 °C, and **f** 400 °C corresponding phase images. In all the images, densely packed grains with well-defined grains are seen [6] (With permission from Elsevier)

ommended for semiconducting thin films [24]. Typically, gold coating (10–20 nm thickness) is performed on the specimen surface by the sputtering technique. Also, it can increase the secondary electron detection, improve edge resolution, and pro-

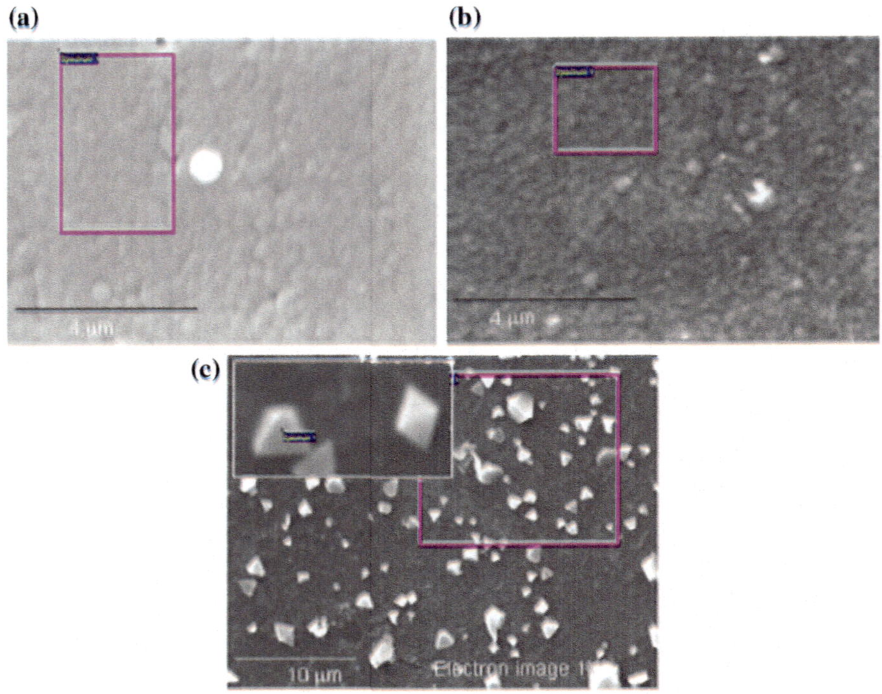

Fig. 3.13 Scanning electron micrograph of **a** Sb_2S_3-5 h deposited film annealed in nitrogen for 30 min (S/Sb—1.2) **b** Sb_2S_3-5 h deposited thin film annealed in air for 15 min **c** Sb_2S_3-0.5 h deposited in air for 30 min which shows the formation of Sb_2O_3 octahedral shaped crystallites (senarmontite) due to annealing in air [20] (With permission from Elsevier)

tect electron beam sensitive samples. Most of the SEMs are equipped with energy dispersive X-ray analysis (EDX) systems which can detect the characteristic X-ray radiations emitted by the sample. Such EDX measurements are used for a semiquantitative analysis of sample composition or elemental detection. Figure 3.13 shows the use of SEM to analyze the surface morphology of chemical bath deposited Sb_2S_3 thin films post-annealed under different conditions [20]. Also, the ratio of compositional analysis (S/Sb) of the different samples (selected regions) is compared.

Another case of surface analysis using SEM is shown in Fig. 3.14. In the figure, antimony oxide nanocrystals, obtained by PLALM of antimony in distilled water [25], show a fibrous morphology.

Scanning electron microscopy is employed to characterize the morphology of tin sulfide (SnS) thin films fabricated from utilizing their nanoparticle colloids and produced by pulsed laser ablation in liquid (PLAL) [26]. Figures 3.15 and 3.16 show the morphologies of these thin films obtained by a spray deposition of the nanocolloids in dimethyl formamide (DMF) and isopropyl alcohol (IA) as liquid media during the ablation process. The films show a porous surface with a kind

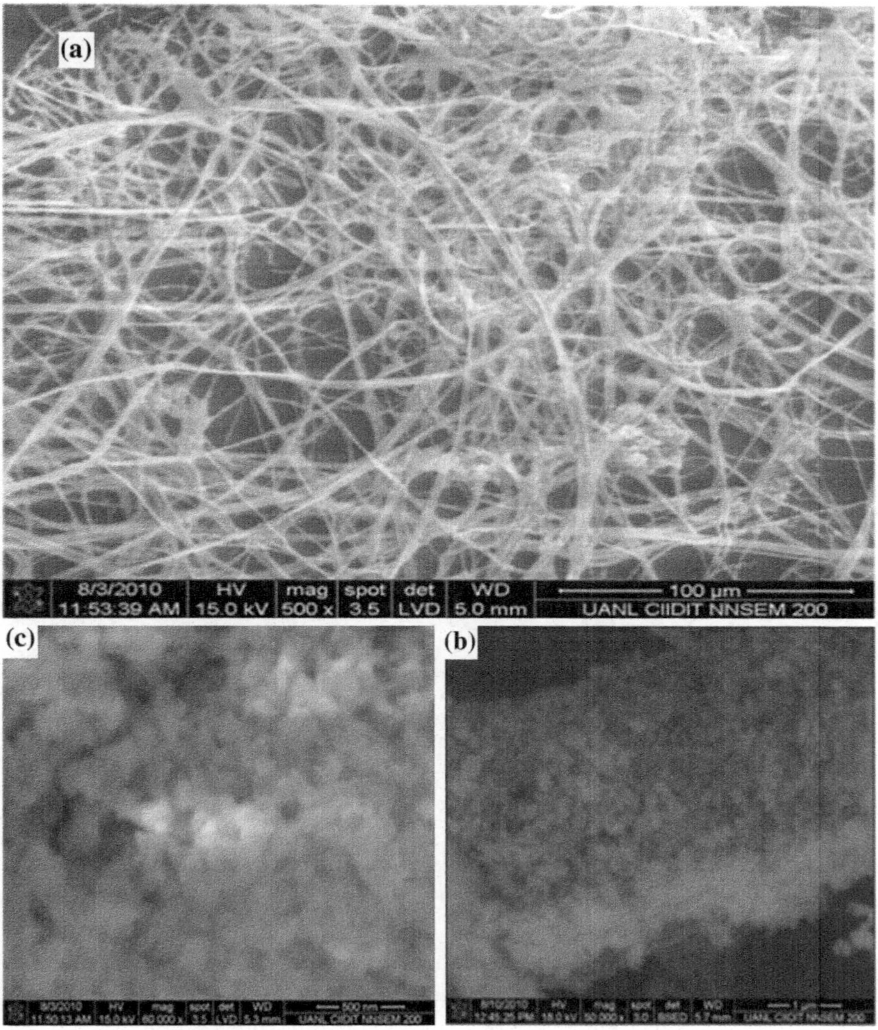

Fig. 3.14 **a** SEM image of antimony oxide nanocrystals obtained by PLALM of antimony in distilled water (×500), **b**, **c** images at higher magnification (×50,000–60,000) of one of the fibers in (**a**) [25] (With permission from Springer)

of layered structure. This morphology will be useful in charge storage, sensor, and photodetector applications of nanomaterials.

The electrical performance of a semiconductor device can also be evaluated using special surface penetration of the electron beam namely voltage contrast and charge collection modes. Figure 3.17 shows the charge collection mode known as electron beam induced current (EBIC) [27]. The figure illustrates the use of EBIC measurements to estimate the current density loss due to incomplete charge collection in

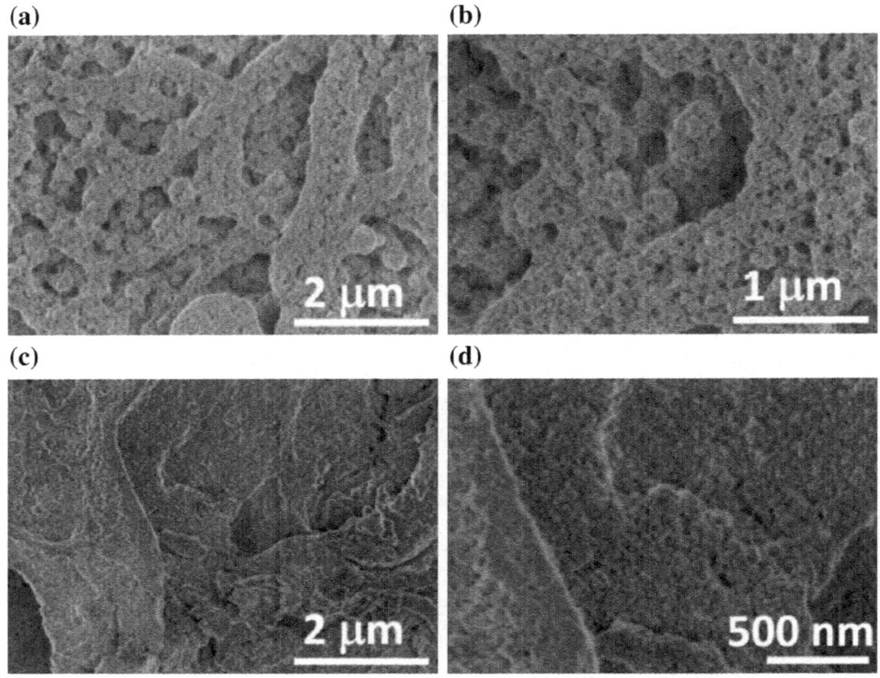

Fig. 3.15 Scanning electron micrographs of SnS thin films **a**, **b** as deposited and **c**, **d** annealed (at 380 °C in vacuum, 1 h) synthesized from SnS nanocolloids in isopropyl alcohol [26] (With permission from John Wiley and Sons)

CdTe solar cells. The images were captured from the device cross section. The measurement accuracy was improved by coating a thin Al_2O_3 film on the device cross section.

Another technique that is utilized to characterize semiconductors is Raman spectroscopy. Raman spectroscopy helps to identify the structure and molecular vibrations of materials. This technique is described in the following section.

3.3.6 Raman Spectroscopy

This technique is a form of vibrational spectroscopy in which Raman bands, corresponding to transitions due to change in the polarizability of a molecule by interaction with light (laser beam, also known as excitation energy), are detected. These transitions can be used to identify the molecule as they provide a "molecular fingerprint" of the material.

The use of Raman spectra for analysis of some semiconductor samples (PbI_2, Sb_2S_3, and CdS) is illustrated in Figs. 3.18, 3.19, and 3.20. In all these cases, the

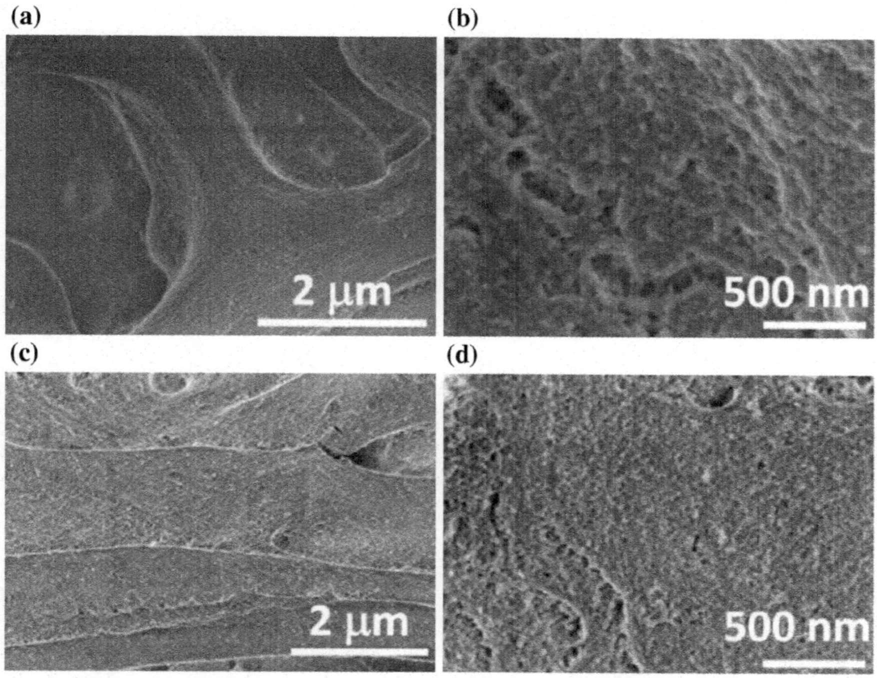

Fig. 3.16 Scanning electron micrographs of SnS thin films **a**, **b** as deposited and **c**, **d** annealed (at 380 °C in vacuum, 1 h) synthesized from SnS nanocolloids in dimethyl formamide [26] (With permission from John Wiley and Sons)

spectra were measured using 532 nm as the excitation wavelength. Figure 3.18 shows the Raman spectra of PbI_2 semiconducting thin films prepared by spin coating. The normal procedure is that the experimental peaks are compared with that of theoretical calculations of the respective vibrational modes of the given molecule. In many cases, the experimental Raman spectra are published in http://rruff.info/ [28]. Examining Fig. 3.18, the spectrum reveals three peaks at wavenumbers 96, 136 and 215 cm^{-1}. All the peaks were identified as those originating from vibrational modes of PbI_2 by comparing the experimental spectra with the reported Raman peaks for this compound [17]. Similarly, the Raman spectral analysis of thin films of antimony sulfide and CdS deposited by laser-assisted chemical bath deposition at different conditions are shown in Figs. 3.19 and 3.20 [22, 29].

The next section deals with the elemental composition and chemical state characterization of semiconductors using XPS.

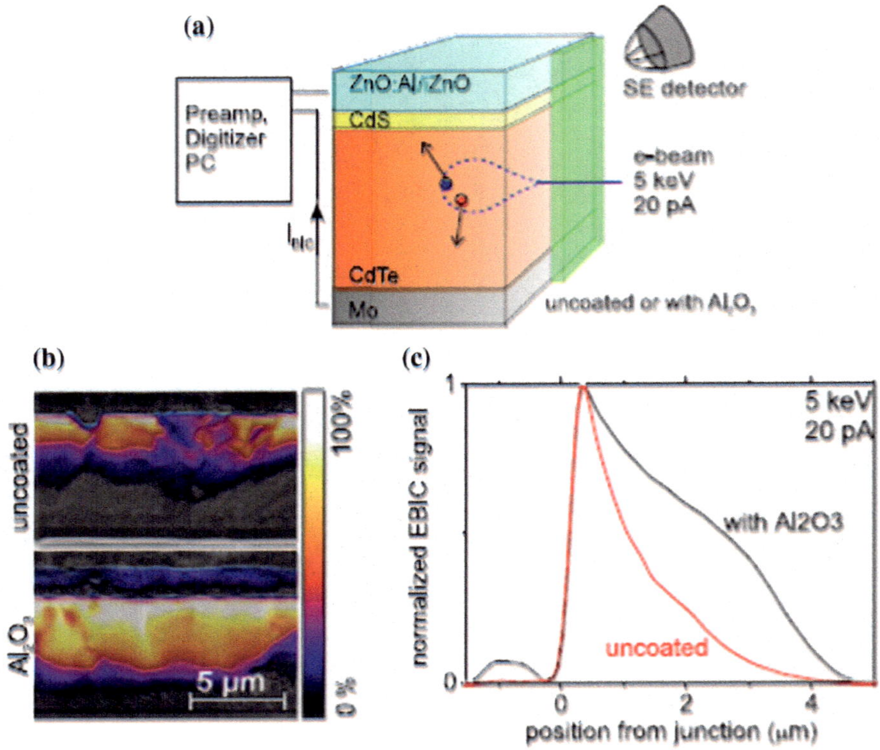

Fig. 3.17 **a** Experimental schematic for EBIC measurements: The electron beam scans the cross section of the electrically connected device, and the induced current IBIC is mapped simultaneously to the secondary electron signal. Measurements were performed without and with a 6 nm Al_2O_3 coating on the cross section. **b** EBIC and SE overlays for CdTe devices without (top) and with (bottom) Al_2O_3 coating, showing that the EBIC signal extends farther and more homogenously towards the back contact in case of Al_2O_3 coating. **c** Normalized EBIC profiles, horizontally averaged over 30 μm, are shown as solid lines for the two cases [27]

3.4 Compositional Properties

3.4.1 X-ray Photoelectron Spectroscopy (XPS)

X-ray photoelectron spectroscopy (XPS) facilitates the analyses of photoelectrons that are emitted from a material by irradiating it with a monochromatic X-ray source (characteristic X-rays from an X-ray source). In the XPS technique, the number of photoelectrons and their kinetic energy (KE) are collected in a defined time interval. In typical XPS spectra, the KE is converted to the respective binding energy (BE) and plotted versus the number of photoelectrons. Binding energy peaks appear in the spectra at discrete energies due to the emission of electrons with specific binding energies from the core levels of atoms that form the material. The position of the

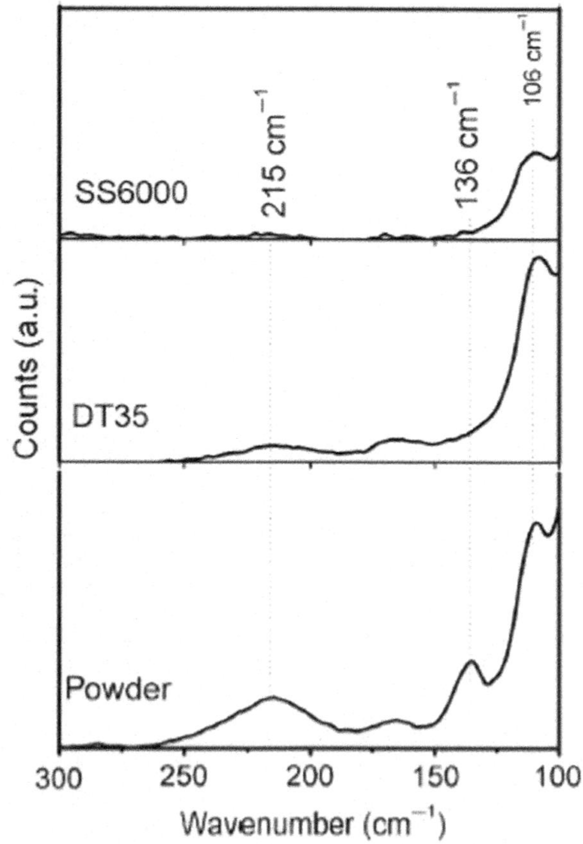

Fig. 3.18 Raman spectra of PbI_2 thin films formed by spin coating (at different spinning speed (SS6000: spin speed 6000 rpm, time: 25 s, DT35 spin speed 7000 rpm, time: 35 s) and powder. PbI_2 (hexagonal) structure) exists in different polytypes in which the common form is $2H^8$, each plane of I and Pb atoms is an identical series of triangles or overlapping hexagons. The unit cell of 2H contains three atoms. There are nine normal modes of vibration for the PbI_2 2H polytype. There are three branches of acoustic modes and six branches of optical modes. The band at 96 cm^{-1} corresponds to the Alg mode, and that at 136 cm^{-1} may be due to the resonant optical excitation of the PbI_2 molecules. The broadband at 215 cm^{-1} was also reported as a characteristic peak of PbI_2 [17] (With permission from Springer)

peaks identifies the type of elements in the material. Also, the peaks show a shift in their binding energy from the respective elemental state when it is in a compound. Hence, from the peak shift, we can identify the chemical states of the elements present in the sample. Peak areas are proportional to the number of orbitals in the analysis volume and are used to quantify the elemental composition knowing the detector sensitivity for each element. Thus, from the XPS peak analysis, we can determine the chemical state of the constituent elements in a semiconducting material [31]. In a typical XPS system, sample surface etching using Ar^+ ions can be achieved, and by

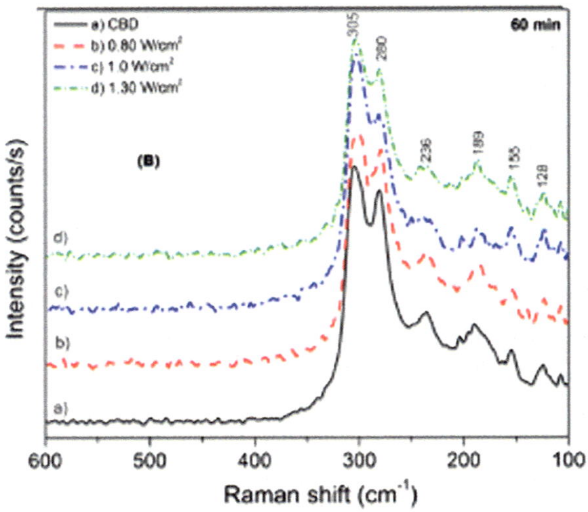

Fig. 3.19 Raman spectra of Sb_2S_3 thin films obtained under normal CBD and LACBD of different laser fluences. Peaks *e* observed at positions of 128, 155, 189, 236, 280, and 305 cm^{-1} for all the samples with slight variations in intensities for different conditions are originating from different vibrational modes of orthorhombic Sb_2S_3. From the theoretical study [30]. There are 30 Raman active modes constituting vibrations of symmetric and antisymmetric Sb-S bonds, and symmetric and antisymmetric S-Sb-S bending vibrations [29] (With permission from Elsevier)

Fig. 3.20 Raman spectra (using 532 nm excitation wavelength) of CdS thin films grown by normal CBD and pulsed laser-assisted CBD (PLACBD) for 20 min under laser fluence of 12, 27 mJ/cm^2 for 532 nm and 50 mJ/cm^2 for 1064 nm. Raman spectrum of bare glass substrate is also included [22] (With permission from Elsevier)

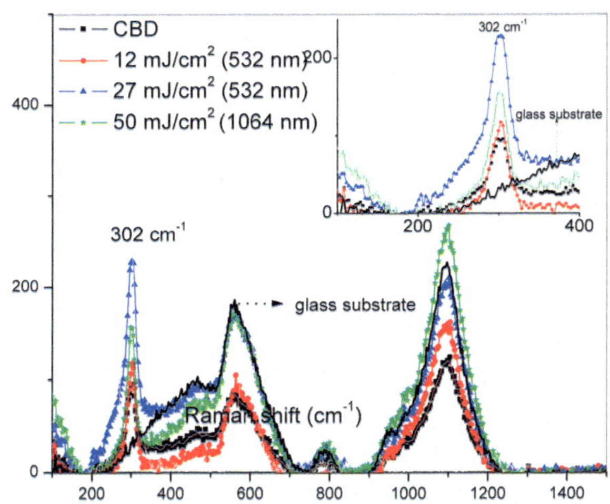

repeating the etching and analysis cycles, depth profile analysis of the composition of a semiconductor can also be accomplished.

Figures. 3.21, 3.22, 3.23, and 3.24 demonstrate the complete analysis of $CuSbS_2$ semiconducting thin films [6]. First, a typical survey spectrum from the sample is recorded to identify the elements as shown in Fig. 3.21. From the survey, the XPS pattern corresponding to the presence of Cu, Sb, and S can be seen. Then, high-

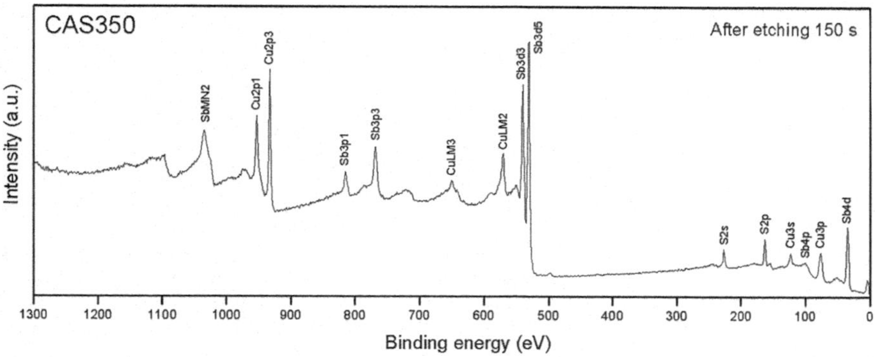

Fig. 3.21 XPS survey spectra of $CuSbS_2$ formed by heating glass/Sb_2S_3/Cu stack layers annealed at 350 °C (after 150 s argon ion etching). From the survey, the XPS pattern corresponding to the presence of Cu, Sb, and S were detected (X-ray source: monochromatized Al Kα radiation ($h\nu$ = 1486.68 eV) [6] (With permission from Elsevier)

resolution core level spectrum for each element (the binding energy region is selected based on the most dominant peak from the respective survey spectrum) is detected separately and plotted to perform a detailed analysis as illustrated in Fig. 3.22. After the peak broadening analysis, the peak features and the values can be compared with the standard/reported spectra [32, 33] for preliminary conclusions and for other research purposes. In this example, the respective peak values suggest the presence of $CuSbS_2$ as a major phase and Sb_2S_3 as a minor phase in the sample. Also, the presence of Sb due to the ion etching effect is noted.

Figure 3.23 illustrates the depth profile Binding Energy (BE) spectra recorded after repeating etching cycles and Fig. 3.24 shows the respective depth profile of composition. Thus, XPS is an effective tool for a complete compositional analysis.

XPS characterization is very useful in the analysis of nanomaterials synthesized by PLAL. The target material was characterized by XPS technique before ablation and the nanoparticles obtained in different liquid media after ablation. The XPS results confirmed that the nanoparticles obtained were of the same elemental composition and chemical state as that of the target.

Figure 3.25 shows the chemical state analysis of tin (Sn) and sulfur (S) present in the target of SnS and the nanoparticles of SnS obtained in different liquid media (ethanol, acetone and distilled water). This is the high-resolution photoelectron spectral analysis which helped to confirm the elemental composition and chemical states of the nanoparticles [34].

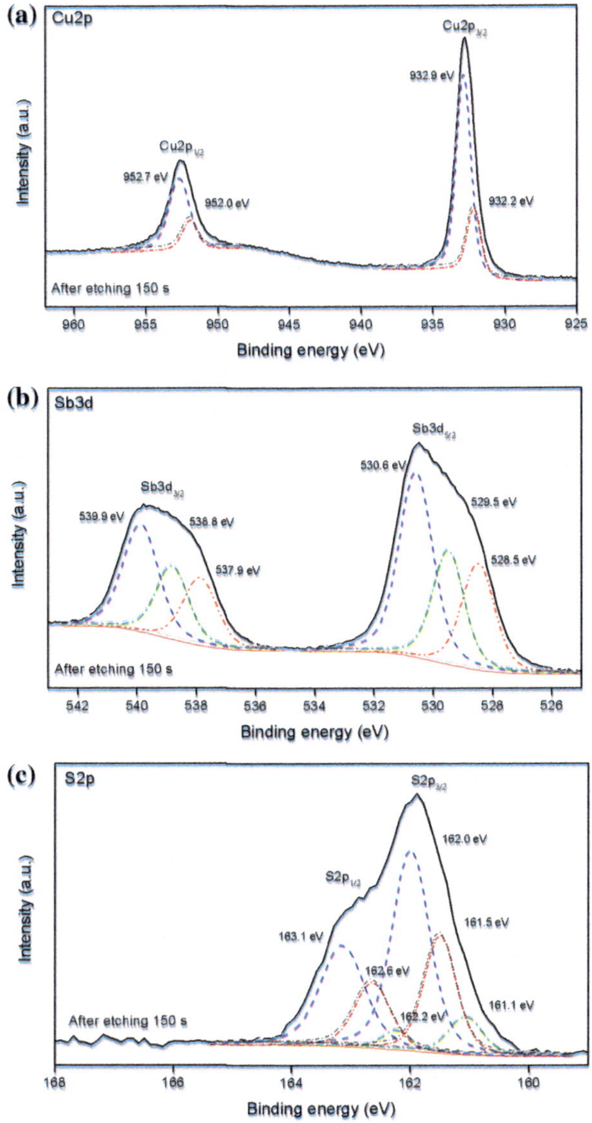

Fig. 3.22 High-resolution spectra of **a** Cu 2p core level, **b** Sb 3d core level, and **c** S 2p core level of $CuSbS_2$ formed by the heating glass/Sb_2S_3/Cu stack layers annealed at 350 °C, after 150 s Ar+ ions etching [6] (With permission from Elsevier). Binding energies (BE) of all the peaks were corrected using C 1s energy at 284.6 eV corresponding to adventitious carbon in addition to the charge compensation by the flood gun associated with the spectrometer. The peaks were de-convoluted using Shirley type background calculation and peak fitting using Gaussian–Lorentzian sum function as shown in the figure. Also, care was taken to keep FWHM same and to maintain the respective intensity ratios of p and d doublets, as demanded by the spectroscopic calculations [31]

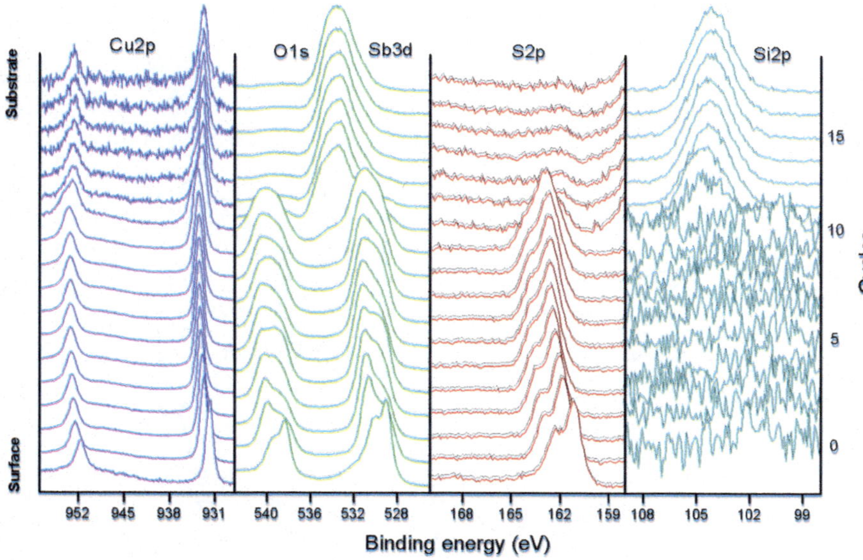

Fig. 3.23 XPS profile montage of $CuSbS_2$ formed by heating glass/Sb_2S_3/Cu stack layers at 350 °C [6] (With permission from Elsevier). Cu 2p, O 1s, Sb 3d, S 2p, and Si 2p core levels were detected throughout the depth of the thin films. Si 2p and O 1s levels were from the glass substrate

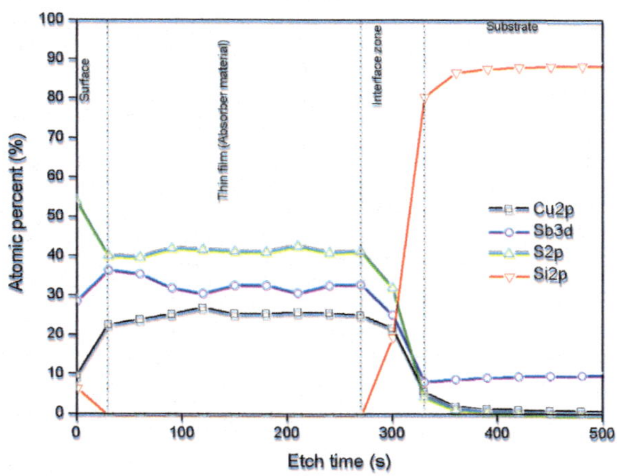

Fig. 3.24 Depth profile for compositional analysis $CuSbS_2$ formed by heating glass/Sb_2S_3/Cu stack layers annealed at 350 °C [6] (With permission from Elsevier). The compositions of copper, antimony, and sulfur are uniform (1:1.2:1.8) throughout the thin film thickness, as noted

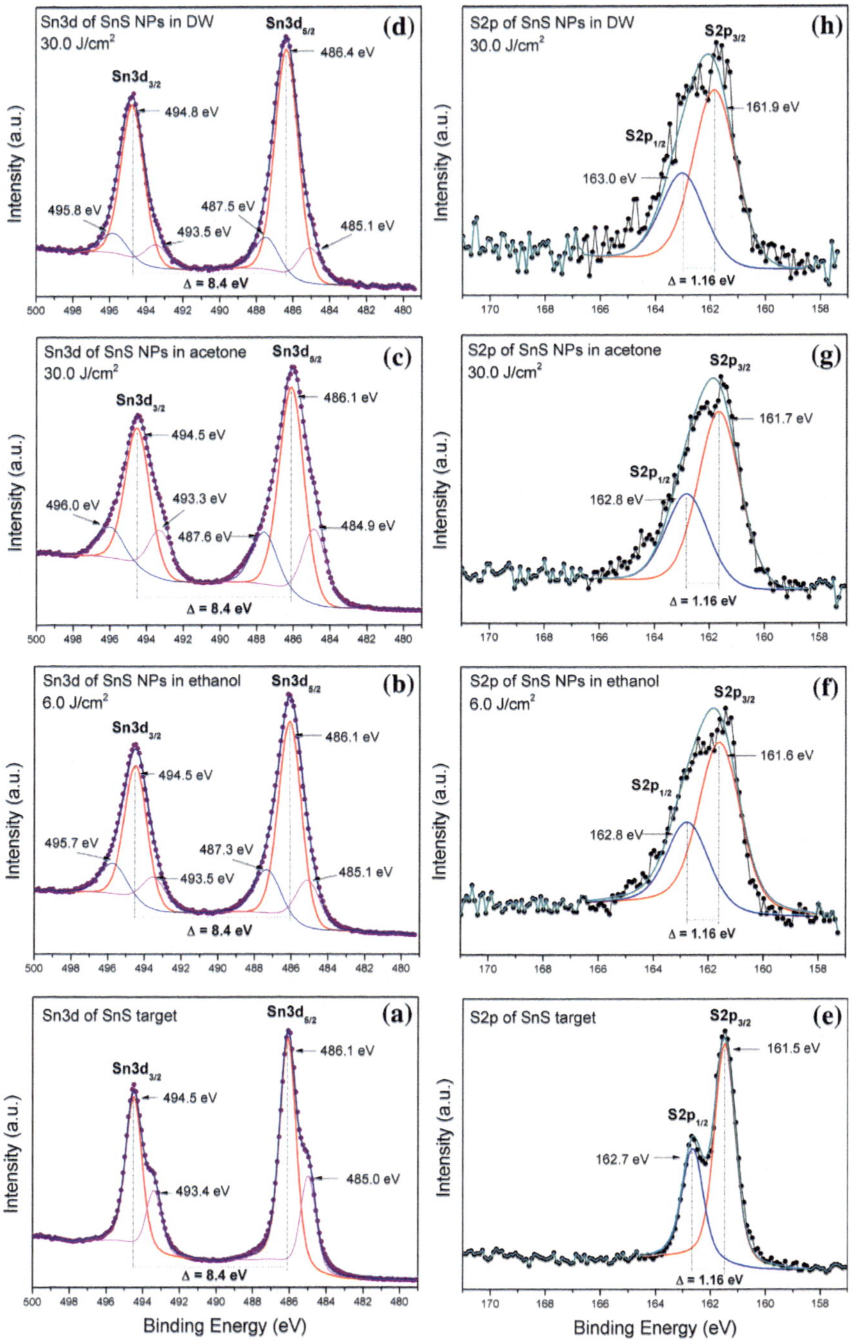

Fig. 3.25 High-resolution Sn 3d and S2p core level spectrum for **a** and **e** SnS target and SnS nanoparticles obtained by ablation; **b** and **f** in ethanol (532 nm) at 6.0 J/cm^2, **c** and **g** in acetone (1064 nm) at 30.0 J/cm^2 and **d** and **h** in distilled water (1064 nm) at 30.0 J/cm^2 [34] (With permission from Springer)

References

1. Valdes LB (1954) Resistivity measurements on germanium for transistors. Proc IRE 42(2):420–427. https://doi.org/10.1109/JRPROC.1954.274680
2. Smits FM (1958) Measurement of sheet resistivities with the four-point probe. Bell Syst Tech J 37(3):711–718. https://doi.org/10.1002/j.1538-7305.1958.tb03883.x
3. García-Ramírez MA, Ghiass AM, Moktadir Z, Tsuchiya Y, Mizuta H (2014) Fabrication and characterisation of a double-clamped beam structure as a control gate for a high-speed non-volatile memory device. Microelectron Eng 114(Supplement C):22–25. https://doi.org/10.1016/j.mee.2013.09.002
4. López-Albarrán P, Navarro-Santos P, Garcia-Ramirez MA, Ricardo-Chávez JL (2015) Dibenzothiophene adsorption at boron doped carbon nanoribbons studied within density functional theory. J Appl Phys 117(23):234301. https://doi.org/10.1063/1.4922452
5. Fox M (2010) Optical properties of solids, 2nd edn. Oxford University Press, Oxford
6. Ornelas-Acosta RE, Shaji S, Avellaneda D, Castillo GA, Das Roy TK, Krishnan B (2015) Thin films of copper antimony sulfide: a photovoltaic absorber material. Mater Res Bull 61:215–225. https://doi.org/10.1016/j.materresbull.2014.10.027
7. Mendivil MI, García LV, Krishnan B, Avellaneda D, Martinez JA, Shaji S (2015) $CuInGaSe_2$ nanoparticles by pulsed laser ablation in liquid medium. Mater Res Bull 72:106–115. https://doi.org/10.1016/j.materresbull.2015.07.038
8. Tiwald TE, Thompson DW, Woollam JA, Pepper SV (1998) Determination of the mid-IR optical constants of water and lubricants using IR ellipsometry combined with an ATR cell. Thin Solid Films 313–314(313):718–721. https://doi.org/10.1016/S0040-6090(97)00984-X
9. Yang C, Williams B, Hulet M, Tiwald T, Miles R, Samuels A (eds) (2011) Optical constants of neat liquid-chemical warfare agents and related materials measured by infrared spectroscopic ellipsometry. Proc SPIE
10. Stenzel O (1996) Das Dünnschichtspektrum. Akademie-Verlag, Berlin, p 35
11. Serway RA (1998) Principles of physics. Saunders College Pub., Fort Worth. ISBN 0-03-020457-71998
12. Griffiths DJ (1999) Electrodynamics. Introduction to electrodynamics, 3rd edn. Prentice Hall, Upper Saddle River, NJ, pp 301–306
13. Giancoli DC (2005) Physics: principles with applications. Pearson Education, Upper Saddle River
14. Elert G (2011) Resistivity of steel. The physics factbook
15. Ohring M (1995) Engineering materials science. Academic Press, London
16. http://www.icdd.com
17. Acuña D, Krishnan B, Shaji S, Sepulveda S, Menchaca JL (2016) Growth and properties of lead iodide thin films by spin coating. Bull Mater Sci 39(6):1453–1460. https://doi.org/10.1007/s12034-016-1282-z
18. Garza D, Grisel García G, Mendivil Palma MI, Avellaneda D, Castillo GA, Das Roy TK et al (2013) Nanoparticles of antimony sulfide by pulsed laser ablation in liquid media. J Mater Sci 48(18):6445–6453. https://doi.org/10.1007/s10853-013-7446-y
19. Cullity BD, Stock SR (2001) Elements of X-ray diffraction, 3rd edn. Pearson, Essex
20. Krishnan B, Arato A, Cardenas E, Roy TKD, Castillo GA (2008) On the structure, morphology, and optical properties of chemical bath deposited Sb_2S_3 thin films. Appl Surf Sci 254(10):3200–3206. https://doi.org/10.1016/j.apsusc.2007.10.098
21. Haugstad G (2012) Atomic force microscopy, understanding basic modes and advanced applications. Wiley, Hoboken
22. Garcia LV, Loredo SL, Shaji S, Aguilar Martinez JA, Avellaneda DA, Das Roy TK et al (2016) Structure and properties of CdS thin films prepared by pulsed laser assisted chemical bath deposition. Mater Res Bull 83:459–467. https://doi.org/10.1016/j.materresbull.2016.06.027
23. Goldstein J, Newbury DE, Joy DC, Lyman CE, Echlin P, Lifshin E et al (2003) Scanning electron microscopy and x-ray microanalysis, 3rd edn. Springer, US

24. http://www.leica-microsystems.com/science-lab/brief-introduction-to-coating-technology-for-electron-microscopy/
25. Mendivil MI, Krishnan B, Sanchez FA, Martinez S, Aguilar-Martinez JA, Castillo GA et al (2013) Synthesis of silver nanoparticles and antimony oxide nanocrystals by pulsed laser ablation in liquid media. Appl Phys A 110(4):809–816. https://doi.org/10.1007/s00339-012-7157-2
26. Johny J, Sepulveda-Guzman S, Krishnan B, Avellaneda DA, Aguilar Martinez JA, Shaji S (2017) Synthesis and properties of tin sulfide thin films from nanocolloids prepared by pulsed laser ablation in liquid. ChemPhysChem https://doi.org/10.1002/cphc.201601186
27. Bissig B, Lingg M, Guerra-Nunez C, Carron R, La Mattina F, Utke I et al. On a better estimate of the charge collection function in CdTe solar cells: Al_2O_3 enhanced electron beam induced current measurements. Thin Solid Films. http://dx.doi.org/10.1016/j.tsf.2016.08.012
28. http://rruff.info/
29. Shaji S, Garcia LV, Loredo SL, Krishnan B, Aguilar Martinez JA, Das Roy TK et al (2017) Antimony sulfide thin films prepared by laser assisted chemical bath deposition. Appl Surf Sci 393:369–376. https://doi.org/10.1016/j.apsusc.2016.10.051
30. Liu Y, Eddie Chua KT, Sum TC, Gan CK (2014) First-principles study of the lattice dynamics of Sb_2S_3. Phys Chem Chem Phys 16(1):345–350. https://doi.org/10.1039/c3cp53879f
31. Pvd Heide (2011) X-ray photoelectron spectroscopy: an introduction to principles and practices. Wiley, Hoboken
32. http://xpssimplified.com/elements/cerium.php
33. Moulder JF, Stickle WF, Sobol PE, Bomben KD (1992) Handbook of X-ray photoelectron spectroscopy: a reference book of standard spectra for identification and interpretation of XPS data. Perkin-Elmer Corporation, Eden Prairie, Minnesota
34. Guillen GG, Mendivil Palma MI, Krishnan B, Avellaneda Avellaneda D, Shaji S (2016) Tin sulfide nanoparticles by pulsed laser ablation in liquid. J Mater Sci: Mater Electron 27(7):6859–6871. https://doi.org/10.1007/s10854-016-4639-6

Chapter 4
Vanadium Oxides: Synthesis, Properties, and Applications

Chiranjivi Lamsal and Nuggehalli M. Ravindra

4.1 Introduction

4.1.1 General Considerations

The state of a material, whether metallic or insulating, provides a clear and fundamental insight into its electric and electronic properties. Transition metal oxides (TMOs) have unique physical, electronic, thermal, optical, chemical, and magnetic properties [1]; they are ideal for the study of insulating and metallic states due to their diverse nature of electrical properties even within the material with similar structure [2]. A large number of TMOs can be categorized under good insulator, metal and yet a third class, which have low resistivity at room temperature similar to conventional metals but a negative temperature coefficient of resistivity (TCR) similar to semiconductor. These materials with "contradictory" properties are insulating/semiconducting at low temperatures but are metallic at higher temperatures above a critical temperature (T_c). Such Insulator–Metal Transitions (IMT) have also been observed in most of the vanadium oxide (VO) compounds over a wide range of temperatures depending on the O/V ratio [3].

The ground state electronic configuration of vanadium is [Ar]$3d^3 4s^2$. Being a d-transition metal, vanadium has different oxidation states that are capable of existing in both single as well as mixed valence states on forming oxides. The vanadium oxides such as VO, V_2O_3, VO_2, and V_2O_5 exist in a single oxidation state whereas many

C. Lamsal
State University of New York, 101 Broad Street, 212 Beaumont Hall, Plattsburgh, NY 12901, USA
e-mail: cl254@njit.edu

N. M. Ravindra (✉)
New Jersey Institute of Technology (NJIT), Newark, NJ, USA
e-mail: nmravindra@gmail.com

M. I. Pech-Canul and N. M. Ravindra (eds.), *Semiconductors*,
https://doi.org/10.1007/978-3-030-02171-9_4

others, for instance, V_3O_5, V_4O_7, V_6O_{11}, V_6O_{13}, V_7O_{13}, V_8O_{15}, etc., remain in mixed (two) valence state. However, these oxides can be categorized under the so-called Magnéli (V_nO_{2n-1}) and Wadsley ($V_{2n}O_{5n-2}$) homologous series. Magnéli phases which remain in mixed valence state can be represented by general stoichiometric formula as [4]

$$V_nO_{2n-1} = V_2O_3 + (n-2)VO_2, \text{ where } 3 \leq n \leq 9$$

In this study, we focus on single valency phases of vanadium oxides: VO_2, V_2O_3, and V_2O_5. Vanadium ions in VO_2 and V_2O_3 have $V^{4+}(d^1)$ and $V^{3+}(d^2)$ electronic structures whereas V_2O_5 has V^{5+} ion with no 3d electrons.

In transition metal oxides, the s-band associated with the transition metal ions and the p-band associated with the oxygen ions are pushed away from the Fermi level by ~±5 eV and only d-orbitals are close to it [2] and are of significant importance. Since the metal d-orbitals form the conduction band, we expect some electrons in this band except with (d^0) configuration. Due to the small width of the d-conduction band in these oxides, electrons are seriously influenced by other interactions [5]. Octahedral coordination geometry in vanadium oxides creates a crystal field (or ligand field) which, in turn, splits the fivefold degenerate d-orbitals into different sets of degenerate orbitals, for instance, low-lying triply degenerate t_{2g} orbitals and higher lying doublet e_g in VO_2 as shown in Fig. 4.1.

The crystal field splitting energy for 3d-series ions in oxides has a value of 1–2 eV [5], which increases as cation oxidation state increases [7]. For spin 1/2 transition metal oxides such as VO_2 to convert into a spin 1/2 insulator, an additional splitting of the higher lying doublet is required along with strong correlations [8]. Electron—electron correlations influence the electronic structure of a material [9].

Within conventional band theory, which treats the electrons as extended plane wave, insulators are defined as materials having completely filled valence band and

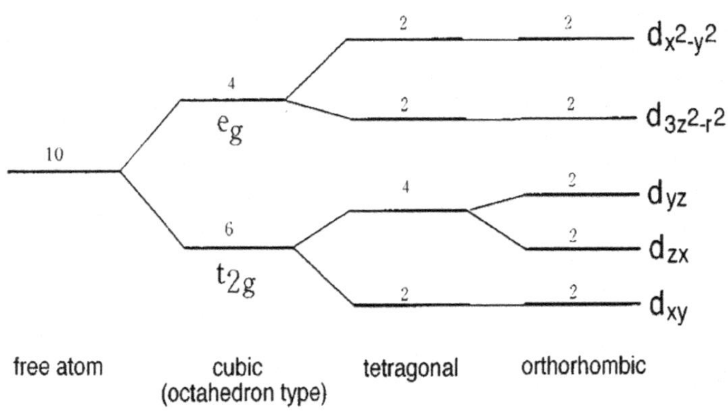

Fig. 4.1 Crystal field splitting of 3d orbitals under cubic, tetragonal, and orthorhombic symmetries. The numbers cited near the levels are the degeneracy including spins [6]

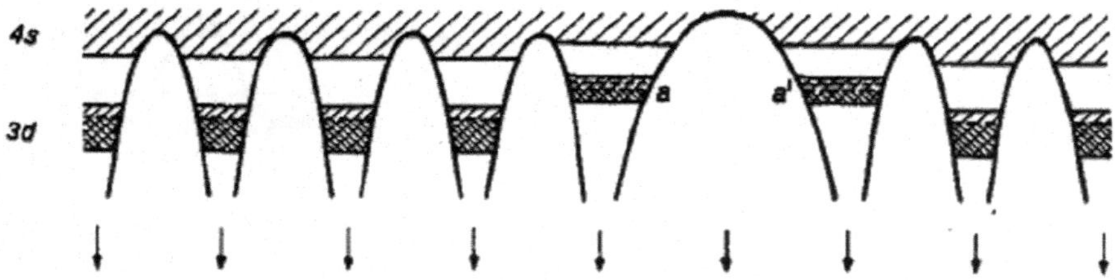

Fig. 4.2 Schematic representation of the 3d and 4s bands in a substance like NiO. The arrows, pointing down, indicate the position of cations except the arrow between *a* and *a*′, which is a vacant lattice point [11]

an empty conduction band separated by a sizeable energy gap [10]. However, the TMOs undergoing IMT have incompletely filled 3d bands. DeBoer and Verwey [11] explained this lack of conductivity, in the insulating phase, as an effect of the potential barrier existing between any two transition metal ions in the crystal as shown in Fig. 4.2.

For the case in which the electronic bandwidth is small, it is possible that the correlation energy is sufficiently larger than the kinetic energy of the electrons and the electrons remain localized. In other words, electrons are considered to be localized if the excited state is short lived as compared to the time that the electron would take to tunnel through the barrier to reach it [12]. For an electron localized around an ion in ionic crystals, it is possible for the electron to be bound in the potential well resulting from the lattice polarization due to its own presence and thus forming a "quasiparticle" known as a polaron. In contrast to large polarons, we call a polaron "small" when the lattice deformation does not extend beyond nearest neighbors. The transfer of such charge carrier occurs only in response to appropriate motions of the neighboring atoms [13]. In fact, in small-polaron model, a self-trapped electron participates in conduction by phonon assistance from site to site in the form of uncorrelated hopping as thermal fluctuation momentarily brings about a configuration equivalent to initial distorted site. Thus, two steps are clearly involved in this process: the polarization arrangement with identical initial and final states and tunneling during the coincidence event. The small-polaron theory has been used to describe the conduction mechanism in TMOs [12].

The transfer of a quantum particle from site to site can be described by two basic mechanisms: hopping and tunneling—which refer, respectively, to the transfer over a barrier due to thermal activation and transfer between two levels of same energy due to the overlap of the wave functions on the sites [14]. A schematic representation of these two mechanisms is shown in Fig. 4.3.

The mechanism of hopping can be explained by Arrhenius law, $\log\sigma\mathbf{T} \sim \mathbf{1/T}$, derived from random walk theory. However, an "equivalent" relation, $\log \sigma \sim \mathbf{1/T}$, has been used frequently in the literature, for the ease of simplicity, to fit with the experimental data. It has been found that the activation energy required to surmount the barrier, the barrier height, obtained using either form is the same within errors in many cases [15]. The corresponding rule for tunneling appears empirically in

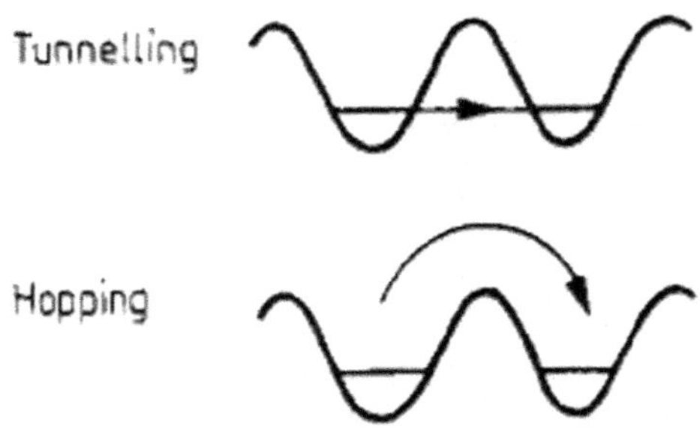

Fig. 4.3 Transfer of a quantum particle across a potential barrier between two identical localized sites [14]

the form, logσ $\sim$ **T**, a prediction of Tredgold model [16] for tunneling of quantum particle through a potential barrier of varying thickness caused by the lattice vibrations.

In transition metal oxides, d-electrons are spatially confined in partially filled orbitals and are considered to be strongly interacting or "correlated" because of Coulombic repulsion between two d-electrons of opposite spin on the same ion. In other words, the two conduction electrons with antiparallel spin at the same bonding site repel each other with strong Coulomb force so as to keep them mutually separated, and hence spatially localized in individual atomic orbitals rather than behaving as delocalized Bloch functions. According to Mott and Peierls [17], electron–electron correlation could be the origin of the insulating behavior observed in TMOs. Obviously, the high potential barriers observed between the two atoms in a TMO are highly opaque for the electrons to pass through. However, the low transparency of the potential barrier cannot merely describe the observed conductivity in TMOs. Peierls [17] noted the solution of the problem as follows: "if the transparency of the potential barriers is low, it is quite possible that the electrostatic interaction between the electrons prevents them from moving at all. At low temperatures, the majority of the electrons are in their proper places in the ions. The minority of the electrons which have crossed the potential barrier find all the other atoms to be occupied, and in order to get through the lattice, they have to spend a long time in ions already occupied by other electrons. This needs a considerable addition of energy and so it is extremely improbable at low temperatures." Hence, low transparency and electron–electron correlation contribute to the electrical properties of TMOs and the conductivity at low temperature is proportional to a "high power" of the initial transparency.

4.1.2 Insulator–Metal Transitions (IMT): Mott, Hubbard, and Peierls Mechanism

Hartree–Fock band approximation, with the inclusion of crystal field splitting, failed to explain the insulating state of TMOs due to the neglect of electronic correlation. Later, Mott [18, 19] showed that the electron–electron correlation can produce insulating states in any material, provided the lattice constant exceeds a critical value.

Mott further suggested that metallic state always exists in these materials as excited state. Conduction in such an insulator is limited due to the formation of pairs of an electron and a hole, an exciton, which are bound to each other via a Coulomb interaction. However, above a critical concentration of excitons, the screened Coulomb interaction becomes so weak that a sharp transition occurs from no free carriers to larger number of carriers. This requires a high-energy state which, in general, is hard to attain at ordinary temperature [2].

Hubbard presented a more quantitative, but still semi-quantitative, description of Mott transition introducing the effects of correlation on the Hamiltonian [20],

$$\mathbf{H} = \sum_{\mathbf{i,j}} \sum_{\sigma} \mathbf{T_{ij} C^{+}_{i\sigma} C_{j\sigma}} + \mathbf{U} \sum_{\mathbf{i}} \mathbf{n_{i\uparrow} n_{i\downarrow}} \tag{4.1.1}$$

where $\boldsymbol{n_{i\sigma}} = \boldsymbol{C^{+}_{i\sigma} C_{i\sigma}}$ is the number operator for an electron in the state $\boldsymbol{i}, \boldsymbol{\sigma}$, $\boldsymbol{T_{ij}}$ is Fourier transform of the Bloch energies $\in (\boldsymbol{k})$, and U is the average intra-ionic Coulomb repulsion. The first term represents the hopping motion of the electrons from atom to atom and the second term describes the repulsion of two electrons on the same atom. This model is simple in that it replaces the long-range Coulomb potential by delta-function repulsion. Two configurations, with and without electrons on the same atom, separated by the onsite Coulomb repulsion U are called the upper and lower Hubbard band in solids. An insulating state is formed if the repulsive term dominates over the hopping term. However, the energy gap due to electronic correlation shrinks continuously as the ratio of bandwidth (E_b) to U increases and becomes zero at a critical ratio when an insulator-to-metal transition occurs. This is not a sharp transition as predicted by Mott and is a consequence of neglecting the interatomic Coulomb term, in Hubbard model, a term responsible for screening effect in Mott transition. It seems feasible that a large number of free carriers can effectively screen U and, thereby, reduce its value. It is doubtful that screening of such an intra-atomic quantity can be significant [2]. At the same time, an insulator–metal transition can be obtained if the effect of neglecting the interatomic Coulomb interaction somehow compensates with an effective screening of U. Even though various explanations [21] have been proposed for the Mott–Hubbard transition, none of them has been clearly proved to prevail. Since the metallic and insulating states do not coexist, a continuous phase transition seems more probable; a discontinuous transition could simply be an artifact of the approximation scheme.

Another mechanism, encountered in the literature, in interpreting the IMT in V-O system is Peierls type [22, 23]. It was Peierls [24] who first pointed out that a one-dimensional (1D) metal coupled to the underlying lattice shows instability at low temperatures. In the absence of electron–electron or electron–phonon interaction, the ground state (at $T = 0$ K) of the coupled system has a periodic array of atoms with lattice constant a. In the presence of an electron–phonon interaction, the ground state is a condensate of electron–hole pairs accompanied by periodic lattice distortion of period $2a$. The condensate is called charge density wave (CDW) characterized by charge density $\boldsymbol{\rho}(\boldsymbol{r})$ associated with the collective mode formed by the electron–hole pairs [25],

$$\rho(\boldsymbol{r}) = \rho_0 + \rho_1 \cos(2\,\boldsymbol{k}_F.\boldsymbol{r} + \varphi) \tag{4.1.2}$$

where, ρ_0 is the unperturbed electron density, φ is the phase of the condensate, and Fermi wave vector is $(\boldsymbol{k}_F) = \frac{\pi}{a}$. CDWs are mainly a 1D phenomenon. However, despite its occurrence in 2D- or 3D-band structures, most of the discussions are based on idealized one-dimensional model.

A schematic representation of Peierls distortion, in 1D metal with half-filled band, is shown in Fig. 4.4. This distortion opens up a gap at the Fermi level and, for small distortion, the configuration of the distorted system is energetically favorable [25]. At finite temperature T, the electron–phonon interaction is screened by the electrons that are excited across the gap and narrow the energy gap [26]. Consequently, the lattice distortion is reduced and a second-order transition occurs at the so-called Peierls temperature (T_p). The material shows metallic properties above T_p but becomes semiconducting below T_p. However, the attempts to interpret IMT in V-O systems, based on Peierls mechanism, have also been criticized [27].

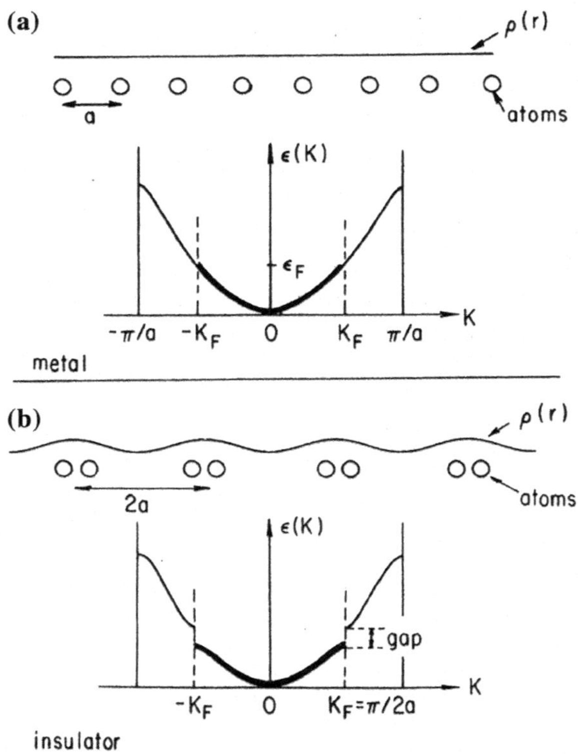

Fig. 4.4 Peierls distortion **a** undistorted metal, **b** Peierls insulator [25]

4.1.3 Literature Review

Correlated electrons are responsible for the extreme sensitivity of materials for small change in external stimuli such as pressure, temperature, or doping [28]. VO_2 is one of the widely studied materials which undergoes IMT at 340 K [29], while V_2O_3 and V_2O_5 show the transitions at 160 K [30] and 530 K [31], respectively. These phase transitions are reversible [32] and are accompanied by drastic change in the crystallographic, magnetic, optical, and electrical properties. During structural transition, atoms undergo displacement with redistribution of electronic charge in the crystal lattice, and hence the nature of interaction changes [33]. Below T_c, VO_2 and V_2O_3 have monoclinic structure [34, 35] and V_2O_5 has orthorhombic structure [36]. At temperatures higher than T_c, they have crystal structures that are different from their low-temperature counterparts [34, 37]. Similarly, the phase transition leads to change in electrical conductivity up to 10 orders of magnitude [38], while optical and magnetic properties show discontinuity. Recently, the phase transition in bulk V_2O_5 has become a controversial issue even though the studies on its thin film show IMT; various transition temperatures have been reported for these materials in the literature [39, 40]. Kang et al. [40] have also concluded that V_2O_5 films undergo IMT without structural phase transition. Furthermore, the precise mechanism of IMT is still a matter of debate [41] and no theoretical understanding has been realized to predict the transition temperature [42].

The vanadium oxides are chromogenic materials and can change their optical properties due to some external stimuli in the form of photon radiation (photochromic), change in temperature (thermochromic), and voltage pulse (electrochromic); the change becomes discontinuous during IMT. Such properties can be exploited to make coatings for energy-efficient "smart windows" [43], and electrical and optical switching devices [44]. Thin films of VO_2 and V_2O_3 have been found to show good thermochromism in the infrared region [45, 46]. While maintaining the transparency to visible light, a smart window modulates infrared irradiation from a low-temperature transparent state to a high-temperature opaque state [47]. The two oxides, VO_2 and V_2O_5, can change their optical properties in a persistent and reversible way in response to a voltage [48]. V_2O_5 exhibits exceptional electrochromic behavior because it has both anodic and cathodic electrochromism, different from VO_2 which has only anodic electrochromism, and is also an integral part in band structure effects [48]. These electrochromic materials have four main applications: information displays, variable reflectance mirrors, smart windows, and variable emittance surfaces.

The V-O systems are widely applicable in technology such as memory devices and temperature sensors [49]. The memory aspect of the material is evidenced from the pronounced hysteresis present in the phase transition [50]. Normally, the range of operation of a device lies outside the hysteresis region. Some bolometric devices are operational within the hysteretic transition [51]. Bolometers are thermal infrared (IR) detectors and can be used in infrared imaging applications such as thermal camera, night vision camera, surveillance, mine detection, early fire detection, medical imaging, and detection of gas leakage. A bolometer requires a material with high-temperature coefficient of resistance (TCR) and a small $1/f$ noise constant [52]. Pure, stoichiometric single crystals of VO_2 and V_2O_5 have high TCR but are difficult to grow. Furthermore, the latent heat involved in IMT is highly unfavorable for the bolometric performance [53]. Since T_c of V_2O_3 is far below room temperature, the resistance, and hence the level of noise is low which makes V_2O_3 a good candidate for the fabrication of efficient microbolometers. Cole et al. [54] have shown that the thin films of all the three oxides, combined together, can produce a desired material with high TCR and optimum resistance for bolometer fabrication.

TMOs, among others, are the best candidates for the cathode materials for rechargeable Li-ion batteries. Due to the layered structure of V_2O_5, it has acceptable ionic storage capacity. The cathode material currently being used, coarse-grained Li_xCoO_2, has a practical energy density <500 Wh/kg which is far less than a theoretical energy density of ~1100 Wh/kg for lithiated V_2O_5 [55]. Panasonic has commercialized the use of V_2O_5 as a cathode material in rechargeable Li-ion batteries for low-energy applications (button cells). The energy density of this product is 100–140 Wh/L and has been found to be insensitive to overcharging [56]. TMOs, being stable at high temperature along with their tunable transport properties and well-established synthesis techniques, are gaining attention for their thermoelectric properties [57]. V_2O_5 is known to have hopping transport mechanism with anisotropic n-type electrical conduction. The Seebeck coefficient varies slowly with temperature indicating the applicability of V_2O_5 as a thermistor [58].

Clearly, phase transition in VO_2 is of high technological interest. IMT occurs near room temperature and T_c can be tuned optically, thermally, electrically [59], and with doping [38]. The phase transition in VO_2 has been used to achieve frequency-tunable metamaterials in the near-infrared range [60, 61]. Recently, Kyoung et al. [62] have extended the study to terahertz range by proposing an active terahertz metamaterial, a gold nano-slot antenna on a VO_2 thin film, which transforms itself from transparent to complete extinct at resonance when the VO_2 film undergoes thermo- or photoinduced phase transition. Cavalleri et al. [34] showed that the phase transition can be photoinduced within hundreds of femtoseconds which can be an underlying principle for an ultrafast switch.

Vanadium dioxide (VO_2) [45, 63] and vanadium sesquioxide (V_2O_3) [64, 65] are the model systems that are used to study IMT in correlated electron systems. V_2O_5 is the most stable among the other two and exhibits highly anisotropic optoelectronic properties [40, 66]. While the study of vanadium oxide systems is an exciting field of research due to its significant technological applications, the phase transition, high sensitivity to microscopic details, and anisotropic nature make the study more difficult. In general, the conventional band theory, which treats the electrons as extended plane waves, can explain the metallic behavior of a material but fails to account for the localized electrons [67]. In transition metal oxides, the d-electrons are partially localized and partially itinerant [68] and, during IMT, an electron changes its behavior from localized to itinerant. Even though the Mott–Hubbard transition (strong electron–electron interactions) and the Peierls mechanisms (electron–phonon interactions) are considered to be responsible for IMT, no general consensus has been reached among the scientific community [69]. Coping with such a transition problem, which involves understanding the "competition" between kinetic (wave-like) and correlation (particle-like) terms in the electronic level, is still an exciting field of research and is the heart of electronic many-body problems.

4.2 First Principles Electronic Structure Methods

Ab Initio calculations are becoming widely popular in studying electronic structures and various physical properties (e.g., ground state energy, dipole moment, polarizability, vibrational frequencies, and nuclear quadrupole moment) of many-electron systems [70–73]. The first principles approaches can be classified into three main categories: the Hartree–Fock (HF), the Density Functional Theory (DFT), and the quantum Monte-Carlo (QMC) approaches [71]. The Hartree–Fock self-consistent method is based on the one-electron approximation in which the motion of each electron, in the effective field of all the other electrons, is governed by a one-particle Schrödinger equation. In this approximation, the "Hartree–Fock" energy of many-electron system having N-electrons can be written as

$$E^{(0)} + E^{(1)} = E^{\mathrm{HF}} \tag{4.2.1}$$

where

$$E^{(0)} = \left\langle \Psi^{(0)} \middle| \hat{H}_0 \middle| \Psi^{(0)} \right\rangle = \sum_i \varepsilon_i \tag{4.2.2}$$

is the expectation value of sum of the one-electron Fock operators and is known as the lowest energy eigenvalue of the unperturbed system. Similarly, the expectation value of the first-order perturbed Hamiltonian $\hat{H}'$ over the unperturbed state $\Psi^{(0)}$ of the system is written as

$$E^{(1)} = \left\langle \Psi^{(0)} \middle| \hat{H}' \middle| \Psi^{(0)} \right\rangle \tag{4.2.3}$$

which is the first-order perturbation energy correction to the unperturbed energy $E^{(0)}$.

The Hartree–Fock wave function satisfies the antisymmetry requirement and it includes the correlation effects arising from the pairs of electrons of the same spin. However, the motions of electrons of the opposite spin remain uncorrelated in this approximation. The methods beyond the HF approximation, which deal with the phenomenon associated with many-electron system, are known as electron correlation methods. One of the approaches to electron correlation is the Møller–Plesset (MP) perturbation method which adds higher excitations to the HF approximation as a non-iterative correction utilizing techniques from many-body perturbation theory [72, 74].

Another first principles approach, to calculate the electronic structure of many-electron systems, incorporating electron correlation, is the Density Functional Theory (DFT). In this theory, exchange correlation energy is expressed, at least formally, as a functional of the resulting electron density distribution, and the electronic states are solved self-consistently as in the HF approximation [71]. In the HF approximation, the exchange interaction is treated exactly but the dynamic correlation, arising due to Coulomb repulsion, between the electrons is neglected. The Density Functional Theory, in principle, is exact but, in practice, both exchange and dynamic correlation effects are treated approximately [75].

4.2.1 Modern Density Functional Theory: Kohn–Sham Approach

The history of Density Functional Theory (DFT) began with the Thomas–Fermi approach in the 1920s [76, 77]. The theory was given subsequent extensions until 1964, when Hohenberg and Kohn [78] proved a theorem that the ground state energy (E) of a many-electron system, from Schrödinger equation, is a unique functional of the electron density, $E[\rho(r,\theta,\varphi)]$. By introducing orbitals, ϕ_i, Kohn and Sham formulated a set of equations called Kohn–Sham (KS) equations

$$\left[-\frac{\hbar^2}{2m}\nabla^2 - \sum_A Z_A \frac{e^2}{|\mathbf{r}-\mathbf{R_A}|} + e^2\int \frac{\rho(\mathbf{r}')}{|\mathbf{r}-\mathbf{r}'|}\mathbf{dr}' + U_{\mathrm{XC}}(\mathbf{r})\right]\phi_i = \varepsilon_i\phi_i \tag{4.2.4}$$

1 2 3 4

where

first term, E_k = Kinetic energy acting on orbital,
second term, V_{nuc} = Nuclear attraction potential acting on orbital,
third term, $V_H[\rho]$ = Columbic interaction of electron in orbital, ϕ_i, with other electrons in the molecule (Hartree potential),
and the fourth term, $U_{\mathrm{XC}}[\rho]$ = exchange electron correlation potential.

Unlike the Schrödinger equation (which depends on N-electrons' $3N$ coordinates), the Kohn–Sham equations depend on the electron density, $\rho(r,\theta,\varphi)$, which, in turn, depend on the orbital,

$$\rho(r,\theta,\varphi) = \sum_i |\phi_i|^2 \tag{4.2.5}$$

Hartree potential includes a self-interaction term since the orbital that is being described in KS is also a part of the electron density, which is unphysical. U_{XC} needs to be constructed in such a way that the self-interaction contained in the Columbic interaction could be removed. Since the true form of exchange–correlation is not yet known, this term needs to be approximated in some way—the major approximation involved in DFT. However, the orbital concept in KS equation gives 99% of the kinetic energy right; thus, the accurate density is obtained except for a small contribution from the exchange–correlation.

The most widely used approximations to exchange correlation within DFT are Local Density Approximation (LDA) and Generalized Gradient Approximation (GGA). LDA was proposed by Kohn and Sham in 1965 based on the concepts that were developed in the Thomas–Fermi–Dirac theory. The main aspect of LDA is to treat a general inhomogeneous electronic system as locally homogeneous. This functional is known exactly but the use of such a uniform electron gas cannot represent the electron density of a molecular system accurately. In order to study the magnetic properties, Local Spin Density Approximation (LSDA) can be used which basically generalizes LDA to include electron spin. Within L(S)DA, wide bands of s and p orbitals are addressed well but the correlations of electrons in narrow bands (d and f bands) are weakly described, in general. GGA, on the other hand, uses the information about local electron density and the local gradient in the electron density [79]. Depending on the way information is used from the electron density and its local gradient, a large number of L(S)DA/GGA functionals have been developed.

In order to solve the KS equation, we need to define the Hartree potential which depends on the electron density. The electron density, in turn, depends on the orbitals, ϕ_i, which are only known after solving the KS equation. In order to solve this "chicken-egg" problem, we adopt an iterative method as indicated in the flowchart (Fig. 4.5) and obtain self-consistent solution (solution obtained from the iterations until the convergence is reached). The closer is the starting wave function to the true function, the lesser are the SCF iterations required in the calculations.

4.2.2 Basis Sets

An isolated system (atoms/molecules/clusters) does not interact with others, and hence no boundary condition is required. A material can be thought of as being made of unit cell by repeating it indefinitely in all directions. In computer simulations, a unit cell along with a set of boundary conditions is used to approximate such a periodic system. The solution of Kohn–Sham equation, for such a periodic system, must satisfy Bloch's theorem; the Kohn–Sham orbital, $\phi_i(\mathbf{r})$, can be written in terms of two factors as

$$\phi_i(\mathbf{r}) = e^{i\mathbf{k}.r} u_i(\boldsymbol{r}) \tag{4.2.6}$$

where $e^{i\mathbf{k}.r}$ is called plane wave and the function $u_i(\boldsymbol{r})$ is such that it is periodic in space with the same periodicity of the unit cell of the system. It is the Bloch theorem that makes the problem easier to solve in reciprocal space than in real space. Further,

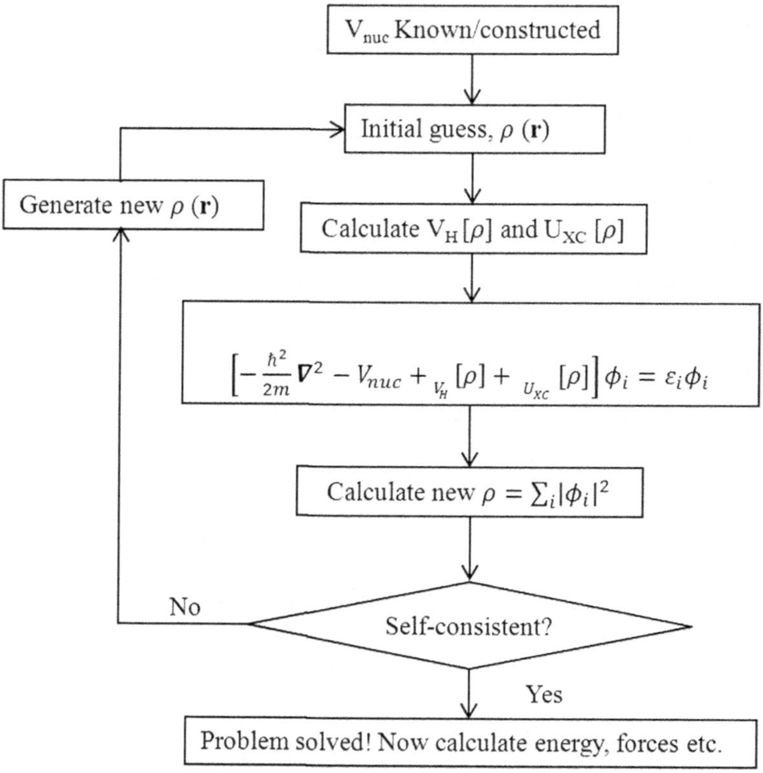

Fig. 4.5 Solving the KS equations for a set of fixed nuclear positions

the periodicity of $u_i(\boldsymbol{r})$ allows us to write it in terms of a set of plane waves as (Fourier expansion of periodic function)

$$u_i(\boldsymbol{r}) = \frac{1}{\sqrt{\Omega}} \sum_G C_{i,G} e^{iG.r} \tag{4.2.7}$$

where Ω is the volume of the unit cell of the lattice, **k** is the wave vector in the first Brillouin zone, and G is the reciprocal lattice vector. Hence, the orbital can be written as

$$\phi_i(\boldsymbol{r}) = \frac{1}{\sqrt{\Omega}} \sum_G C_{i,G} e^{i(\boldsymbol{k}+\boldsymbol{G}).\boldsymbol{r}} \tag{4.2.8}$$

The most natural choice of basis sets for studying periodic system would be plane wave basis sets. Equation (4.2.8) consists of infinite number of plane waves in the sum and in real calculation, we need to truncate it to some finite number.

With the orbital of the form (4.2.8), the kinetic energy is written as

$$E = \frac{\hbar^2}{2m}|k + G|^2 \tag{4.2.9}$$

Since solutions with lower energies are more physically important, we can truncate the infinite sum (4.2.8) to a value E_{cut} such that $\frac{\hbar^2}{2m}|k + G|^2 \leq E_{\text{cut}}$. (It is just a tradition to express this cutoff in energy units.)

4.2.3 *Pseudopotential*

For the true nuclear potential, wave functions corresponding to the core electrons exhibit sharp peaks as shown in Fig. 4.6. In order to address this fast spatial variation, we need a large number of plane waves. In other words, the plane waves associated with tightly bound core electrons in atoms oscillate in the short length scale (also, valence wave functions have wiggles: $n - 1$ nodes near nucleus, n being principle quantum number) and involve large energy cutoffs. Since the chemical and physical properties of a material mainly depend on less-tightly-bound electrons of the outer shells rather than tightly bound core electrons, core electrons can be treated as a big system with the nucleus. Hence, instead of using exact potential, we use an effective potential called pseudopotential [i.e., to replace potentials 1 through 4 in Eq. (4.2.4)] which is pre-calculated in an atomic environment and kept frozen during the calculation. In other words, the use of pseudopotential avoids the situation of taking into account the core electrons in the plane wave expansion, which would otherwise require higher cutoff energy. Pseudopotentials are used in electronic structure calculations with great success, reasonable accuracy, and a much more time efficient manner.

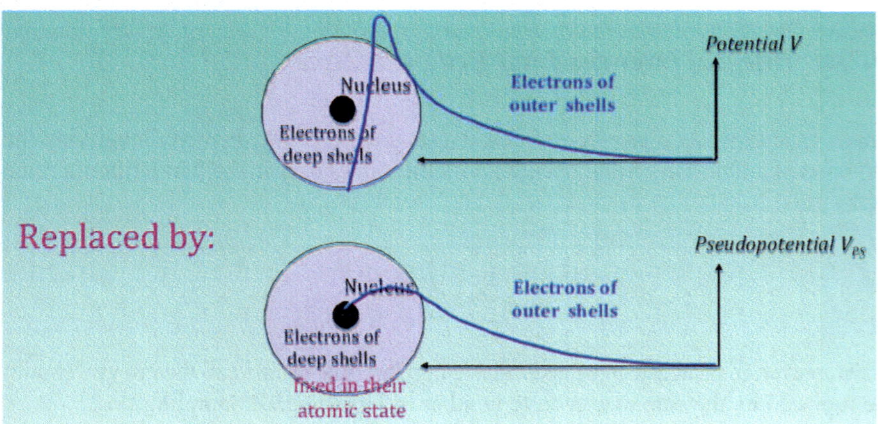

Fig. 4.6 A typical comparison of exact potential with pseudopotential [80]

All-electron (exact) potential and pseudopotential are the same beyond a particular distance from the nucleus called cutoff or core radius. A particular pseudopotential defines the atom, number of electrons treated as valence/core electrons, type of exchange–correlation functional, and minimum energy cutoff required in a calculation. Since the pseudopotentials are not unique, we are free to choose a form that simplifies the calculation and the interpretation of the resulting electronic structure [81]. Pseudopotentials can be soft and hard depending on the cutoff energies considered; hard potentials require high cutoff energies while the soft ones require low cutoff energies. The mostly used pseudopotentials are norm-conserving and ultrasoft pseudopotentials. The norm-conserving pseudopotential was first introduced by Hamann, Schlüter, and Chiang [82] based on Ab Initio atomic calculations. The norm-conservation criterion generally produces hard pseudopotential. However, it is not always necessary that the norm of pseudo-wave and all-electron wave functions coincide [83]. On the other hand, the ultrasoft pseudopotentials (USPPs) which require substantially lower cutoff energies are based on the work of Vanderbilt [84]. In practice, the cutoff radius in USPP is chosen to be very large, well beyond the maximum of the radial wave function; sharp peaks in the pseudo-wave function are replaced by smoother wave functions. However, it has a disadvantage of requiring a large number of empirical parameters during the construction process.

Blöchl introduced another frozen core approach that generalizes both the linear augmented-plane-wave (LAPW) method and the pseudopotential method "in a natural way" [85]. Later, Kresse and Joubert [86] implemented the Projector Augmented Wave (PAW) method in the plane wave calculations. Even though all electrons are not treated explicitly in the PAW formalism, it is sometimes referred to as an all-electron method in a sense that the valence electronic states are kept orthogonal to the core wave functions. Some of the disadvantages of USPPs can be avoided with the PAW method.

4.2.4 Integral Over the First Brillouin Zone

Many quantities such as electron density, total energy, etc., involve integrating the ith band of a function, $f_i(\boldsymbol{k})$, over the possible values of **k** in the first Brillouin Zone (BZ) as

$$\bar{f_i} = \frac{1}{\Omega_{\mathrm{BZ}}} \int_{\mathrm{BZ}} \mathrm{d}\boldsymbol{k}\, f_i(\boldsymbol{k}) \tag{4.2.10}$$

However, in principle, we need infinite number of **k**-points and the integral should be replaced by the sum over a finite number of k-points: BZ Sampling.

$$\bar{f_i} = \frac{1}{N_k} \sum_k f_i(\boldsymbol{k}) \tag{4.2.11}$$

The sum over the k-points in the Brillouin Zone can be reduced to the sum in the irreducible part of the Brillouin zone (IBZ), by employing all symmetry operations of the Bravais lattice to individual k-point. (This will significantly reduce the computational time.) We can calculate the "weight", w_k, for each k-point in the IBZ by the following relation:

$$w_k = \frac{\text{Number of symmetry connected}\, k - \text{points}}{\text{Total number of}\ k - \text{points in the BZ}} \tag{4.2.12}$$

Then, the sum over the BZ becomes sum over the IBZ

$$\bar{f_i} = \sum_{k}^{\text{IBZ}} w_k f_i(\boldsymbol{k}) \tag{4.2.13}$$

There are mainly two commonly used approximations for the sum over the irreducible part of the BZ: (Linear) tetrahedron method and special k-points approaches. Special k-points scheme has now been widely used in the calculations, which are Chadi–Cohen k-points, Cunningham k-points, and equidistance or Monkhorst–Pack grids schemes.

When the general shape of a function, to be integrated, is unknown, the most straightforward approach is to construct equidistant grid. Monkhorst and Pack (MP) [87] developed a method for obtaining sets of special points in the Brillouin zone. The points in the k-space are written as a linear combination of lattice vectors in k-space: $\boldsymbol{k} = n_1\boldsymbol{b}_1 + n_2\boldsymbol{b}_2 + n_3\boldsymbol{b}_3$, where $(q_1 \times q_2 \times q_3)$ grid has the coordinates: $n_i = \frac{2p_i - q_i - 1}{2q_i}$, $p_i = 1, \ldots, q_i; i = 1, 2, 3$.

4.2.5 Metals—Fermi Surface Sampling

In metals, the Fermi surface separates the occupied electronic states from the unoccupied ones. In other words, the functions to be integrated in k-space show a discontinuity at the Fermi surface as indicated by "Θ-function"/step function

$$\bar{f_i} = \sum_{k}^{IBZ} w_k f_i(\boldsymbol{k}) \Theta(\varepsilon_i(\boldsymbol{k}) - \varepsilon_F) \tag{4.2.14}$$

where the step function is defined as $\Theta(\varepsilon_i(\boldsymbol{k}) - \varepsilon_F) = \begin{matrix} 1, & \text{if}\, \varepsilon_i(\boldsymbol{k}) < \varepsilon_F \\ 0, & \text{if}\, \varepsilon_i(\boldsymbol{k}) > \varepsilon_F \end{matrix}$. This imposes a serious complication in calculating integrals, and hence special efforts have to be made to address this issue; otherwise, a very large number of k-points are required to reach the convergence in the calculations. On the other hand, the Fermi level is adjusted to satisfy the normalization condition: $\sum_{\boldsymbol{k} \in \boldsymbol{BZ}} N_k w_k = N$, during self-consistency; for k-points that are closer to the Fermi surface, the coarse grid can lead to the highest occupied bands entering or exiting the sums from one iteration step to the next. This introduces instability in the finite sum during the self-consistent procedure.

The remedy to this problem is to replace the step function by a smoother function $f(\{\varepsilon_i(\boldsymbol{k})\})$ which allows partial occupancies at the Fermi level. This, basically, "smears out" the discontinuity by forcing the function being integrated to be continuous, and hence such approaches are called smearing methods. One of the commonly used smearing functions is the Fermi–Dirac (FD) function

$$f(\{\varepsilon_i(\boldsymbol{k})\}) = \frac{1}{e^{\frac{(\varepsilon_i(k)-\varepsilon_F)}{\sigma}} + 1}; \quad \sigma = k_B T \tag{4.2.15}$$

The broadening energy parameter, σ, should be adjusted to avoid instability in the SCF convergence procedure, and hence regarded as one of the convergence parameters. Clearly, the *F–D* function approaches step function in the limit $T \rightarrow$ zero. However, the broadening parameter cannot be related to the electronic Fermi temperature unless the system is at really finite electronic temperature (finite temperature extension of Kohn–Sham theory by Mermin [88]), in general, it is just a technical issue. Several other smearing functions have been introduced to approximate the step function such as Gaussian smearing function, method of Methfessel–Paxton, and "cold" smearing of Marzari–Vanderbilt. Besides the smearing techniques, a method of tetrahedron is also used in which BZ is divided into tetrahedra, the function is interpolated within these tetrahedra and then integration is performed.

4.2.6 ABINIT

ABINIT is a full Ab Initio, free-to-use, simulation package based on Density Functional Theory, pseudopotentials, and plane waves. In such plane wave based electronic structure calculations, Fast Fourier transforms (FFTs) are used, which are proved to be an efficient algorithm to transform functions (wave functions/electron densities) from real space to their reciprocal space counterparts. ABINIT mainly computes charge density, total energy, and electronic structure of a periodic system based on Kohn–Sham density functional approach. In the usual ground state (GS) calculation or structural relaxation, the potential has to be determined self-consistently. Several choices for the selection of the algorithm for SCF are possible in ABINIT. In our calculations, using ABINIT (version 7.6.3) [89, 90], we chose the default integer for self-consistent-field cycle: iscf = 17 which refers to Pulay mixing of the density [91], the algorithm for accelerating convergence in SCF procedure. ABINIT requires parameters such as atomic species and their position within a particular structure as its input variables—the atoms are placed in a unit cell, which is built by taking the symmetries of the system into consideration. Band occupation scheme has to be specified during the calculations, which is basically the smearing technique: "cold smearing" of Marzari (bump minimization) [92] is chosen for metallic case (occopt = 4) but default value (occopt = 1) is taken elsewhere. The temperature of smearing is taken to be 0.01 Hartree (tsmear = 0.01 Ha) as suggested, for d-band metals, in ABINIT website (the convergence calculation with respect to this parameter should be checked; large value leads to the convergence at wrong value and small value requires a large number of k-points). Monkhorst–Pack grid was chosen for Brillouin zone sampling and a convergence test was performed to determine the density of k-mesh. Another convergence test was also performed to truncate the plane wave expansion, which was found to be 40 Hartree (i.e., ecut = 40 Ha).

PAW potential used in our calculations are those with the smallest pseudopotential radii, generated using a program called AtomPaw [93, 94], for both vanadium (V) and oxygen (O), obtained from Case Western Reserve University, which were well tested with the previously confirmed data; the choice of valence bands was such that no "ghost" or "phantom" bands appeared in the band structures; minimum volume of the unit cell, bulk modulus, and the derivative of the bulk modulus were examined [95]. In oxygen, electrons in the first shell ($n = 1$) were treated as core electrons while in vanadium, the electrons in first and second shells ($n = 1$ and 2) were treated as core electrons. Approximation to exchange–correlation used in the calculations is Local Density Approximation (LDA), with the functional of Perdew and Wang (PW92) [96]. We have an advantage of using LDA in our calculations in that they have the tendency of having smaller radii for their pseudopotential spheres which is consistent with the relatively tightly packed vanadium oxide (VO_x: $x = 1.5$, 2 and 2.5) structures.

4.2.7 *Vanadium Oxides: Symmetries and Structure*

A crystal structure can be constructed uniquely from its lattice defined by lattice vectors, $\mathbf{R_i}$, and its basis defined by lattice basis vectors, $\mathbf{r_i}$. A basis vector in the unit cell can be expressed as a linear combination of the lattice vectors. For a unit cell having p number of atoms, we can write

$$\mathbf{r_i} = x_i\mathbf{R}_1 + y_i\mathbf{R}_2 + z_i\mathbf{R}_3; i = 1, 2, 3, \ldots, p \tag{4.2.16}$$

where relative coordinates x_i, y_i, z_i (with $|x_i|$, $|y_i|$, $|z_i| < 1$) are used to locate atoms in the unit cell. Here, the lattice and basis are connected to each other by the symmetry elements of the crystal structure. In ABINIT, the unit cell of a crystal structure is defined using acell/rprim (rprim, in general, defines the unit lattice vectors which are then scaled by acell variable to obtain $\mathbf{R}_i$); alternately, acell/angdeg input variables can also be used. Using the Wyckoff positions, we can locate the atoms in the unit cell in terms of the relative coordinates.

For small change in external stimuli such as pressure, temperature, or doping [28], several vanadium oxides undergo insulator-to-metal phase transition (IMT) that are accompanied by structural change and this leads to the change in number of formula unit (z) in the primitive unit cell. Below the phase transition temperature T_c, vanadium dioxide exhibits monoclinic (insulating) structure and its primitive cell contains four formula units (z =4; 12 atoms). In its metallic phase, above T_c, it acquires rutile structure and contains two formula units (z = 2; 6 atoms). For temperature below T_c, V_2O_3 has monoclinic structure with 20 atoms in its unit cell (z = 4), while above T_c, the structure changes to trigonal corundum ("corundum" was derived from Sanskrit word "Kuruvinda", meaning ruby) with 10 atoms in the primitive unit cell. V_2O_5 has simple orthorhombic structure and its bulk phase does not undergo phase transition (i.e., remains semiconducting at all temperatures). The primitive unit cell contains two formula units (z =2; 14 atoms). Due to its layered structure and weak bonds between layers, few layers thick (~nm) or even a monolayer of V_2O_5 is possible to extract. IMT has been observed in the film of V_2O_5 but without structural change. Since graphene (single layer of graphite) exhibits interesting properties unlike its bulk counterpart, researchers are curious whether analogous useful and exciting properties can be observed in the single layer of V_2O_5.

4.2.7.1 Vanadium Dioxide (VO_2)

Vanadium oxide, below T_c, can exist in two phases: stable phase/M_1 phase (β = 122.62°) and metastable phase/M_2 phase (β = 91.88°). The metastable phase is normally induced by either doping or stress. Since we are mainly focused on the temperature induced phase transition, we will study the most stable phase and its primitive unit cell is defined by the following lattice parameters [97]:

$$a = 5.7517\text{Å}; b = 4.5378\text{Å}; c = 5.3825\text{Å}; \beta = 122.646° \tag{4.2.17}$$

The M_1 phase of stoichiometric VO_2 is characterized by a simple monoclinic lattice with space group $P2_1/c\left(C_{2h}^5, \text{No.}14\right)$ [98]. Vanadium atom and the two types of oxygen atoms occupy the Wyckoff position (4e): $\pm(x, y, z), \pm\left(x, \frac{1}{2} - y, \frac{1}{2} + z\right)$ where x, y, z are shown in Table 4.1 [98]. With this definition of the crystal structure of VO_2 and "cut3d" utility of ABINIT, XCrySDen [99] (a program for displaying Crystalline Structure and Densities under X-window environment) was used to visualize the crystal structure, which is shown in Fig. 4.7.

Similarly, at temperature above T_c, VO_2 has the rutile structure and its primitive unit cell is defined by the following lattice parameters:

Table 4.1 Wyckoff parameters of the stable phase of VO_2: simple monoclinic structure

Atom	Wyckoff position	Parameters		
		x	y	z
V	4e	0.23947	0.97894	0.02646
O_1	4e	0.10616	0.21185	0.20859
O_2	4e	0.40051	0.70258	0.29884

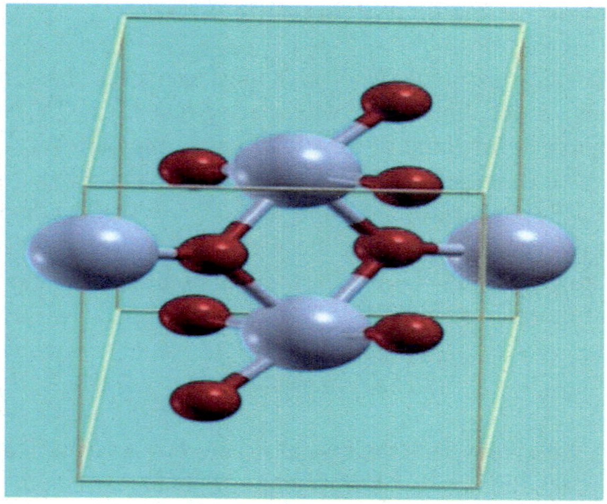

Fig. 4.7 Monoclinic structure of stable phase of VO_2: V atoms (grey) and O atoms (red) are shown

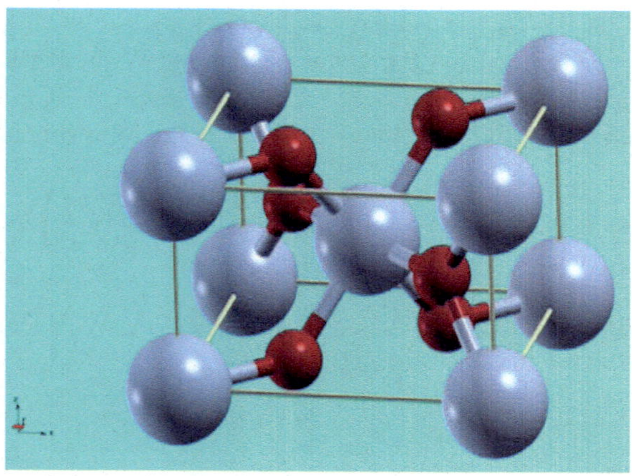

Fig. 4.8 Rutile phase of VO_2: V atoms (grey) and O atoms (red) are shown

$$\mathbf{a_R} = a_R(1, 0, 0); \mathbf{b_R} = a_R(0, 1, 0); \mathbf{c_R} = c_R(0, 0, 1) \tag{4.2.18}$$

with $a_R = 4.5546$ Å and $c_R = 2.8514$ Å. Simple tetragonal lattice has space group $P4_2/\text{mnm}\left(D_{4h}^{14}, \text{No.}136\right)$, and accordingly vanadium atoms occupy the Wyckoff position (2a): $(0, 0, 0), (\frac{1}{2}, \frac{1}{2}, \frac{1}{2})$ and oxygen atoms occupy the Wyckoff position (4f): $(x, x, 0), (-x, -x, 0), \left(-x + \frac{1}{2}, x + \frac{1}{2}, \frac{1}{2}\right), \left(x + \frac{1}{2}, -x + \frac{1}{2}, \frac{1}{2}\right)$, with $x = 0.3001$ [98]. With this data, the crystal structure of VO_2 (Rutile) is visualized, which is shown in Fig. 4.8.

4.2.7.2 Vanadium Sesquioxide (V_2O_3)

Vanadium sesquioxide (V_2O_3), in its insulating phase, has inner centered (body centered) monoclinic lattice with space group $I2/a\left(C_{2h}^6, \text{No.15}\right)$, and accordingly vanadium and first kind of oxygen atoms occupy the Wyckoff position (8f): $\pm(x, y, z)$, $\pm\left(\frac{1}{2}+x, -y, z\right)$, $\pm\left(\frac{1}{2}+x, \frac{1}{2}+y, \frac{1}{2}+z\right)$, and $\pm\left(x, \frac{1}{2}-y, \frac{1}{2}+z\right)$ with $(x, y, z)_V =$ (0.3438, 0.0008, 0.2991) and $(x, y, z)_{O1} =$(0.407, 0.845, 0.652) while the other kind of oxygen atoms occupy the Wyckoff position (4e): $\pm\left(\frac{1}{4}, y_{O2}, 0\right)$, $\pm\left(\frac{3}{4}, \frac{1}{2}+y_{O2}, \frac{1}{2}\right)$ with $y_{o2} = 0.191$ [100]. The unit cell of the monoclinic lattice is defined by following lattice parameters [37]:

$$a = 7.255\,\text{Å}; b = 5.002\,\text{Å}; c = 5.548\,\text{Å}; \beta = 96.75^{\circ} \tag{4.2.19}$$

With these data, the crystal structure was visualized which is shown in Fig. 4.9.

Metallic phase of V_2O_3 has trigonal corundum structure with space group $R\bar{3}C\left(D_{3d}^6, \text{No.167}\right)$. Vanadium atoms, located in the trigonal lattice, are located at the Wyckoff positions (4c): $\pm(x, y, z)$, $\pm\left(\frac{1}{2}+x, \frac{1}{2}+y, \frac{1}{2}+z\right)$ with $x = y = z = 0.3463$ and oxygen atoms occupy the Wyckoff positions (6e): $\pm\left(x, \frac{1}{2}-y, \frac{1}{4}\right)$, $\pm\left(\frac{1}{4}, y, \frac{1}{2}-z\right)$, $\pm\left(\frac{1}{2}-x, \frac{1}{4}, z\right)$ with $x = y = z = 0.56164$ [100]. The primitive unit cell of the trigonal lattice is defined by the following lattice parameters:

$$\mathbf{a_t} = a_t\left(1/2, -\frac{\sqrt{3}}{2}, \frac{c_t}{a_t}\right); \mathbf{b_t} = a_t\left(1/2, \frac{\sqrt{3}}{2}, \frac{c_t}{a_t}\right); \mathbf{c_t} = a_t(-1, 0, c_t/a_t) \tag{4.2.20}$$

with $a_t = 2.85875\,\text{Å}$ and $c_t = 4.66767\,\text{Å}$. With these specifications, the crystal structure is visualized as shown in Fig. 4.10.

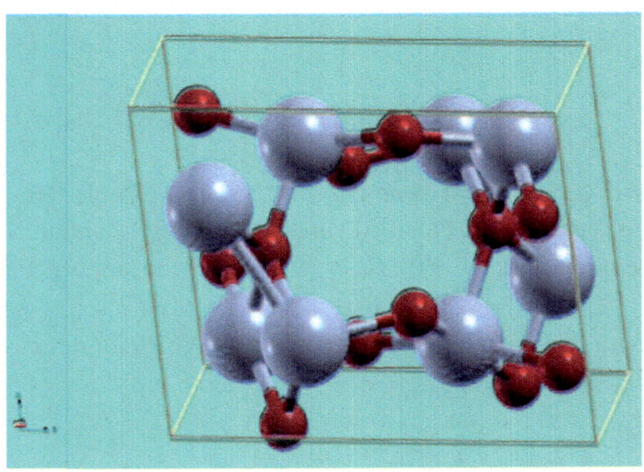

Fig. 4.9 Insulating phase of V_2O_3: V atoms (grey) and O atoms (red) are shown

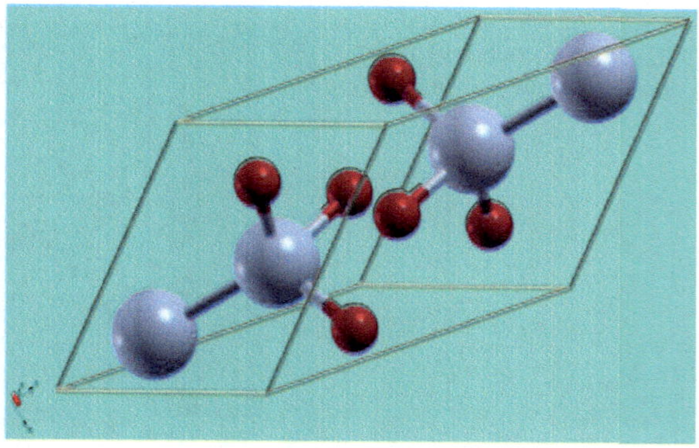

Fig. 4.10 Metallic phase of V_2O_3: V atoms (grey) and O atoms (red) are shown

4.2.7.3 Vanadium Pentoxide (V_2O_5)

The crystal structure of vanadium pentoxide consists of simple orthogonal lattice with space group Pmmn$\left(D_{2h}^{13}, \text{No.59}\right)$ and lattice constants for the primitive unit cell are given by [101]

$$a = 11.512\,\text{Å}; b = 3.564\,\text{Å}; c = 4.368\,\text{Å} \tag{4.2.21}$$

The structure consists of vanadium atom and three different types of oxygen atoms; vanadium and two types of oxygen atoms occupy the Wyckoff positions (4f): (x, 1/4, z), ($-x$ +1/2, 1/4, z), ($-x$, 3/4, $-z$), (x +1/2, 3/4, $-z$) and the third type of oxygen atom occupies the Wyckoff positions (2a): (1/4, 1/4, z), (3/4, 3/4, $-z$) [102–105], where x, y, z are given in Table 4.2 [106]. Based on the crystal structure parameters, the unit cell was visualized as is shown in Fig. 4.11.

Table 4.2 Wyckoff parameters of V_2O_5: simple orthorhombic lattice

Atom	Wyckoff positions	Parameters		
		x	y	z
V	(4f)	0.10118	0.25	−0.1083
O_1	(4f)	0.1043	0.25	−0.469
O_2	(4f)	−0.0689	0.25	0.003
O_3	(2a)	0.25	0.25	0.001

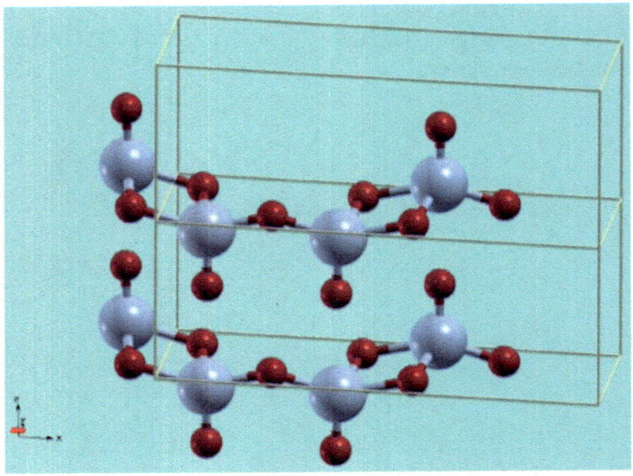

Fig. 4.11 Crystal structure of V_2O_5: V atoms (grey) and O atoms (red) are shown

4.2.8 *Optimization of Unit Cell*

Structure optimization in ABINIT can be simulated, at 1 atmosphere and 0 K, using its input variable "optcell" (=2) which optimizes both cell shape and dimensions when "ions" are allowed to move using "ionmov" variable. Since the symmetry of the system is taken into account, effectively relevant degrees of freedom are optimized. In order to avoid the discontinuities (as a result of abrupt change in the number of plane waves with cell size), suitable value of "ecutsm" has to be used. Keeping in mind that larger sphere of plane waves may be necessary during the optimization, proper value of "dilatmx" should be provided in the input file. In our calculation, dilatmx = 1.14 was used, allowing the lattice parameters to change up to 14% from input values. Using ionmov = 2, full variable cell relaxation was performed using the Broyden–Fletcher–Goldfarb–Shanno (BFGS) minimization procedure.

The result of the structural optimization is given in Table 4.3. The differences between the original and the corresponding relaxed parameters are close to each other within 5%, except the monoclinic structure of V_2O_3 and V_2O_5 in *c* direction, which shows slightly more deviation but lies within 10%: which were defined to be a little further than their relaxed positions. Since the overall changes were very small, unit cells were defined with reasonable accuracy.

4.2.9 *Convergence Studies*

The variables that mainly affect the convergence of ground state energy calculations in ABINIT are: density of k-points in the first Brillouin zone and kinetic energy cutoff for plane waves. Figure 4.12 shows the variation of total energy, at gamma point, of vanadium oxides (per unit cell used in the calculations) as a function of

Table 4.3 Lattice constants for bulk VO_x

	Before relaxation	After relaxation
V_2O_5 (simple orthorhombic)		
Lattice parameter (Angstrom)	$a = 11.512$; $b = 3.564$; $c = 4.368$	$a = 11.695$; $b = 3.546$; $c = 3.951$
Angle (degree)	$\alpha = \beta = \gamma = 90$	$\alpha = \beta = \gamma = 90$
V_2O_3 (trigonal corundum)		
Lattice parameter (Angstrom)	$a = b = c = 5.4735$	$a = b = c = 5.4673$
Angle (degree)	$\alpha = \beta = \gamma = 53.7843$	$\alpha = \beta = \gamma = 51.0410$
V_2O_3 (monoclinic)		
Lattice parameter (Angstrom)	$a = 7.255$; $b = 5.002$; $c = 5.548$	$a = 7.431$; $b = 4.781$; $c = 5.166$
Angle (degree)	$\alpha = \gamma = 90; \beta = 96.75$	$\alpha = \gamma = 90; \beta = 100.3311$
VO_2 (Rutile)		
Lattice parameter (Angstrom)	$a = b = 4.5546$; $c = 2.8514$	$a = b = 4.5479$; $c = 2.7328$
Angle (degree)	$\alpha = \beta = \gamma = 90$	$\alpha = \beta = \gamma = 90$
VO_2 (monoclinic)		
Lattice parameter (Angstrom)	$a = 5.7517$; $b = 4.5378$; $c = 5.3825$	$a = 5.541$; $b = 4.527$; $c = 5.286$
Angle (degree)	$\alpha = \gamma = 90; \beta = 122.646$	$\alpha = \gamma = 90; \beta = 121.63$

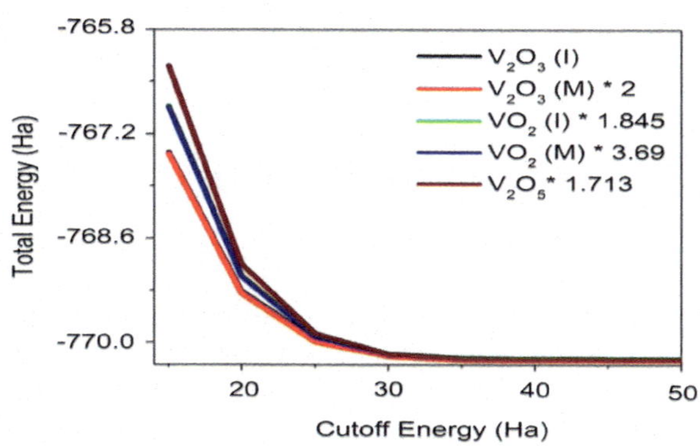

Fig. 4.12 Variation of total energy, of vanadium oxides, with plane wave cutoff; scaled by multiplication factors as shown

cutoff energy. In our calculations, we have considered the value of the plane wave cutoff to be 40 Hartree.

Similarly, convergence tests were performed to determine the optimum number of k-points required in the BZ; the number of k-points depends on whether the system is metallic or not and, of course, the precision of the calculations. The metallic system usually requires more k-points than the insulating/semiconducting systems. In the present study, ground state calculations were performed with Monkhorst–Pack grid having different k-point densities represented by $2 \times 2 \times 2$, $4 \times 4 \times 4$, $6 \times 6 \times 6$, and

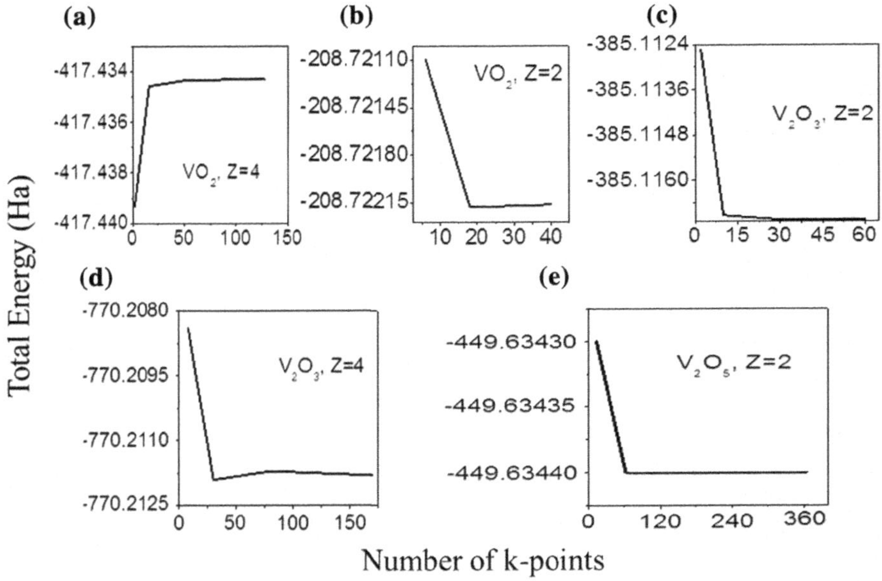

Fig. 4.13 Variation of total energy, of vanadium oxides, with number of *k*-points

$8 \times 8 \times 8$ grids as shown in Fig. 4.13. $6 \times 6 \times 6$ grid is found to be appropriate for all the systems except V_2O_5, for which $4 \times 4 \times 4$ is sufficient.

4.2.10 BoltzTraP: Calculations of Boltzmann Transport Properties

BoltzTraP [107] is a program for calculating the semi-classical transport coefficients based on Boltzmann transport equations. The code performs smooth Fourier expansion of band energies, maintaining the space group symmetry. In the present study, the electron density from Self-Consistent Field (SCF) calculation was used as an input for Non-Self-Consistent (NSCF) calculation to obtain the electronic structure on much finer grid (as compared to the SCF grid). The purpose of NSCF calculation, here, was to effectively reduce the time of transport coefficient calculations. Smooth highly resolved electronic bands, thus obtained, were processed using the BoltzTraP code. Transport properties were calculated using band energies with constant relaxation time approximation. Since the calculations are performed at 0 K and 1 atmospheric pressure, the temperature dependency in energy band is ignored. However, the electronic properties at higher temperature are simulated by applying the Fermi distribution over electronic states.

$$f_0(\varepsilon) = \frac{1}{\left(e^{(\varepsilon-\varepsilon_F)/K_BT}\right)+1} (\text{equilibrium } F - D \text{ function}) \tag{4.2.22}$$

The main transport properties studied are the following: (a) electrical conductivity: Ohm's law, in the absence of magnetic field and thermal gradient, (b) Seebeck coefficient (thermopower), and (c) thermal (electronic) conductivity which are, respectively, given by Eqs. (4.2.23), (4.2.25), and (4.2.26) [108]. Clearly, the transport coefficients are function of temperature (T) and chemical potential (μ).

$$\sigma_{\alpha\beta}(T;\mu) = \frac{1}{\Omega}\int \sigma_{\alpha\beta}(\varepsilon)\left[-\frac{\partial f_\mu(T;\varepsilon)}{\partial\varepsilon}\right]\mathrm{d}\varepsilon \tag{4.2.23}$$

$$\sigma_{\alpha\beta}(\varepsilon) = \frac{e^2}{N}\sum_{i,k} T_{i,k} v_\alpha(i,k) v_\beta(i,k)\delta(\varepsilon-\varepsilon_{ik}) \tag{4.2.24}$$

$$S_{\alpha\beta}(T;\mu) = \frac{1}{eT\Omega\sigma_{\alpha\beta}(T;\mu)}\int \sigma_{\alpha\beta}(\varepsilon)(\varepsilon-\mu)\left[-\frac{\partial f_\mu(T;\varepsilon)}{\partial\varepsilon}\right]\mathrm{d}\varepsilon \tag{4.2.25}$$

$$K^0_{\alpha\beta}(T;\mu) = \frac{1}{e^2T}\int \sigma_{\alpha\beta}(\varepsilon)(\varepsilon-\mu)^2\left[-\frac{\partial f_\mu(T;\varepsilon)}{\partial\varepsilon}\right]\mathrm{d}\varepsilon \tag{4.2.26}$$

where e is the electronic charge, Ω is the volume of unit cell, N is number of k-points sampled, $v(k)$ is the band velocity, $\varepsilon(k)$ is band energy, and $\tau(k)$ is relaxation time. Due to the presence of delta function like factor, BoltzTraP considers transport of electrons in a narrow energy range and the relaxation time is practically same for such range [108].

4.3 Electronic Properties of Vanadium Oxides

4.3.1 General Considerations

In mean-field approximation, every electron in a crystal, with $>10^{23}$ electrons/cm^3, experiences the same average potential $V(\mathbf{r})$ and the Schrödinger equation for each electron is given by

$$\left(\frac{p^2}{2m} + V(\mathbf{r})\right)\Phi_n(\mathbf{r}) = E_n\Phi_n(\mathbf{r}) \tag{4.3.1}$$

where the one-electron Hamiltonian operates on wave function $\Phi_n(\mathbf{r})$ to yield the energy of the electron in an eigenstate n (each of which can be occupied, at most, by two electrons of opposite spin—Pauli's exclusion principle). Using Bloch's theorem, the eigenstate of one-particle Hamiltonian can be written as the product of plane waves and a lattice periodic function

$$\Phi_{nk}(\mathbf{r}) = e^{ik.r}\mathrm{u}_{nk}(\mathbf{r}) \tag{4.3.2}$$

Thus, the wave function is indexed with a quantum number n and the wave vector **k**. A plot of the electron energies in (4.3.1) versus k is known as the electronic band structure or "spaghetti plot". Each value of k has a discrete spectrum of states, labeled by band index n, the energy band numbering. The number of bands, in a band structure diagram is equal to the number of atomic orbitals in the unit cell [109]. The overlap integral (overlap between the interacting orbitals) determines the width of the band or bandwidth or band dispersion, which is the difference in energy between the lowest and highest points in a band. The greater the overlap between neighboring unit cells, the greater is the bandwidth and vice versa.

The wave vector k can take any value within the Brillouin zone. All the points in a Brillouin zone can be classified using the symmetry of the reciprocal lattice. Symmetric points or Lifschitz points [110], also called special/specific high-symmetry points, are those points which remain fixed or transform into an equivalent one under a symmetry operation of the Brillouin zone. These points play a specific role in solid-state physics: (a) if two k-vectors can be transformed into each other due to some set of symmetry elements, electronic energies at those k-vectors must be identical; (b) "wave functions can be expressed in a form such that they have definite transformation properties under symmetry operations of the crystal" [111]. Similarly, we can define symmetric lines and planes in the Brillouin zone. Customarily, high-symmetry points and lines inside the Brillouin zone are denoted by Greek letters while those on the surface are denoted by Roman letters. The center of a Brillouin zone is always denoted by the Greek letter Γ.

The behavior of electrons in a solid can be studied microscopically from its electronic band structure [111]. One can extract important information about a material such as its stability, transport coefficients, optical properties, and intra as well as inter-molecular bonding interactions from the band structure diagrams. Since electronic band structure of a material is dependent on the crystal structure, it (band structure) can be considered as a link between crystal structure and physical properties.

Experiments have been performed to investigate the electronic structure of vanadium oxides through photoemission and X-ray absorption spectroscopy. The studies have revealed that the electronic structure of these three oxides of vanadium, VO_2, V_2O_3, and V_2O_5, is characterized by a strong hybridization between the 2p (O) and 3d (V) bands. It has been found that the hybridization energy exceeds both the Coulomb repulsion energy of two 3d electrons (U_{dd}) and the energy required to transfer an electron from a ligand orbital (e.g., 2p (O)) to 3d (V) orbital, i.e., the charge transfer energy [112]. It is the hybridization that determines the shape of the valence band [113, 114].

4.3.2 Electronic Band Structure of Bulk V_2O_5

Figure 4.14 shows the electronic band structure and density of states of bulk V_2O_5 calculated by ABINIT. The variable "acell" was set to the theoretical value of 21.7545, 6.7350, 8.2543 Bohr [101] and 76 bands were computed by solving Kohn–Sham equation for uniformly spaced k-points, at least 12 k-points along each symmetry line, using the variable "ndivsm = 12". Occupation option was chosen such that all k-points have the same number of bands and the same occupancies of the band. The electron density from SCF calculation (data set number 1) was used as an input for NSCF calculation (data set number 2), and hence the potential used for Kohn–Sham equation does not vary during the k-point scanning. The variable "tolwfr = 10^{-12}" was used as a tolerance criterion for non-self-consistent calculations in ABINIT, which is tolerance on wave function squared residual. However, not all bands can be converged within the specified tolerance. By default, two upper bands constitute such "buffer" bands. Since buffer allows reaching the convergence of important non-buffer bands, the number of bands can be increased, for instance, by two so as to achieve the same convergence for all bands. The following high-symmetry points, specified in reduced coordinates within the first Brillouin zone, were chosen in the calculations: Γ(0, 0, 0), X(0.5, 0, 0), S(0.5, 0.5, 0), Y(0, 0.5, 0), Γ(0, 0, 0), Z(0, 0, 0.5), U(0.5, 0, 0.5), R(0.5, 0.5, 0.5), T(0, 0.5, 0.5) and Z(0, 0, 0.5) as indicated in Fig. 4.15. XCrySDen [99] was used for the visualization of Brillouin Zone.

The Fermi energy obtained at the end of SCF calculations, in bulk V_2O_5, is 1.62 eV; the origin is shifted by this amount as shown by the horizontal dotted line in the band structure plot. The Fermi level corresponds to the highest occupied band. The number of valence electron bands (as opposed to core electron bands considered in the pseudopotential) is equal to one half the number of the "valence electrons" in the unit cell (i.e., $2 \times 13 + 5 \times 6 = 56$ but only 30 are shown in Fig. 4.14 and other 26 bands, not shown, range from -62.1 to -15.4 eV) and the rest, $18 + 2 = 20$, bands

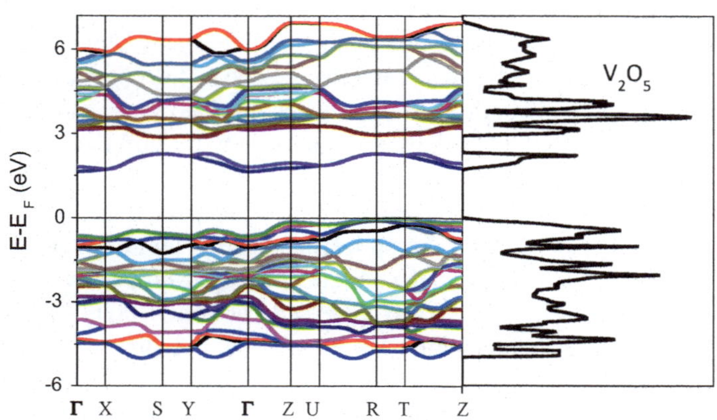

Fig. 4.14 Electronic band structure and density of states for bulk V_2O_5

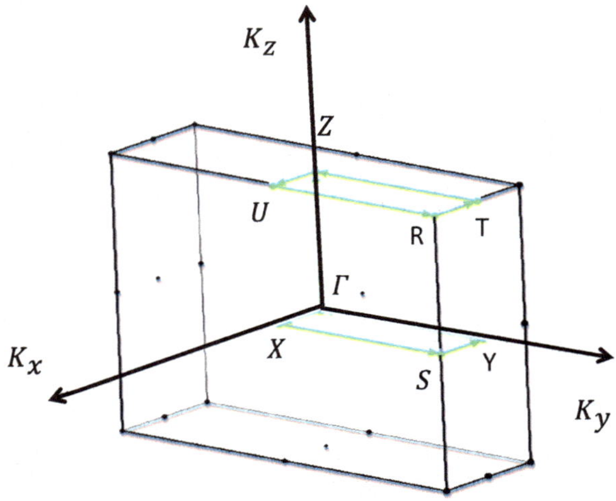

Fig. 4.15 Definition of high-symmetry points and lines in BZ for bulk V_2O_5

correspond to the conduction band. Sometimes, the two bands appearing around 2 eV are also called intermediate bands. These (intermediate) bands, with narrow bandwidth 0.65 eV (≈0.75 eV [115]), separated from higher conduction band by 0.58 eV (≈0.6 eV [115]), can make significant contribution to transport phenomena or mediate optical transitions from valence to higher conduction bands. Also, not shown in the figure are the 4s like bands which lie at a much higher energy. Bands appearing at −62.1 and −15.4 eV correspond to V3s and O2s, respectively. Some other bands appearing at −37.1 eV correspond to V3p. Valence-top bands (5 eV wide ≈5.1 eV [115]) comprise mainly O2p states (total of 30: p orbitals consist of three states per oxygen atom and there are ten oxygen atoms in the unit cell) plus some contribution from V3d states. Similarly, conduction-bottom bands (5.3 eV wide) consist of empty V3d states (total of 20: d-orbitals consist of five states per vanadium atom and there are four vanadium atoms in the unit cell) hybridized with O2p states, the bonding oxygen orbitals with higher energy (as compared to anti-bonding counterpart). Triply degenerate t_{2g} and doublet e_g states, separated from each other by the crystal field, are located at the lower edge of the conduction band and higher energies, respectively. All the degeneracies involved in t_{2g} and e_g states are lifted eventually because the octahedral coordination in V_2O_5 suffers strong deviation [106]. The density of electronic states (DOS) for V_2O_5 is shown on the right side of Fig. 4.14. As can be seen in Fig. 4.14, the most bonding part of the d-band is split-off from the d-band spectrum and the optical bandgap is the difference in energy between the top of the O2p band and the bottom of the split-off part of the V3d band [117]. An indirect bandgap of 1.7 eV is observed between T-point of the valence band to the Γ-point of the conduction band which is smaller than the experimental value of 2.2 eV [106]. It is to be noted here that this result is consistent with the reported bandgap (1.74 eV) of Eyert and Höck [106]. This underestimation of the bandgap is an inherent problem in determining bandgap using the DFT formalism.

Table 4.4 Band structure of V_2O_5: comparison table

	Present calculation	Comparison with remark	(Remark)
Bandgap	1.7 eV (indirect: $\Gamma - T$)	1.74 eV (indirect: $\Gamma - T$)	Similar calculation [106]
		1.6 eV (indirect: $\Gamma - R$)	Similar calculation [95]
		2.2 eV (indirect)	Experiment [106]
V3s bands	4 bands (~appearing at −62.1 eV)	N/A	Not shown in figure
V3p bands	12 bands (~appearing at −37.1 eV)	N/A	Not shown in figure
O2s bands	10 bands (~appearing at −15.4 eV)	N/A	Not shown in figure
Valence bands (mainly O2p)	30 bands (width: 5 eV)	30 bands (width: 5.1 eV)	Similar calculation [115]
Conduction bands (mainly V3d)	20 bands (width: 5.3 eV)	20 bands	Similar calculation [106]
Intermediate bands	~2 eV (width: 0.65 eV)	~2 eV (width: 0.75 eV)	Similar calculation [115]
		Width: 0.45 eV	Experiment [116]
	0.58 eV below higher band	0.6 eV below higher band	Similar calculation [115]
		0.35 eV below higher band	Experiment [116]
"Valence electrons" considered by PPs	$56 \times 2 = 112$ e	–	–
Number of bands calculated	76 bands	–	–

For convenience, the analyses of band structures of bulk V_2O_5 are summarized in Table 4.4 and are compared with the literature.

It is also seen from Fig. 4.14 that the bands are dispersive to a varying extent. Bands are relatively flat (narrow bandwidth) along Γ-X while larger dispersions (wide bandwidth) are observed along Γ-Y and Γ-Z directions, for instance. This is a clear indication of anisotropies of the crystal structure. Disperse bands are indication of hybridization: the more the bands are dispersive, the stronger is the hybridization and vice versa [118]. In general, the DOS is proportional to the inverse of the slope of the "band structure". Conduction bands below 4.7 eV are relatively flat which is also confirmed by the peaks in the DOS in the corresponding energy range. Bands along X-S, S-Y, U-R, and R-T are doubly degenerate throughout the energy range and the rest are nondegenerate.

4.3.3 Electronic Band Structure of Bulk VO_2

Figure 4.16 shows the electronic band structure and the DOS of bulk VO_2 in its metallic rutile phase. The following high-symmetry points, specified in reduced coordinates within the first Brillouin zone, were chosen in the calculations: Γ(0, 0, 0), X(0, 0.5, 0), R(0 0.5 0.5), Z(0, 0, 0.5), Γ(0 0 0), R(0, 0.5, 0.5), A(0.5, 0.5, 0.5), Γ(0, 0, 0), M(0.5, 0.5, 0), A(0.5, 0.5, 0.5) and Z(0, 0, 0.5) as indicated in Fig. 4.17. As usual, the origin was shifted by 7.47 eV, the Fermi energy, as shown by the horizontal dotted line in the band structure plot. Out of the computed 35 bands, the first 24 $[(1 \times 13+2 \times 6) - 1=24]$ were fully occupied while the rest were either partially occupied or unoccupied. The Fermi level passes through partially occupied bands with indices 25–28, where the band with index 25 corresponds to the valence electrons (as opposed to core electrons considered in the pseudopotential). This is a clear indication of finite overlapping of valence and conduction bands, a factor that characterizes metallic behavior. We mainly observed seven groups of bands but the first three groups lying around −63, −38, and −18 eV, not shown in the figure, are due to V3s, V3p, and O2s states, respectively. These are less important in our discussion because they are very tightly bound. The rest of the four groups of bands lie below (one), at (one) and above (two) the Fermi energy. Out of the four groups, the first one lying between −7.6 and −1.6 eV (energy range = 6 eV [119]) consists of 12 bands and are mainly due to O2p states. The second group which starts at 6.8 eV is due to V4s states. The third and fourth groups, consisting of six and four bands, in the ranges from −0.6 to 2.0 eV and from 2.2 to 5.1 eV, respectively, are mainly from V3d states and some contribution from O2p states as a result of the p–d hybridization. The groups of bands discussed above can be more clearly understood from the following picture of crystal field splitting.

It is known that nearly octahedral crystal field partially lifts the degeneracy of the d-orbitals yielding two sets of bands: t_{2g} and e_g^{σ} (total of 10 bands, the sum of the bands in the third and the fourth groups mentioned in the previous paragraph).

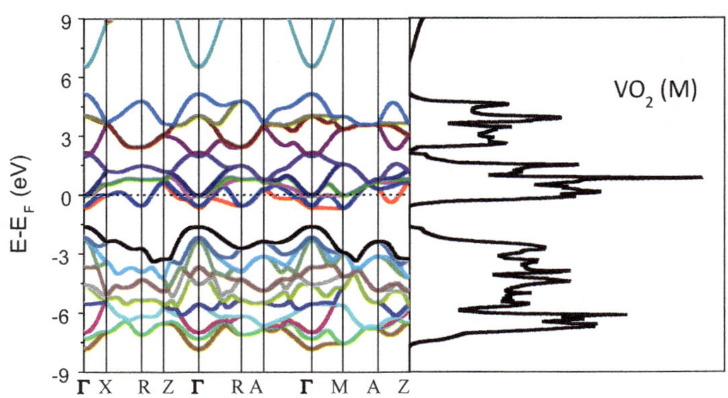

Fig. 4.16 Electronic band structure and density of states for bulk VO_2 rutile structure

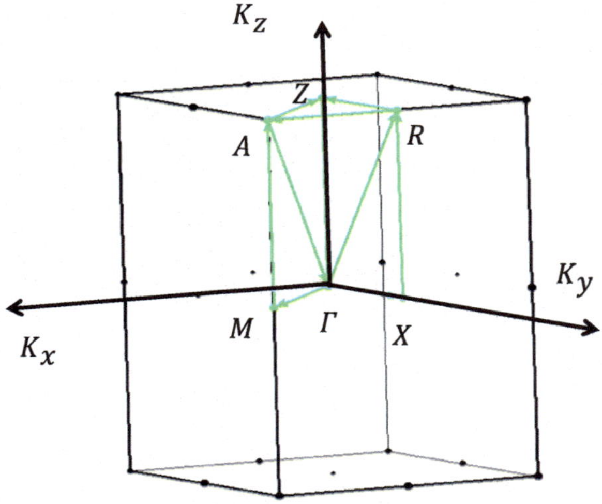

Fig. 4.17 Definition of high-symmetry points and lines in BZ for VO_2 rutile structure

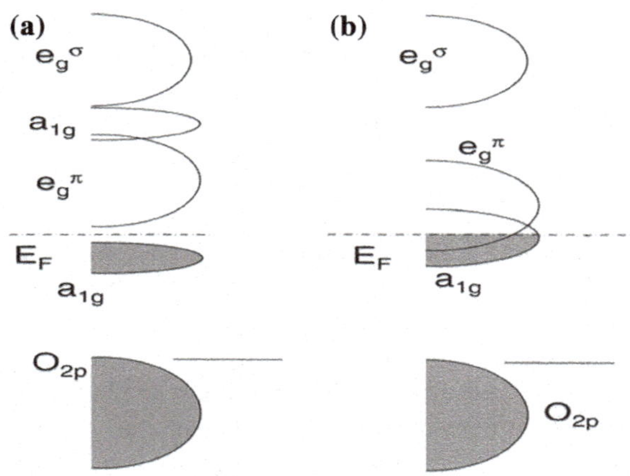

Fig. 4.18 Schematic energy diagrams of the V3d and O2p states at the **a** insulating and **b** metallic phases of VO_2 [119]

Further, the tetragonal crystal field splits triply degenerate t_{2g} states, containing the single d-electron, into an a_{1g} state and e_g^{π} doublet (which are also called $d_{\|}$ and π^*, respectively, in the literature) [23] as shown in Fig. 4.18b. The Fermi level passes through partially occupied e_g^{π} and a_{1g} bands. Octahedral distortions can be measured in terms of $t_{2g} - e_g$ configuration mixing [98]. In the case of insulating, monoclinic (M_1) phase, a_{1g} further splits into lower and upper a_{1g} bands as shown in Fig. 4.18a, which are, respectively, called lower and upper Hubbard band, in the Mott picture while bonding (lower) and anti-bonding (upper) bands in the Peierls model [119]. This latest splitting has been explained by metal–metal pairing in the vanadium chains [100].

Figure 4.19 shows the electronic band structure and density of states of bulk VO_2 in its monoclinic (M_1) insulating phase. The following high-symmetry points, specified in reduced coordinates within the first Brillouin zone, were chosen in the

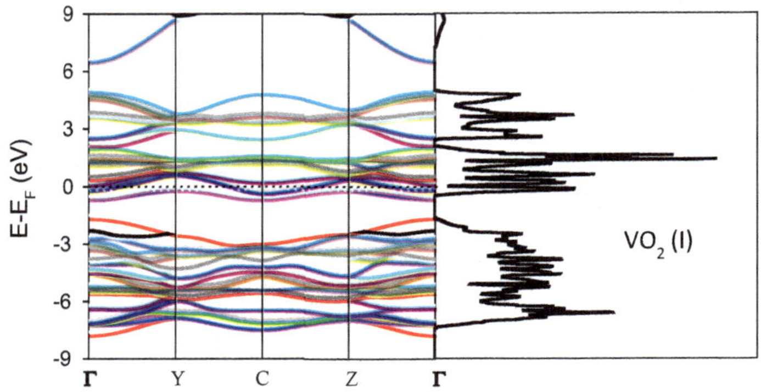

Fig. 4.19 Electronic band structure and density of states for bulk VO_2 monoclinic (M_1) phase

calculations: Γ(0, 0, 0), Y(0, 0, 0.5), C(0, 0.5, 0.5), Z(0, 0.5, 0), Γ(0, 0, 0) as indicated in Fig. 4.20. As usual, the origin was shifted by 7.44 eV, the Fermi energy, as shown by the horizontal dotted line in the band structure plot. Out of 69 bands that were computed, the first 50 $[2 \times (1 \times 13 + 2 \times 6) = 50]$ are fully occupied; the band with indices 51 and 52 are partially occupied and the rest are unoccupied. The number of band groups in the insulating phase is exactly the same as in the metallic phase. They appear in similar locations as in the metallic phase, and hence can be interpreted with same arguments. One difference is that the number of bands in each group is doubled in the insulating phase, which is consistent with the fact that the number of formula units in the real space primitive unit cell doubles during the phase transition. The fact that electronic states are present at the Fermi level indicates the shortcomings of LDA in reproducing the observed optical gap of 0.6 eV [3]. However, some differences in the band properties can be observed in the insulating phase: (a) number of bands crossing the Fermi level is less, with smaller amount of overlapping between the valence and conduction bands and (b) bands corresponding to "valence electrons" considered in the pseudopotential are fully occupied, unlike in the metallic phase, as shown in Fig. 4.21. For convenience, the analyses of band structures of bulk VO_2 are summarized in Table 4.5 and are compared with the literature.

4.3.4 Electronic Band Structure of Bulk V_2O_3

Figure 4.22 shows the electronic band structure and density of states of bulk V_2O_3 in its metallic corundum phase. The following high-symmetry points, specified in reduced coordinates within the first Brillouin zone, were chosen in the calculations: Γ(0, 0, 0), L(0, −0.5, 0), Z(0, 0, 0.5), F(0.5, 0, 0.5), Γ(0, 0, 0), and Z(0, 0, 0.5) as indicated in Fig. 4.23. As usual, the origin was shifted by 9.2 eV, the Fermi energy, as shown by the horizontal dotted line in the band structure plot. Out of 61 computed bands, the first 42 were fully occupied while the rest were either partially occupied

Table 4.5 Band structure of VO_2: comparison table

	Present calculation			Comparison with remark	(Remark)
Bandgap	M	0		0	Metal
	I	0		0.6 eV	Experiment [120]
				0	Similar calculation [121]
V3s bands	M	2 bands	Appearing at ~−63 eV	N/A	Not shown in figure
	I	4 bands		N/A	Not shown in figure
V3p bands	M	6 bands	Appearing at ~−38 eV	N/A	Not shown in figure
	I	12 bands		N/A	Not shown in figure
O2s bands	M	4 bands	Appearing at ~−18 eV	N/A	Not shown in figure
	I	8 bands		N/A	Not shown in figure
Valence bands (mainly O2p)	M	12 bands	Width: 6 eV	Width: 6 eV	Experiment [119]
				Width: 5.4 eV	Similar calculation [98]
	I	24 bands	Width: 6 eV		
Conduction bands (mainly V3d)	M	Two groups: 6 bands: t_{2g} 4 bands: e_g	From −0.6 to 2.0 eV & from 2.2 to 5.1 eV	From −0.6 to 2.0 eV & from 2.0 to 5.5 eV	Similar calculation [98]
	I	20 bands	Similar location as in metallic case		
Upper conduction bands (V4s)	M	2 bands	Starting at 6.8 eV	N/A	
	I	4 bands			
"Valence electrons" considered by PPs	M	50 electrons	–	–	–
	I	100 electrons	–	–	–
Number of bands calculated	M	35 bands	–	–	–
	I	69 bands	–	–	–

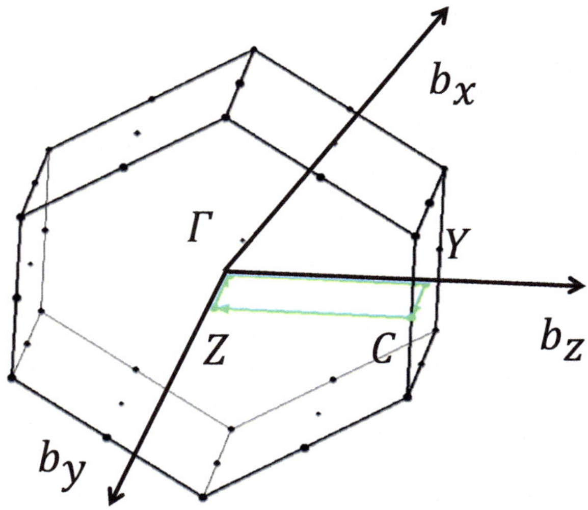

Fig. 4.20 Definition of high-symmetry points and lines in BZ for VO_2 monoclinic (M_1) phase

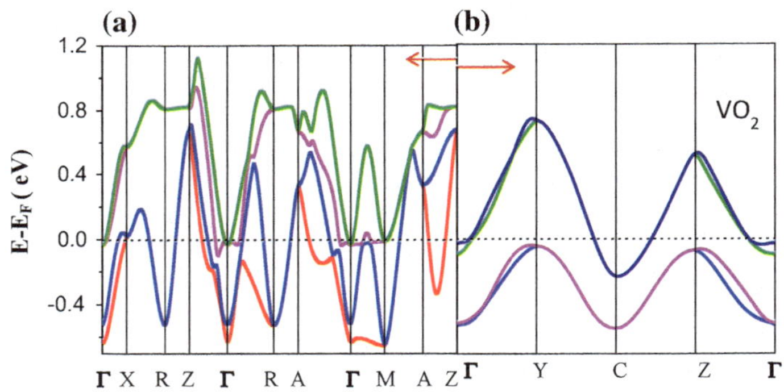

Fig. 4.21 **a** Metallic phase and **b** insulating phase of bulk VO_2 at Fermi level

or unoccupied. The Fermi level passes through partially occupied bands with indices 43–47, where the bands with indices 43 and 44 correspond to the valence electrons (as opposed to core electrons considered in the pseudopotential). This is a clear indication of finite overlapping of valence and conduction bands, a property that indicates metallic behavior. The V3s, V3p, and O2s states appear in similar locations as in VO_2 but are not important for our purpose because they are tightly bound. By excluding these groups of bands, we can now see, mainly, four groups of bands as in VO_2. The first one lying between -8.0 and -3.6 eV (energy range $\approx$4.0 eV [122]) consists of 18 bands and are mainly due to O2p states. The second group which starts at 4 eV, consistent with the work of Mattheiss [122], is due to V4s states. The third and fourth groups consisting of 12 and 8 bands, in the ranges from -1.23 to 1.4 eV and from 1.94 to 3.7 eV, respectively, are mainly from V3d states and some contribution from O2p states as a result of p–d hybridization. The magnitude of the p–d hybridization is the same as that found for VO_2 [100]. The groups of bands,

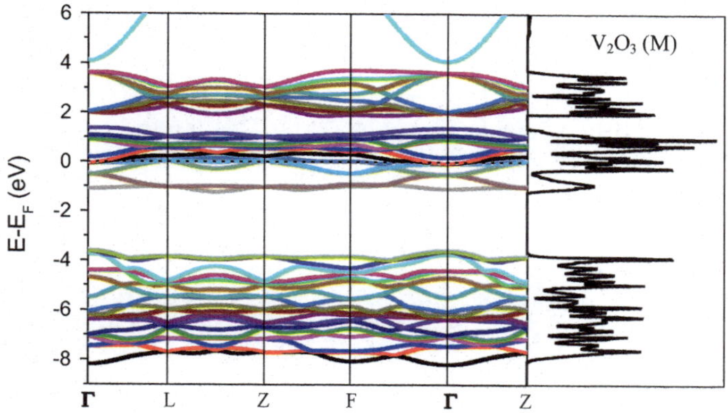

Fig. 4.22 Electronic band structure and density of states for bulk V_2O_3 metallic phase

Fig. 4.23 Definition of high-symmetry points and lines in BZ for V_2O_3 metallic phase

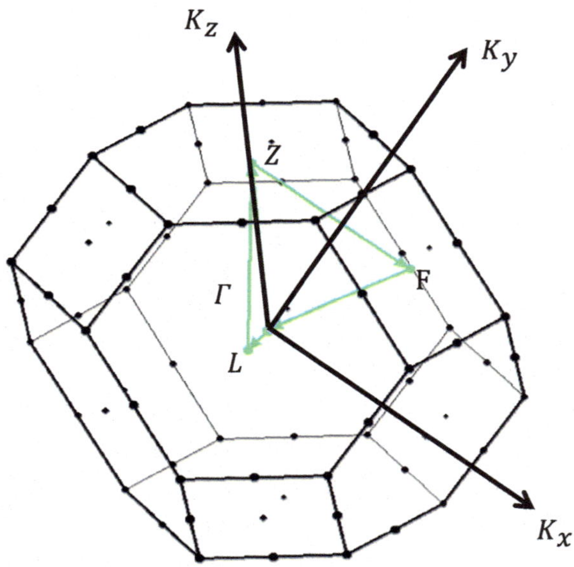

discussed above, can be better understood from the following picture of crystal field splitting (Fig. 4.24).

As in the case of VO_2, the crystal field due to the oxygen octahedra (VO_6) splits the d-orbitals into two sets of bands: t_{2g} (lower band) and e_g^{σ} (upper band). The upper band e_g^{σ} is twofold degenerate and consists of d_{xz} and d_{yz} bands. This oxygen octahedron is not perfect but has trigonal distortion. The influence of non-cubic arrangement of distant metal ions in the lattice, combined with the trigonal distortion, splits the threefold degenerate t_{2g} orbitals into a_{1g} singlet ($d_{3z^2-r^2}$) and twofold degenerate e_g^{π} bands $\left(d_{xy}, d_{x^2-y^2}\right)$ [123] as shown in Fig. 4.24b. In the case of monoclinic phase, the lower symmetry of the crystal field (due to structural change during phase transition)

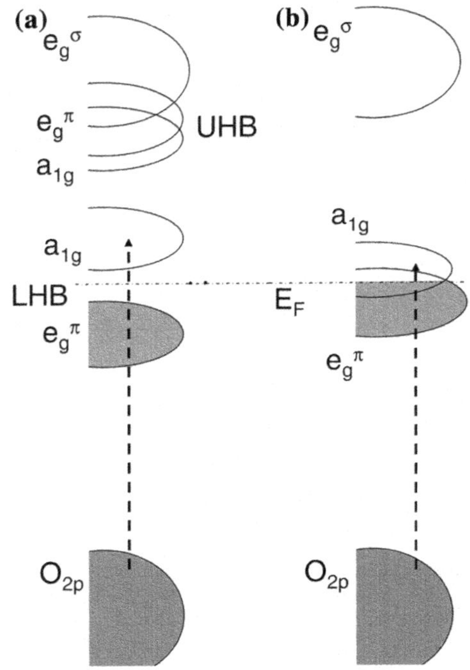

Fig. 4.24 Schematic energy diagrams of the V3d and O2p states at the **a** insulating and **b** metallic phases in V_2O_3 [119], LHB and UHB are lower and upper Hubbard bands, respectively

further lifts the degeneracy between two e_g^π bands as shown in Fig. 4.24a [119, 123, 124]. In this model, the antiferromagnetic insulating (AFI) phase remains in Hund's rule (maximum spin) configuration $e_g^{\uparrow\uparrow}$, leaving a_{1g} band empty (putting $V3d^2$ ions in the $S = 1$ state). This is contrary to the one-band Hubbard model, where one V electron enters bonding a_{1g} orbital state and the other enters the doubly degenerate e_g^π state ($S = 1/2$ on each V site) [124].

Figure 4.25 shows the electronic band structure and density of states of bulk V_2O_3 in its monoclinic insulating phase. The following high-symmetry points, specified in reduced coordinates within the first Brillouin zone, were chosen in the calculations: Γ(0, 0, 0), Y(0.5, 0, 0), C(0.5, 0.5, 0), Z(0, 0.5, 0), and Γ(0, 0, 0), as indicated in Fig. 4.26. As usual, the origin was shifted by 9.2 eV, the Fermi energy, as shown by the horizontal line at zero energy. Out of the 121 bands that were computed, the first 88 $[2 \times (2 \times 13 + 3 \times 6) = 88]$ are fully occupied; the band with indices 89, 90, and 91 are partially occupied and the rest are unoccupied. The number of band groups in the insulating phase is exactly the same as in the metallic phase. They appear in similar locations as in the metallic phase, and hence can be interpreted with the same arguments. One difference is that the number of bands in each group is doubled in the insulating phase, which is consistent with the fact that the number of formula units in the real space primitive unit cell doubles during the phase transition. The fact that electronic states are present at the Fermi level indicates the shortcomings of LDA in reproducing the observed optical gap of 0.66 eV [125]. However, some differences in the band properties can be clearly seen in the insulating phase: (a) number of bands

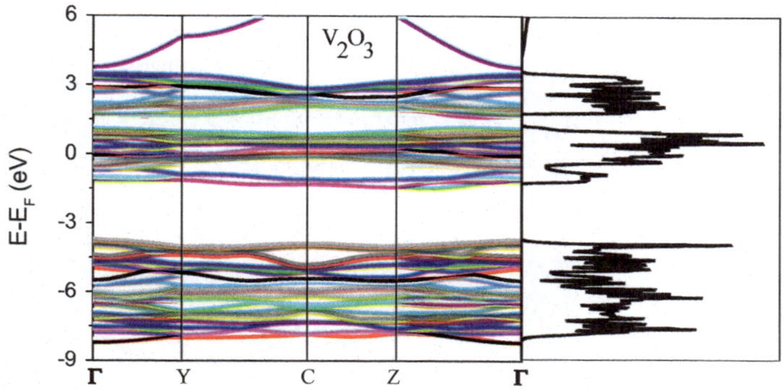

Fig. 4.25 Electronic band structure and density of states for bulk V_2O_3 in insulating phase

Fig. 4.26 Definition of high-symmetry points and lines in BZ for V_2O_3 insulating phase

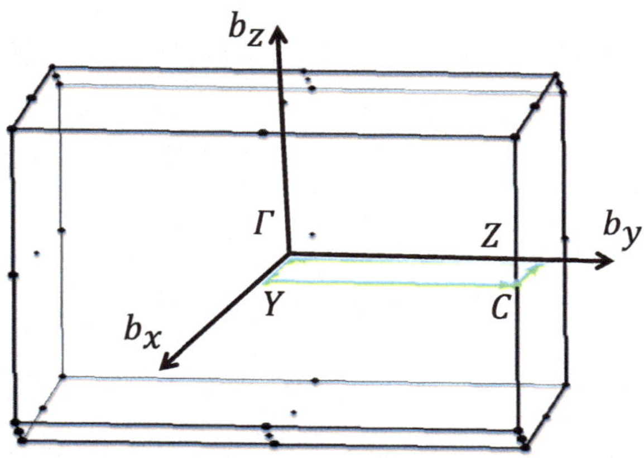

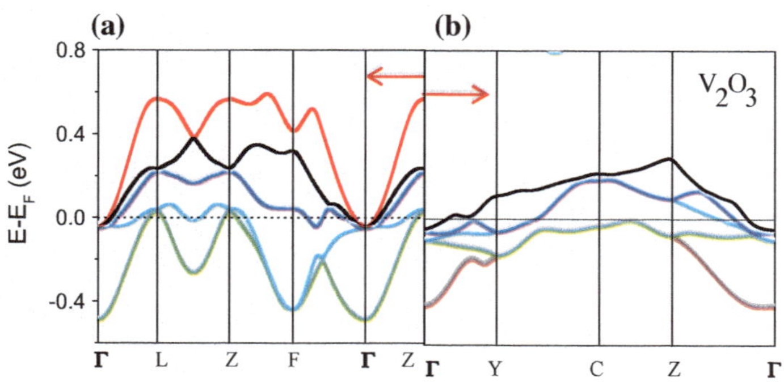

Fig. 4.27 **a** Metallic phase and **b** insulating phase of V_2O_3 at Fermi level

crossing the Fermi level is less, (b) bands corresponding to "valence electrons" considered in the pseudopotential are fully occupied, unlike in the metallic phase as shown in Fig. 4.27. For convenience, the analyses of band structures of bulk V_2O_3 are summarized in Table 4.6 and are compared with the literature.

Table 4.6 Band structure of V_2O_3: comparison table

	Present calculation			Comparison with remark	(Remark)
Bandgap	M	0		0	Metal
	I	0		0.66 eV	Experiment [124]
				0	Similar calculation [126]
V3s bands	M	4 bands	Appearing at similar location as in VO_2	N/A	Not shown in figure
	I	8 bands		N/A	Not shown in figure
V3p bands	M	12 bands	Appearing at similar location as in VO_2	N/A	Not shown in figure
	I	24 bands		N/A	Not shown in figure
O2s bands	M	6 bands	Appearing at similar location as in VO_2	N/A	Not shown in figure
	I	12 bands		N/A	Not shown in figure
Valence bands (mainly O2p)	M	18 bands	Width: 4.4 eV	Width: 4.0 eV	Calculation [122]
	I	36 bands	Width: 4.4 eV		
Conduction bands (mainly V3d)	M	Two groups: 12 bands: t_{2g} 8 bands: e_g	From −1.2 to 1.4 eV & from 1.9 to 3.7 eV		
	I	40 bands	Similar location as in metallic case		
Upper conduction bands (V4s)	M	4 bands	Starting at 4.1 eV	N/A	Starting at 4 eV [122]
	I	8 bands			
"Valence electrons" considered by PPs	M	88 electrons	–	–	–
	I	176 electrons	–	–	–
Number of bands calculated	M	61 bands	–	–	–
	I	121 bands	–	–	–

4.3.5 Summary

In this section, we have summarized the findings of our LDA one-electron, Kohn–Sham band-structure calculations of bulk vanadium oxides in both insulating and metallic phases. Electronic band structures depend on the crystal structure and contribute to stability, bonding interactions, and physical properties such as transport and optical properties. Band structures, obtained in our study, are consistent with similar studies in the literature. Band structures are found to be influenced by crystal field and the strong hybridization between 2p (O) and 3d (V) bands.

Bands in V_2O_5 are remarkably dispersive to different extent, along various high-symmetry lines in the Brillouin zone (BZ). This implies that the crystal structure is highly anisotropic. A group of conduction bands (with narrow bandwidth, called intermediate bands), lying close (0.6 eV apart) to higher conduction bands observed in V_2O_5, seem to play an important role in a variety of device applications. An indirect gap of 1.7 eV is observed between T and Γ points of the BZ, which is close to the value of 1.74 eV reported in the literature [106].

Similar calculations are performed in both metallic and insulating phases of bulk vanadium dioxide (VO_2) and sesquioxide (V_2O_3). In the metallic phase, we see finite overlapping of valence and conduction bands. Bands in the insulating phases appear in similar locations as in corresponding metallic phases. The number of bands is doubled in the insulating phases (as compared to corresponding metallic counterpart), which is consistent with the fact that the number of formula units of the unit cell, in the calculations, is doubled in each of the oxides as the materials transition from metallic to insulating phase. Even though an optical gap is not observed in the insulating phase (shortcomings of LDA), we have noticed some differences in the band structures that characterize phase—whether insulating or metallic. In the insulating phase, (a) number of bands crossing the Fermi level is less and (b) bands corresponding to "valence electrons" considered in the pseudopotential are fully occupied, unlike in the metallic phase.

4.4 Transport Properties of Vanadium Oxides

4.4.1 General Considerations

Boltzmann Transport Equation (BTE) is a semi-classical transport equation: quantum mechanics is embedded in the electronic band structure but the equation of motion for electrons is written in momentum space that looks like the Newton's equation of motion

$$\frac{\mathrm{d}\left(\hbar\vec{k}\right)}{\mathrm{d}t} = -\nabla_r E_c(\vec{r}) = -q\vec{E}(\vec{r}) \tag{4.4.1}$$

where $\vec{E}(\vec{r})$ =electric field and $E_c(\vec{r})$ = bottom of conduction band. Here, the force is due to the electric field in the sample. The solution to Eq. (4.4.1) yields the time-dependent momentum coordinate. Similarly, the velocity of an electron is obtained from the band structure

$$\overrightarrow{v_g}(t) = \frac{1}{\hbar}\nabla_k E\left[\vec{k}(t)\right] \tag{4.4.2}$$

and its solution gives the position of the particle with time. With these two solutions, the electrons can be tracked in phase space (this proves the semi-classical nature of the approach because both momentum and position of a particle cannot be specified quantum mechanically). There are some assumptions considered in the Boltzmann approach such as the variation of the bottom of the conduction band with space is slow so that quantum mechanical reflections are not considered. This is a single particle approach but many-body effects are treated. We include the effects of scattering (collision) but we consider the scattering to be very short in time duration such that the positions of electrons do not change during those events. It is the scattering or source term, in the second-order, linear, nonhomogeneous differential Eq. (4.4.3), that imposes difficulty in finding an analytical solution. In other words, these scattering phenomena are complicated, and hence some approximation techniques need to be adopted to obtain analytical solutions. One of the widely used approximations is the relaxation time approximation (RTA) which considers a constant averaged (momentum) relaxation time, τ_m, for the electrons. The basic idea in RTA is that the effect of collision is to take a small perturbation and to react in a negative way to try to pull the system back to equilibrium. In other words, if the system is perturbed, it will decay back to equilibrium as $\sim e^{-t/\tau_m}$. Within RTA, the BTE can be written as

$$\vec{v}.\nabla_r f - q\vec{E}.\nabla_p f = -\frac{\delta f(\vec{p})}{\tau_m} \tag{4.4.3}$$

where $f = f(r, p, t)$ is the occupation function which, in equilibrium, is given by

$$f_0(\varepsilon) = \frac{1}{e^{(\varepsilon-\varepsilon_F)/K_BT}+1}(\text{equilibrium } F - \text{D function}) \tag{4.4.4}$$

The solution of BTE within RTA can be used to derive thermoelectric (TE) transport parameters, which surprisingly yields accurate results for various systems [127]. In this study, the "BoltzTraP" code [107] is used which implements the Boltzmann theory to calculate the thermoelectric properties under constant relaxation time and rigid band approach [128]. The main transport properties calculated in this study are electrical conductivity (σ), Seebeck coefficient (S), and thermal (electronic) conductivity (K) as given by Eqs. (4.4.5)–(4.4.7), respectively. Since the Seebeck coefficient is independent of τ_m, it can be tested against the experimental data to obtain the relatively realistic comparison.

$$J_e = \sigma E(\text{In the absence of magnetic field, } B \text{ and thermal gradient, } \nabla T) \tag{4.4.5}$$

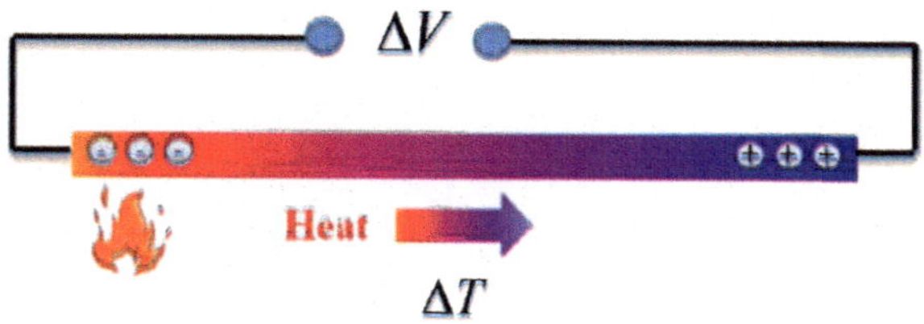

Fig. 4.28 Voltage due to heat flow

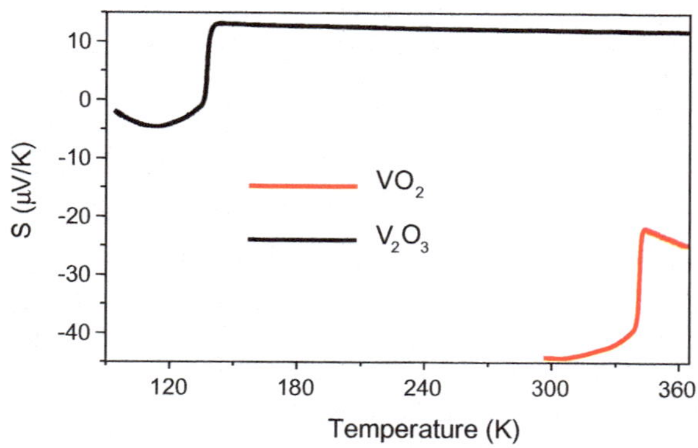

Fig. 4.29 Seebeck coefficient versus temperature for VO_2 (red) and V_2O_3 (black) before and after phase transition [131, 132]

$$S = -\Delta V/\Delta T(-\text{ve sign, only when } \Delta V \& \nabla T \text{ are opposite}) \quad (4.4.6)$$

$$J_Q = -K(\nabla T)(\text{In the absence of electric field, ie } J_e = 0) \quad (4.4.7)$$

However, these transport coefficients, in a crystal, are tensors given by the relations (4.2.23), (4.2.25), and (4.2.26), respectively.

Thermoelectric effect refers to conversion of thermal energy into electrical energy as shown in Fig. 4.28. Good thermoelectric performance of a device is understood as having low thermal conductivity but high electrical conductivity as well as high Seebeck coefficient. These transport coefficients depend largely on the electronic structure of materials, specifically, electronic states near the Fermi level [129]. Moreover, the electron–electron correlation effect, present in vanadium oxides, is of immense interest for properties related to the details of electronic structure [9]. Since the layered structures of oxide materials demonstrate interesting thermoelectric behavior [130], we are mainly interested in the thermoelectric properties of V_2O_5 which consists of layered unit cells. V_2O_5 is the most stable than the other two (vanadium dioxide and vanadium sesquioxide) and easy to produce using simple, inexpensive, and nontoxic sol–gel deposition technique. Among the three oxides, only V_2O_5 exhibits thermoelectric (TE) properties [57] while VO_2 and V_2O_3 show little TE response (small value of Seebeck coefficient) as shown in Fig. 4.29. In order to study the effect of phase transition on the thermoelectric properties, the Seebeck coefficient, electrical conductivity, and thermal (electronic) conductivity of vanadium dioxide, VO_2 are calculated.

Table 4.7 Non-SCF grids used in the calculations

	Number of k-points in IBZ
V_2O_5	500
VO_2 (Metallic)	550
VO_2 (Insulating)	595

Even though "BoltzTraP" works well for a large number of systems, it is computationally expensive because it requires high k-point mesh for convergence. In order to avoid the computational expense, a self-consistent (SCF) calculation was performed for the theoretical structure (before relaxation) with fewer (but well-converged) k-point density. The electron density obtained in the SCF calculation was then fed to non-SCF calculation to obtain the electronic structure on a finer mesh of k-points as shown in Table 4.7. Both calculations use the cutoff energy to be 40 Hartree, a converged value with corresponding tolerance in energy of 0.1 meV for all the systems studied. In our study, the x-axis is defined to be parallel to the crystallographic a-axis (of conventional cell); y-axis is orthogonal to x-axis and lies in the a–b plane while z-axis is orthogonal to both the x- and y-axis. In other words, the coordinate axes are the same as crystallographic axes if the latter one is an orthogonal system. We have studied the direction-dependent (along the three axes) transport properties as well as their averaged (trace of a 3×3 tensor quantity) value.

4.4.2 Transport Properties of Bulk V_2O_5

Figure 4.30 shows the variation of Seebeck coefficient with temperature from 10 to 700 K along the three crystallographic axes a, b, and c, at the value of chemical potential $(\mu) = 1.619$ eV ($\approx$1.624 eV, the Fermi energy). Clearly, the Seebeck coefficient is anisotropic. The absolute value of Seebeck coefficient, $|S|$, along a, is higher throughout the range beyond $T = 150$ K, as compared to the other two. The maximum (absolute) value of average Seebeck coefficient is found to be 246.54 μV/K at 300 K. This value is consistent with the absolute value of 258 μV/K obtained in similar calculation [114] and an absolute value of 218 μV/K measured experimentally at room temperature [133]. Maxima and crossovers are also seen in the variation of the Seebeck coefficient and have been attributed to the (a) "Specific details" of the band structure and (b) direction-dependent electron–phonon interactions [134].

Figure 4.31 shows the temperature-dependent variation of electrical and thermal (electronic) conductivities along the three crystallographic axes, in terms of constant relaxation time. Figure 4.32 shows the comparison of the calculated averaged conductivity $\left(\sigma/\tau \times 10^{-18}\right)$ with the available experimental value of conductivity for V_2O_5 films (both as-deposited and post-annealed at 773 K) [133]. Assuming that the behavior of thin film does not differ drastically with its bulk counterpart of V_2O_5 [135], we conclude that RTA, within the Boltzmann theory, cannot repro-

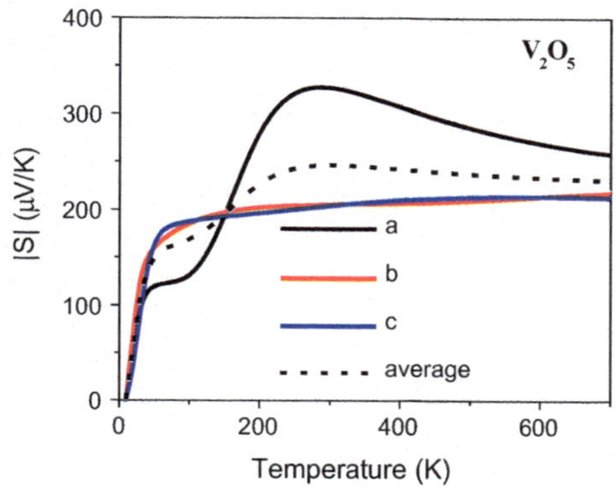

Fig. 4.30 Seebeck coefficient versus temperature for bulk V_2O_5

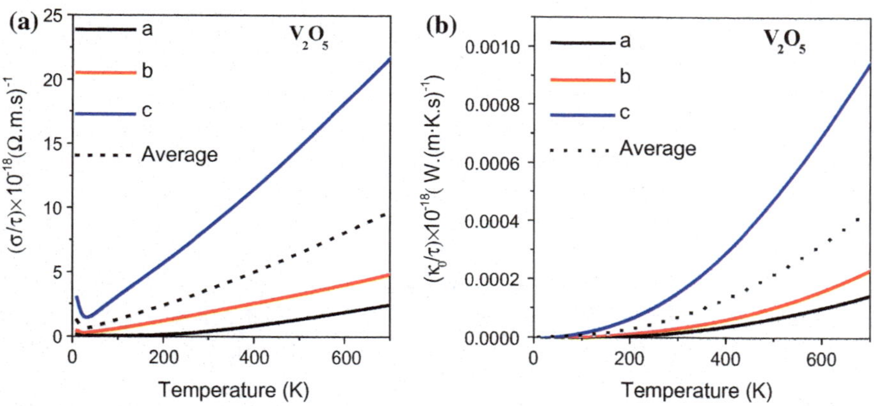

Fig. 4.31 **a** Electrical and **b** thermal (electronic) conductivities of V_2O_5

duce the experimental results of the electrical conductivity with desired accuracy. In other words, the rate of variation of electrical conductivity is much higher than that predicted by the BTE. However, it can predict the anisotropic behavior as clearly manifested in Fig. 4.31a in the following order: $\sigma_a < \sigma_b < \sigma_c$ and $k_a^{el} < k_b^{el} < k_c^{el}$. With attosecond momentum relaxation time ($\tau \sim 10^{-18}$s), as seen from Fig. 4.32, the thermal (electronic) conductivity, calculated in the Boltzmann framework, at 300 K is 6.7944×10^{-5} W/(m · K), which is ~3 orders of magnitude smaller than the thermal conductivity value of 0.45 W/(m · K), observed in the experiment [133]. Hence, the contribution of phonons might be taken into account to understand the thermal conductivity of bulk V_2O_5. With this analysis, we will mainly focus on the Seebeck coefficient of VO_2 in the following discussion.

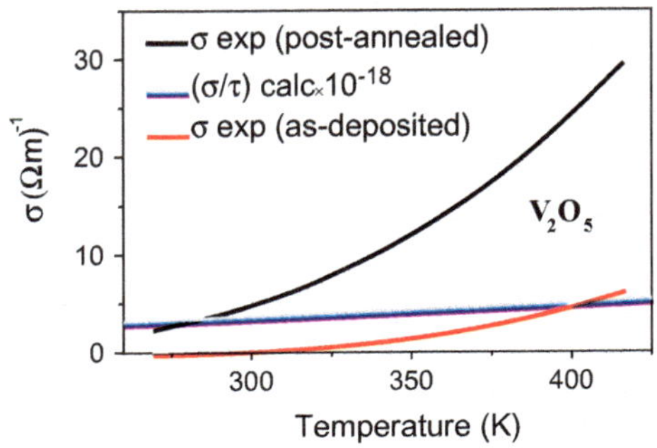

Fig. 4.32 Calculated electrical conductivity of V_2O_5 compared with experiment [133]

4.4.3 Transport Properties of Bulk VO_2

Figure 4.33 shows the temperature-dependent variation of Seebeck coefficient of bulk VO_2 in its metallic rutile phase, at the value of chemical potential $(\mu) = 7.51$ eV ($\approx$7.47 eV the Fermi energy). In VO_2, there are two distinct directions for electric polarization: **E** $\perp$ *c* and **E** $\parallel$ *c* axis which is also manifested by the Seebeck coefficient in Fig. 4.33. The absolute value of Seebeck coefficient increases with temperature. The absolute value of Seebeck coefficient, parallel to the rutile *c*-axis, is slightly lower than its perpendicular counterpart. Berglund and Guggenheim [131] have measured Seebeck coefficients at 348 K as 23.1 ± 0.2 μV/K and 21.1 ± 0.2 μV/K (absolute values) in the directions perpendicular and parallel to the rutile *c*-axis, respectively. The corresponding BoltzTraP (absolute) values are 227 and 205 μV/K, which are of one order of magnitude higher. This is due to the insufficient k-points used in our calculation. A rule of thumb is that the number of *k*-points in the calculation should be more than $16 \times 10^6 / V_{puc}$, where V_{puc} is the volume of the primitive unit cell [107]. Figure 4.34 shows the temperature-dependent variation of electrical and thermal (electronic) conductivities along the three crystallographic axes, in terms of constant relaxation time. The anisotropic features are observed from both the plots, but the electrical conductivity seems to be relatively reluctant to the directional variation as compared to the electronic component of the thermal conductivity.

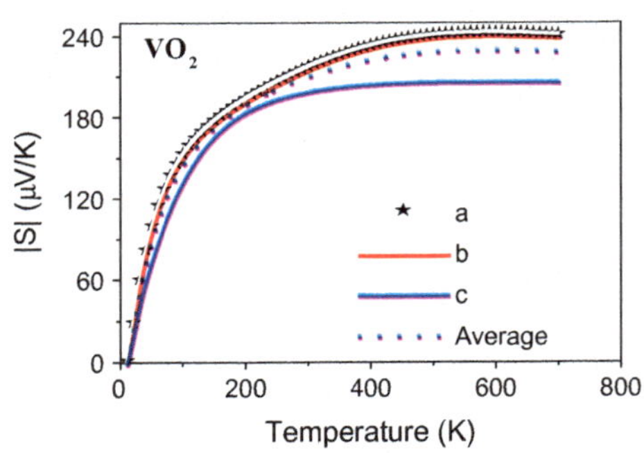

Fig. 4.33 Seebeck coefficient versus temperature for bulk VO_2 in rutile phase

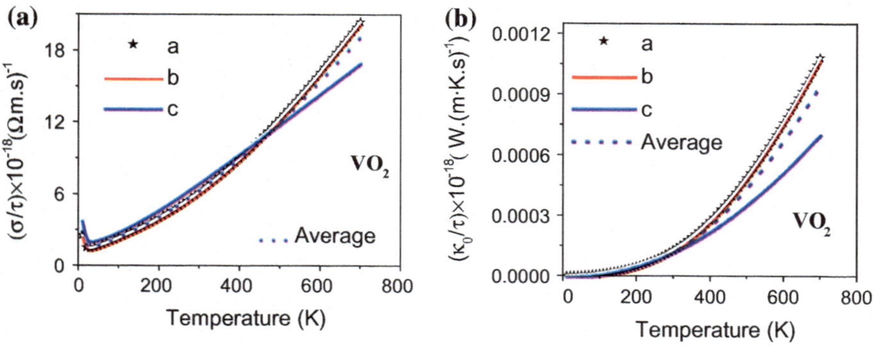

Fig. 4.34 **a** Electrical and **b** thermal (electronic) conductivities of bulk VO_2 in high-temperature phase

Figure 4.35 shows the temperature-dependent variation of Seebeck coefficient of bulk VO_2 in its monoclinic phase, at the value of chemical potential $(\mu) = 7.51$ eV ($\approx$7.44 eV, the Fermi energy). There is no noticeable anisotropy in the Seebeck coefficient. The Seebeck coefficient decreases with temperature. The absolute value of Seebeck coefficient at 298 K is found to be 223 μV/K, whose magnitude is about five times greater than that measured in an experiment (43 μV/K, the absolute value) [131]. The Seebeck coefficient in insulating phase deviates less from the experimentally measured values than that in the metallic phase. This relatively reduced value can be attributed to the reduction in BZ volume, where the k-point mesh is relatively denser. (The volumes of primitive unit cells, in real space, are 59.149834 (Metallic phase) and 118.289910 Å^3 (Insulating phase) with a ratio of 1:2, which will be flipped in reciprocal space). Figure 4.36 shows the temperature-dependent variation of electrical and thermal (electronic) conductivities along the three crystallographic axes, in terms of the constant relaxation time. Figure 4.37 shows the comparison of temperature-dependent variations of Seebeck coefficient in the low- and high-temperature phases with the experimental data; Seebeck coefficient in the metallic and insulating phases are scaled by factors 216/22.3 and 223/43, respectively. It is seen that the Seebeck coefficient, at experimental critical temperature of 340 K, changes by 18.9 μV/K which lies within 10% of the observed discontinuity of 17.3 μV/K during the phase transition [131]. "Kohn–Sham–Boltzmann" approach can predict phase transition in VO_2 with reasonable accuracy.

4.4.4 *Summary*

Among the three oxides of our interest, in the present study, only V_2O_5 exhibits thermoelectric (TE) properties. It is the layered structure, stability, and easiness to prepare by inexpensive and nontoxic approaches, which make V_2O_5 an interest-

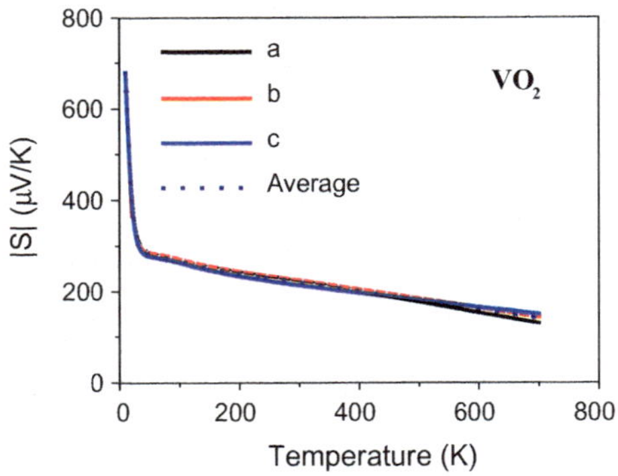

Fig. 4.35 Seebeck coefficient versus temperature for bulk VO_2 in monoclinic phase

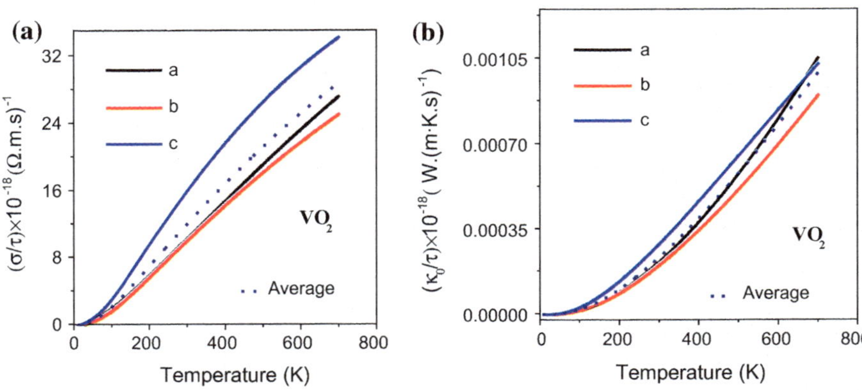

Fig. 4.36 **a** Electrical and **b** thermal (electronic) conductivities of VO_2 in low-temperature phase

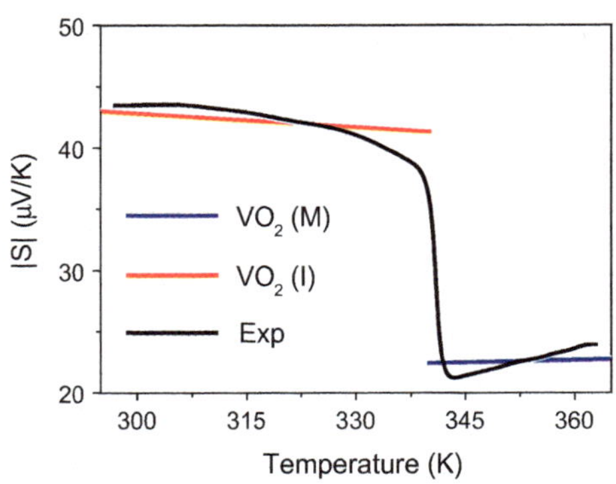

Fig. 4.37 Seebeck coefficient in low- and high-temperature phases of bulk VO_2: "Kohn–Sham–Boltzmann" prediction of phase transition in VO_2

ing thermoelectric material. Seebeck coefficient, electrical conductivity and thermal (electronic) conductivity are studied as a function of temperature at a fixed value of chemical potential close to the Fermi energy using Kohn–Sham band structure approach coupled with Boltzmann transport equations. All the transport parameters have correctly reproduced the highly anisotropic electrical conduction that has been observed in V_2O_5. Maxima and crossovers are also seen in the temperature-dependent variation of Seebeck coefficient which can be the consequences of "specific details" of the band structure and anisotropic electron–phonon interactions. Comparisons of the averaged electrical conductivity with that of as-deposited as well as post-annealed V_2O_5 films have shown the value of momentum relaxation time to be ~10^{-18} s.

For understanding the effect of phase transition on transport properties, we have also calculated the thermoelectric properties of vanadium dioxide, VO_2, for both metallic and insulating phases. Seebeck coefficient and thermal (electronic) conductivity in metallic phase show similar and more pronounced anisotropic feature as compared to the electronic part of the thermal conductivity. However, in the insulating phase, Seebeck coefficient does not exhibit noticeable anisotropy. The absolute value of Seebeck coefficient increases monotonically with temperature in the metallic phase while it decreases monotonically with temperature in the insulating phase. Seebeck coefficient, at experimental critical temperature of 340 K, is found to change by 18.9 μV/K which lies within 10% of the observed discontinuity of 17.3 μV/K during the phase transition. "Kohn–Sham–Boltzmann" approach can predict phase transition in VO_2 with reasonable accuracy.

4.5 Optical Properties of Vanadium Oxides

4.5.1 General Considerations

The optical property of a material originates from the response of electrons to perturbation due to the incident radiation and transition between electronic states. The two optical parameters, namely, frequency-dependent complex refractive index $\tilde{n}(\omega)$ and dielectric function $\in(\omega)$ are related to the electronic structure and band structure of the solid. The bandgap calculation and absorption edge estimation have been of immense interest in research due to their applications in the design of optical, electronic, and optoelectronic devices. Bandgaps of VO_2 and V_2O_5, at room temperature, have been reported as 0.6 eV [3] and 2.3 eV [136], respectively, while an energy gap of 0.66 eV is found in V_2O_3 at 70 K [125]. In this study, we have mainly analyzed the spectral dependence of the complex dielectric function $\in(\omega)$ of both bulk and thin film of the V-O systems deposited on Al_2O_3 substrates, based on the data available in the literature. The observed peaks in the corresponding spectra have been interpreted and compared as a function of structure, polarization, and temperature. The complex dielectric function, $\in(\omega)$, is related to the complex refractive index by the following equations:

$$\epsilon_1 = n^2 - k^2 \tag{4.5.1}$$

$$\epsilon_2 = 2nk \tag{4.5.2}$$

where $n(\omega)$ and $k(\omega)$ are frequency (ω) dependent refractive index (n) and extinction coefficient (k), respectively.

Since the dielectric function is a complicated function of frequency [137], we have used the Penn model, a simplified model of a semiconductor or insulator [138, 139] to account for the average isotropic energy gap in terms of the long-wavelength electronic dielectric constant in the nondispersive region. Van Vechten's [140] extension of Penn model to d-electrons has also been implemented to account for the energy gap and ionicity of the bonds have been calculated using the empirical theory developed by Phillips [141]. Also, the sum rule has been applied to the V-O system to describe the effective number of electrons participating in the optical transitions.

4.5.2 Review of Optical Spectra

Above T_c, VO_2 has a tetragonal crystal structure with two distinct directions for electric polarization. The lower symmetry monoclinic structure is energetically favorable for the crystal phase below T_c, and hence higher degree of anisotropic behavior is expected with three distinct directions for electric polarization. However, the "domain" pattern [142], observed in this low-temperature phase, reduces the degree of anisotropy, and hence electric vector (**E**) sees only two independent directions. Therefore, the optical properties have been studied with $\mathbf{E} \perp a$-axis for monoclinic phase and $\mathbf{E} \perp c$-axis for tetragonal phase and their parallel counterparts. Anisotropic character of V_2O_3 is rarely taken into consideration since the experimental study of its electrical and optical properties show very small directional dependence [143, 144]. However, V_2O_5 is highly anisotropic [115].

By definition, the dielectric function of an insulator or semiconductor quantifies the dielectric polarization, which in turn is described classically by the oscillation of a spring connecting a pair of electric charges generated by an external electric field. Resonant oscillation of the spring, followed by light absorption, can be observed when the frequency of the incident radiation matches with the oscillating frequency of the spring. In other words, ϵ_2, which is proportional to the amount of light absorbed in the medium, shows a peak corresponding to the resonance frequencies of the spring. Since the region of interest for incident photons lies within the infrared to the vacuum ultraviolet range, we will analyze, in essence, the atomic and electronic polarization. On the other hand, refraction or absorption of light in a medium can be completely determined by the complex refractive index, $(n + ik)$, as well. Clearly, the real part (n) controls the speed of light in the medium while the extinction coefficient (k) signifies absorption and modulates the amplitude of the electromagnetic radiation in the medium.

Figures 4.38, 4.39, 4.40, and 4.41 show the variations in optical properties such as $\in_1$, $\in_2$, n, k, and R of the bulk and thin film of VO_2, V_2O_3, and V_2O_5 with photon energy at different temperatures and polarizations of electric field. It is evident from the figures that the optical parameters show strong variation with energy of incident photons from infrared to vacuum ultraviolet range (up to 12 eV). The value of $\in_1$ at temperature higher than T_c decreases with frequency at the lower end of the spectrum and becomes negative while the $\in_2$-E spectra show the corresponding exponential increase with decrease in frequency. This can be attributed to the free carrier absorption or Drude tail of the metallic [145] phase and can further be justified by the rapid increase in reflectivity with decrease in frequency below $\omega < \omega_p$, the plasma frequency at which $\in_1$ becomes zero. The Drude absorption feature can also be observed in the k-E spectra at high-temperature metallic phase as evidenced in Fig. 4.40b. The anisotropy is manifested from the amplitude, width, and energy position of the corresponding structure in the optical spectra for polarizations parallel to the crystallographic axis a, b, and c. By comparing the peaks in the $\in_2$-E spectra of the insulating phase, for instance, of all the three oxides, V_2O_5 shows high anisotropic behavior; most of the peaks are more sharply peaked in V_2O_5 than those seen in VO_2 and V_2O_3. Similarly, unlike in the high-temperature phase, the absorption peaks in the $\in_2$-E spectra, at low-temperature phase, are relatively sharper. The temperature dependence of the spectral variation of the optical properties is highly manifested at the lower end of the spectrum. These changes in the infrared region during IMT are due to the onset of free carrier dominated absorption, a characteristic of metallic phase [142]. However, no remarkably high qualitative difference is observed in the optical spectra between different temperatures at the higher frequency.

Figure 4.38 shows the comparison of the reflectivity spectra and dielectric function—both real and imaginary part—of vanadium dioxide in the energy range of 0.25–5.0 eV as a function of temperature below and above T_c. Figure 4.38a, b show $\in_1$-E, $\in_2$-E and R-E spectra of bulk single crystal of VO_2 for polarization $\mathbf{E} \parallel a$-axis and $\mathbf{E} \perp a$-axis, respectively, while Fig. 4.38c shows the optical spectra for 1000 Å polycrystalline thin film of VO_2 deposited on Al_2O_3 substrate. The bandgap absorption, as expected in the $\in_2$-E spectra, cannot be observed which might be due to stoichiometric impurity and other imperfections in the samples. Comparison of Fig. 4.38a, b indicates the direction dependence of optical properties; the absorption and reflectivity peaks, observed in the low-temperature phase, are higher for the polarization $\mathbf{E} \perp a$-axis as compared to the parallel counterparts. The contribution of atomic polarization to the dielectric function, which is indicated by the first resonance peak located in the infrared region, at 300 K for electric field $\mathbf{E} \perp a$-axis is higher than that for the polarization $\mathbf{E} \parallel a$-axis. A small peak appearing near 0.6 eV in the $\in_1$-E spectra in the metallic phase at 355 K, as seen in Fig. 4.38b, is absent for the polarization $\mathbf{E} \parallel a$-axis. It means that the anisotropy is significant in the infrared region. No significant difference in structural feature between the bulk and thin film spectra is seen. The structures in the spectra below 2.0 eV at high-temperature phase have been described as a result of metallic free carrier dominated absorption [142]. However, significant peaks can be seen at energies above 2.5 eV in both phases as indicated in Table 4.8 and are explained in terms of (direct) inter-band transitions,

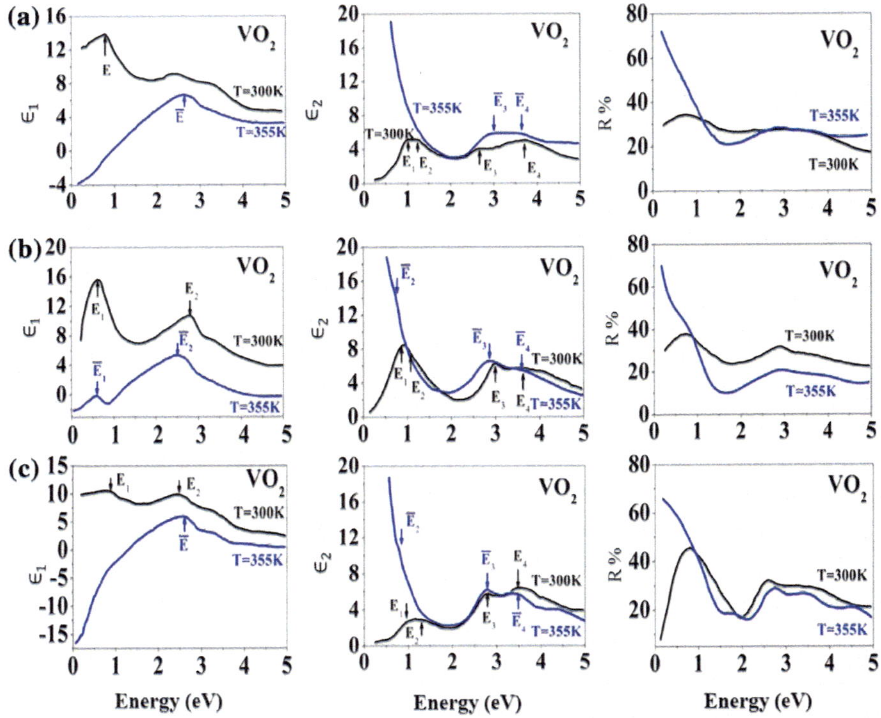

Fig. 4.38 Variation of ϵ_1, ϵ_2 and R with photon energy at temperatures 300 and 355 K [142] for bulk VO_2 with two polarizations of electric field **a E** ∥ *a*-axis, **b E** ⊥ *a*-axis, and **c** a 1000 Å thin film of VO_2 on Al_2O_3 substrate

Table 4.8 Photon energies corresponding to the peaks and shoulders as seen in the ϵ_2-E spectra of VO_2 [142]

E (eV)→		Bulk								Film on Al_2O_3			
		E ∥ *a*-axis				**E** ⊥ *a*-axis							
		E_1	E_2	E_3	E_4	E_1	E_2	E_3	E_4	E_1	E_2	E_3	E_4
ϵ_2	300 K	1.0	1.3	2.6	3.6	0.85	1.3	3.0	3.7	1.0	1.3	2.8	3.5
	355 K	–	–	3.0	3.6	–	0.75	2.9	3.6	–	0.85	2.8	3.5

i.e., the transitions between the 2p (O) and 3d (V) bands which are separated approximately by 2.5 eV [142]. A shoulder appearing relatively distinct near 0.7 eV in the ϵ_2-E spectra of high-temperature phase, in Fig. 4.38b, has been interpreted as inter-band transition within the 3d bands [142].

Figure 4.39 shows the reflectivity spectra and dielectric function—both real and imaginary part—of vanadium sesquioxide (V_2O_3) from infrared to vacuum ultraviolet range (up to 10.0 eV) as a function of temperature below and above T_c. Figure 4.39a shows the ϵ_1-E and ϵ_2-E spectra of bulk V_2O_3 calculated using density

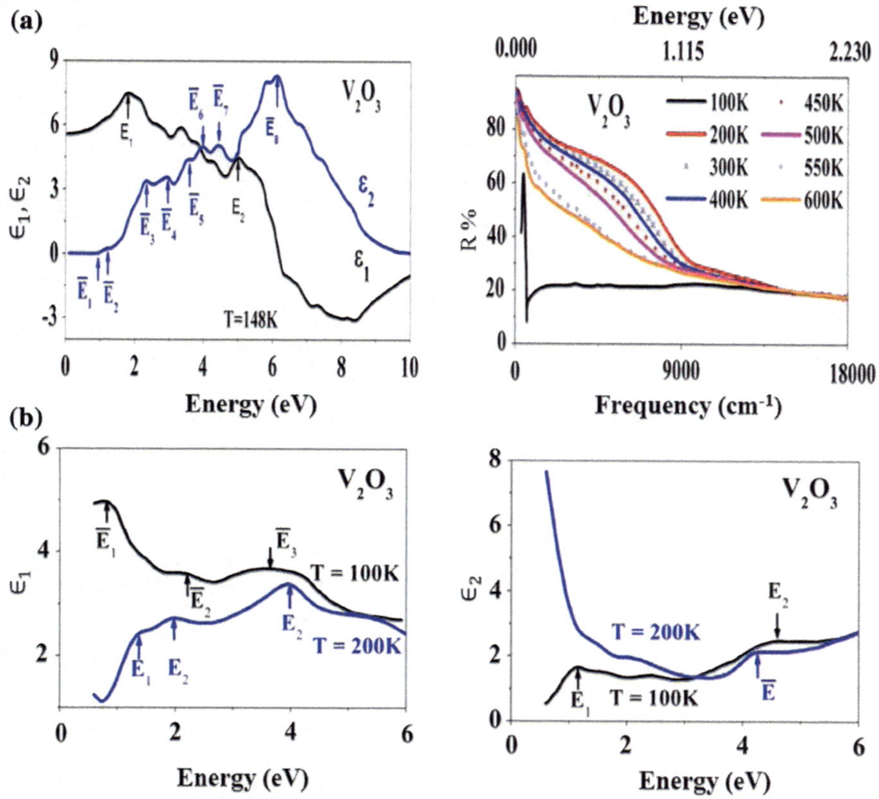

Fig. 4.39 Variation of $\in_1$, $\in_2$ and R with photon energy for **a** bulk V_2O_3 at different temperatures [64, 146], **b** thin film of V_2O_3 on Al_2O_3 substrate [119]

functional theory [146] and R-E spectra of single crystal of V_2O_3 at near-normal incidence [64]. The absorption, in bulk V_2O_3 at temperature of 148 K, starts at 1 eV as seen in the $\in_2$-E spectra of Fig. 4.39a. It means that the low-temperature insulating phase is transparent to infrared radiation. Other absorption peaks appear at near infrared, visible, and near ultraviolet regions as indicated in Table 4.9. The highest but wider peak centered at 6.1 eV covers the range from 5 to 9 eV, which is an indication of strong absorption in the ultraviolet region. The infrared reflectivity spectra of a single crystal of V_2O_3, measured in the temperature range of 100 to 600 K, show two distinct behaviors. Unlike the R-E spectra at the temperature of 100 K, the reflectivity at low-frequency edge increases with decrease in frequency at all temperatures above 200 K. This can be attributed to the high-temperature metallic behavior. However, the temperature-dependent variations at higher temperatures are relatively insignificant, and it seems that the significant changes in reflectivity occur within a few degrees of T_c. On the other hand, the high-frequency reflectivity tails merge with each other, irrespective of the temperature. Figure 4.39b shows the dielectric

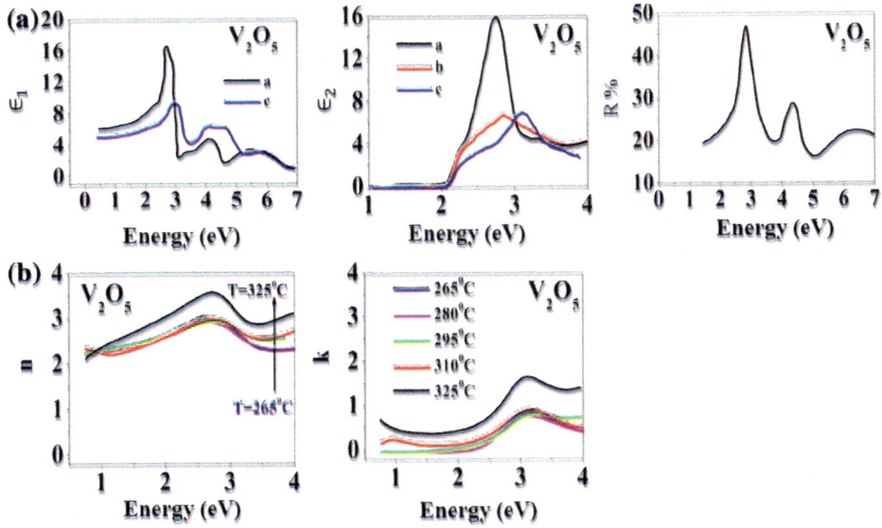

Fig. 4.40 Variation of $\in_1$ [116], $\in_2$[115], R [116] n, and k [40] with photon energy for **a** bulk V_2O_5 at polarization **E** ∥ a (black), **E** ∥ b (red), and **E** ∥ c (blue) **b** thin film of V_2O_5 on Al_2O_3 substrate

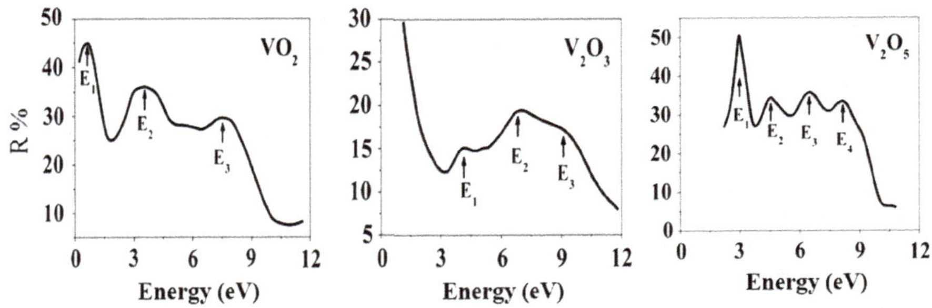

Fig. 4.41 Variation of R with photon energy for bulk VO_2, V_2O_3, and V_2O_5 at 298 K [147]

function of a 75 nm polycrystalline film of V_2O_3 deposited on Al_2O_3 substrate at temperatures below and above T_c [119]. Table 4.9 lists the absorption peaks observed in the $\in_2$-E spectra. Clearly, fewer structures are seen in the case of experimental spectra of the thin film as compared to the calculated spectra of the bulk V_2O_3. This can be partially attributed to the optical property calculations combined with other parameters such as temperature difference and possibly surface effects. However, the peaks in the insulating phase, existing at three energy locations viz. E_2, E_3, and E_7 (see Table 4.9), seem to refer to major optical transitions.

Figure 4.40 shows the dielectric function, reflectivity spectra, and both the real and imaginary parts of the refractive index of vanadium pentoxide (V_2O_5) from infrared to near-vacuum ultraviolet range (up to 7.0 eV) as a function of temperature below and above T_c. Figure 4.40a shows the $\in_1$-E, $\in_2$-E and R-E spectra of bulk

Table 4.9 Photon energies corresponding to the peaks and shoulders as seen in the $\in_2$-E spectra of V_2O_3 [119, 146]

Energy (eV)→		E_1	E_2	E_3	E_4	E_5	E_6	E_7	E_8
Bulk	148 K	1.0	1.2	2.3	3.0	3.6	4.0	4.5	6.1
Film on Al_2O_3	100 K	–	1.2	2.4	–	–	–	4.6	–
	200 K	–	–	2.0	–	–	–	4.3	–

single crystal of V_2O_5 for polarizations parallel to crystallographic axis a, b, and c. The band-edge absorption is visible around 2.0 eV from the $\in_2$-E spectra, which cannot be described by a unique inter-band optical transition but can only be partially attributed to the direct forbidden transitions ($\mathbf{k} \neq 0$, where **k** is the wave vector) [115]. Beyond the intrinsic edge toward higher energy, the peaks represent the absorption and correspond to electronic transitions from filled 2p (O) to empty 3d (V) states [116]. It can be seen from the $\in_2$-E and R-E spectra for polarization vector $\mathbf{E} \parallel a$-axis, in Fig. 4.40a, that the first sharp absorption occurs at around 2.8 eV, whereas the second and third peaks appear at 4.3 and 6.4 eV, respectively. The dielectric function shows a very high anisotropy in the range 2.2–3.3 eV (visible region) as noticed in the $\in_1$-E and $\in_2$-E spectra; the spectra in $\mathbf{E} \parallel a$-axis deviates most from the other two. Clearly, the anisotropy depends on the spatial distribution of electron wave functions and it is possible that the 3d-orbitals directed along the a-axis are relatively more localized, as indicated by the narrow intense peak in the $\in_2$-E spectra for $\mathbf{E} \parallel a$-axis, forming a wider conduction band. Figure 4.40b shows the frequency-dependent refractive index and extinction coefficient of a polycrystalline thin film of α-V_2O_5 deposited on Al_2O_3 substrate. The measurements were taken from 0.75 to 4.0 eV at various temperatures ranging from 265 to 325 °C with an increment of 15 °C. Both n-E and k-E spectra exhibit significant temperature-dependent change over the entire energy range indicating the phase transition. The k-E spectra shows a shift in the absorption edge from 1.5 eV to less than 0.75 eV as the temperature is increased from 280 to 295 °C [40]. The sharp absorption observed in the k-E spectra, at around 3.0 eV, is close to the corresponding absorption peak located at 2.8 eV in bulk V_2O_5.

4.5.3 Application of Penn Model

Band structure of a material is related to its R-E spectra ($\in_2$-E spectra). By definition, an intensity maximum in R (or $\in_2$) in the R-E spectra (or $\in_2$-E spectra) represents a maximum number in the optically induced electronic transitions in the material [148]. The energy corresponding to the peak should, therefore, correspond to a band-to-band energy difference or a bandgap. Since this is a macroscopic gap [139], it should be related to the high-frequency dielectric constant $\in_\infty$ (=n^2), where n is the refractive index.

Several models [139, 149, 150] have been proposed to interpret the frequency and wave-vector dependence of the dielectric function. All these models have, however, been proposed for elemental semiconductors. Extrapolation of the applicability of these models to amorphous semiconductors, [151] and narrow and wide gap materials, including alkali halides [140, 152, 153], has been carried out with reasonable success. Here, we demonstrate the applicability of one such model to the three oxides of vanadium. For a model semiconductor, the high-frequency dielectric constant is given by [139]

$$\varepsilon_\infty = 1 + \left(\hbar\omega_p/E_p\right)^2\left[1 - \left(E_p/4E_F\right) + \frac{1}{3}\left(E_p/4E_F\right)^2\right] \tag{4.5.3}$$

where E_p is the Penn gap [139] and E_F is the Fermi energy given by [152]

$$E_F = 0.2947\left(\hbar\omega_p\right)^{4/3} \tag{4.5.4}$$

with the valence electron plasmon energy given by [154] $\hbar\omega_\mathrm{p} = 28.8\left[(N_\upsilon\rho/W)^{1/2}\right]$, W is the molecular weight and N_υ is the number of valence electrons per molecule calculated using

$$N_\upsilon = \mathrm{M}a + N(8 - b) \tag{4.5.5}$$

for a compound $A_M B_N$, where $a(b)$ is the number of valence electrons per atom of type A(B) and M(N) is the atomic fraction of element A(B). Equation (4.5.3) can be rewritten as

$$\varepsilon_\infty = 1 + \left(\hbar\omega_p/E_p\right)^2 S_0 \tag{4.5.6}$$

where S_0 represents the terms inside the square brackets in Eq. (4.5.3). Since the most significant variation occurs in the expression before S_0, Penn neglects the smaller terms containing E_g/E_F and thus approximates the value of S_0 as 1 [155]. This is true for materials with bandgaps in the commonly occurring range where $E_g/E_F = 0.3$ [150]. However, Grimes and Cowley [150] found that the value of S_0 is only weakly dependent on the bandgap and that a value of 0.6 is a fairly good representation of S_0. Thus, with this slightly more accurate value for S_0, the energy gap can be determined using appropriate values of the dielectric constant. In Table 4.10, the peak energies from the reflectivity data of the V-O systems, at temperature of 298 K, are summarized.

At room temperature, VO_2 and V_2O_5 are in insulating phase while V_2O_3 is in metallic phase with very low density of states at the Fermi level [147]. It is believed that IMT is governed by the change in the 3d band structure [156, 157]. In Penn model, the effect of the d-band is to increase the number of valence electrons per molecule, N_υ. Van Vechten [140] has considered, in detail, the effect of d-electrons

Table 4.10 Peak energies from reflectivity data of the V-O systems at temperature of 298 K

Energy (eV)→	E_1	E_2	E_3	E_4
VO_2	0.7	3.6	7.4	–
V_2O_3	4.2	6.8	9.0	–
V_2O_5	2.97	4.51	6.5	8.1

on the dielectric properties of materials. In Table 4.11, we also evaluate the Penn gap incorporating the d-electron contribution to N_υ as indicated within the parenthesis.

It is important to note, here, that the effective valence–conduction bandgap, for the material consisting of different atoms in the unit cell, can be separated into homopolar (E_h) and heteropolar part (C) as introduced by Phillips [158]. Accordingly, we write $E_p^2 = E_h^2 + C^2$ and introduce a parameter, the Phillips ionicity, defining the ionic character in bonds as, $f_{\rm i} = C^2/(E_h^2 + C^2)$, where E_h is related to the static dielectric constant (ε_0) by [141]

$$\varepsilon_0 = 1 + \left(\hbar\omega_p/E_h\right)^2 S_0 \tag{4.5.7}$$

In order to study the Penn gap, which is the macroscopic gap accounting for all the possible optically induced electronic transitions in the material, we rely on the reflectivity data, measured at room temperature, extended for longer range of photon energy as shown in Fig. 4.41 [147]. The size of the single crystal samples of V_2O_5 and V_2O_3, used in these measurements, were $10 \times 10 \times 5$ mm^3 each, whereas that of VO_2 was $7 \times 5 \times 5$ mm^3. The measurements were performed for the polarization **E** $\parallel$ *a*-axis. The energies corresponding to maxima in intensities seen in Fig. 4.41 are listed in Table 4.10.

We can clearly see three major peaks for VO_2 crystal as summarized in Table 4.10. The first peak, appearing at 0.7 eV, in the insulating phase, corresponds to the shoulder appearing in the ϵ_2-E spectra of the high-temperature phase, in Fig. 4.38b, having the origin of transition, i.e., from occupied to empty states within the d-band [142, 147]. The other two peaks, at 3.6 and 7.4 eV, correspond to the transition from 2p (O) to 3d (V) band. The fact that the 3d-bandwidth in V_2O_3 is around 3 eV [157], which is the largest of all three V-O system [147], and the electronic transitions start at 4.2 eV implies that no transition occurs within the d-band in V_2O_3. Also, the transitions are mostly to the 4s, 4p bands of vanadium only after 10 eV [147], the observed peaks between 4.2 eV and 9.0 eV can be referred to the transition from 2p (O) to 3d (V) band. We can see four major peaks for V_2O_5 crystal as indicated in Table 4.10; these refer to the transition from 2p (O) to 3d (V) band, where the first peak at 2.97 eV shows the highest optical transition and is attributed to excitonic transitions [147]. Since the first peak appears at relatively high energy, no transition occurs within the d-band in V_2O_5.

The results of the calculations based on Penn model are presented in Table 4.11. Also, listed in Table 4.11 are the values of the bandgap energy (E_g), zero-frequency (ϵ_0) and high-frequency (ϵ_∞) dielectric constants, Phillips Ionicity (f_i), average

Table 4.11 Properties of the V-O systems; the parenthesis value of $\overline{E}$ is the arithmetic average of all the energies corresponding to the peaks and shoulders in the R–E spectra

V-O system	Mol. Wt (W)	ρ (g/cc)	E_g (eV)	N_v	$\hbar\omega_p$ (eV)	E_F (eV)	ϵ_∞	E_p (eV)	$\overline{E}$ (eV)	ϵ_0	f_i	C (eV)	E_h (eV)
VO_2	82.94	4.68 [161]	0.60 [3]	6	16.56	12.44	9.7 [162]	4.4 (4.6)	3.9 (4.79)	25.9 [162]	0.65	3.55	2.60
V_2O_3	149.88	4.98 [143]	0.66 [125]	10	16.60	12.48	5.0 [163]	6.4 (6.7)	6.6 (6.30)	15.0 [164]	0.71	5.43	3.44
V_2O_5	181.88	3.36 [165]	2.30 [136]	14	14.64	10.55	4.0 [166]	6.5 (6.8)	7.36 (5.52)	13.8 [116]	0.77	5.73	3.17

homopolar (E_h) and heteropolar (C) energy gaps, Fermi energy (E_F), and arithmetic average of all the energies corresponding to the peaks in the R-E spectra ($\overline{E}$). Using the values of E_F and E_p listed in Table 4.11, we have evaluated the value of S_0 and it is found to be 0.88 which is more than our approximation but still less than unity. It can be seen from Table 4.11 that the calculated value of E_p for the single crystal of V_2O_3 is close to $\overline{E}$. E_p of VO_2 and V_2O_5 are also seen to be in good accord with the corresponding values of $\overline{E}$; the difference between E_p and $\overline{E}$ for V_2O_5 is relatively higher as compared to that of VO_2. These relative deviations are consistent with the degree of anisotropy of the three V-O systems. It is important to note, here, that the E_p values of VO_2 and V_2O_3 are closer to the value in the parenthesis (the average of all the energies corresponding to the peaks and shoulders in the R-E spectra, $\overline{E}$). This indicates that an isotropic, nearly free electron model such as the Penn model seems to be valid in explaining the energies corresponding to the peaks in the reflectivity spectra of these vanadium oxides. It is to be noted, here, that such a procedure of comparing the calculated values of E_p with the average of the energies corresponding to the peaks in the R-E spectra was proposed by Phillips [141]. Examining the ionicity, we see that the V-O systems are more than 65% ionic. V_2O_3 and VO_2 follow the general trend that low oxidation states of vanadium oxides are more ionic and undergo IMT [159]. However, V_2O_5 is highly ionic but is consistent with the fact that V_2O_5 is more ionic than VF_5 [160].

4.5.4 Sum Rule

At this stage, it would be worthwhile to look into the number of electrons participating in the optical transitions. Most of the electrons in the material are core electrons and are tightly bound to the atomic nuclei. If we consider that the core electrons are excited at high enough frequencies, the sum rule can be written as [167]

$$\frac{2\pi^2 Nne^2}{m} = \int_0^\infty \omega \in_2 (\omega)\mathrm{d}\omega \tag{4.5.8}$$

where m is the mass of a free electron; e, the electronic charge; N, the number of atoms per unit volume (atom density); ω, the angular frequency of light; and n, the total number of electrons per atom. However, the electrons contributing to the optical properties of solids are conduction and valence electrons, and hence, the core states can be neglected. Further assuming that other absorptive processes such as phonon excitation are not overlapping with electronic excitation [168], the effective number of electrons per atom participating in the optical transitions over a given frequency range is approximated by

$$n_{\mathrm{eff}}(\omega_0) = \frac{m}{2\pi^2 Ne^2} \int_0^{\omega_0} \omega \in_2 (\omega)\mathrm{d}\omega \text{ (In terms of frequency)}$$

$$n_{\mathrm{eff}}(\omega_0) = \frac{(4\pi \in_0)m}{2\pi^2 N\hbar^2} \int_0^{E_0} E \in_2 (E)\mathrm{d}E \text{ (In terms of Energy and SI system)} \tag{4.5.9}$$

where $\in_0$ is the permittivity of free space and $n_{\mathrm{eff}}(\omega_0)$ is the effective number of electrons per atom governed by polarization of electron shells, contributing to optical transitions below a frequency ω_0. Since we are interested in calculating the effective number of electrons per formula unit, we define N as the number of vanadium ions per formula unit per unit volume.

Figure 4.42 shows the variation of n_{eff} and its slope with photon energy, which were calculated numerically using Eq. (4.5.9) for all the three V-O systems at temperatures below and above T_c. The effective number of electrons shows a clear temperature-dependent variation with photon energy below and above T_c. The calculated n_{eff} for the insulating phase is zero below certain photon energy but varies with photon energy. The rate of change of n_{eff} with respect to energy of incident photons, referred to as slope, is not constant throughout the frequency range and shows significant variation.

The n_{eff} for both the bulk and thin film of VO_2, corresponding to the insulating phase (300 K), is almost zero below 0.6 eV which then varies with the energy of incident photons. The slope for bulk phase, at 300 K, initially increases until 1.2 eV, remains fairly constant from 1.2 to 2.1 eV, rises to a maximum at 3.8 eV, with a small shoulder in between, and finally decreases with increase in photon energy. The region of the shoulder may refer to the transition from 2p (O) to 3d (V) states and is small when compared to its expected value of unity for the absorption due to the one "extra" d-electron per formula unit. While there are some differences in the magnitude between the slope of the bulk and film of VO_2, their energy-dependent variations show similar pattern. The slope for the high-temperature phase of VO_2

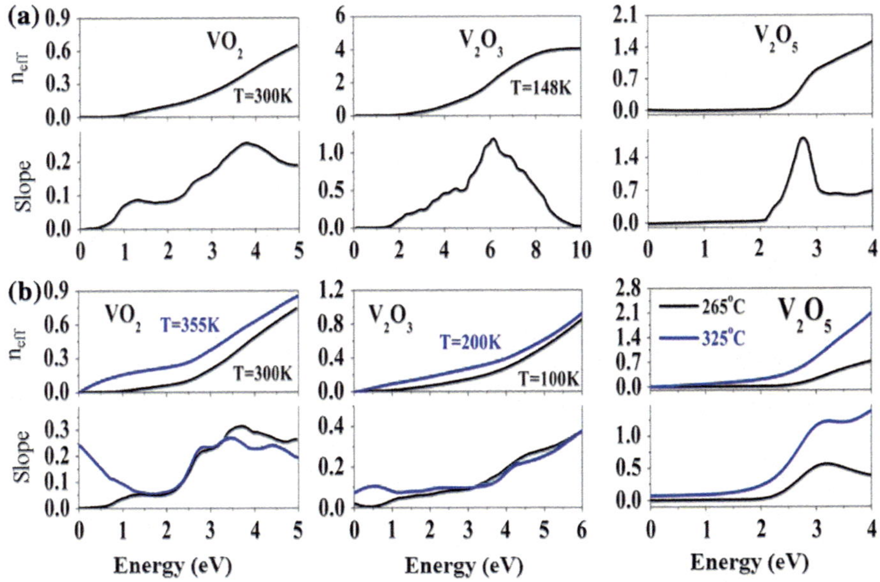

Fig. 4.42 Variation of n_{eff} with photon energy calculated using Eq. (5.9) along with its slope with respect to the energy for **a** bulk at polarization **E** ∥ *a* [115, 142, 146] and **b** film of VO_2, V_2O_3, and V_2O_5 on Al_2O_3 substrate [40, 119, 142]

initially decreases and reaches minimum at 1.75 eV and then rises until 2.5 eV, and finally shows a slight rise and fall alternatively as shown in Fig. 4.42b.

The slope for V_2O_3 in Fig. 4.42a shows the highest peak, besides other small structures, at around 6.1 eV which indicates a strong absorption in the ultraviolet region. This peak is consistent with the corresponding peak in the ϵ_2-E spectra observed in Fig. 4.42a. On comparing Figs. 4.42a, b, we see that the n_{eff} for both the bulk and film of V_2O_3 in its insulating phase show similar trend until 3 eV. However, after 3 eV, the n_{eff} in the film of V_2O_3 deviates considerably from its bulk counterpart, which in fact shows saturation near a value of four electrons per formula unit at the end of the ultraviolet spectrum. Assuming that the density functional theory [146] correctly predicts the optical properties of V_2O_3 in the photon energy range of 0–10 eV, this saturation can be attributed mainly to absorption due to the two d-electrons per vanadium ion combined with some contribution due to the transition from 2p (O) to 3d (V) states. However, there is a remarkable difference in the n_{eff} between the bulk and film of V_2O_3 and may require further study to make a definite conclusion to interpret the difference.

The n_{eff} for both the bulk and film of insulating phase of V_2O_5 is almost zero below 2.2 eV and is consistent with the observed absorption band edge. An abrupt change in the slope near the peak region, as seen in Fig. 4.42a, is characteristic of inter-band transition. This peak is consistent with the peak in Fig. 4.40a in the ϵ_2-E spectra observed at polarization **E** ∥ *a*-axis. The n_{eff} in bulk phase of V_2O_5 shows a value of

1.49 at photon energy of 4.0 eV. Since V_2O_5 does not have any d-electron in its V^{5+} ion, this should be the contribution due to the transition from 2p (O) to 3d (V) states. The value of n_{eff} corresponding to this transition is higher in V_2O_5 as compared to the other two oxides and can be attributed to the higher number of oxygen atoms per formula unit. A similar interpretation can be made for the film of V_2O_5 on Al_2O_3 substrate. However, the variation of n_{eff} with energy at two different temperatures below and above T_c appears to show more consistent pattern at sufficiently high photon energy in both VO_2 and V_2O_3 while a divergence pattern can be easily seen from Fig. 4.42b for the corresponding variation in V_2O_5. Comparison of n_{eff} of bulk at room temperature and film of V_2O_3 at 265 °C shows different pattern of variations with photon energy which may be partly attributed to the difference in temperature and the highly anisotropic nature of V_2O_5. This may be due to the fact that studies [3, 40] of optical property do not pertain to the same crystallographic axis besides ambient conditions and other aspects of the experiment such as the quality of the crystal and analysis procedures. This conclusion is consistent with the perspective of Kang et al. [40], where assertion has been made that the structural phase transition in V_2O_5 does not occur and that V_2O_5 film undergoes an IMT at a critical temperature of 280 °C instead of 257 °C [31] as reported in most of the literature.

4.5.5 Summary

Vanadium oxides, which consist of strongly correlated d-electrons, are extremely sensitive to the external stimulus such as temperature and undergo insulator–metal transitions (IMT) at a particular temperature depending on the *O*/*V* ratio. Vanadium oxides are widely used in technology in which devices make use of their properties such as IMT, high-temperature coefficient of resistance (TCR), and a small 1/*f* noise constant. In this study, we have analyzed the optical properties such as ϵ_1, ϵ_2, *n*, *k*, and *R* of bulk and film of VO_2, V_2O_3, and V_2O_5 deposited on Al_2O_3 substrates, based on the data available in the literature. The observed peaks in the corresponding optical spectra have been interpreted and compared as a function of structure, polarization, and temperature. The anisotropy is significant in the infrared region for VO_2 and in the visible region for V_2O_5. Penn model leads to an explanation of the energies corresponding to the peaks in the R-E spectra of the single crystal of the V-O systems at room temperature. E_p values for VO_2 and V_2O_5 are close to the average of the energies corresponding to the peaks ($\overline{E}$) while their values are even closer in V_2O_3, clearly reflecting the degree of anisotropy in the order of $V_2O_3 < VO_2 < V_2O_5$. The vanadium–oxygen bonds are highly ionic and undergo IMT at T_c as a function of the oxidation state of the vanadium ion, i.e., the transition temperature increases with oxidation states of the vanadium atom. Optical transitions and the effective number of electrons participating in these processes are described from the ϵ_2-E spectra and its numerical integration using the well-known sum rule. The results of these calculations show that the optical transitions from valence to conduction bands, including the transition from 2p (O) to 3d (V) bands and the inter-band transitions, occur within the d-bands. The optical spectra have no indication of the transition occurring from occupied to empty states within the d-band for V_2O_3 and V_2O_5 systems but the intra-band transition seems to occur in VO_2. The change in n_{eff} with respect to the energy of the incident photons is also calculated and it is found that this change is consistent with the peaks observed in the ϵ_2-E spectra.

4.6 Vanadium Oxides: Synthesis/Deposition

Preparation and characterization of single crystals of vanadium oxides have been of immense interest since Morin [30] discovered that certain oxides undergo semiconductor-to-metal, widely known as insulator-to-metal (IMT), transitions. Despite the fact that the bulk crystals of VO_2 and V_2O_3 are vulnerable to fracture, especially as these materials undergo phase change, and that different configurations are compatible with technologically varying devices, numerous efforts have been made to deposit thin films of these oxides in the past [169]. As near-ambient critical temperature (T_c) can be tuned optically, thermally, electrically [59], and with doping [38], IMT in VO_2 is of high technological interest and significant research has been performed in the literature [53, 54, 170–192]. However, thin films of stoichiometric VO_2 and nanoparticles cannot be easily prepared as deposition techniques require exact physical conditions to yield films with the proper composition. Beside thin films, nanoparticles of VO_2 last longer in the phase transition cycle as compared to bulk crystals [193].

Vanadium oxides exist in various phases such as VO, V_2O_3, VO_2, V_2O_5, V_3O_5, V_4O_7, V_6O_{11}, V_6O_{13}, V_7O_{13}, V_8O_{15}, etc. However, these oxides can be categorized under the so-called Magnéli (V_nO_{2n-1}) and Wadsley ($V_{2n}O_{5n-2}$) homologous series. Magnéli phases which remain in mixed (two) valence state can be represented by the general stoichiometric formula as [4]

$$V_nO_{2n-1} = V_2O_3 + (n-2)VO_2, \text{ where } 3 \leq n \leq 9$$

Their crystal structure constitutes typical dioxide-like and sesquioxide-like regions and undergo IMT with transition (critical) temperature as shown in Table 4.12 [194].

Particularly interesting single valency phases of vanadium oxides are VO_2, V_2O_3, and V_2O_5 which undergo IMT at 340 K [29], 160 K [30], and 530 K [31], respectively. It is the existence of more than 20 stable phases [195], with various order and disordered defects, that make the synthesis of vanadium oxides particularly challenging. For instance, while vanadium oxides synthesized under certain conditions yield a mixture of different phases, further adjustment of the physical parameters might possibly yield almost single phase but it becomes more nonstoichiometric [196] as shown in the phase diagram (Fig. 4.43). Since the existence of mixed phases and/or variability in stoichiometry influence the film performance, pure stoichiometric growth of sample is desired in order to maximize the effect of phase transition

Table 4.12 Phase transition temperature of Magnéli phases

Parameter (n)	V_nO_{2n-1}	Transition temperature (K)
2	V_2O_3	160
3	V_3O_5	430
4	V_4O_7	238
5	V_5O_9	135
6	V_6O_{11}	170
7	V_7O_{13}	Metallic
8	V_8O_{15}	68
9	V_9O_{17}	79
∞	VO_2	340

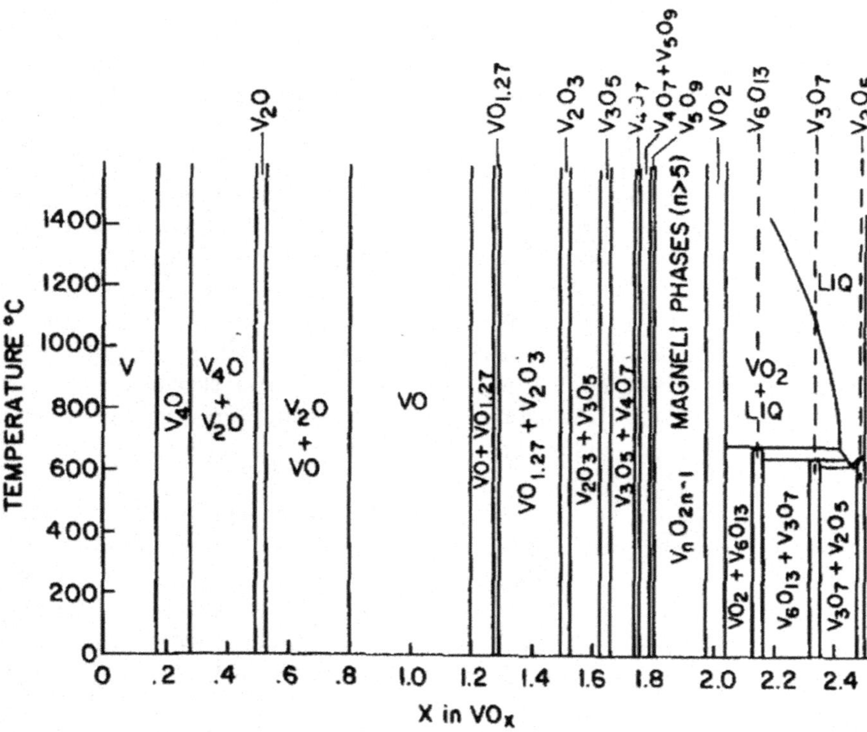

Fig. 4.43 A rough phase diagram of vanadium–oxygen system [197]

in vanadium oxide. In spite of these challenges in growing both the bulk and thin film of VO_2, several methods have been used such as a variety of sputtering methods, reactive evaporation, Pulsed Laser Deposition (PLD), sol–gel deposition, and Metal-Organic Chemical Vapor Deposition (MOCVD) methods [193].

4.6.1 Sol–Gel Methods

Typically, sol–gel method consists of sequential hydrolysis and polycondensation of a precursor in a liquid intermediate [198]. The method starts with the formation of homogeneous sol, in which solid particles are immersed in a liquid in the form of colloidal suspension. By definition, solids particles of diameters varying from 1 to 100 nm are considered to be colloids [199]. After polycondensation, a porous network is formed in a liquid—this wet structure is called a gel. The term “gel”, in general, means the following [199]:

(a) orderly lamellar structure,
(b) completely disordered polymeric network (covalent),
(c) mostly disordered polymer network (physical aggregation), and
(d) particularly disordered structure.

Sol–gel technique is extensively used for preparing thin film oxides of various properties [200–216]. These include: alkoxides hydrolysis, oxide hydrolysis, peroxide-assisted hydrolysis, and cation exchange [217].

Sol–gel methods have been used extensively by researchers in the past [198, 209, 218–226] to grow thin films of vanadium dioxide. It is a simple, low-cost method and allows high-purity deposition over a large area. Being a low-temperature technique, chemical composition in the product can be finely tuned. Even a small amount of dopant gets finely distributed in the product, and hence this method is widely used for doping metal into VO_2 to change its phase transition behavior. Sol comprises a solvent and molecular precursor such as vanadium oxyacetylacetone $VO(AcAc)_2$ or vanadium acetylacetonate $V(AcAc)_4$ or chemical precursor such as metal salt or metal alkoxides [227–229]. When the sol is condensed to form a gel, it is further reduced by annealing, or oxidized in a neutral environment at approximately 600 °C to form VO_2 network [198].

Ningyi et al. [230] melted the V_2O_5 powder and poured into DI water to make a brownish V_2O_5 solution and prepared highly oriented V_2O_5 thin films on SiO_2/Si substrates using sol–gel method. Then, by heating at a temperature above 400 °C, under a pressure below 2 Pa in air, VO_2 films were obtained. They described the reduction process as $V_2O_5 \rightarrow V_3O_7 \rightarrow V_4O_9 \rightarrow V_6O_{13} \rightarrow VO_2$. It is worthwhile to note, here, that the phase sequence of single valency vanadium oxides is $VO \rightarrow V_2O_3 \rightarrow VO_2 \rightarrow V_2O_5$ [231]. However, the films obtained in this way were in mixed phase and not highly oriented. Wang et al. [225] prepared thermochromic VO_2 films by annealing two different precursor gels, namely, V-H_2O_2 ($V_2O_5.2.2H_2O$) and V_2O_5–H_2O_2 ($V_2O_5.1.8H_2O$) at 750 °C in vacuum. Since their process did not involve an intermediate reduction step, it can be considered to be a one-step annealing process for the formation of VO_2 films. In XRD (Fig. 4.44), their samples showed only one phase, VO_2, for both the samples after annealing at 750 °C.

Partlow et al. [222] used sol–gel method to deposit thin polycrystalline films of VO_2 and V_2O_3 on various substrates. They used a pale yellow commercial vanadium oxide isopropoxide ($VO(OC_3H_7))_3$ to prepare precursor solution and deposited the

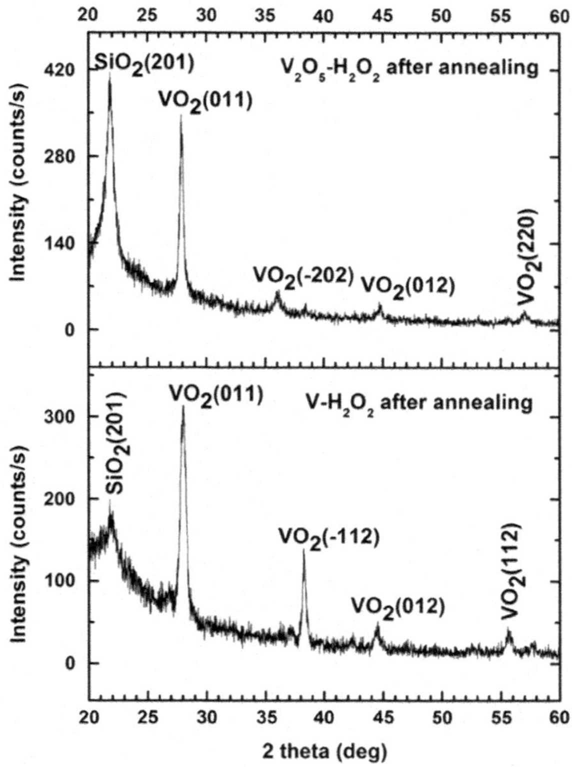

Fig. 4.44 XRD spectra of V-H_2O_2 and V_2O_5–H_2O_2 precursor gels after annealing at 750 °C [225]

films by dipping or spinning. After removing the organic components, the film was first converted to V_2O_5 and then reduced to V_2O_3 or VO_2 using appropriate reducing environment. It is the pyrolysis conditions, rather than starting material, that determine the final phase of vanadium oxide; they used the equivalent vanadium oxide *n*-propoxide and obtained similar results. V_2O_5 films were reduced to VO_2 by heating in a CO–CO_2 (50/50 by percentage) mixture above 400 °C. On the other hand, reduction of V_2O_5 to V_2O_3 was achieved by heating the sample in dry H_2 at a temperature between 350 and 850 °C, with preferable temperature of above 700 °C for better crystallinity. It is believed that proper heat treatments can reduce any V_xOy material to VO_2 or V_2O_3. It appears that sol–gel method is very effective in producing high-quality thin film switches [222] of VO_2 and V_2O_3.

4.6.2 Sputtering Methods

Sputtering is one of the most commonly used Physical Vapor Deposition (PVD) processes in which materials from a target are deposited onto a substrate in the form

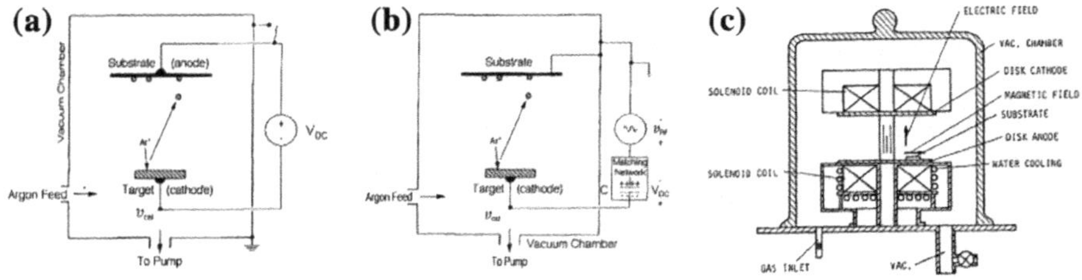

Fig. 4.45 **a** A DC sputtering apparatus; **b** an RF sputtering apparatus; **c** a magnetron sputtering apparatus [193]

of layers of atoms or molecules, in a vacuum chamber, by bombarding the target with ionized gas. Fuls et al. [232] invented reactive (ion-beam) sputtering method to grow VO_2 thin films. Since sputtering shows good reproducibility [233], it has been widely used to prepare the film of VO_x [234–238]. Commonly used sputtering methods are Direct Current (DC), Radio Frequency (RF), and magnetron sputtering [193]. Choi et al. [239] formed a thermochromic substrate in a three step process: coating pure vanadium on a glass substrate, seed layer formation by heat treatment, then coating VO_2 thin film on the seed layer by sputtering. The target consisted, preferably, of vanadium (metal) or vanadium oxide (mixture of VO_2, V_2O_3, and V_2O_5 and is conducting). DC sputtering deposition methodology was used, unlike for other nonconducting vanadium oxides, Radio Frequency (RF) sputtering deposition is required. The main difference between DC and RF sputtering is the power source used in the process. In DC sputtering, the ions from the ionized gas are deposited on the target if the target is not conducting, and hence the process becomes ineffective. On the other hand, if AC power is used, the ions accumulated on the (nonconducting) target surface during the half cycle of AC will be removed on changing its polarity. Typically, the deposition rate for DC sputtering method is almost five times faster than that of RF sputtering [239].

Similarly, magnetron sputtering is a plasma coating process in which magnetic fields are used to keep plasma in front of the target thereby intensifying the ion bombardment. Magnetron sputtering is characterized by a smooth and uniform layer but at the cost of a relatively slow deposition rate. Ping and Sakae [240] deposited thermochromic single phase VO_2 films on glass substrates by RF magnetron sputtering with (a) vanadium target cooled with water, (b) argon–oxygen gas discharge, (c) RF power of 200 W, and (d) sputtering pressure of 1.5 Pa. With precise control of oxygen flow, the film was deposited at low substrate temperature of 300 °C. A typical DC sputtering, RF sputtering, and a magnetron sputtering apparatus are shown in Fig. 4.45.

4.6.3 Reactive Evaporation

Reactive evaporation is a method in which metal atoms are evaporated and deposited using a partial pressure of a reactive gas present in the chamber. In order to form the vanadium oxides, oxygen is channeled into a vacuum chamber where it reacts with metallic vanadium vapor. In the past, several researchers have used reactive evaporation methods to prepare vanadium oxide films [241–246]. Nyberg and Buhrman [245] reported that optical contrast can be improved during IMT when a VO_2 film is deposited by reactive evaporation. This is attributed to improved stoichiometry of the films thus produced. Leisenberger et al. [247] grew vanadium oxides films on Pd (111) using reactive evaporation. By studying from sub-monolayer to 15 monolayer coverages, they found VO/VO_2 like behavior at lower coverage while thicker layers consisted of V_2O_3. Surnev et al. [246] produced ultrathin vanadium oxides on Pd (111) using reactive evaporation method. Their study showed that the morphology of the low coverage phases depends on the oxygen pressure and substrate temperature. Lovis and Imbihl [248] studied ultrathin vanadium oxides (VO_x) deposited on Rh(111) surface using reactive evaporation, in which a highly pure (>99.8%) vanadium rod was evaporated in the ambience of 2×10^{-7} mbar oxygen partial pressure at 675 K temperature. The VO_x layers thus produced were subjected to H_2 + O_2 reaction, where the layers underwent self-organization giving rise to stripe or island pattern (depending on the vanadium coverage) due to condensation of V and O. These patterns are nonequilibrium structures and are characterized by memory effects [248].

4.6.4 Pulsed Laser Deposition

Pulsed Laser Deposition (PLD) is also one of the most commonly used Physical Vapor Deposition (PVD) processes in which materials from a target are deposited onto a substrate in the form of layers of atoms or molecules, in a vacuum chamber, by focusing a high-power pulsed laser beam on the target. The laser beam used in the process ablates (i.e., melts, evaporates, and ionizes) the material from the target and forms a highly luminous plasma plume. When the plume moves away from the target, the ablated material condenses onto the substrate. A schematic of PLD method is shown in Fig. 4.46 [193].

Borek et al. [249] deposited oriented grains of VO_2 on (012) sapphire substrates for the first time in 1993 using the PLD method. They used a KrF laser of wavelength 248 nm on pure vanadium target in a vacuum chamber, partially filled with an Ar–O atmosphere. The partial pressure of oxygen in the chamber was found to be the determining factor for optimizing the oxygen content of vanadium oxide phases. Oxygen partial pressure of 10% in an argon atmosphere was found to be optimum for growing the VO_2 films. IMT shown by the film thus produced was characterized by the sharp change in resistivity. Since then, many researchers have used PLD method

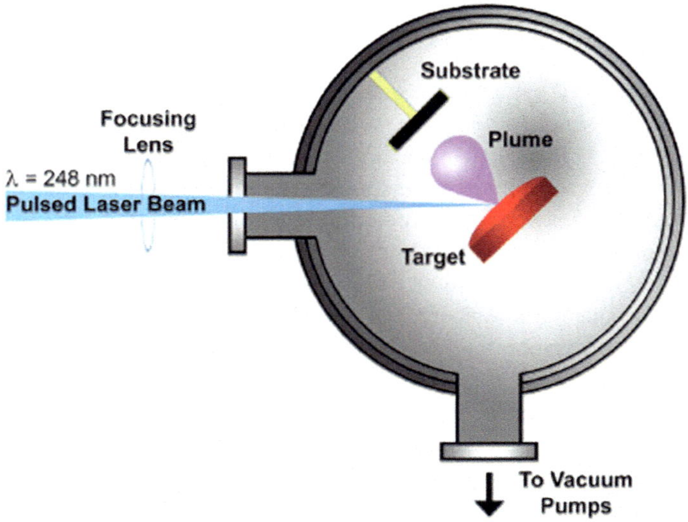

Fig. 4.46 Schematics of pulsed laser deposition device [193]

to grow vanadium oxides [53, 174, 250–258]. Rajendra Kumar et al. [53] used PLD to deposit vanadium oxide thin films at room temperature using neodymium-doped yttrium aluminum garnet (Nd:YAG) laser at 532 nm. Pure V_2O_5 powder (99.999%) was used as a target and various laser fluences were used for the deposition. A laser fluence of 1.4 J/cm^2 was determined to be the optimum for using vanadium oxide films for microbolometer applications. Huotari et al. [257, 259] used PLD technique to deposit vanadium oxide thin films using Lambda Physik Compex 201 Excimer laser (308 nm). At 400 °C substrate temperature, different oxygen partial pressures of 1.0×10^{-2} and 1.5×10^{-2} mbar yielded pure V_2O_5 thin films and films with mixed phase structure of orthorhombic V_2O_5 and triclinic V_7O_{16}, respectively. In studying the films as resistive gas sensors for NH_3, the mixed-phase films showed better gas (NH_3) sensing than the pure V_2O_5 films. Madiba et al. [260] studied the effects of gamma irradiations on vanadium dioxide thin films deposited using reactive pulsed laser deposition technology. In their study, they found that long-range crystal structure remains unaffected by the gamma radiations up to 100 kGy but creates some crystal defects, disordered region, and atomic displacement. Electrical resistivity during IMT of gamma irradiated sample also remains unaffected.

4.6.5 *Chemical Vapor Deposition*

Chemical Vapor Deposition (CVD) is a method for producing high-quality, high-performance thin films, mostly used in the semiconductor industry. In typical CVD, one or more volatile precursors flow into a chamber containing the heated wafer (substrate) and then react and/or decompose on the hot surface resulting in the desired deposit. The process is accompanied by chemical byproducts, which along with the unused precursor gas flows out of the chamber. Depending on the nature of the material, application types, epitaxial versus polycrystalline or amorphous film, process enhancement methods and derivatives of CVD technology, and many variants of CVD can be seen in practice. There is a significant amount of literature available for CVD [261, 262]; a Google search of "chemical vapor deposition" shows more than a million hits. Most of the vanadium oxides are grown using organometallic precursors, and hence the CVD process, in this case, is sometimes called metal-organic chemical vapor deposition (MOCVD).

Groult et al. [263] grew vanadium oxide thin films using atomic layer chemical vapor deposition (ALCVD) technology. Main advantage of using ALCVD is to control film thickness at atomic layer level with high precision and grow uniform and compact thin films. They deposited the vanadium oxide thin films on Si (100) substrate from triisopropoxyvanadium oxide [$VO(OC_3H_7)_3$] and water precursors, after 300 and 1000 ALCVD cycles, heat treated at 250 and 500 °C, respectively, and studied the evolution in morphology and structure by thermal annealing. It was concluded that the films annealed at temperature below 400 °C had a granulous but homologous surface. The transition from an amorphous to a crystalline vanadium oxide thin film, characterized by typical elongated V_2O_5 plates, was observed at temperatures starting at 400 °C. Su et al. [264] used chemical vapor deposition methods to prepare vanadium oxides of different morphologies such as V_2O_5 rods, VO_2 blocks, and VO_x microspheres with V^{5+} and V^{4+} cations using vanadium oxyacetylacetone $VO(AcAc)_2$ as a vanadium precursor at different deposition temperatures 500, 150, and 450–250 °C, respectively. Malarde et al. [265] used an optimized atmospheric-pressure chemical vapor deposition (APCVD) method to grow a thermochromic VO_2 thin film for smart window applications. In their study, vanadium chloride (VCl_4, 99%) was used as a source of vanadium (vanadium precursor), ethyl acetate (EtAc, $C_4H_8O_2$, 99.8%) was used as a source of oxygen (oxygen precursor), and float glass coated with a 50 nm SiO_2 barrier layer was used as the substrate. Researchers in the past have used water as an oxygen precursor in growing the VO_2 films from VCl_4 but the films thus produced are inhomogeneous and/or porous because of "unclean" reaction of VCl_4 with water. Because EtAc reacts smoothly with VCl_4 and has a desired high vapor pressure, it was found to be an excellent alternative to water, resulting in homogeneous and smooth thin films. At a constant temperature of 550 °C, the optimum molar flow ratio of VCl_4/EtAc was determined to be 2:1. The single layer of monoclinic VO_2 thin films synthesized in this study showed low solar transmittance of 7.8–12%, as required for better thermochromic performance [117].

4.7 Simulation of Spectral Emissivity of Vanadium Oxides (VO_x) Based Microbolometer Structures

4.7.1 Introduction

Materials at nonzero temperature emit radiation and the radiated energy varies as the fourth power of the absolute temperature (Stefan–Boltzmann law). At room temperature, the emission is mostly in the infrared (IR) region which ranges from 0.75 to 1000 μm [266]. The other major source of thermal radiation is Cosmic Microwave Background (CMB) radiation, a remnant of the hot Big Bang theory for the origin of the universe (Nobel prize in 1978, Penzias and Wilson of Bell Labs). The microwave ranges in the region between 103 and 106 μm. Detection of dangers in advance, superior situation awareness and the use of precise weapon on time are the main requirements for both military (defense) and nonmilitary (civilian) security applications. Infrared detectors, sensitive in both short- and long-wavelength infrared region, fulfill most of these requirements [117].

The atmosphere allows infrared (IR) transmission in the following region [266]: 0.79–1.7 μm (near infrared, near IR), 2–6 μm (mid-wave infrared, MWIR), and 8–14 μm (Long Wave Infrared, LWIR). As can be seen from Fig. 4.47, the transmission is affected primarily due to the IR scattering and absorption by various atmospheric gases such as H_2O, CO_2, O_3, and O_2. According to Wien's displacement law, a sensor such as the human eye can see an object in the presence of radiation within certain range of electromagnetic (EM) spectrum and a temperature source around it, with certain minimum temperature (for example: industrial fluorescent light bulbs for eye) as shown in Table 4.13. Usually 8–14 μm window is a choice for high-performance thermal imagers because of the following reasons: (a) smoke and mist particles are small as compared to these wavelengths and IR can easily transmit through them; (b) it is highly sensitive to objects at ambient temperature; and (c) scattering by gas molecules is very small in the region above 2 μm [267]. However, at high altitude in the atmosphere, where the temperature is far below −35 °C, the IR detection is still a challenge to the current technology.

In order to detect radiation, a photodetector, a device which converts the absorbed photons into a measurable form is used. There are mainly two types of detectors that can convert the absorbed photons into a measurable form: photon detectors and thermal detectors. A photon detector is an optoelectronic device which gives rise to an electrical output signal when the energy distribution of electrons changes as a result

Table 4.13 Minimum temperature of sources required to "See" an object in different spectral regions [268]

	Eye	Near IR	SWIR	LWIR
λ (μm)	0.38–0.72	0.79–1.7	2–6	8–14
Temp (°C)	525	275	−25	−35

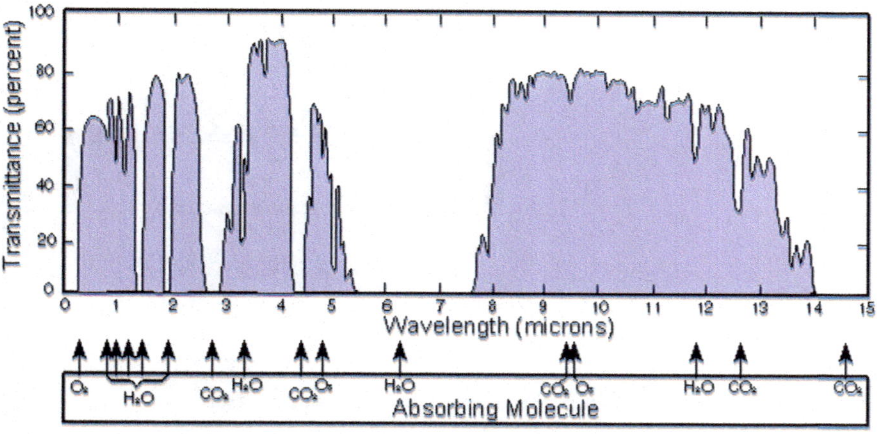

Fig. 4.47 IR transmission in earth's atmosphere [266]

of the interaction of radiation with either free or bound charge carriers in a material. Interaction can be either internal or external. In internal interaction, photons either interact with charge carriers (bound or free) or produce a localized excitation of an electron to a higher energy state [269]. However, in external interaction, electrons are emitted as a result of Einstein's photoelectric effect. On the other hand, thermal detectors absorb the photon energy and convert it into heat which, in turn, affects the physical or electrical parameters such as electrical conductivity, thermoelectric voltage, and pyroelectric voltage. Hence, thermal detectors do not depend on the nature of the photon or spectral content of the radiation but depend on radiant power; the spectral response of a sensing material is determined by the emissivity of the surface. Since heating and cooling are slower processes compared to the interaction between photons and electrons, thermal response is relatively slower than spectral response. Typically, thermal effects occur in millisecond time scale while the effects due to photons are observed on micro- or nano-second time scale.

Photon detectors are used in IR detector technology due to two major performance parameters: excellent signal-to-noise ratio and fast response time. Since the energy of incident photons is comparable to the average thermal energies ($K_B T$) of atoms of the sensing element [270], the noise due to thermal charge carriers is inevitable, and hence these photon detectors require cryogenic cooling to 77 K or below [1]. Cooling mechanism, included in the photodetectors, makes the device not only heavy, bulky, and inconvenient but also expensive. Furthermore, photon detectors lack in broad band response, i.e., they exhibit selective wavelength-dependent response to incident radiation. The difficulty in operating photon detectors with appropriate spectral response in the IR region [271] is their other drawback. Thermal detectors such as thermocouples, bolometers, thermopiles, and pyroelectric detectors are interesting because they are rugged, reliable, light, and inexpensive and they can be operated at room temperature. Most of the thermal detectors are passive devices since they do

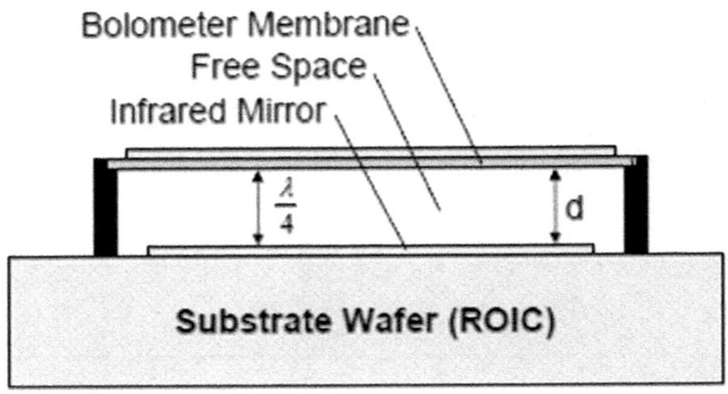

Fig. 4.48 Schematic of a microbolometer pixel structure [266]

not require bias and, most importantly, they provide flatter spectral response. In this section, a discussion of one of the thermal IR detectors, the bolometer, is presented.

The bolometer consists of a sensing element having a strong temperature coefficient of resistance [TCR, $\alpha = \frac{1}{R}\frac{dR}{dT}$] so that a small temperature change, caused by the incident radiation, can be measured. Commonly used bolometer sensing elements are: VO_x, SiO_2, and amorphous silicon (α − Si) [266]. Thermal bolometric detectors can be fabricated on thermally isolated hanging membranes by utilizing micro-electro-mechanical systems (MEMS) technology [272]. A "monolithic" structure, patented and fabricated at Honeywell, by silicon micromachining is termed as a microbolometer which consists of a two-level structure with a gap (d) of ~2.5 μm between them as shown in Fig. 4.48 (Texas Instruments developed the pyroelectric detector arrays). The upper layer is a square-shaped silicon nitride (Si_3N_4) plate, of side 50 μm and thickness 0.5 μm, suspended over an underlying silicon readout integrated circuit (ROIC) substrate. Encapsulated in the center [267] of Si_3N_4 bridge is 500 Å of polycrystalline VO_x—a popular thermistor. A reflective layer of Al coated on top of Si wafer increases the absorption effectively at 10 μm wavelength due to the formation of a quarter-wave resonant cavity.

Due to the development of MEMS technology, uncooled infrared bolometers have now maintained the performance levels of cooled infrared photon detectors [52]. The performance of a thermal detector can be divided into two steps: raising the temperature of a sensing material by input radiation and using the temperature-dependent variation of a particular property of the material as its response. The second step, involving the use of material property, depends on the type of thermal detector and, for a bolometer, TCR is utilized. TCR is related to the voltage responsivity, a widely used parameter to specify the performance of a bolometer, as

$$R_v = \frac{\alpha I R_0 \Delta T}{P_i} = \frac{\alpha\ I\ R_0 \epsilon}{G\left(1+\omega^2\tau^2\right)^{1/2}} \quad (4.7.1)$$

where P_i is the incident power, G is thermal conductance, ω is frequency of sinusoidal excitation, τ is thermal time constant, ΔT is the rise in temperature, ϵ is emissivity, R_0 is resistance at ambient temperature, and I is bias current. The complete characterization of a microbolometer requires the understanding of both electrical and optical properties. The focus of the present study is the simulation of room temperature spectral emissivity of a single pixel industry standard VO_x-based microbolometer.

4.7.2 Results and Discussion

The Multi-Rad, a copyrighted software, was developed at Massachusetts Institute of Technology (MIT) to study the radiative properties of silicon-related layered materials [273–275]. It implements thin film optics in the form of the matrix method of multilayers [273] and assumes the layers to be optically smooth, parallel to each other, optically isotropic (no variation in azimuthal direction), and the dimension in question is much larger than the wavelength of the incident radiation (no edge effects).

A generic layered structure is shown in Fig. 4.49 [276]. The analyses in this section follows the earlier approach [276]. There are N-layer interfaces (circled) and $N+1$ "layers" (squared), including the unbounded transparent media on each side of the actual stack. The terms A_i and B_i are the amplitudes of the forward and backward propagating electric field vectors on the left side of the interface, i. The prime notation on A'_{N+1} and B'_{N+1} indicates that these are the amplitudes on the right side of interface N. Light is incident on interface 1, with an angle of incidence $\theta = \theta_1$.

The central equation of the multilayer theory relates the amplitudes on the left side of interface 1 with those on the right side of interface N

$$\begin{pmatrix} A_1 \\ B_1 \end{pmatrix} = \left[\prod_{i=1}^{N} P_i D_i^{-1} D_{i+1}\right]\begin{bmatrix} A'_{N+1} \\ B'_{N+1} \end{bmatrix} = \begin{bmatrix} m_{11} & m_{12} \\ m_{21} & m_{22} \end{bmatrix}\begin{bmatrix} A'_{N+1} \\ B'_{N+1} \end{bmatrix} \quad (4.7.2)$$

where P_i is the propagation matrix, D_i is the dynamical matrix, and m_{ij} is an element of the transfer function matrix. The propagation matrix accounts for the effect of absorption and interference within a layer i bounded by two interfaces.

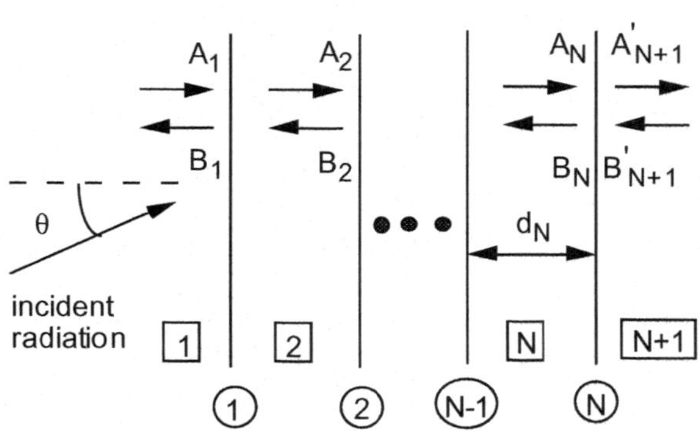

Fig. 4.49 Notation for matrix method of multilayers

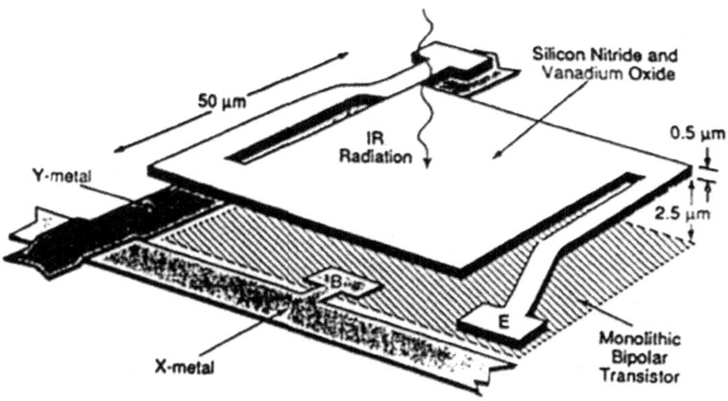

Fig. 4.50 A microbolometer pixel structure [278]

Reflectance, transmittance, and emittance of a multilayer stack can be studied in the spectral range of 0.4–20 μm at different thickness and angle of incidence. Radiation at a given wavelength is treated as coherent; so interference effects are taken into account [277]. The details of the modeling and the approach to the simulation have been described in earlier studies [276].

Emissivity, $\varepsilon(\lambda, \theta, \phi, T)$, is the ratio of energy radiated from a material surface to that radiated from a blackbody at the same temperature (T), wavelength (λ), and viewing condition (θ, ϕ). It also depends on the property of a material and its surface roughness. According to Kirchhoff's law, emissivity is equal to absorptivity for an object in thermodynamic equilibrium which can be clearly conceptualized from the notion that an object absorbing the entire incident light will emit more radiation. Radiation penetrates certain thickness of the specimen before being absorbed, and hence opacity is not only a material property but it depends on thickness as well. Emissivity also depends on its thickness, for instance, thinner sample is characterized by lower emissivity. Emissivity is usually measured (experimentally) at a direction normal to the surface. In this study, the point in question is singular with respect to ϕ, and hence $\theta = 0$ and $0 \leq \phi \leq 2\pi$. The present study analyzes the spectral emissivity of the VO_x-based microbolometer structure [Fig. 4.50] under conditions of normal incidence using Multi-rad. The microbolometer structure, considered in this study, was originally developed by Honeywell, Inc. [278].

In the Honeywell microbolometer structure, the bolometer sensing element has been chosen to be VO_x, a standard sensing material with x equal to 1.8, and is used in large-scale production [279]. The thin films (500 Å) of mixed oxides of vanadium (VO_2, V_2O_3, and V_2O_5) result in a family of materials having desired properties for bolometric operation such as high TCR and well-defined electrical properties with good fabrication capability [54]. As mixed oxides of vanadium, the VO_x can be realized as

$$xVO_2 + yV_2O_3 + zV_2O_5 = VO_{1.8} \tag{4.7.3}$$

Equation (4.7.3) yields the two relations

$$x + 2y + 2z = 1$$
$$2x + 3y + 5z = 1.8$$

which after solving yields the following relation:

$$y(V_2O_3) + (1.4 - 4 \times y)(VO_2) + (y - 0.2)(V_2O_5) = VO_{1.8} \quad (4.7.4)$$

This leads to a set of inequalities for nonzero amount of all three oxides as $y > 0$; $(1.4 - 4 \times y) > 0$; $y - 0.2 > 0$, i.e., $0.2 < y < 0.35$. Since $x = 1.8$ in VO_x is nearly equal to 2, we assume that other oxides (V_2O_3 and V_2O_5) are formed during VO_2 deposition process, due to their smaller heat of enthalpy (enthalpy of formation, ΔH, for V_2O_5, V_2O_3, and VO_2 are −1557, −1219, and −713 cal/mole, respectively [280]). By choosing $y = 0.2025$, a value close to 0.2, relation (4.7.4) can be written as

$$0.2025(V_2O_3) + 0.59 \times (VO_2) + 0.0025(V_2O_5) = VO_{1.8} \quad (4.7.5)$$

On the other hand, the value of y can be chosen in such a way that the maximum possible amount of V_2O_5 is obtained in the mixture, consistent with the lowest enthalpy of all the three vanadium oxides. By choosing the value of $y = 0.3475$, the following relation is obtained:

$$0.3475(V_2O_3) + 0.01 \times (VO_2) + 0.1475(V_2O_5) = VO_{1.8} \quad (4.7.6)$$

However, by considering the heat of formation, ΔH, and assuming that different oxides are formed with a probability which varies linearly with ΔH, the following relation is obtained:

$$y : (1.4 - 4 \times y) : (y - 0.2) = 1219 : 713 : 1557 \quad (4.7.7)$$

The nonnegative value of y, obtained by solving Eq. (4.7.7), is 0.3 and leads to the following relation:

$$0.3(V_2O_3) + 0.2 \times (VO_2) + 0.1(V_2O_5) = VO_{1.8} \quad (4.7.8)$$

Accordingly, 0.05 μm thick VO_x layer in 50 μm × 50 μm pixel, in Fig. 4.50, can be realized in the following three combinations:

(a) $t(VO_2) = 4.7589 \times 10^{-4}$ μm, $t(V_2O_3) = 0.0281$ μm, $t(V_2O_5) = 0.0214$ μm [based on relation (4.7.7)]—consistent with lowest ΔH of V_2O_5;
(b) $t(VO_2) = 0.0313$ μm, $t(V_2O_3) = 0.0183$ μm, $t(V_2O_5) = 4.0551 \times 10^{-4}$ μm [based on relation (4.7.5)]—with the assumption that the two oxides, V_2O_3 and V_2O_5, are formed during the deposition process of VO_2;

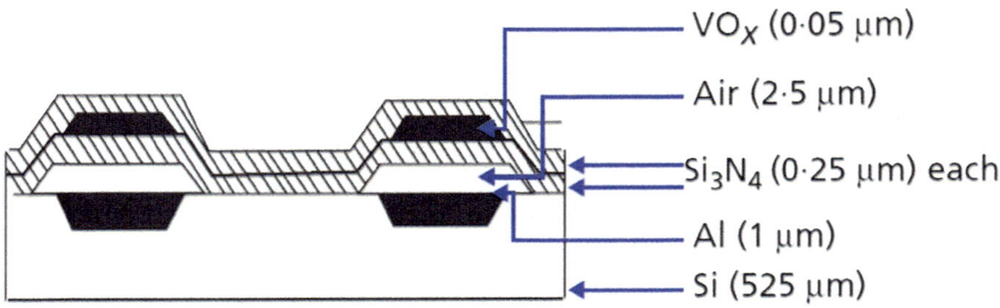

Fig. 4.51 Simulated structure of a microbolometer pixel [278, 281]

(c) $t(VO_2) = 0.0099$ μm, $t(V_2O_3) = 0.0251$ μm, $t(V_2O_5) = 0.0150$ μm [based on relation (4.7.8)]—with the assumption that formation probability of the three oxides is linear with enthalpy.

For clarification and simplicity, we will label them as combinations (a), (b), and (c), respectively, in the following discussion. The use of thickness (and composition) to four decimal place is due to the capability of Multi-rad.

Materials, in Multi-rad, are defined in terms of their real and imaginary parts of their refractive indices; the air gap is defined as $n = 1$ and $k = 0$ throughout; the wavelength-dependent optical constants, n and k, of Al, VO_2, V_2O_3 and V_2O_5 have been taken from the literature [222, 282, 283]. Since typical microbolometer pixels are fabricated on an industry standard substrate of 4 inch diameter [284], wafer thickness is taken as 525 μm [285], with 1 μm thick Al layer "deposited" on the substrate. The simulated structure of the microbolometer pixel is shown in Fig. 4.51.

Room temperature emissivity of VO_2/Si, V_2O_3/Si, and Si, simulated using Multi-Rad, and their comparison with experimental data are presented in Fig. 4.52. As can be seen in the figure, the simulated values of emissivity of these structures are in good agreement with the experimental data. It can be further noted that V_2O_3/Si exhibits high emissivity that is consistent with the metallic behavior of V_2O_3 at room temperature. The emissivity spectra of VO_2/Si follows the wavelength-dependent emissivity of Si. Thus, having established the validity of the application of Multi-Rad in simulating the emissivity of vanadium oxides/Si successfully, the study is extended to simulate the emissivity of V_2O_5/Si. Similar to VO_2/Si, V_2O_5/Si also follows the spectral emissivity of Si. It should be noted that VO_2 and V_2O_5 are insulators at room temperature.

Figure 4.53 shows the calculated emissivity of each of the VO_x constituents, at 30 °C, in the wavelength range of 2.4–20 μm, for different thicknesses of the three combinations defined by relations (4.7.6), (4.7.5), and (4.7.8). As can be seen in the figure, the emissivity of each oxide at different thickness, corresponding to the combinations (a) through (c), is calculated and compared with a situation where 50 nm thick VO_x is either VO_2 or V_2O_3 or V_2O_5. The emissivity of V_2O_3 (metallic phase [286]) shows almost flat spectral response throughout the wavelength range considered in all the combinations. However, the emissivity of the insulating phase

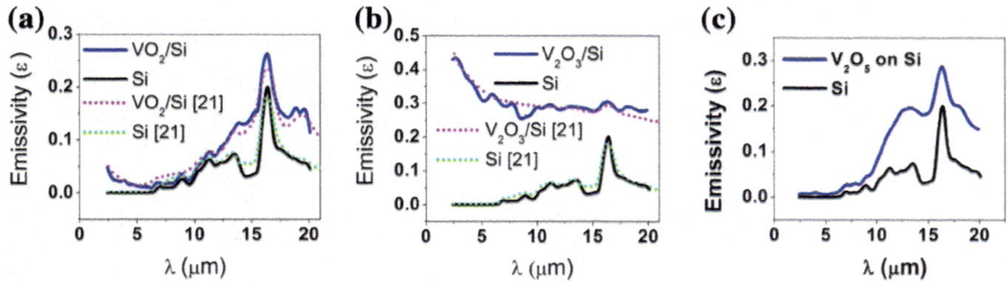

Fig. 4.52 Comparison of simulated emissivity of VO_2/Si, V_2O_3/Si, and Si with experiments [222]; only simulated emissivity is presented for V_2O_5/Si

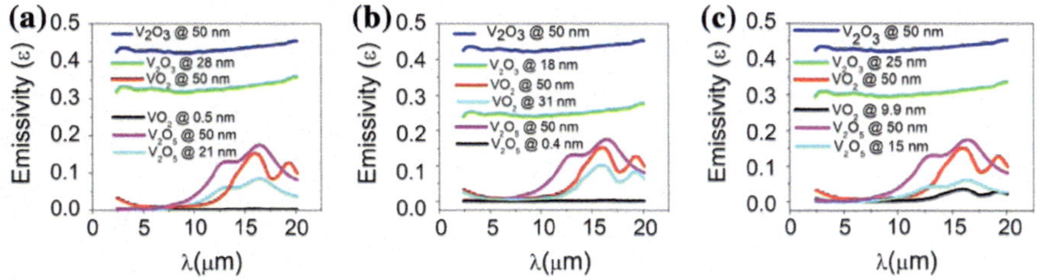

Fig. 4.53 Room temperature (30 °C) emissivity of each of VO_x constituents for different thicknesses of the three combinations defined by relations (4.7.5), (4.7.6), and (4.7.8)

Table 4.14 Emissivity peaks for each V-O system of three different combinations

VO_2	Wavelength (λ)		$\lambda_1 = 15.9$ μm	$\lambda_2 = 19.2$ μm
	Emissivity peak for thickness (t)	$t_1 = 0.0099$ μm	0.035	0.029
		$t_2 = 0.0313$ μm	0.103	0.085
		$t_3 = 0.05$ μm	0.154	0.129
V_2O_5	Wavelength (λ)		$\lambda_1 = 13.2$ μm	$\lambda_2 = 16.5$ μm
	Emissivity peak for thickness (t)	$t_1 = 0.0150$ μm	0.044	0.061
		$t_2 = 0.0214$ μm	0.061	0.085
		$t_3 = 0.05$ μm	0.132	0.177

[286] of both VO_2 and V_2O_5 exhibit wavelength-dependent variation with two major peaks as shown in Table 4.14. Emissivity spectra of VO_2 have peaks at wavelengths 15.9 and 19.2 μm, whereas those of V_2O_5 have peaks at wavelengths 13.2 and 16.5 μm. Without implying generality, wavelength $\lambda = 16.5$ μm corresponding to the highest value of emissivity, listed in Table 4.14, will be considered as a reference value for comparison as seen in the following discussion.

It can also be seen from Fig. 4.53 that the qualitative features of emissivity of the three oxides, at two different thicknesses, are generally similar for all the combinations, the deviation being $\leq$7.03% (excluding films of thickness less than 1 nm for

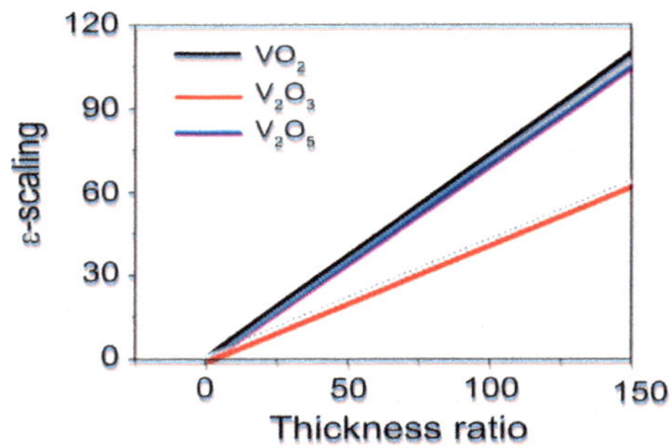

Fig. 4.54 Variation of ε-scaling with thickness ratio for each vanadium oxide layer

Table 4.15 Thickness dependence of emissivity for each V-O system

		VO_2	V_2O_3	V_2O_5
Combination (*a*)	ε-scaling	73.5000	1.3009	2.0824
	Thickness ratio	100.0000	1.7794	2.3364
Combination (*b*)	ε-scaling	1.5000	1.7027	88.5000
	Thickness ratio	1.5974	2.7322	125.0000
Combination (*c*)	ε-scaling	4.4545	1.3912	2.9016
	Thickness ratio	5.0505	1.9920	3.3333
ε at $\lambda = 16.5$ μm for 0.05 μm thick film		0.1470	0.4410	0.1770

which the emissivity is close to zero). Since the industry uses 0.05 μm thick film of VO_x (Fig. 4.51), all the calculated emissivity of the vanadium oxides have been normalized with respect to the emissivity of 0.05 μm thick film of VO_x. Emissivity is a volume effect. In order to correlate the film thickness with emissivity, the parameters: "thickness ratio", defined as the ratio of 0.05 μm to the thickness of a single oxide film and "ε-scaling", defined as the ratio of emissivity at different thicknesses to emissivity at thickness of 0.05 μm, at $\lambda = 16.5$ μm, are compared. Variation of ε-scaling with thickness ratio for each V-O system is shown in Fig. 4.54 and summarized in Table 4.15. It can be seen from this table and the figure that ε-scaling varies linearly with thickness ratio as ε-scaling = (0.73 × thickness ratio) + 0.55 (for VO_2); ε-scaling = (0.42 × thickness ratio) + 0.55 (for V_2O_3), and ε-scaling = (0.70 × thickness ratio) + 0.50 (for V_2O_5). This indicates that the slope of this variation is less than one for each of the V-O systems. Furthermore, due to the metallic phase of V_2O_3 at room temperature [286], the slope of the ε-scaling versus thickness ratio for V_2O_3 differs from that of the insulating phase of the other oxides at room temperature. The emissivity of 0.05 μm thick film at 16.5 μm varies as $V_2O_3 > V_2O_5 > VO_2$ such that the emissivity for V_2O_3 is more than twice that for VO_2 or V_2O_5.

Figure 4.55 shows the emissivity of VO_x in the wavelength range of 2.4–20 μm, with all six possible layer stacking of its constituents of the three combinations defined by relations (4.7.5), (4.7.6), and (4.7.8). The emissivity change within the entire wavelength range is <0.085. As can be seen in Fig. 4.55, a closer examination

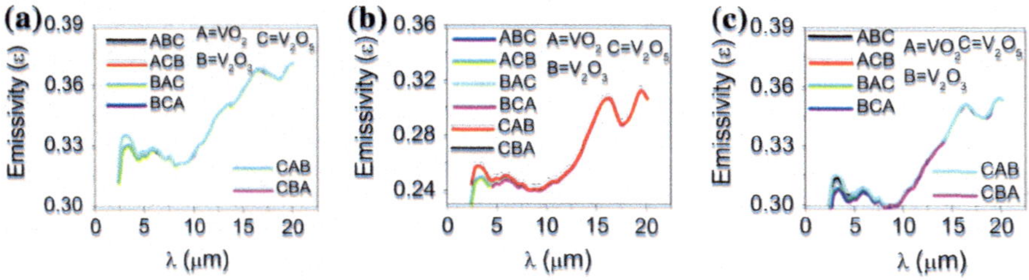

Fig. 4.55 Emissivity of VO_x with all six possible layer stacking of its constituents of the three combinations defined by relations (4.7.5), (4.7.6), and (4.7.8)

within the framework of specified "variation" shows that the wavelength-dependent emissivity of all possible layer stacking completely overlaps, especially in the range of 8–14 μm, for all combinations (a) through (c). In other words, stacking pattern of layers of the three different vanadium oxides has no effect on the emissivity in the region of interest (LWIR). This is an indication that the combined structure of the thin films of vanadium oxides behave as a mixed system instead of thin isolated layered system. In the present simulation, however, the thin film layer of V_2O_3 is stacked on the VO_2 layer with V_2O_5 layer on top of V_2O_3.

Figure 4.56 shows the evolution in emissivity of the microbolometer pixel structure [278] in the wavelength range of 8–14 μm, as the individual component layers are introduced on top of silicon substrate in the specific pattern of the three combinations defined by relations (4.7.5), (4.7.6), and (4.7.8). It can be seen that the wavelength-dependent emissivity for the aluminum–silicon system is minimal. In other words, due to the highly reflecting nature of Al coating on silicon substrate, the emissivity is practically zero over the entire spectral range. Furthermore, the addition of Si_3N_4 overlayer on air/Al/Si has no contribution to the emissivity. The two configurations, without and with Si_3N_4 on top of VO_x/Si_3N_4/Air/Al/Si, hereafter labeled as configuration 1 (Config. 1) and configuration 2 (Config. 2), respectively, show high emissivity throughout the LWIR range; the wavelength-dependent variation of emissivity, $(\varepsilon - \lambda)$, in the shorter wavelength of the LWIR range seems to be linear but tends to saturate in the longer wavelengths. It is also apparent that the role of Si_3N_4 overlayer on top of configuration 1, combination (b), is to linearize the spectral emissivity. The arithmetic and weighted averages of emissivity for the two configurations, in all the three combinations (a) through (c), as defined below, have been presented in Table 4.16.

$$\text{Arithmetic mean} = \frac{\sum \varepsilon_i}{\sum i} \text{ and Weighted arithmetic mean} = \frac{\sum \varepsilon_i \times \lambda_i}{\sum \lambda_i}$$

The configuration 2 ($Si_3N_4/VO_x/Si_3N_4$/Air/Al/Si) is the structure that is being utilized in the current uncooled VO_x-based microbolometer technology, whose average emissivity is reported to be 0.8 [287] in the wavelength range of 8–14 μ.

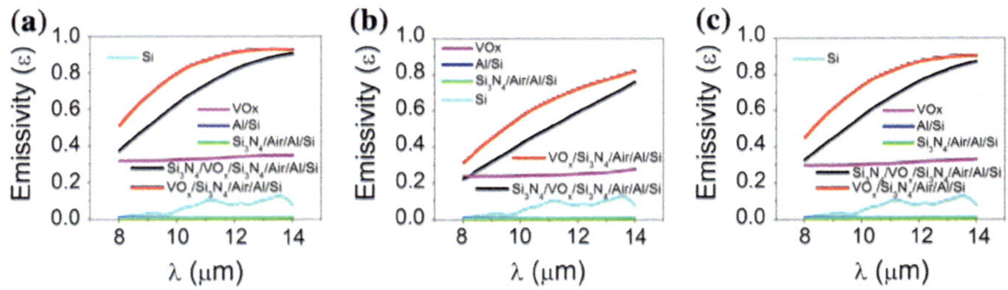

Fig. 4.56 Evolution in emissivity of a microbolometer pixel structure [278] with the addition of individual constituent layers on top of silicon substrate in the specific order of the three combinations defined by relations (4.7.5), (4.7.6), and (4.7.8)

Table 4.16 Emissivity comparison for the two configurations in all three combinations (*a*) through (*c*)

Emissivity	Combination (*a*)		Combination (*b*)		Combination (*c*)	
	Config. 1	Config. 2	Config. 1	Config. 2	Config. 1	Config. 2
Arithmetic average	0.8197	0.6979	0.6249	0.4973	0.7707	0.6431
Weighted average	0.8393	0.7252	0.6496	0.5240	0.7925	0.6711

Calculations in Table 4.16 show that this Config. 2, with Combination (*a*), has the weighted average value of emissivity of 0.73. This represents about 8.7% deviation from the reported average emissivity of 0.8 [287] for the "VO_x-based industry microbolometer". It is interesting to note that the calculated emissivity for Config. 1 is consistently higher than that of Config. 2 and is irrespective of the combination (*a*), (*b*), or (*c*). The overall influence of the top Si_3N_4 layer seems to reduce the spectral emissivity of the bolometer structure while at the same time providing the much desired linear performance of the bolometer (Fig. 4.56b).

4.7.3 Summary

The analysis of an industry standard VO_x-based microbolometer [278], with x equal to 1.8 [279], has been performed in the above study. This analysis is based on modeling room temperature spectral emissivity using Multi-Rad, a simulation package that utilizes thin film optics in the form of matrix method of multilayers [273] and assumes the layers to be optically smooth and parallel to each other. The study of spectral emissivity of individual component of bare VO_x, i.e., VO_2, V_2O_3, and V_2O_5, with different thickness, shows that the wavelength-dependent variation of emissivity, $(\varepsilon - \lambda)$, scales with thickness almost linearly. Unlike the insulating phase (VO_2 and V_2O_5), the metallic phase (V_2O_3) of same thickness shows appreciable emissivity throughout the spectral range and its emissivity is less sensitive to thickness variation. The $(\varepsilon - \lambda)$ of bare VO_x, in the wavelength range of 8–14 μm, with all six possible layer stacking of its constituents for all three combinations, shows that the stacked layers of the thin films of VO_2, V_2O_3, and V_2O_5, of total thickness of 500 Å, behave as a mixed system.

The evolution in the $(\varepsilon - \lambda)$ of the microbolometer pixel structure [278] has been studied in detail by considering the constituent layers, one layer at each step, on top of the silicon substrate in the specific stacking order of the three specific combinations of VO_2, V_2O_3, and V_2O_5 layers. Si_3N_4 overlayer does not change the spectral emissivity of Al/Si while it decreases the spectral emissivity of the VO_x/Si_3N_4/Air/Al/Si system. Calculations show that the Si_3N_4 layer provides the much desired linear performance of the VO_x-based bolometer.

4.8 Applications of Vanadium Oxides

Vanadium oxides are used in technology in which devices make use of their properties such as Insulator–Metal Transition (IMT), high-temperature coefficient of resistance (TCR) and a small $1/f$ noise constant. Rajendra Kumar et al. [53] used Pulsed Laser Deposition technique to deposit vanadium oxide thin films at room temperature using neodymium-doped yttrium aluminum garnet (Nd:YAG) laser with 532 nm. A laser fluence of 1.4 J/cm^2 was determined to be the optimum one for using vanadium oxide films for microbolometer applications. Bolometers are thermal infrared (IR) detectors and can be used in infrared imaging applications such as thermal camera, night vision camera, surveillance, mine detection, early fire detection, medical imaging, and detection of gas leakage. Furthermore, the vanadium–oxygen system (VO_x) has relatively low processing temperature, and hence it is also integrated into microelectromechanical system (MEMS) fabrication [288, 289].

Vanadium oxides are applicable in technology such as memory devices and temperature sensors [49]. The memory aspect of the material is evidenced from the pronounced hysteresis present in the phase transition [50]. Normally, the range of operation of a device lies outside the hysteresis region. Some bolometric devices are operational within the hysteretic transition [51].

Transition Metal Oxides (TMOs), among others, are the best candidates for the cathode materials for rechargeable Li-ion batteries. Due to the layered structure of V_2O_5, it has good ionic storage capacity. The cathode material currently being used, coarse-grained Li_xCoO_2, has a practical energy density <500 Wh/kg which is far less than a theoretical energy density of ~1100 Wh/kg for lithiated V_2O_5 [55]. Panasonic has commercialized the use of V_2O_5 as a cathode material in rechargeable Li-ion batteries for low-energy applications (button cells).

TMOs, being stable at high temperature along with their tunable transport properties and well-established synthesis techniques, are gaining attention for their thermoelectric properties [57]. V_2O_5 is known to have hopping transport mechanism, with anisotropic n-type electrical conduction. The Seebeck coefficient varies slowly with temperature indicating the applicability of V_2O_5 as a thermistor [58].

The vanadium oxides are chromogenic materials and can change their optical properties due to some external stimuli in the form of photon radiation (photochromic), change in temperature (thermochromic), and voltage pulse (electrochromic); the change becomes discontinuous during IMT. Such properties can be exploited to make coatings for energy-efficient "smart windows" [43], and electrical and optical switching devices [44]. Thin films of VO_2 and V_2O_3 have been found to show good thermochromism in the infrared region [45, 46]. While maintaining the transparency to visible light, a smart window modulates infrared irradiation from a low-temperature transparent state to a high-temperature opaque state [47]. The two oxides, VO_2 and V_2O_5, can change their optical properties in a persistent and reversible way in response to a voltage [48]. V_2O_5 exhibits exceptional electrochromic behavior because it has both anodic and cathodic electrochromism, different from VO_2 which has only anodic electrochromism, and is also an integral part in band structure effects [48]. These electrochromic materials have four main applications: information displays, variable reflectance mirrors, smart windows, and variable emittance surfaces.

Malarde et al. [265] used an optimized Atmospheric-Pressure Chemical Vapor Deposition (APCVD) method to grow a thermochromic VO_2 thin film for smart window applications. In their study, vanadium chloride (VCl_4, 99%) was used as a source of vanadium (vanadium precursor), ethyl acetate (EtAc, $C_4H_8O_2$, 99.8%) was used as a source of oxygen (oxygen precursor), and float glass coated with a 50 nm SiO_2 barrier layer was used as the substrate. Researchers in the past have used water as an oxygen precursor in growing the VO_2 films from VCl_4 but the films thus produced are inhomogeneous and/or porous because of "unclean" reaction of VCl_4 with water. Because EtAc reacts smoothly with VCl_4 and has a desired high vapor pressure, it was found to be an excellent alternative to water, resulting in homogeneous and smooth thin films. At a constant temperature of 550 °C, the optimum molar flow ratio of VCl_4/EtAc was determined to be 2:1. The single layer of monoclinic VO_2 thin films synthesized in this study showed low solar transmittance of 7.8–12%, as required for better thermochromic performance [117].

Since TMOs can show charge transfer pseudocapacitance, they can be used as electrode materials in supercapacitors, which has potential applications in hybrid electric vehicles, memory backup systems, and electronic devices [290]. Huang et al. [291] synthesized hollow flowers-like vanadium oxide (V_6O_{13}) of good electrochemical performance which can be a possible electrode material for supercapacitors. Jung and Um [292] studied vanadium oxide thin films on metallic bipolar plates which can be applicable as self-heating sources for enhancing cold-start operation of fuel cells at subzero ambient temperatures. Huotari et al. [257, 259] deposited vanadium oxide thin films and studied gas sensing properties. They discovered that the coexisting phases of vanadium oxide, V_2O_5 and vanadium oxide nanotubes, are good candidates for gas sensors. Biette et al. [293] prepared macroscopic fibers of vanadium oxides and found them to be effective in sensing alcohol vapors.

Wu et al. [294] synthesized a low-cost vanadium pentoxide (V_2O_5) by a simple method and introduced it into Dye-Sensitized Solar Cells (DSCs) as counter electrode (CE). They found the material to be efficient for photovoltaic performance. Chen et al. [295] studied vanadium oxide–carbon nanotube composite and found the composite to be a potential catalyst in partial oxidation of *n*-butane.

Clearly, phase transition in VO_2 is of high technological interest. IMT occurs near room temperature and T_c can be tuned optically, thermally, electrically [59], and with doping [38]. The phase transition in VO_2 has been used to achieve frequency-tunable metamaterials in the near-infrared range [60, 61]. Kyoung et al. [62] extended the study to terahertz range by proposing an active terahertz metamaterial, a gold nano-slot antenna on a VO_2 thin film, which transforms itself from transparent to complete extinct at resonance when the VO_2 film undergoes thermo or photoinduced phase transition. Cavalleri et al. [34] showed that the phase transition can be photoinduced within hundreds of femtoseconds which can be an underlying principle for an ultrafast switch.

References

1. Darling RB, Iwanaga S (2009) Structure, properties, and MEMS and microelectronic applications of vanadium oxides. Sadhana 34:531–542
2. Adler D (1968) Mechanisms for metal-nonmetal transitions in transition-metal oxides and sulfides. Rev Mod Phys 40:714–736
3. Parker JC et al (1990) Optical properties of the vanadium oxides VO_2 and V_2O_5. J Am Ceram Soc 73(11):3206–3208
4. [cond-mat.str-el], a.c.-m
5. Henrich VE, Cox PA (1996) The surface science of metal oxides. Cambridge University Press, Cambridge
6. Imada M, Fujimori A, Tokura Y (1998) Metal-insulator transitions. Rev Mod Phys 70:1039–1263
7. Woodward PM et al (2006) Metal oxides: chemistry and applications. CRC Press, Boca Raton, Florida
8. Available from: http://www.cpfs.mpg.de/~kasinath/
9. Wu C, Feng F, Xie Y (2013) Design of vanadium oxide structures with controllable electrical properties for energy applications. Chem Soc Rev 42(12):5157–5183
10. Lalena JN, Cleary DA (2005) Principles of inorganic materials design. Wiley, Hoboken, New Jersey
11. de Boer JH, Verwey EJW (1937) Semi-conductors with partially and with completely filled 3d-lattice bands. Proc Phys Soc 49:59–71
12. Adler D, Brooks H (1967) Theory of semiconductor-To-metal transitions. Phys Rev 155:826–840
13. Emin D (1982) Small polarons. Phys Today 35:34–40
14. Hurd CM (1985) Quantum tunnelling and the temperature dependent DC conduction in low-conductivity semiconductors. J Phys C: Solid State Phys 18:6487–6499
15. Bruce PG (ed) (1997) Solid state electrochemistry. Cambridge University Press, United Kingdom
16. Tredgold RH (1962) On very low mobility carriers. Proc Phys Soc 80:807–810
17. Mott NF, Peierls R (1937) Discussion of the paper by de Boer and Verwey. Proc Phys Soc 49:72–73

18. Mott NF (1949) The basis of the electron theory of metals, with special reference to the transition metals. Proc Phys Soc. Section A 62:416–422
19. Mott NF (1961) The transition to the metallic state. Phil Mag 6:287–309
20. Hubbard J (1964) Electron correlations in narrow energy bands. III. An improved solution. Proc R Soc Lond Ser A 281:401–419
21. Gebhard F (1997) The Mott metal-insulator transition: models and methods, Issue 137. Springer, New York
22. Wentzcovitch RM, Schulz WW, Allen PB (1994) VO_2: Peierls or Mott-Hubbard? A view from band theory. Phys Rev Lett 72:3389–3392
23. Biermann S et al (2005) Dynamical singlets and correlation-assisted peierls transition in VO_2. Phys Rev Lett 94:026404
24. Peierls RE (1955) Quantum theory of solid. Oxford University, New York/London
25. Gruner G (1988) The dynamics of charge-density waves. Rev Mod Phys 60:1129–1181
26. Kuper CG (1955) On the thermal properties of Fröhlich's one-dimensional superconductor. Proc R Soc Lond. Series A, Math Phys Sci 227:214–228
27. Rice TM, Launois H, Pouget JP (1994) Comment on "VO_2: Peierls or Mott-Hubbard? A view from band theory". Phys Rev Lett 73:3042
28. Kotliar G, Vollhardt D (2004) Strongly correlated materials: insights from dynamical mean-field theory. Phys Today 57:53–59
29. Zylbersztejn A, Mott NF (1975) Metal-insulator transition in vanadium dioxide. Phys Rev B 11:4383–4395
30. Morin F (1959) Oxides which show metal-to-insulator transitions at the neel temperature. Phys Rev Lett 3:34–36
31. Nadkarni GS, Shirodkar VS (1983) Experiment and theory for switching in Al/V_2O_5/Al devices. Thin Solid Films 105:115–129
32. Lamb DM (2009) Semiconductor to metallic phase transitions from vanadium and titanium oxides induced by visible light. Missouri State University, Springfield
33. Batista C, Ribeiro RM, Teixeira V (2011) Synthesis and characterization of VO_2-based thermochromic thin films for energy-efficient windows. Nanoscale Res Lett 6:301
34. Cavalleri A et al (2001) Femtosecond structural dynamics in VO_2 during an ultrafast solid-solid phase transition. Phys Rev Lett 87:237401
35. Wessel C et al (2012) Electronic structure and thermodynamics of V_2O_3 polymorphs. J Comput Chem 33:2102–2107
36. Chain EE (1991) Optical properties of vanadium dioxide and vanadium pentoxide thin films. Appl Opt 30:2782–2787
37. Dernier PD, Marezio M (1970) Crystal structure of the low-temperature antiferromagnetic phase of V_2O_3. Phys Rev B 2:3771–3776
38. Chudnovskiy F, Luryi S, Spivak B (2002) Switching device based on first-order metalinsulator transition induced by external electric field. In: Luryi S, Xu JM, Zaslavsky A (eds) Future trends in microelectronics: the nano millennium, pp 148–155. Wiley Interscience
39. Blum R-P et al (2007) Surface metal-insulator transition on a vanadium pentoxide (001) single crystal. Phys Rev Lett 99:226103
40. Kang M et al (2011) Metal-insulator transition without structural phase transition in V_2O_5 film. Appl Phys Lett 98:131907
41. Liu W-T et al (2011) Intrinsic optical properties of vanadium dioxide near the insulator–metal transition. Nano Lett 11:466–470
42. Pergament AL (2011) Metal-insulator transition temperatures and excitonic phases in vanadium oxides. ISRN Condens Matter Phys Article ID 605913
43. Granqvist CG (1985) Spectrally selective coatings for energy efficiency and solar applications. Phys Scr 32:401–407
44. Simic-Milosevic V et al (2008) Local band gap modulations in non-stoichiometric V_2O_3 films probed by scanning tunneling spectroscopy. Phys Rev B 77:125112
45. Kana JBK et al (2008) Thermochromic VO_2 thin films synthesized by rf-inverted cylindrical magnetron sputtering. Appl Surf Sci 254:3959–3963

46. Thomas MS, DeNatale JF, Hood PJ (1997) High–temperature optical properties of thin–film vanadium oxides–VO_2 and V_2O_3. Mater Res Soc Symp Proc 479:161–166
47. Zhang Z et al (2011) Solution-based fabrication of vanadium dioxide on F: SnO_2 substrates with largely enhanced thermochromism and low-emissivity for energy-saving applications. Energy Environ Sci 4:4290–4297
48. Granqvist CG (1995) Handbook of inorganic electrochromic materials. Elsevier Science, Amsterdam, Holland
49. Gupta A et al (2009) Semiconductor to metal transition characteristics of VO_2 thin films grown epitaxially on Si (001). Appl Phys Lett 95:111915
50. Driscoll T et al (2009) Phase-transition driven memristive system. Appl Phys Lett 95:043503
51. de Almeida LAL et al (2000) Thermal dynamics of VO_2 films within the metal–insulator transition: Evidence for chaos near percolation threshold. Appl Phys Lett 77:4365
52. Niklaus F et al (2008) Performance model for uncooled infrared bolometer arrays and performance predictions of bolometers operating at atmospheric pressure. Infrared Phys Technol 51:168–177
53. Kumar RTR et al (2003) Pulsed laser deposited vanadium oxide thin films for uncooled infrared detectors. Sens Actuators A 107:62–67
54. Cole BE, Higashi RE, Wood RA (1998) Monolithic two-dimensional arrays of micromachined microstructures for infrared applications. Proc IEEE 86:1679–1686
55. *Interfaces, Phenomena and Nanostructures in Lithium Batteries: Proceedings*. 2001, New Jersey: The Electrochemical Society, Inc
56. Kiehne HA (2003) Battery technology handbook. Marcel Dekker Inc., New York
57. Walia S et al (2013) Transition metal oxides—thermoelectric properties. Prog Mater Sci 58(8):1443–1489
58. El Mandouh ZS, Selim MS (2000) Physical properties of vanadium pentoxide sol gel films. Thin Solid Films 371(1):259–263
59. Driscoll T et al (2009) Memory metamaterials. Science 325:1518–1521
60. Driscoll T et al (2008) Dynamic tuning of an infrared hybrid-metamaterial resonance using vanadium dioxide. Appl Phys Lett 93:024101
61. Dicken MJ et al (2009) Frequency tunable near-infrared metamaterials based on VO_2 phase transition. Opt Express 17:18330–18339
62. Kyoung JS et al (2011) Active terahertz metamaterials: nano-slot antennas on VO_2 thin films. Phys Status Solidi C 8:1227–1230
63. Hilton DJ et al (2007) Enhanced photosusceptibility near T_c for the light-induced insulator-to-metal phase transition in vanadium dioxide. Phys Rev Lett 99:226401
64. Baldassarre L et al (2008) Quasiparticle evolution and pseudogap formation in V_2O_3: an infrared spectroscopy study. Phys Rev B 77:113107
65. Perucchi A et al (2009) Optical properties across the insulator to metal transitions in vanadium oxide compounds. J Phys: Condens Matter 21:323202
66. Losurdoa M et al (2001) Spectroscopic ellipsometry investigation of V_2O_5 nanocrystalline thin films. Thin Solid Films 384:58–64
67. Mott NF (1964) Electrons in transition metals. Adv Phys 13:325–422
68. Gunnarsson O (1976) Band model for magnetism of transition metals in the spin-density-functional formalism. J Phys F: Met Phys 6:587–606
69. Aurelian C et al (2010) Exploiting the semiconductor-metal phase transition of VO2 materials: a novel direction towards tuneable devices and systems for RF-microwave applications. In: Mukherjee M (ed) Advanced microwave and millimeter wave technologies semiconductor devices circuits and systems, pp 35–56. InTech
70. Levine IN (ed) (1999) Quantum chemistry, 5th edn. Prentice-Hall, Upper Saddle River, New Jersey
71. Ohno K, Esfarjani K, Kawazoe Y (1999) Computational materials science. Springer, Berlin
72. Szabo A, Ostlund NS (1989) Modern quantum chemistry. Dover Publications, New York
73. Lamsal C, Mishra DR, Ravindra NM (2014) Equilibrium configuration of $(H_2O)_n$, for $n = 1-3$. Nanomaterials Energy 3(4):129–138

74. Foresman JB, Frisch Æ (1999) Exploring chemistry with electronic structure methods. Gaussian, Inc. Pittsburgh, Pennsylvania
75. Thijssen J (2001) Computational physics. Cambridge University Press, Cambridge
76. Thomas LH (1927) The calculation of atomic fields. Math Proc Camb Philos Soc 23(05):542–548
77. Fermi E (1927) Un Metodo Statistico per la Determinazione di alcune Priopriètà dell'Atomo. Rend Accad Naz Lincei 6:602–607
78. Hohenberg P, Kohn W (1964) Inhomogeneous electron gas. Phys Rev 136(3B):B864–B871
79. Sholl D, Steckel JA (2011) Density functional theory: a practical introduction. Wiley, Hoboken
80. Torrent M (2008) International school on vibrational spectroscopies: a hands-on introduction to ABINIT. Queretaro, Mexico, March 30–April 5, 2008
81. Martin RM (2004) Electronic structure: basic theory and practical methods. Cambridge University Press, Cambridge
82. Hamann DR, Schlüter M, Chiang C (1979) Norm-conserving pseudopotentials. Phys Rev Lett 43(20):1494–1497
83. Kohanoff J (2006) Electronic structure calculations for solids and molecules: theory and computational methods. Cambridge University Press, Cambridge
84. Vanderbilt D (1990) Soft self-consistent pseudopotentials in a generalized eigenvalue formalism. Phys Rev B 41(11):7892–7895
85. Blöchl PE (1994) Projector augmented-wave method. Phys Rev B 50(24):17953–17979
86. Kresse G, Joubert D (1999) From ultrasoft pseudopotentials to the projector augmented-wave method. Phys Rev B 59(3):1758–1775
87. Monkhorst HJ, Pack JD (1976) Special points for Brillouin-zone integrations. Phys Rev B 13(12):5188–5192
88. Mermin ND (1965) Thermal properties of the inhomogeneous electron gas. Phys Rev 137(5A):A1441–A1443
89. Gonze X et al (2009) ABINIT: first-principles approach to material and nanosystem properties. Comput Phys Commun 180(12):2582–2615
90. Gonze X (2005) A brief introduction to the ABINIT software package. In: Zeitschrift für Kristallographie, p 558
91. Pulay P (1980) Convergence acceleration of iterative sequences. The case of scf iteration. Chem Phys Lett 73(2):393–398
92. http://www.abinit.org
93. Holzwarth NAW, Tackett AR, Matthews GE (2001) A projector augmented wave (PAW) code for electronic structure calculations, Part I: Atompaw for generating atom-centered functions. Comput Phys Commun 135(3):329–347
94. http://users.wfu.edu/natalie/papers/pwpaw/man.html
95. Stewart G (2012) Electronic band structure of bulk and monolayer V2O5. Department of Physics, Case Western Reserve University
96. Perdew JP, Wang Y (1992) Accurate and simple analytic representation of the electron-gas correlation energy. Phys Rev B 45(23):13244–13249
97. Longo JM et al (1970) A refinement of the structure of VO2. Acta Chem Scand 24:420–426
98. Eyert V (2002) The metal-insulator transitions of VO2: a band theoretical approach. Ann Phys 11(9):650–704
99. Kokalj A (1999) XCrySDen—a new program for displaying crystalline structures and electron densities. J Mol Graph Model 17(3–4):176–179
100. Schwingenschlögl U (2003) The interplay of structural and electronic properties in transition metal oxides. Department of Physics, University of Augsburg, 86135 Augsburg, Germany
101. Enjalbert R, Galy J (1986) A refinement of the structure of V2O5. Acta Crystallogr Sect C 42(11):1467–1469
102. Aroyo MI et al (2011) Crystallography online: Bilbao crystallographic server. Bul Chem Commun 43(2):183–197
103. Aroyo Mois I et al (2006) Bilbao crystallographic server: I. Databases and crystallographic computing programs. Zeitschrift für Kristallographie 221:15

104. Aroyo MI et al (2006) Bilbao crystallographic server. II. Representations of crystallographic point groups and space groups. Acta Crystallogr Sect A 62(2):115–128
105. http://www.cryst.ehu.es/cgi-bin/cryst/programs/nph-wp-list
106. Eyert V, Höck KH (1998) Electronic structure of V_2O_5: role of octahedral deformations. Phys Rev B 57(20):12727–12737
107. Madsen GKH, Singh DJ (2006) BoltzTraP. A code for calculating band-structure dependent quantities. Comput Phys Commun 175(1):67–71
108. Reshak AH, Khan SA, Auluck S (2014) Thermoelectric properties of a single graphene sheet and its derivatives. J Mater Chem C 2(13):2346–2352
109. Hoffmann R (1988) Solids and surfaces: a chemist's view of bonding in extended structures. VCH Publishers
110. Izyumov IUA, Syromyatnikov VN (1990) Phase transitions and crystal symmetry. Springer, Berlin
111. Yu P, Cardona M (2010) Fundamentals of semiconductors: physics and materials properties. Springer, Berlin
112. Zimmermann R et al (1998) Strong hybridization in vanadium oxides: evidence from photoemission and absorption spectroscopy. J Phys: Condens Matter 10:5697–5716
113. Jakub G et al (2009) Relative stability of low-index V_2O_5 surfaces: a density functional investigation. J Phys: Condens Matter 21(9):095008
114. Yu C et al (2012) Thermoelectric transport in V_2O_5 thin films. J Phys: Conf Ser 395(1):012016
115. Parker JC et al (1990) Optical properties of vanadium pentoxide determined from ellipsometry and band-structure calculations. Phys Rev B 42:5289–5293
116. Mokerov VG et al (1976) Optical properties of vanadium pentoxide in the region of photon energies from 2 eV to 14 eV. Opt Spectrosc 40:58–61
117. Lamsal C, Ravindra NM (2014) Vanadium oxides for energy and security applications. In: Spectroscopic techniques for security, forensic and environmental applications. Nova Science Publishers, Hauppauge, NY
118. Yu B et al (2012) Predictions of high-pressure structural, electronic and thermodynamic properties of α-Si_3N_4. Acta Metall Sinica (Engl Lett) 25(2):131–140
119. Qazilbash MM et al (2008) Electrodynamics of the vanadium oxides VO_2 and V_2O_3. Phys Rev B 77:115121
120. Continenza A, Massidda S, Posternak M (1999) Self-energy corrections in VO_2 within a model GW scheme. Phys Rev B 60(23):15699–15704
121. Kozhevnikov AV, Anisimov VI, Korotin MA (2007) Calculation of the electronic structure of the vanadium dioxide VO_2 in the monoclinic low-temperature phase M 1 using the generalized transition state method. Phys Metals Metall 104(3):215–220
122. Mattheiss LF (1994) Band properties of metallic corundum-phase V_2O_3. J Phys: Condens Matter 6(32):6477
123. Ezhov SY et al (1999) Orbital occupation, local spin, and exchange interactions in V_2O_3. Phys Rev Lett 83(20):4136–4139
124. Saha-Dasgupta T et al (2009) Electronic structure of V_2O_3: Wannier orbitals from LDA-NMTO calculations. arXiv:0907.2841
125. Thomas GA et al (1994) Observation of the gap and kinetic energy in a correlated insulator. Phys Rev Lett 73:1529–1532
126. Held K et al (2001) Mott-Hubbard metal-insulator transition in paramagnetic V_2O_3: an LDA+DMFT(QMC) study. Phys Rev Lett 86(23):5345–5348
127. Xu B, Verstraete MJ (2013) First principles explanation of the positive Seebeck coefficient of lithium. arXiv:1311.6805
128. Reshak AH (2014) Fe2MnSixGe1-x: influence thermoelectric properties of varying the germanium content. RSC Adv 4(74):39565–39571
129. Ernst A et al (2005) Weakly dispersive band near the fermi level of GaMnAs due to Mn interstitials. Phys Rev Lett 95(23):237207
130. Sotojima K et al (2007) Thermoelectric properties and phase transition of $(Zn_xCu_{2-x})V_2O_7$. J Jpn Soc Powder Powder Metall 54(5):356–361

131. Berglund CN, Guggenheim HJ (1969) Electronic properties of VO_2 near the semiconductor-metal transition. Phys Rev 185:1022–1033
132. Austin IG, Turner CE (1969) The nature of the metallic state in V_2O_3 and related oxides. Phil Mag 19(161):939–949
133. Santos R et al (2013) Thermoelectric properties of V_2O_5 thin films deposited by thermal evaporation. Appl Surf Sci 282:590–594
134. Chumakov Y et al (2013) Ab initio calculations and measurements of thermoelectric properties of V_2O_5 films. J Electron Mater 42(7):1597–1603
135. Surnev S, Ramsey MG, Netzer FP (2003) Vanadium oxide surface studies. Prog Surf Sci 73(4–8):117–165
136. Scanlon DO et al (2008) An ab initio study of reduction of V_2O_5 through the formation of oxygen vacancies and Li intercalation. J Phys Chem C 112:9903–9911
137. Cardona M (1970) Optical properties and electronic density of states. J Res Natl Bur Stand-A. Phys Chem 74A:253–265
138. Cardona M (1965) Infrared dielectric constant and ultraviolet optical properties of solids with diamond, zinc blende, wurtzite, and rocksalt structure. J Appl Phys 36:2181–2186
139. Penn DR (1962) Wave-number-dependent dielectric function of semiconductors. Phys Rev 128:2093–2097
140. Vechten JAV (1969) Quantum dielectric theory of electronegativity in covalent systems. I. Electronic dielectric constant. Phys Rev 182:891–905
141. Phillips JC (1973) Bonds and bands in semiconductors. Academic, New York
142. Verleur HW, Barker AS, Berglund CN (1968) Optical properties of VO_2 between 0.25 and 5 eV. Phys Rev 172:788–798
143. Zhuze VP, Andreev AA, Shelykh AI (1969) The Hall effect in V_2O_3 single crystals in the metallic conductivity region. Sov Phys Solid State 10:2914–2916
144. Andrianov GO et al (1973) The optical absorption by the frenkel exciton in V_2O_3. Phys Status Solidi B 60:79–84
145. Fujiwara H (2007) Spectroscopic ellipsometry: principles and applications. Wiley, Hoboken
146. Guang-Huaa L, Xiao-Yan D, Chun-Hai W (2011) Electronic and optical properties of monoclinic and rhombohedral vanadium sesquioxide. Chin J Struct Chem 30:638–643
147. Shin S et al (1990) Vacuum-ultraviolet reflectance and photoemission study of the metal-insulator phase transitions in VO_2, V_6O_{13}, and V_2O_3. Phys Rev B 41:4993–5009
148. Ravindra NM, Weeks RA, Kinser DL (1987) Optical properties of GeO_2. Phys Rev B 36:6132–6134
149. Breckenridge RA, Robert J, Shaw W, Sher A (1974) Improved calculations of the complex dielectric constant of semiconductors. Phys Rev B 10:2483–2489
150. Grimes RD, Cowley ER (1975) A model dielectric function for semiconductors. Can J Phys 53:2549–2554
151. Ravindra NM, Narayan J (1986) Optical properties of amorphous silicon and silicon dioxide. J Appl Phys 60:1139
152. Ravindra NM et al (1981) Model based studies of some optical and electronic properties of narrow and wide gap materials. Infrared Phys 21:369–381
153. Kumar A, Ravindra NM (1982) Effective charges and ionicity in alkali halides: an analysis. Phys Rev B 25:2889–2892
154. Jackson JD (1978) Classical electrodynamics. Wiley, New York
155. Ravindra NM, Ganapathy P, Choi J (2007) Energy gap–refractive index relations in semiconductors—an overview. Infrared Phys Technol 50:21–29
156. Goodenough JB (1971) The two components of the crystallographic transition in VO_2. J Solid State Chem 3:490–500
157. Guo Q et al (1999) Ordered binary oxide films of V_2O_3 (0001) on Al_2O_3. J Vac Sci Technol A 17:1887–1892
158. Phillips JC (1968) Dielectric definition of electronegativity. Phys Rev Lett 20:550–553
159. Rao KJ et al (2006) Structural and electrochemical behaviour of sputtered vanadium oxide films: oxygen non-stoichiometry and lithium ion sequestration. Bull Mater Sci 29(5):535–546

160. Malati MA (1999) Experimental inorganic/physical chemistry: an investigative, integrated approach to practical project work. Woodhead Publishing, England
161. Donnay JDH (1963) Crystal data, determinative tables, 2nd edn. American Crystallographic Association, Buffalo, USA
162. Barker AS Jr, Verleur HW, Guggenheim HJ (1966) Infrared optical properties of vanadium dioxide above and below the transition temperature. Phys Rev Lett 17:1286–1289
163. Zhuze VP, Lukirskii DP, Startsev GP (1971) Reflectivity of the metallic phase of V_2O_3. Sov Phys Solid State 13:260–261
164. Neuberger M (1971) Data compilation on vanadium oxides, p 61. Hughes Aircraft Company, Culver City, California
165. Kennedy TN, Hakim R, Mackenzie JD (1967) Preparation and properties of crystalline and amorphous vanadium pentoxide. Mater Res Bull 2:193–201
166. Kenny N, Kannewurf CR, Whitmore DH (1966) Optical absorption coefficients of vanadium pentoxide single crystals. J Phys Chem Solids 27:1237–1246
167. Phillip HR, Ehrenreich H (1962) Optical properties of Ag and Cu. Phys Rev 128:1622–1629
168. Smith DY (1985) Handbook of optical constants of solids ed by Palik ED. Academic Press, Orlando, Florid
169. Rozgonyi GA, Polito WJ (1968) Preparation of thin films of vanadium (Di-, Sesqui-, and Pent-) oxide. J Electrochem Soc 115(1):56–57
170. Becker MF et al (1994) Femtosecond laser excitation of the semiconductor-metal phase transition in VO_2. Appl Phys Lett 65(12):1507–1509
171. Chen S et al (2003) A novel structural VO_2 micro-optical switch. Opt Quant Electron 35(15):1351–1355
172. Soltani M et al (2007) Micro-optical switch device based on semiconductor-to-metallic phase transition characteristics of W-doped VO_2 smart coatings. J Vac Sci Technol A: Vac Surf Films 25(4):971–975
173. Soltani M et al (2004) Effects of Ti–W codoping on the optical and electrical switching of vanadium dioxide thin films grown by a reactive pulsed laser deposition. Appl Phys Lett 85(11):1958–1960
174. Soltani M et al (2004) Optical switching of vanadium dioxide thin films deposited by reactive pulsed laser deposition. J Vac Sci Technol A: Vac Surf Films 22(3):859–864
175. Lopez R et al (2004) Switchable reflectivity on silicon from a composite VO_2–SiO_2 protecting layer. Appl Phys Lett 85(8):1410–1412
176. Chen S et al (2004) Smart VO_2 thin film for protection of sensitive infrared detectors from strong laser radiation. Sens Actuators A 115(1):28–31
177. Bong-Jun K et al (2012) Laser-assisted control of electrical oscillation in VO_2 thin films grown by pulsed laser deposition. Jpn J Appl Phys 51(10R):107302
178. Qian Y et al (2013) Oxygen pressure manipulations on the metal–insulator transition characteristics of highly (0 1 1)-oriented vanadium dioxide films grown by magnetron sputtering. J Phys D Appl Phys 46(5):055310
179. Théry V et al (2017) Structural and electrical properties of large area epitaxial VO_2 films grown by electron beam evaporation. J Appl Phys 121(5):055303
180. Walden RH (1970) Two switching devices utilizing VO<inf>2</inf>. IEEE Trans Electron Devices 17(8):603–612
181. Kivaisi RT, Samiji M (1999) Optical and electrical properties of vanadium dioxide films prepared under optimized RF sputtering conditions. Sol Energy Mater Sol Cells 57(2):141–152
182. Sobhan MA et al (1996) Thermochromism of sputter deposited WxV1–xO2 films. Sol Energy Mater Sol Cells 44(4):451–455
183. Soltani M et al (2006) Thermochromic vanadium dioxide smart coatings grown on Kapton substrates by reactive pulsed laser deposition. J Vac Sci Technol A: Vac Surf Films 24(3):612–617
184. Manning TD et al (2002) Intelligent window coatings: atmospheric pressure chemical vapour deposition of vanadium oxides. J Mater Chem 12(10):2936–2939
185. Roach WR (1971) Holographic storage in VO_2. Appl Phys Lett 19(11):453–455

186. Jorgenson GV, Lee JC (1986) Doped vanadium oxide for optical switching films. Sol Energy Mater 14(3):205–214
187. Chivian J et al (1985) An improved scan laser with a VO<inf>2</inf> programmable mirror. IEEE J Quant Electron 21(4):383–390
188. Jerominek H, Picard F, Vincent D (1993) Vanadium oxide films for optical switching and detection. Opt Eng 32(9):2092–2099
189. Duchene J, Terraillon M, Pailly M (1972) R.F. and D.C. reactive sputtering for crystalline and amorphous VO_2 thin film deposition. Thin Solid Films 12(2):231–234
190. Hood PJ, DeNatale JF (1991) Millimeter-wave dielectric properties of epitaxial vanadium dioxide thin films. J Appl Phys 70(1):376–381
191. Rini M et al (2005) Photoinduced phase transition in VO_2 nanocrystals: ultrafast control of surface-plasmon resonance. Opt Lett 30(5):558–560
192. Lakes RS et al (2001) Extreme damping in composite materials with negative-stiffness inclusions. Nature 410(6828):565–567
193. Joyeeta N, Haglund JRF (2008) Synthesis of vanadium dioxide thin films and nanoparticles. J Phys: Condens Matter 20(26):264016
194. http://etd.library.vanderbilt.edu/available/etd-04202011-182358/unrestricted/Dissertation_Joyeeta.pdf, Accessed 18 May 2017
195. Nag J (2011) The solid-solid phase transition in vanadium dioxide thin films: synthesis, physics and application. Vanderbilt University, Nashville
196. Abdullaev MA, Kamilov IK, Terukov EI (2001) Preparation and properties of stoichiometric vanadium oxides. Inorg Mater 37(3):271–273
197. Griffiths CH, Eastwood HK (1974) Influence of stoichiometry on the metal-semiconductor transition in vanadium dioxide. J Appl Phys 45(5):2201–2206
198. Seyfouri MM, Binions R (2017) Sol-gel approaches to thermochromic vanadium dioxide coating for smart glazing application. Sol Energy Mater Sol Cells 159:52–65
199. Hench LL, West JK (1990) The sol-gel process. Chem Rev 90(1):33–72
200. Livage J (1991) Vanadium pentoxide gels. Chem Mater 3(4):578–593
201. Pelletier O et al (2000) A detailed study of the synthesis of aqueous vanadium pentoxide nematic gels. Langmuir 16(12):5295–5303
202. Berezina O et al (2015) Vanadium oxide thin films and fibers obtained by acetylacetonate sol–gel method. Thin Solid Films 574:15–19
203. Fontenot CJ et al (2000) Vanadia gel synthesis via peroxovanadate precursors. 1. In situ laser Raman and 51V NMR characterization of the gelation process. J Phys Chem B 104(49):11622–11631
204. Fontenot CJ et al (2001) Vanadia gel synthesis via peroxovanadate precursors. 2. Characterization of the gels. J Phys Chem B 105(43):10496–10504
205. Livage J et al (1997) Optical properties of sol-gel derived vanadium oxide films. J Sol-Gel Sci Technol 8(1):857–865
206. Lee C-Y et al (2011) Synthesis and characterization of sodium vanadium oxide gels: the effects of water (n) and sodium (x) content on the electrochemistry of $NaxV_2O_5$[middle dot]nH_2O. Phys Chem Chem Phys 13(40):18047–18054
207. Livage J, Pelletier O, Davidson P (2000) Vanadium pentoxide sol and gel mesophases. J Sol-Gel Sci Technol 19(1):275–278
208. Carn F et al (2005) Tailor-made macroporous vanadium oxide foams. Chem Mater 17(3):644–649
209. Guzman G, Morineau R, Livage J (1994) Synthesis of vanadium dioxide thin films from vanadium alkoxides. Mater Res Bull 29(5):509–515
210. Chaput F et al (1995) Synthesis and characterization of vanadium oxide aerogels. J Non-Cryst Solids 188(1):11–18
211. Wu YF et al (2013) Spectroscopic analysis of phase constitution of high quality VO_2 thin film prepared by facile sol-gel method. AIP Adv 3(4):042132
212. Augustyn V, Dunn B (2010) Vanadium oxide aerogels: Nanostructured materials for enhanced energy storage. C R Chim 13(1–2):130–141

213. Li H et al (2011) High-surface vanadium oxides with large capacities for lithium-ion batteries: from hydrated aerogel to nanocrystalline VO2(B), V6O13 and V2O5. J Mater Chem 21(29):10999–11009
214. Mege S et al (1998) Surfactant effects in vanadium alkoxide derived gels. J Non-Cryst Solids 238(1–2):37–44
215. Leventis N et al (2008) Polymer nanoencapsulated mesoporous vanadia with unusual ductility at cryogenic temperatures. J Mater Chem 18(21):2475–2482
216. Luo H et al (2008) Synthesis and characterization of the physical, chemical and mechanical properties of isocyanate-crosslinked vanadia aerogels. J Sol-Gel Sci Technol 48(1):113–134
217. Fears TM (2015) Synthesis and characterization of vanadium oxide nanomaterials. Doctoral Dissertations. 2443. http://scholarsmine.mst.edu/doctoral_dissertations/2443
218. Greenberg CB (1983) Undoped and doped VO_2 films grown from $VO(OC_3H_7)_3$. Thin Solid Films 110(1):73–82
219. Takahashi Y et al (1989) Preparation of VO_2 films by organometallic chemical vapour deposition and dip-coating. J Mater Sci 24(1):192–198
220. Speck KR et al (1988) Vanadium dioxide films grown from vanadium tetra-isopropoxide by the sol-gel process. Thin Solid Films 165(1):317–322
221. Deki S, Aoi Y, Kajinami A (1997) A novel wet process for the preparation of vanadium dioxide thin film. J Mater Sci 32(16):4269–4273
222. Partlow DP et al (1991) Switchable vanadium oxide films by a sol-gel process. J Appl Phys 70:443–452
223. Lu S, Hou L, Gan F (1993) Preparation and optical properties of phase-change VO_2 thin films. J Mater Sci 28(8):2169–2177
224. Yin D et al (1996) High quality vanadium dioxide films prepared by an inorganic sol-gel method. Mater Res Bull 31(3):335–340
225. Wang N et al (2013) Simple sol–gel process and one-step annealing of vanadium dioxide thin films: synthesis and thermochromic properties. Thin Solid Films 534:594–598
226. Béteille F, Mazerolles L, Livage J (1999) Microstructure and metal-insulating transition of VO_2 thin films. Mater Res Bull 34(14–15):2177–2184
227. Gao Y et al (2012) Nanoceramic VO_2 thermochromic smart glass: a review on progress in solution processing. Nano Energy 1(2):221–246
228. Babkin EV et al (1987) Metal-insulator phase transition in VO_2: influence of film thickness and substrate. Thin Solid Films 150(1):11–14
229. Pan M et al (2004) Properties of VO_2 thin film prepared with precursor $VO(acac)_2$. J Cryst Growth 265(1–2):121–126
230. Ningyi Y, Jinhua L, Chenglu L (2002) Valence reduction process from sol–gel V_2O_5 to VO_2 thin films. Appl Surf Sci 191(1–4):176–180
231. Cui J, Da D, Jiang W (1998) Structure characterization of vanadium oxide thin films prepared by magnetron sputtering methods. Appl Surf Sci 133(3):225–229
232. Fuls EN, Hensler DH, Ross AR (1967) Reactively sputtered vanadium dioxide thin films. Appl Phys Lett 10(7):199–201
233. de Castro MSB, Ferreira CL, de Avillez RR (2013) Vanadium oxide thin films produced by magnetron sputtering from a V2O5 target at room temperature. Infrared Phys Technol 60:103–107
234. Wang H, Yi X, Chen S (2006) Low temperature fabrication of vanadium oxide films for uncooled bolometric detectors. Infrared Phys Technol 47(3):273–277
235. Shishkin NY et al (2005) Doped vanadium oxides phase transitions vapors influence. Sens Actuators B: Chem 108(1–2):113–118
236. Chen S et al (2006) Vanadium oxide thin films deposited on silicon dioxide buffer layers by magnetron sputtering. Thin Solid Films 497(1–2):267–269
237. Cho C-R et al (2006) Current-induced metal–insulator transition in VOx thin film prepared by rapid-thermal-annealing. Thin Solid Films 495(1–2):375–379
238. Chen X, Dai J (2010) Optical switch with low-phase transition temperature based on thin nanocrystalline VOx film. Optik Int J Light Electron Opt 121(16):1529–1533

239. Choi Y et al (2014) Thermochromic substrate and method of manufacturing the same. Google Patents
240. Ping J, Sakae T (1994) Formation and thermochromism of VO_2 films deposited by rf magnetron sputtering at low substrate temperature. Jpn J Appl Phys 33(3R):1478
241. Case FC (1987) Reactive evaporation of anomalous blue VO_2. Appl Opt 26(8):1550–1553
242. Case FC (1988) The influence of substrate temperature on the optical properties of ion-assisted reactively evaporated vanadium oxide thin films. J Vac Sci Technol A: Vac Surf Films 6(3):2010–2014
243. Case FC (1990) Low temperature deposition of VO2 thin films. Journal of Vacuum Science & Technology A: Vacuum, Surfaces, and Films 8(3):1395–1398
244. Case FC (1991) Improved VO_2 thin films for infrared switching. Appl Opt 30(28):4119–4123
245. Nyberg GA, Buhrman RA (1987) High optical contrast in VO_2 thin films due to improved stoichiometry. Thin Solid Films 147(2):111–116
246. Surnev S et al (2000) Growth and structure of ultrathin vanadium oxide layers on Pd(111). Phys Rev B 61(20):13945–13954
247. Leisenberger FP et al (1999) Nature, growth, and stability of vanadium oxides on Pd(111). J Vac Sci Technol A: Vac Surf Films 17(4):1743–1749
248. Lovis F et al (2011) Self-organization of ultrathin vanadium oxide layers on a Rh(111) surface during a catalytic reaction. Part II: A LEEM and spectromicroscopy study. J Phys Chem C 115(39):19149–19157
249. Borek M et al (1993) Pulsed laser deposition of oriented VO_2 thin films on R-cut sapphire substrates. Appl Phys Lett 63(24):3288–3290
250. Kim DH, Kwok HS (1994) Pulsed laser deposition of VO_2 thin films. Appl Phys Lett 65(25):3188–3190
251. Maaza M et al (2000) Direct production of thermochromic VO_2 thin film coatings by pulsed laser ablation. Opt Mater 15(1):41–45
252. Suh JY et al (2004) Semiconductor to metal phase transition in the nucleation and growth of VO_2 nanoparticles and thin films. J Appl Phys 96(2):1209–1213
253. Pauli SA et al (2007) X-ray diffraction studies of the growth of vanadium dioxide nanoparticles. J Appl Phys 102(7):073527
254. Lopez R, Feldman LC, Haglund RF (2004) Size-dependent optical properties of VO_2 nanoparticle arrays. Phys Rev Lett 93(17):177403
255. Chae BG et al (2004) Fabrication and electrical properties of pure VO_2 phase films. J Korean Phys Soc 44(4):884–888
256. Kumar RTR et al (2003) Study of a pulsed laser deposited vanadium oxide based microbolometer array. Smart Mater Struct 12(2):188
257. Huotari J et al (2015) pulsed laser deposited nanostructured vanadium oxide thin films characterized as ammonia sensors. Sens Actuators B: Chem 217:22–29
258. Masina BN et al (2015) Phase-selective vanadium dioxide (VO_2) nanostructured thin films by pulsed laser deposition. J Appl Phys 118(16):165308
259. Huotari J et al (2013) Gas sensing properties of pulsed laser deposited vanadium oxide thin films with various crystal structures. Sens Actuators B: Chem 187:386–394
260. Madiba IG et al (2017) Effects of gamma irradiations on reactive pulsed laser deposited vanadium dioxide thin films. Appl Surf Sci 411:271–278
261. http://onlinelibrary.wiley.com/journal/10.1002/(ISSN)1521-3862 Online ISSN: 1521-3862. Chemical vapor deposition
262. Hawkins DT (ed) (1981) Chemical vapor deposition 1960–1980, a bibliography. IFI/ Plenum, New York
263. Groult H et al (2006) Two-dimensional recrystallisation processes of nanometric vanadium oxide thin films grown by atomic layer chemical vapor deposition (ALCVD) evidenced by AFM. Appl Surf Sci 252(16):5917–5925
264. Su Q et al (2009) Formation of vanadium oxides with various morphologies by chemical vapor deposition. J Alloy Compd 475(1–2):518–523

265. Malarde D et al (2017) Optimized atmospheric-pressure chemical vapor deposition thermochromic VO_2 thin films for intelligent window applications. ACS Omega 2(3):1040–1046
266. Niklaus F, Vieider C, Jakobsen H (2008) MEMS-based uncooled infrared bolometer arrays: a review. Proc SPIE 6836:68360D
267. Rogalski A (2011) Infrared detectors. CRC Press, Boca Raton
268. Naoum JA et al (2003) dynamic infrared system level fault isolation. In: 29th international symposium for testing and failure analysis. ASM International, Santa Clara, California
269. Art J (2006) Photon detectors for confocal microscopy. In: Pawley JB (ed) Handbook of biological confocal microscopy. Springer Science + Business Media, LLC, New York
270. Hyseni G, Caka N, Hyseni K (2010) Infrared thermal detectors parameters: semiconductor bolometers versus pyroelectrics. Wseas Trans Circuits Syst 9:238–247
271. Stotlar SC () Infrared detector. In: Waynant R, Ediger M (ed) Electro-Optics Handbook, pp 17.1–17.24. McGraw-Hill, New York
272. Bhan RK et al (2009) Uncooled infrared microbolometer arrays and their characterisation techniques. Appl Opt 59:580–589
273. Yeh P (1988) Optical waves in layered media. Wiley, New York
274. Hebb JP, Jensen KF (1996) The effect of multilayer patterns on temperature uniformity during rapid thermal processing. J Electrochem Soc 143:1142–1151
275. Howell JR, Siegel R, Menguc MP (1992) Thermal Radiation Heat Transfer. Hemisphere Publishing Corporation, Washington D. C
276. Ravindra NM et al (2003) Modeling and simulation of emissivity of silicon-related materials and structures. J Electron Mater 32:1052–1058
277. Hebb J (1997) Pattern effects in rapid thermal processing. In: Mechanical engineering. Massachusetts Institute of Technology
278. Rogalski A (2003) Infrared detectors: status and trends. Prog Quantum Electron 27:59–210
279. Fieldhouse N et al (2009) Electrical properties of vanadium oxide thin films for bolometer applications: processed by pulse dc sputtering. J Phys D Appl Phys 42:055408
280. Melnik V et al (2012) Low-temperature method for thermochromic high ordered VO_2 phase formation. Mater Lett 68:215–217
281. Lamsal C, Ravindra NM (2014) Simulation of spectral emissivity of vanadium oxides (VO_x)-based microbolometer structures. Emerg Mater Res 3:194–202
282. http://refractiveindex.info/?group=METALS&material=Aluminium (Date: 11/20/2013: 1:25 pm)
283. Folks WR et al (2008) Spectroscopic ellipsometry of materials for infrared micro-device fabrication. Phys Status Solidi C 5:1113–1116
284. Rogalski A (2002) Comparison of photon and thermal detector performance. In: Henini M, Razeghi M (ed) Handbook of Infra-red Detection Technologies. Elsevier Science Inc., New York
285. http://www.semiwafer.com/products/silicon.htm (Date: 11/19/2013: 5:21 pm)
286. Lamsal C, Ravindra NM (2013) Optical properties of vanadium oxides-an analysis. J Mater Sci 48:6341–6351
287. Richwine R et al (2006) A comprehensive model for bolometer element and uncooled array design and imaging sensor performance prediction. Proc SPIE 6294:62940F
288. Wang S et al (2017) Atomic layer-deposited titanium-doped vanadium oxide thin films and their thermistor applications. J Electron Mater 46(4):2153–2157
289. Venkatasubramanian C, Horn MW, Ashok S (2009) Ion implantation studies on VOx films prepared by pulsed dc reactive sputtering. Nucl Instrum Methods Phys Res Sect B 267(8–9):1476–1479
290. Chen L-M et al (2009) Investigations on capacitive properties of the AC/V_2O_5 hybrid supercapacitor in various aqueous electrolytes. J Alloy Compd 467(1–2):465–471
291. Huang Z et al (2011) Synthesis of vanadium oxide, V_6O_{13} hollow-flowers materials and their application in electrochemical supercapacitors. J Alloy Compd 509(41):10080–10085
292. Jung H-M, Um S (2013) Electrical and thermal transport properties of vanadium oxide thin films on metallic bipolar plates for fuel cell applications. Int J Hydrogen Energy 38(26):11591–11599

293. Biette L et al (2005) Macroscopic fibers of oriented vanadium oxide ribbons and their application as highly sensitive alcohol microsensors. Adv Mater 17(24):2970–2974
294. Wu K et al (2016) Vanadium oxides (V_2O_5) prepared with different methods for application as counter electrodes in dye-sensitized solar cells (DSCs). Appl Phys A 122(9):787
295. Chen X-W et al (2007) Carbon nanotube-induced preparation of vanadium oxide nanorods: application as a catalyst for the partial oxidation of n-butane. Mater Res Bull 42(2):354–361

Chapter 5
Graphene: Properties, Synthesis, and Applications

Sarang Muley and Nuggehalli M. Ravindra

5.1 Introduction

5.1.1 Objective

The objective of this chapter is to study and analyze the electronic, optical, and thermoelectric properties of graphene and graphene nanoribbons, as a function of number of layers, doping, chirality, temperature, and lattice defects. Some aspects related to the methods of synthesis of graphene are described.

5.1.2 Background

Coal, a form of carbon, has been the driving force for the industrial revolution. In recent years, nanostructured form of carbon has become a significant part of another technological and scientific revolution in the field of nanotechnology. The field of nanoscale science has been significantly molded by research on carbon nanostructures. Different structures of sp^2 hybridized carbon have been investigated and most important ones are the soccer ball structure called fullerene (C_{60}), the single atom thick planar form of carbon called graphene, the "rolled up" tubular sheets of graphene termed as carbon nanotubes (as shown in Fig. 5.1) [1].

S. Muley
DW National Standard Stillwater LLC, 3602 N Perkins Road, Stillwater, OK 74075, USA
e-mail: sarang.muley@gmail.com

N. M. Ravindra (✉)
Interdisciplinary Program in Materials Science & Engineering, New Jersey Institute of Technology, Newark, NJ, USA
e-mail: nmravindra@gmail.com

M. I. Pech-Canul and N. M. Ravindra (eds.), *Semiconductors*,
https://doi.org/10.1007/978-3-030-02171-9_5

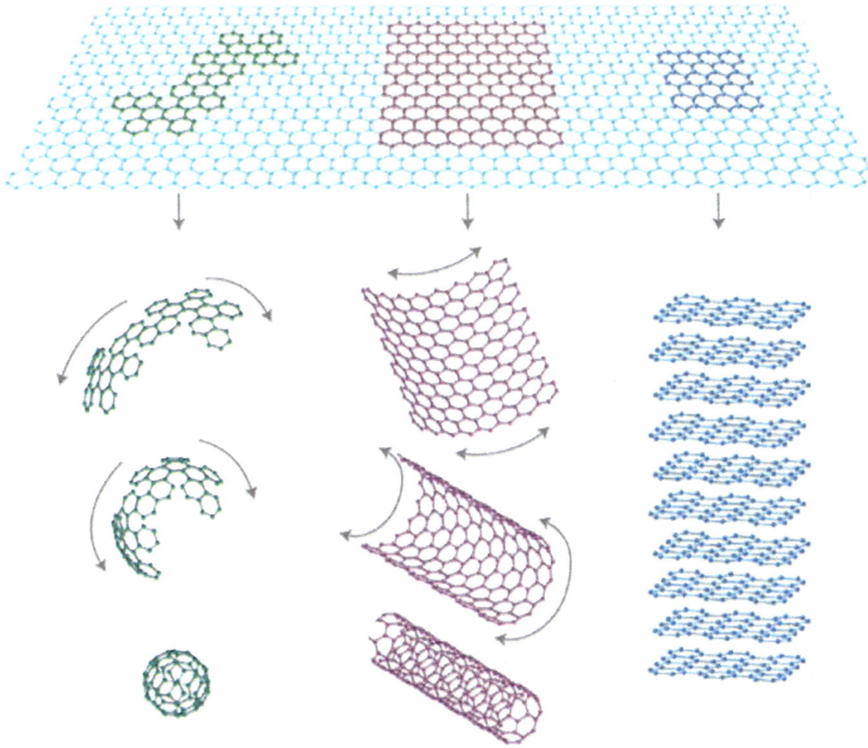

Fig. 5.1 Graphene is a 2D building material for carbon materials of all other dimensionalities. It can be wrapped up into 0D buckyballs, rolled into 1D nanotubes or stacked into 3D graphite [1]

Graphene can be considered as the building block for the other carbon allotropes such as fullerenes and carbon nanotubes. The research on the different nanostructures of carbon has been shown to be "self-enhancing" as this field is highly interconnected and many research groups are working together with a view to bridge the gap between laboratory and industry as well as to use the significant properties of the novel materials in a vast variety of applications [2]. Discovery of fullerenes in 1985 [3, 4] paved the way for the incidental discovery of carbon nanotubes in 1990s [5] and the experimental feasibility of graphene in 2004 [6].

Fibrous carbon materials with uniquely high strength-to-weight ratio such as carbon fiber reinforced composites and graphene have been applied in various equipment for sports (such as tennis rackets) as well as orthopedics [7, 8]. These nanostructures are reported to exhibit exceptional mechanical properties, such as Young's modulus higher than 1 TPa in case of CNTs [9] as well as in graphene [10] and fullerene. Advances in the knowledge of the various properties of these nanostructures provide motivation for further research and exploring possibilities for practical applications and production of these materials in various industries.

Theoretically, the electronic and optical absorption properties of graphene have been published in 1947 by Wallace [11]. However, in October 2004, monocrystalline and highly stable graphitic films were successfully fabricated using the mechanical exfoliation of graphite with a scotch tape, under ambient conditions [6, 12]. Films were found to be semimetallic with a small overlap of the valence and conduction bands and shown to have strong ambipolar electric field effect. Scientists who isolated the graphene, Andre Geim and Konstantin Novoselov, were awarded the Nobel Prize in Physics in 2010 [13]. Earlier, graphene was considered to be inseparable and thermodynamically unstable to exist as a single layer [14]. The stability of graphene is attributed to its strong covalent planar bonds [1]. As shown in Fig. 5.2, graphene is about 0.34 nm thick and it is composed of carbon atoms arranged hexagonally in a honeycomb structure, with sp^2 bonds, which are about 0.14 nm long [15, 16]. Carbon atoms have a total of six electrons; two electrons in the inner shell and four electrons in the outer shell. The four outer shell electrons, in an individual carbon atom, take part in chemical bonding; but it is known that each carbon atom in planar structure of graphene is bonded to three carbon atoms on the two-dimensional plane, so that one electron is free for electronic conduction in the third dimension. These free electrons are called pi (π) electrons. They are located above and below the graphene sheet and are highly mobile. In graphene, these pi orbitals are known to overlap and help in enhancing the carbon–carbon bonds. Multilayer graphene with less than 10 layers is sometimes referred to as Few-Layer Graphene (FLG). The properties of graphene with more than 10 layers are the same as bulk graphite [1, 17]. Carbon atoms in a layer of graphene are covalently bonded and Van der Waals interaction exists between the neighboring layers of graphene.

Fig. 5.2 Graphene's honeycomb lattice [16]

Graphene is known to be optically transparent, between 70 and 90%, the transparency being dependent on its thickness. It has high electronic and thermal conductivities as well as excellent transport properties [18]. With these extraordinary properties, it has been reported to be used for applications such as touchscreens, fuel cells [19], batteries, sensors [20, 21], solar cells [22], supercapacitors [23], nanocomposites, wear resistant films, Organic Photovoltaics (OPV) as well as transparent displays and flexible electronics [24–27].

The main obstacle has been to commercially produce graphene. The key challenge is the synthesis and processing of bulk quantities of graphene sheets. Graphene is known to form irreversible agglomerates or even restacking to form graphite through interlayer Van der Waals interaction, unless they are well separated from each other. It is not commercially viable in making OPVs with tiny flakes of graphene using scotch tape; therefore, alternative techniques such as epitaxial growth and copper foil technique have been developed [28, 29]. With advances in these techniques, it is important to completely understand and study the structure and properties of graphene and its derivatives.

Graphene has been studied within the context of condensed physics phenomena [30, 31], and it is a material that has been suggested to replace silicon due to its excellent electron mobility (about 100 times greater than silicon), large mean free path [6], as well as the ability to modify its electrical properties by doping and chirality [32]. There has been a plethora of theoretical and experimental researches to investigate the electronic, optical, mechanical, chemical, and thermoelectric properties of graphene. Scientifically important phenomena such as half-integer quantum Hall effect and Berry's phase [33], the breakdown of the Born–Oppenheimer approximation [34], and confirmation of the existence of massless Dirac Fermions [33], have been observed in graphene.

In the attempts to fabricate single-layer graphene, various top-down approaches have been utilized such as mechanical exfoliation [1, 6], liquid phase exfoliation of graphene [35] as well as bottom-up approaches such as epitaxial growth of graphene on SiC substrate [36] or metal substrates [37–39], Chemical Vapor Deposition [40], and substrate free gas synthesis [41]. All these approaches have paved the way for further research on graphene.

As a thin-film, graphene is grown on metallic substrates and the growth phenomenon has been studied since the 1970s. Single-layer graphite growth was also reported on various transition metal substrates [42–44]. Even before these experiments, the separation of graphene layers in graphite, in the form of graphite intercalation compounds, exfoliated graphite and so-called graphene oxide (GO), has been studied [45]. Recently, Nina et al. [46] have reported that intercalated graphite can be readily exfoliated in dimethylformamide to obtain suspensions of crystalline single- and few-layer graphene sheets. The concept of reversible intercalation of graphite by using nonoxidizing bronsted acids can narrow the path for commercialization of graphene. This research has revisited decades-old belief that graphite intercalation must involve host–guest charge transfer, resulting in partial oxidation, reduction or covalent modification of graphene sheets. Majority of the experimental research has been focused on the development of synthetic routes leading to mass production of well-defined sheets [19].

The term "graphene" was proposed by Boehm et al. in 1986, to describe a single atomic sheet of graphite [47]. Mechanical exfoliation route enables one to obtain very high crystallinity and purity samples, which have been used to explore the transport properties of graphene. Following this route of making graphene, various papers have reported unique properties of graphene, in contrast to bulk graphite. Novoselov et al. [1, 6, 12, 48], Nair et al. [18, 49, 50] and Berger et al. [51] are some of the scientists who have reported the same.

The feasibility of making graphene commercially and the possibility to tailor its electronic properties makes it a promising material for the electronics industry. It is known that the optical and thermoelectric properties are a function of the electronic and structural properties of a material. This has led to the research on improving these properties by various techniques, especially chemical doping, isotopic substitution, isoelectronic impurities, and hydrogen adsorption [52]. The effects of layers and edges of graphene on its properties have been explored to understand their significance in applications as a layer in multilayered configuration of devices. The need for thermal management of devices in view of their nano-sizes is the driving force for such a material with enhanced thermoelectric properties.

5.2 Literature Review

In this section, a brief theoretical background of the electronic, optical, mechanical, and thermoelectric properties of graphene, along with the relevant research in scientific literature, is presented. Table 5.1 presents the properties of graphene and its comparison with other materials.

Table 5.1 Properties of graphene and comparison with other materials

Property	Value	Comparison with other materials	References
Breaking strength	42 N/m	More than 200 times greater than steel	[53]
Elastic limit	~20%	<1% for steel	[25]
Carrier mobility at room temperature	200,000 $cm^2 V^{-1} s^{-1}$	More than 100 times higher than Si	[54]
Thermal conductivity	~5000 W/mK	More than 10 times higher than copper	[55]
Maximum current density	>10^8 A/cm	~100 times larger than Cu	[56]
Optical absorption coefficient	2.30%	~50 times higher than GaAs	[18]

5.2.1 Electronic Properties

One of the most important applications of graphene is in electronics. As discussed in the previous section, the studies of electronic structure of graphene date back to the year 1947. Most of the major manufacturers in the field of semiconductor fabrication have vested huge interest in analyzing the properties of graphene for its applications in electronic circuits. This is because of the fact that silicon is approaching its fundamental size limits in device miniaturization.

A linked and important property that differentiates graphene is the high mobility of charge carriers in excess of 200,000 cm^2/Vs. This shows that nearly ballistic transport is observed in the submicron regime. The challenging fact is that this value of mobility is true only for large-scale graphene, which is a gapless material. Due to zero bandgap, it is not possible to turn off the device completely without high leakage current. This hampers the prospects of using graphene in making Field-Effect Transistors (FETs) for applications in logic circuits. Various techniques used by researchers in order to open up the bandgap in graphene include quantum confinement in one direction giving rise to graphene nanoribbons (GNRs) (as shown in Fig. 5.3) [57–59], application of strain [60, 61] and use of bilayer graphene (BLG) [62, 63]. Jin et al. [64] have studied the surface modulation of graphene field-effect transistors on periodic trench structure made by carbonized poly(methylmethacrylate). Graphene nanoribbons have been found to exhibit two achiral structures: armchair and zigzag nanoribbons. Both armchair and zigzag GNRs are known to demonstrate bandgaps above 200 meV for widths less than 10 nm [65]. Zigzag GNRs belong to mid-gap semiconductors and antiferromagnetic configuration which is possible to make experimentally. Armchair GNRs possess nonmagnetic configurations which have width-dependent energy gaps [66]. Recent experiments have shown the fabrication of flat and curved GNRs by various techniques such as unzipping carbon nanotubes, e.g., wet chemical technique using acid reactions, a catalytic approach using metal nano-clusters as scalpels, as well as a physicochemical method using argon plasma treatment [67].

Bandgaps, to the extent of a few hundreds of milli-electron volts, have been achieved with BLG by application of an electric field that is perpendicular to the bilayer [68]. The gap in the stacked bilayer graphene arises due to the formation of pseudospins between the layers, and hence makes it possible to electrically induce a bandgap [69].

However, the predicted high mobility could not be achieved for large-area graphene. This is explained by the fact that, as the bandgap opens up, it becomes more parabolic, and so the effective mass increases and the mobility decreases. Thus, the bandgap leads to reduced mobility of GNRs. Apart from this, GNRs have rough edges which wipe out the bandgap opening, thus raising further challenges in fabricating them. Since the bandgap is inversely proportional to the width of GNRs, a bandgap of 0.5 eV is required for room temperature operation of transistors and this will require the width of the GNR to be less than 5 nm which is difficult to fabricate with much accuracy.

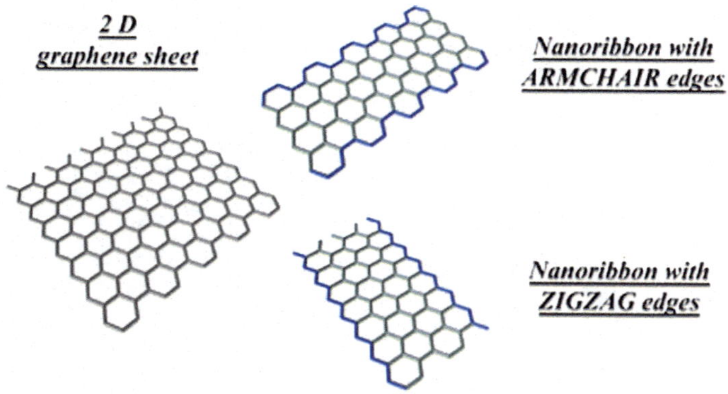

Fig. 5.3 Zigzag and armchair edges in monolayer graphene nanoribbons. The edge structure and the number of atomic rows of carbon atoms normal to the ribbon axis determine the electronic structure and ribbon properties [59]

As discussed earlier, carbon atoms in graphene are arranged in a honeycomb lattice with two atoms per unit cell, as shown in Fig. 5.2. The electronic band structure of graphene can either be described using tight binding approximation (TBA) or the similar Linear Combination of Atomic Orbitals (LCAO), which is used more commonly in chemistry. The two atoms in graphene which make up two nonequivalent sub-lattices are bonded by trigonal σ bonds. These σ bonding sp^2 orbitals are formed by the superposition of the s, p_x, and p_y orbitals of atomic carbon, whereas the p_z orbital remains non-hybridized. The hybridized orbital is geometrically trigonal and planar. This is the reason why each carbon atom within graphite has three nearest neighbors in the graphite sheet. There is an overlap of p_z-orbitals of neighboring carbon atoms and distributed π-bonds are formed above and below each graphite sheet. This leads to the presence of delocalized electron π bonds, similar to the case of benzene, naphthalene, anthracene, and other aromatic molecules. In this regard, graphene can be considered to have an extreme size of planar aromatic molecules.

The geometry of sp^2 hybridized carbon is shown in Fig. 5.4 [70]. The covalently bonded in-plane σ bonds are found to be primarily responsible for the mechanical strength of graphene and other sp^2 carbon allotropes. The σ electronic bands are completely filled and they have an energy separation larger than the π bands and hence the effects of σ bands on the electronic behavior of graphene can be neglected in a first approximation. The out-of-plane π-bond is primarily responsible for its peculiar electronic and optical properties. It should be understood that, in a real sample of graphene, the layer is not strictly a 2D crystal; it is found to be rippled when suspended [71] or it adheres to the corrugation of its supporting substrate [72]. In these cases, a mixing of the σ and π orbitals occurs, which needs to be taken into consideration when calculating the electronic properties of graphene [73]. One of the simplest known evaluations of the band structure and, therefore, its electronic properties is obtained by examination of the π bands in a tight binding approximation. The first reference to the band structure calculation was published by Wallace in 1947

[11]. The 2D nature of graphene allows plotting the relationship in the entire first Brillouin zone (as shown in Fig. 5.5).

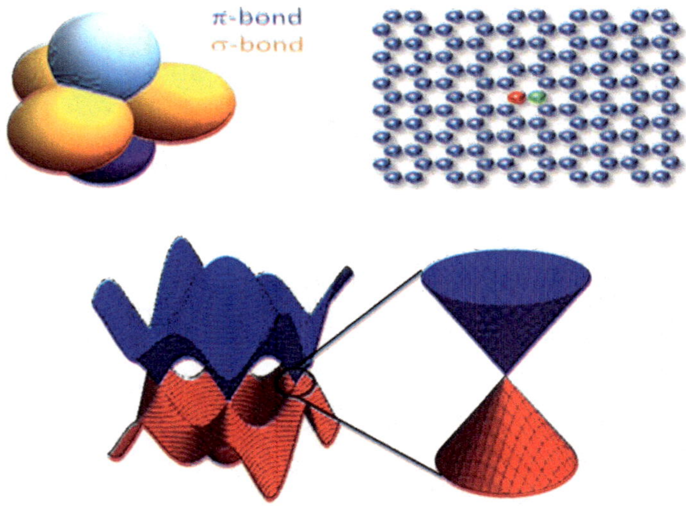

Fig. 5.4 Geometry of sp^2 hybridized carbon atom. Each of the two equivalent carbon atoms within one unit cell (red and green) contributes one cosine-shaped band to the electronic structure. These bands cross exactly at the Fermi level, where they form a Dirac cone with a linear electronic dispersion. Valence and conduction bands are shown in red and blue, respectively [70]

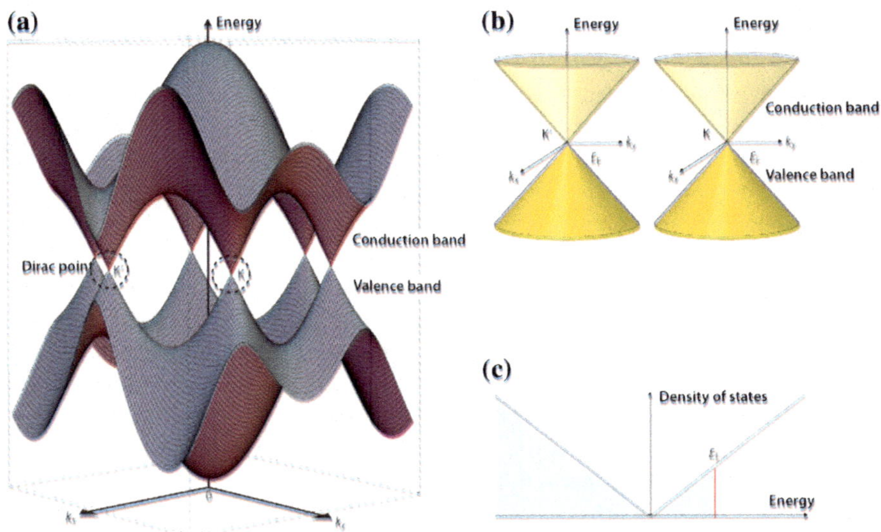

Fig. 5.5 **a** Energy bands near the Fermi level in graphene. The conduction and valence bands cross at points K and K′. **b** Conical energy bands in the vicinity of the K and K′ points. **c** Density of states near the Fermi level with Fermi energy E_F [75]

In the case of graphene, the bottom of the conduction band and the top of the valence band is not at the Γ point as in the case of most of the metals and semiconductors, but it is present at another high symmetry point at the boundary of the first Brillouin zone, at the so-called K-points, as shown in Fig. 5.5. The valence and conduction bands meet but they do not overlap. Density of states is null at the K-points themselves. This is the reason for graphene to be known as a zero bandgap semiconductor or semimetal. The first Brillouin zone contains two nonequivalent K-points called K and K′.

The most interesting aspect of graphene physics is that the band structure and physical properties of this material can be influenced by nanostructuring, functionalizing, mechanically straining, etc., yielding new physics to be studied and further explored. The Dirac points, K and K′, are the most important points in the structure of graphene. They are located at the corners of the Irreducible Brillouin Zone (IBZ) and one can define their positions in momentum space [74].

Figure 5.6 shows the band structure of pure graphene sheet [76]. It can be found that the bandgap location of graphene is different from that of a typical semiconductor. In graphene, the conduction and valence bands coincide at a conical point, known as a Dirac point. The energy–momentum plot shows the quasi-particles in the material behaving like massless Dirac Fermions [77].

The unique band structure of graphene allows higher mobility of electrons than in other materials. Electronic transport in a medium with negligible electrical resistivity is called ballistic transport, which is possible in very pure and defect-free graphene. Multichannel ballistic transport has also been reported in the case of multi-walled carbon nanotubes [78]. Electrons are capable of ballistic movement over long distances in graphene [77]. The velocity of electrons in graphene is a maximum at the Fermi velocity, which is about 1/300 of the speed of light. This allows graphene to be an excellent conductor. Doping enables changing the position of the Fermi levels in the band structure of graphene at room temperature [79]. Richter et al. [80] have reported the robust 2D electronic properties of rotationally stacked turbostratic graphene microdisks, where twisting of intermediate layers leads to electronic decoupling resulting in charge transport properties which retain the 2D nature of graphene. This reduced dimensionality leads to charge carrier localization which has been observed at lower temperatures.

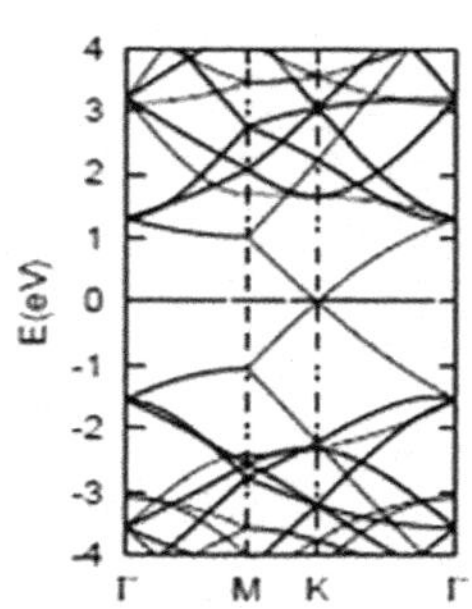

Fig. 5.6 Band structure of pure graphene sheet [76]

5.2.1.1 Electronic Density of States (DOS) of Graphene

Figure 5.7 shows the theoretical DOS of graphene [74] indicating its semimetallic nature. Zero bandgap in graphene is completely different than in the case of diamond, which is a wide bandgap semiconductor.

5.2.1.2 Effects of Doping on Electronic Structure in Graphene

Recently, there have been numerous attempts to fabricate graphene-based devices by engineering its bandgap by doping. Investigations on doped graphene nanoribbons [81, 82] indicate that doping with nitrogen or boron can make it possible to obtain n-type or p-type semiconducting graphene. It is shown experimentally that nitrogen doping of graphene [81] tends to move the Dirac point in the band structure of graphene below the Fermi level, E_F, and an energy gap is found to appear at high-symmetric K-point. Sharma et al. [83] have reported the effect of substitution of boron and nitrogen on the electronic structure of graphene. It was reported that the band structure and Density of States of B-/N-doped graphene depend on the choice of cell parameters.

Ci et al. [84] synthesized a novel two-dimensional nanomaterial in which a few carbon atoms on a graphene sheet are replaced by equal number of boron and nitrogen atoms. The concentration of the dopant atoms was controlled by keeping same B/N ratio. This novel nanomaterial was found to be semiconducting with a very small bandgap. Synthesis of similar BNC materials has been reported by Panchakarla et al. [85]. The electronic properties of nitrogen and boron-doped armchair edged graphene nanoribbons (AGNR) is also reported [86]. Nitrogen is found to introduce an impurity level above the donor level, while boron introduces an impurity level below that

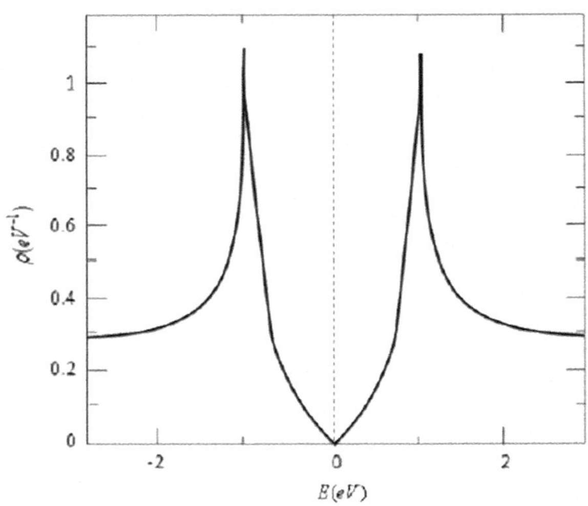

Fig. 5.7 Electronic Density of States (DOS) of graphene [74]

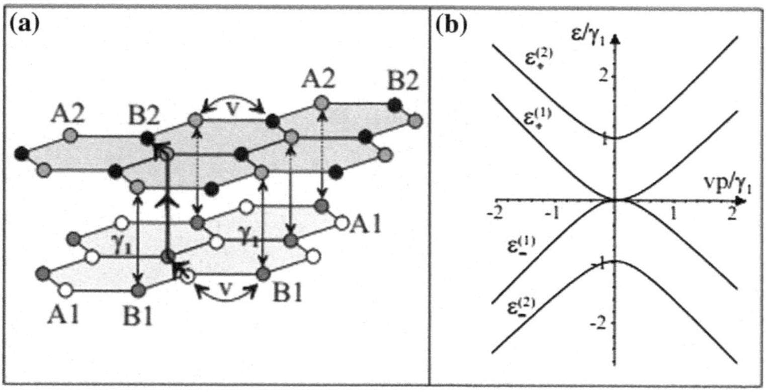

Fig. 5.8 **a** Schematic figure of bilayer graphene containing four sites of unit cell. **b** Electronic band structure of bilayer graphene [95]

of the acceptor level. This is different from single-wall Carbon Nanotubes (CNTs), in which, the impurity level is neither donor nor acceptor in their systems. In CNTs, the donor and acceptor levels are derived mainly from the lowest unoccupied orbital and the highest occupied orbital. Cao et al. [87] have reported superior boron, nitrogen, and iron ternary-doped carbonized graphene oxide-based catalysts for oxygen reduction in microbial fuel cells. Nitrogen-doped graphene oxide has been reported to be effective in the removal of boron ions from seawater by Chen et al. [88].

Theoretically, the introduction of energy gap in graphene is shown by oxidation of mono-vacancies in graphene [89], hetero-bilayers of graphene/boron nitride [90, 91], F-intercalated graphene on SiC substrate [92], and bilayer graphene-BN heterostructures [93]. Experimentally, the substitutional doping in carbon of boron nitride nanosheets, nanoribbons, and nanotubes has been reported [94]. It is reported experimentally that the sp^2 hybridized BNC nanostructure, with equal number of boron and nitrogen atoms, can also open finite bandgap. The band structure of bilayer graphene is as shown in Fig. 5.8 [95].

Fujii et al. [96] studied the role of edge geometry and chemistry in the electronic properties of graphene nanostructures. This paper has presented scanning probe microscopic as well as first-principles characterization of graphene nanostructures. It is clear that the challenges in making chemically doped graphene and different edge geometries, experimentally, have been overcome to a large extent. One of the challenges of experimentally grown graphene is the grain boundaries which are known to affect the electronic and optoelectronic properties of graphene [97]. In fact, grain boundaries present localized states, which have been proven to be crucial in regard to the electronic, magnetic, and mechanical properties that depend on the atomic line junctions. These localized states allow for decoration of line defects with adsorbates, hence opening a novel route for nanosensor applications [98].

Denis et al. [99] have studied the electronic properties and chemical reactivity of triple-doped graphene systems by means of DFT calculations. They have observed

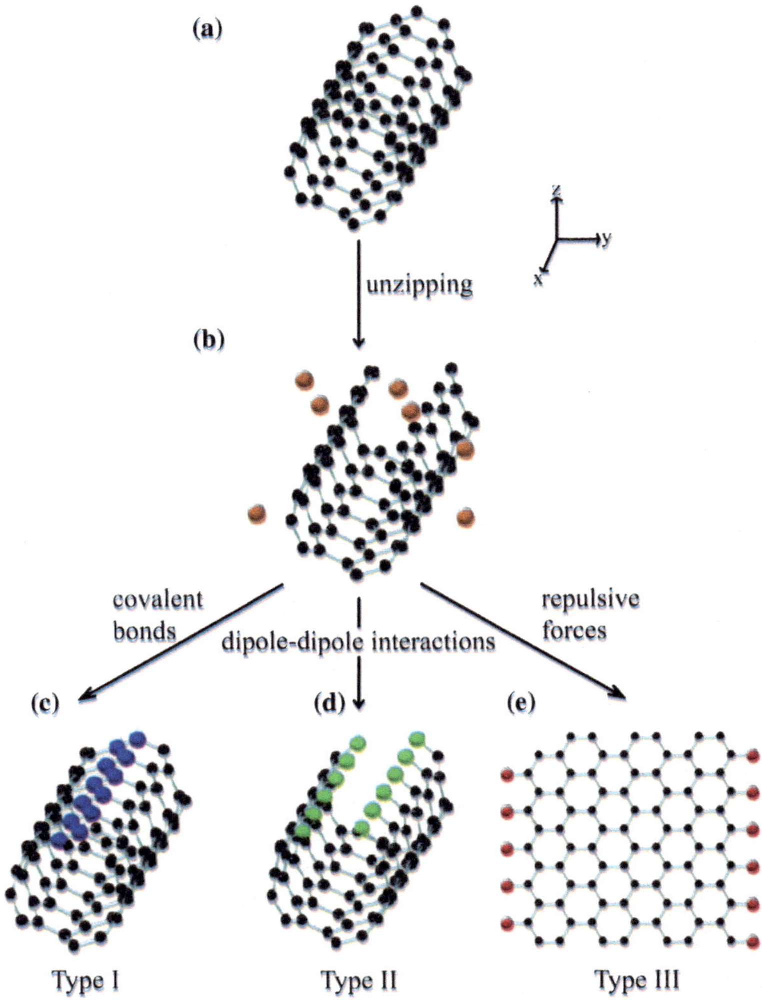

Fig. 5.9 **a** A single-wall carbon nanotube is used as starting material. **b** The unzipped nanotube is put in an environment of decorating atoms. **c–e** Three possible geometric structures are formed by three types of decorating atoms with different edge–edge interactions [66]

that, in order to increase the reactivity of graphene, Al, P, and S should be combined with BN motifs. Chang et al. [66] have investigated the geometric and electronic properties of edge decorated graphene nanoribbons using DFT. Three stable geometric structures have been demonstrated as shown in Fig. 5.9. These structures have been found to have high free carrier densities, whereas a few are semiconductors due to zigzag-edge-induced antiferromagnetism.

Sanjuan et al. [100] have presented the correlation of geometry of graphene with its mechanical, electronic, and chemical properties. Chen et al. [101] have studied the electronic properties of graphene, supported on (0001) SiO_2 surface, using DFT. The electronic properties have been shown to be dependent on the underlying substrate surface properties as well as on the percentage of hydrogen passivation. By application of methyl for passivation of oxygen-terminated SiO_2, it is possible to improve the charge carrier mobility of graphene by further reducing the interaction of graphene sheet with oxygen-terminated SiO_2. External electric field can also aid in modulating the charge transfer between the graphene and the SiO_2 surface. Hexagonal BN (h-BN) is a widely used substrate for graphene devices. Kretinin et al. [102] have reported the electronic properties of graphene encapsulated with different two-dimensional atomic crystals such as molybdenum, tungsten disulphides, h-BN, mica, bismuth strontium calcium copper oxide, and vanadium pentoxide. They have noted that devices made by encapsulating graphene with molybdenum or tungsten disulphides are found to have high carrier mobilities of ~60,000 $cm^2 V^{-1} s^{-1}$. Encapsulation with other substrates (such as mica, bismuth strontium calcium copper oxide, and vanadium pentoxide) results in exceptionally bad quality of graphene devices with carrier mobilities ~1000 $cm^2 V^{-1} s^{-1}$. The differences have been attributed to the self-cleansing occurring at the interfaces of graphene, h-BN, and transition metal dichalcogenides. This cleansing process is not found to take place on atomically flat oxide surfaces.

Motivated by the current experimental and theoretical reports, we have investigated the effect of doping of boron and nitrogen on the electronic properties of the graphene systems using first-principles electronic structure calculations based on density functional theory. Majidi [103] has used DFT to study the electronic properties of porous graphene, graphene-like, α-graphyne, graphene-like, and graphyne-like BN sheets.

5.2.2 *Optical Properties of Graphene*

Research on graphene has shown its unique optical properties [104], including its strong coupling with light [18], high-speed operation [105], and gate-variable optical conductivity [106]. These are extremely useful for addressing the future needs of the electro-optic (EO) modulators. It is found that, in the optical range, graphene has a constant index of refraction of 2.6 in an ultrawide band of wavelengths ranging from ultraviolet up to near-infrared and it shows constant absorption of 2.3%.

The rise in interest of graphene in photonics and optoelectronics is clear from its applications ranging from solar cells and light-emitting devices as well as in touch screens, photodetectors, and ultrafast lasers. These applications in nano-photonics are explored due to the unique combination of its optical and electronic properties [107].

Due to its visual transparency, graphene has immense potential as transparent coatings. Optical absorption of graphene is anisotropic for light polarization being parallel/perpendicular to the axis normal to the sheet. Experimentally, it is reported that as compared to graphite, the optical and energy loss spectra of graphene show a redshift of absorption bands and $\pi + \sigma$ electron plasmon and disappearance of bulk plasmons [108, 109]. Optical properties are the prominent characteristics that differentiate graphene from graphite. Ebernil et al. [110] have reported that π and $\pi + \sigma$ surface plasmon modes in free-standing single sheets of graphene are present at 4.7 and 14.6 eV, respectively, which are found to be present at 7 and 25 eV in case of bulk graphite. This redshift is reported to decrease as the number of layers of graphite reduces.

Among the many areas in which graphene has prominent applications, an important one is for fabricating sensors due to the sensitivity of its electronic structure to adsorbates [20]. Low energy loss electron spectroscopy provides a way of detecting changes in the electronic structure that is highly spatially resolved.

5.2.2.1 Linear Response: The Kubo Formula

An entirely interacting electronic system is considered here. It is described by the Hamiltonian H, in Eq. (5.2.1), under the action of an external time-dependent field [111]:

$$\mathbf{H_{tot}} = \mathbf{H} + \mathbf{H_{ext}(t)} = \mathbf{H} + \sum_{\mathbf{i}} \varnothing^{\mathbf{ext}}(\mathbf{r_i}, \mathbf{t}) = \mathbf{H} + \int \mathbf{dr}\boldsymbol{\rho}(\mathbf{r})\varnothing^{\mathbf{ext}}(\mathbf{r}, \mathbf{t}) \quad (5.2.1)$$

The density operator is defined as in Eq. (5.2.2):

$$\boldsymbol{\rho}(\mathbf{r}) = \sum_{\mathbf{i}} \boldsymbol{\delta}(\mathbf{r} - \mathbf{r_i}) \quad (5.2.2)$$

The induced density is defined in Eq. (5.2.3):

$$\boldsymbol{\rho}^{\mathbf{ind}}(\mathbf{r}, \mathbf{t}) = \langle \psi(\mathbf{t}) | \boldsymbol{\rho}(\mathbf{r}) | \psi(\mathbf{t}) \rangle - \langle \rho \rangle \quad (5.2.3)$$

In the linear response regime, the external field is assumed to be weak, so that one can expand the exact time-dependent ground state in the field, at the first order, as per Eq. (5.2.4) with external Hamiltonian in the picture.

$$|\psi(t) > \equiv \mathrm{e}^{-iHt} |\psi_I(t) > \approx |\psi\rangle + i \int_{-\infty}^{t} \mathrm{d}t' H_I^{\mathrm{ext}}(t') |\psi\rangle \quad (5.2.4)$$

In accordance with the Kubo formula:

$$\rho^{\mathrm{ind}}(r, t) = \int_{-\infty}^{t} \mathrm{d}t' \int_{-\infty}^{\infty} \mathrm{d}r' \chi(rr', t - t') \varnothing^{\mathrm{ext}}(r', t') \quad (5.2.5)$$

where the response function is defined as:

$$\chi_{\rho\rho}(\mathbf{rr}', \mathbf{t}) \equiv -\mathbf{i}\langle[\rho_{\mathrm{I}}(\mathbf{r}, \mathbf{t}), \rho_{\mathrm{I}}(\mathbf{r}')]\rangle = -i\langle[\delta_{\rho_{\mathrm{I}}(\mathbf{r},\mathbf{t})}, \delta_{\rho_{\mathrm{I}}(\mathbf{r}')}]\rangle \tag{5.2.6}$$

5.2.2.2 Dielectric Function and Energy Loss Spectra

In determining the optical properties, a calculation of the dielectric function, $\varepsilon(q, \omega)$ is generally reported, as a function of the frequency, ω, and the momentum transfer, q. If one assumes that the light polarization is parallel to the electric field momentum q, the cross section for optical absorption, $\sigma(\omega)$, i.e., the optical absorption spectrum, is then proportional to the imaginary part of the macroscopic dielectric function, as shown in Eq. (5.2.7).

$$\boldsymbol{\sigma}(\boldsymbol{\omega}) \propto \mathbf{Im}(\in_{\mathbf{M}} (\mathbf{q} \to 0, \boldsymbol{\omega})) \tag{5.2.7}$$

where

$$\in_M^{-1}(\omega) \equiv 1 + \lim_{q\to 0} \frac{4\pi}{|q|^2} \langle\langle \chi(q, \omega) \rangle\rangle \tag{5.2.8}$$

The limit $q \to 0$ is taken because the momentum carried by a photon is vanishingly small compared to the crystal momenta of a bulk material.

One of the important quantities, which can be measured experimentally, is the energy loss function. The loss function, $\Gamma(q, \omega)$, is related to the imaginary part of the inverse dielectric function:

$$\boldsymbol{\Gamma}(\mathbf{q}, \boldsymbol{\omega}) \propto \mathbf{Im}\left(\frac{1}{\boldsymbol{\varepsilon}(\mathbf{q}, \boldsymbol{\omega})}\right) \tag{5.2.9}$$

Contrary to the absorption cross section, the loss function is defined for finite momentum transfer q. One can measure the momentum transfer in the Electron Energy Loss Spectroscopy (EELS) through the deflection of the electron beam. Equation (5.2.9) is only valid for angular-resolved EELS on bulk materials and not for spatially resolved EELS on isolated nano-objects.

5.2.2.3 Literature Review of Graphene and Doped Graphene

It is known that graphene is optically transparent in the visible spectrum. Hence, in order to use it in optoelectronic applications, it needs to be tailored in order to absorb specific wavelength region of the spectra.

Optical properties of graphite and graphene have been studied by Sedelnikova et al. [112] and Marinopoulos et al. [113, 114]. Marinopoulos et al. reported the absorption spectrum for different values of c/a ratio of graphite and compared it with BN sheet. Eberlein et al. [110] performed the plasmon spectroscopy of graphene in conjunction with ab initio calculations of the low loss function. Numerous reports have been published regarding the surface plasmons in graphite and carbon nanotubes. The studies of SWCNTs (Single-Wall CNTs), for radius $r \rightarrow \infty$, can be applicable for free-standing single graphene sheets. The E field of a fast moving particle is elongated along its direction of travel; therefore, when it passes perpendicular through a foil of graphene, mainly, the out-of-plane mode with momentum $\hbar q$ parallel to E should be excited. These modes are reported to be forbidden in single-layer graphene, while they have a weak intensity in graphite. Although momentum transfer is close to zero, q is known to have considerable in-plane component. Sedelnikova et al. [112] have reported the effects of ripples on the optical properties of graphene.

It can be observed that the peaks at 4 eV, in Fig. 5.10, in single boron-doped and nitrogen-doped graphene is lesser in intensity and there are well-defined changes in the peak in the broader plateau. Since individual boron or nitrogen doping does not induce bandgap in graphene, it is found that the peak at 4 eV is not changed. However, it is found that, by increasing doping of boron or nitrogen in graphene, the high-intensity peak at 4 eV decreases in intensity, indicating the reduction in absorption. The peak intensity and position are not found to change for out-of-plane polarization. Hence, one can note that the 4 eV peak is of principal importance while considering tailoring of optical properties of graphene. However, this is not the case for in-plane modes in bilayer graphene with AB stacking and graphite. Thus, any loss in absorption below 10 eV, due to these plasmons, can be attributed to the presence of adsorbates on graphene. This characteristic of graphene makes it highly sensitive to adsorbates.

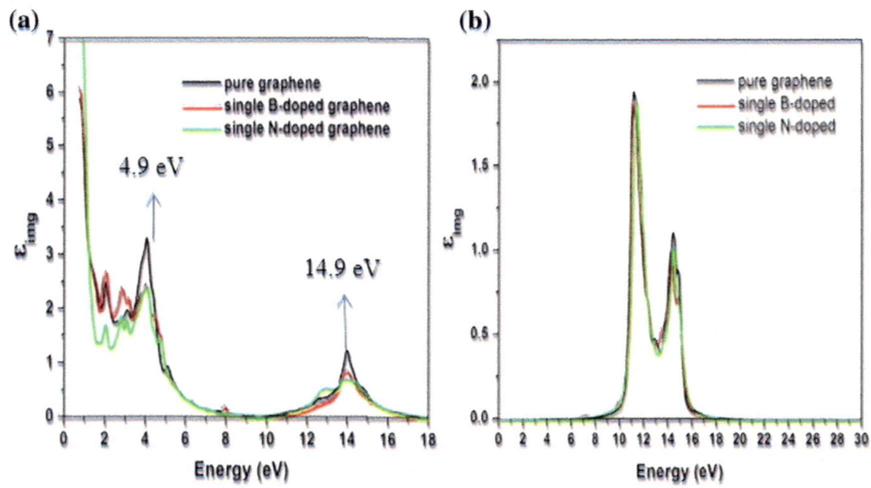

Fig. 5.10 Comparison of the imaginary part of the dielectric function of pure graphene with that of single boron and nitrogen atom-doped graphene sheet for E⊥c (**a**) and E||c (**b**), based on simulations. Peaks observed at 4.9 and 14.9 eV originate from $\pi \rightarrow \pi^*$ and $\sigma \rightarrow \sigma^*$ interband transitions, respectively [107]

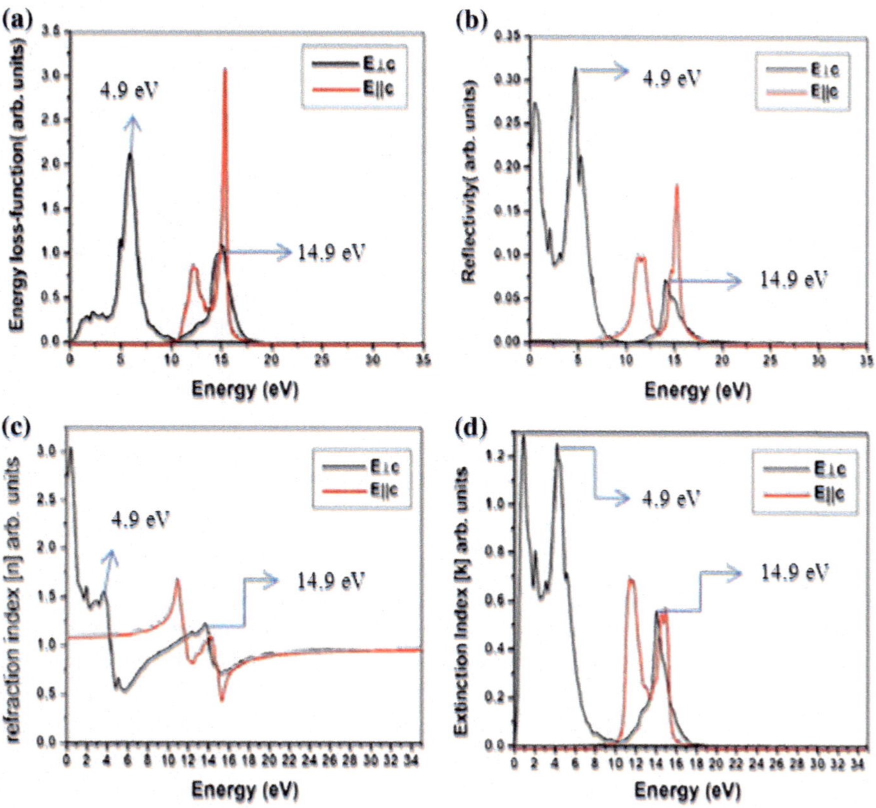

Fig. 5.11 Simulation of the electron energy loss function (**a**), reflectivity (**b**), refractive index (**c**), and extinction coefficient (**d**) of pure graphene for E⊥c and E||c [107]. Peaks observed at 4.9 and 14.9 eV originate from $\pi \rightarrow \pi^*$ and $\sigma \rightarrow \sigma^*$ interband transitions, respectively [107]

Figure 5.10 shows the imaginary part of the dielectric function of graphene, both pristine and doped graphene, for E⊥c and E||c [107].

The absorption spectrum for E⊥c in graphene shows a significant peak at energies up to 5 eV and another peak structure of broader energy range which starts at about 10 eV and has a weak intensity peak at 14 eV. These peaks originate from $\pi \rightarrow \pi^*$ and $\sigma \rightarrow \sigma^*$ interband transitions, respectively, in accordance with the previous interpretations by Marinopoulos et al. [107]. The experimental value of the high-intensity peak is at 4.6 eV, as reported by Eberlein et al. [110]. The imaginary part of the dielectric function shows a singularity at zero frequency for E⊥c. This makes graphene to exhibit optically metallic property for E⊥c, while for E||c, graphene shows semiconductor properties. Electron energy loss function, reflectivity, refractive index, and extinction coefficient are presented in Fig. 5.11. For parallel polarization of light with respect to the plane of graphene sheet, reflectivity is more and transition

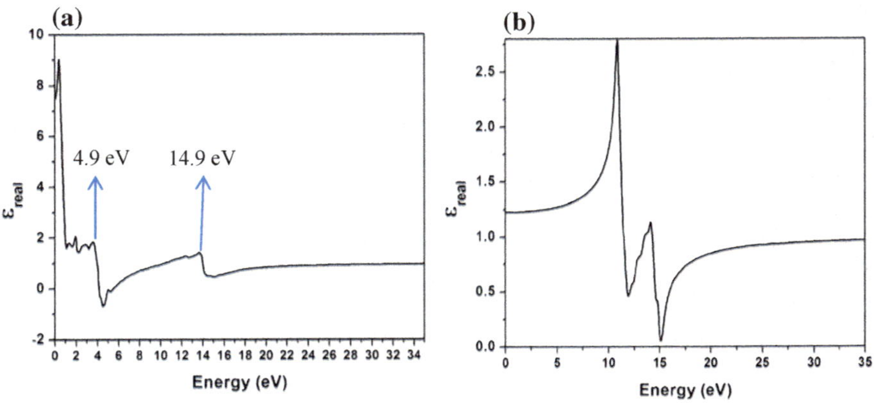

Fig. 5.12 Simulated real part of the dielectric function of pure graphene for E⊥c (**a**) and E||c (**b**). Peaks observed at 4.9 and 14.9 eV originate from $\pi \rightarrow \pi^*$ and $\sigma \rightarrow \sigma^*$ interband transitions, respectively [107]

is less in this energy range. In the energy range of 10–15 eV, for light polarization perpendicular to graphene sheet, the reflectivity is on the higher side.

The imaginary part of the dielectric constant of single-layer graphene, for out-of-plane polarization (E||c), consists of two prominent peaks at about 11 and 14 eV, as presented by Marniopoulos et al. [107]. This polarization spectrum varies with respect to graphite. Graphite has a weak intensity peak in the energy range of 0–4 eV. This peak is absent in the case of single-layer graphene. The real part of the dielectric function of pure graphene is plotted in Fig. 5.12. The value of the static dielectric constant (value of dielectric function at zero energy), in case of E⊥c, is reported to be 7.6, while it is 1.25 in the case of E||c.

Corato et al. [115] presented the optical properties of bilayer graphene nanoflakes theoretically, in order to explore the role of π–π interactions. They considered two different types of π-stacking with varying electronic gap or ionization potential. Their results indicate a redshift and broadening of lowest excitations. Thus, one can expect overall broadening of the optical absorption in ensemble of flakes. In heterogeneous ensemble of flakes, there is a possibility of the presence of low-energy excitations with considerable charge transfer character, which can be demonstrated by proper exploitation of chemical edge functional.

Chernov et al. [116] have discussed the optical properties of graphene nanoribbons encapsulated in single-walled carbon nanotubes and reported the photoluminescence in both visible and infrared spectral range. These photoluminescent peaks are found to be resonant and their position is dependent on the geometrical structure of the ribbon.

Hong et al. [117] have reported the thermal and optical properties of free-standing flat and stacked single-layer graphene in aqueous media. They demonstrated that stacked graphene structures thermalize rapidly than flat graphene and display non-equilibrated electron and phonon temperatures upon excitation. Hot electron luminescence, characteristic of graphene, is found to depend on Fermi energy and the morphology of graphene. Thus, it has been interpreted that the morphology of graphene structure could affect its optical and thermal properties.

Graphene is also known to be an extraordinary material for THz devices as its THz properties have an origin in the band structure [118]. In the low THz regime, graphene is found to exhibit linear response, where its optical conductivity follows Drude-like behavior and is mainly governed by its Fermi level. Based on this phenomenon, many THz devices have been proposed. In the high THz regime, graphene is found to exhibit third-order nonlinear response which can be orders of magnitude larger than what can be achieved in other frequency ranges such as IR or visible.

The experimental measurements of the optical properties of graphene, in the long wavelength range, are very limited in the literature. The electronic and optical properties of 2D materials have been discussed in detail by Marco et al. [119].

5.2.2.4 Emissivity (an Infrared Optical Property): Significance and Basics

Rapid thermal processing (RTP) is a widely used processing technique for the manufacture of silicon and other semiconductor devices. The short process times, high-temperature ramp rates, and very high temperatures are essential in RTP [120]. During RTP, factors such as the temperature uniformity across the surface of the wafer, process reproducibility, and accuracy are core requirements to its successful operation. Temperature uniformity across the wafer is affected by design parameters such as wafer patterning, temperature accuracy, and uniformity of irradiation. Temperature accuracy depends on the technique used for the measurement. The use of thermocouples in temperature measurement is highly intrusive and the delicate thermocouple wires make handling of wafers especially difficult and pose problems in sealing vacuum chambers [121].

In this regard, pyrometers are the most suitable choice of temperature measurement techniques [122]. Pyrometers are used to measure the amount of radiation emitted within narrow window of wavelength. However, for obtaining accurate temperature measurements using these devices, the knowledge of some key optical properties of the material being analyzed, is essential. Spectral emissivity is defined as the ratio of the radiation emitted by a given substrate to that of a black body under the same conditions of temperature, wavelength, angle of incidence, and direction of polarization [123]. Emissivity is a number between 0 and 1. The wavelength, transmittance, absorptivity, absorption coefficient, reflectivity, etc., of the materials are taken into consideration [124] in designing pyrometers. Ratio pyrometry is a technique of radiometric method which can eliminate the like terms from ratios of measured signals. Multi-wavelength Imaging Pyrometers (MWIP) are designed to obtain profiles of temperature, remotely, of targets of unknown wavelength-dependent emissivity [123]. MWIP incorporates least-squares fit of the signal taken from the radiometric model of an infrared (IR) camera and is based on the simultaneous measurement of emissivity and temperature. The selection of the emissivity model is a decisive factor for the accuracy of the least-squares-based MWIP technique for measurement of temperature [125, 126].

The significant interest in radiative properties of materials is in applications such as process monitoring and control of materials, noncontact temperature sensors, pyrometry, infrared detectors including bolometers, night vision, etc. [124]. However, these properties are not readily available in the literature and the results presented in this study can be helpful in various applications including thermal management of high power electronic devices. The need for thermal management in electronics has led to the development of alternate materials, techniques of manufacturing, and designs, in order to have a higher lifecycle of the electronic devices; cost factor is also an important consideration.

Radiative properties of graphene are of significant interest in applications such as process monitoring and control, noncontact temperature sensors, pyrometry, and infrared detectors. These require knowledge of spectral emissivity. Knowledge of emissivity as well as the relation of resistance as a function of temperature for multilayered configuration, with graphene as one of the layers, can be used to fabricate noncontact sensing devices such as Hot Electron Bolometer (HEB). We have studied various configurations of graphene-based bolometers and proposed alternative configurations for achieving better ratio of change in resistance with respect to temperature, i.e., for improving sensitivity.

5.2.2.5 Significance of Multilayered Structures and Their Optical Properties

In industrial scenario, one needs to control or reduce friction, wear, and corrosion of components, in order to extend the life of the device. This also leads to conservation of scarce material resources, saving of energy as well as improvising safety in engineering applications. Nano-coatings such as thin films and engineered surfaces have been developed and applied in industry for decades. Nano-coatings can be used to enhance the surface-related characteristics such as optical, magnetic, electronic, and catalytic properties.

In phenomenon of reflection of light in a thin film, both the top and bottom boundaries have to be taken into consideration. In a thin layer, the incident light will be reflected back at the secondary boundary, followed by the transmission through the first boundary. This leads to a path difference between the two waves (incident light wave and reflected wave from the inner surface). This phase difference results in interference of the waves. Therefore, the reflectance is a function of the thickness of the films and the wavelength of light.

5.2.2.6 Device Studies

A bolometer is a device which responds to increase in temperature (on interaction with the incident thermal radiation) by changing its resistance [123, 127]. When an electron system is coupled with power, electrons as well as phonons are driven out of thermal equilibrium, leading to the creation of hot electrons. Hence, such a bolometer is called Hot Electron Bolometer (HEB) [128]. It is essentially a sensitive thermometer. It can be used in conjunction with a spectroscope to measure the ability of some chemical compounds to absorb wavelengths of infrared radiation. One can obtain important information about the structure of the compounds with this technique.

Graphene has the ability to absorb light from mid-infrared to ultraviolet with nearly equal strengths. Hence, it has found applications in optical detectors. Graphene is particularly well suited for HEBs due to its small electron heat capacity and weak coupling of electrons and phonons, which causes large induction of light with small changes in electron temperature. The small value of electronic specific heat of graphene makes it possible for faster response times, higher sensitivity, and low noise equivalent power [123]. At low temperatures, usually in the cryogenic range, electron–phonon coupling in metals is very weak. The usual operating range of operation of HEBs is cryogenic. Graphene-based HEBs can be used at higher operating temperatures due to lower electron–phonon scattering even at room temperature along with the highest known mobilities of charge carriers at room temperature [48, 127, 129].

5.2.3 Mechanical Properties

The past few decades have witnessed an exponential increase in research on nanomaterials; examples include two-dimensional structures and carbon nanotubes. Nanomaterials are materials with a characteristic length less than 100 nm. At the nanoscale, increased ratio of surface area to volume can drastically change the mechanical behavior and properties of a material. This phenomenon in novel nanomaterials has been unexplored at large in view of a large number of innovations in materials science. Mechanical characterization of nanomaterials is performed by indentation testing on nanoscale thin films. In a traditional indentation test, such as the Vickers hardness test, the hardness is evaluated by division of the force applied by a pyramidal tip by the projected indentation area after unloading. This approach is easier at the macroscale. However, at the nanoscale, it requires significantly more accurate equipment. The measured area is known to have errors from effects such as elastic recovery and pile-up [130].

In 1986, Doerner and Nix proposed a methodology to determine the hardness and elastic modulus from load–displacement curves as obtained by nanoindentation technique [131]. This method was further refined in 1992 by Oliver and Pharr in order to determine the elastic modulus of thin films from the linear portion of the force–displacement unloading curve [130, 132]. The reduced elastic modulus, E_r, contact stiffness, S, and area are shown to be related by the Eq. (5.2.10).

$$S = \frac{2\sqrt{A_c}}{\sqrt{\pi}} * E_r \tag{5.2.10}$$

S can be calculated from the linear portion of the load–displacement unloading curve, as shown in Fig. 5.13 [130].

Reduced modulus can be easily calculated by knowing the area. Hardness is related to force and indented area in accordance with the following equation:

$$H = \frac{F}{A_c} \tag{5.2.11}$$

where H is hardness, F is maximum force applied, and A_c is the contact area at the maximum load. Nanohardness machines record the force applied, but the contact area is difficult to determine at the nanoscale [130]. The elastic modulus and hardness require the determination of the accurate contact area. The Oliver–Pharr technique includes the creation of an "area function" for the indenter. This area function is characteristic of each indenter, and it relates the contact area to the depth of penetration of the tip. The accurate indentation area corresponds to the projected contact area at maximum load, as shown in Fig. 5.14. This aspect is a variation from the classical approach of hardness testing, in which the projected area of the residual indentation is considered, after unloading.

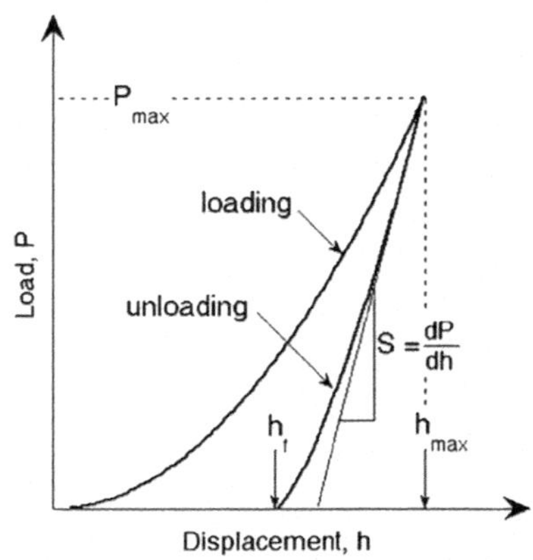

Fig. 5.13 Typical load–displacement curve. Notice the linear portion of the unloading curve, where S is found [130]

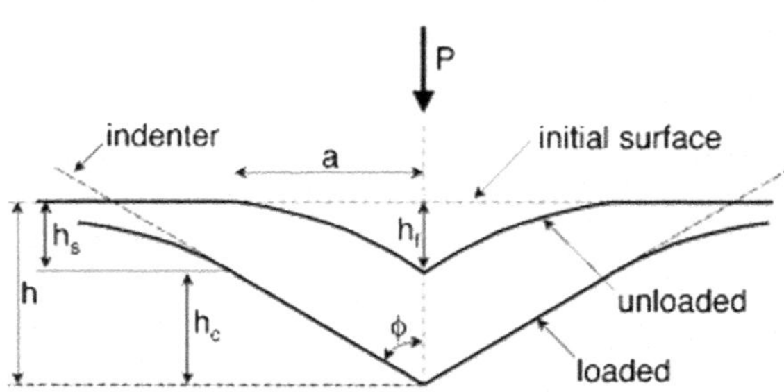

Fig. 5.14 Schematic of indentation with sink-in deformation. The schematic shows multiple indentation depths at maximum loading and after unloading. The correct contact area is related to the depth h_c [130]

5.2.3.1 Literature Review

Graphene is reported to have an elastic modulus of the order of 1 TPa and intrinsic strength (130 GPa) [133]. However, the mechanical properties of graphene need to be fully understood in view of its proposed applications in nanodevices (sensors, resonators, etc.). For graphene, being a two-dimensional material, its in-plane tensile behavior is the most important mechanical behavior. The elastic modulus E, the Poisson's ratio ν, and the intrinsic strength are the most fundamental mechanical properties. The values of elastic modulus and intrinsic strength have been obtained using free-standing indentation based on Atomic Force Microscopy (AFM). Lee et al. [53] used instrumented indentation and reported the value of elastic modulus as 1.0 TPa for single-layer graphene, while Frank et al. [134] have measured the value of E as ~0.5 TPa for stack of graphene nanosheets ($n < 5$). Zhang and Pan [135] reported the elastic modulus of monolayer of graphene as 0.5 TPa, bilayer graphene as 0.89 TPa, and observed reduction in the values with increasing number of layers of graphene. Lee et al. [136], using Raman spectroscopy, reported the values of elastic modulus of monolayer and bilayer graphene as 2.4 and 2.0 TPa, respectively.

Force friction microscopy (FFM) showed that there is monotonical decrease in the frictional force of monolayer graphene with increasing thickness. It is reported that the graphene's frictional force is roughly twice that of bulk graphite, and the tip–graphene adhesion is found to be constant [137–139]. Interlayer shear strength of graphene has been determined by performing AFM on a corrugated substrate and it is determined to be greater than or equal to 5.6 MPa, which is orders of magnitude less than its tensile yield stress [140]. This research prompted the application of graphene as a thin film and led to the determination of mechanical properties of graphene as suspended sheet and in polymer composites [24]. However, its out-of-plane mechanical properties, such as hardness, are relatively less explored. In the applications using graphene as an electrode, it is important to understand and quantify the load and impact that graphene can withstand. A protocol was developed recently in the Sansoz lab to evaluate the mechanical properties, especially the hardness, of thin films by using Atomic Force Microscopy (AFM) nanoindentations [141].

Cao [133] has discussed the recent advances in the atomistic studies of mechanical properties of graphene, with a focus on the in-plane tensile stress response, geometric characteristics, and free-standing indentation response of graphene. Numerical analysis can offer link between scientific research and engineering applications such as nanosensors, nanotransistors, and others. Three basic tensile loading modes, such as uniaxial tensile stress/strain and biaxial tension are applied based on displacement control. Graphene has a sixfold symmetric lattice. The in-plane orientation of graphene has been typically stated using chirality angle θ. θ varies from 0 to 30°, where $\theta = 0°$ and 30° correspond to zigzag and armchair directions, respectively. It is reported that irrespective of the magnitude of tensile deformation, graphene is isotropic with regards to elastic modulus and the Poisson's ratio [133].

The intrinsic stress on graphene cannot be measured directly using AFM; it can be evaluated by performing inverse analysis of experimental data using Finite Element Modeling (FEM). Breaking indentation force from FEM needs to be compared to the experimental counterpart. Based on this approach, Lee et al. [53] reported the Piola–Kirchhoff (second PK) stress as 42 ± 4 N/m and the Lagrange strain as 0.25. However, in this approach, it is assumed that the bending stiffness of graphene is to be neglected and the difference between mechanical behavior (calculated by considering the second and third-order nonlinear terms of elastic modulus) and the true behavior in free-standing indentation is neglected.

DFT studies can also help to determine the values of the 2nd PK stress and Lagrange strain. Liu et al. [142] have determined the phonon spectrum of monolayer graphene under uniaxial tensile stress with DFT simulation, and it was proposed that the maximum stress will correspond to the first occurrence of phonon instability. Based on this, they reported the values of 2nd PK stress and Lagrange strain to be 31.03 N/m and 0.3 for zigzag direction, while for armchair direction, the values are 30.3 N/m and 0.21, respectively. Wei et al. [143] used DFT simulations and predicted the ductile nature of graphene by plotting the stress–strain behavior. In the case where the tensile strain is larger than Lagrange strain, graphene is still found to support a stress which is lower than the tensile stress, especially in the zigzag direction.

Figure 5.15 shows that the failure strain of graphene is much higher than the Lagrange strain along zigzag and armchair directions [133]. Thus, graphene is found to be anisotropic with regards to the fracture resistance. Graphene exhibits much stronger failure strain along zigzag direction than the armchair direction.

The literature reports variations in the failure strains in graphene using DFT simulations and MD (Molecular Dynamics) methods. This is attributed to different C–C bond behavior. The C–C bond behavior, in DFT simulations, is dependent on the behavior of electrons explicitly calculated in the simulations. This type of behavior is a superposition of both the electrostatic attraction between the nucleus and electrons, as well as, repulsion between the electrons. Thus, the bonds show nonlinear

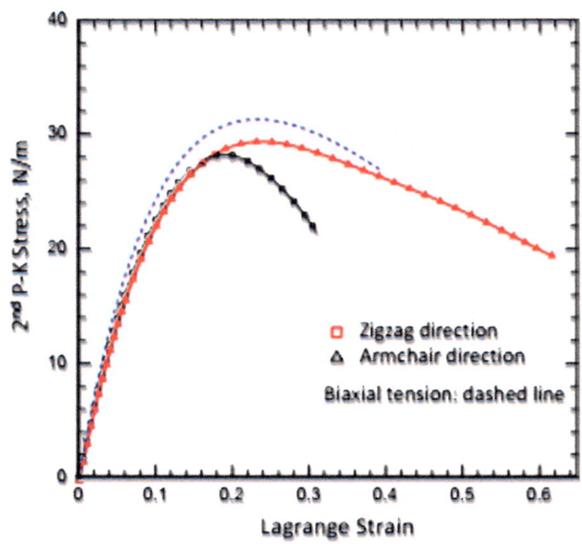

Fig. 5.15 Simulation-based relationship between the second P–K stress and Lagrange strain of graphene determined under uniaxial/biaxial stress tension [133]

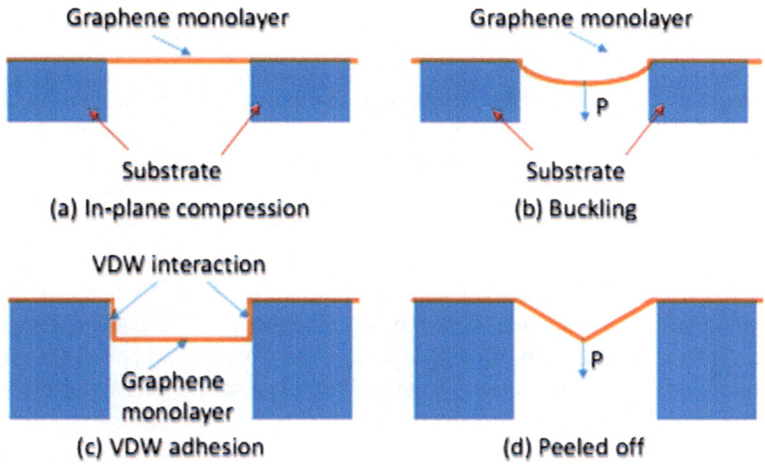

Fig. 5.16 Schematic of the creating mechanism of the true boundary condition of graphene in free-standing indentation: **a** in-plane compression; **b** buckling; **c** adhesion by the vdW interaction between substrate wall and graphene; and **d** peeled off by the indentation load [133]

behavior without cutoff feature, which is the typical characteristic of the bonds in MD methods.

It is reported that the elastic modulus is highly overestimated in the free-standing indentation test mainly due to vdW (Van der Waals) interaction between indenter tip and graphene, known as the vdW effect. The vdW force between graphene and side wall is not large enough to maintain a clamped boundary condition of graphene in free-standing indentation tests.

Figure 5.16 shows the reported mechanism for large size suspended graphene in free-standing indentation [133]. One can find that, for a 2D material, the mechanical properties are a function of the thickness, vdW interaction with indenter tip, substrate and/or within graphene layers, geometric defects (vacancies). These factors affect the mechanical deformation of graphene under different loading conditions/modes. Using the conventional continuum theory, the measured mechanical properties of graphene will be different, as the factors mentioned earlier, are not accounted for properly. Hence, numerical simulations, at the atomistic scale, play a significant role in the understanding of the mechanical properties of 2D materials.

5.3 Computational Methods

In this section, various computational techniques used to model the electronic, optical, mechanical, and thermoelectric properties are discussed.

5.3.1 Density Functional Theory (DFT)

The main principle of the ab initio techniques is to solve the Schrodinger's equation. This single equation is sufficient to describe arbitrary systems accurately. When applied to many-body problems, the analytical solution is highly impractical. Hence, there is an intense need to look for approximations.

In a short span of time, after Schrodinger's equation was published, the first rudimentary predecessor of DFT was developed by Thomas and Fermi [144]. In this approach, a multi-electronic system based on Fermi–Dirac statistics, assuming the behavior of the system as a homogeneous electron gas, is modeled. The model considers interacting electrons moving in an external potential and provides a highly oversimplified relation between the potential and its electronic density. This theory is quite useful for describing some qualitative trends, such as total energy of atoms. However, there is a major drawback of not being able to predict the chemical bonding of atoms in the system.

Another most successful attempt for dealing with many-electron systems is the Hartree–Fock (HF) hypothesis, developed in 1930 by Hartree, Fock, and Slater [145]. In this approximation, a multi-electron wavefunction HF can be calculated as an antisymmetric combination (known as the Slater determinant) of wavefunctions i of N electrons (known as spin-orbitals) composing it, including the Pauli Exclusion Principle.

By definition, the HF theory is only an approximation since it is assumed that all-electron wavefunctions have a particular shape. Hence, in cases where accuracy is a concern or when there are strong electron–electron interactions, it is inferior as compared to other methods. However, at present, it is widely used in case of periodic systems.

Over the past few decades, DFT has been the most successful and widely used method in the field of Computational Condensed Matter Physics [146]. The DFT describes a many-body interacting system via its particle density and not via its many-body wavefunction. This reduces the 3 N degrees of freedom of the N-body system to only three spatial coordinates through its particle density. Hohenberg and Kohn tried to formulate the DFT as an exact theory for many-body systems [147]. DFT is based on the Hohenberg–Kohn theorems. These theorems basically state that the properties of the ground state of a many-electron system are determined uniquely by the electron density (first theorem) and that this quantity can be calculated by using a variational principle (second theorem). The ground state energy is, hence, a function of the density. Along with the energy, all the physical properties of the system are also a function of the density. This enormously simplifies the problem of a many-electron system.

The Hohenberg–Kohn formulation applies to a system of interacting particles in an external potential $V_{\text{ext}}(r)$, including any problem of electrons and fixed nuclei, where the Hamiltonian is given by Eq. (5.3.1) [148],

$$\hat{\mathrm{H}} = -\frac{\hbar^2}{2m_e} + \sum \nabla_{i^2} + \sum V_{\mathrm{ext}}(r_i) + \frac{1}{2}\sum_{i \neq j} \frac{\mathrm{e}^2}{(\check{\mathrm{r}}_i - \check{\mathrm{r}}_j)} \tag{5.3.1}$$

The Hohenberg–Kohn theorems are as given below:

(a) **Theorem I**: For any system of interacting particles, in an external potential $V_{\mathrm{ext}}(r)$, the potential $V_{\mathrm{ext}}(r)$ is uniquely determined, except for a constant, by the ground state particle density $n_0(r)$.
(b) **Theorem II**: A universal functional for the energy $E[n]$ in terms of the density $n(r)$ can be defined. This is valid for any external potential $V_{\mathrm{ext}}(r)$. For any given external potential $V_{\mathrm{ext}}(r)$, exact ground state energy of the system is the global minimum value of this functional, and the density $n(r)$ that minimizes the functional is the exact ground state density $n_0(r)$.

Hohenberg–Kohn theorem has made it possible to use the ground state density to calculate the electronic properties of the system. But it does not provide a way of finding the ground state energy. Kohn–Sham (KS) equations provide a method to determine the ground state energy. In order to derive the KS equations, one needs to consider the ground state energy as the function of charge density. Kohn–Sham derived a set of single-particle Schrodinger equations given by [149],

$$\left[\frac{-\hbar^2}{2m}\nabla^2 - V_{\mathrm{eff}}(r)\right]\Psi_i(r) = \varepsilon_i \Psi_i(r) \tag{5.3.2}$$

where ε_i are Kohn–Sham eigenvalues, $\psi_i(r)$ are Kohn–Sham single-particle orbitals.
The effective potential is given by [150],

$$V_{\mathrm{eff}}(r) = V_{\mathrm{ext}}(r) + V_{\mathrm{Hartree}}(r) + V_{xc}(r) \tag{5.3.3}$$

where $V_{\mathrm{eff}}(r)$ represents the electron–ion interaction [148]. $V_{\mathrm{Hartree}}(r)$ represents the Hartree potential, i.e., the classical electrostatic interaction. $V_{xc}(r)$ is the exchange-correlation potential.

$V_{\mathrm{Hartree}}(r)$ and $V_{xc}(r)$ are given by Eqs. (5.3.4) and (5.3.5), respectively.

$$V_{\mathrm{Hartree}}(r) = \int \frac{n(r')}{|r - r'|}\mathrm{d}r' \tag{5.3.4}$$

$$\mathbf{V_{xc}(r)} = \frac{\boldsymbol{\delta}\mathbf{E_{xc}}}{\boldsymbol{\delta}\mathbf{n(r)}} \tag{5.3.5}$$

where E_{xc} is the exchange-correlation energy. $V_{xc}(r)$ represents the many-body effect. Approximation for the E_{xc} [148] is stated in Eq. (5.3.6) in Sect. 5.3.1.1.

5.3.1.1 LDA (Local Density Approximation)

LDA is a class of approximations to exchange-correlation energy functionals (E_{xc}) in DFT. For spin-unpolarized systems, LDA for E_{xc} is given in Eq. (5.3.6):

$$E_{xc}^{\mathrm{LDA}}[\rho] = \int \rho(r) \in_{xc} (r)\mathrm{d}r \tag{5.3.6}$$

where ρ is the electronic density and $\in_{\mathrm{xc}}$ is the exchange-correlation energy per particle of a homogeneous electron gas of charge density ρ. In LDA, the functional depends only on the electronic density at the coordinates, where the functional is evaluated. This approximation can be used to find Eigen function and eigenvalues of the Hamiltonian. It is commonly used along with plane wave basis set. LDA considers a functional whose functional derivative is taken with respect to density at that point only. It is an approximation to the exchange-correlation, which depends on the value of the electronic density at each point. Local approximations to the exchange-correlation energy are derived from the homogeneous electron gas model (such as the Jellium model). The exchange functional can also be expressed as the energy of interaction between the electron density and the Fermi coulomb hole charge distribution. LDA is synonymous with functionals based on the HEG approximations, which are applied to realistic systems such as large molecules and solids. The expression for energy and potential is given below [148]:

$$E_{xc}(n) = \int E_{xc}[n]n\left(\left(\check{\mathrm{r}}\right)\right)\mathrm{d}^3r \tag{5.3.7}$$

where E_{xc} = exchange-correlation energy per electron. The exchange-correlation energy is composed of two terms: exchange term and the correlation term, linearly as in Eq. (5.3.8).

$$E_{xc}[n] = E_x + E_c \tag{5.3.8}$$

where E_x is the exchange term and E_c is the correlation term. The exchange term takes a simple analytical form for the HEG, while only limiting terms are precisely known for the correlation density. This leads to various approximations for E_c.

The exchange-correlation potential, corresponding to the exchange-correlation energy, is stated in Eq. (5.3.9).

$$\mathbf{V_{xc}(r)} = \frac{\boldsymbol{\delta}\mathbf{E_{xc}(r)}}{\boldsymbol{\delta}\mathbf{n}} \tag{5.3.9}$$

In finite systems, the LDA potential is known to decay asymptotically with exponential form. This is slightly erroneous. The true exchange-correlation potential decays much slower in a Coulombic way. The LDA potential cannot support the Rydberg series and such states are too high in energy. LDA does not provide an accurate description of electron-rich anions. Therefore, LDA is unable to bind an additional electron and predicts the anionic species to be stable erroneously [151].

5.3.1.2 LDA+U

This approximation is an improved form of LDA. In case of transition metal or rare earth metal ions which have strongly correlated system, the LDA approximation is not sufficient to describe the electronic properties of the system. If one applies LDA to a transition metal compound, it will provide the metallic electronic structure with partial d-band, which is erroneous. Several approaches have been worked out for improving LDA to include the differences in the strongly correlated electron–electron interaction. The Hamiltonian of LDA for the case, described in Sect. 5.3.1.1, can be improved by using the calculated self-energy in a consistent procedure.

The orbital-dependent potential, taking LDA+U approximation, exhibits both the upper and lower Hubbard bands with difference of coulomb parameter U. LDA+U approximation shows that the information obtained is not sensitive to a particular form of localized orbitals. LDA+U theory can be described as the Hartree–Fock theory for localized states (orbital of rare earth metal). By using the Hubbard U term, a correction term to Hamiltonian of LDA, there is a large increase in the number of calculations for the electronic structure. The Hubbard parameter term U relates the single-particle potential to the magnetic order parameter. For impurity systems, high T_c superconductors, Mott insulators, transition metals, the LDA+U approach is highly accurate. Delocalized s, p electrons can be described by using an independent one-electron potential. For localized d and f electrons, LDA, including Hubbard-like term, can be used instead of averaged coulomb energy [148].

5.3.1.3 Quantum Espresso

It is an integrated suite of computer codes, including the plane wave pseudopotential DFT code PWSCF. This suite is used for electronic structure calculations and materials modeling at the nanoscale [148]. The software is released under the GNU General Public License. The full Quantum ESPRESSO distribution contains the following core packages for the calculation of electronic structure properties within Density Functional Theory (DFT), using a Plane Wave basis set and pseudopotentials:

- PWSCF (PW): Plane Wave Self-Consistent Field,
- CP (CPV): Car–Parrinello Molecular Dynamics.

It is based on Density Functional Theory (electron–ion interaction), plane waves, and pseudopotentials (both norm conserving and ultrasoft for electron–electron interactions).

In DFT, for a given periodic system, by determining the electronic states, one can evaluate the thermal, optical, and magnetic properties of solids, equations of state, electron density distributions, and cohesive energies of the system.

DFT has become a very widely used method in computational physics and computational chemistry research for determining the electronic structures of many-body systems, such as atoms and molecules. It has been particularly applied in the condensed state. PWSCF is a method used to calculate band structures by expanding wavefunctions into plane waves. Various auxiliary packages are also included along with the codes. PWgui is graphical user interface, which produces input data files for PWSCF. Several additional packages which can use the PWSCF data as input for post-processing have been included.

5.3.1.4 Pseudopotential

It is a modified effective potential term instead of coulombic potential term in Schrödingers' equation for core electrons. The information about the type of exchange-correlation functional and the type of pseudopotential can be found from the literature [150]. Pseudopotential is an effective potential constructed in order to replace the atomic all-electron potential such that the core states are eliminated and valence electrons are described by pseudo-wavefunctions with significantly lesser nodes. In pseudopotential, the Kohn Sham's radial equation is considered. It contains the contribution from valence electrons.

Pseudopotentials with larger cutoff radius are considered to be softer. However, they are found to be less accurate in different environments. Thus, the motivation in using pseudopotentials is to reduce the size of basis sets, reduce the number of electrons, and include relativistic (and other) effects. There are two types of pseudopotentials, i.e., ultrasoft pseudopotential and norm conserving pseudopotential used in modern plane wave electronic structure codes. These methods make it possible to consider basis sets with significantly lower cut-off (the frequency of the highest Fourier mode). These basis sets can be used to describe the electron wavefunctions. Thus, proper numerical convergence can be achieved with less computing resources [152].

5.3.1.5 XCrysDen

It is a molecular and crystalline-structure visualization program for input and output files generated by PWSCF. Its principal function is to serve as a property analyzer program. It can run on most UNIX platforms, without any specific hardware or software requirements. Special efforts have been made to allow appropriate display of 3D iso-surfaces and 2D contours. XCrysDen is also a graphical user interface for some other software codes such as CRYSTAL. It can perform real-time operations such as rotation and translation. It can be used for measuring distances, angles, dihedrals for a given crystal lattice.

5.3.2 *Emissivity Calculations*

Multi-Rad software is the modeling tool which has been used in our simulations of the radiative properties of thin-film stacks. In Multi-Rad, the thin-film optics is implemented in the form of the matrix method of multilayers. The model assumes the following: (a) the layers are optically smooth and parallel, as well as the materials are isotropic; and (b) constancy of the properties in azimuthal direction. For a given multilayer stack, it is possible to calculate the radiative properties as a function of wavelength and angle of incidence, at a specific temperature [153]. In Multi-Rad, a material can be defined using its refractive indices, n, and extinction coefficients, k. Most of the parameters in this study have been based on data from Refs. [154–156].

The matrix method of multilayers can predict the reflectance and transmittance of a multilayer stack for a specific wavelength and angle of incidence. For a specific wavelength, the coherent radiation is considered. This enables consideration of the interference effects. Most important assumptions of this theory are that the surface dimension on which the radiation is incident is much larger than the wavelength of the incident radiation, i.e., there are no edge effects [157]. The detailed description of the matrix method of multilayers can be obtained in Ref. [153].

The spectral absorptance is a directional property and it can be calculated by subtracting the reflectance and transmittance from unity, and the spectral emittance can be calculated by assuming Kirchhoff's law on a spectral basis:

$$\alpha_{\lambda,\theta} = \varepsilon_{\lambda,\theta} = 1 - \mathsf{R}_{\lambda,\theta} - \mathsf{T}_{\lambda,\theta} \tag{5.3.10}$$

where the subscripts are introduced to indicate the spectral and directional properties, respectively. Spectral directional reflectance and transmittance are denoted, respectively, as $\mathsf{R}_{\lambda,\theta}$ and $\mathsf{T}_{\lambda,\theta}$. They are calculated from the average of s and p wave properties. If it is assumed that a given emitter is in local thermodynamic equilibrium, Kirchhoff's law can be considered as valid on a spectral basis. It is characterized by a single temperature. In cases where there are high gradients of temperature across the emitting wafer, or where the electrons and phonons are not in local thermal equilibrium, Kirchhoff's law is not valid.

5.3.3 *Theory of Atomistic Simulation*

Atomistic simulation refers to a suite of computational techniques which are used to model the interaction and configuration of a system of atoms. In this work, the term "atomistic simulation" will pertain to molecular dynamics. Detailed and comprehensive reviews of "atomistic simulation" can be found in books by Allen and Tildesley [158] and Haile [159].

Atomistic simulations are commonly classified into two categories: equilibrium and nonequilibrium [159]. In equilibrium atomistic simulations, the system is completely isolated from its surroundings with a fixed number of atoms, volume, and constant total energy. These boundary conditions are considered to be corresponding to the microcanonical (NVE) ensemble in statistical mechanics [160]. In nonequilibrium atomistic simulations, the system is allowed to interact with the surrounding environment through either thermal or physical constraints (such as a thermostat or an applied force). Depending on the equations of motion which describe the system of atoms, these calculations may correspond to the canonical, either NVT [Ensemble with a constant number of particles (N) of the system, Volume (V) and Temperature (T)] or NPT [Ensemble with a constant number of particles (N) of the system, Pressure (P) and Temperature (T)], in statistical mechanics. Many different methods exist to specify the interaction between the atomic system and the environment. All of them are considered as nonequilibrium MD methods.

In the atomistic framework, each atom is represented as a point mass in space while the interatomic potential provides a model for the potential energy of a system of atoms. Commonly, the total potential energy of the system is written solely as a function of the positions of the atomic nuclei. This simplification avoids having to specifically account for the motion and interaction of the individual electrons. Since interatomic forces are conserved, the force on a given atom, F^i, is related to the interatomic potential, U, through the gradient operator, i.e.,

$$F^i = -\frac{\partial U(r_N)}{\partial r^i} \tag{5.3.11}$$

where "r" is the atomic position vector. In this work, superscripts denote variables assigned to individual atoms, while subscripts denote variables associated with sets of atoms, directions or at specific timesteps. Thus, $\boldsymbol{r}_N$ represents the position vectors for the system of N atoms while "r^i" is the atomic position vector for the ith atom. One of the inherent limitations of the atomistic method is the extremely high expense of computational resources. This makes it essential to limit the systems to relatively small numbers of atoms. The studies of nanoscale surface effects that are related to length scale are extremely important. The goal of this work is to examine the atomic scale behavior which can represent a bulk sample with micro or nanoscale grain structure. Thus, periodic boundary conditions are used in many of these calculations for eliminating the influence of free surface effects. As shown in Fig. 5.17, we consider a two-dimensional example of using periodic boundary conditions in the atomistic framework. The primary cell is outlined with solid lines and it represents a small portion of the material. The atoms which lie within this cell are explicitly modeled using atomistic methods. The bordering "image" cells, which are shown with dashed lines in Fig. 5.17, represent the infinite repetition of the primary cell in two dimensions. With this method, it is possible to model an infinite amount of material in each direction.

Suppose blue atom in the primary cell moves to a point outside of this region during the simulation, as shown with a solid blue arrow in Fig. 5.17. The image of this atom will be reflected back into the primary cell on the opposite side with the same momentum, as shown with the dashed blue arrow. Note that the atoms which

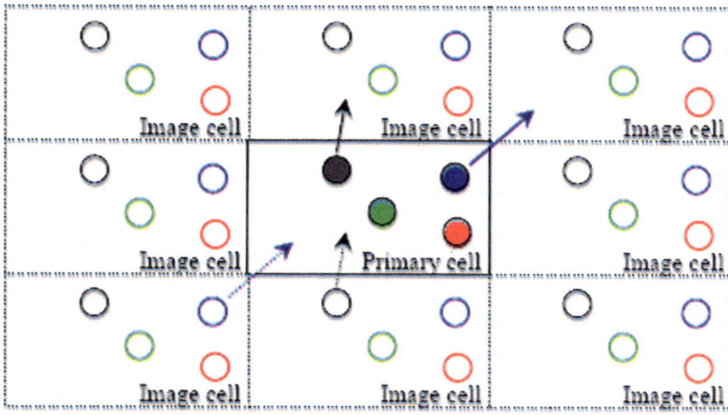

Fig. 5.17 Two-dimensional illustration of periodic boundary conditions in the atomistic framework

lie near the borders of the primary computational cell can interact with neighbor atoms across the periodic boundary. While periodic boundary conditions remove the effects of free surfaces, they are imparting image constraints on the system which must be taken into consideration when simulating defect behavior with long-range interactions.

The atomistic code, used in this study, is classical molecular dynamics code LAMMPS, which stands for Large-scale Atomic/Molecular Massively Parallel Simulator. LAMMPS was written by Steve Plimpton at Sandia National Laboratories/Albuquerque, NM. LAMMPS has potentials for soft materials (biomolecules, polymers) and solid-state materials (metals, semiconductors) as well as coarse-grained systems. It can be used to model atoms or, more generically, as a parallel particle simulator at the mesoscale or even up to the continuum levels [161].

LAMMPS runs on single processor or in parallel using message passing techniques. It performs a spatial decomposition of the simulation domain. The code is designed to be easy to modify or extend for better functionalities. LAMMPS is distributed as an open source code under the terms of the GPL. LAMMPS is distributed by Sandia National Laboratories, a US Department of Energy laboratory. The code is written in C++. This MD code is capable of performing molecular dynamics simulations in the microcanonical (NVE), constant volume canonical (NVT) ensembles and also the isothermal–isobaric ensemble NPT. The latest stable version of LAMMPS is 28 June 2014 and has been used in simulations performed in this study.

5.3.3.1 Molecular Dynamics

In the molecular dynamics method, the evolution of atomic positions is described using Newton's second law of motion,

$$p^i = F^i = mv^i \tag{5.3.12}$$

$$r^i = \frac{dr^i}{dt} = \frac{p^i}{m} \tag{5.3.13}$$

where m denotes the mass, p^i is the momentum, and v^i is the velocity of the ith atom. The most widely used method to solve the above equations in molecular dynamics is the "Velocity-Verlet finite-difference" algorithm [162]. This algorithm has been applied because its form is exactly time reversible. This allows the equations of motion to be propagated forward in time without iteration and symplectic, i.e., the volume in phase space is conserved. Thus, one can ensure stability and convergence of simulation for a bigger atomic system.

While the study of material behavior in isolated systems has merit, the vast majority of problems in mechanics and materials science require consideration of interaction of the system with the surrounding environment (nonequilibrium molecular dynamics). One way to accomplish this, in molecular dynamics, is to introduce the concept of an extended system [163]. Essentially, Newton's equations of motion are augmented and coupled to additional differential equations which can describe the relationship between the system and the environment. Commonly, molecular dynamics calculations are performed at a constant temperature or pressure (or both).

Of course, an accurate evaluation of the thermodynamic quantities makes it essential to consider the size of the atomistic system. The use of periodic boundary conditions in the atomistic framework serves as an effective way to approximate the thermodynamic limit.

5.3.3.2 Input File Parameters in MD Simulation Using LAMMPS

(a) **Initialization**

One has to define the initial configuration of the system in accordance with the literature. The dimensionality, boundary condition (periodic or fixed), atomic positions, timesteps, unit cells, and simulation box size are the parameters that need to be set for the material being analyzed.

(b) **Force Field Implementation**

Interactive potential needs to be defined in a system. A suitable empirical potential has been chosen as a function of time. Tersoff potential has been chosen to model graphene. This potential has been shown to be successful in describing atomic interactions for carbon.

(c) **Prescribing Ensemble and Running Simulation**

Before running an actual simulation, a thermalization process needs to be performed so that the system is in thermal equilibrium—the system is in thermal equilibrium at minimum energy. The ensemble is essential to perform this operation. In order to equilibrate the system, microcanonical ensemble (NVT) is designed using Nose–Hoover thermostat. It performs time integration on Nose-Hoover style and is based on non-Hamiltonian equations of motion which are designed to generate positions and velocities. This is achieved by adding some dynamic variables which are coupled to the particle velocities (thermostatting). When used correctly, the time-averaged temperature of the particles will match the specified target values.

5.3.3.3 Thermal Transport in Graphene

In recent years, electronic devices are becoming smaller. The interface between layers of materials is becoming increasingly important in the determination of their thermal properties. This is especially true in case of nanoscale materials in which the interface thermal resistance significantly affects their overall thermal conductivity. As the thermal management in electronic devices is a prime concern, the role of materials which are widely used in electronic devices such as carbon nanostructures, silicon-based materials, and the interface between these materials is becoming increasingly important. In this study, the method of molecular dynamics simulations is used for the calculations of thermal conductivity of graphene and its derivatives.

The thermal conductivity of graphene is difficult to be determined experimentally [1]. Hence, its thermal conductivity has been mostly predicted from theoretical methods. Experiments performed by Alexander et al. [164] show the thermal conductivity of graphene to be in the range of ~4840 to ~5300 W/mK. It is even higher than CNT. Several models have been successfully tested to investigate the thermal conductivity of graphene and graphite. Nika et al. [165] determined the thermal conductivity from phonon dispersion and they found the values to be in the range of 2000–5000 W/mK with varying flake size of graphite. Lan et al. [166] determined the thermal conductivity through Green–Kubo function and found the values of ~3410 W/mK. Baladin et al. [167] performed experiments to calculate the thermal conductivity of single-layer graphene as ~4200 W/mK, while graphite has a thermal conductivity of ~2000 W/mK. NEMD technique is applied in this simulation of thermal conductivity.

NEMD simulation has been implemented in this study to obtain the temperature gradient resulting from the swapping of kinetic energy. The thermal conductivity, κ, is then calculated from the Fourier law:

$$\kappa = \frac{-J}{(\Delta T\,.\,\lambda)} \tag{5.3.14}$$

where ΔT is the gradient of the temperature, J is the heat flux from the heat bath to the system, and λ is the total kinetic energy transfer between slabs. λ is defined as:

$$\lambda = \frac{\sum \text{transfers}\ \frac{m}{2}\left(v_h^2 - v_c^2\right)}{2tL_xL_y\langle \partial T/\partial z\rangle} \tag{5.3.15}$$

The subscript "h" refers to the hot particles and subscript "c" refers to the cold particles of identical mass "m". The velocities of these particles are interchanged. $L_x \times L_y$ is the cross-sectional area of simulation box. The factor of two in the denominator is used to denote heat flow in two directions of the slabs, effectively doubling the area of heat flux. The term T/z is the temperature gradient obtained from the average of the ensemble.

Equation (5.3.15) is used to determine the thermal conductivity of graphene and its derivatives. This Müller-Plathe technique is deployed to apply heat to the system. This is different from the Green–Kubo method [168]. This technique is sometimes called as the reverse nonequilibrium molecular dynamics (R-NEMD). This is because the usual NEMD approach is to impose a temperature gradient on the system and measure the response as the resulting heat flux. In the Müller-Plathe method, the heat flux is imposed, and the temperature gradient is the system's response. The schematic diagram of the Müller-Plathe method is shown in Fig. 5.18.

The basic principle of the Müller-Plathe method is that the complete system is to be divided into slabs along the axial direction, and the temperature of each slab is calculated by the following statistics:

$$T_k = \frac{1}{3n_k k_B} \sum_{i \in_k}^{n_k} m_i v_i^2 \tag{5.3.16}$$

In this technique, the first layer is assumed to be the cold slab, while layer $N/2$ (middle layer) is simulated as a hot slab. The hottest atom having maximum kinetic energy exchanges its energy with the adjacent atoms till the heat energy reaches the end of the cold slab (one with minimum kinetic energy). The temperature gradient between the atoms and the distribution of kinetic energy of atom is very broad. Hence, the hottest atom at cold slab always has higher kinetic energy than coldest atom at hot slab. The linear momentum and energy of the system are conserved, if the mass of the swapping atoms remains the same. However, the angular momenta of the atoms are not conserved. This is not a problem since the introduction of periodic boundary to the system allows the angular momentum to be neglected.

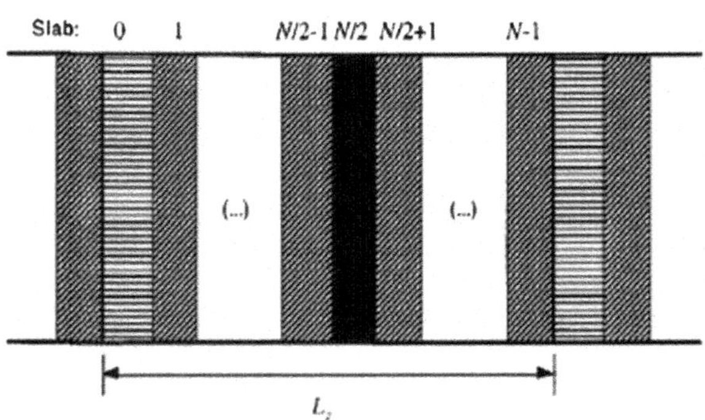

Fig. 5.18 Müller-Plathe method, SWNT has been differentiated into several slabs [168]

The Müller-Plathe algorithm is relatively easy to implement, as well as to maintain the system linear momentum; the total energy is conserved, and does not require an external heat bath. However, the Müller-Plathe method assumes that the cold slab and hot slab exchange of atomic mass are equal.

5.4 Modeling and Simulation

This section has been divided into four subsections based on the properties of graphene and its functional derivatives.

5.4.1 Electronic Properties

Graphene is known to have relaxed 2D honeycomb structure (Fig. 5.2) and the doped graphene will be assumed to have a similar structure, unless violated by energy minimization considerations. The optimizations of the lattice constants and the atomic coordinates are made by the minimization of the total energy. A sample of graphene nanostructure analyzed has been shown in Fig. 5.19.

The following steps and methodology have been used for simulating the various graphene-based nanostructures:

1. Setting up structure by inputting coordinates. VMD (Visual Molecular Dynamics) software [169] has been used for generating the input structure of graphene nanoribbons. An open source code "latgen" has been used for obtaining coordinates of graphene nanosheets and its derivatives.
2. Performing self-consistent (SCF) simulation of graphene nanostructure to obtain minimum total energy configuration.
3. Performing non-SCF calculation by varying energy cutoff, Monkhorst-Pack grid parameters, and lattice constant.

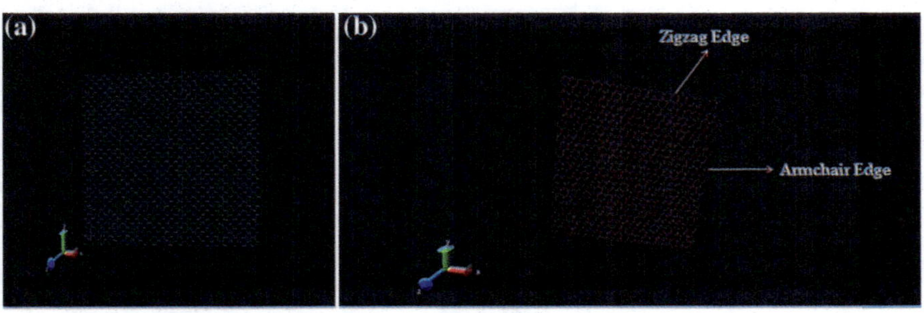

Fig. 5.19 Simulated structure of **a** graphene nanosheet and **b** graphene nanoribbon

4. Repeating the simulation till obtaining the lowest energy configuration in terms of energy, k-points as well as lattice constant.
5. Kinetic energy cutoff for wavefunctions has been chosen as 70 Ry, while the kinetic energy cutoff for charge density and potential has been set as 840 Ry (as a rule, it should be 8–12 times the kinetic energy cutoff for wavefunctions). It is noted that the total energy of the system is minimum at this cutoff.
6. Smearing with degauss 0.03 has been used in considering the semimetallic nature of graphene.
7. Energy convergence parameter has been set as 1.0E−8 Ry, i.e., the system converges once it reaches this energy level.
8. Path for plotting band structure has been chosen as G, M, K, G.

Figure 5.19 shows the simulated graphene nanostructures. These structures have been modeled using VMD software.

Once the verification of the model is established by comparison with the literature (experimental or simulated), the model is then extended to simulate the properties of graphene as function of orientation, number of layers, doping, vacancies, etc.

5.4.1.1 Band Structures

A. **Undoped Graphene**

As seen in Fig. 5.20, the Dirac point is located on K and the Fermi level is present at energy 0 eV. Thus, it is observed that undoped graphene behaves as a semimetal.

B. **Boron-Doped Graphene**

Upon boron doping, it is observed that the Fermi level is shifted below the Dirac point in graphene. The boron atoms adjust themselves to the surrounding host carbon atoms. When graphene sheet is doped with boron atoms, the boron atoms also

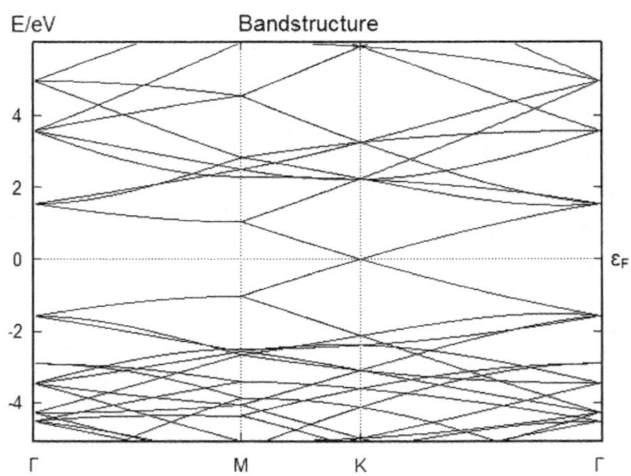

Fig. 5.20 Band structure of undoped graphene

undergo sp^2 hybridization. Due to similar size of carbon and boron atoms, there is no significant distortion in the structure of graphene, except for change in adjoining bond length. Doping induces bandgap in graphene. The variations of band structure are shown in Fig. 5.21. Due to electron deficient nature of boron, it is noted that the Fermi level moves about 1–2 eV below the Dirac point. This is evident from the band structures and shows that boron is a p-type dopant in graphene.

C. **Nitrogen-Doped Graphene**

Figure 5.22 shows the band structure trends by nitrogen doping of graphene. Figure 5.23 shows the trends of bandgap as a function of % nitrogen doping in graphene. It is evident from the study that nitrogen is an n-type dopant in graphene, as the Fermi level moves significantly above the Dirac point with increasing nitrogen doping. This also shows the electron-rich character of nitrogen as a dopant for graphene. It is known that the bandgap results due to breaking of symmetry in graphene sub-lattices. Hence, increasing the dopant concentration in graphene is expected to modify the bandgap in all the cases. However, it is found that the nature of bandgap versus % nitrogen doping is not uniform and it has a peak at about 8% nitrogen doping in graphene.

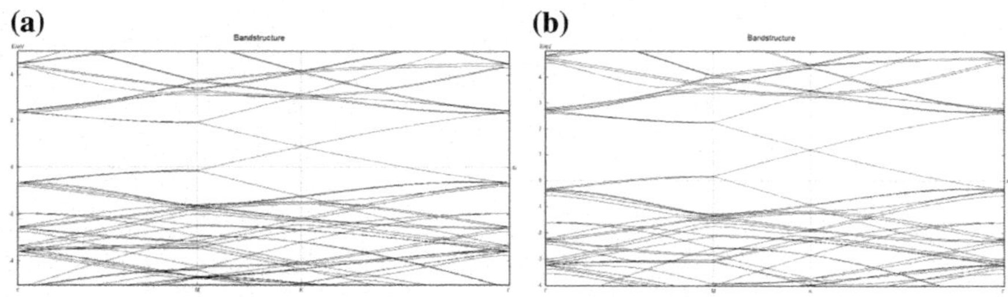

Fig. 5.21 Band structure of **a** 2% boron-doped graphene and **b** 4% boron-doped graphene

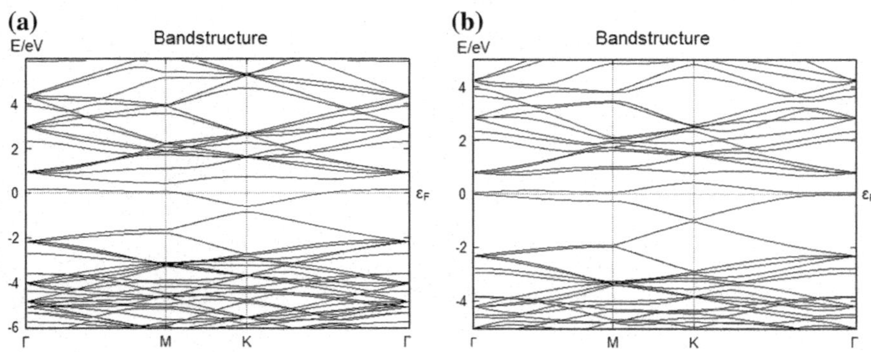

Fig. 5.22 Band structure of **a** 2% nitrogen-doped graphene and **b** 4% nitrogen-doped graphene

5.4.1.2 Trends in Density of States (DOS)

A. Pure Single-Layer Graphene

Figure 5.24 shows changes in the DOS of graphene computed using energy dispersion [74]. It is seen that the DOS is nil in the proximity of charge neutrality point (CNP) for graphene. Its zoomed-in view is shown in the figure on the right.

Figure 5.25 shows the density of states calculated for pristine graphene. It is found to exhibit the peculiar nature of graphene with nil DOS at the CNP, as compared with the literature.

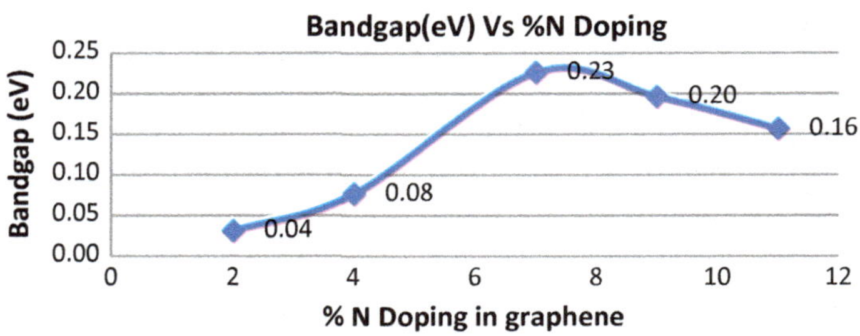

Fig. 5.23 Bandgap versus % nitrogen doping in graphene

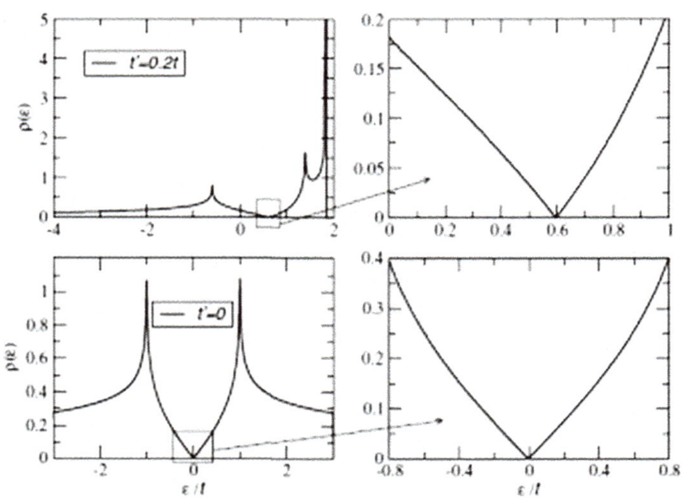

Fig. 5.24 DOS per unit cell as a function of energy (in units of t) computed from energy dispersion, with two different values of t' and the zoomed-in view of density of states close to the neutrality point of one electron per site [74]

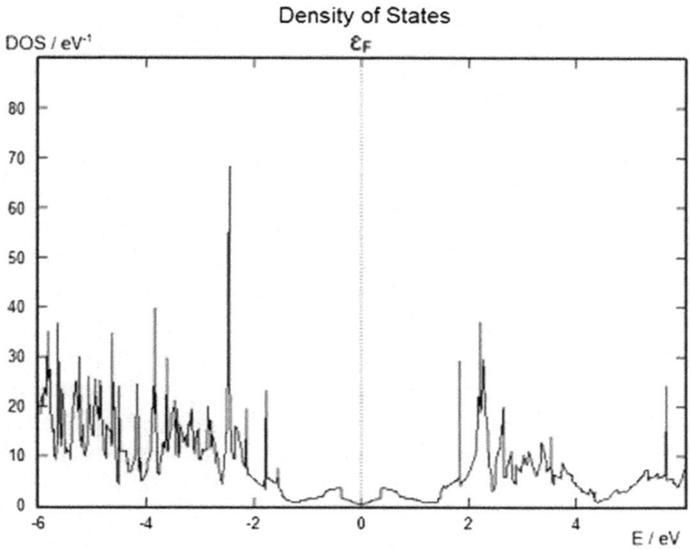

Fig. 5.25 DOS for pristine single-layer graphene, based on simulations

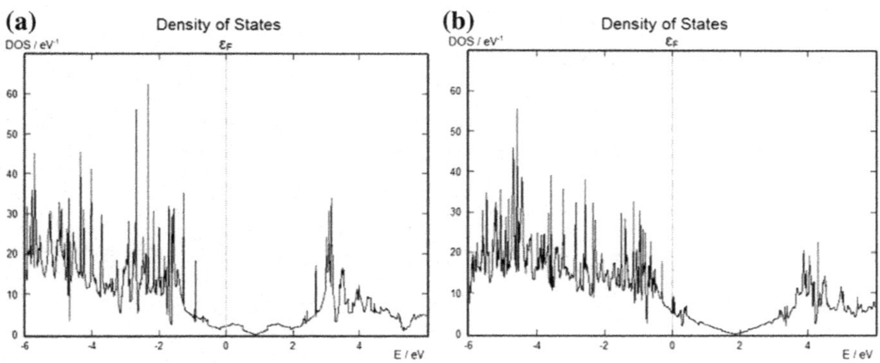

Fig. 5.26 DOS of **a** 2% boron-doped graphene and **b** 11% boron-doped graphene

B. Boron-Doped Graphene

Figure 5.26 shows the variations in the DOS for varying % boron doping. The shift in the Dirac point from the Fermi level can be observed. The point with nil DOS is found to shift from about 0.7 eV for 2% boron to about 1.8 eV for 11% boron doping in graphene. Similar studies have also been performed for nitrogen doping in graphene.

5.4.2 Optical Properties

5.4.2.1 Pure Graphene

Figure 5.27 shows the trends in the dielectric constant from simulations. This is comparable with the literature, as in Fig. 5.12. Figure 5.27a shows the real part, with peaks at about 4.9 and 14.6 eV, which is also predicted experimentally. Figure 5.27b shows the imaginary part of the dielectric constant for pristine graphene. It also exhibits peaks at 4.9 and 14.6 eV while it displays a valley between the two peaks.

Figure 5.28 shows the refractive index and extinction coefficient for pristine graphene. Peaks are observed at the same positions, as in the previous cases. It can be observed that the dielectric constant of graphene has an anisotropic nature.

Figure 5.29 shows the comparison of the imaginary part of the dielectric constant for layered graphene configuration. It is observed that the curves have the peaks of higher intensity with increasing number of layers. This can be understood from the increase in absorptance with increase in number of layers of graphene.

Figure 5.30 shows the variation of refractive index (top) and optical absorption (bottom) spectra of graphene monolayer as a function of wavelength. It can be observed that it is independent of wavelength in the range of 300–1000 nm.

5.4.2.2 Boron-Doped Graphene

The static dielectric constant of graphene increases significantly with increase in boron doping. The variation is seen in Fig. 5.31.

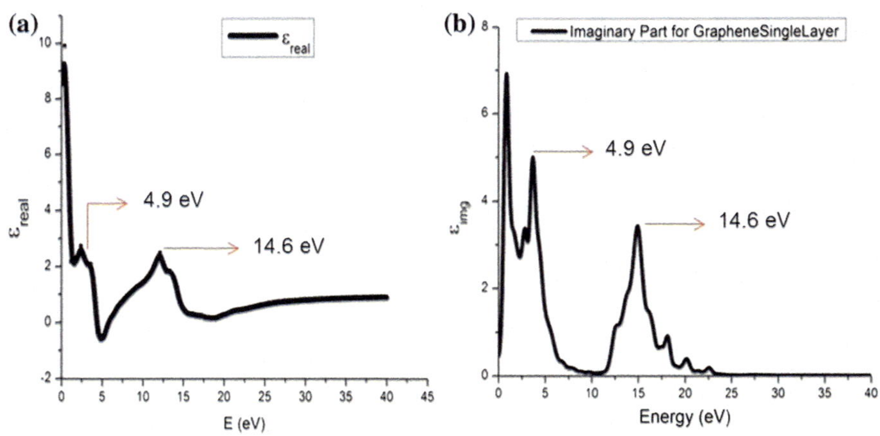

Fig. 5.27 Optical spectra of pristine graphene showing **a** real and **b** imaginary component of the dielectric constant, from our simulations. Peaks are observed at about 4.9 and 14.6 eV corresponding to $\pi \rightarrow \pi^*$ and $\sigma \rightarrow \sigma^*$ interband transitions, respectively

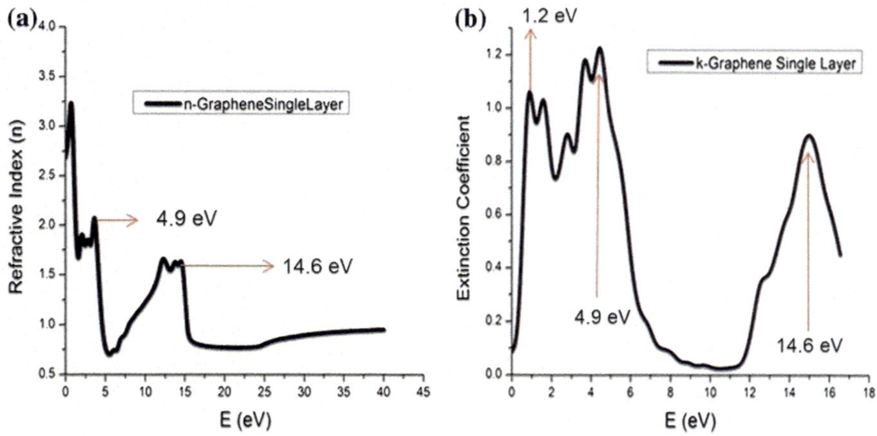

Fig. 5.28 Simulated **a** refractive index and **b** extinction coefficient of pristine graphene. Peaks are observed at about 4.9 and 14.6 eV corresponding to $\pi \rightarrow \pi^*$ and $\sigma \rightarrow \sigma^*$ interband transitions, respectively

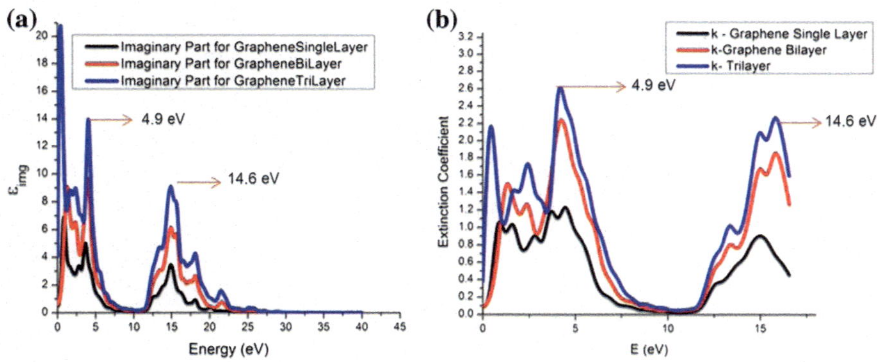

Fig. 5.29 Comparison of **a** simulated imaginary parts of dielectric constant and **b** simulated extinction coefficients for single, bilayer, and trilayer graphene. Peaks are observed at about 4.9 and 14.6 eV corresponding to $\pi \rightarrow \pi^*$ and $\sigma \rightarrow \sigma^*$ interband transitions, respectively

5.4.2.3 Nitrogen-Doped Graphene

The variation of static dielectric constant with nitrogen doping in graphene is shown in Fig. 5.32.

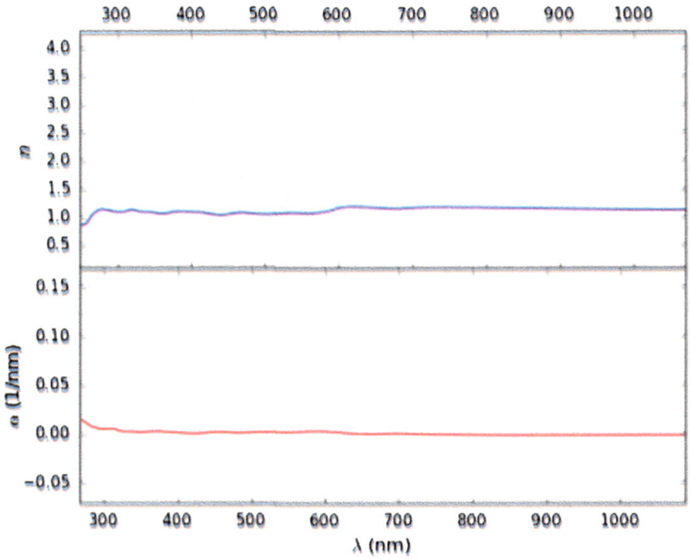

Fig. 5.30 Refractive index and optical absorption spectra of graphene monolayer as a function of wavelength from 300 to 1000 nm, based on simulations

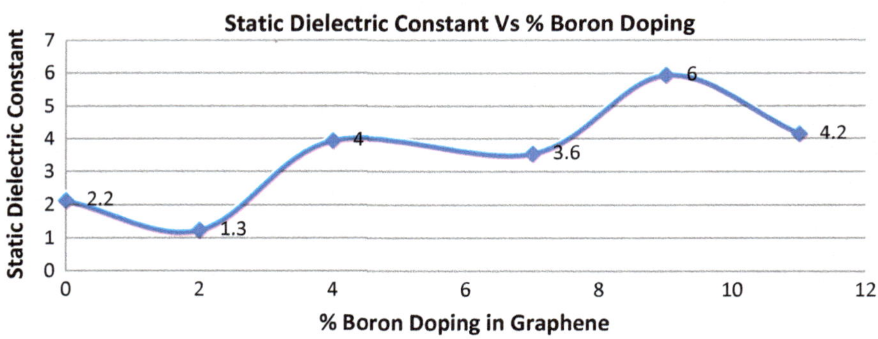

Fig. 5.31 Static dielectric constant versus % boron doping in graphene, based on simulations

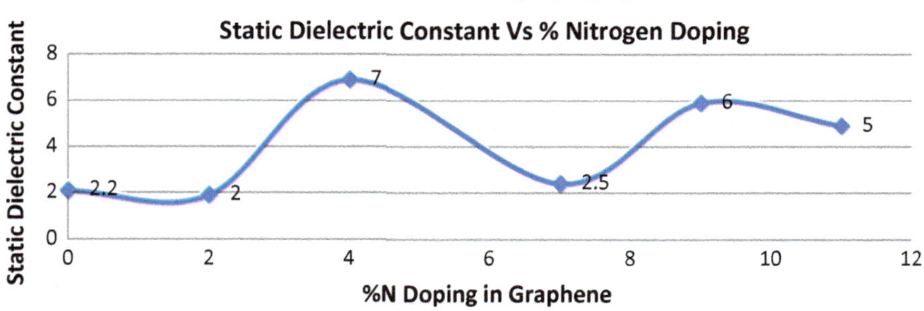

Fig. 5.32 Static dielectric constant versus % nitrogen doping, based on simulations

5.4.3 Mechanical Properties

Graphene has been reported to exhibit extremely high elastic modulus and hardness. In this study, the mechanical properties of graphene have been evaluated as a function of layers.

The simulation model consists of circular graphene sheet of diameter 16 nm. The simulation cell is relaxed until a slight energy drift is observed. After achieving equilibrium configurations, nanoindentation is performed. Here, the circular sheet of atoms, surrounding the graphene sheet, has been fixed. Indenter is considered to be spherical and has diamond-like properties. The diameter of the indenter is considered as 25 Å. The force constant of the indenter is 1 keV/Å.

A detailed analysis of Oliver–Pharr method to determine the elastic modulus from the load versus displacement curves is presented by Kan et al. [170]. The Oliver–Pharr method begins by fitting the unloading portion of load–displacement curve to power-law relation as shown below:

$$P = \alpha\, m\left(h - h_f\right) \tag{5.4.1}$$

where "α" and "m" are the fitting parameters. Originally, h_f is meant as the final depth after completion of unloading. However, practically when Oliver–Pharr method is used, h_f becomes only a fitting parameter.

The slope of unloading curve at the maximum indentation depth is known as "Contact Stiffness" (S). The contact depth of spherical indentation, h_c, can be determined using the Oliver–Pharr method in accordance with the following equation:

$$h_c = h_m - 0.75\frac{P_m}{S} \tag{5.4.2}$$

where h_m is the maximum indentation depth and P_m is the load at maximum indentation depth. The contact area, A_c, can be computed directly from the contact depth h_c and the radius of the indenter tip R:

$$A_c = \pi\left(2Rh_c - h_c^2\right) \tag{5.4.3}$$

The contact stiffness, S, and the contact area, A_c, can be used to calculate the reduced modulus from the following equation:

$$E_r = \frac{\sqrt{\pi}}{2\beta}\frac{S}{\sqrt{A_c}} \tag{5.4.4}$$

where β is a dimensionless correction factor. It accounts for the deviation in stiffness from the lack of axisymmetry of the indenter tip with $\beta = 1.0$ for axisymmetric indenters, $\beta = 1.012$ for a square-based Vickers indenter, and $\beta = 1.034$ for a triangular Berkovich punch [171]. For spherical indentations, β is taken as unity in this work.

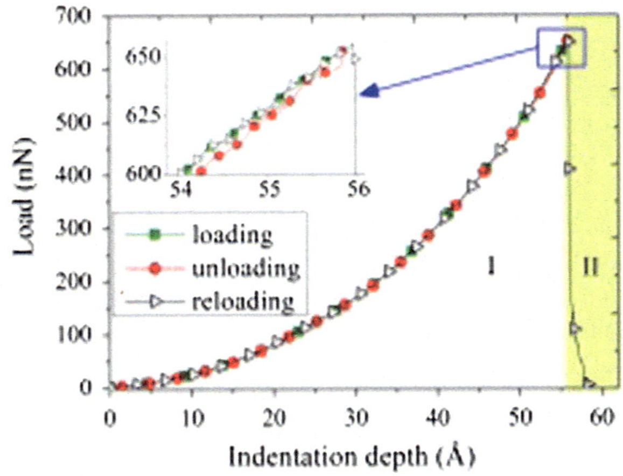

Fig. 5.33 Load versus indentation depth with maximum indentation depth smaller than critical indentation depth [172]

Wang et al. [172] have used molecular dynamics simulations to emulate the nanoindentation experiments for some single-layer rectangular graphene films with four clamped edges. The obtained typical load versus displacement curves are shown in Fig. 5.33. The effects of indenter radii, loading speeds, and aspect ratios are discussed. The Young's modulus of single-layer graphene films has been found to be 1 TPa and its yield strength is reported to be 200 GPa.

Graphene film is ruptured at a critical indentation depth. Figure 5.34 shows the loading–unloading–reloading process with depth less than the maximum indentation depth. The indenter considered is diamond, so that there is no atomic configuration of the indenter. Researchers have used energy function as described by adaptive intermolecular reactive empirical bond-order potentials to describe the interatomic interactions in carbon atoms of graphene layer. In performing MD simulations, canonical (NVT) ensemble is used and temperatures are controlled to within 0.01 K. In order to control the interatomic thermal fluctuations, Nose-Hoover method is used with a timestep of 1 fs. Energy minimization is performed initially and the system is allowed to relax to the lowest energy configuration before indentation.

The formation of dislocations in graphene has been described in detail in Refs. [173, 174]. Researchers have mainly focused on the tensile deformation and shear deformation [175]. Most of the references in the literature utilize Density Functional Theory (DFT), Tight Binding Molecular Dynamics (TBMD), and ab initio total energy calculations. Studies show the creation of dislocations and defects, followed by their effects on the properties of graphene. Very few experimental investigations, to study dislocation activities, have been published on graphene, till date [172]. Warner et al. have reported the observation of dislocations in graphene using HRTEM. Edge dislocations are shown to result in substantial deformation of atomic structure of graphene with bond compression or elongation of $\pm 27\%$, along with shear strain and rotations of lattice.

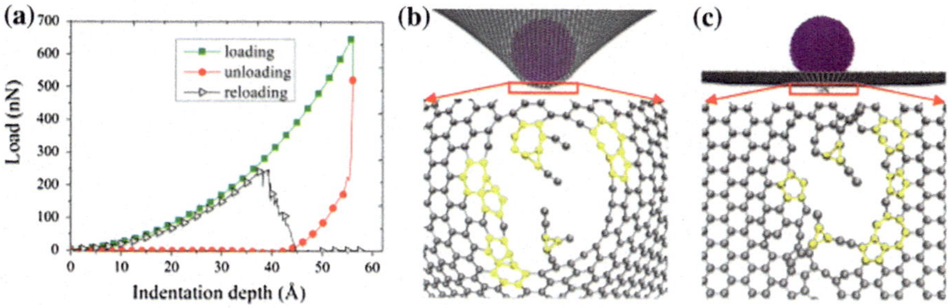

Fig. 5.34 Loading–unloading–reloading process with the maximum indentation depth smaller than the critical indentation depth. **a** Load–displacement curve. **b** Local atom configuration when the loading process is finished. **c** Local atom configuration when the unloading process is finished, from literature [172]

5.4.4 Thermal Conductivity Calculations

Thermal conductivity calculations have been performed with LAMMPS software. Following steps are implemented for the simulation:

1. Periodic boundary conditions have been considered in the direction of width and length.
2. Tersoff potential has been used to describe the interatomic interactions for C, B, and N atoms.
3. Neighboring atom cutoff distance for force calculations is considered as 2 Å.
4. Conjugate gradient algorithm, used for energy minimization, is performed to minimize the energy of system up to the levels of 1.0E−6 eV.
5. Initially, Gaussian distribution is created by setting temperature of system to 300 K and the system is described as NPT ensemble for equilibration, which is done up to 5000 timesteps with each timestep of 0.001 fs.
6. NVE ensemble is simulated by thermostating the system using Berendsen thermostat.
7. Müller-Plathe technique is used to calculate the thermal conductivity of a given material. Each simulation is executed for about 2,000,000 timesteps. Sample graphs of temperature profile and kinetic energy, swapped as a function of time, are shown in Figs. 5.35 and 5.36.

Once the verification of the model is established by comparison with the literature (experimental or simulated), the model is then extended to simulate the properties of graphene as function of orientation, number of layers, doping, vacancies, etc. The comparison with the literature along with the description of the models will be explained in detail in Sect. 5.5.

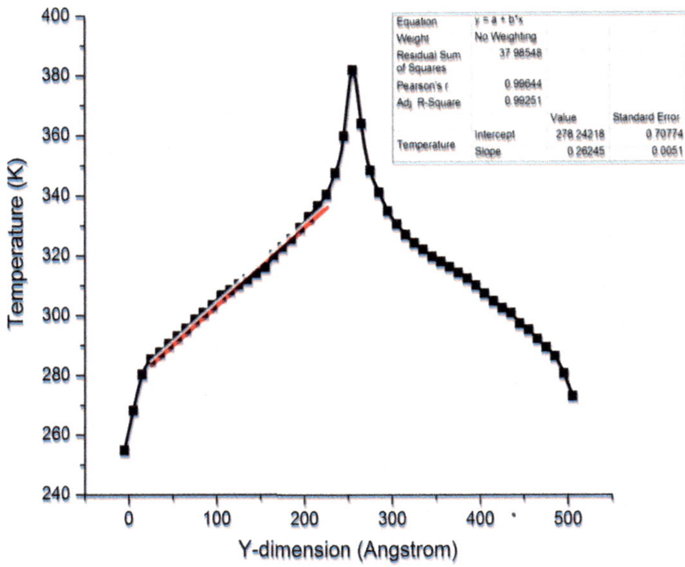

Fig. 5.35 Temperature versus size of simulation box in *Y* direction, based on simulations

Fig. 5.36 Kinetic energy versus time, based on simulations

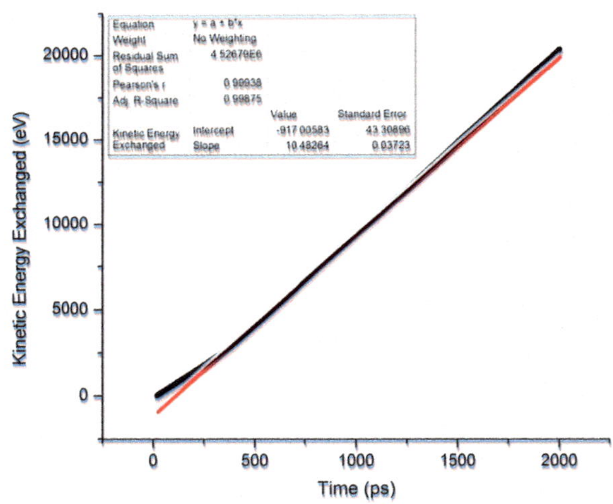

5.4.4.1 Thermal Conductivity Calculations for Graphene Nanosheets and Nanoribbons

Figure 5.37 shows the thermal conductivity of single-layer pristine graphene, armchair and zigzag graphene nanoribbons as a function of their length. It is observed that zigzag graphene nanoribbons have the highest thermal conductivity with respect to the armchair graphene nanoribbons as well as pristine graphene nanosheets. These results are consistent with the theory of Nika et al. [176] on small graphene flakes.

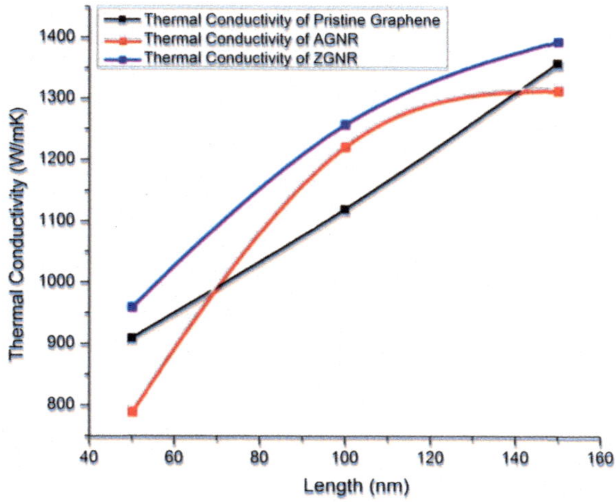

Fig. 5.37 Simulated thermal conductivity of pristine graphene, armchair graphene nanoribbons, and zigzag graphene nanoribbons as a function of length

The important role of phonon scattering by graphene edges on the temperature dependence of thermal conductivity has also been discussed [177].

The edge in the GNRs can decrease the thermal conductivity, which can be attributed to two reasons [178]. First, compared with that of graphene, there appear two edge-localized phonon modes in the low-energy region for the GNRs, i.e., the transverse acoustic mode, and the lowest lying optical mode [179]. The edge-localized phonons can interact with other low-energy phonons and thus reduce their PMFP (Phonon Mean Free Path) edge effect. This would remarkably reduce the low-energy phonon contribution to the thermal conductivity, which is very substantial and significant for thermal transport. Second, the boundary scattering at the edges of GNRs also reduces the thermal conductivity.

Figure 5.38a shows the variations of thermal conductivity for graphene and GNRs as a function of the width for a constant 50 nm length in each case. Figure 5.38b shows the variations of thermal conductivity for AGNR and ZGNR as a function of width for larger values of width. It is found that there is a sudden drop in the thermal conductivities initially from 0.5 to 1.5 nm widths while the thermal conductivity remains constant beyond 2 nm. The constancy of thermal conductivity at higher widths is also evident in Fig. 5.38b. The thermal conductivity is also found to be constant for increasing widths in case of AGNR. In the case of ZGNR, it is observed that the thermal conductivity first increases up to about 30 nm but decreases with further increase in width of ZGNR. As the width of GNR increases, the transport becomes more ballistic and is found to approach its limit. The heat flux is known to have preferential flow along the direction of width reduction in asymmetric graphene nanoribbons [180]. This is explained from the phenomena that, with increasing widths of GNRs, the total number of phonon modes increases, while the number of edge-localized phonon modes does not change. Hence, the effect of edges is found to reduce with further increase in width of GNRs. The energy gap between various phonons is also known to reduce with increase in the width of GNRs. This leads to higher possibility

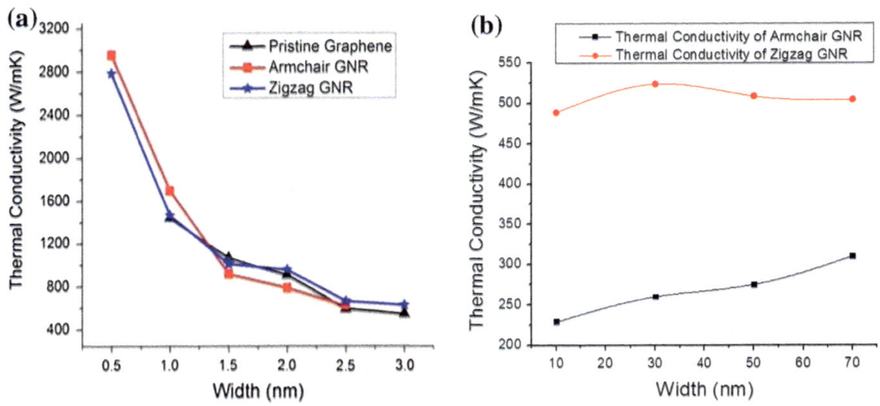

Fig. 5.38 Thermal conductivity of pristine graphene, armchair graphene nanoribbons, and zigzag graphene nanoribbons as a function of their widths in the range of **a** 0.5–3 nm and **b** 10–70 nm with a constant length of 50 nm, based on simulations

of phonon related Umklapp process and reduction in the thermal conductivity. These phenomena govern the variations in the values of thermal conductivity of asymmetric GNRs with varying widths.

Boron and nitrogen are known to be effective in modifying the electronic and optical properties of graphene making it suitable for optoelectronic devices. This makes it essential to understand the behavior of the thermal conductivity of doped graphene structures.

The effect of doping pristine graphene, AGNR, and ZGNR with boron is shown in Fig. 5.39. It is observed that there is a drastic reduction (~50%) in the thermal conductivity of graphene structures on doping with boron up to about 1%. Such a reduction has also been reported by NEMD techniques for doping of graphene with nitrogen by Mortazavi et al. [181]. It is noted that there is a reduced chirality dependence upon increase in dopant concentration.

Nitrogen is known to have atomic size comparability with carbon and forming strong bonds with carbon atoms. Doping with nitrogen has been shown to improve biocompatibility and sensitivity of carbon nanotubes [182]. Figure 5.40 shows the thermal conductivity trends in pristine graphene, AGNR, and ZGNR as a function of % nitrogen doping. It is found to exhibit a drastic reduction in thermal conductivity, as in the case of boron doping in graphene structures.

The temperature dependence of thermal conductivity of graphene is shown in Fig. 5.41. In graphene structures, a reduction in the thermal conductivity is observed at higher temperatures. The results presented here would be significantly useful in applications involving large temperature variations and gradients. It is noted that the increase in temperature leads to reduction in chirality dependence. At about 400 K, all the graphene nanostructures are found to have the same value of thermal conductivity.

Depending on the procedure used during synthesis of graphene or experiments to modify its properties by doping by chemical reduction, graphene samples are

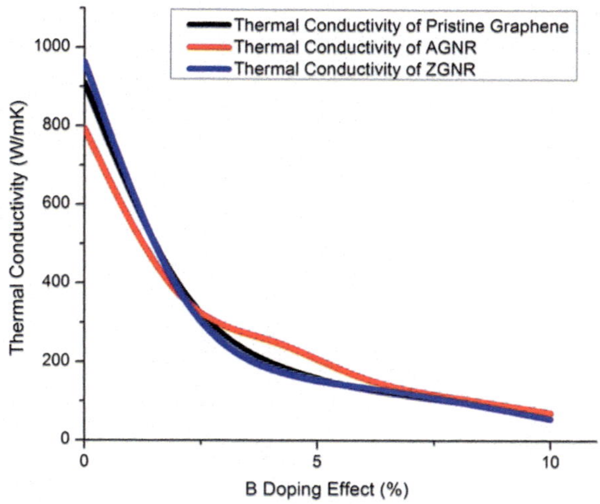

Fig. 5.39 Effect of boron doping on the thermal conductivity of pristine graphene, armchair graphene nanoribbons, and zigzag graphene nanoribbons, based on simulations

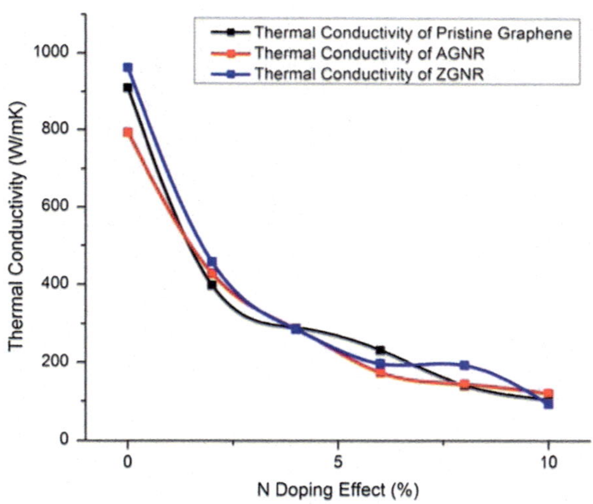

Fig. 5.40 Effect of nitrogen doping on the thermal conductivity of pristine graphene, armchair graphene nanoribbons, and zigzag graphene nanoribbons, based on simulations

inevitably expected to have single vacancies, Stone–Wales defects, grain boundaries, etc. [183]. We have simulated the thermal conductivity of pristine graphene, AGNR and ZGNR structures with varying concentration of vacancies up to about 10%, using NVE ensembles and periodic boundary conditions using MD. It is noticed that there is a drastic reduction in the thermal conductivity values. Principally, the bonds surrounding the vacancy defects should have higher stiffness (stiffness of bonds comparable to double or triple bonds) due to less coordination than the original graphene structures. Also, it is shown by Klemens et al. [184] that the relaxation time due to point defect scattering is inversely proportional to vacancy concentration and DOS. This leads to higher mean free path with increase in concentration of vacancies in the structure. Thus, a drastic reduction in thermal conductivity is observed with

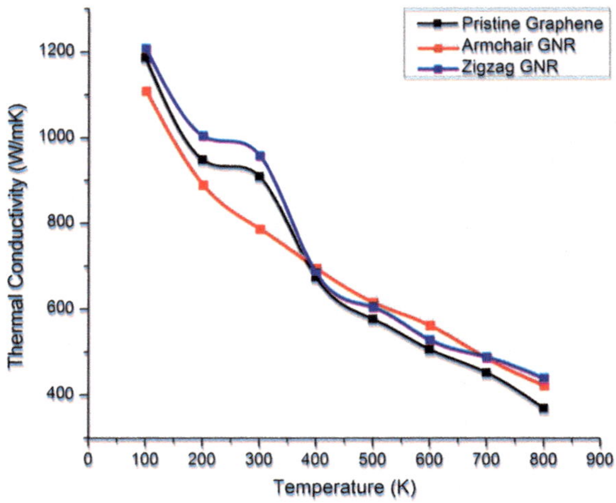

Fig. 5.41 Simulated temperature dependence of thermal conductivity of pristine graphene, armchair GNR, and zigzag GNR in the range of 100–800 K

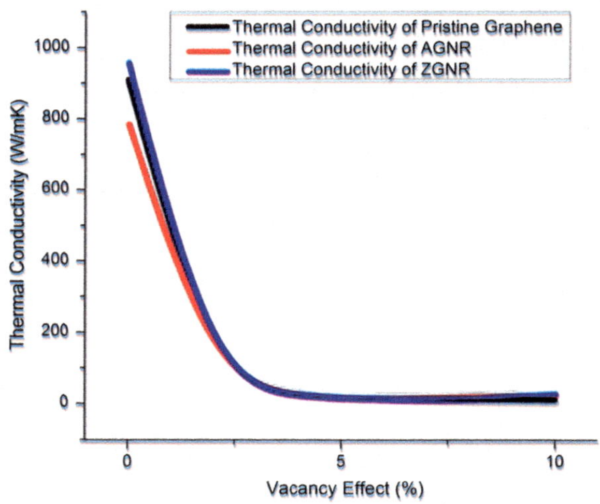

Fig. 5.42 Effect of vacancies on thermal conductivity of pristine graphene, armchair GNR, and zigzag GNR, based on simulations

increasing vacancy concentration. A detailed study of graphene with defects along with the corresponding phonon DOS has been presented by Zhang et al. [183] using NEMD simulations. The effect of vacancies on thermal conductivities of graphene nanostructures is presented in Fig. 5.42.

In principle, reduction in thermal conductivity of graphene structure is expected with increasing number of layers. These trends have been shown in Fig. 5.43. The interlayer interactions are found to be responsible for the substantial reduction of thermal conductivity in graphene. Pristine graphene is found to have the lowest thermal conductivity and ZGNR has highest thermal conductivity with increasing number of layers, indicating the dominance of chirality dependence in graphene nanostructures. Applications in thermal management for semiconductor devices and circuits have been proposed for multilayer graphene-based materials [185].

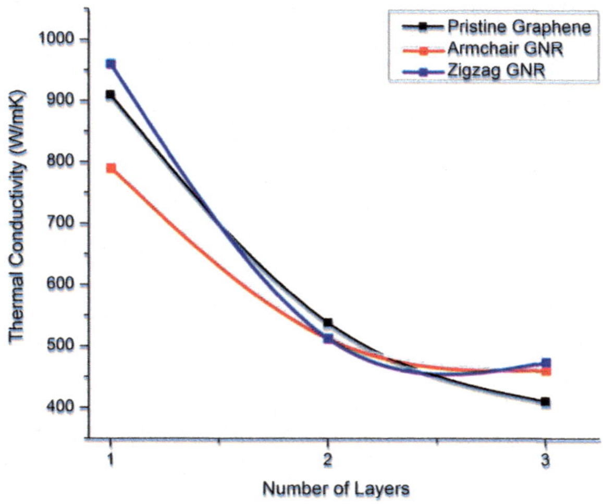

Fig. 5.43 Effect of number of layers (up to 3) of pristine graphene, armchair GNR, and zigzag GNR on thermal conductivity, based on simulations

Ghosh et al. [176] have experimentally found that graphene has thermal conductivities in the range of 3000–5000 W/m K depending on the specific sizes, which vary from 1 to 5 μ.

5.4.5 Transport Parameter Calculations

The transport parameter calculations, in this study, have been performed using Boltztrap Software, as explained in Sect. 5.3. Boltztrap uses the electronic band structure calculations performed using Quantum Espresso for post-processing and evaluating transport parameters. All the transport parameters are found to be interdependent as a function of chemical potential and temperature. This leads to the need for evaluation and comparison of each of these parameters.

Chemical doping is known to change the electronic band structure. Thus, there are changes in transport parameters as a function of % doping (p- or n-type). In order to improve the thermoelectric properties, graphene needs to have bandgap. This objective is achieved by using GNRs by appropriate patterning. GNRs exhibit bandgap dependence on ribbon width. AGNRs are semiconductors and have the bandgap inversely proportional to their widths. Bandgap is found to be dependent on the chirality of graphene (armchair/zigzag). Also, attempts to reduce phonon-induced lattice thermal conductivity of graphene have made researchers work on disordered graphene structures. It is found that such structures have improved Seebeck coefficient. However, the increase in bandgap in such disordered GNRs is found to lead to reduction in electrical conductivity. Thus, the overall thermoelectric performance of such structures is affected and needs to be optimized [186].

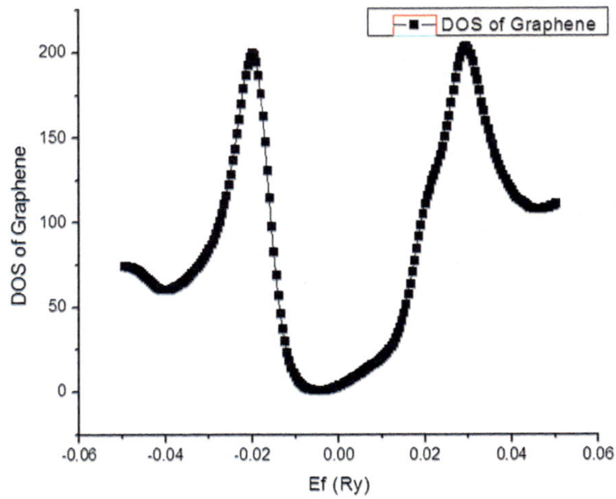

Fig. 5.44 Simulated DOS for single-layer graphene as a function of chemical potential

Zigzag GNRs are found to be metallic with very low Seebeck coefficient. The transport in ZGNRs is independent of line edge roughness in the first conduction plateau around the Fermi level [187]. The thermoelectric performance of ZGNRs has been reported to be improved in the presence of extended line defects by Hossein et al. [188] using quantum mechanical nonequilibrium Green's function simulations.

Figure 5.44 shows the plot of the density of states (DOS) of pristine graphene as a function of chemical potential. This is found to be in accord with the DOS plots in the literature, as shown in Fig. 5.7. It is known that graphene has a peculiar electronic density of states with nil DOS at the point of connection between the conduction band and the valence band (called Dirac Point), as discussed in Sect. 5.2.

Figure 5.45 shows the comparison of the DOS of 0, 4, 8, and 12% boron-doped armchair GNR. A finite displacement of the Dirac point to the positive side of the energy, indicating p-doping, is observed. Also, there is a finite bandgap near the Dirac point.

Figure 5.46 shows the comparison of electrical conductivity of armchair GNR with different % boron doping as a function of chemical potential. The peak electrical conductivity reduces as a function of boron doping. This is attributed to the changes in band structure of GNRs with doping. The electrical conductivity is found to be a function of temperature. Boron or nitrogen doping causes increase in electrical conductivity in the low-temperature region, while it causes reduction of electrical conductivity in the high-temperature region [189].

Figure 5.47 shows the variations of Seebeck coefficient as a function of chemical potential for AGNR and ZGNR with boron doping. The Seebeck coefficient peaks at $\pm k_BT$ and is found to nearly vanish around the bandgap. Seebeck coefficient curves are symmetric for AGNRs, which can be attributed to the symmetrical distribution of first conduction channels. There is a change in the sign of Seebeck coefficient across the charge neutrality point (CNP) as majority carrier density changes from electrons to holes. This is also reported in the literature [52, 190]. 8% boron-doped ZGNRs are

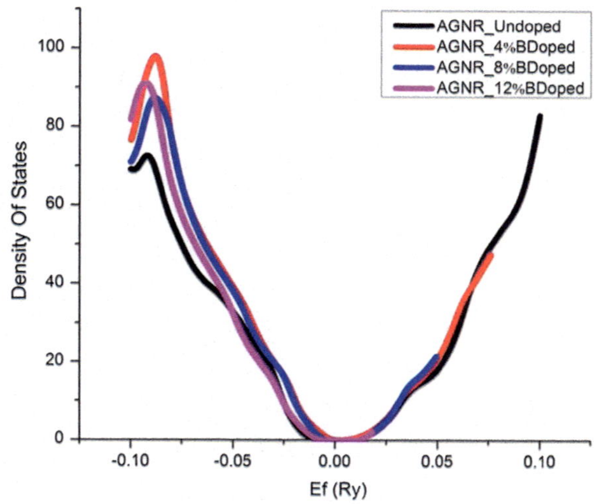

Fig. 5.45 Simulated DOS of AGNR and 4, 8, and 12% boron-doped AGNR

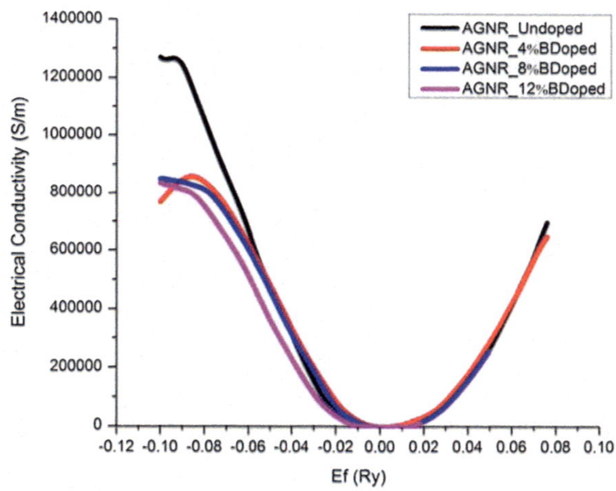

Fig. 5.46 Simulated electrical conductivity comparison of AGNR with 0, 4, 8, and 12% boron doping

not found to be symmetrical around Fermi level, as shown in Fig. 5.47. Thermopower for undoped AGNR is ~200 μV/K. Thermopower ~600 μV/K is observed in case of 12% boron doping of AGNR. The peak thermopower for boron-doped ZGNR is lesser than the corresponding boron-doped AGNR to the extent of ~300 μV/K.

The variation in peak Seebeck coefficient (TEP) with temperature for AGNR with boron doping is shown in Fig. 5.48. The peak value reduces with increase in temperature from 30 to 800 K almost linearly at higher temperatures.

Seebeck coefficient as a function of temperature for ZGNR doped with boron at Fermi level is shown in Fig. 5.49. The peak value of the Seebeck Coefficient reduces with increase in temperature from 50 to 800 K almost linearly at higher temperatures. This is indicative of the thermoelectric generation mechanism being diffusive TEP

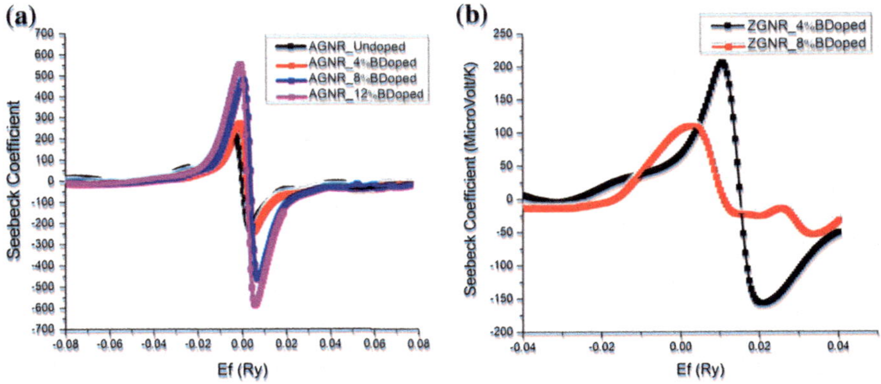

Fig. 5.47 Thermopower (Seebeck coefficient) μV/K versus chemical potential (Ry) for **a** AGNR with 0, 4, 8, and 12% B-doping and **b** ZGNR with 4 and 8% B-doping at 300 K, based on simulations

Fig. 5.48 Peak Seebeck coefficient as a function of temperature for AGNR with 0, 4, 8, and 12% boron doping, based on simulations

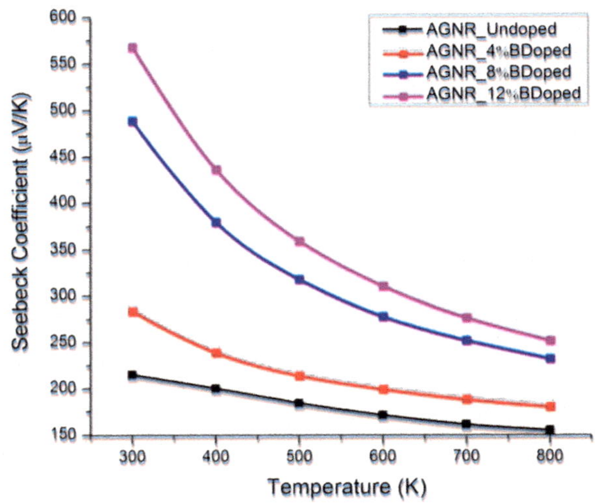

(ThermoElectric Power), with the absence of phonon drag component, as reported in the literature [190].

The effect of doping on the Hall resistivity of AGNR is shown in Fig. 5.50. It exhibits similar variations as observed in the Seebeck coefficient. The maximum Hall resistivity is noted at 12% boron doping.

The electronic thermal conductivity is found to exhibit a behavior similar to electrical conductivity for AGNR. The peak value is found to be ~10 W/mK as shown in Fig. 5.51, which is extremely small as compared to phonon-induced lattice thermal conductivity. It reduces with increase in % boron doping in AGNR. This reduction is attributed to the disturbed lattice coordination in the regions around the doped atoms.

Nitrogen doping in AGNR also leads to variation in the transport parameters. The trends for variations in Seebeck coefficient is as shown in Fig. 5.52.

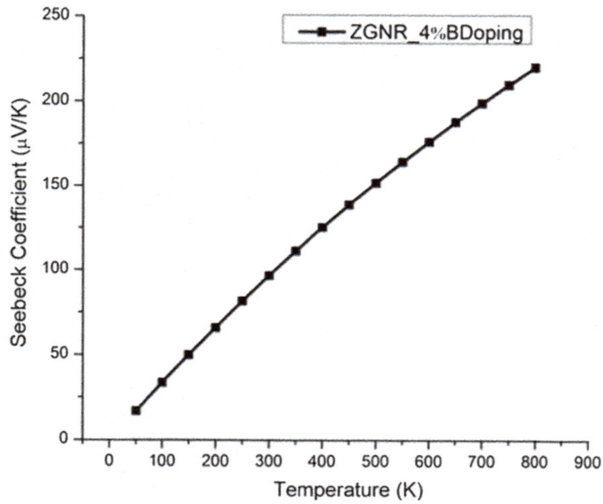

Fig. 5.49 Simulated Seebeck coefficient as a function of temperature for 4% boron-doped ZGNR at the Fermi level

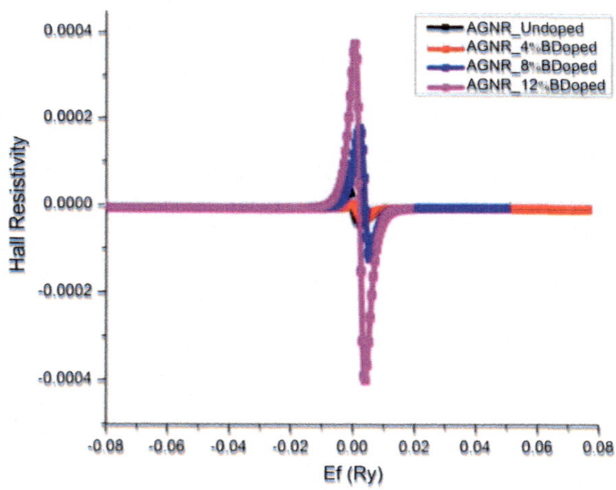

Fig. 5.50 Comparison of simulated Hall resistivity (m^3/C) versus chemical potential for AGNR with 0, 4, 8, and 12% boron doping

Peak thermopower is observed at ~8% nitrogen doping in AGNR. There is a substantial increase in the thermopower as compared to undoped AGNR. Variation of Hall resistivity for AGNR with nitrogen doping is shown in Fig. 5.53. It is found to exhibit behavior similar to the Seebeck coefficient.

The variation in the electronic thermal conductivity for undoped AGNR and 4% nitrogen-doped AGNR is shown in Fig. 5.54 in the range of −0.07 to +0.07 Ry.

The peak values of the electronic thermal conductivity reduce with increase in nitrogen doping. This is comparable with boron doping in AGNR and the reason for the reduction is attributed to changes in the lattice coordination leading to corresponding changes in phonon DOS.

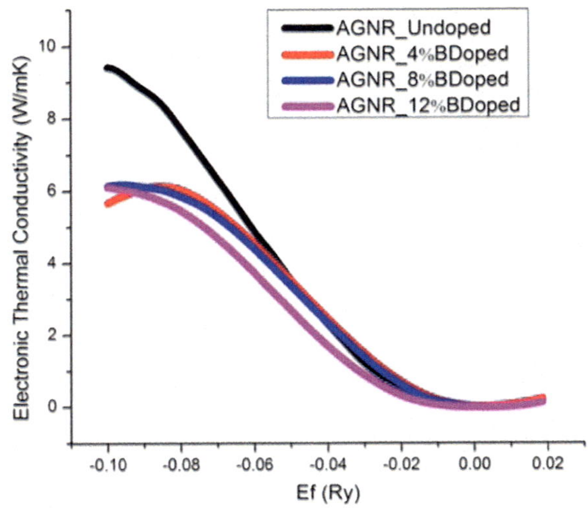

Fig. 5.51 Comparison of simulated electronic thermal conductivities of AGNR with 0, 4, 8, and 12% boron doping as a function of chemical potential (Ry)

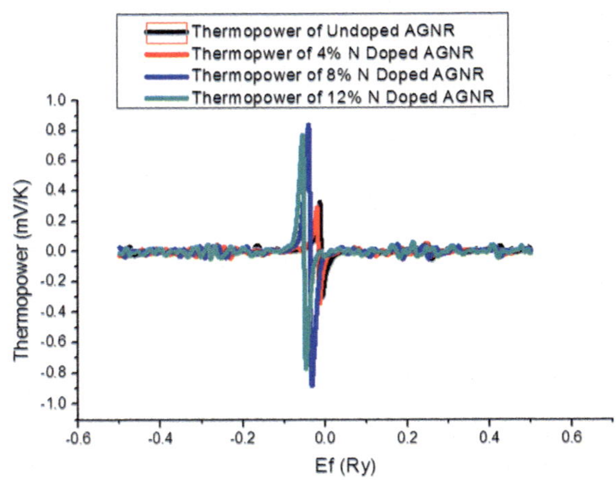

Fig. 5.52 Simulated thermopower of AGNR (mV/K) with 0, 4, 8, and 12% nitrogen doping as a function of chemical potential (Ry)

The corresponding Fermi energy variations for boron doping in AGNR and ZGNR are shown in Fig. 5.55. P-doping leads to increase in the value of Fermi level for each case.

Figure 5.56 shows ZT calculated for single-layer graphene as a function of length with a constant width of 2 nm. It is found to decrease with increase in length, as the thermal conductivity increases as a function of length in this regime.

Figure 5.57a, b show the comparison of ZT values for similar lengths of armchair and zigzag GNR and the effects of boron and nitrogen doping at room temperature. It is found that the values of thermal conductivity are much lower for AGNR w.r.t. ZGNR, hence leading to higher values of ZT. The peak ZT value observed is about 1.6, which is much higher than conventional thermoelectric materials such as Bi_2Te_3. Even higher values of ZT have been reported for GNRs with optimized geometries.

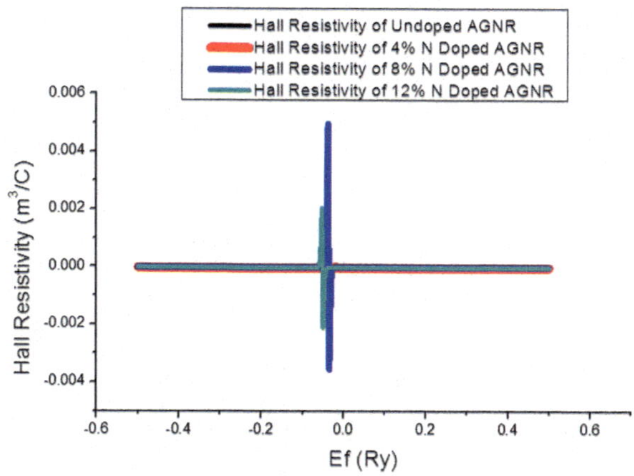

Fig. 5.53 Comparison of simulated Hall resistivity for AGNR with 0, 4, 8, and 12% nitrogen doping as a function of chemical potential (Ry)

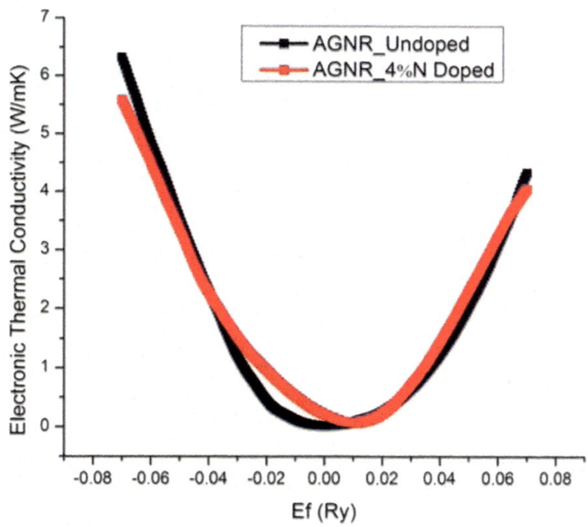

Fig. 5.54 Simulated electronic thermal conductivity of AGNR with 0 and 4% nitrogen doping as a function of chemical potential (Ry)

In attempts to reduce the phonon-induced lattice thermal conductivity, ZGNRs with vacancy defects have been simulated. The vacancies have been randomly distributed in the structure. Firstly, the relaxation is performed till the energy of the system reaches 1E−10 eV. This is followed by self-consistent calculation with cut-off for energy as 70 Ry. The peak thermopower is found to increase with increasing concentration of vacancies.

Figure 5.58 shows the behavior of Seebeck coefficient as a function of temperature for ZGNRs with about 4, 8.3 and 12.5% vacancies. ZGNRs with 12.5% vacancies are found to have enhanced Seebeck coefficient. Seebeck coefficient is noted to increase slightly with temperature for ZGNR with 4% and 8.5% vacancies. ZGNR with 12.5% vacancies is found to exhibit reduction in Seebeck coefficient with increase in temperature from 300 to 800 K.

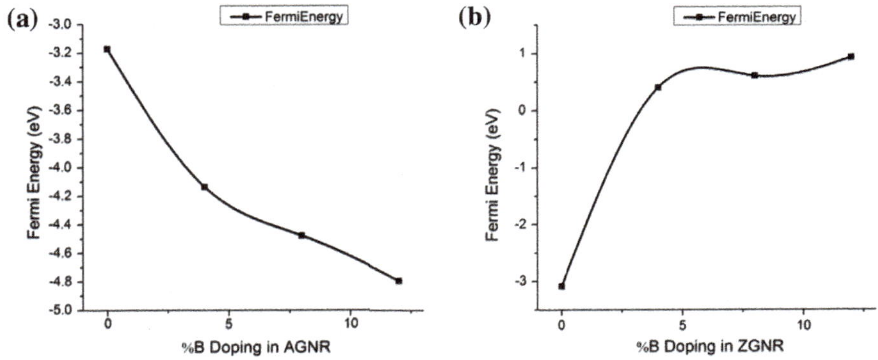

Fig. 5.55 Fermi energy trends for boron-doped **a** AGNR and **b** ZGNR, based on simulations

Fig. 5.56 ZT values as a function of length of graphene

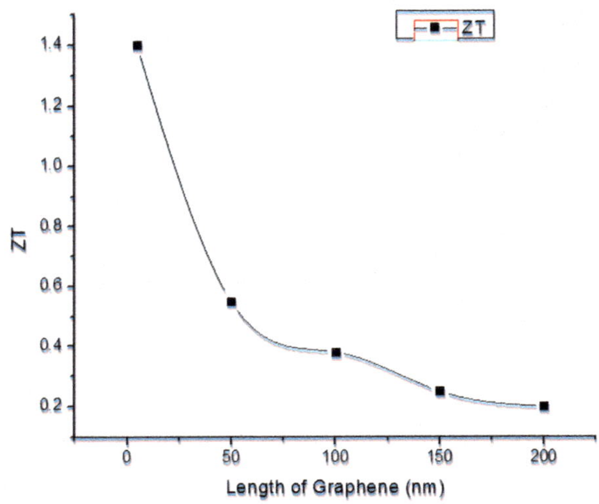

Bahamon et al. have investigated the electrical properties of ZGNRs that included extended line defects (ELD-ZGNRs) along the length of the nanoribbon [191]. It was noted that the extended line defects break the electron-hole energy symmetry in the GNRs. It introduces an additional electron band around the Fermi level. In this way, the asymmetry in the density of states and transmission function is achieved which improves the Seebeck coefficient. This structure has been experimentally realized recently by Lincoln and Mark [192].

Figure 5.59 shows the improvement in ZT for ZGNR with increase in % vacancy. This is as a result of increased thermopower and reduced value of thermal conductivity for the defect structure of ZGNR. Thus, inducing vacancies in ZGNR is an important technique for improvement of thermoelectric performance of graphene nanostructures.

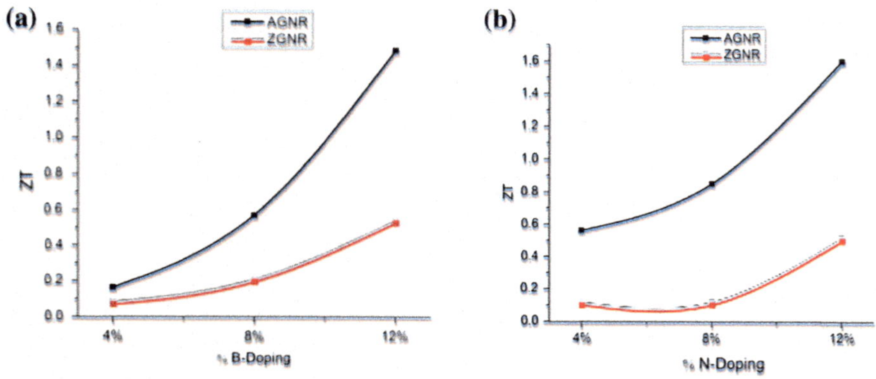

Fig. 5.57 ZT as a function of **a** % boron doping and **b** % nitrogen doping in armchair GNR and zigzag GNR

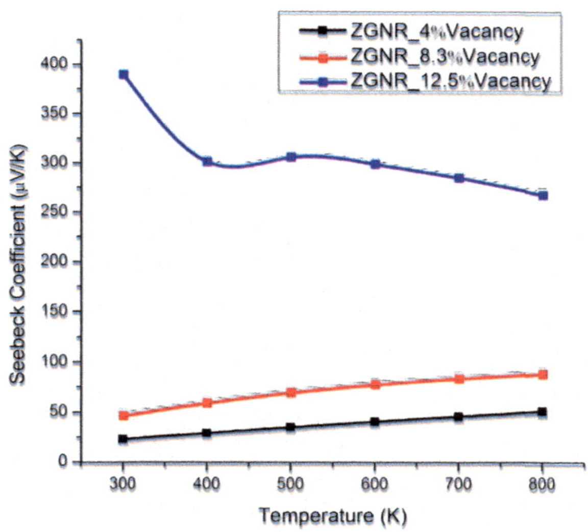

Fig. 5.58 Simulated Seebeck coefficient as a function of temperature for ZGNR with 4, 8.3, and 12.5% vacancies

5.4.6 *Emissivity Calculations*

Three case studies have been presented in this work depending on the application of graphene in specific areas [123]. These are:

I. Bulk materials
II. Multilayers
III. Device applications

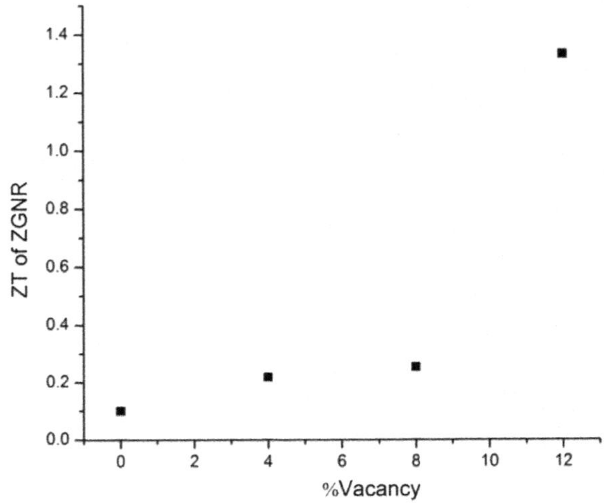

Fig. 5.59 ZT as a function of % vacancy in zigzag GNR, based on simulations

5.4.6.1 Bulk Materials Analysis

In this section, the optical properties, mainly emissivity, transmittance, and reflectance spectra of carbon-like materials: diamond, graphite, and graphene are presented. These materials are of significant importance in electronics.

As discussed in the literature, carbon like materials are finding significant applications in electronics due to their excellent physical, mechanical, electronic, and electrical properties. As observed in Fig. 5.60a, emissivities of bare substrates of carbon allotropes, natural diamond, and graphite for different thicknesses, in the wavelength range of 0.4–20 μm, have been compared. The emissivity increases with increasing thickness of diamond in the range of 500–5000 μm from 0 to ~0.6. For the same range of thickness, i.e., 500–5000 μm, absorptance in graphite is found to follow a single trend rising from 0.3 at 0.43 μm to 0.48 at 1.7 μm and then the emissivity values decrease linearly to ~0.11 at 20 μm. The emissivity of diamond for a thickness of 5000 μm is found to be a maximum, showing the trend of increasing linearly with wavelength from about 0 at 0.4 μm to 0.57 at 4 μm wavelength.

Figure 5.60b shows emissivity versus wavelength for graphene as well as few layers of graphene (FLG) at room temperature. It is found to be almost constant with respect to wavelength in the range of 1–2 μm (temperature in °C).

Single-layer graphene is found to absorb ~2.3% of the incoming IR radiation, theoretically as well as experimentally, which is attributed to the interband absorption in a wide range of wavelengths spanning from the visible to infrared [193]. The values obtained from our calculations are found to be ~2.5% (as can be seen in the emissivity plot in Fig. 5.60b) for single-layer graphene and are found to increase to ~20% with increasing number of layers to 10. Graphene has a very low reflectivity, and most of the incident electromagnetic waves are found to be transmitted (about 97%).

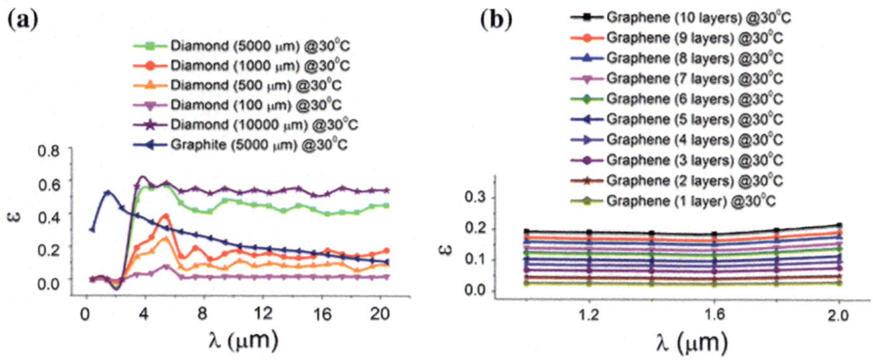

Fig. 5.60 **a** Simulated emissivity versus wavelength for diamond and graphite. **b** Simulated emissivity versus wavelength for graphene up to 10 layers (temperature in °C) [123]

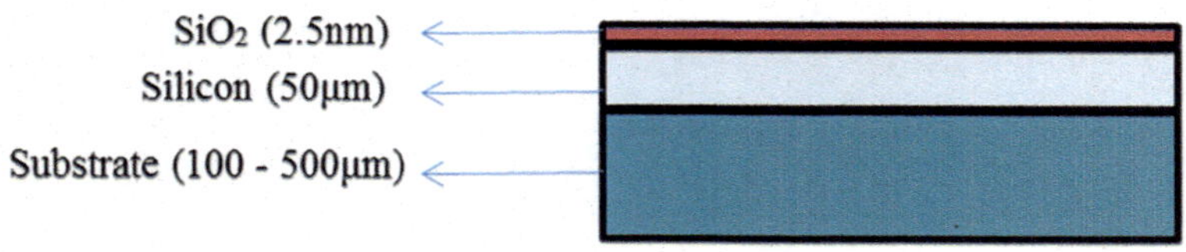

Fig. 5.61 Multilayered structure simulated for silicon on graphene/graphite [123]

5.4.6.2 Multilayered Structures

It is essential to obtain accurate values of the temperature in specific spectral range for multilayered structures. This leads to application of noncontact temperature sensing devices such as pyrometers. Accurate values of wavelength and temperature-dependent emissivity of a given material or structure are essential in order to obtain temperatures using pyrometers. Here, the trends in optical properties in the IR wavelength range in case of silicon on graphene, silicon on graphite, and graphene on silicon have been presented. A multilayered structure, as shown in Fig. 5.61, has been considered in the simulations of the optical properties of silicon on graphene and silicon on graphite. The variation in the optical properties with changing substrate thickness from 100 to 500 μm has been shown.

SiO_2/Si/Graphene

This multilayered structure is simulated as shown in Fig. 5.61, with substrate material taken as graphene (1–4 layers) at room temperature and compared with silicon(50 μm) and SiO_2(2.5 nm)/Si (50 μm). The simulated emissivity as a function of wavelength for this structure is shown in Fig. 5.62a. It is observed that the effect of graphene is to increase the values of emissivity at a given wavelength and its value stays linear over the entire wavelength range.

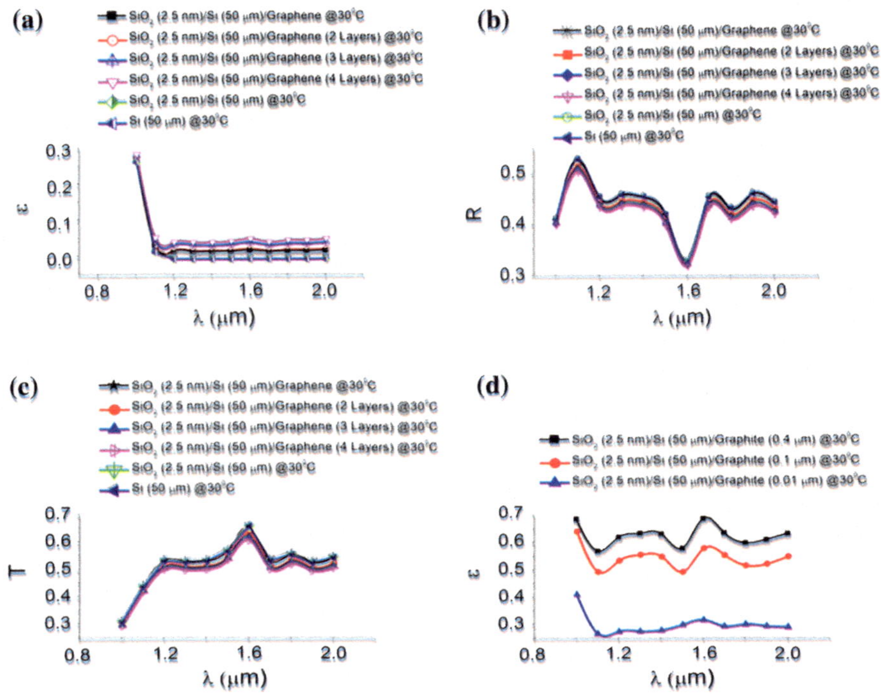

Fig. 5.62 **a** Emissivity, **b** reflectance and **c** transmittance as a function of wavelength in the wavelength range of 1–2 μm for SiO_2/Si/graphene structure, **d** emissivity as a function of wavelength in the wavelength range of 1–2 μm for SiO_2 (2.5 nm)/Si (50 μm)/graphite (0.4, 0.1, and 0.01 μm) (temperature in °C) [123]

The transmittance is found to decrease slightly with increasing number of layers of graphene as shown in Fig. 5.62c. As shown in Fig. 5.62b, a sharp decrease in average reflectance from 0.45 to 0.32, at 1.6 μm wavelength, is observed. Comparison of emissivity of substrate as graphite (Fig. 5.62d) versus graphene (Fig. 5.62a) indicates that the values of emissivity are much higher in the case of graphite substrate than in the case of graphene at room temperature. The emissivity changes slightly for 0.01 μm thick graphite. In Fig. 5.62d, there are specific features like flat plateaus corresponding to emissivity of 0.65 for the case of graphite substrate (0.4 μm thickness). Also, a valley at about 1.5 μm wavelength and a peak at about 1.6 μm wavelength are observed for the graphite substrate.

SiO_2/Si/Graphite

The multilayered structure, as shown in Fig. 5.61, has been simulated. Graphite is the substrate layer with varying thicknesses of 0.01, 0.1, and 100 μm. Based on our simulations, above ~0.3 μm thickness of graphite, the values of emissivity are

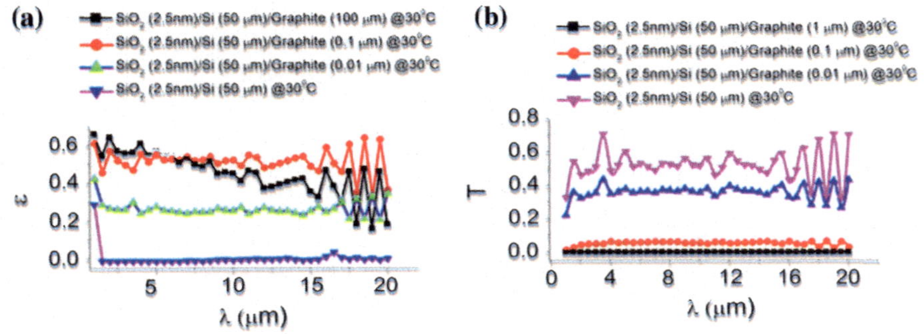

Fig. 5.63 Emissivity of SiO_2 (25 Å)/silicon (50 μm)/graphite (1 μm) (temperature in °C) [123]

constant and independent of thickness. The values of emissivity are found to increase with increase in thickness of graphite from 0.01 to 0.1 μm. The emissivity of graphite decreases above wavelength of ~5 μm up to 20 μm, for thicknesses above 0.3 μm. Some of these trends are as shown in Fig. 5.63.

Graphene/SiO_2/Si

The structure with top layer of graphene (1–10 layers thick)/SiO_2(300 nm)/Si(50 μm) has been simulated. Results of emissivity and transmittance as a function of wavelength are as shown in Fig. 5.64.

It is evident from Fig. 5.64 that the emissivity increases with increasing number of layers of graphene from 1 to 10 layers almost linearly from ~0.02 to 0.2, respectively. It is observed that the emissivity is almost constant for a particular structure within the wavelength range of 1.2–2 μm. The corresponding transmittance versus wavelength plots show a constant decrease in transmittance for a given wavelength as the number of layers of graphene are increased, with a peak at about 1.6 μm wavelength and almost constant transmittance from 1.7 to 2 μm.

5.4.6.3 Device Studies

A multilayered bolometer device configuration, as shown in Fig. 5.65, has been simulated. A bolometer is a device that responds by changing its resistance due to increase in temperature on interaction with the incident thermal radiation [127]. When power is coupled to electron system, electrons are driven out of thermal equilibrium along with phonons, creating hot electrons and hence such a bolometer is called the Hot Electron Bolometer (HEB) [128]. It is essentially a sensitive thermometer. It can be used with a spectroscope to measure the ability of some chemical compounds to absorb wavelengths of infrared radiation, by which one can obtain important information about the structure of the compounds.

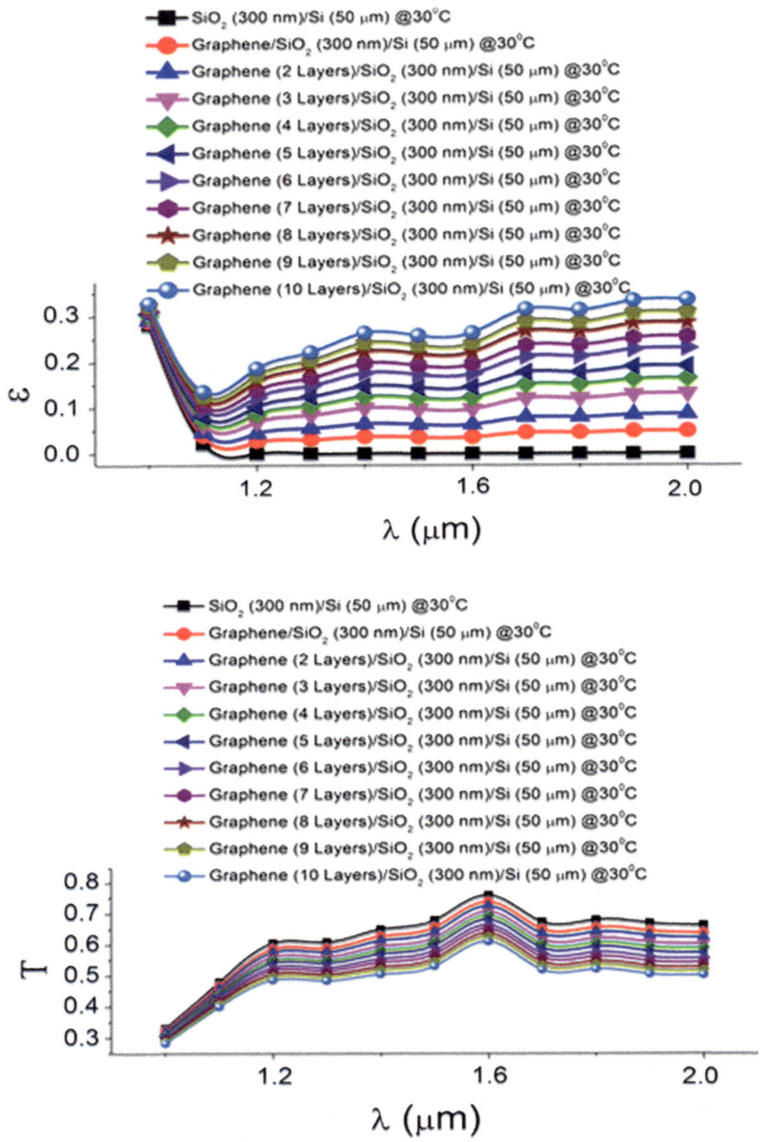

Fig. 5.64 Emissivity and transmittance of SiO_2 (25 Å)/silicon (50 μm)/graphite (1 μm) (temperature in °C) [123]

Due to its ability to absorb light from mid-infrared to ultraviolet with nearly equal strengths, graphene has found applications in optical detectors. Graphene is typically well suited for HEBs due to its small electron heat capacity and weak coupling of electrons and phonons, which causes large light-induced changes in electron temperature. Small electronic specific heat makes it possible for faster response times,

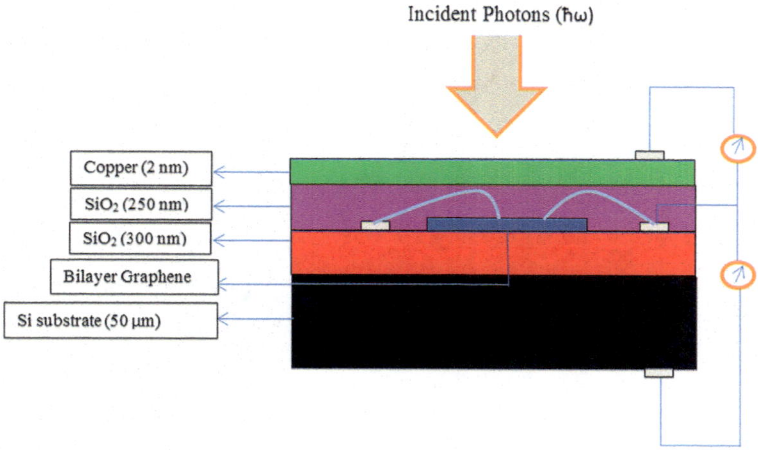

Fig. 5.65 Bolometer device configuration [194]

higher sensitivity, and low noise equivalent power. At low temperatures, usually in the cryogenic range, electron–phonon coupling in metals is very weak. The usual range of operation of HEBs is cryogenic, while graphene-based HEBs can be used at higher temperatures due to its low electron–phonon scattering even at room temperature and its highest known mobilities of charge carriers at room temperature [48, 127, 129].

However, the resistance of pristine graphene is weakly sensitive to electron temperature. Various approaches have been attempted to address this issue in the literature. The first one is a dual-gated bilayer graphene (DGBLG) bolometer [194], which has temperature-dependent resistance as well as weak electron–phonon coupling. Light absorption by DGBLG causes electrons to heat up due to their small electron specific heat, while the weak coupling of electrons and phonons helps in creating bottleneck in heat path, decoupling the electrons from phonon path. Good light sensitivity causes a change of resistance in the sample, which can then be converted to a detectable electrical signal. The second approach, proposed in the literature, is to use disordered graphene instead of pristine graphene. Disordered graphene has been shown to exhibit highly temperature-dependent resistance. Also, graphene film is separated from the electrical contacts by a layer of boron nitride, which acts as a tunneling barrier to increase the contact resistance and hence thermal resistance, resulting in better thermal isolation [127, 194].

It is to be noted that, in order to obtain higher responsivity of a device, one needs to essentially increase the absorptance in the device. Various methods for improving this characteristic are as follows: using multilayered graphene, surface plasmonics enhancement or microcavity, with the latter two introducing selectivity of wavelengths [127]. Two types of graphene-based hot electron bolometer devices have been simulated on the basis of these concepts.

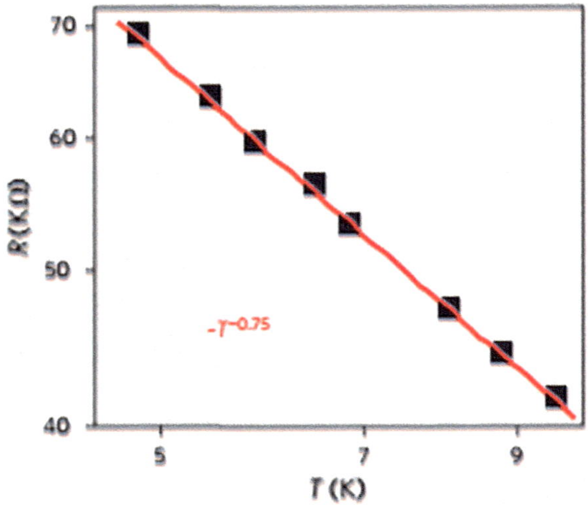

Fig. 5.66 Resistance as a function of temperature for bilayer graphene bolometer [127]

Case I

In this case, we have simulated the bolometer structure, as shown in Fig. 5.65, by changing the number of layers of graphene and studying their corresponding emissivity (or absorption) to develop understanding of its responsivity [194].

Figure 5.66 shows the behavior of resistance with temperature for a bilayer graphene bolometer [127].

As shown in the emissivity plot in Fig. 5.67, it is observed that, for silicon(50 μm) and SiO_2(0.3 μm)/Si(50 μm), the curves follow similar trends, with the latter having emissivity (or absorptance) (higher by 0.1) than the former. The emissivity (or absorptance) is found to increase for each added layer in the case of the bolometer configuration. The increase in emissivity (or absorptance) is found to be highest with copper layer of 2 nm on the top, in the range of wavelength from 0.8 to 2 μm for the bolometer structure. This shows improvement in the responsivity of the device.

It is clear that the influence of copper layer on the top is to increase the absorptance to about 20%, as compared to about 10% for the multilayered structure with the corresponding reduced transmittance for the structure. Also, copper has high value of temperature coefficient of resistance (about 4.29×10^{-3}/°C) [195] and hence can be considered as a material that enhances the responsivity of the device by detecting smallest change in temperature. Also, due to the layer of bilayer graphene (BLG), the device is expected to have higher speed of response, as BLG has very less electron–phonon interaction and very high mobility of electrons even at room temperature.

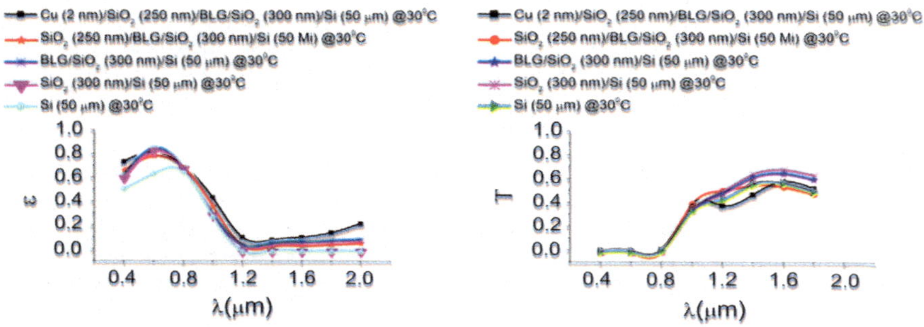

Fig. 5.67 Emissivity (or absorptance) and transmittance as a function of wavelength for the various layers in the proposed bolometer configuration (temperature in °C)

Case II

Graphene has been shown to be an excellent material for electronic applications based on its electron transport properties. However, there are some limitations of graphene-based devices such as induction of surface optical phonons on graphene that are in contact with substrate materials (commonly SiO_2/Si), reduction in carrier mobilities than its freely standing form, surface roughness, and inhomogeneity of charge carriers [196].

Hence, a novel approach of suppressing surface dangling bonds and surface charge traps has been proposed by using hexagonal boron nitride (h-BN) as a substrate for graphene, as h-BN has strong ionic bonding in hexagonal lattice structure. It is known that there is about 1.7% lattice mismatch between h-BN and graphene and hence there is little electronic coupling with graphene. This approach has been shown to endow the device with higher electron mobilities as well as electron-hole charge inhomogeneity.

In this case, the bolometer structure, as shown in Fig. 5.68, has been simulated, based on the research by Han et al. [127] and the multilayered structure considered by Wang et al. [196]. The optical properties of (a) pristine graphene (instead of disordered graphene) and (b) varying layer thickness of graphene and hexagonal boron nitride (h-BN) on SiO_2/Si substrate, have been simulated. The evolution in their emissivity and transmittance, as a function of wavelength (1–2 μm), has been presented.

Figure 5.69 shows the resistance versus temperature plot for this bolometer [127]. It is clear from this figure that pristine graphene layer has a very small change in resistance with temperature. Therefore, increasing the thickness of graphene layer can be considered as an alternate approach to increase the absorption of the device and hence improving its responsivity. Also, varying the thickness of BN will lead to change in emissivity of the device and hence the responsivity.

Figure 5.70a shows that the emissivity has an increasing trend with increasing thickness of BN layer from 20 to 2000 nm. The corresponding optical transmittance is found to decrease with increasing thickness of BN layer as shown in Fig. 5.70b.

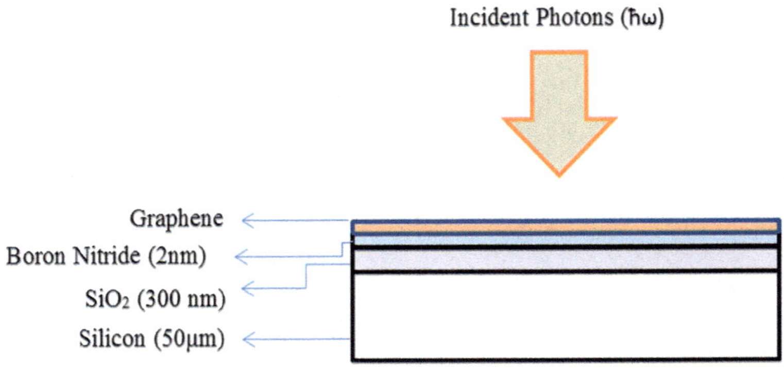

Fig. 5.68 Bolometer device structure with multilayered configuration graphene/BN /SiO_2/Si [127]

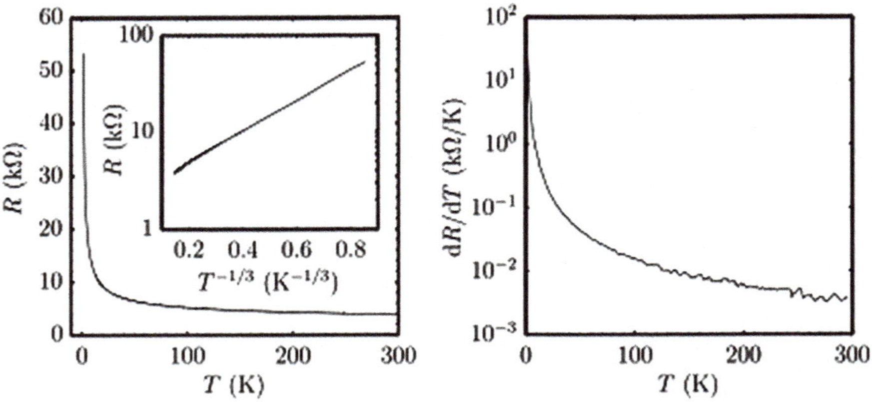

Fig. 5.69 Temperature dependence of resistance of graphene nanoribbon-based bolometer device [127]

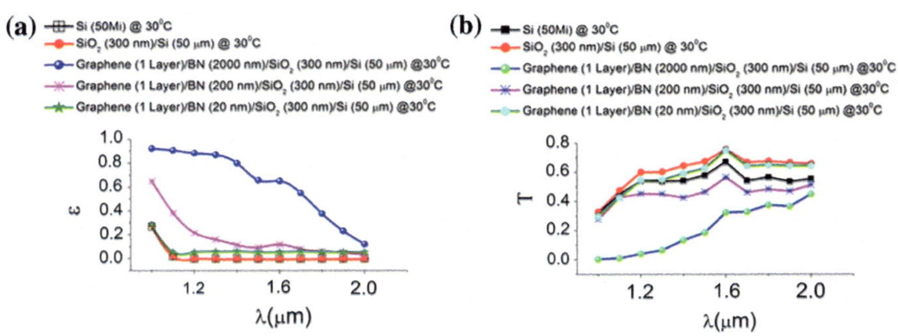

Fig. 5.70 Effect of variation of BN thickness on the **a** emissivity and **b** transmittance of bolometer structure configuration (temperature in °C) [123]

5.5 Results and Discussion

In this section, analyses of the results presented in the previous section are presented with a view to compare them with the available literature. The modeling techniques, used in our calculations, are presented along with those in the literature.

5.5.1 *Summary of Methods Used for Simulations*

Table 5.2 summarizes the computational methods used for simulating the thermal, mechanical, electronic, optical, and thermoelectric properties of graphene and its derivatives. The software tools namely LAMMPS (Large-Scale Atomic/Molecular Massively Parallel Simulator), Quantum Espresso, YAMBO, and Boltztrap are available under GNU public license. These software have been installed in a Linux operating system.

5.5.2 *Electronic Properties*

Figure 5.71a shows the simulated electronic properties of pure graphene nanosheets, based on the DFT calculations by Rani et al. [76] using VASP (Vienna Ab initio Simulation Package) code. Figure 5.71b shows the simulated electronic properties of pure graphene nanosheets, in accordance with our DFT calculations using Quantum Espresso.

Table 5.2 Synopsis of various simulation methods used in this study

Simulation methods	Molecular dynamics (MD)	Density functional theory (DFT)	Linear response: The Kubo formula
Parameters	(a) Tersoff /Airebo Potentials for simulating atomic interactions (b) NVT thermostat for equilibration and NVE ensemble for NEMD (c) Periodic boundary condition	(a) Quantum mechanical calculations (b) Energy and k-point convergence ($20 \times 20 \times 1$) (c) Periodic boundary conditions (d) Ballistic transport regime	(a) It takes input from electronic band structure calculations by Quantum Espresso (b) Dielectric matrix calculation within random phase approximation (RPA)
Software used	LAMMPS	Quantum Espresso	YAMBO
Properties	Thermal and mechanical	Electronic	Optical

5.5.2.1 Analysis of Models for Electronic Properties

Graphene has been shown to have two-dimensional honeycomb structure. The results of the electronic band structure calculations from the literature [76] are shown in Fig. 5.71. This analysis has been performed using geometry optimizations and electronic band structure calculations by using the VASP (Vienna Ab initio Simulation Package) code [197] based on density functional theory (DFT). This approach is based on an iterative solution of the Kohn–Sham equation [149] of the density functional theory in a plane wave set with the projector-augmented wave pseudopotentials. In the calculations, the Perdew–Burke–Ernzerhof (PBE) [198] exchange-correlation (XC) function of the generalized gradient approximation (GGA) has been used. The cutoff energy for plane waves was set to 400 eV. The lattice constant optimizations are made by the minimization of the total energy of the structure.

The 5×5 supercell (consisting of 50 atoms) has been used to simulate the isolated sheet and the sheets have been separated by larger than 12 Å along the perpendicular direction to avoid interlayer interactions. The Monkhorst-Pack scheme has been adopted for sampling the Brillouin zone. The structures have been fully relaxed with a gamma-centered $7 \times 7 \times 1$ k-mesh. During these processes, except for the band determination, partial occupancies have been treated using the tetrahedron methodology with Blöchl corrections. For band structure calculations, the partial occupancies for each wavefunction were determined by applying the Gaussian smearing method with a smearing of 0.01 eV. For geometry optimizations, the internal coordinates were relaxed until the Hellmann–Feynman forces were less than 0.005 Å.

Figure 5.72 shows the band structure of pristine graphene and the corresponding density of states, based on our simulations. The geometry of the structure and their coordinates are generated using VMD (Visual Molecular Dynamics) software.

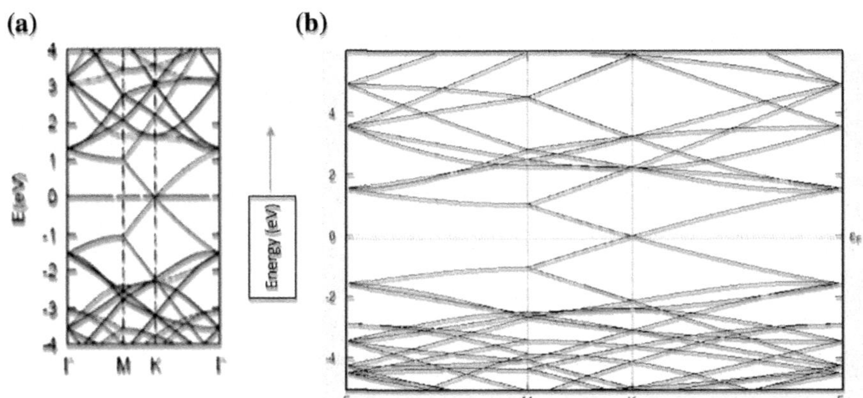

Fig. 5.71 **a** Simulated band structure of pure graphene nanosheet, from literature [76]. **b** Electronic band structure of pristine graphene, simulated in this work

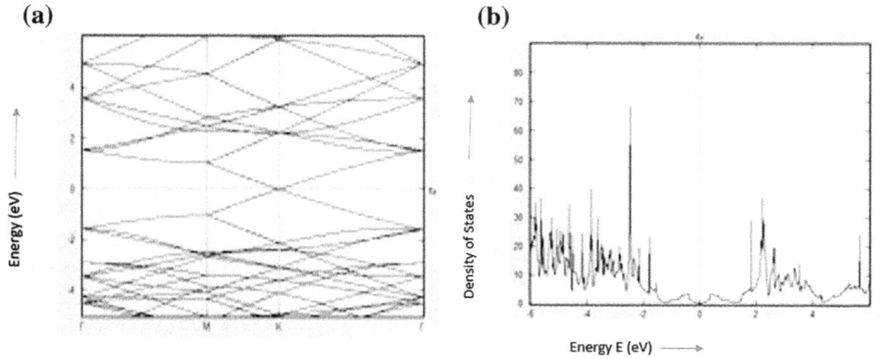

Fig. 5.72 **a** Simulated band structure of pure graphene. **b** Corresponding Density of States (DOS) for pure graphene

The structural parameters, i.e., lattice constant and coordinates are optimized by relaxation of the structure by using Quantum Espresso code based on DFT. Self-consistent (SCF) equations, proposed by the Kohn–Sham approach, have been iteratively solved for density functional theory in plane wave set. The Perdew–Burke–Ernzerhof (PBE) exchange-correlation (XC) function of the generalized gradient approximation (GGA) is applied. The non-SCF calculations have been performed by varying energy cutoff, Monkhorst-Pack grid parameters, and lattice constant. The optimized parameters obtained from non-SCF calculations are used to perform further simulations. Kinetic energy cutoff for wavefunctions has been chosen as 70 Ry, while the kinetic energy cutoff for charge density and potential has been set as 840 Ry (as a rule should be 8–12 times kinetic energy cutoff for wavefunctions). It is noted that the total energy of the system is minimum at this cutoff. Smearing with degauss 0.03 has been used considering the semimetallic nature of graphene. Energy convergence parameter has been set as 1.0E−8 Ry, i.e., the system converges once it reaches this energy level.

It can be observed that the Dirac point is at nil energy (0 eV) at K-point. This is reported in the literature as in Fig. 5.71. Also, the DOS plot for pristine graphene is shown to have its minima at the Dirac point, as in Fig. 5.72b.

Further, doping of graphene nanosheet is carried out and its band structure as well as the corresponding DOS is plotted as a function of concentration of dopants. Boron is a p-type dopant in graphene and it has a diameter that is very close to carbon atoms. Hence, upon doping of graphene with boron, the boron atom undergoes sp^2 hybridization and there is no significant distortion in the structure of the graphene sheet. The corresponding band structure for boron-doped graphene is shown in Fig. 5.73. However, there is change of adjoining bond lengths of boron atoms. It is observed that, with about 2% boron doping, the lattice parameter is 2.5794 Å as compared with about 2.455 Å for relaxed undoped graphene. As shown in Fig. 5.74, there is a shift in the Fermi level below the Dirac point and the linear dispersion near the Dirac point is still intact. This shift is attributed to the electron deficient nature of boron in the

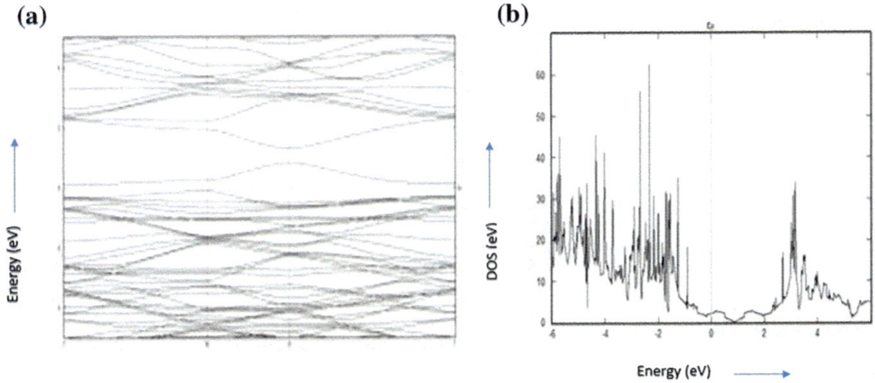

Fig. 5.73 **a** Simulated band structure of 4% boron-doped graphene. **b** Simulated DOS for 4% boron-doped graphene, from calculations

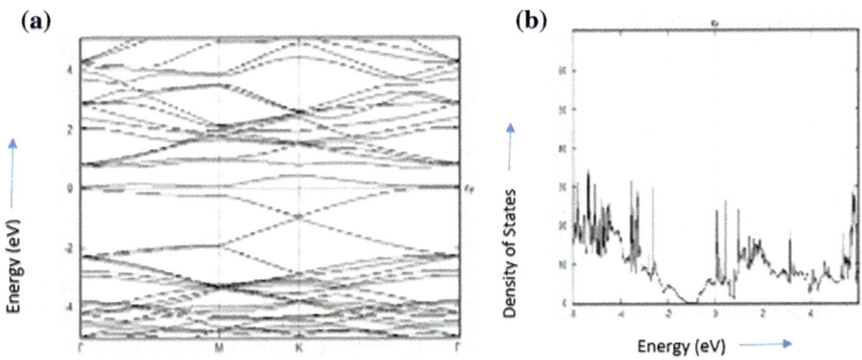

Fig. 5.74 **a** Simulated band structure of 4% nitrogen-doped graphene. **b** Simulated DOS for 4% nitrogen-doped graphene, from calculations

carbon lattice. Another important observation is the presence of bandgap in boron-doped structure, to the extent of about 0.14 eV on 2% boron doping in graphene. Thus, graphene changes its semimetallic nature to a semiconductor.

When graphene is doped with nitrogen atoms, there is a similar behavior as observed in B-doping. However, in this case, the bond length of C–N bonds is decreased. This results in the lattice parameter of about 4% nitrogen-doped graphene sheet as 2.4338 Å. Thus, there is a reduction in the lattice parameter of graphene nanosheets with N-doping. Due to the electron-rich nature of nitrogen atoms in graphene lattice, one observes that the Fermi level shifts by 0.7 eV above the Dirac point. Due to the broken symmetry of graphene sub-lattice on doping, there is a bandgap of about 0.14 eV with about 2% nitrogen doping.

Electrons in bilayer graphene exhibit an unusual property of being chiral quasi-particles, which is characterized by Berry phase 2π [199]. The low-energy Hamiltonian of a bilayer describes chiral quasi-particles with a parabolic dispersion and Berry

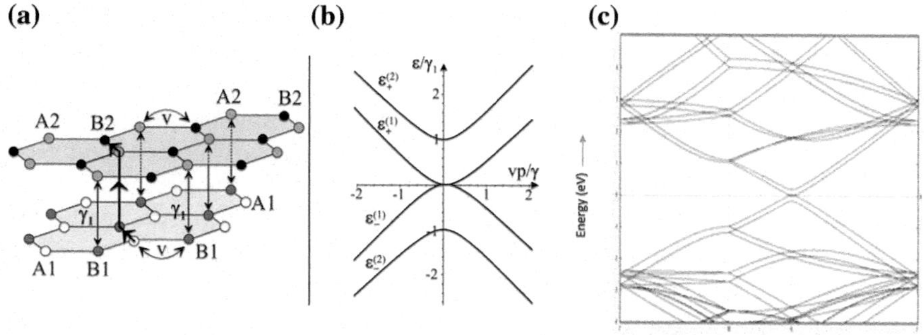

Fig. 5.75 **a** Schematic of the bilayer lattice containing four sites in the unit cell: A1 (white circles) and B1 (gray) in the bottom layer, and A2 (gray) and B2 (black) in the top layer [199]. **b** Schematic of the low-energy bands near the K-point obtained by taking into account intralayer hopping with velocity v, B1A2 interlayer coupling $\gamma 1$, A1B2 interlayer coupling γ_3 [with $v_3/v = 0.1$] and zero layer asymmetry Δ, as per literature [199]. **c** Simulated band structure of bilayer graphene, from our calculations

phase 2π. This is confirmed by quantum Hall effect (QHE) and Angle-Resolved Photoemission Spectroscopy (ARPES) measurements. The asymmetry between the on-site energies in the layers leads to a tunable gap between the conduction and valence bands.

As discussed in Ref. [199] and shown in Fig. 5.75a, bilayer graphene is considered to consist of two coupled hexagonal lattices with inequivalent sites A1B1 and A2B2 on the bottom and top graphene sheets, respectively. They are arranged in accordance with Bernal (A2–B1) stacking. Every B1 site in the bottom layer lies directly below an A2 site in the upper layer, but sites A1 and B2 do not lie directly below or above a site in the other layer. Tight binding model of graphite has been applied by considering the Slonczewski–Weiss–McClure parameterization. The inversion symmetric pristine bilayer graphene can be seen as a zero bandgap semiconductor, as shown in Fig. 5.75b.

Our calculated results of the band structure of pristine bilayer graphene are presented in Fig. 5.75c. The simulations have been performed using bilayer graphene structure similar to the one explained in the literature. The results are calculated using Quantum Espresso with implementation of DFT and plane wave sets. Similar to literature, it can be observed that pristine bilayer graphene is a zero bandgap semiconductor, with a possibility to modulate its bandgap by functionalization and doping.

5.5.3 Optical Properties

Optical properties have been simulated by performing band structure calculations using Quantum Espresso along with a post-processing tool, YAMBO [111]. Yambo is a FORTRAN/C code for many-body calculations in solid-state and molecular physics. Yambo relies on the Kohn–Sham wavefunctions generated by the DFT public code: Quantum Espresso.

Application of graphene in nano-photonics has been explored because of the unique combination of its optical and electronic properties. Transparency of graphene in the visible range of energy has led to its application as transparent coatings. Optical absorption of graphene is anisotropic due to difference in properties with light polarization along parallel and perpendicular to axis normal to the sheet. Experiments have indicated that, as compared to graphite, there is redshift of absorption bands $\pi + \sigma$ electron plasmon and disappearance of bulk plasmons. As discussed in Sect. 5.3, Eberlein et al. [110] have shown that π and $\pi + \sigma$ surface plasmon modes in free-standing single sheets are present at 4.7 and 14.6 eV. These values exhibit substantial redshift from the corresponding values in graphite.

5.5.3.1 Analysis of Models Used for Evaluating Optical Properties

As discussed in literature, in Ref. [107], VASP (Vienna Ab initio Simulation Package) based on DFT has been used to simulate the properties of graphene. The self-consistent Kohn–Sham equations of the density functional theory in a plane wave set, with the projector-augmented wave pseudopotentials, have been applied. The Perdew–Burke–Ernzerhof (PBE) exchange-correlation (XC) function of the generalized gradient approximation (GGA) has been adopted in the calculations. The plane wave cutoff energy has been set to 400 eV. The 4×4 supercell (consisting of 32 atoms) has been used to simulate the isolated sheet. The graphene sheets have been separated by larger than 12 Å along the perpendicular direction in order to avoid interlayer interactions. The Monkhorst-Pack scheme has been used for sampling the Brillouin zone. The structures have been fully relaxed with a gamma-centered $7 \times 7 \times 1$ k-mesh. The partial occupancies have been treated using the tetrahedron methodology with Blöchl corrections. For geometry optimizations, the coordinates have been relaxed until the Hellmann–Feynman forces were less than 0.005 Å. Dielectric function $\varepsilon(\omega)$ has been calculated in the energy interval from 0 to 25 eV within the random phase approximation (RPA).

Figure 5.76a shows the real part of the dielectric constant for pristine graphene. Results have been simulated as described above. Figure 5.76b shows the real part of the dielectric constant for pristine graphene simulated in this study. Quantum Espresso, with implementation of DFT and plane wave sets, has been used for the band structure calculations. We have considered graphene sheet of width 2 nm and length 2 nm with periodic boundary conditions. The width of the vacuum layer, above the graphene layer, is assumed to be 13 Å in order to ensure that there is no interlayer interaction. The Monkhorst-Pack scheme has been used for sampling the Brillouin zone and it is found that $13 \times 13 \times 1$ is the K-point mesh for which the energy of the system is a minimum. Dielectric function $\varepsilon(\omega)$ has been calculated in the energy interval from 0 to 40 eV within the random phase approximation (RPA). It is found that there are two major features—one peak at around 5 eV corresponding

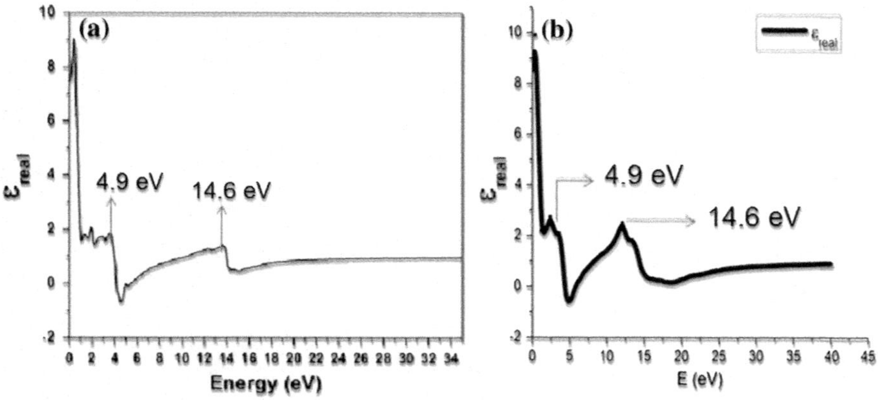

Fig. 5.76 **a** Simulated real part of dielectric function of pure graphene for $E\perp c$, from literature [107]. **b** Simulated real part of dielectric function of pure graphene, from calculations. Peak at around 5 eV is due to $\pi \rightarrow \pi*$ and another peak at about 14.6 eV corresponds to $\sigma \rightarrow \sigma*$ interband transitions

to $\pi \rightarrow \pi^*$ and the another peak at about 14.6 eV corresponding to $\sigma \rightarrow \sigma^*$ interband transitions.

Figure 5.77a shows the imaginary part of the dielectric constant evaluated in Ref. [107]. It is observed that the plot for E⊥c consists of a very significant peak at small frequencies (up to 5 eV) and also another peak structure of broader frequency range which starts from about 10 eV and has a weak intensity peak at 14 eV. Figure 5.77b shows the imaginary part of the dielectric constant in accordance with our DFT simulations with Quantum Espresso in combination with post-processing tool—YAMBO. The procedure adopted for modeling and analysis is explained in the previous section.

Once the verification of the model is established by comparison with the literature (experimental or simulated), the model is then extended to simulate the properties of graphene and zigzag GNR, as a function of number of layers. The results are presented in Fig. 5.78a, b. It is observed that there is a rise in intensity of peaks at 4.9 and 14.6 eV. Other features such as shoulder in the range from 5 to 10 eV are constant irrespective of the number of layers. Above 27.5 eV, the imaginary part of the dielectric constant shows nil value.

Figure 5.79a shows the trend in the refractive index as a function of energy from Ref. [107]. Refractive index follows similar trends as it has been evaluated from the corresponding dielectric functions. It shows minima positions at the maximum positions in the absorption spectra, as expected. Figure 5.79b shows the trend of refractive index as a function of energy from our calculations for the case of E⊥c. Similar trends are observed in the literature.

Figure 5.80a shows the trends for refractive indices for armchair GNR as a function of energy for single-layer, bilayer, and trilayer armchair GNR. The peaks are present at the same positions, i.e., 4.9 and 14.6 eV.

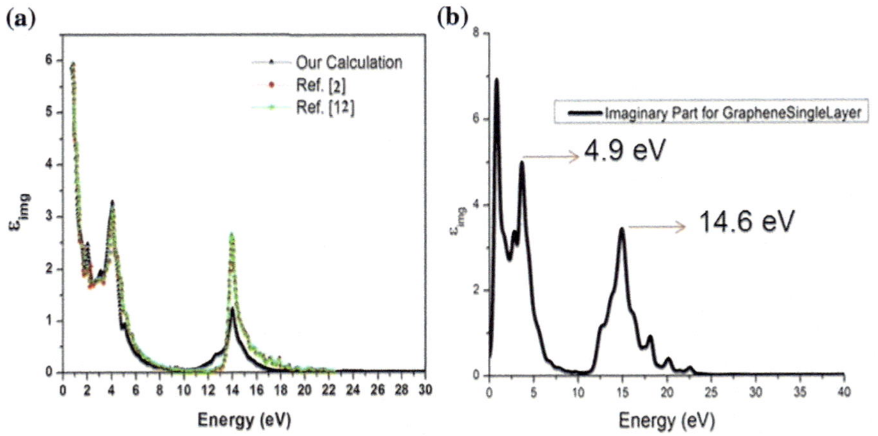

Fig. 5.77 **a** Simulated imaginary part of the dielectric function of pure graphene for light polarization perpendicular to the plane of graphene sheet ($E \perp c$), from literature [107] and **b** simulated imaginary part of dielectric function of pure graphene for light polarization perpendicular to the plane of graphene sheet ($E \perp c$), from our calculations

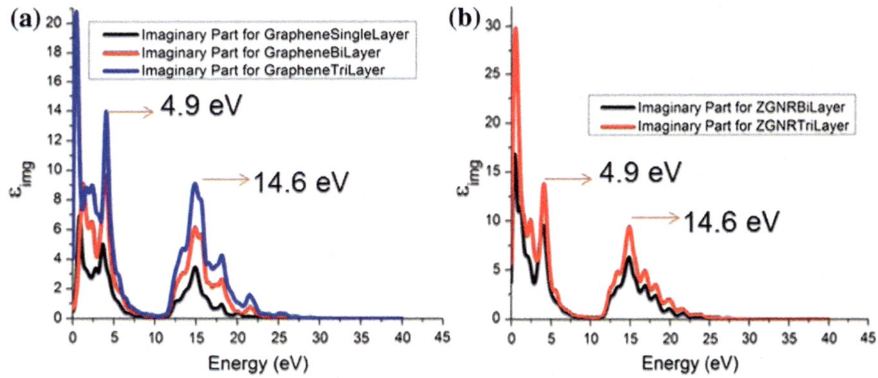

Fig. 5.78 **a** Simulated imaginary part of the dielectric constant for single-layer, bilayer, and trilayer graphene, from our calculations. **b** Simulated imaginary part of the dielectric constant for bilayer and trilayer zigzag GNRs, from our calculations

There is similar observation of rise in intensity of peaks at 4.9 and 14.6 eV. It can be seen that the shoulder locations are found at the same positions in the lower energy region from 5 to 15 eV, while beyond 15 eV, there is a slight rise in all the portions of the curves for refractive indices. Figure 5.80b shows the trends observed for zigzag bilayer and trilayer GNRs. It can be observed that the curves follow similar trends as reported in earlier cases.

Figure 5.81a shows the extinction coefficient (k) as a function of energy presented in Ref. [107]. It is observed that there is an additional peak at about 1.2 eV in the curve for E⊥c, along with those at 4.9 eV and 14.6 eV, as reported in all the previous

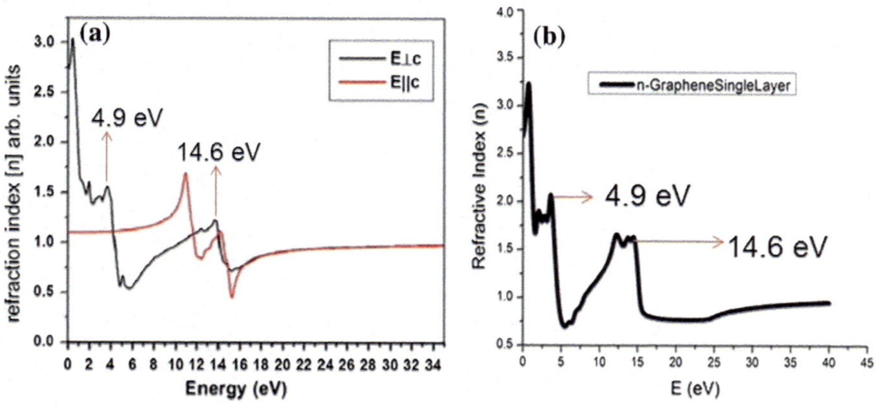

Fig. 5.79 **a** Simulated refractive index of graphene for light polarization perpendicular to the plane of graphene sheet (E⊥c) as well as (*E*||*c*), from literature. **b** Simulated refractive index of graphene for light polarization perpendicular to the plane of graphene sheet (*E*⊥*c*), from our calculations

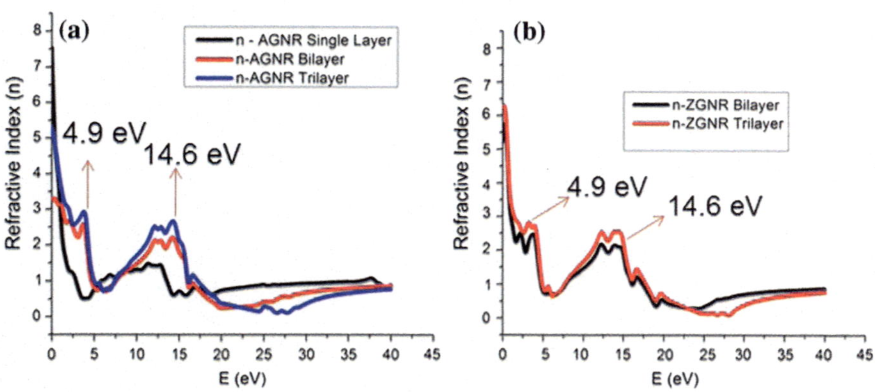

Fig. 5.80 **a** Simulated refractive indices of single-layer, bilayer, and trilayer armchair GNRs. **b** Simulated refractive indices of bilayer and trilayer zigzag GNRs, from our calculations

results. Figure 5.81b shows the extinction coefficient (*k*) as a function of energy simulated from our calculations. Similar peaks are observed at 1.2, 4.9 and 14.6 eV for the orientation E⊥c.

Figure 5.82a shows the trends in extinction coefficient for single-layer, bilayer, and trilayer graphene as a function of energy. A rise in intensity is observed at all the portions of the curve except the shoulder between 8 and 12 eV. Figure 5.82b shows the trends in extinction coefficient for bilayer and trilayer zigzag GNRs as a function of energy.

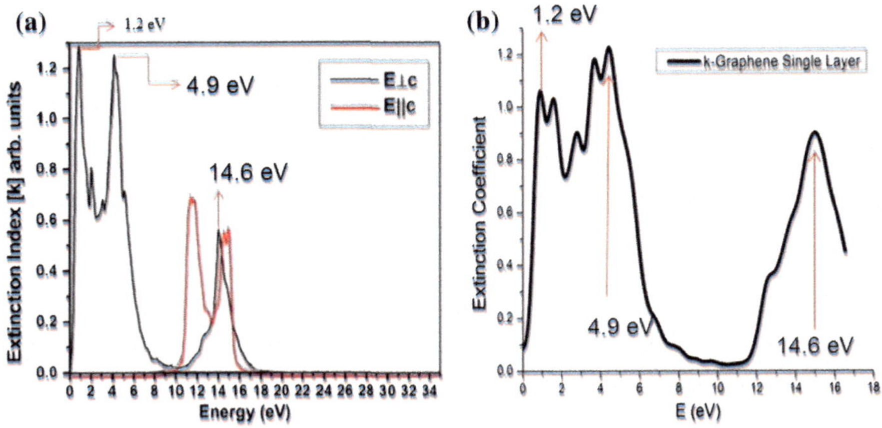

Fig. 5.81 **a** Simulated extinction coefficient as a function of energy for single-layer graphene for light polarization perpendicular to the plane of graphene sheet ($E \perp c$) as well as ($E \| c$), from literature [107]. **b** Simulated extinction coefficient as a function of energy for single-layer graphene for light polarization perpendicular to the plane of graphene sheet ($E \perp c$), from our calculations

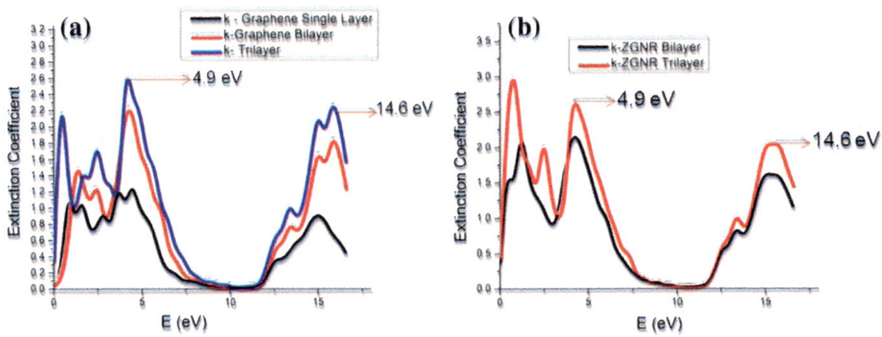

Fig. 5.82 **a** Simulated extinction coefficient for single-layer, bilayer, and trilayer graphene as a function of energy, from our calculations. **b** Simulated extinction coefficient for bilayer and trilayer zigzag GNRs, from our calculations

5.5.3.2 Experimental Observations of Optical Properties of Graphene

Kravets et al. [200] have demonstrated the optical transparency of two-dimensional system with a symmetric electronic spectrum by a fine structure constant. They have measured the ellipsometric spectra and extracted the optical constants of a graphene layer.

A reconstruction of the electronic dispersion relation near the K-point using optical transmission spectra has been reported. Spectroscopic ellipsometry analysis of graphene deposited on amorphous quartz substrates has been reported in Fig. 5.83 [200, 201]. A pronounced peak in the ultraviolet absorption at 4.6 eV is observed because of a van Hove singularity in the DOS of graphene. The peak has been found

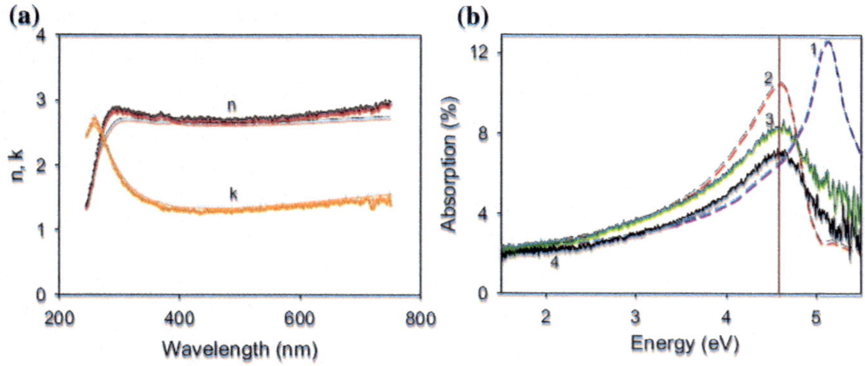

Fig. 5.83 Experimental measurement of variable angle spectroscopic ellipsometry of graphene on amorphous quartz substrate—**a** reconstructed optical constants of graphene are shown, and **b** absorption spectra of single-layer graphene. Solid curves 3 and 4 are experimental data. Dashed curves 1 and 2 are calculations from Ref. [201]

to be asymmetric and downshifted by 0.5 eV. The downshift is attributed to possible excitonic effects. The symmetric peak at 5.2 eV (curve 1) is expected by noninteracting theory, whereas interaction effects should result in asymmetric peak downshifted to 4.6 eV (curve 2) [200].

Based on the analysis of the optical properties of graphene, Wu et al. [202] have explored the sensitivity of graphene-based optical biosensor. Experimental measurements by Nair et al. [18] on light transmission through suspended graphene membranes show that the transparency of graphene is a universal constant and is independent of the wavelength.

On the basis of measurements, the dielectric function ε or complex refractive index n of graphene, in the visible range, has been obtained within the framework of Fresnel coefficient calculations:

$$\boldsymbol{n} = \mathbf{3.0} + \boldsymbol{i}\frac{\boldsymbol{C_1}}{\mathbf{3}}\boldsymbol{\lambda_0} \tag{5.5.1}$$

where the constant $C_1 \approx 5.446\,\mu\text{m}^{-1}$ is implied by the opacity measurement by Nair [18], and λ_0 is the vacuum wavelength. In order to validate this experimental model, Wu et al. [202] used full-wave electromagnetic field simulation in frequency domain using CST MICROWAVE STUDIO® 2009. In the calculations, the thickness of graphene $d = L \times 0.34$ nm (where L is the number of graphene layers) is sandwiched between two vacuum blocks. The light transmittance through monolayer graphene is about 97.7%, which is related to the fine structure constant α by $\pi\alpha = 2.3\%$, as shown in Fig. 5.84. Thus, monolayer of graphene would absorb 2.3% of the incident light. The simulated transmittance is also shown to follow the same trends and it is shown to absorb 2.3% of the incident light for each added layer of graphene. The measured optical spectra of graphene enable one to use the complex refractive index

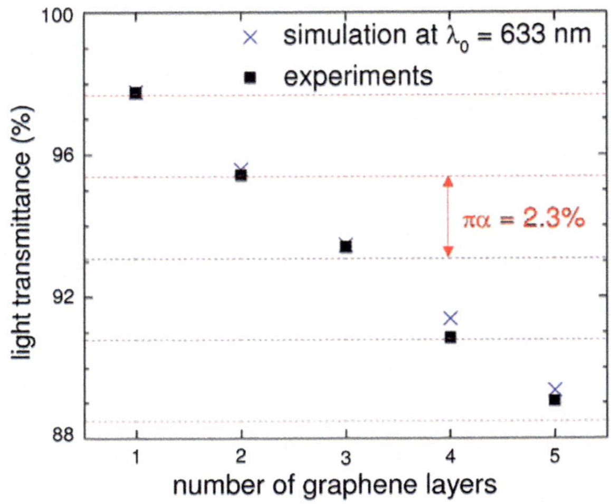

Fig. 5.84 Simulated transmittance of light at $\lambda_0 =$ 633 nm (crosses) and measured transmittance of white light (squares) [18] as a function of the number of graphene layers, from Ref. [202]. The dashed lines correspond to an intensity reduction by $\pi\alpha = 2.3\%$ with each added layer, where α is the fine structure constant [202]

for prediction of the optical behavior of graphene for surface plasmon resonance biosensing.

5.5.3.3 Experimental Measurements of Emissivity of Graphite

In most of the applications of thermography, relative temperature variations are of interest. Variation of measured intensity can be interpreted in terms of temperature variations or emissivity change of the surface with respect to its surroundings. Typical areas of application of emissivity measurements are oxidation of metals and semiconductors or erosion of materials on reentry of vehicles from space. Knowledge of emissivity is the first step for determination of accurate radiation temperatures.

The method of comparison of radiation is known to be a well-suited approach for measuring emissivities that are essential for temperature measurements. The sample surface radiance is measured by using calibrated detector and the surface temperature can be estimated independently. Temperature difference between sample surface and black body can be calculated using heat conduction equation. The accuracy in the determination of the temperature difference depends on the thermal conductivity of the material, which should be either known or measured [203].

For detectors with linear operation, the emittance ε can be evaluated as the ratio of detector signals measured from the sample (U_s) to the signal measured at the black body (U_{BB}) at the same temperature, from Eq. (5.5.2).

$$\boldsymbol{\varepsilon} = \frac{\boldsymbol{U}_s(\boldsymbol{T}_s)}{\boldsymbol{U}_{\mathbf{BB}}(\boldsymbol{T}_s)} \tag{5.5.2}$$

The radiation from the sample consists of contribution from the direct radiation as well as reflected radiation. The contribution from reflection can be either calculated or determined experimentally. The samples in this study are 15 mm in diameter and 3–6 mm thick. The sample is heated using electron gun and the vacuum in the chamber is $<0.7 \times 10^{-4}$ mbar. Linear pyrometer is used for the measurements in the lower wavelength range and a thermal detector is used for the determination of emissivity in the higher wavelength >1.3 μm range.

Spectral emissivity has been measured at temperatures in the range of 1100–2000 K and is presented in Fig. 5.85a. No systematic temperature dependence has been noted. The rate of decrease of emissivity, at longer wavelengths, is reported to be much higher for composite samples as compared with pure graphite. The reason for the lower value of emissivity for composite samples is attributed to the extreme surface smoothness of fiber bundles. In contrast to the temperature-independent nature of spectral emissivity, the total normal emissivity increases with increasing temperature. In accordance with Wien's displacement law, lower wavelength corresponds to higher radiation intensity. Hence, the total (integrated) emissivity increases with temperature if the spectral emissivity increases with reduction in wavelengths. Figure 5.85b shows the experimental trends of total normal emissivity as a function of temperature.

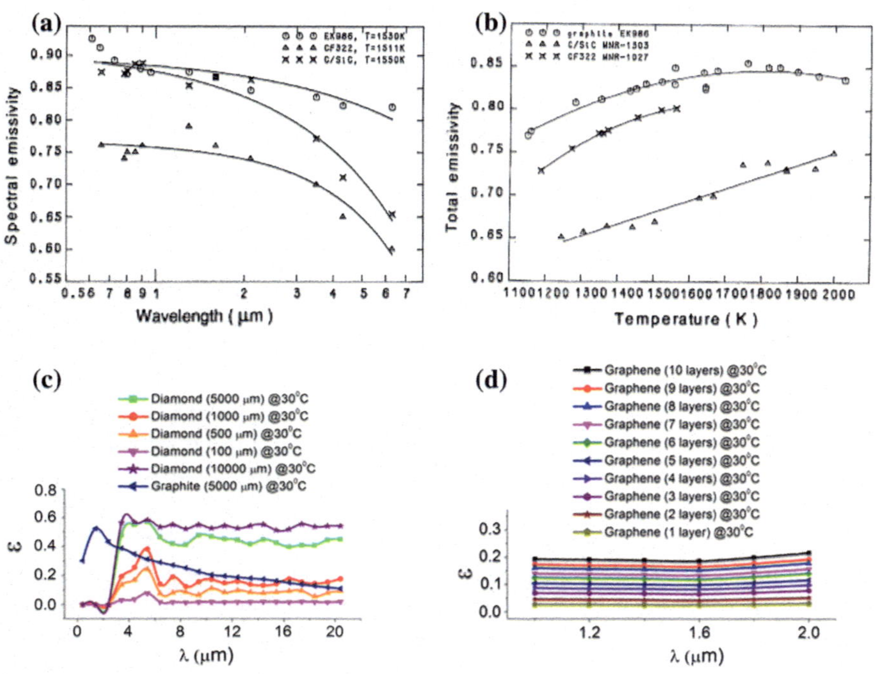

Fig. 5.85 **a** Experimental normal spectral emissivity of graphite (EK986) carbon/carbon (CF322) and carbon/silicon carbide (C-SiC) composites at higher temperature, from literature [203]. **b** Total normal emissivity of graphite (EK986), carbon/carbon (CF322), and carbon/ silicon carbide (C-SiC) composites [203]. **c** Simulated normal emissivity versus wavelength for diamond and graphite [123]. **d** Emissivity versus wavelength for graphene up to 10 layers (temperature in °C) [123]

Figure 5.85c shows the simulated trends of spectral emissivity of 5000 μm thick graphite at room temperature. The simulated trends of spectral emissivity for various thicknesses of natural diamond, at room temperature, are also included for comparison. It is observed that the values of emissivity of graphite at room temperature are much lower than the reported experimental values for higher temperatures. In the case of natural diamond, the emissivity is found to be increasing with increase in thickness at room temperature. Emissivity of natural diamond is found to be almost independent of wavelength for thicknesses below 1000 μm, in the wavelength range of 6–20 μm. For higher thicknesses >1000 μm of natural diamond, there are some variations in the emissivity values.

Figure 5.85d shows the simulated trends of emissivity for graphene (1–10 layers) in the wavelength range of 0.4–2.0 μm. It is observed that the emissivity increases with increasing number of layers of graphene, independent of the wavelength. Increase in the emissivity of graphene with increase in thickness is attributed to the higher absorption with increasing number of layers.

5.5.4 Mechanical Properties

Graphene is reported to have intrinsic strength exceeding any other material as well as other carbon allotropes. Hence, there is a motivation for its application in carbon fiber reinforcements in advanced composites. However, the high theoretical values of its mechanical properties, reported in the literature, are very difficult to realize experimentally. Recent experimental studies have made possible use of single-layer graphene in applications.

Nanoindentation, based on Atomic force microscopy (AFM), has been used to determine the mechanical properties of graphene [53]. Advantages of AFM for testing the properties of graphene over CNT are as follows: precise definition of sample geometry, 2D structure is less sensitive to the presence of defects, and sheet is clamped around the entire hole circumference, which is different from CNTs. Figure 5.86 shows the nanoindentation study of 5 × 5 mm array of circular wells patterned on Si substrates with 300 nm SiO_2 epilayer formed by nanoimprint lithography and reactive ion etching. Flakes of graphene have been mechanically deposited on the substrate. Graphene has been shown to adhere to the vertical wall of the hole of 2–10 nm diameter due to van der Waals attraction force to the substrate.

Prior to the start of indentation, graphene membranes have been scanned using AFM in noncontact mode. AFM tip has been placed within 50 nm of the center. Mechanical testing has been performed at a constant displacement rate, followed by reversal of load. The repetition of such cycles for several times for each film tested showed no hysteresis. This indicates the elastic behavior of graphene film around the periphery of the well. Followed by studies of elastic behavior of graphene films, the samples were subjected to indentation till failure. Using ab initio values for in-plane stiffness and flexural rigidity, it has been reported that the energy from bending the

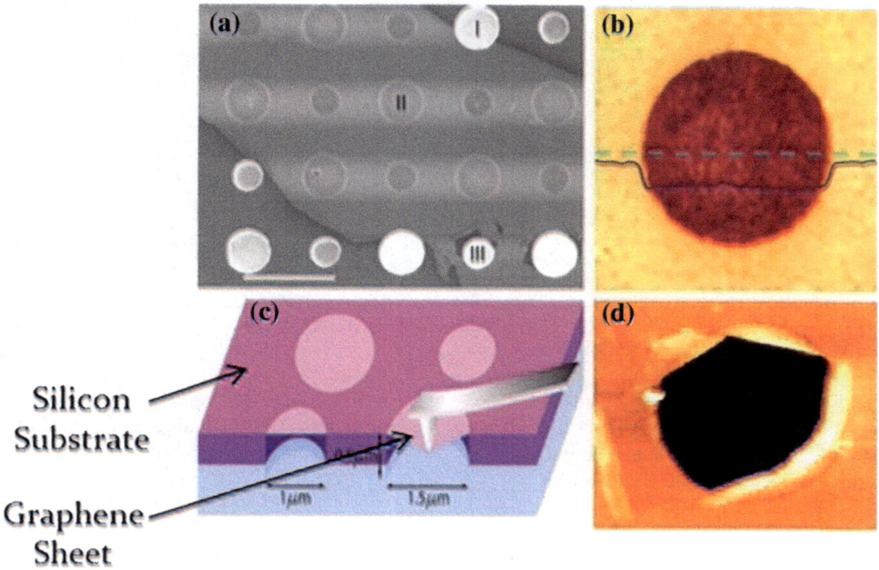

Fig. 5.86 Images of suspended graphene membranes. **a** Scanning electron micrograph of a large graphene flake spanning an array of circular holes 1 and 1.5 μm in diameter. Area I shows a hole partially covered by graphene, area II is fully covered, and area III is fractured from indentation. Scale bar, 3 μm. **b** Noncontact mode AFM image of one membrane, 1.5 μm in diameter. The solid blue line is a height profile along the dashed line. The step height at the edge of the membrane is about 2.5 nm. **c** Schematic of nanoindentation on suspended graphene membrane. **d** AFM image of a fractured membrane [53]

graphene membranes is about three orders of magnitude less than the energy from in-plane strain. Images of suspended graphene membranes are shown in Fig. 5.86.

The resulting force versus displacement curve for loading/unloading is shown in Fig. 5.87a. Using numerical simulations and molecular dynamics simulations, it has been shown that the elastic response of graphene nanosheets is nonlinear. Force versus displacement behavior has been approximated in accordance with Eq. (5.5.3).

$$F = \sigma_0^{2D}\left(\pi a^2\right)\left(\frac{\delta}{a}\right) + E^{2D}\left(q^3 a\right)\left(\frac{\delta}{a}\right)^3 \qquad (5.5.3)$$

where F is the applied force, δ is the deflection at the center point, σ_0^{2D} is the pretension in the film, ν is Poisson's ratio [taken as 0.165, the Poisson's ratio for graphite in the basal plane], and $q = 1/(1.05 - 0.15\nu - 0.16\nu^2) = 1.02$ is a dimensionless constant. The solid line in Fig. 5.87a shows the least-squares curve fit of one set of experimental data, based on Eq. (5.5.3), taking σ_0^{2D} and E^{2D} as free parameters. The closeness of the fit has been considered to validate the appropriateness of this model.

Figure 5.87b shows the typical breaking curves for different indenter tip radii and well diameters. The graphene film has been reported to be hanging around the edge

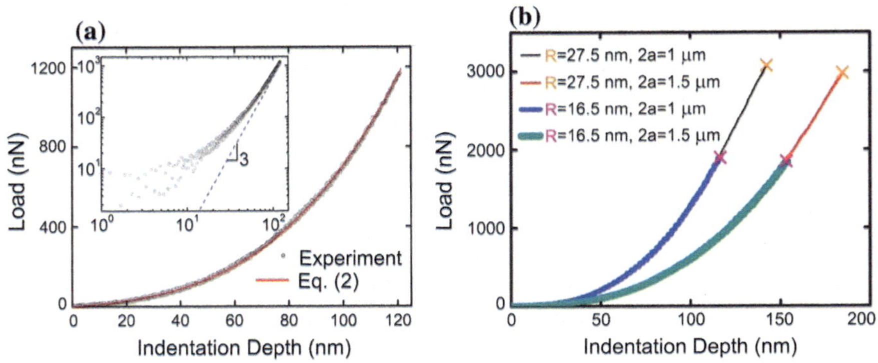

Fig. 5.87 **a** Loading/unloading curve and curve fitting to Eq. (5.5.3). The curve approaches cubic behavior at high loads (inset). **b** Fracture test results. Four typical tests, with different tip radii and film diameters; fracture loads are indicated by × marks. Breaking force depends strongly on tip radius but not on sample diameter [53]

of the hole without significant sign of slippage or irreversible deformation prior to catastrophic failure. This indicates the fracture in the graphene film initiated at the indentation point. The indentation forces in this process have been sufficient to break Si AFM tips. However, the diamond tips used in this experiment have been confirmed by TEM to have no damage.

Wang et al. [172] have performed molecular dynamics study on nanoindentation experiments for single-layer rectangular graphene films with four clamped edges. Load versus displacement curves have been presented and the effects of indenter radii, loading speeds, and aspect ratios of graphene film on the mechanical properties have been discussed. Young's Modulus and strength of single-layer graphene film have been determined and the values are 1.0 TPa and 200 GPa, respectively. Graphene film is shown to rupture at a critical indentation depth, as shown in Fig. 5.88.

A spherical diamond indenter has been used to simulate the nanoindentation (Fig. 5.88) [172]. The upper ball in Fig. 5.88a is diamond indenter. It is considered to be rigid, so that there are no changes in the atomic configuration of the indenter during the molecular dynamics simulation. The lower layer is single-layer of rectangular graphene.

The interatomic interaction of carbon atoms in the graphene layer has been described by AIREBO (Adaptive Intermolecular Reactive Empirical Bond Order) potential. AIREBO potential has been shown to consider multi-body potential effects and local atomic circumstance effects. AIREBO introduces long-range interaction and torsion term. This is in contrast to the Tersoff–Brenner potential. Hence, AIREBO potential is found to be more accurate in the evaluation of Young's modulus as well as breaking and reforming of bonds in carbon atoms of graphene layer. The cutoff parameter of AIREBO potential has been considered as 2 Å, which helps to avoid the influence of nonphysical explanations with improper cutoff on fracture mechanics. Canonical ensemble (NVT) has been considered during molecular dynamics simu-

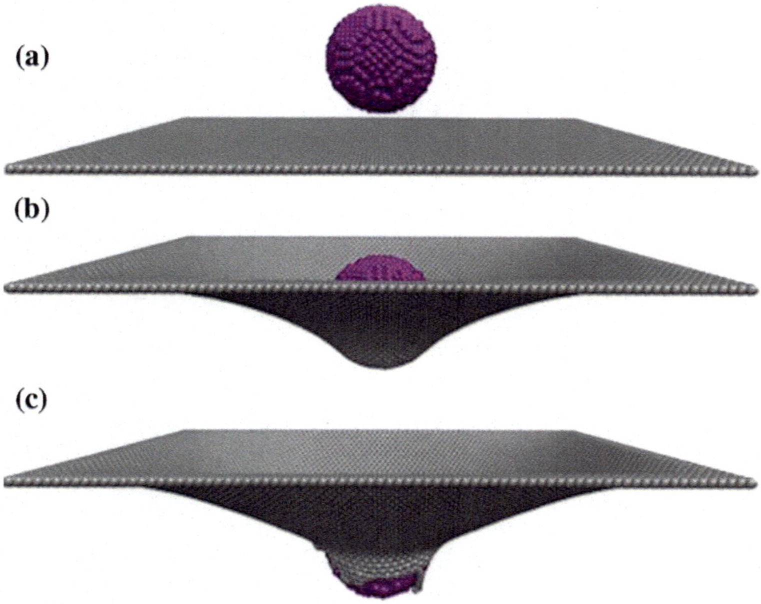

Fig. 5.88 Atomic configuration of the system model during the nanoindentation experiment. **a** The origin model, **b** the state during the loading process, and **c** at rupture state, from literature [172]

lations and the temperature has been maintained at 0.01 K. Nosé-Hoover thermostat has been used for avoiding the complex effects of atomic thermal fluctuations and timestep has been set as 0.001 ps. Initially, the system has been relaxed and equilibration has been maintained to keep the system at lowest energy state during the simulation.

Figure 5.89a shows the effect of various indenter speeds and Fig. 5.89b shows the effect of various diameters of indenter on the load versus displacement plots. It is found that, when the loading speed is higher than the critical value, with the increase in speed, the maximum load increases rapidly; simultaneously, the critical indentation depth decreases rapidly. The higher is the indenter loading speed, the lesser is the time required for the indenter to pass through the graphene sheet. This leads to higher load and lower indentation depth than the ones at lower speed. Figure 5.89b shows that the radii of the spherical indenter affect the indentation depth. It is noted that critical indentation depth increases with increase in the indenter radius.

Figure 5.90a shows our simulation of the indentation of a rectangular single-layer graphene sheet. We have considered a molecular dynamics simulation with two different interatomic potentials describing interactions of carbon atoms in graphene layer, one with AIREBO potential and other with Tersoff–Brenner potential. The indenter is considered to be spherical rigid type with force constant of 1000 eV/Å. Initially, the system is relaxed and the structure is equilibrated during the simulation. The atoms at the edges of the graphene sheet (about 10 Å on each edge) are kept

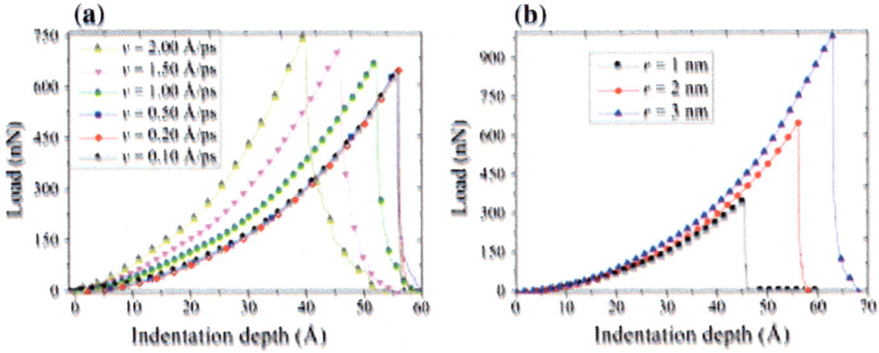

Fig. 5.89 Comparison of load versus indentation depth for different parameters. **a** The indenter is loaded at different loading speeds between 0.10 and 2 Å/ps. **b** The indenter is loaded with different indenter radii of 1, 2, and 3 nm, from literature [172]

fixed during the simulation in order to provide physical support to the system during indentation. The timestep is considered as 1 fs and the cutoff distance for the AIREBO potential is considered to be 2.5 Å. We have considered NVT ensemble (canonical) and the system temperature is maintained at 300 K during the simulation. Berendsen thermostat has been applied to maintain the constant temperature of the system. As noted in the literature, we found that AIREBO potential has more accuracy than Tersoff–Brenner type interatomic potential for calculating the indentation force. The higher values of indentation force for the AIREBO potential are attributed to the inclusion of additional long-range interaction term and torsion in the case of AIREBO potential, which is absent for Tersoff–Brenner type potential.

Figure 5.90b shows the effects of variation of speeds of the diamond indenter on the force versus displacement curves. It is noted that the higher is the indentation speed, the higher is the slope of the curve. This is consistent with the observation in the literature. Figure 5.90c shows the effect of variation of indenter diameter on the force versus displacement curves. It is found that the higher is the indenter radius for the same aspect ratio of the graphene sheet, the higher is the slope of the force versus indentation depth curves. This indicates that the system will show higher stiffness with higher indenter radius. However, the elastic moduli, evaluated from the force versus indentation depth curves, are found to be in the range of 0.9–1.2 TPa, which are also noted from various AFM nanoindentation experiments in the literature. Thus, our model is verified with the reported literature. The nonlinear elastic response of graphene is evident from the curves in Fig. 5.90.

Figure 5.91 shows the comparison of nanoindentation behavior of single-layer and bilayer graphene. The simulation methodology is as explained in earlier cases. It can be noted that the values of indentation force are much higher (almost double) in the case of bilayer graphene as compared to single-layer graphene. It is seen that the values of indentation force are rapidly increasing with increasing depth of

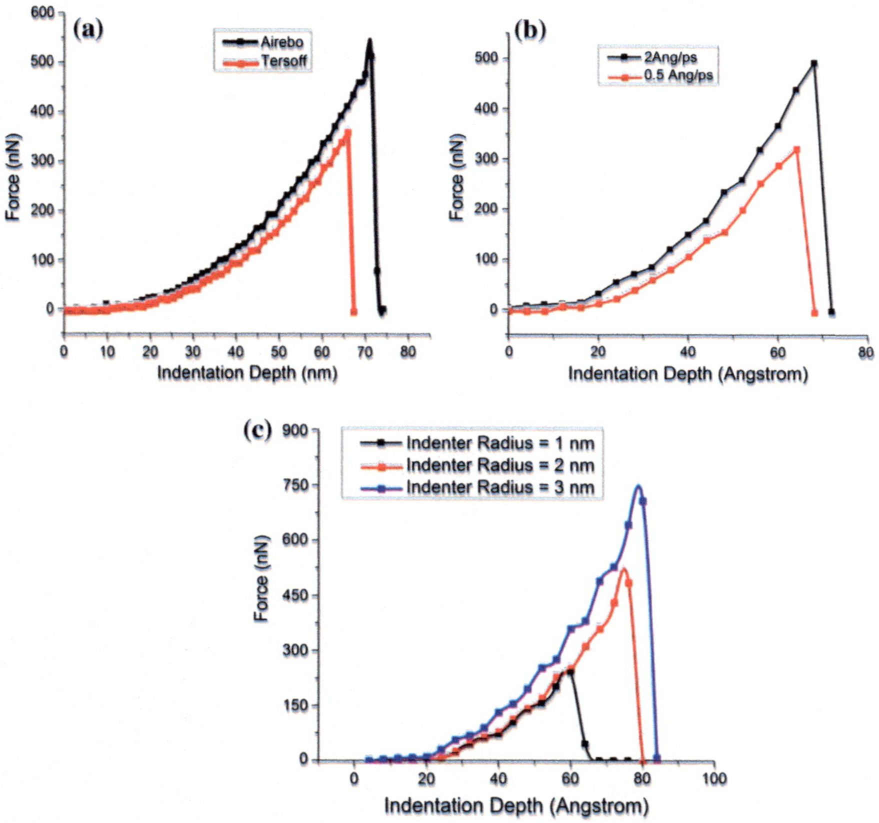

Fig. 5.90 **a** Comparison of simulated force versus indentation depth using Tersoff–Brenner and AIREBO interatomic potentials, from our calculations. **b** Simulated force exerted by indenter as a function of indentation depth for variation in speed of indenter, and **c** simulated force exerted by indenter as a function of indentation depth for variation of indenter radius, from our calculations

indentation. Higher stiffness values of bilayer graphene are evident from the curves in Fig. 5.91.

Zandiatashbar et al. [204] have studied the effects of defects on the intrinsic strength and stiffness of graphene. It is reported that even with high density of sp^3 defects in graphene, the two-dimensional elastic modulus is maintained. The defective graphene in sp^3 regime is noted to have breaking strength ~14% smaller than its pristine counterpart. In contrast to this, the mechanical properties of graphene have been reported to have significant drop in the vacancy defect regime. They have provided a mapping between the Raman spectra of defective graphene sheets and its mechanical properties.

In experimental studies, $1 \times 1\ cm^2$ array of circular wells, with diameters ranging from 0.5 to 5 μm, have been patterned on Si chip with a 300-nm SiO_2-capping layer, by means of photolithography and reactive ion etching. Suspended membranes have

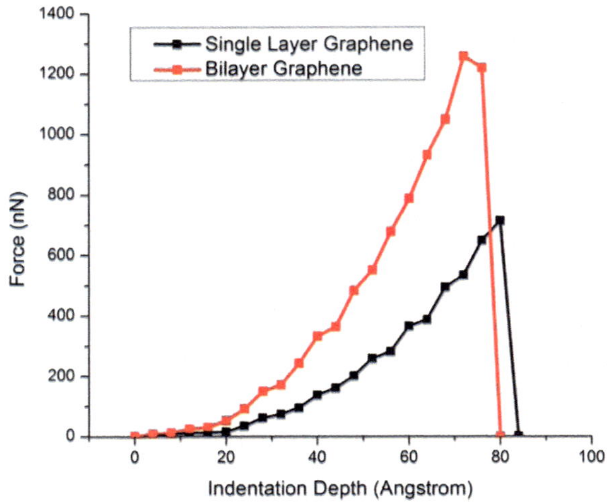

Fig. 5.91 Simulated force versus indentation depth trends for single-layer and bilayer graphene, from our calculations

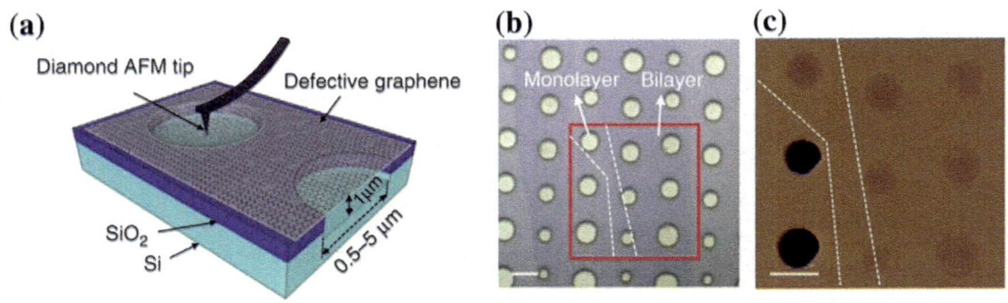

Fig. 5.92 **a** Schematic representation of Atomic Force Microscopy (AFM) nanoindentation test on suspended graphene sheets with defects. Graphene sheet is suspended over a hole with diameters ranging from 0.5 to 5 μm and depth of ~1 μm. **b** Optical micrograph of exfoliated graphene sheets suspended over holes. White dashed line indicates the boundary of each layer. **c** Noncontact mode AFM image of suspended graphene sheet obtained from the red square box region marked in (**b**). Scale bars, 3 μm [204]

been created by mechanical exfoliation of graphene on the patterned substrate. Elastic stiffness and breaking strength have been evaluated using AFM nanoindentation with a diamond tip indenter, as shown in Fig. 5.92.

Defects in graphene sheet have been induced using a tabletop oxygen plasma etcher. Some other methods, reported in the literature, for inducing defects in graphene are weak oxidation by ion bombardment, oxygen plasma or ultraviolet irradiation to etch graphene. The etch rate in oxygen plasma has been reported to be ~9 layers per minute at a chamber pressure of ~215 mTorr. This rate is much faster as compared to the other approaches. Raman spectroscopy has been used to characterize defective graphene sheets.

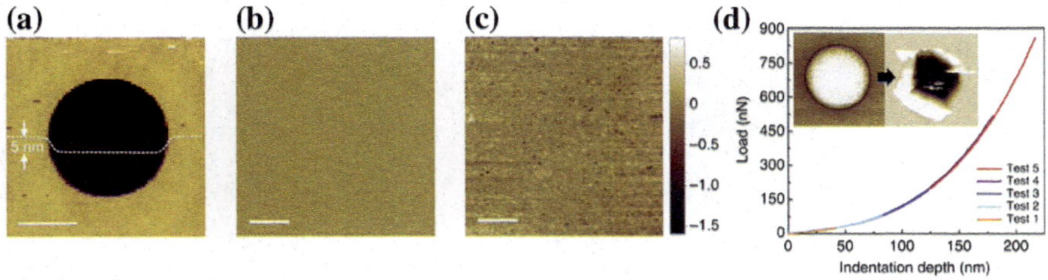

Fig. 5.93 **a** AFM image of a graphene sheet fully covering a hole. High-resolution AFM images of suspended graphene sheet **b** before and **c** after oxygen plasma exposure of 55 s. The plasma treatment leaves the surface pock-marked with a multitude of nanopores that are several nm in size (the dark spots in the image represent the nanopores). **d** Typical force versus displacement curves of AFM nanoindentation test for defective graphene exposed to oxygen plasma for 30 s. Tests are repeated at increasing indentation depths until the sample breaks. The curves fall on top of each other (no hysteresis), which indicates no significant sliding or slippage between the graphene membrane and the substrate. The AFM images in the inset of d show a graphene sheet before and after fracture. Scale bars, 1 μm (**a**); 100 nm (**b**, **c**) [204]

Figure 5.93a, b, c show the images of AFM nanoindentation of the graphene sheet at various stages. Figure 5.93d is the typical force versus indentation depth curve for a defective graphene sheet. It has been reported that the breaking stress shows higher sensitivity to defects than the elastic stiffness, irrespective of the type of defects. Principal finding of this study has been to show that graphene can maintain a large fraction of its pristine strength and stiffness even in the presence of sp^3 type defects.

Figure 5.94 shows the aberration-corrected high-resolution TEM (AC-HRTEM) image of defective graphene lattice with different oxygen plasma exposure times. These images confirm that, in sp^3 defect regime, sp^3 point defects in the form of oxygen adatoms are generated. With higher plasma exposure, carbon atoms are etched from the lattice. This leads to the formation of nanocavities or nanopores in the lattice.

5.5.5 *Thermal Conductivity*

The continuing progress in the electronics industry has led to miniaturization of circuit components. This is leading to challenges in the thermal management in electronic devices, circuits, and systems and hence the immense interest in the thermal properties of materials that are used in nanostructured form in electronic components and circuits. The room temperature thermal conductivity of carbon-based allotropes has a very wide range—of over five orders of magnitude—least for amorphous form of carbon to the highest in carbon nanotubes and graphene. A field of special interest has been the size dependence of thermal conductance of graphene and its derivatives [164].

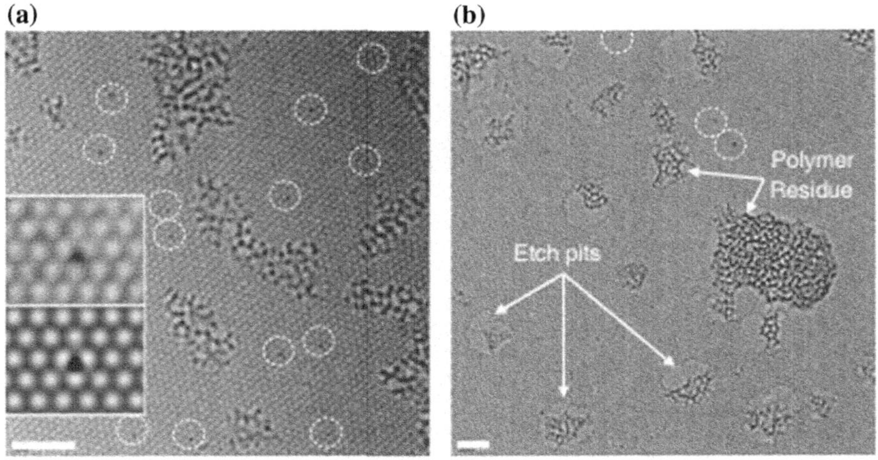

Fig. 5.94 AC-HRTEM characterization of defect structures. Images of a typical graphene sheet in **a** sp^3-type and **b** vacancy-type defect regime. Polymer residue associated with the transfer process onto the TEM grid is indicated by arrows. The defective graphene of the vacancy-type defect regime contains an abundance of nanocavities (that is, etch pits), while the defective graphene of the sp^3-type defect regime shows a contrasting absence of such cavities. The black dots circled with dashed lines in (**a**) and (**b**) are oxygen adatoms. The insets of (**a**) show the experimentally obtained TEM image (upper) and the corresponding simulated image (lower) of oxygen atoms bonded to carbon forming sp^3 point defects. Scale bars, 2 nm (**a, b**) [204]

Heat conduction in carbon materials is usually dominated by phonons, even in case of graphite (having metallic properties). This is attributed to the strong covalent sp^2 bonding which results in efficient heat transfer by lattice vibrations. The thermal conductivity of both suspended as well as supported graphene has been studied by molecular dynamics simulations. Length dependence has been reported in the case of suspended single-layer graphene. The thermal conductivity of supported single-layer graphene has been found to be independent of its length. Figure 5.95 shows the simulated thermal conductivity of suspended single-layer zigzag graphene as a function of its length at room temperature [205].

In this study, MD simulations have been performed using LAMMPS package. Tersoff potential with optimized potential parameters has been used in the simulations to describe C–C interatomic interactions. Graphene interlayer interactions have been modeled using van der Waal (vdW) type Lennard-Jones potential. Nonequilibrium molecular dynamics (NEMD) simulations with Langevin heat bath have been used to study thermal transport in graphene structures.

Figure 5.96 shows our calculated results of the thermal conductivity for pristine graphene nanosheets as well as armchair and zigzag GNRs, as a function of their lengths, with NEMD simulations. All the structures, considered for these simulations, have a width of 2 nm. It is seen that κ_p increases as a function of length for the suspended graphene structures, as reported in the literature. As the sample size increases, more low-frequency acoustic phonons can be excited and contribute

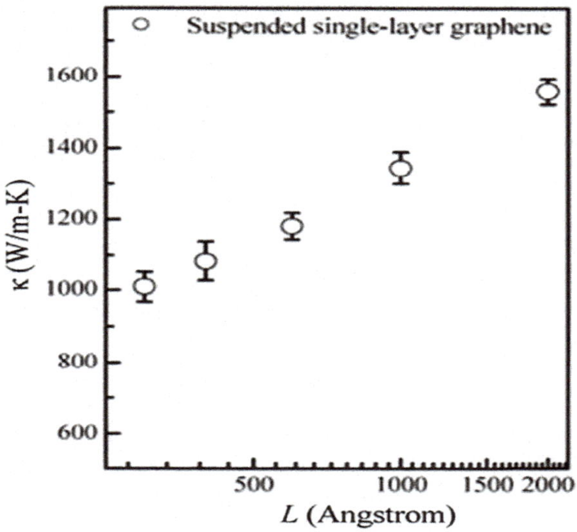

Fig. 5.95 Simulated thermal conductivity κ of suspended (empty circle) and single-layer graphene versus the length L at room temperature, from literature. Here, zigzag graphene with fixed width $W = 52$ Å is used [205]

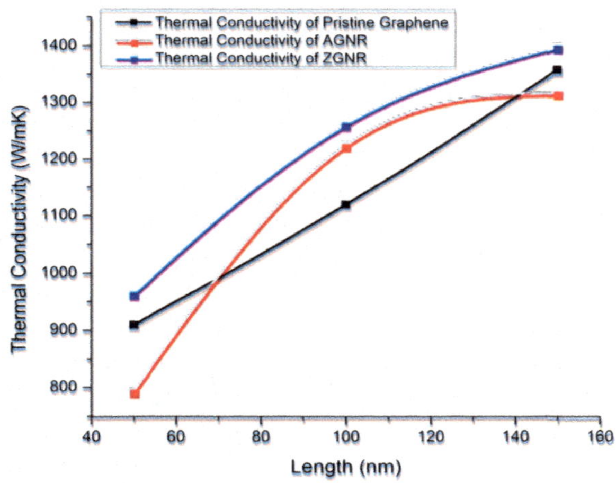

Fig. 5.96 Simulated thermal conductivity of pristine graphene, armchair GNR, and zigzag GNR, from our calculations

to thermal conduction, resulting in a length-dependent behavior. Extremely long-wavelength low-frequency acoustic phonons have ballistic transport mechanism.

The variations in widths of single-layer graphene are known to have significant effect on their thermal conductivity values. Figure 5.97a shows the width dependence of thermal conductivity as a function of number of layers of graphene [206].

A decreasing trend in thermal conductivity is observed with increase in width from 1.7 to 6.6 nm for graphene. This is explained from the phenomena that, with increasing widths of GNRs, the total number of phonon modes increases, while the number of edge-localized phonon modes does not change.

Tersoff–Brenner potential has been used in these simulations for describing C–C bonding. Nosé–Hoover thermostat is used for controlling temperature during the

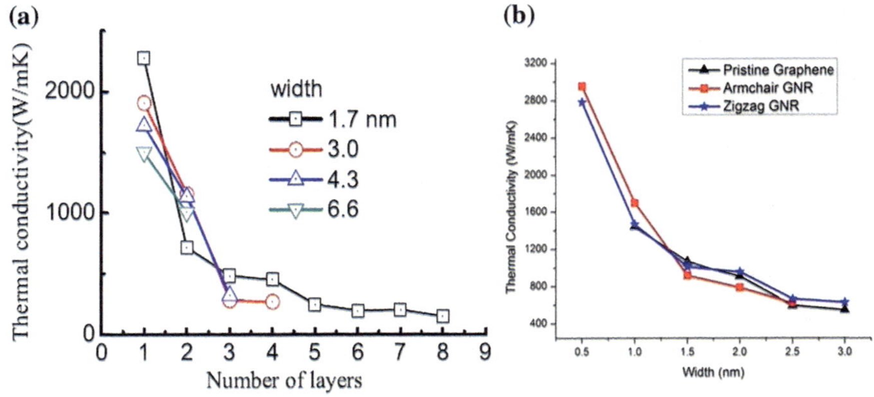

Fig. 5.97 **a** Layer-dependent simulated thermal conductivity for various widths of zigzag GNRs at 325 K, from literature [206]. **b** Effect of increasing width on thermal conductivity of pristine graphene and GNRs, from our calculations

simulations. Room temperature (RT) κ_p decreases monotonically with number of layers in few-layer graphene.

Figure 5.97b shows our simulated results of width dependence of κ_p at room temperature. In our simulations, the length of graphene as well as GNRs is considered as 50 nm in all the cases with variations in width. The reason for decrease in κ_p with increase in width up to 3 nm is attributed to the constant number of edge-localized phonon modes and increasing Umklapp scattering effects with higher widths.

Figure 5.98a shows the trends in κ_p for higher widths of zigzag and armchair GNRs [207]. The increase in κ_p with increase in width is attributed to the increase in the number of phonon modes. However, κ_p gets saturated with increase of width beyond a certain limit as the energy gap between different phonons reduces with increasing width. This leads to higher probability of Umklapp scattering.

Figure 5.98b shows our simulated results of the thermal conductivity of armchair and zigzag GNRs with widths beyond 10 nm up to 60 nm. The simulation parameters have been kept constant for all the calculations. Length of graphene and GNRs is maintained as 50 nm and width is varied from 10 to 60 nm. It is observed that similar trends occur with increase in width of graphene and GNRs. The reason for this behavior has been explained previously and is consistent with the findings in the literature.

Figure 5.99a shows the simulated effect of nitrogen doping on armchair and zigzag GNRs as a function of concentration of nitrogen in the structure, from Ref. [181] by Mortazavi et al. All the simulations of thermal conductivity have been performed using NEMD implemented in LAMMPS. Bonding interactions between C–C and C–N atoms have been modeled using optimized Tersoff potentials. The atoms at the corner of graphene structures are fixed during simulation. It is observed that doping of ~1% nitrogen in graphene results in considerable decrease in κ_p as well as reduction in chirality dependence of κ_p.

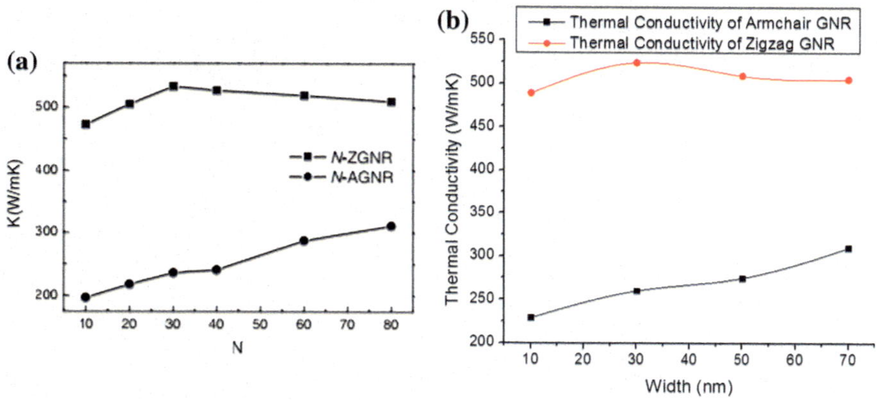

Fig. 5.98 **a** Simulated thermal conductivity of *N* -AGNR and *N* -ZGNR with variation of *N*, where the length of GNRs is fixed to be 11 nm. The ZGNR's thermal conductivity increases first and then decreases with *N* increasing, while the AGNR's thermal conductivity monotonously increases with *N*, from literature [207]. **b** Simulated thermal conductivity of AGNR and ZGNR as a function of width, from our calculations

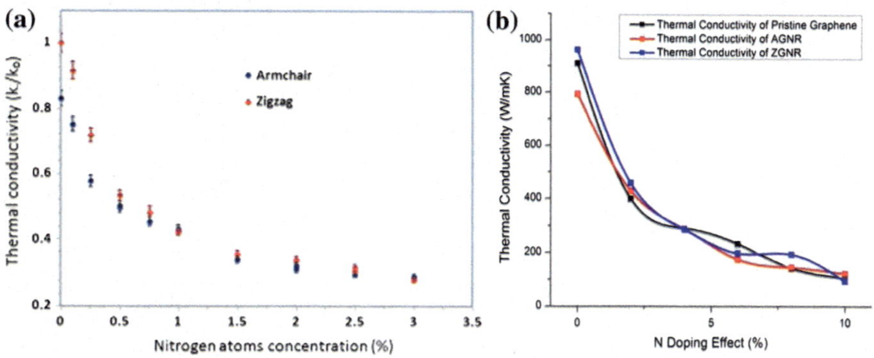

Fig. 5.99 **a** Simulated effect of nitrogen atom concentrations on the normalized thermal conductivity of single-layer graphene along the armchair and zigzag chirality directions, from literature [207]. **b** Simulated thermal conductivity of pristine single-layer graphene and armchair/zigzag GNRs, from our calculations

Figure 5.99b shows the trends for thermal conductivity as a function of % nitrogen for pristine graphene nanosheets as well as armchair and zigzag GNRs, from our simulations. NEMD simulations have been performed in LAMMPS at room temperature (300 K). Lengths and widths of all the graphene structures are maintained as 50 and 2 nm, respectively. Tersoff–Brenner potentials are used to describe C–C and C–N bonds and the cutoff in each case is 0.25 nm. Berendsen thermostat is used for maintaining constant temperature during the simulations. It is observed that there is a drastic reduction in the thermal conductivity with about 2.5% nitrogen doping and the values have been found to be consistent with the literature. The reason for this decrease is reduction in lattice symmetry and broadening of peaks in DOS in the

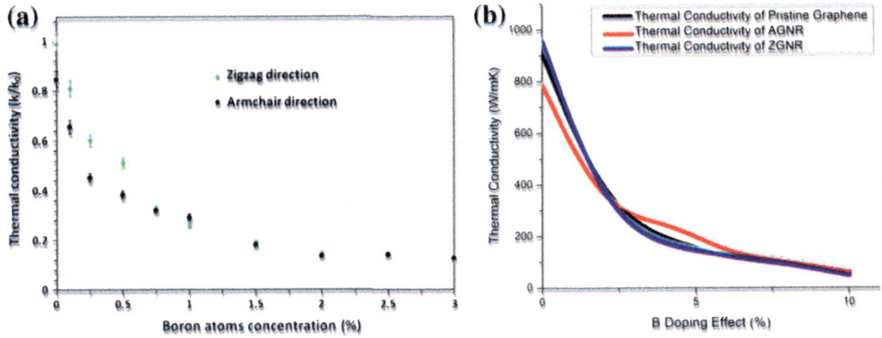

Fig. 5.100 **a** Simulated effect of substitutions of boron atoms on the thermal conductivity of single-layer graphene along the armchair and zigzag chirality directions, from literature [208]. **b** Comparison of simulated thermal conductivity of pristine graphene and armchair/zigzag GNRs, from our calculations

frequency range around 50 THz [181]. These phonon modes are considered to be dominant heat-carrying modes and contribute to heat transport in graphene.

Mortazavi et al. [208] have presented their studies of boron doping in armchair and zigzag GNRs by NEMD simulations implemented in LAMMPS. Interatomic interactions for C–C and C–B have been modeled using optimized Tersoff potential parameters. As shown in Fig. 5.100a, it is observed that about 0.75% boron doping results in a drastic reduction in its thermal conductivity. The reasons for the reduction in thermal conductivity, with increase in dopant concentration, are similar to the ones in case of nitrogen doping in graphene.

Figure 5.100b shows our simulated results of thermal conductivity for pristine graphene nanosheets and armchair and zigzag GNRs as a function of % boron doping at room temperature, using NEMD simulations implemented in LAMMPS. It is observed that the trends are similar to the reported literature as in Fig. 5.100a. The reduction in thermal conductivity of graphene structures is higher with boron doping. There is reduced chirality dependence of thermal conductivity with increase in % boron doping in graphene.

Figure 5.101a shows the trends for thermal conductivity as a function of temperatures up to 700 K. The presented data is a collection of experimental results from various references presenting similar trends in the thermal conductivity calculations [164]. It is observed that increasing temperature leads to lower values of thermal conductivity of graphene structures. The higher is the temperature, more are the collision of phonons. This leads to lower values of thermal conductivity with increasing temperature beyond 100 K. The increase in thermal conductivity with increase of temperatures up to 100 K is attributed to increase in mobility of phonons.

Figure 5.101b shows the results of our simulations of thermal conductivity of pristine graphene nanosheets and nanoribbons, using NEMD simulations implemented in LAMMPS. In all the cases, the size of graphene structures is maintained as 50 nm × 2 nm ($L \times W$). It is observed that similar reducing trends occur with increasing

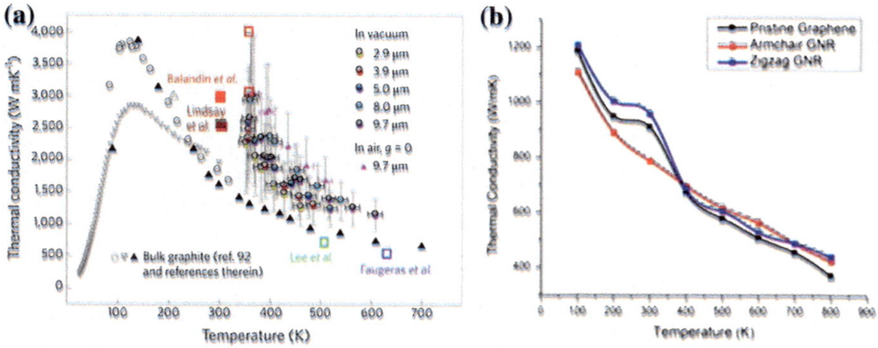

Fig. 5.101 **a** Experimental thermal conductivity of graphene as a function of temperature. Experimental data points are indicated by empty rectangular boxes. The filled red and brown boxes are theoretical data points. These two sets of points are for different graphene flake sizes—3 and 5 μm, respectively. Setting $L = 3$ μm would give $K \approx 2500$ W/mK. The experimental results from different research groups obtained for graphene by the Raman optothermal technique are in agreement within the experimental uncertainty of the method [164]. **b** Simulated thermal conductivity as a function of temperature for pristine graphene and armchair/zigzag GNRs, from our calculations

temperature from 100 to 800 K, as shown in Fig. 5.101b. Reason for this reduction is attributed to higher Umklapp scattering as well as collision of phonons at high temperature.

Figure 5.102a shows the reduction in simulated thermal conductivity of pristine graphene with varying concentration of vacancies, from Ref. [183] by Zhang et al. In these simulations, REBO (Reactive Empirical Bond Order) potentials are used to model C–C interatomic interactions. Green–Kubo method is employed in MD simulations. It is based on linear response theory. Green–Kubo method has been known to have an advantage of being devoid of artificial thermostat perturbation. The presence of thermostat could have influence on the thermal conductivity and heat flux. NVE ensemble (microcanonical) has been considered for isolated pristine graphene sheet at room temperature (300 K). Zhang et al. have observed that even 0.42% vacancy concentration in graphene can cause significant reduction in its thermal conductivity. At about 8.75% vacancy concentration, its thermal conductivity can be reduced to ~3.08 ± 0.31 W/mK.

The reduction in thermal conductivity has been attributed to two important factors. First one is the broadening of phonon mode peaks around 15 THz, where the valley-shaped curves are almost flattened by the broadening of nearby peaks with increasing concentration of vacancies. Broadening of phonon modes shows reduction of lifetime of those modes leading to lower mean free path. Thus, the thermal conductivity reduces. The second is an average increase of DOS for low-frequency modes below 15 THz. This leads to reduction in relaxation time and corresponding mean free path which causes lower thermal conductivity values.

Figure 5.102b shows our simulated thermal conductivity values of pristine graphene as well as armchair and zigzag GNRs with varying concentration of vacan-

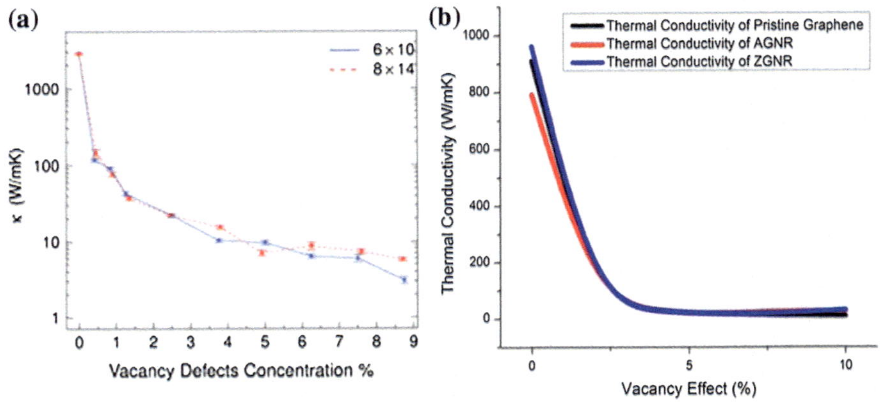

Fig. 5.102 **a** Simulated thermal conductivity of graphene as a function of vacancy defect concentration (at 300 K) using molecular dynamics simulations with the Green–Kubo method. The solid blue (dark gray) line and the dashed red (gray) line correspond to PB sizes of 6 × 10 and 8 × 14, respectively [183]. **b** Simulated thermal conductivity of pristine graphene, armchair/zigzag GNR as a function of % vacancies at 300 K using molecular dynamics simulations with NEMD-based Müller-Plathe technique, from our calculations

cies, using NEMD simulations implemented in LAMMPS. Initially, the system is relaxed and equilibration is maintained in the structure for about 50,000 timesteps with NPT ensemble. Timestep is considered to be 1 fs in the simulation. NEMD simulation is performed on NVE ensemble and Müller-Plathe technique is simulated, as explained in Sect. 5.3. Periodic boundary conditions are considered for each case and size of graphene is 50 nm × 2 nm. Similar trends have been observed and drastic reduction in thermal conductivity is found, which is consistent with the literature.

Figure 5.103a shows the trends of simulated thermal conductivity of graphene as well as armchair and zigzag GNRs, from Ref. [206] by Zhong et al. It is observed that there is decrease in thermal conductivity in each case due to cross-plane coupling of low-energy phonons. The cross-plane coupling is absent in the case of single-layer graphene. Hence, the mode of thermal transport in single-layer graphene is ballistic. In the presence of cross-plane coupling, particles at the interface between the layers are subject to collision leading to phonon scattering. This leads to lower values of thermal conductivity for few-layer graphene (FLG).

Figure 5.103b shows our results for NEMD simulation of pristine graphene as well as armchair and zigzag GNRs, as a function of number of layers from 1 to 3 at room temperature. Modeling parameters have been considered to be same as in previous cases. The size of graphene in each case is 50 nm × 2 nm. Similar trends are observed as the ones reported in the literature. An important observation is the reduced chirality dependence of κ_p of GNRs with increasing number of layers.

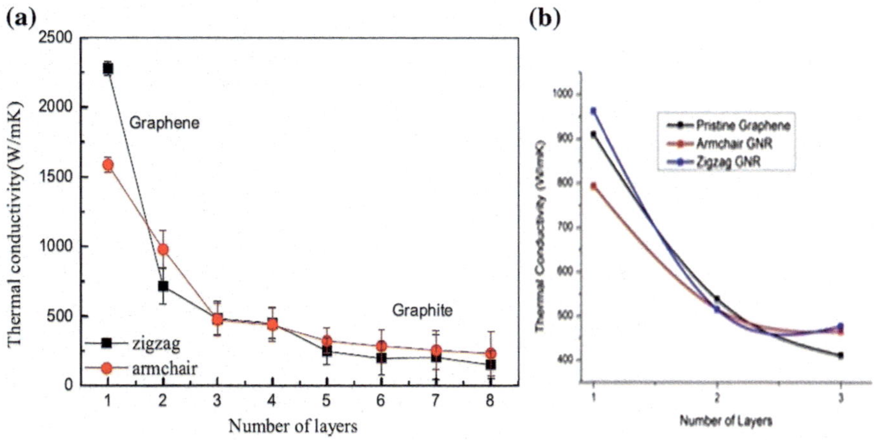

Fig. 5.103 **a** Simulated layer-dependent thermal conductivity of few-layer graphene. The one-layer nanoribbons refer to graphene and the five to eight layers nanoribbons are similar to ultra-thin graphite, from literature [206]. **b** Simulated layer-dependent thermal conductivity of pristine graphene and armchair/zigzag GNRs 1–3 layers, from our simulations

5.6 Synthesis Techniques

For graphene to become cost-effective and to be utilized in a variety of applications, it is essential to be manufacturable on a large scale. In the literature, various techniques have been used to separate the single-layer graphene from graphite—mechanical or chemical. Similarly, there have been various techniques to chemically reduce Graphene Oxide (GO) to reduced graphene oxide (RGO). This reduction has been attractive due to its low-cost scalability. GO is the precursor of RGO.

The synthesis of GO was first published by Brodie in 1859. In this technique, equal weight of graphite is to be mixed with 3 equal weights of $KClO_3$ and needs to be reacted to fuming HNO_3 at 60 °C for 4 days.

Cooper et al. [209] have presented non-oxidative route of producing few-layer graphene via the electrochemical intercalation of tetraalkylammonium cations into pristine graphite. Highly Oriented Pyrolytic Graphite (HOPG) has been found to be more advantageous amongst different forms of graphite but the source electrode set up poses difficulties to the procedure and needs sonication. Few-layer graphene flakes (2 nm thickness) have been reported to be formed directly using graphite electrode. However, flake diameters from this source are shown to be typically small (100–200 nm). For solvent-based method, graphite rod does not require secondary physical processing of resulting dispersion.

Sasha et al. [210] have reported the synthesis of graphene-based nanosheets via chemical reduction of exfoliated graphite oxide. In this experiment, colloidal suspension of exfoliated graphene oxide sheets is obtained by reduction with hydrazine hydrate in water. This results in aggregation and subsequent formation of high surface area carbon material consisting of thin graphene-based sheets. Pei et al. [211] have presented detailed studies on reduction of graphene oxide.

Abeladin et al. [212] have synthesized large-area high-quality graphene from different liquid alcohols (methanol, ethanol, and propanol) precursors by chemical vapor deposition on copper foils in a tube furnace. They have performed a systematic investigation with various conditions of growth. Alcohol vapor exposure times of 5 min and average temperature of 850 °C have been found to yield continuous graphene monolayer films. Raman spectroscopy has been used to observe growth of graphene layer. X-Ray Photoelectron Spectroscopy has been used to show that the oxygen moieties found in the source molecules have no measurable doping or oxidation effect in the synthesized graphene. Using Raman spectroscopy, it has been shown that transfer of graphene films to insulating substrates are of high quality.

Liu et al. [213] have investigated the synthesis of high-quality monolayer and bilayer graphene on copper using chemical vapor deposition. It has been shown that the growth of graphene on copper is determined by process parameters during growth as well as quality of copper substrate and pretreatment of the substrate. Microtopography of copper surface has been found to strongly affect the uniformity of graphene growth while the number of synthesized layers of graphene at low-pressure conditions is determined by the purity of copper film. Also, a minimum partial pressure of hydrocarbon is essential for graphene to cover the surface of copper during the growth. They have reported a new mode of growth resulting in tetragonal shaped graphene domain. This is different from the previously known lobe structure (for monolayer) or hexagonal (for few-layer) graphene. HRTEM (High-Resolution Transmission Electron Microscopy) has shown nonideal nature of CVD graphene structure for the first time which indicates important cause of electron/hole mobility degradation which is typically observed in CVD graphene.

Dong et al. [214] have synthesized large-sized thin films of single- and few-layered graphene by chemical vapor deposition on copper foils under atmospheric pressure using ethanol or pentane as precursor. By using various characterization techniques such as confocal Raman, TEM, and Scanning Tunneling Microscopy (STM), the major portion of the obtained films have been found to be hexagonal graphene lattice. Electrical measurements and optical microscopy have been used to confirm the continuity of the films. It has been reported that CVD-grown graphene, with ethanol as precursor, exhibits lower defect density, higher electrical conductivity and Hall mobility than that grown with pentane as precursor. The liquid precursor-based atmospheric pressure CVD synthesis has been presented as a simple, cost-effective, and safe thin-film graphene growth method.

Wu et al. [215] have presented wafer-scale synthesis of graphene by chemical vapor deposition and its application in hydrogen sensing. Graphene with large area has been synthesized on copper foils by chemical vapor deposition under ambient pressure. $4'' \times 4''$ graphene film has been transferred onto a $6''$ Si wafer with a thermally grown oxide film. 4 mm $\times$ 3 mm size graphene film with 1 nm palladium film deposited for hydrogen detection has been fabricated. This has been used to detect hydrogen in air with concentrations in the range of 0.0025–1%. Pd decorated graphene films show high sensitivity, fast response, and recovery. They can be used for multiple cycles. Mechanism of hydrogen detection by these films has also been discussed in the literature.

Liao et al. [216] have presented different strategies of graphene-dielectric integration which is significantly important for the development of graphene transistors and new generation of electronics. Physical Vapor Deposition (PVD) can be used for direct deposition of dielectric materials on graphene, which often leads to defects in the monolayer of carbon lattice.

Kuilla et al. [217] have presented recent advances in the modification of graphene and fabrication of graphene-based polymer nanocomposites. Modification of graphene/graphene oxide and the utilization of these materials in the fabrication of nanocomposites with different polymer matrices have been explored.

5.7 Applications of Graphene

Continuous depletion of fossil fuel and finite sources of energy has been a motivation for increased research and development in the field of nanotechnology. Developing renewable and more efficient energy conversion and storage systems represents one of the most important and viable approaches to the demands on energy. The structure and functions of materials and devices have significant effect on energy conversion efficiency and storage.

Due to the flexibility to tune the morphology and structure of graphene and its derivatives, it has been important templates for the structural and functional properties. Quan et al. [218] have reviewed the functional applications of materials obtained from graphene-templated approaches for potential applications in energy storage and conversion.

Palla et al. [219] have used bandgap engineered graphene and hexagonal boron nitride for resonant tunneling diode. Doubled-barrier resonant tunneling diode (DBRTD) has been modeled by taking advantage of single-layer hexagonal graphene and hexagonal boron nitride (h-BN). Peak-to-valley ratio of approximately 8 (3) for p-type (n-type) DBRTD with quantum well of 5.1 nm (4.3 nm) at a barrier width of 1.3 nm has been achieved for zero bandgap graphene at room temperature.

Ali and Garcia [220] have discussed the necessity of irregularities for the chemical applications of graphene. In most of the applications such as sensors, energy storage, electrochemical systems, catalysis, etc., of graphene, the superior properties are usually because of reactivity of intrinsic defects and dangling bonds. There has been a gradual shift to the use of graphene nanoribbons and quantum dots instead of seamless graphene. Reduction of reactive sites makes graphene more well-defined structure-wise but practically less useful. In order to design graphene for practical applications, one needs to understand the role of structural irregularities. The research in the manipulation of structure of graphene has led to useful applications in electrochemistry, luminescence, and catalysis.

Tang et al. [221] have presented the functionalization of graphene mainly due to its semimetallic property. Changing properties of graphene from semimetal to semiconductor has been a topic of constant research. CVD growth techniques of monolayer and double layer graphene have been developed. The large-area doped graphene films have been fabricated to modify the structure-related optical and electrical properties of graphene. A novel technique has been invented named "Tang-Lau method" for making Graphene Oxide (GO). Growths of graphene quantum dots using microwave assisted hydrothermal method as well as "Soft-Template method" have been developed; preparations of Cl-, S-, and K-doped graphene quantum dots by hydrothermal methods have also been invented. Novel room temperature photodetectors have been developed for covering the detection bands from UV, Vis, and NIR.

Shi et al. [222] have discussed the role of surface engineering in improving in vitro stability and enriching and the functionality of graphene-based nanomaterials. This can enable single/multi-modal imaging (e.g., magnetic resonance imaging, positron emission tomography, etc.) and cancer therapy (e.g., photodynamic therapy, photothermal therapy, etc.). Qu et al. [223] have synthesized a versatile microscale-thick graphene foil via simple chemical vapor deposition on nickel foil. This distinctive free-standing graphene foil has lightweight, less defects and high electrical conductivity (4149 S cm^{-1}) unlike the previously reported self-supporting graphene films.

Xu et al. [224] have synthesized large-area high-quality graphene using polystyrene by atmospheric pressure chemical vapor deposition on copper foils in a short time period. These films of graphene on glass substrates show high optical transmittance and electrical conductivity. Magnetic transport studies have demonstrated that the as-grown monolayer graphene exhibits high carrier mobility of 3395 cm^2 V^{-1} s^{-1}. Kojima et al. [225] have developed magnetic field sensor, with high precision and significant usability, of graphene for automotive applications. Sensitivity of Hall sensor is directly proportional to its carrier mobility, which in the case of graphene is extremely high as compared to Si, GaAs, and InSb. Graphene Hall sensors are expected to be having very high sensitivity, which can enable sensing of Earth's magnetic field. Also, low-temperature dependence of carrier mobility due to ballistic transport in graphene can lead to increased usability of graphene.

Marini et al. [226] have theorized saturable graphene absorption which is a non-perturbative nonlinear optical phenomenon that can play a vital role in the generation of ultrafast light pulses. It has been reported that modulation depth of saturable absorption in graphene can be controlled through externally applied voltage. These results have practical significance and applications in the development of graphene-based optoelectronic devices, as well as mode-locking and random lasers. Bystrov et al. [227] have used computational molecular modeling of graphene/graphene oxide (G/GO) and polyvinylidene fluoride (PVDF) ferroelectric polymer composite nanostructures using semiempirical quantum approximation PM3 in HyperChem. Modeling of GO-methane-hydrates nanostructures based on hexagonal ice model shows that, after relaxation, the system keeps a stable deformed state. This can serve in gas hydrates storage and separation applications. Also, these modeled composites can be used as multifunctional molecular units. Ruhl et al. [228] have reviewed the availability of techniques for the production of graphene on dielectric substrates.

Dubey et al. [229] have reviewed advances and ability of graphene and its related forms to induce stem cells differentiation into osteogenic lineages. Graphene and its derivatives (GO and reduced GO) have gained attention in biomedical applications due to their extraordinary properties such as high surface area, excellent mechanical strength, and ease of functionalization.

5.8 Conclusions

Graphene is found to be an excellent material for applications in optoelectronics in view of its extraordinary electronic, optical, mechanical, and thermoelectric properties.

Graphene is a zero bandgap material. Modification of electronic band structure of graphene has been shown to be possible by means of chemical doping. Boron and nitrogen are effective p- and n-type dopants, respectively, in graphene. The band structure analysis shows that n- and p-type of dopants shift the Dirac point of graphene on the negative or positive side of energy, respectively. The armchair graphene nanoribbons are shown to have negligible bandgaps depending on their size while zigzag graphene nanoribbons are known to exhibit metallic nature. The metallic character of zigzag nanoribbons is attributed to the high density of edge states at the Fermi energy. The chirality dependence of electronic properties of graphene, leading to its metallic or semiconducting properties for zigzag and armchair nanoribbons, could present a possibility of full carbon-based electronic devices, where semiconducting tubes could be used as channels and metallic ones as interconnects. However, the lack of control on chirality is found to be a challenging aspect in engineering their electronic properties and industrial applications. Introducing impurities and functional groups has been shown to be an effective way of controlling the properties of graphene nanoribbons.

Graphene is shown to be optically transparent in the visible range. Both the real and imaginary parts of the dielectric constants of graphene are found to be varying with the number of layers. It is noted that there is an increase in the peak intensity of the dielectric functions with increase in number of layers of graphene. Variations in the static dielectric constant are found to be dependent on the concentration of p- or n-type dopants in graphene and GNRs.

A comparison of the wavelength and thickness-dependent optical properties, mainly emissivity and transmittance, has been presented for various multilayered configurations with graphene as one of the component layers. Such structures have been found to be useful as hot electron bolometers. Graphite is found to have emissivity that is independent of thickness. Emissivity of graphene remains low and increases with increase in number of layers. For BN-based hot electron bolometer, the emissivity is found to increase with increase in BN layer thickness. The effect of graphene is to increase the emissivity of the bolometer structure. A comparison of the emissivity of graphite and graphene show that graphite exhibits higher emissivity.

Graphene has been reported to have an extremely high elastic modulus of the order of 1 TPa. The values of elastic modulus are found to be varying slightly depending on the indenter diameter, i.e., the Young's modulus increases with increase in diameter.

The values of indentation force depend on the type of interatomic potential considered in the simulation. Airebo potential has values of indentation force ~1.5 times higher than the Tersoff potential, for the same indentation depth. The higher is the indentation velocity, the greater is the indentation force recorded at the same indentation depth. Bilayer graphene is found to provide higher resistance to indentation. Hence, bilayer graphene has higher values of indentation force at the same values of indentation depth. Load versus indentation depth curves are found to be useful tools in the evaluation of the mechanical properties. Atomistic simulations are successful in predicting the mechanical properties which are comparable to the experimental findings.

Single-layer graphene and GNRs are found to exhibit increase in phonon-induced lattice thermal conductivity (κ_p) with increase in length, while κ_p is found to increase with decrease in width. The sudden drop in κ_p is observed with increase in width from 0.5 to 1.5 nm. κ_p is noted to remain constant for higher widths in case of AGNRs. In case of ZGNRs, κ_p increases up to about 30 nm but decreases further. Transport is found to become more ballistic with increasing widths of GNRs. Thus, the role of edges reduces with increasing width. κ_p is found to reduce drastically ~50% with doping of B or N. Increasing dopant concentration also marks reducing chirality dependence of thermal conductivity. Boron doping leads to higher reduction in κ_p as compared to nitrogen doping. Higher temperatures cause reduction in κ_p for temperatures in the range of 100–800 K. At about 400 K, all the graphene nanostructures i.e., pristine graphene and GNRs are found to have similar values of κ_p. The presence of vacancies, to the extent of ~4%, causes about ~70% reduction in the thermal conductivity with increasing number of layers of graphene and GNRs indicating interlayer interactions. Room temperature κ_p of ZGNR is found to be higher than AGNR, while AGNR is reported to have higher κ_p at higher temperatures. The electronic thermal conductivity (κ_e) is found to increase with temperature. The rate of increase of κ_e is faster in bilayer graphene as compared to single-layer graphene. High thermal conductivity of graphene leads to its potential application as ballistic field-effect transistors (FETs).

Graphene is found to have extremely high electrical conductivities ~10^6–10^8 S/m, depending on the applied potentials and temperature. The electrical conductivity of GNRs is also a function of its chirality, i.e., armchair and zigzag. The peak electrical conductivity of graphene and GNRs decreases with increasing doping concentration. This is attributed to the corresponding changes in the electronic band structures upon doping. Seebeck coefficients are found to be symmetrical around the charge neutrality point. Peak Seebeck coefficients are found to reduce with increase in temperatures, mostly linearly at higher temperatures. For a given chemical potential, away from the Fermi energy, it is found that Seebeck coefficients increase linearly with temperature. This shows the diffusive thermoelectric generation mechanism in the absence of phonon drag components. Bandgaps reduce with increasing widths for AGNRs and hence Seebeck coefficients reduce. Similar effects have been observed in case of ZGNRs. Seebeck coefficients (TEP) are observed to be higher for AGNRs with respect to ZGNRs. TEP for AGNRs are symmetrical about the charge neutrality point (CNP) with the change in majority charge carriers from electrons to holes.

Seebeck coefficient is found to increase with increase in concentration of B or N dopants. Peak values of TEP for AGNR are observed at 12% p-doping while, for ZGNR, it is observed at about 4% doping of p-type and at 8% n-type doping. Hall resistivity is observed to follow similar trends as the Seebeck coefficient for graphene on a given substrate.

Power factor is found to increase with temperature in the range of 100–500 K. $\sigma_{graphene}$ is highest at $\pm k_B T$, while it reduces to nil at the Fermi level.

The Figure of Merit (ZT) is found to increase with decreasing length of graphene in the range of 200–25 nm with constant width of 2 nm. The highest values observed are ~1.4 for graphene at about 25 nm length. The reason for this improvement is the decrease in lattice-induced thermal conductivity for lower lengths of graphene. The effect of doping p- or n-type dopants in AGNR and ZGNR is to improve ZT values. It is found that ZT increases at a higher rate in AGNR than ZGNR with increasing concentration of dopants. The values of ZT are found to be higher for AGNR than ZGNR. Thus, AGNR is more suitable for thermoelectric device applications. The ZT values for modified geometries or doped graphene nanostructures are noted to be higher than conventional thermoelectrics such as Bi_2Te_3. The thermopower is found to improve with increase in concentration of vacancies in ZGNRs. This improvement is attributed to breaking of electron-hole energy symmetry with induced defects. Asymmetry in DOS leads to increased thermopower. This also leads to improved Figure of Merit (ZT) in ZGNR with respect to the corresponding pristine ZGNR (without vacancies). The reason for this improvement is attributed to reduced thermal conductivities and a minimal reduction in power factor.

References

1. Geim AK, Novoselov KS (2007) The rise of graphene. Nat Mater 6(3):183–191
2. Tsang ACH, Kwok HYH, Leung DYC (2017) The use of graphene based materials for fuel cell, photovoltaics, and supercapacitor electrode materials. Solid State Sci 67:A1–A14
3. Talbot C (1999) Fullerene and nanotube chemistry: an update. Sch Sci Rev 81:37–48
4. Birkett PR et al (1995) Holey fullerenes! a bis-lactone derivative of fullerene with an eleven-atom orifice. J Chem Soc, Chem Commun 18:1869–1870
5. Dresselhaus MS, Dresselhaus G, Eklund PC (1996) Chapter 2—Carbon materials. In: Eklund MS, Dresselhaus G, Dresselhaus PC (eds) Science of fullerenes and carbon nanotubes. Academic Press, San Diego, pp 15–59
6. Novoselov KS et al (2004) Electric field effect in atomically thin carbon films. Science 306(5696):666–669
7. Luo H et al (2014) Preparation of three-dimensional braided carbon fiber-reinforced PEEK composites for potential load-bearing bone fixations. Part I. Mechanical properties and cytocompatibility. J Mech Behav Biomed Mater 29:103–113
8. Chung DDL (2000) Thermal analysis of carbon fiber polymer-matrix composites by electrical resistance measurement. Thermochim Acta 364(1–2):121–132
9. Yakobson B, Avouris P (2001) Mechanical properties of carbon nanotubes. In: Dresselhaus M, Dresselhaus G, Avouris P (eds) Carbon nanotubes. Springer, Berlin, pp 287–327
10. Ranjbartoreh AR et al (2011) Advanced mechanical properties of graphene paper. J Appl Phys 109(1):014306 (6 p)

11. Wallace PR (1947) The band theory of graphite. Phys Rev 71(9):622–634
12. Novoselov KS et al (2005) Two-dimensional atomic crystals. Proc Natl Acad Sci USA 102(30):10451–10453
13. Gerstner E (2010) Nobel Prize 2010: Andre Geim & Konstantin Novoselov. Nat Phys 6(11):836
14. Landau LD (1937) Zur Theorie der phasenumwandlungen II. Phys Z Sowjetunion 11:26–35
15. Cahn RW, Harris B (1969) Newer forms of carbon and their uses. Nature 221(5176):132–141
16. Intriguing state of matter previously predicted in graphene-like materials might not exist after all, 2013 [11/13/2014]. Available from: http://phys.org/news/2013-05-intriguing-state-previously-graphene-like-materials.html
17. Partoens B, Peeters FM (2006) From graphene to graphite: electronic structure around the K point. Phys Rev B 74(7):075404
18. Nair RR et al (2008) Fine structure constant defines visual transparency of graphene. Science 320(5881):1308
19. Georgakilas V et al (2012) Functionalization of graphene: covalent and non-covalent approaches, derivatives and applications. Chem Rev 112(11):6156–6214
20. Muley SV, Ravindra NM (2014) Graphene–environmental and sensor applications. In: Hu A, Apblett A (eds) Nanotechnology for water treatment and purification. Springer International Publishing, pp 159–224
21. Yin PT et al (2013) Prospects for graphene-nanoparticle-based hybrid sensors. Phys Chem Chem Phys 15(31):12785–12799
22. Wang JT-W et al (2013) Low-temperature processed electron collection layers of graphene/TiO_2 nanocomposites in thin film perovskite solar cells. Nano Lett 14(2):724–730
23. El-Kady MF et al (2012) Laser scribing of high-performance and flexible graphene-based electrochemical capacitors. Science 335(6074):1326–1330
24. Stankovich S et al (2006) Graphene-based composite materials. Nature 442(7100):282–286
25. Kim KS et al (2009) Large-scale pattern growth of graphene films for stretchable transparent electrodes. Nature 457(7230):706–710
26. Liu Q et al (2009) Polymer photovoltaic cells based on solution-processable graphene and P3HT. Adv Funct Mater 19(6):894–904
27. Bae S et al (2010) Roll-to-roll production of 30-inch graphene films for transparent electrodes. Nat Nano 5(8):574–578
28. Tzalenchuk A et al (2010) Towards a quantum resistance standard based on epitaxial graphene. Nat Nano 5(3):186–189
29. Li X et al (2009) Large-area synthesis of high-quality and uniform graphene films on copper foils. Science 324(5932):1312–1314
30. Katsnelson MI, Novoselov KS (2007) Graphene: new bridge between condensed matter physics and quantum electrodynamics. Solid State Commun 143(1–2):3–13
31. Simon J, Greiner M (2012) Condensed-matter physics: a duo of graphene mimics. Nature 483(7389):282–284
32. Chahardeh JB (2012) A review on graphene transistors. Int J Adv Res Comput Commun Eng 1(4)
33. Novoselov KS et al (2005) Two-dimensional gas of massless Dirac fermions in graphene. Nature 438(7065):197–200
34. Pisana S et al (2007) Breakdown of the adiabatic Born-Oppenheimer approximation in graphene. Nat Mater 6(3):198–201
35. Zhang D et al (2012) Preparation, characterization, and application of electrochemically functional graphene nanocomposites by one-step liquid-phase exfoliation of natural flake graphite with methylene blue. Nano Res 5(12):875–887
36. Yoon TL et al (2013) Epitaxial growth of graphene on 6H-silicon carbide substrate by simulated annealing method. J Chem Phys 139(20):204702
37. Yao Y et al (2011) Controlled growth of multilayer, few-layer, and single-layer graphene on metal substrates. J Phys Chem C 115(13):5232–5238

38. Lizzit S et al (2012) Transfer-free electrical insulation of epitaxial graphene from its metal substrate. Nano Lett 12(9):4503–4507
39. Voloshina E, Dedkov Y (2012) Graphene on metallic surfaces: problems and perspectives. Phys Chem Chem Phys 14(39):13502–13514
40. Zhang Y, Zhang L, Zhou C (2013) Review of chemical vapor deposition of graphene and related applications. Acc Chem Res 46(10):2329–2339
41. Dato A et al (2008) Substrate-free gas-phase synthesis of graphene sheets. Nano Lett 8(7):2012–2016
42. Koo Y et al (1972) Photocatalyst nanomaterials for environmental challenges and opportunities. Nature 238:37–38
43. Hawthorne MF, Owen DA (1971) Chelated biscarborane transition metal derivatives formed through carbon-metal sigma bonds. J Am Chem Soc 93(4):873–880
44. Blakely JM, Kim JS, Potter HC (1970) Segregation of carbon to the (100) surface of nickel. J Appl Phys 41(6):2693–2697
45. Ebert LB (1976) Intercalation compounds of graphite. Annu Rev Mater Sci 6(1):181–211
46. Kovtyukhova NI et al (2014) Non-oxidative intercalation and exfoliation of graphite by Brønsted acids. Nat Chem (advance online publication)
47. Boehm HP, Setton R, Stumpp E (1994) Nomenclature and terminology of graphite intercalation compounds (IUPAC Recommendations 1994). Pure Appl Chem 66:1893
48. Morozov SV et al (2008) Giant intrinsic carrier mobilities in graphene and its bilayer. Phys Rev Lett 100(1):016602
49. Nair RR et al (2011) Spin-half paramagnetism in graphene induced by point defects. Nat Phys 8:199–202
50. Nair RR et al (2013) Dual origin of defect magnetism in graphene and its reversible switching by molecular doping. Nat Commun 4:1–6
51. Berger C et al (2006) Electronic confinement and coherence in patterned epitaxial graphene. Science 312(5777):1191–1196
52. Tan X et al (2012) Optimizing the thermoelectric performance of zigzag and chiral carbon nanotubes. Nanoscale Res Lett 7:116
53. Lee C et al (2008) Measurement of the elastic properties and intrinsic strength of monolayer graphene. Science 321(5887):385–388
54. Chen J-H et al (2008) Intrinsic and extrinsic performance limits of graphene devices on SiO2. Nat Nano 3(4):206–209
55. Ghosh S et al (2010) Dimensional crossover of thermal transport in few-layer graphene. Nat Mater 9(7):555–558
56. Murali R et al (2009) Breakdown current density of graphene nanoribbons. Appl Phys Lett 94(24):243114
57. Xu G et al (2010) Enhanced conductance fluctuation by quantum confinement effect in graphene nanoribbons. Nano Lett 10(11):4590–4594
58. Son Y-W, Cohen ML, Louie SG (2006) Energy gaps in graphene nanoribbons. Phys Rev Lett 97(21):216803
59. Jia X et al (2011) Graphene edges: a review of their fabrication and characterization. Nanoscale 3(1):86–95
60. Giavaras G, Nori F (2010) Graphene quantum dots formed by a spatial modulation of the Dirac gap. Appl Phys Lett 97(24):243106
61. Guinea F, Katsnelson MI, Geim AK (2010) Energy gaps and a zero-field quantum Hall effect in graphene by strain engineering. Nat Phys 6(1):30–33
62. Allen MT, Martin J, Yacoby A (2012) Gate-defined quantum confinement in suspended bilayer graphene. Nat Commun 3:934
63. Żebrowski DP, Wach E, Szafran B (2013) Confined states in quantum dots defined within finite flakes of bilayer graphene: coupling to the edge, ionization threshold, and valley degeneracy. Phys Rev B 88(16):165405
64. Jin JE et al (2016) Surface modulation of graphene field effect transistors on periodic trench structure. ACS Appl Mater Interfaces 8(28):18513–18518

65. Chen Y-C et al (2013) Tuning the band gap of graphene nanoribbons synthesized from molecular precursors. ACS Nano 7(7):6123–6128
66. Chang SL et al (2014) Geometric and electronic properties of edge-decorated graphene nanoribbons. Sci Rep 4:6038
67. Cano-Márquez AG et al (2009) Ex-MWNTs: graphene sheets and ribbons produced by lithium intercalation and exfoliation of carbon nanotubes. Nano Lett 9(4):1527–1533
68. Mak KF et al (2009) Observation of an electric-field-induced band gap in bilayer graphene by infrared spectroscopy. Phys Rev Lett 102(25):256405
69. Xu X et al (2014) Spin and pseudospins in layered transition metal dichalcogenides. Nat Phys 10(5):343–350
70. Massless Dirac Carriers in Graphene, 2014 [05/09/2014]. Available from: http://www.mpsd.mpg.de/mpsd/en/research/cmdd/ohg-ued/research/graphene
71. Bao W et al (2009) Controlled ripple texturing of suspended graphene and ultrathin graphite membranes. Nat Nano 4(9):562–566
72. Cho D-H et al (2013) Effect of surface morphology on friction of graphene on various substrates. Nanoscale 5(7):3063–3069
73. Chattopadhyaya M, Alam MM, Chakrabarti S (2012) On the microscopic origin of bending of graphene nanoribbons in the presence of a perpendicular electric field. Phys Chem Chem Phys 14(26):9439–9443
74. Castro Neto AH et al (2009) The electronic properties of graphene. Rev Mod Phys 81(1):109–162
75. Ando T (2009) The electronic properties of graphene and carbon nanotubes. NPG Asia Mater 1:17–21
76. Rani P, Jindal VK (2013) Designing band gap of graphene by B and N dopant atoms. RSC Adv 3(3):802–812
77. Preuss P (2008) Surprising graphene: honing in on graphene electronics with infrared synchrotron radiation. http://www.lbl.gov/publicinfo/newscenter/pr/2008/ALS-graphene-electrons.html
78. Li HJ et al (2005) Multichannel ballistic transport in multiwall carbon nanotubes. Phys Rev Lett 95(8):086601
79. de La Fuenta J (2013) Properties of graphene. http://www.graphenea.com/pages/graphene-properties#.UpkBPGSxPnU
80. Richter N et al (2017) Robust two-dimensional electronic properties in three-dimensional microstructures of rotationally stacked turbostratic graphene. Phys Rev Appl 7(2):024022
81. Li X et al (2009) Simultaneous nitrogen-doping and reduction of graphene oxide. Science 324:768–771
82. Martins TB et al (2007) Electronic and transport properties of boron-doped graphene nanoribbons. Phys Rev Lett 98(19):196803
83. Sharma R et al (2017) Investigation on effect of boron and nitrogen substitution on electronic structure of graphene. FlatChem 1:20–33
84. Ci L et al (2010) Atomic layers of hybridized boron nitride and graphene domains. Nat Mater 9(5):430–435
85. Panchakarla LS et al (2009) Synthesis, structure, and properties of boron- and nitrogen-doped graphene. Adv Mater 21(46):4726–4730
86. Yu SS et al (2006) Nature of substitutional impurity atom B/N in zigzag single-wall carbon nanotubes revealed by first-principle calculations. IEEE Trans Nanotechnol 5(5):595–598
87. Cao C et al (2017) Superiority of boron, nitrogen and iron ternary doped carbonized graphene oxide-based catalysts for oxygen reduction in microbial fuel cells. Nanoscale 9(10):3537–3546
88. Chen F et al (2017) Nitrogen-doped graphene oxide for effectively removing boron ions from seawater. Nanoscale 9(1):326–333
89. Kaloni TP et al (2011) Oxidation of monovacancies in graphene by oxygen molecules. J Mater Chem 21(45):18284–18288

90. Fan Y et al (2011) Tunable electronic structures of graphene/boron nitride heterobilayers. Appl Phys Lett 98(8):083103
91. Kaloni TP, Cheng YC, Schwingenschlogl U (2012) Electronic structure of superlattices of graphene and hexagonal boron nitride. J Mater Chem 22(3):919–922
92. Cheng YC et al (2011) Origin of the high p-doping in F intercalated graphene on SiC. Appl Phys Lett 99(5):053117
93. Ramasubramaniam A, Naveh D, Towe E (2011) Tunable band gaps in bilayer graphene–BN heterostructures. Nano Lett 11(3):1070–1075
94. Wei X et al (2011) Electron-beam-induced substitutional carbon doping of boron nitride nanosheets, nanoribbons, and nanotubes. ACS Nano 5(4):2916–2922
95. McCann E, Abergel DS, Fal'ko VI (2007) The low energy electronic band structure of bilayer graphene. Eur Phys J Spec Top 148(1):91–103
96. Fujii S et al (2014) Role of edge geometry and chemistry in the electronic properties of graphene nanostructures. Faraday Discuss
97. Huang PY et al (2011) Grains and grain boundaries in single-layer graphene atomic patchwork quilts. Nature 469(7330):389–392
98. Ayuela A et al (2014) Electronic properties of graphene grain boundaries. New J Phys 16:083018
99. Denis PA, Ullah S, Sato F (2017) Triple doped monolayer graphene with boron, nitrogen, aluminum, silicon, phosphorus and sulfur. ChemPhysChem 18:1864–1873
100. Pacheco Sanjuan AA et al (2014) Graphene's morphology and electronic properties from discrete differential geometry. Phys Rev B 89(12):121403
101. Chen K et al (2012) Electronic properties of graphene altered by substrate surface chemistry and externally applied electric field. J Phys Chem C 116(10):6259–6267
102. Kretinin AV et al (2014) Electronic properties of graphene encapsulated with different two-dimensional atomic crystals. Nano Lett 14(6):3270–3276
103. Majidi R (2016) Electronic properties of porous graphene, α-graphyne, graphene-like, and graphyne-like BN sheets. Can J Phys 94(3):305–309
104. Bonaccorso F et al (2010) Graphene photonics and optoelectronics. Nat Photon 4(9):611–622
105. Xia F et al (2009) Ultrafast graphene photodetector. Nat Nano 4(12):839–843
106. Wang F et al (2008) Gate-variable optical transitions in graphene. Science 320(5873):206–209
107. Rani P, Dubey GS, Jindal VK (2014) DFT study of optical properties of pure and doped graphene. Physica E 62:28–35
108. Falkovsky LA (2008) Optical properties of graphene. J Phys Conf Ser 129(1):012004
109. Cheng JL, Salazar C, Sipe JE (2013) Optical properties of functionalized graphene. Phys Rev B 88(4):045438
110. Eberlein T et al (2008) Plasmon spectroscopy of free-standing graphene films. Phys Rev B 77(23):233406
111. Marini A et al (2009) Yambo: an ab initio tool for excited state calculations. Comput Phys Commun 180(8):1392–1403
112. Sedelnikova OV, Bulusheva LG, Okotrub AV (2011) Ab initio study of dielectric response of rippled graphene. J Chem Phys 134(24):244707
113. Marinopoulos AG et al (2004) Ab initio study of the optical absorption and wave-vector-dependent dielectric response of graphite. Phys Rev B 69(24):245419
114. Marinopoulos AG et al (2004) Optical absorption and electron energy loss spectra of carbon and boron nitride nanotubes: a first-principles approach. Appl Phys A 78(8):1157–1167
115. De Corato M et al (2014) Optical properties of bilayer graphene nanoflakes. J Phys Chem C
116. Chernov AI et al (2013) Optical properties of graphene nanoribbons encapsulated in single-walled carbon nanotubes. ACS Nano 7(7):6346–6353
117. Hong T et al (2014) Thermal and optical properties of freestanding flat and stacked single-layer graphene in aqueous media. Appl Phys Lett 104(22):223102
118. Yang K, Arezoomandan S, Sensale-Rodriguez B (2013) The linear and non-linear THz properties of graphene. Int J Terahertz Sci Technol 6(4):223–233
119. Bernardi M et al (2016) Optical and electronic properties of two-dimensional layered materials

120. Wolf S, Tauber RN (1986) Crystalline defects, thermal processing, and gettering. In: Wolf S, Tauber RN (eds) Silicon processing for the VLSI era, Volume 1—Process technology. Lattice Press, Sunset Beach, CA, pp 36–72
121. Borca-Tascuic T, Achimov DA, Chen G (1998) Difference between wafer temperature and thermocouple reading during rapid thermal processing. Mater Res Soc Symp Proc 525:103–108
122. Wagner J, Boebel FG (1996) Temperature measurement at RTP facilities—an overview. Mater Res Soc Symp Proc 429:303–308
123. Muley SV, Ravindra NM (2014) Emissivity of electronic materials, coatings, and structures. JOM 66:616–636
124. Ravindra NM et al (2001) Emissivity measurements and modeling of silicon-related materials: an overview. Int J Thermophys 22(5):1593–1611
125. Kosonocky WF et al (1994) Multiwavelength imaging pyrometer. In: Infrared detectors and focal plane arrays III, Proceedings SPIE, 1994, vol 2225, pp 26–43
126. Kaplinsky MB et al (1997) Recent advances in the development of a multiwavelength imaging pyrometer. Opt Eng 36(11):3176–3187
127. Han Q et al (2013) Highly sensitive hot electron bolometer based on disordered graphene. Sci Rep 3(3533):6
128. Gu D (2007) Terahertz imaging system using hot electron bolometer technology. In: Electrical Engineering. University of Massachusetts Amherst, Ann Arbor, p 174
129. Kubakaddi SS (2009) Interaction of massless Dirac electrons with acoustic phonons in graphene at low temperatures. Phys Rev B 79(7):075417
130. Oliver WC, Pharr GM (2004) Measurement of hardness and elastic modulus by instrumented indentation: advances in understanding and refinements to methodology. J Mater Res 19(01):3–20
131. Doerner MF, Nix WD (1986) A method for interpreting the data from depth-sensing indentation instruments. J Mater Res 1(04):601–609
132. Oliver WC, Pharr GM (1992) An improved technique for determining hardness and elastic modulus using load and displacement sensing indentation experiments. J Mater Res 7(06):1564–1583
133. Cao G (2014) Atomistic studies of mechanical properties of graphene. Polymers 6(9):2404–2432
134. Frank IW et al (2007) Mechanical properties of suspended graphene sheets. J Vac Sci Technol B 25(6):2558–2561
135. Zhang Y, Pan C (2012) Measurements of mechanical properties and number of layers of graphene from nano-indentation. Diam Relat Mater 24:1–5
136. Lee J-U, Yoon D, Cheong H (2012) Estimation of Young's modulus of graphene by Raman spectroscopy. Nano Lett 12(9):4444–4448
137. Lee C et al (2009) Elastic and frictional properties of graphene. Physica Status Solidi (b) 246(11–12):2562–2567
138. Lee H et al (2009) Comparison of frictional forces on graphene and graphite. Nanotechnology 20(32):325701 (6 p)
139. Lee C et al (2010) Frictional characteristics of atomically thin sheets. Science 328(5974):76–80
140. Scharfenberg S et al (2011) Probing the mechanical properties of graphene using a corrugated elastic substrate. Appl Phys Lett 98(9):091908
141. Sansoz F, Gang T (2010) A force-matching method for quantitative hardness measurements by atomic force microscopy with diamond-tipped sapphire cantilevers. Ultramicroscopy 111(1):11–19
142. Liu F, Ming P, Li J (2007) Ab initio calculation of ideal strength and phonon instability of graphene under tension. Phys Rev B 76(6):064120
143. Wei X et al (2009) Nonlinear elastic behavior of graphene: ab initio calculations to continuum description. Physical Review B 80(20):205407

144. Morgan III J (2006) Thomas-Fermi and other density-functional theories. In: Drake G (ed) Springer handbook of atomic, molecular, and optical physics. Springer New York, pp 295–306
145. Adachi H, Mukoyama T, Kawai J (2006) Hartree-Fock-Slater method for materials science: the DV-X Alpha method for design and characterization of materials. Springer, Berlin
146. Dinh PM (2005) Condensed matter theories. Nova Science Publishers, New York
147. Hohenberg P, Kohn W (1964) Inhomogeneous electron gas. Phys Rev 136(3B):B864–B871
148. Kotochigova S et al (2014) Atomic reference data for electronic structure calculations. NIST: Physical Measurement Laboratory, NIST
149. Kohn W, Sham LJ (1965) Self-consistent equations including exchange and correlation effects. Phys Rev 140(4A):A1133–A1138
150. Narasimhan S (2011) The Self Consistent Field "(SCF) loop and some relevant parameters for quantum—ESPRESSO 2221(5)
151. Perdew JP, Zunger A (1981) Self-interaction correction to density-functional approximations for many-electron systems. Phys Rev B 23(10):5048–5079
152. Bachelet GB, Hamann DR, Schlüter M (1982) Pseudopotentials that work: from H to Pu. Phys Rev B 26(8):4199–4228
153. Ravindra NM et al (2003) Modeling and simulation of emissivity of silicon-related materials and structures. J Electron Mater 32(10):1052–1058
154. Palik ED (1998) Handbook of optical constants of solids. In: Palik ED (ed). Academic Press, Waltham MA, pp 275–798
155. Zaitsev AM (2001) Refraction. In: Zaitsev AM (ed) Optical properties of diamond: a data handbook. Springer, Berlin, pp 1–9
156. Weber JW, Calado VE, van de Sanden MCM (2010) Optical constants of graphene measured by spectroscopic ellipsometry. Appl Phys Lett 97(9):091904 (4 p)
157. Hebb JP, Jensen KF (1996) J Electrochem Soc 143(3):1142–1151
158. Allen MP, Tildesley DJ (1987) Computer simulation of liquids. Clarendon Press, Oxford
159. Haile JM (1992) Molecular dynamics simulation: elementary methods. Wiley, Clemson, USA, p 489
160. Greiner W, Neise L, Stöcker H (1995) Thermodynamics and statistical mechanics. Classical theoretical physics. Springer, New York
161. Plimpton S (1995) Fast parallel algorithms for short-range molecular dynamics. J Comput Phys 117(1):1–19
162. Swope WC et al (1982) A computer simulation method for the calculation of equilibrium constants for the formation of physical clusters of molecules: application to small water clusters. J Chem Phys 76(1):637–649
163. Andersen HC (1980) Molecular dynamics simulations at constant pressure and/or temperature. J Chem Phys 72(4):2384–2393
164. Balandin AA (2011) Thermal properties of graphene and nanostructured carbon materials. Nat Mater 10(8):569–581
165. Nika DL et al (2009) Phonon thermal conduction in graphene: role of Umklapp and edge roughness scattering. Phys Rev B 79(15):155413
166. Lan J et al (2009) Edge effects on quantum thermal transport in graphene nanoribbons: tight-binding calculations. Phys Rev B 79(11):115401
167. Balandin AA et al (2008) Superior thermal conductivity of single-layer graphene. Nano Lett 8(3):902–907
168. Müller-Plathe F (1997) A simple nonequilibrium molecular dynamics method for calculating the thermal conductivity. J Chem Phys 106(14):6082–6085
169. Humphrey W, Dalke A, Schulten K (1996) VMD—visual molecular dynamics. J Mol Graph 14(1):33–38
170. Kan Q et al (2013) Oliver-Pharr indentation method in determining elastic moduli of shape memory alloys—a phase transformable material. J Mech Phys Solids 61(10):2015–2033
171. King RB (1987) Elastic analysis of some punch problems for a layered medium. Int J Solid Structures 23(12):1657–1664

172. Wang W et al (2014) Nanoindentation experiments for single-layer rectangular graphene films: a molecular dynamics study. Nanoscale Res Lett 9(41):1–8
173. Kiselev SP, Zhirov EV (2013) Molecular dynamics simulation of deformation and fracture of graphene under uniaxial tension. Phys Mesomech 16(2):125–132
174. Zhao H, Min K, Aluru NR (2009) Size and chirality dependent elastic properties of graphene nanoribbons under uniaxial tension. Nano Lett 9(8):3012–3015
175. Carpio A, Bonilla LL (2008) Periodized discrete elasticity models for defects in graphene. Phys Rev B 78(8):085406
176. Ghosh S et al (2008) Extremely high thermal conductivity of graphene: prospects for thermal management applications in nanoelectronic circuits. Appl Phys Lett 92(15):151911
177. Hu J et al (2009) Frontiers of characterization and metrology for nanoelectronics. In: AIP Conference Proceedings No. 1173. AIP, New York, p 135
178. Xu X et al (2014) Length-dependent thermal conductivity in suspended single-layer graphene. Nat Commun 5:3689
179. Yamamoto T, Watanabe K, Mii K (2004) Empirical-potential study of phonon transport in graphitic ribbons. Phys Rev B 70(24):245402
180. Yang N, Zhang G, Li B (2009) Thermal rectification in asymmetric graphene ribbons. Appl Phys Lett 95(3):033107
181. Mortazavi B et al (2012) Nitrogen doping and curvature effects on thermal conductivity of graphene: a non-equilibrium molecular dynamics study. Solid State Commun 152(4):261–264
182. Carrero-Sánchez JC et al (2006) Biocompatibility and toxicological studies of carbon nanotubes doped with nitrogen. Nano Lett 6(8):1609–1616
183. Zhang H, Lee G, Cho K (2011) Thermal transport in graphene and effects of vacancy defects. Phys Rev B 84(11):115460
184. Klemens PG, Pedraza DF (1994) Thermal conductivity of graphite in the basal plane. Carbon 32(4):735–741
185. Subrina S, Kotchetkov D (2008) Simulation of heat conduction in suspended graphene flakes of variable shapes. J Nanoelectron Optoelectron 3:1–21
186. Areshkin DA, Gunlycke D, White CT (2006) Ballistic transport in graphene nanostrips in the presence of disorder: importance of edge effects. Nano Lett 7(1):204–210
187. Ouyang Y, Guo J (2009) A theoretical study on thermoelectric properties of graphene nanoribbons. Appl Phys Lett 94(26):093104
188. Karamitaheri H et al (2012) Engineering enhanced thermoelectric properties in zigzag graphene nanoribbons. J Appl Phys 111(5):093104
189. Mousavi H, Moradian R (2011) Nitrogen and boron doping effects on the electrical conductivity of graphene and nanotube. Solid State Sci 13(8):1459–1464
190. Sankeshwar NS, Kubakaddi SS, Mulimani BG (2013) Thermoelectric power in graphene. In: Aliofkhazraei DM (ed) Advances in graphene science. InTech, pp 217–271
191. Bahamon D, Pereira A, Schulz P (2011) Third edge for a graphene nanoribbon: a tight-binding model calculation. Phys Rev B 83(7)
192. Carr LD, Lusk MT (2010) Defect engineering: graphene gets designer defects. Nat Nano 5(5):316–317
193. Avouris P (2010) Graphene: electronic and photonic properties and devices. Nano Lett 10(11):4285–4294
194. Jun Y et al (2012) Dual-gated bilayer graphene hot-electron bolometer. Nat Nanotechnol 7(7):472–478
195. Resistivity, Conductivity and Temperature Coefficients for some Common Materials, 2014 [01/06/2014]. Available from: http://www.engineeringtoolbox.com/
196. Wang M et al (2013) A platform for large-scale graphene electronics—CVD growth of single-layer graphene on CVD-grown hexagonal boron nitride. Adv Mater 25(19):2746–2752
197. Kresse G, Furthmüller J (1996) Efficient iterative schemes for ab initio total energy calculations using a plane-wave basis set. Phys Rev B 54(16):11169–11186
198. Perdew JP, Burke K, Ernzerhof M (1996) Generalized gradient approximation made simple. Phys Rev Lett 77(18):3865–3868

199. McCann E, Abergel DSL, Fal'ko VI (2007) Electrons in bilayer graphene. Solid State Commun 143(1–2):110–115
200. Kravets VG et al (2010) Spectroscopic ellipsometry of graphene and an exciton-shifted van Hove peak in absorption. Phys Rev B 81(15):155413
201. Yang L et al (2009) Excitonic effects on the optical response of graphene and bilayer graphene. Phys Rev Lett 103(18):186802
202. Wu L et al (2010) Highly sensitive graphene biosensors based on surface plasmon resonance. Opt Express 18(14):14395–14400
203. Neuer G (1992) Emissivity measurements on graphite and composite materials in visible and infrared spectral range. In: EETI (ed) Quantitative infrared thermography (QIRT) 92. QIRT Archives, Paris
204. Zandiatashbar A et al (2014) Effect of defects on the intrinsic strength and stiffness of graphene. Nat Commun 5:3186
205. Chen J, Zhang G, Li B (2013) Substrate coupling suppresses size dependence of thermal conductivity in supported graphene. Nanoscale 5(2):532–536
206. Zhong W-R et al (2011) Chirality and thickness-dependent thermal conductivity of few-layer graphene: a molecular dynamics study. Appl Phys Lett 98(11):113107 (4 p)
207. Guo Z, Zhang D, Gong X-G (2009) Thermal conductivity of graphene nanoribbons. Appl Phys Lett 95(16):163103
208. Mortazavi B, Ahzi S (2012) Molecular dynamics study on the thermal conductivity and mechanical properties of boron doped graphene. Solid State Commun 152(15):1503–1507
209. Cooper AJ et al (2014) Single stage electrochemical exfoliation method for the production of few-layer graphene via intercalation of tetraalkylammonium cations. Carbon 66:340–350
210. Stankovich S et al (2007) Synthesis of graphene-based nanosheets via chemical reduction of exfoliated graphite oxide. Carbon 45(7):1558–1565
211. Pei S, Cheng H-M (2012) The reduction of graphene oxide. Carbon 50(9):3210–3228
212. Guermoune A et al (2011) Chemical vapor deposition synthesis of graphene on copper with methanol, ethanol, and propanol precursors. Carbon 49(13):4204–4210
213. Liu W et al (2011) Synthesis of high-quality monolayer and bilayer graphene on copper using chemical vapor deposition. Carbon 49(13):4122–4130
214. Dong X et al (2011) Growth of large-sized graphene thin-films by liquid precursor-based chemical vapor deposition under atmospheric pressure. Carbon 49(11):3672–3678
215. Wu W et al (2010) Wafer-scale synthesis of graphene by chemical vapor deposition and its application in hydrogen sensing. Sens Actuators B Chem 150(1):296–300
216. Liao L, Duan X (2010) Graphene–dielectric integration for graphene transistors. Mater Sci Eng R Rep 70(3):354–370
217. Kuilla T et al (2010) Recent advances in graphene based polymer composites. Prog Polym Sci 35(11):1350–1375
218. Quan Q et al (2017) Graphene and its derivatives as versatile templates for materials synthesis and functional applications. Nanoscale 9(7):2398–2416
219. Palla P et al (2016) Bandgap engineered graphene and hexagonal boron nitride for resonant tunnelling diode. Bull Mater Sci 39(6):1441–1451
220. Eftekhari A, Garcia H (2017) The necessity of structural irregularities for the chemical applications of graphene. Mater Today Chem 4:1–16
221. Tang L et al (2017) Functionalization of graphene by size and doping control and its optoelectronic applications
222. Shi S et al (2014) Surface engineering of graphene-based nanomaterials for biomedical applications. Bioconjug Chem 25(9):1609–1619
223. Qu L et al (2017) A versatile graphene foil. J Mater Chem A
224. Junqi X et al (2017) Large-area, high-quality monolayer graphene from polystyrene at atmospheric pressure. Nanotechnology 28(15):155605
225. Kojima E et al (2017) Magnetic field sensor of graphene for automotive applications. SAE International

226. Marini A, Cox JD, García de Abajo FJ (2017) Theory of graphene saturable absorption. Phys Rev B 95(12):125408
227. Bystrov VS et al (2017) Graphene/graphene oxide and polyvinylidene fluoride polymer ferroelectric composites for multifunctional applications. Ferroelectrics 509(1):124–142
228. Ruhl G et al (2017) The integration of graphene into microelectronic devices. Beilstein J Nanotechnol 8:1056–1064
229. Dubey N et al (2015) Graphene: a versatile carbon-based material for bone tissue engineering. Stem Cells Int 2015:12

Chapter 6
Transition Metal Dichalcogenides Properties and Applications

Nuggehalli M. Ravindra, Weitao Tang and Sushant Rassay

6.1 Background

Two-Dimensional (2D) materials, sometimes referred to as mono or single-layer materials, are crystalline materials which consist of a single layer of atoms. Since the discovery of graphene, as the first 2D material in 2004, about 700 2D materials have been predicted. In this context, it must be emphasized that this area of research, 2D Materials, represents the largest growing field in condensed matter physics, materials science and engineering and applications today. This is reflected by several journals that are dedicated to this area of research. These journals include the following: *2D Materials* [1], FlatChem [2], Graphene [3], Graphene Technology [4], and npj 2D Materials and Applications [5], in addition to journal such as Carbon [6]. The American Chemical Society has put together a collection of papers, from various journals such as *ACS Nano, ACS Photonics, Chemistry of Materials, and Nano Letters* [7], that represent the growing opportunities in 2D Materials, including and beyond graphene.

Research on other 2D materials has taken place due to the zero bandgap of graphene that has limited its use as a semiconductor, although significant efforts have been made to be able to dope graphene and fabricate electronic devices with reasonable success. In this context, it must be pointed out that the Journal of Materials Research has recently put together a focused issue on porous carbon and carbonaceous materials for energy conversion and storage [8]. The discovery of graphene has directed scientists and researchers about new emerging physical properties when a bulk crystal of a macroscopic dimension is thinned down to an atomic layer. Transition Metal Dichalcogenides (TMDCs) are such atomically thin semiconductors having the type MX_2; they are anticipated to have significant potential in electronic,

N. M. Ravindra (✉) · W. Tang · S. Rassay
New Jersey Institute of Technology, Newark, NJ 07102, USA
e-mail: nmravindra@gmail.com

M. I. Pech-Canul and N. M. Ravindra (eds.), *Semiconductors*,
https://doi.org/10.1007/978-3-030-02171-9_6

optoelectronic and energy storage applications, which makes them more preferred materials over graphene.

In addition to the Transition Metal Dichalcogenides (TMDCs), 2D Materials include the following: bismuthene, boron nitride (hexagonal), borophene, germanene, graphane, graphene, graphyne, phosphorene, silicene, stanene, etc. Every effort is being made in research labs today to investigate a variety of bulk materials in the 2D form. While the advantages in utilizing 2D materials can be attributed to the obvious decrease in material consumption and the ability to make them flexible, while perhaps retaining their mechanical properties, the challenges remain in the fundamental understanding of other material properties while addressing the larger picture of the ability to manufacture these materials, at low costs, on a large scale.

In the public domain, the available research literature is vast on this topic. In this chapter, only a very small fraction of the topics are covered. In particular, this chapter focuses on MoS_2, WS_2, $MoSe_2$, and WSe_2, which are four important members of the TMDC class of materials. The structural, electronic, optical, and electrical properties of these materials, and some applications are discussed.

6.2 Physical Properties of MoS_2 and WS_2

The layered Transition Metal Dichalcogenides (TMDCs), i.e., MX_2 are typical 2D semiconductors with atomic scale thickness. M refers to a transition metal atom such as Mo or W, while X refers to a chalcogen atom. Monolayer TMDCs present distinctive properties from bulk TMDCs and thus draw enormous interest.

The TMDC monolayers are direct bandgap semiconductors while bulk TMDCs exhibit indirect bandgap with a smaller bandgap. The lack of inversion center also allows high degree of freedom for charge carriers or K-valley index and thus leads to valleytronics [9]. In bulk TMDCs, the TMDC layers are combined with each other, layer by layer, with van der Waals force, and the weak interaction between layers influences the properties of the bulk TMDCs significantly. TMDCs can also combine with other 2D materials to form van der Waals heterostructure devices through weak van der Waals force [10].

6.2.1 Monolayer Structure

The bulk TMDCs are stacked by monolayer TMDCs through van der Waals force. Thus, in order to understand the structure of bulk TMDCs, the structure of monolayer TMDCs need to be studied first.

As shown in Fig. 6.1, the monolayer TMDCs have interior layered structure. One layer of transition metal atoms is sandwiched between two layers of chalcogen atoms. The bonds between the transition metal atom and transition metal atom, chalcogen atoms and chalcogen atoms, and transition metal atoms and chalcogen atoms are all

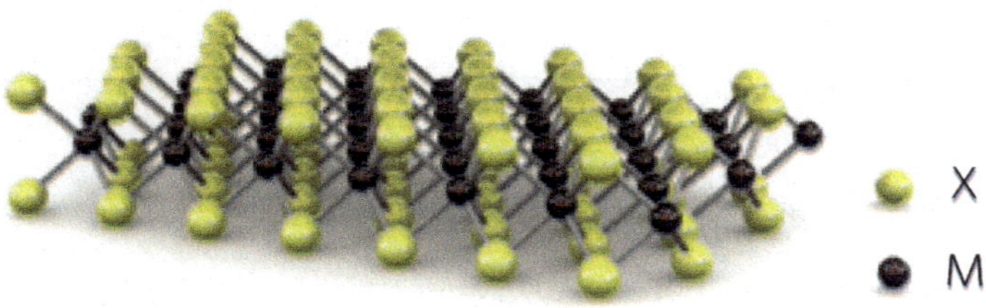

Fig. 6.1 Structure of monolayer TMDCs; M represents transition metal atoms while X stands for chalcogen atoms [11]

covalent bond, and thus, much more stable compared with the weak van der Waals force between bulk layers.

Both the MoS_2 monolayer and WS_2 monolayer have a thickness of 0.65 nm [12]. The ultra-thin thickness of the two materials enables more light to pass through the structure and thus reduces the optical absorptance. Unlike the most commonly studied 2D material, graphene, these materials have intrinsic direct bandgap, and are thus promising in fabricating new transistor channel as an ideal switch material [13]. It should be noted that the direct bandgap of both the materials are sensitive to strain [14, 15]. For example, the direct bandgap in WS_2 can be transformed to indirect bandgap with only 1% strain [15]. The electronic properties of these materials will be further discussed in Sect. 6.3.

6.2.2 Bulk Structure

In bulk form, both MoS_2 and WS_2 are inorganic compounds. They are classified as metal dichalcogenides and have a similar structure.

Figure 6.2 represents the structure of bulk TMDCs, which includes MoS_2 and WS_2. Bulk TMDCs contain three types of structures: 1T, 2H, and 3R. The three structures constitute different combination of single layers that are held by weak van der Waals force. It should be noted that bulk TMDCs often consist of all these three structures, and are not formed by only one of the structures. The 2H structure is the most abundant structure in the bulk crystal [16]. In the 2H structure, the metal atoms stay in the center of a trigonal prismatic coordination sphere which consists of one metal atom and six chalcogen atoms, and each chalcogen atom is bonded with three metal atoms. Because the layers are held together by weak van der Waals force, bulk TMDCs can be exfoliated to form single or few-layer structure/s.

Table 6.1 presents the physical properties of bulk MoS_2 and WS_2. Bulk MoS_2 crystal is a silvery black solid and bulk WS_2 crystal is dark gray. They both exhibit dry lubricant properties. The monolayer TMDCs can be produced from the bulk crystal by Chemical Vapor Deposition (CVD), liquid exfoliation, mechanical exfoliation, and Molecular-Beam Epitaxy (MBE); mechanical exfoliation provides much cleaner,

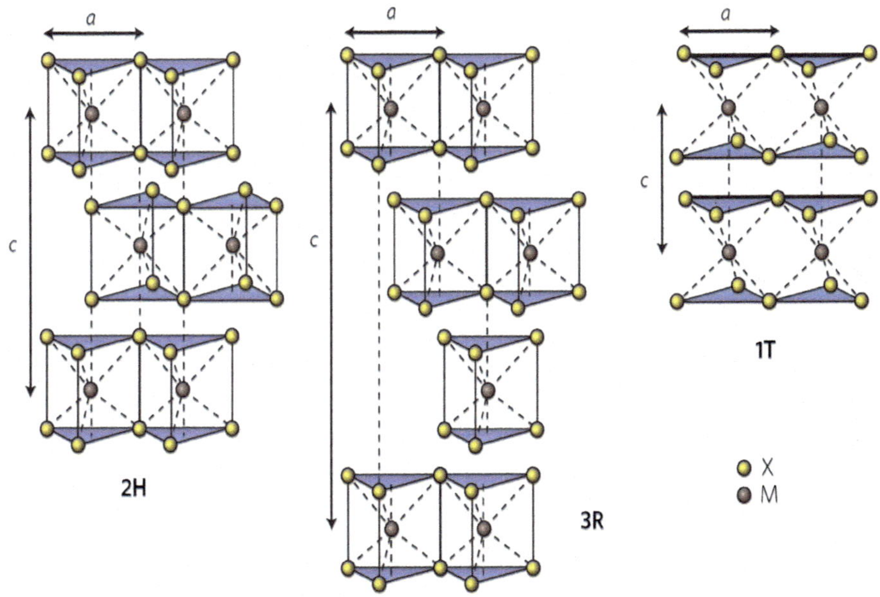

Fig. 6.2 Structure of bulk TMDCs; M represents transition metal atoms while X constitutes chalcogen atoms [11]

Table 6.1 Physical properties of bulk MoS_2 and WS_2

	Molybdenum disulfide	Tungsten disulfide
Chemical formula	MoS_2	WS_2
Molar mass	160.07 g/mole [17]	247.98 g/mole [17]
Density	5.06/cm^3 [17]	7.5 g/cm^3 [17]
Melting point	1185 °C decompose	1250 °C decompose
Crystal structure	hP6, space group P6 3/mmc, No 194 (2H) hR9, space group R3 m, No 160 (3R)	
Lattice constants	$a = 0.3160$ nm, $c/a = 3.89$ (2H) [18]	3.154 nm, $c/a = 3.920$ (2H) [18]
Solubility	Insoluble in water [17]	Slightly soluble in water [17]

more pristine, and higher quality structures, which are more suitable for fundamental studies and potential applications [14].

6.3 Electronic Properties of MoS_2 and WS_2

In this section, a brief overview of the electronic properties of MoS_2 and WS_2, with a focus on the band structure of these two materials, as function of number of layers, is presented. The temperature dependence of the bandgap of MoS_2 and WS_2 monolayers are investigated by MATLAB simulations.

6.3.1 Band Structure of MoS_2 and WS_2

The band structure of materials is critical in understanding the properties of materials and identifying possible applications. The reduction in the number of layers causes significant changes in the band structure of both the materials. The special band structure of monolayer MoS_2 and WS_2 has drawn the attention of scientists.

6.3.1.1 Electronic Band Structure

The electronic band structure (or band structure) of a solid describes the range of energies that the electrons within the solid can possess, and the range of energy states that they may not possess. Due to quantum mechanical wave functions, the electrons in the solid can only possess certain range of energy in certain locations in the band structure.

The repetitiveness of the periodic lattice, which is described by the Brillouin zone, allows utilizing the band structure of a lattice to represent the band structure of the entire solid. The band theory has been utilized to explain many important physical properties and is essential in designing all the electronic and optoelectronic devices.

6.3.1.2 Brillouin Zone

Brillouin zone is often used in mathematics and solid-state physics to describe the primitive cell in reciprocal space. The reciprocal lattice is broken into Brillouin zones as the Bravais lattice is broken into cells. The solution to the Bloch wave in a periodic medium can be represented by their behavior in a single Brillouin zone.

MoS_2 and WS_2 are both comprised of hexagonal lattice, which is presented in Fig. 6.3 to give a better understanding of the electronic band structure of MoS_2 and WS_2.

6.3.1.3 Band Structure of MoS_2 and WS_2

In 2012, Kumar and Ahluwalia [21] reported a thorough investigation of the electronic structure of several TMDCs utilizing first principle calculations. They showed that their results are in good agreement with the experimental measurements [21].

Figure 6.4 shows the band structure of MoS_2, as function of the number of layers. As can be seen in the figure, bulk MoS_2 has an indirect bandgap of 0.75 eV. However, with the decreasing number of layers, the value of the indirect bandgap increases. As

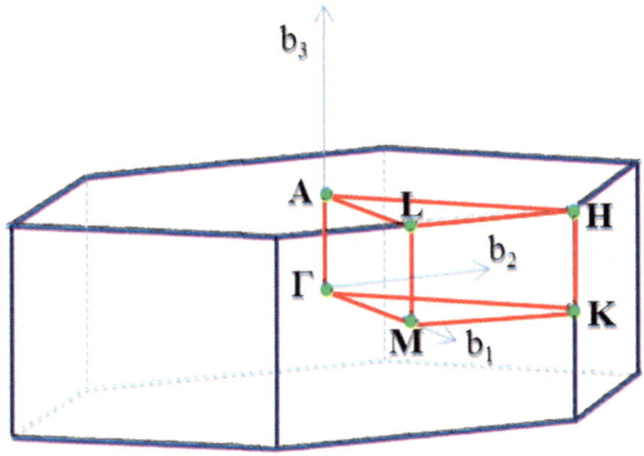

Fig. 6.3 The Brillouin zone and special point of the hexagonal lattice system for MoS_2 and WS_2. A is the center of a hexagonal face, H is the corner point, K is the middle of an edge joining two rectangular faces, L is the middle of an edge joining a hexagonal and a rectangular face, M is the center of a rectangular face, and Γ is the center of the Brillouin zone [20]

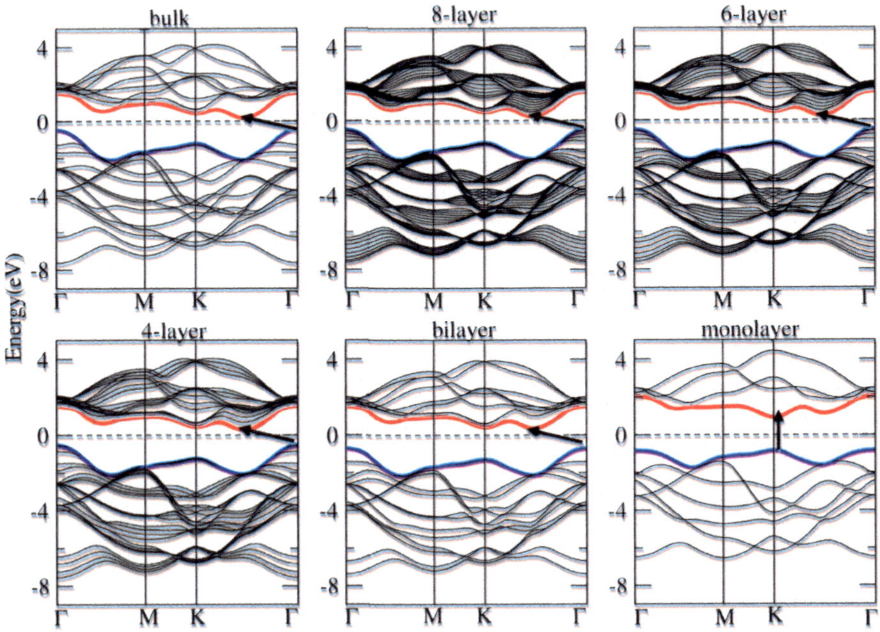

Fig. 6.4 Band structure of MoS_2 as function of number of layers. The arrow represents the approximate location of the bandgap. The symbols in the figure are illustrated in Fig. 6.3 [21]

the number of layers reaches one, the location of the bandgap shifts and the monolayer MoS_2 becomes a direct bandgap semiconductor with a bandgap of 1.89 eV. The corresponding blue shift is 1.14 eV.

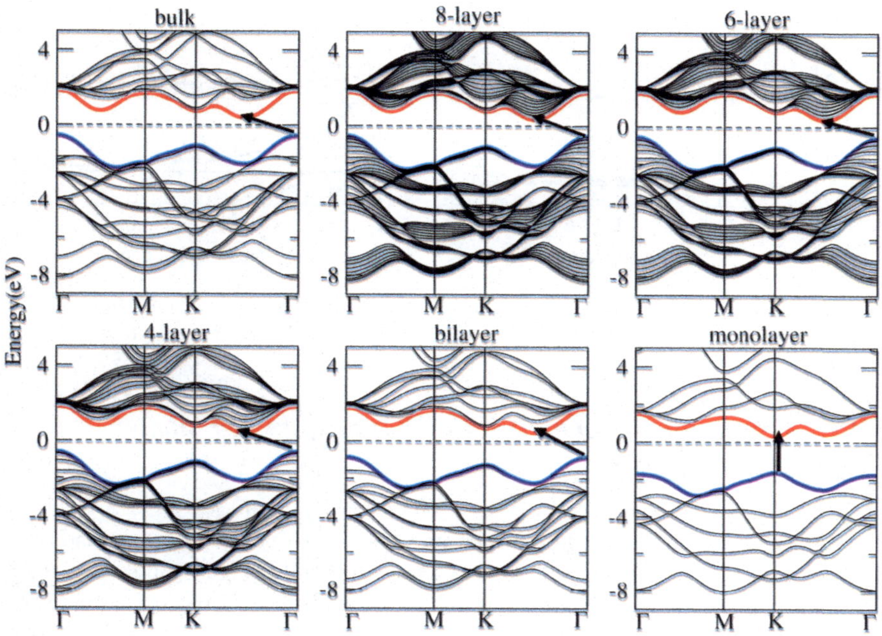

Fig. 6.5 Band structure of WS_2 as function of number of layers. The arrow represents the transition from the highest valence band to the lowest conduction band. The symbols in this figure are illustrated in Fig. 6.3 [21]

The value of blue shift is even larger than the value of the indirect bandgap of the bulk MoS_2. This indicates that there will be a significant change in the properties of the monolayer MoS_2. The location of the bandgap also changes; both the maximum valence band and the minimum conduction band are located at the K point in the Brillouin zone (the middle of an edge joining the two rectangular faces) in monolayer MoS_2. The reduced number of layers causes the bandgap of MoS_2 to shift from the infrared region toward visible region and switch the type of the bandgap from indirect to direct. These changes lead to applications of monolayer MoS_2 for optoelectronic devices.

Figure 6.5 shows the band structure of WS_2 as function of the number of layers. Generally, the change in band structure of WS_2 is similar to that of MoS_2. Different from MoS_2, the value of the indirect bandgap in bulk WS_2 is 0.89 eV, while the value of the direct bandgap in monolayer WS_2 is 2.05 eV. The value of the blue shift is 1.16 eV.

The variation in the value of bandgaps of MoS_2 and WS_2, as function of number layers, is summarized in Fig. 6.6. It should be noted that the value of the direct bandgap at K point does not vary significantly with the number of layers, while the indirect bandgap varies drastically with the number of layers from one layer to four layers.

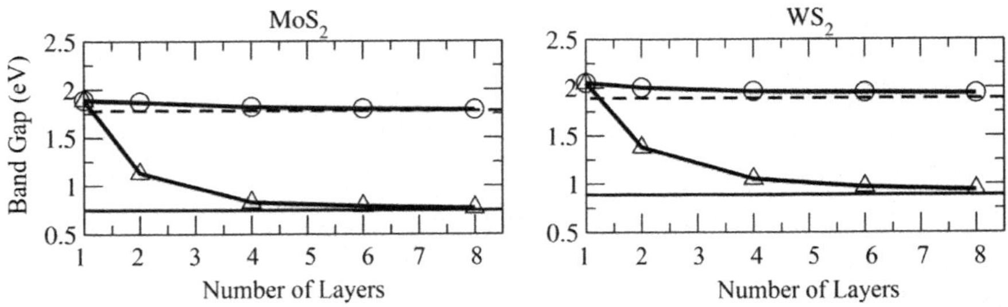

Fig. 6.6 The value of direct and indirect bandgap in MoS_2 and WS_2. The solid lines with triangles are for indirect bandgap, while the solid lines with circles are for direct bandgap at K point [21]

6.3.2 *Temperature Dependence of Bandgap in Monolayer MoS_2 and WS_2*

The properties of semiconductor devices are very sensitive to temperature, and thus, the knowledge of the influence of temperature on the bandgap is essential. Generally, the energy gap of a semiconductor tends to decrease with increase in temperature [22]. As the thermal energy increases, the amplitude of the atomic vibrations increases leading to increase in the atomic spacing [23]. Therefore, an increased interatomic spacing decreases the potential seen by the electrons in the material, which in turn reduces the energy gap [23]. The relationship between bandgap and temperature can be expressed by Eq. 6.1 which is formulated by O'Donnell and Chen [24]; the equation serves to be a direct replacement of Varshini's equation [25].

$$E_g(T) = \left\{E_g(0) - S\langle\hbar\omega\rangle[(\cot h\langle\hbar\omega\rangle/2kT) - 1]\right\} \tag{6.1}$$

$E_g(0)$ is the bandgap of the semiconductor at 0 K, S is a dimensionless coupling constant, and $\langle\hbar\omega\rangle$ is an average phonon energy. This equation has been used explicitly due to poor fitting results obtained from using full theoretical treatments [23, 24].

Table 6.2 shows the fitting parameters of monolayer MoS_2 and monolayer WS_2 based on Eq. 6.1. By introducing these parameters into Eq. 6.1, the variation in the bandgap with temperature are plotted in Fig. 6.7.

As shown in Fig. 6.7, the variation in bandgap versus temperature is generally a straight line except for the initial low-temperature component. The bandgap decreases significantly with the increase in temperature. In order to have a better understanding

Table 6.2 Fitting parameters of E_g for monolayer MoS_2 and WS_2

Material	$E_g(0)$ (eV)	S	$\langle\hbar\omega\rangle$ (meV)	References
MoS_2	1.86	1.82	2.25	[26]
WS_2	2.08	2.47	1.30	[27]

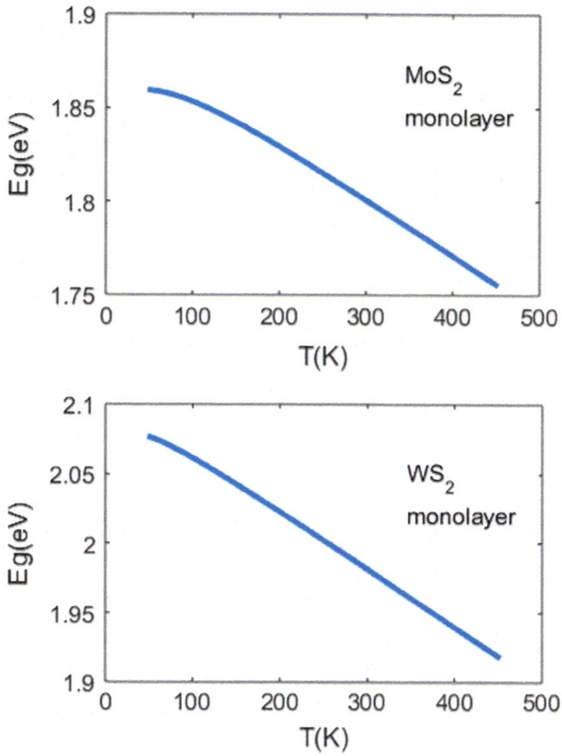

Fig. 6.7 Temperature dependence (T) of bandgap (E_g) for monolayer MoS_2 and monolayer WS_2

of the relationship between bandgap and temperature, (dE_g/dT) versus T is also plotted in Fig. 6.8. As can be seen in this figure, (dE_g/dT) tends to saturate with temperature at room temperature for monolayers of MoS_2 and WS_2.

6.4 Optical Properties of MoS_2 and WS_2

In recent years, there has been a large volume of research conducted on the optical properties of TMDCs; however, much of the research revolves around the spectral reflectance, differential reflectance, differential transmittance, spectral absorptance and absorbance.

These studies are usually based on experimental research and very less or, in some cases, no simulations are carried out. Despite the intense research carried out on the optical properties of TMDCs, none of the studies give definite values of refractive indices and extinction coefficients. Furthermore, the results of most of the studies involving reflectance, absorptance, and transmittance calculations are not in accord

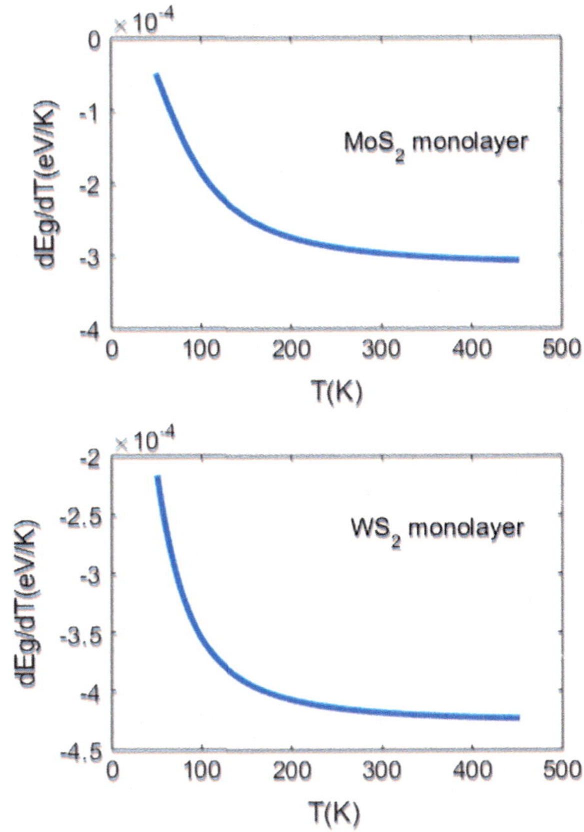

Fig. 6.8 $\left(\frac{dE_g}{dT}\right)$ as function of temperature T of monolayer MoS_2 and WS_2

with each other and there is a large spread in the obtained data of the optical properties in the literature.

In this Section, the optical properties such as refractive index, extinction coefficient, absorptance, reflectance and transmittance of MoS_2 and WS_2, in the energy range of 1.50–3.00 eV, are studied by MATLAB. The optical bandgap is modeled utilizing MATLAB. In order to give a better perspective of future applications, the optical properties of these materials on selected substrates (gold, silicon, and fused silica) are also studied utilizing MATLAB.

6.4.1 *Optical Properties of Suspended MoS_2 and WS_2*

The fundamental optical constants, i.e., the refractive index (n) and the extinction coefficient (k) and their spectral dependence are critical to the optical properties of materials. In this section, these values of the optical constants (n and k) and optical properties (Reflectance R, Transmittance T, and Absorptance A) of MoS_2 and WS_2 are simulated by MATLAB under conditions of normal incidence. The optical bandgaps of the monolayers are also determined to compare with their electronic bandgap.

6.4.1.1 Dielectric Constants of Suspended MoS_2 and WS_2

As an introduction, in electromagnetism, dielectric constants are known as relative permittivity. When an electric field is applied to a material, the charges in the material will be displaced due to the electric force induced by the electric field. However, the separation of charged particles generates an electric field which attenuates or resists the total electric field. Here, the permittivity is a measurement of the resistance of a material when an electric field is formed; it reflects how an electric field affects and is affected by a material. Relative permittivity (ε_r), or the dielectric constant, is the ratio of the permittivity of the material (ε) to the vacuum permittivity ($\varepsilon_0 = 8.85419 \times 10^{-12}$ F/m). It is a dimensionless quantity expressed by Eq. 6.2:

$$\varepsilon_r = (\varepsilon/\varepsilon_0) \tag{6.2}$$

In real situations, the applied field is always time-dependent with changing direction and strength in certain frequency. The permittivity becomes complex in these situations given by Eq. 6.3.

$$\varepsilon_r = (\varepsilon_1 - i * \varepsilon_2) \tag{6.3}$$

ε_1 is the real part of the complex dielectric constant, while ε_2 is the imaginary part of the complex dielectric constant.

Li et al. [28] have determined the dielectric constants for monolayer MoS_2 and WS_2 at room temperature from experimental reflectance spectra by a constrained Kramers–Kronig analysis; we have utilized these values to calculate the photon energy-dependent refractive index and extinction coefficient [29, 30]. As reported by Mukherjee et al. [31], the values of the dielectric constants, determined by Li et al. [28], were found to be better than the set of values reported by other authors and were hence chosen to evaluate the optical properties in this work.

Figure 6.9 presents the results of the dielectric constants of MoS_2 and WS_2 in both monolayer and bulk. It should be noted that the A B points in the ε_2—E spectra actually represent the A–B splitting of monolayer MoS_2 and monolayer WS_2. The values of the A–B splitting, determined in this method, are in agreement with the modeling results that are obtained utilizing the three-band tight-binding model [32].

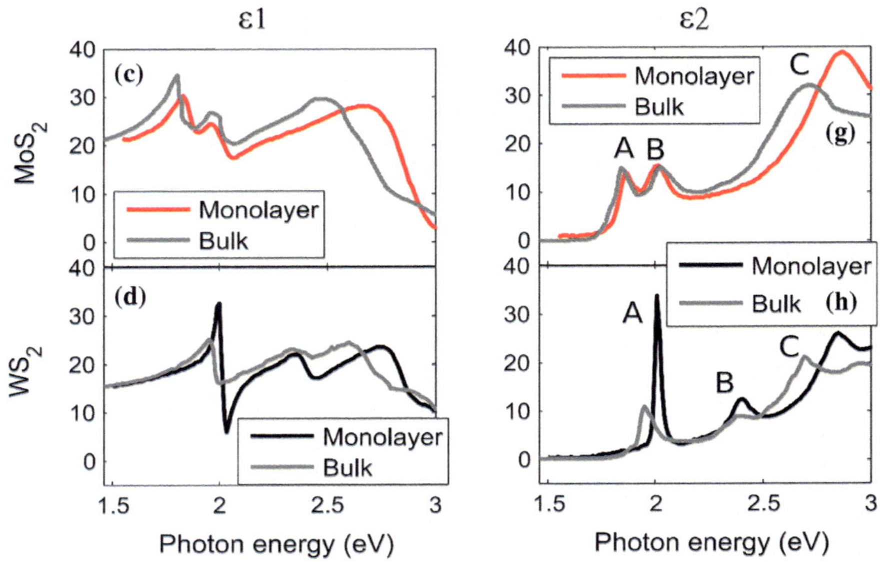

Fig. 6.9 Real (ε_1) and imaginary (ε_2) components of the dielectric constants of monolayer and bulk MoS_2 and WS_2, A, B, and C are the labels of the peaks in ε_2—E spectra [28]

6.4.1.2 Complex Refractive Index of Suspended MoS_2 and WS_2

When light propagates through a media, refraction and absorption will occur. The complex refractive index describes both the refraction and absorption. The complex refractive index can be expressed by the following equation:

$$n = (n - ik) \tag{6.4}$$

n on the left side of the equation is the complex refractive index; n on the right side of the equation is the real part of the complex refractive index; and k on the right side of the equation is the imaginary part of the refractive index, also called the extinction coefficient or attenuation index.

The real and imaginary parts of the complex dielectric constant, ε_1 and ε_2, are related to the refractive index and extinction coefficient by the following two equations:

$$\varepsilon_1 = \left(n^2 - k^2\right) \tag{6.5}$$

$$\varepsilon_2 = (2nk) \tag{6.6}$$

Here, n and k are the refractive index and extinction coefficient of the material, respectively.

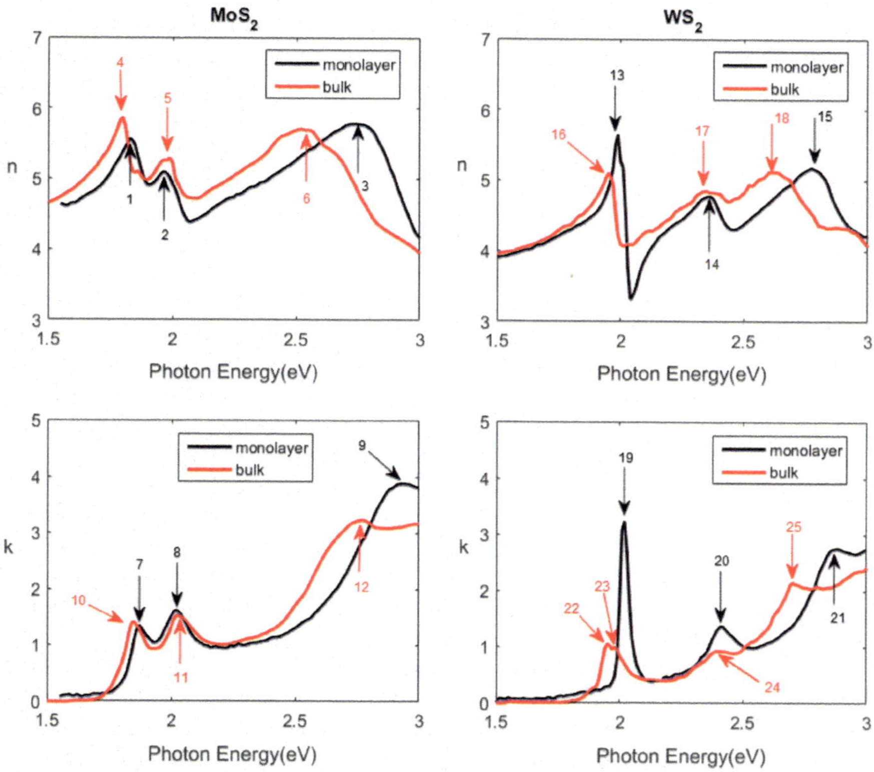

Fig. 6.10 Real component of the refractive index (*n*) and extinction coefficient (*k*) of suspended monolayer and bulk MoS_2 and WS_2

By solving the sets of equations, (6.4) to (6.6), multiple values of *n* are obtained; by considering only the real and positive value of *n*, the value of extinction coefficient *k* is determined.

Figure 6.10 shows the results of *n* and *k* for MoS_2 and WS_2 in both monolayer and bulk.

In order to get a better understanding of Fig. 6.10, the significant features in the $(n - E)$ and $(k - E)$ spectra are presented in Table 6.3.

6.4.1.3 Reflectance, Transmittance, and Absorptance of Suspended MoS_2 and WS_2

When a beam of light penetrates a medium, the energy of the beam will be reflected by the surface of the medium, transmitted through the medium and absorbed by the medium. The reflectance (*R*), transmittance (*T*), and absorptance (*A*) correspond, respectively, to the above three phenomena. They represent the fraction of the incident

Table 6.3 Energies corresponding to features of $n(E)$ and $k(E)$ for suspended MoS_2 and WS_2

Material	No.	E (eV)	n	No.	E (eV)	k
MoS_2 monolayer	1	1.83	5.563	7	1.87	1.341
	2	1.97	5.089	8	2.02	1.607
	3	2.74	5.781	9	2.93	3.879
MoS_2 bulk	4	1.8	5.868	10	1.85	1.3411
	5	1.99	5.294	11	2.03	1.524
	6	2.52	5.714	12	2.77	3.237
WS_2 monolayer	13	1.99	5.648	19	2.02	3.237
	14	2.36	4.772	20	2.41	1.364
	15	2.77	5.176	21	2.88	2.747
WS_2 bulk	16	1.95	5.112	22	1.95	1.056
	17	2.34	4.857	23	1.98	1.002
	18	2.62	5.133	24	2.4	0.933
				25	2.7	2.154

electromagnetic power that is reflected, transmitted and absorbed by the medium. This also means that their sum total will be 100%.

These fractions are determined by the incident energy, the incident angle, and the properties (or the complex refraction index, surface morphology, and thickness) of the medium. It should be noted that, in this study, all the simulations are carried out for normal incidence and at room temperature. The range of the incident energy is from 1.50 to 3.00 eV (wavelength 826.7–413.3 nm).

The reflectance, transmittance, and absorptance are related to the complex refractive index and thickness of the medium by the following equations.

$$R = \frac{(n-1)^2 + k^2}{(n+1)^2 + k^2} \tag{6.7}$$

$$\lambda = \left(\frac{hc}{E}\right) \tag{6.8}$$

$$\alpha = \left(\frac{4\pi k}{\lambda}\right) \tag{6.9}$$

$$T = (1-R)\left(e^{-\alpha t}\right) \tag{6.10}$$

$$A = 1 - R - T \tag{6.11}$$

α in Eqs. 6.9 and 6.10 is the absorption coefficient, which describes the absorbing ability of a material at that energy, in a given thickness. λ, in Eqs. 6.8 and 6.9 is the wavelength of the incident light. t in Eq. 6.10 is the thickness of the material. R is reflectance, T is transmittance, and A is absorptance. The reflectance is independent of the thickness of the medium; it is a surface property. However, transmittance, and

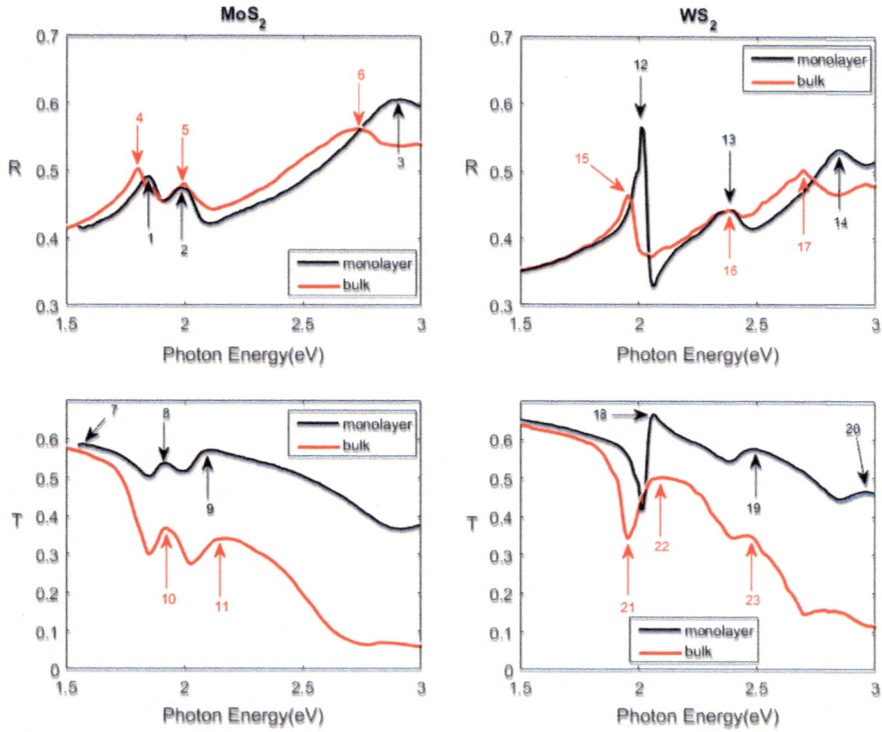

Fig. 6.11 Simulated reflectance (R) and transmittance (T) of monolayer and bulk MoS_2 and WS_2

absorptance are closely related to the thickness of the medium as well as the surface morphology.

In these simulations, the thickness of monolayer MoS_2 and WS_2 are considered to be 0.65 nm [12]. The thickness of bulk MoS_2 and WS_2 are taken as 20 nm, which seem to be the optimal value [33]. Based on the above equations, the reflectance, transmittance, and absorptance of monolayer and bulk MoS_2 and WS_2 are simulated and presented in Fig. 6.11.

The simulated reflectance and transmittance spectra of monolayer and bulk MoS_2 and WS_2, under conditions of normal incidence and at room temperature, are presented in Fig. 6.11. For all four monolayer and bulk TMDCs, the two lowest energy peaks in the reflectance spectra correspond to the excitonic features that are associated with the interband transitions in the K (K′) point in the Brillouin zone [18]. The two significant peaks in Fig. 6.11 can be attributed to the splitting of the valence band by spin–orbit coupling, including the transitions near the Γ point [34, 35].

The maximum value of Reflectance (R) and the corresponding energy (E) for monolayer MoS_2 and WS_2 are as follows: 60.5% (2.91 eV) and 56.5% (2.01 eV), respectively. Similarly, the maximum value of R for bulk MoS_2 and WS_2 are as follows: 50.7% (1.8 eV) and 50.6% (2.69 eV), respectively.

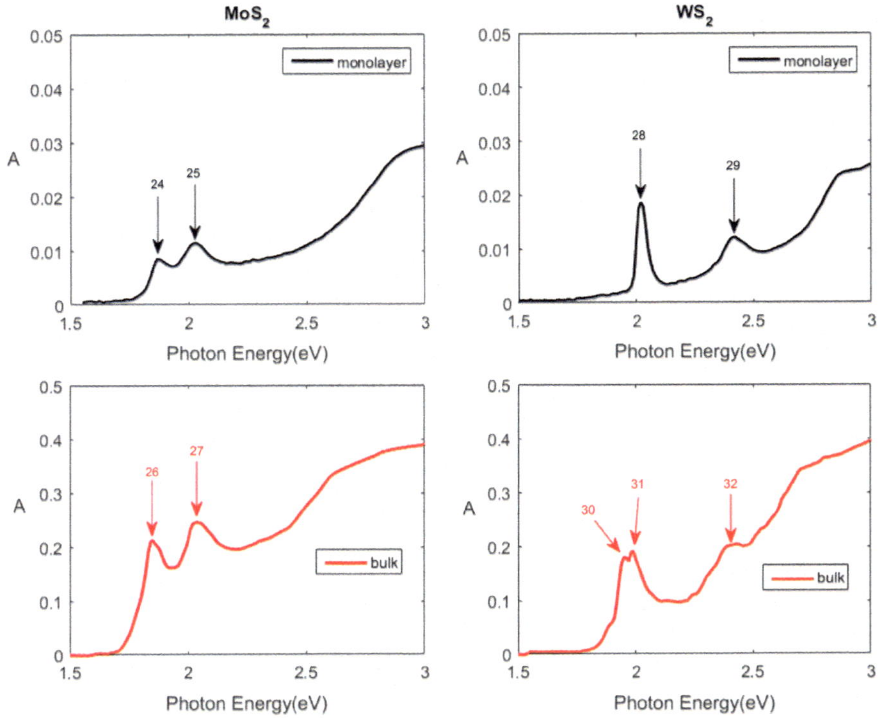

Fig. 6.12 Simulated absorptance (A) of monolayer and bulk MoS_2 and WS_2

For the simulated transmittance spectra, the maximum values of Transmittance (T) and the corresponding energy (E) for monolayer MoS_2 and WS_2 are as follows: 57% (2.11 eV) and 66.5% (2.06 eV), respectively. Similarly, the maximum value of T for bulk MoS_2 and WS_2 are 37.6% (1.92 eV) and 51.9% (1.79 eV), respectively.

The simulated absorptance spectra of monolayer and bulk MoS_2 and WS_2, under conditions of normal incidence and at room temperature, are shown in Fig. 6.12. The spectra of monolayer and bulk are separately shown due to the large change in the magnitude of absorptance value of monolayer and bulk. This change is reasonable due to the increase in the number of layers which leads to increase in absorptance of the incident light by the material.

Within the range of photon energy considered in this study, as observed from Fig. 6.12, the maximum values of Absorptance (A) in the energy range for monolayer MoS_2 and WS_2 are as follows: 2.93 and 2.55%, respectively. Similarly, the maximum value of A for bulk MoS_2 and WS_2 are as follows: 39.1 and 39.7%, respectively. The location of the peaks, in either case, remains relatively similar; these peaks are the A and B exciton absorption peaks which originate from the spin-split direct gap transitions at the K point of the Brillouin zone.

All the values of energy corresponding to the features, present in Figs. 6.11 and 6.12, are listed in Table 6.4.

Table 6.4 Energies corresponding to features of *R*, *T*, and *A* for monolayer and Bulk MoS_2 and WS_2

Material	No.	*E* (eV)	*R*	No.	*E* (eV)	*T*	No.	*E* (eV)	*A*
MoS_2 monolayer	1	1.85	0.491	7	1.57	0.585	24	1.87	0.00848
	2	1.99	0.474	8	1.92	0.537	25	2.03	0.01135
	3	2.91	0.605	9	2.11	0.57			
MoS_2 bulk	4	1.8	0.507	10	1.92	0.376	26	1.85	0.214
	5	1.99	0.485	11	2.17	0.349	27	2.04	0.247
	6	2.73	0.566						
WS_2 monolayer	12	2.01	0.565	18	2.06	0.665	28	2.02	0.0185
	13	2.38	0.446	19	2.49	0.576	29	2.41	0.0121
	14	2.85	0.531	20	2.96	0.466			
WS_2 bulk	15	1.95	0.469	21	1.95	0.35	30	1.95	0.181
	16	2.38	0.446	22	2.1	0.51	31	1.99	0.191
	17	2.69	0.506	23	2.45	0.359	32	2.43	0.205

6.4.1.4 Optical Bandgap of Monolayer and Bulk MoS_2 and WS_2

In semiconductor science and technology, the bandgap is a fundamental property of a material since it determines its applications. For monolayer MoS_2 and WS_2, since their direct bandgaps are in the visible range, it makes them promising candidates in a variety of applications in optoelectronics. The bandgap refers to the minimum energy difference between the bottom of the conduction band and the top of the valence band.

In addition to the ability to determine the bandgap from temperature dependent measurements of the electrical conductivity, there are generally two other methods for estimating the bandgap of semiconductors. One can measure the bandgap optically by excitation spectroscopy in which the charge state of the material is not changed. The bandgap, measured in this method, is called as the optical bandgap. One can also measure the bandgap by utilizing photoemission spectroscopy.

Utilizing the absorptance versus energy spectra, as described in Sect. 6.4.1.3, the optical bandgap of a material can be determined. By plotting certain powers of the absorption coefficient versus photon energy, one can normally confirm the value of the bandgap, and its nature—whether or not it is direct. If a plot of α^2 versus photon energy forms a straight line, it can normally be inferred that the bandgap is direct, measurable by extrapolating the straight line to $\alpha = 0$ axis. On the other hand, if a plot of $\alpha^{1/2}$ versus photon energy forms a straight line, it can normally be inferred that the bandgap is indirect, measurable by extrapolating the straight line to $\alpha = 0$ axis.

The monolayer MoS_2 and WS_2 are examined for the direct semiconductor case. The square of the absorption coefficient versus energy for monolayer MoS_2 and WS_2 is presented in Fig. 6.13.

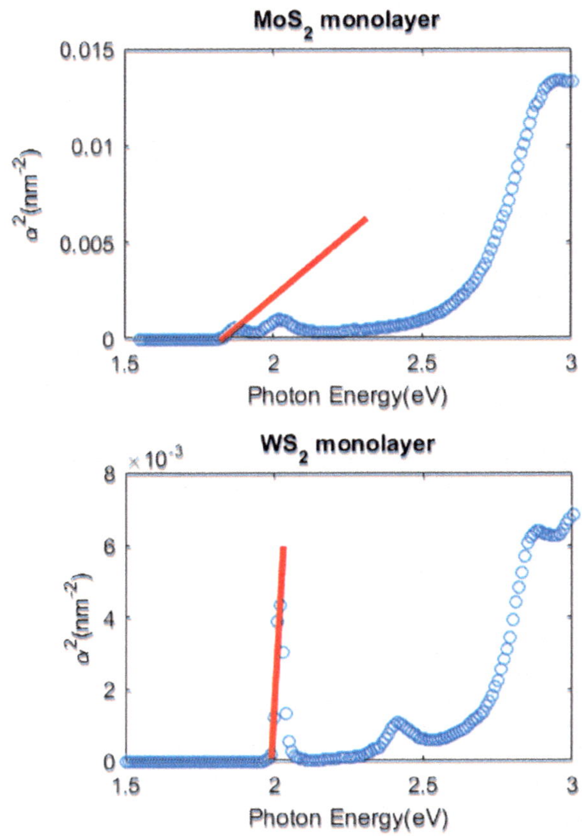

Fig. 6.13 Square of the absorption coefficient versus energy for monolayer MoS_2 and WS_2; the red line is the extrapolating straight line

As can be seen in Fig. 6.13, a straight line behavior is observed near the first peak for the monolayer MoS_2 and WS_2; this shows that the monolayer MoS_2 and WS_2 have a direct bandgap. The values of the optical bandgap of monolayer MoS_2 and WS_2 are obtained by solving the equations to straight lines. The optical bandgaps of monolayer MoS_2, WS_2 are 1.82 and 1.98 eV, respectively. The calculated optical bandgaps are slightly smaller than the electronic bandgaps (in Sect. 6.3.2) within a reasonable range; this is due to the additional energy absorbed by an electron while undergoing a transition from the valence band to the conduction band; there is a difference in the Coulomb energies of the two systems (excitation spectroscopy and tunneling spectroscopy) which, therefore, causes a change in the optical bandgaps and electrical bandgaps.

The analyses of the $\alpha - E$ spectra for bulk MoS_2 and WS_2 has not been possible due to the limited data of the optical properties available in the literature; the bandgap for bulk TMDCs is obviously much smaller than 1.50 eV (Fig. 6.6), which corresponds to the minimum of the photon energy range considered in this study.

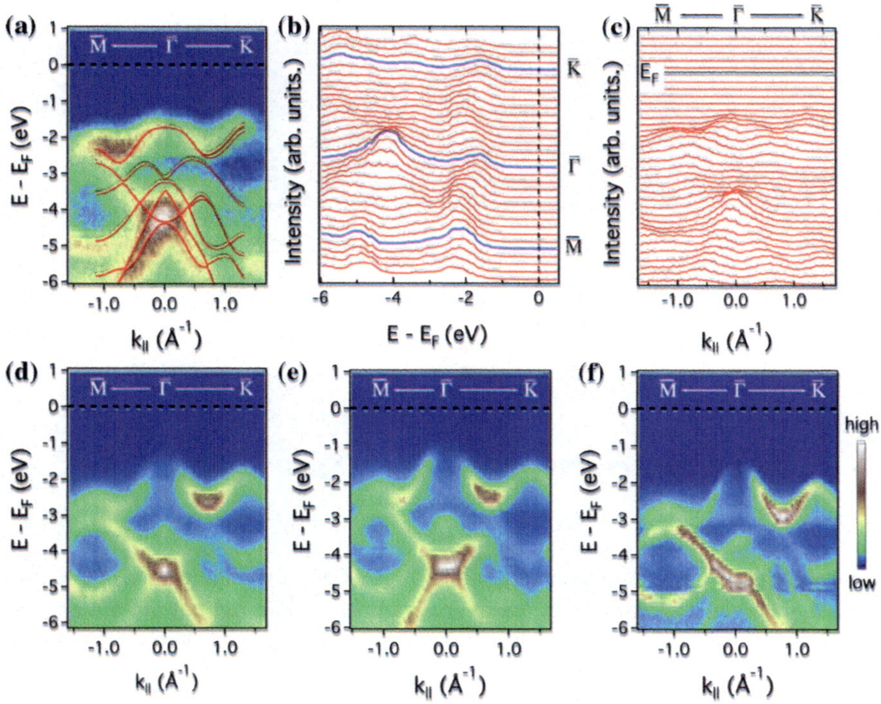

Fig. 6.14 **a** ARPES band map of exfoliated monolayer MoS_2 along the high symmetry lines. Density Functional Theory (DFT)-based calculations are overlaid on it for comparison. **b**, **c** Corresponding Energy Distribution Curves (EDCs) and Momentum Distribution Curves (MDCs). **d–f** ARPES band maps of exfoliated bilayer, trilayer, and bulk MoS_2, respectively [36]

6.4.1.5 Angle-Resolved Photoemission Spectroscopy of MoS_2

In 2013, Jin and coworkers [36] studied the band structure of MoS_2 using Angle-Resolved Photoemission Spectroscopy (ARPES). Some of the highlights of their results are presented in Figs. 6.14 and 6.15.

Figures 6.14 and 6.15 show the evolution in the band structure of MoS_2 as a function of number of layers, as obtained from Angle-Resolved Photoemission Spectroscopy (ARPES). From Fig. 6.15e, it can be seen that, with increasing number of layers, the energy difference between Valence Band Maximum (VBM) of $\overline{K}$ and $\overline{\Gamma}$ decreases with increasing number of layers, transitioning from monolayer to bulk. This phenomenon is due to the quantum confinement as the number of layers changes. The VBM at $\overline{K}$ (derived from the localized in-plane Mo $d_{x^2-y^2}/d_{xy}$ orbitals) is not likely to be affected by quantum confinements. But the VBM at $\overline{\Gamma}$ (derived from the rather delocalized out-of-plane Mo d_{z^2} orbitals and S_{p_z} orbitals) is lowered in energy when the interlayer interactions increase with increasing number of layers [36].

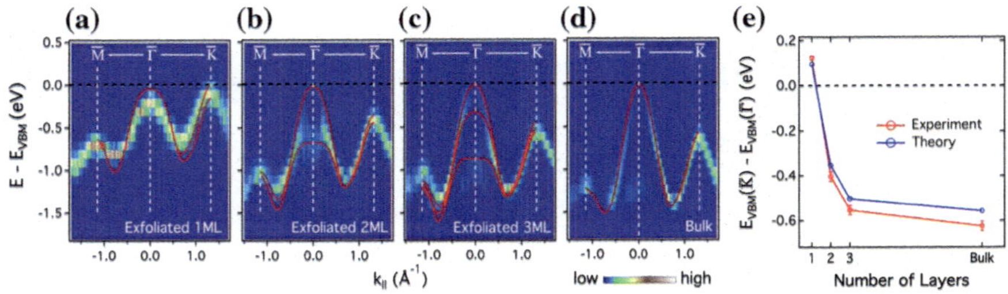

Fig. 6.15 **a–d** 2D curvature intensity plots of the low energy valence band of exfoliated monolayer, bilayer, trilayer, and bulk MoS_2, respectively. Red curves are the corresponding DFT calculated bands. **e** Thickness dependence of the energy difference between Valence Band Maximum (VBM). The theoretical and experimental results are plotted together for comparison [36]

6.4.2 *Optical Properties of MoS_2 and WS_2 on Selected Substrates*

Optical properties of monolayer and bulk TMDCs on a representative semiconductor (Si), metal (Au), and insulator(SiO_2) substrates could contribute to understand and promote the applications of this multilayer system in areas such as coatings, electronics, optoelectronics, sensors, circuits, and systems.

6.4.2.1 Simulation Method

In order to simulate the optical properties of the double layer, the complex dielectric constants of the substrates are required.

Once the optical properties of the substrates are obtained, the simulation of the reflectance, transmittance, and absorptance of the double layer can be performed by using the following equations:

$$R = \frac{(n_1 - n_2)^2 + k_1^2}{(n_1 + n_2)^2 + k_2^2} \tag{6.12}$$

$$\alpha_1 = \frac{4\pi k_1}{\lambda} \tag{6.13}$$

$$\alpha_2 = \frac{4\pi k_2}{\lambda} \tag{6.14}$$

$$T = (1 - R) \times e^{-\alpha_1 t_1} \times e^{-\alpha_2 t_2} \tag{6.15}$$

$$A = 1 - R - T \tag{6.16}$$

In the above equations, n_1 and k_1 are the optical constants of the top layer, n_2 and k_2 are the optical constants of the substrate. α_1 and α_2 are the absorption coefficients of the top layer and substrate, respectively. t_1 and t_2 are the thickness of the upper

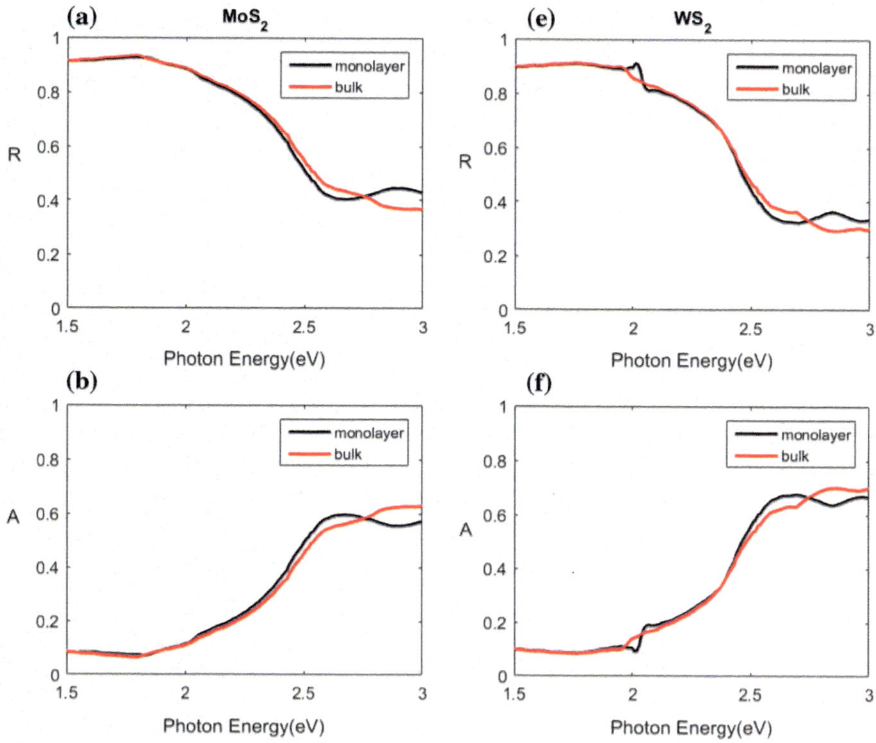

Fig. 6.16 Simulated reflectance and absorptance spectra of monolayer and bulk MoS_2 and WS_2 on gold substrate

layer and substrate, respectively. *R*, *T*, and *A* are the reflectance, transmittance, and absorptance of the double layer structure.

6.4.2.2 MoS_2 and WS_2 on Gold Substrate

The thickness of the gold substrate is chosen as 10 μm. The thicknesses of the monolayers are chosen as 0.65 nm and the thicknesses of the bulk are chosen as 20 nm.

Figure 6.16 represents the simulated reflectance and absorptance spectra of monolayer and bulk MoS_2 and WS_2 on gold substrate. The transmittance in the selected energy range is almost zero and thus is not shown. This is due to the opaqueness of the gold substrate in the wavelength range considered.

It is observed that the reflectance tends to decrease with increase in photon energy, while the absorptance of MoS_2 and WS_2 on gold substrate increases with the increase in photon energy. The reflectance and absorptance spectra are complementary to each other due to the negligible transmittance. The change in the number of layers of MoS_2 and WS_2 does not necessarily cause a significant change in the reflectance and absorptance spectra. No sharp peaks are observed in both the spectra. Maximum values of reflectance and transmittance for monolayer and bulk MoS_2 and WS_2 on gold substrate remain approximately the same, with an average maximum reflectance of ~90% and average maximum absorptance of ~70%.

6.4.2.3 MoS_2 and WS_2 on Silicon Substrate

The thickness of the silicon substrate is chosen as 650 μm. This silicon wafer thickness represents the industry standard. The thicknesses of the monolayers are chosen as 0.65 nm and the thicknesses of the bulk are chosen as 20 nm.

The transmittance in the selected energy range is almost zero due to the opaqueness of the silicon substrate. In Fig. 6.17, it can be observed that the reflectance has a relatively low value as compared to that of the gold substrate (average maximum value of ~90%). However, a sharp peak is observed in the reflectance spectra of monolayer WS_2 on silicon. The absorptance of the monolayer and bulk MoS_2 and WS_2 on silicon substrate is relatively high compared to that of gold substrate.

6.4.2.4 MoS_2 and WS_2 on Fused Silica Substrate

The thickness of the fused silica substrate is chosen as 650 μm. The thicknesses of the monolayers are chosen as 0.65 nm and the thicknesses of the bulk are chosen as 20 nm.

The simulated reflectance and transmittance spectra of monolayer and bulk MoS_2 and WS_2 on fused silica substrate is presented in Fig. 6.18. It is observed that the transmittance of TMDC/fused silica is considerably high which is due to the transparent nature of fused silica and TMDCs.

As observed in Fig. 6.18, the maximum value of Reflectance (R) for monolayer MoS_2 and WS_2 are 48.06 and 43.59%, respectively. Similarly, the maximum value of R for bulk MoS_2 and WS_2 are 43.48 and 34.70%, respectively.

The maximum value of Transmittance (T) for monolayer MoS_2 and WS_2 are 72.56 and 80.26%, respectively. Similarly, the maximum value of T for bulk MoS_2 and WS_2 are 72.48 and 78.64%, respectively.

Figure 6.19 presents the simulated absorptance spectra of monolayer and bulk MoS_2 and WS_2 on fused silica substrate. The maximum value of Absorptance (A) for monolayer MoS_2 and WS_2 are as follows: 3.83 and 3.26%, respectively. Similarly, the maximum value of A for bulk MoS_2 and WS_2 are 50.33 and 50.40%, respectively.

The absorptance spectra, shown in Fig. 6.19, are very similar to that in Fig. 6.12; however, the values differ from each other by a small factor. The trend in the change in the value of the monolayer and bulk TMDCs are the same.

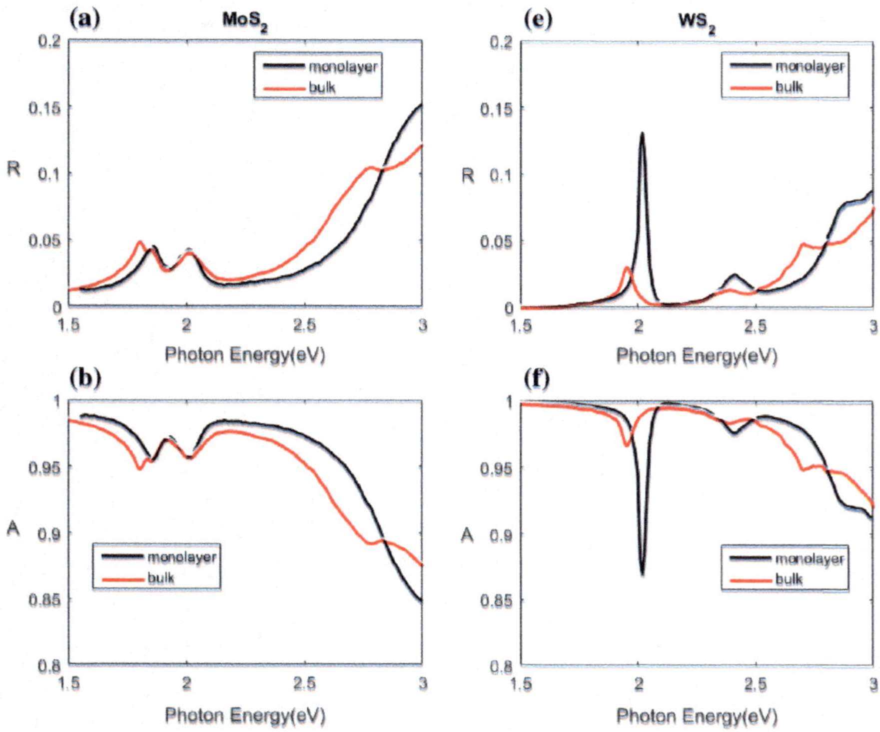

Fig. 6.17 Simulated reflectance and absorptance spectra of monolayer and bulk MoS_2 and WS_2 on silicon substrate

6.5 Electrical Properties of MoS_2 and WS_2

In Sect. 6.4, a brief introduction to the optical properties of the monolayer and bulk MoS_2 and WS_2 were presented. In this Section, the electrical properties of MoS_2 and WS_2 will be discussed. The temperature dependence of the electrical properties such as sheet conductivity and field-effect mobility are considered to study the conduction mechanism. This will help to understand the performance of electronic devices that are formed by monolayers of MoS_2 and WS_2.

6.5.1 Carrier Mobility

Charge carrier mobility in bulk MoS_2, at room temperature, has been studied by Fivaz and Mooser [37]. Their results show that the charge carriers can attain mobility in the range of 200–500 cm^2/Vs [37]. Recently, experiments have been carried out on monolayer MoS_2 at room temperature. It should be noted that the mobility decreases

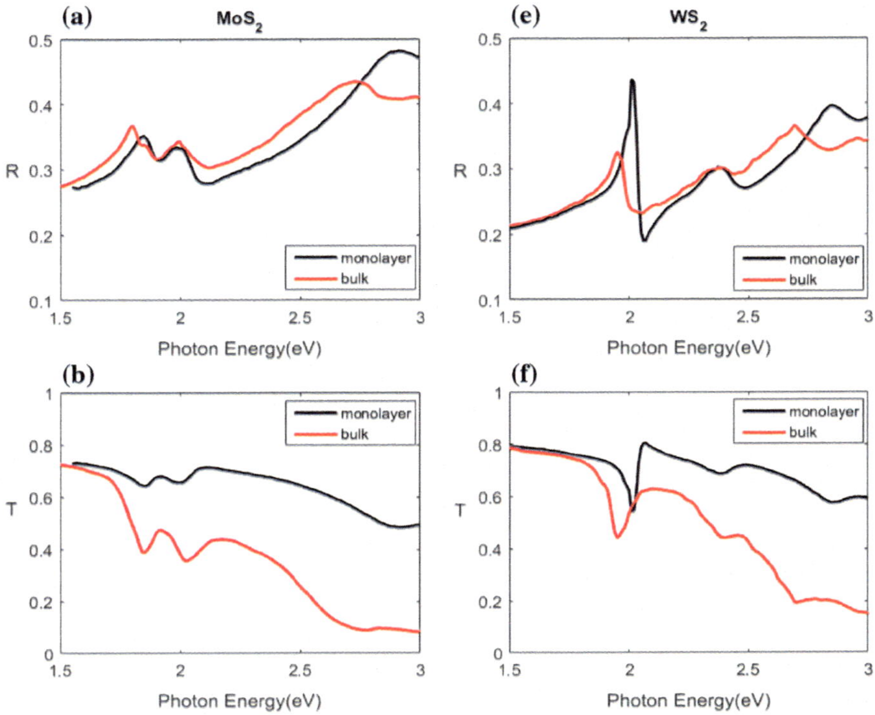

Fig. 6.18 Simulated reflectance and transmittance spectra of monolayer and bulk MoS_2 and WS_2 on fused silica substrate

to 0.1–10 cm^2/Vs [38, 39]. The main reason for this decrease is the charge traps that are present at the interface between MoS_2 and the substrate [40].

In order to increase the mobility of charge carriers in MoS_2-based Field-Effect Transistors (FETs), at room temperature, the understanding of the conduction mechanism is very important. In 2013, Radisavljevic and Kis reported their study on the mobility measurements in monolayer MoS_2, based on Hall effect [41]. In their study, the mobility of charge carriers in MoS_2-based FETs are measured at different temperatures and the results are fitted to the theoretical equations to study the current conduction mechanism.

Figure 6.20 presents the structure of MoS_2-based FET which is commonly fabricated and studied in device research.

The results in Fig. 6.21 represent Hall measurements. In Fig. 6.21a, the variation in conductance with back-gate voltage, for various device temperatures, is shown. For each back-gate voltage, the corresponding charge concentration can be calculated using the following equations:

$$n_{2d} = \left(C_{ox1} * \Delta V_{bg}/e\right) \tag{6.17}$$

$$C_{ox1} = (\varepsilon_0 \varepsilon r_1/d_{ox1}) \tag{6.18}$$

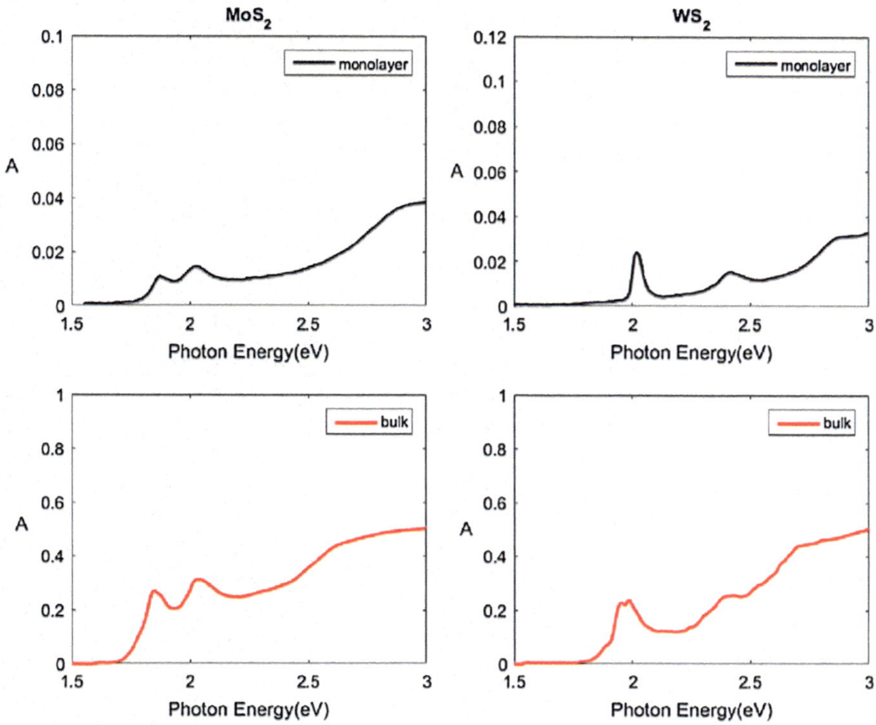

Fig. 6.19 Simulated absorptance spectra of monolayer and bulk MoS_2 and WS_2 on fused silica substrate

$$\Delta V_{\mathrm{bg}} = \left(V_{\mathrm{bg}} - V_{\mathrm{bg,th}}\right) \tag{6.19}$$

$V_{\mathrm{bg,th}}$ is the threshold voltage for each device and is close to its pinch-off voltage estimated from the conductance curves. For instance, the highest back-gate voltage of 40 V corresponds to a charge concentration of $n_{2d} \sim 3.6 \times 10^{12}\ \mathrm{cm}^{-2}$.

As shown in Fig. 6.21b, for each V_{bg}, in a certain temperature range, the conductance follows Eq. 6.20:

$$G = \left[G_0 * \exp\left(\frac{-E_g}{kT}\right)\right] \tag{6.20}$$

Equation 6.20 is valid for the condition when electron transport is thermally activated; good agreement of the experimental data with Eq. 6.20 shows that the electron transport in these temperature ranges is thermally activated. At lower temperatures, the variation in conductance is weakened for all V_{bg}. This is due to hopping through localized states which becomes a dominant mechanism of electron transport, and the system approaches the localized regime.

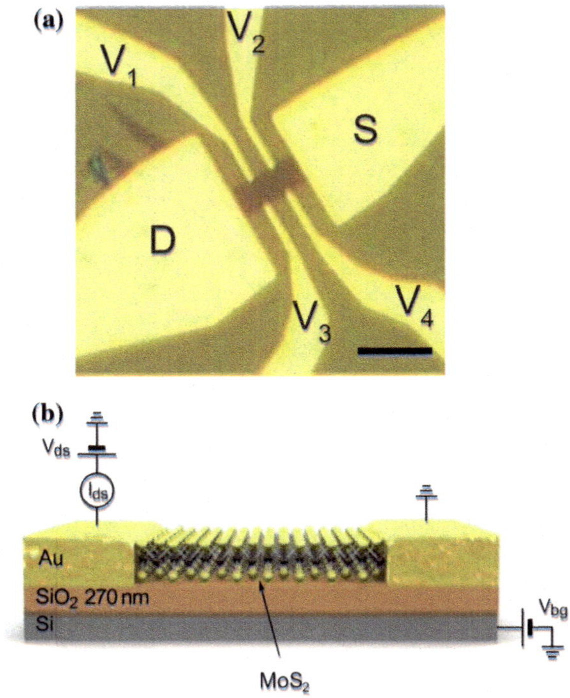

Fig. 6.20 Structure of MoS_2-based FET. **a** Optical image. **b** Cross section view. V_{ds} is drain-source voltage. I_{ds} is drain-source current. V_{bg} is back-gate voltage [41]

In Fig. 6.21c, the mobility dependence on temperature is presented. The results are extracted from Fig. 6.21a, b in the range of 30–40 V of back-gate voltage, V_{bg}. The following equation for mobility, μ, is utilized in the calculation:

$$\mu = \left[\mathrm{d}G/\mathrm{d}V_{bg}\right] * \left[L_{12}/(WC_{ox1})\right] \tag{6.21}$$

A sharp peak corresponding to 18 cm^2/Vs is observed at 200 K in Fig. 6.21c; subsequently, the mobility decreases with increase in temperature. This shows that electron–phonon scattering becomes the dominant mechanism in influencing the temperature dependence of mobility. The decreasing component of the curve can be fitted in accordance with $\mu \sim T^{\gamma}$, $\gamma \approx -1.4$, which is in agreement with the theoretical predictions of $\gamma \approx -1.69$ [42].

WS_2 is an important member of the TMDC class of materials. In 2011, Liu et al. performed a theoretical study of all the TMDCs [43]. Among all these materials, WS_2 shows the highest mobility which results from its reduced effective mass [43]. A recent report estimates that monolayer WS_2-based transistors exhibit 44 cm^2/(Vs) mobility at room temperature [44]. Similar to MoS_2, the temperature dependence studies of the electrical properties of WS_2 is critical for understanding the current conduction mechanism which is fundamental to improving its performance. In 2014, temperature-dependent mobility studies of WS_2-based devices were performed by Ovchinnikov et al. [45]. In this study, the mobility of WS_2-based FET, at different

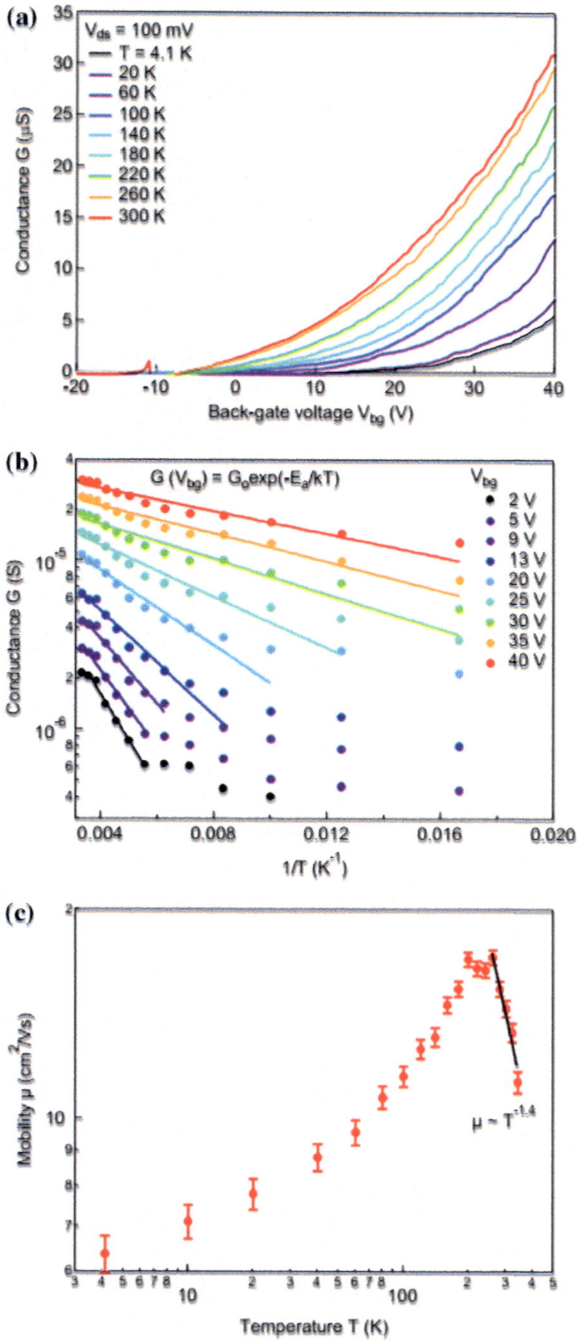

Fig. 6.21 Temperature dependence of electrical properties of monolayer MoS_2. **a** Conductance G versus back-gate voltage at different temperatures. **b** Conductance G versus temperature at different back-gate voltages. **c** Mobility as function of temperature [41]

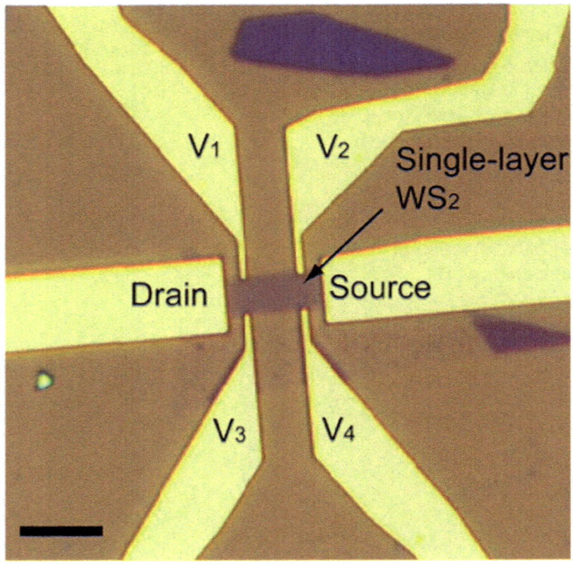

Fig. 6.22 Optical image of WS_2-based FET [45]

temperatures, were measured, and the results were fitted to the theoretical equations to study the conduction mechanism.

The current conduction mechanism in MoS_2 and WS_2-based FETs are similar.

Figure 6.22 shows the optical image of WS_2-based FET. The structure is similar to that of MoS_2 FET. Si is chosen as the back-gate material and a 270 nm thick SiO_2 is chosen as the oxide on Si.

In Fig. 6.23a, b, the variation in sheet conductivity with gate voltage, as a function of temperature, is shown. Sheet conductivity and conductance can be transformed to each other by utilizing the equation:

$$G = (\sigma * t) \tag{6.22}$$

As can be seen in Fig. 6.23a, when V_g is below a certain threshold value, the sheet conductivity increases with increase in temperature. However, when V_g is greater than the threshold value, the sheet conductivity decreases with increase in temperature. This indicates a transition from insulator regime to conductor regime.

Figures 6.21 and 6.23 show that there are differences in the temperature dependence of the electrical properties of MoS_2 and WS_2. In Fig. 6.23c, the mobility decreases with increasing temperature for WS_2, which is different from that of MoS_2. In MoS_2, there emerges a sharp peak in the change in mobility with temperature, while in WS_2, no peak is observed.

This demonstrates a significant difference in the conduction mechanism of the two materials at low temperatures. At higher temperatures, WS_2 shows a power-law dependence; experiments shows that $\mu \sim T^{\gamma}$, $\gamma \approx -0.73$.

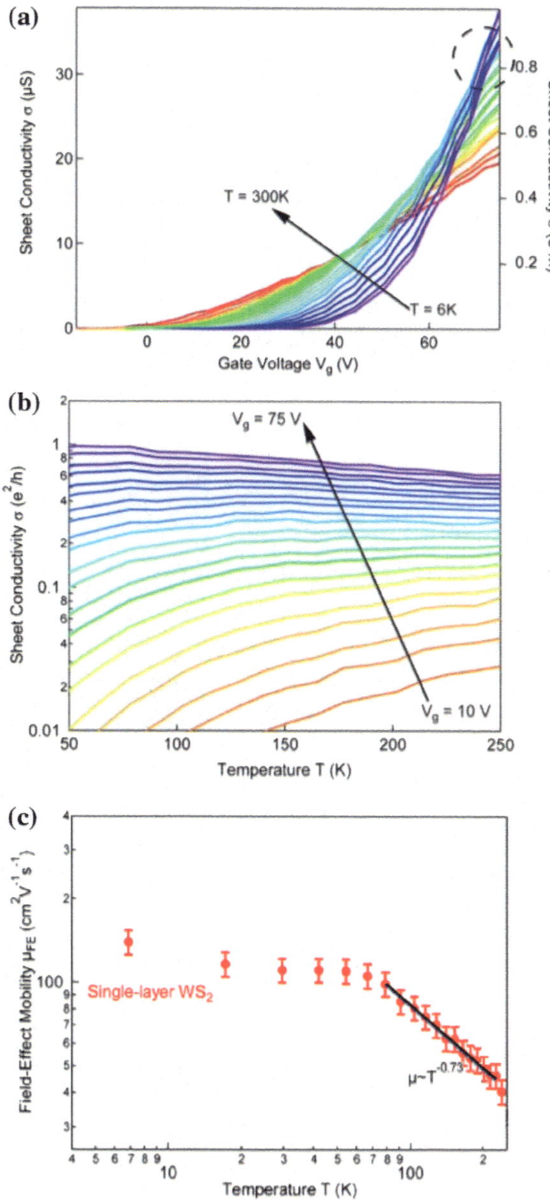

Fig. 6.23 Temperature dependence of electrical properties of monolayer WS_2. **a** Sheet resistivity versus back-gate voltage at different temperatures. **b** Sheet resistivity versus temperature at different back-gate voltage. **c** Mobility dependence on temperature [45]

6.6 Applications of MoS_2 and WS_2

6.6.1 Introduction

The advancements in fabrication methods such as exfoliation and synthetic techniques such as Chemical Vapor Deposition and Molecular-Beam Epitaxy, have made it possible to fabricate ultra-thin monolayer TMDCs, including MoS_2 and WS_2. These two materials can be classified as semiconductors with stability in air. Their direct bandgap in the visible energy range makes these materials promising in digital electronics and optoelectronics. There have been many reports on the applications of these materials in devices such as Field-Effect Transistors (FETs), solar cells, photodetectors, etc. In this section, a brief overview of these applications, the architecture, operating principles and the physics of electronic and optoelectronic devices will be presented.

6.6.2 Digital Electronics

MoS_2 and WS_2 are promising material candidates for applications in future digital electronics such as FETs, inverters, logic gates, and junctions and heterostructures.

6.6.2.1 FET

The structures and functions of FETs have been discussed in Sect. 6.5. These materials have initially been used in FETs only in the past decade [46]. Initially, the field-effect mobility of monolayer MoS_2 was found to be a lot lower than that in graphene [39]. However, in 2011, Radisavljevic et al. have reported a monolayer MoS_2 top-gated with a relatively high mobility, large on/off ratios and low subthreshold swings at room temperature [38]. These reports have inspired interest for further studies. The large bandgap (>1 eV) and ultra-thin top gate makes it possible to achieve small cutoff currents and large switching ratios.

Recent studies have focused on the difference in the theoretically simulated and experimentally measured values of charge carrier mobility in devices based on these TMDCs. Theoretical studies predict that the intrinsic mobility at room temperature in monolayer MoS_2 is 300–400 cm^2/Vs at high carrier densities of $10^{13}/cm^2$ [42]. This value is five times greater than the largest experimentally reported values at 300 K [47].

The sensitivity of carrier mobility to the local dielectric environment has inspired studies on the electron transition mechanism in TMDCs. These studies have been carried out by temperature-dependent mobility measurements. With the improvement in the sample quality and device processing, it is observed that phonon scattering is dominant for $T > 100$ K and charged impurity scattering is dominant for $T < 100$ K [41,

47, 48]. As for theoretical predictions, the influence of remote interfacial phonons from the oxide dielectric have been considered recently, and the results show a better agreement with experimental studies [49]. The lack of considering defects and impurities also lead to overestimating the mobility values.

The contact resistance plays an important role in determining the conductance of FET devices. Since MoS_2 and WS_2 are ultra-thin 2D materials, the substrates significantly influence the performance of the corresponding devices. Theoretical prediction and experimental studies on contact resistance have been carried out [50–53] in the literature. Generally, these studies found that low work function metals such as Ti and Sc form lower resistance contacts than high work function metals which lead to higher drain currents [50]. Nevertheless, even with the optimal contact approach, the contact resistance still dominates the charge transport with reduced channel dimensions.

It should be noted that these flexible 2D materials make it possible to fabricate flexible FETs on flexible substrates [54]. The influence of strain on the properties of these materials has been carried out [55]. Figure 6.24 presents an illustration of the fabrication of flexible FET. These initial studies demonstrate that ultra-thin FETs on flexible substrates remain a desirable option under mechanical bending [56, 57].

The promise of MoS_2 and WS_2-based high-performance FET has inspired researchers to perform further studies on their applications to more complex, functional digital circuitry, inverters, and logic gates [58].

6.6.2.2 Junctions and Heterostructures

In addition to semiconducting TMDC-based devices and circuits, these materials present possibilities for special device geometry-based junctions and heterostructures [59, 60]. Such heterojunction devices have been widely developed for crystalline III–V semiconductors for high-frequency applications (Fig. 6.25).

Epitaxy is the most common method for the growth of III–V semiconductor heterostructures. This limits the materials to have relatively similar lattice constants. The weak van der Waals bonding between TMDCs can relax this limit and thus enhance the possibilities for different materials to form different heterostructures [60].

6.6.3 Optoelectronics

In Sect. 6.3, it was shown that monolayer MoS_2 and WS_2 exhibit direct bandgap in the visible region of the electromagnetic spectrum and have large binding energies. These two materials also show strong photoluminescence. These properties make them promising candidates for applications in optoelectronics including photodetectors, photovoltaics, and light-emitting devices.

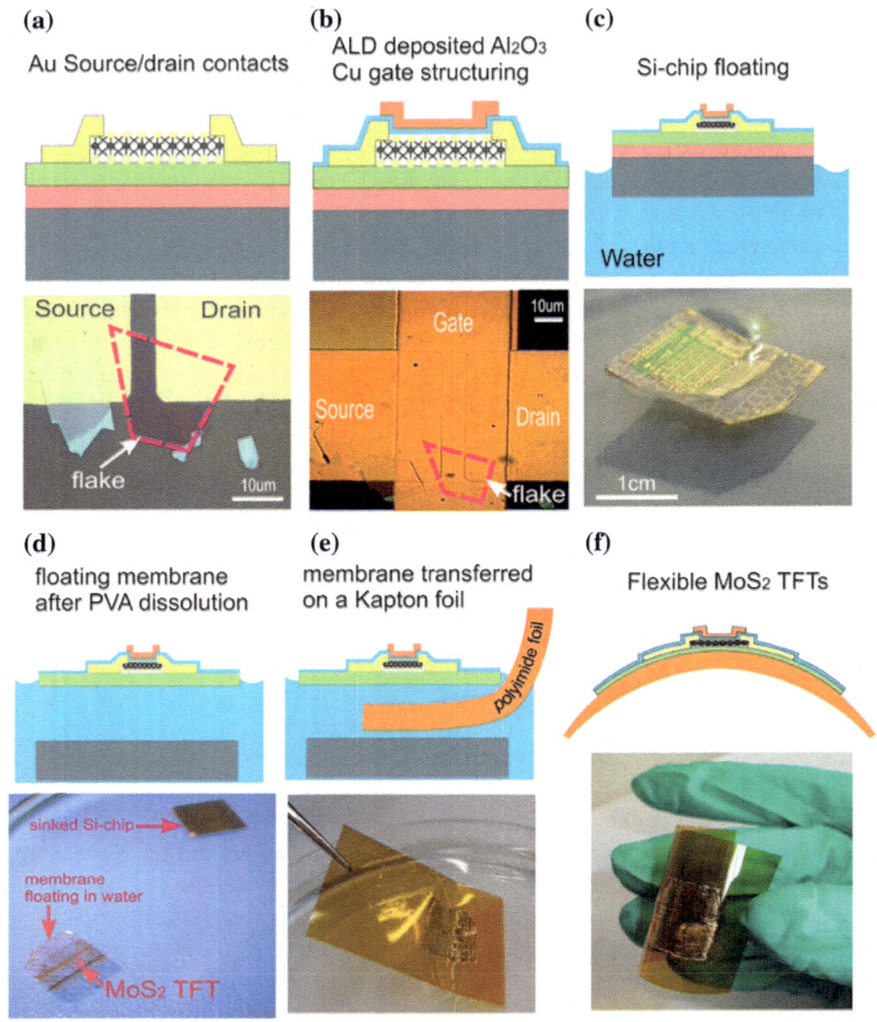

Fig. 6.24 Illustration of the fabrication of flexible FET [56]

6.6.3.1 Photodetectors

When the energy of the incident photon is larger than the bandgap of a semiconductor, electron hole pairs (or excitons) or free carriers, which depend on the exciton binding energy in the semiconductor, are created. Bound excitons can be separated by an applied or built-in electric field and thus generate a photocurrent. The two most common types of photodetectors are phototransistors and photodiodes.

Phototransistors are types of bipolar transistors with their base lead removed and replaced with a light-sensitive area. These devices have only two terminals. When

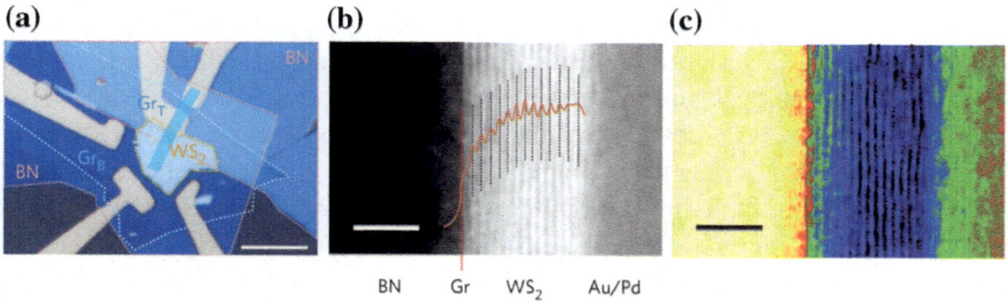

Fig. 6.25 Illustration of Graphene–WS_2 heterotransistor. **a** Optical image (scale bar, 10 mm). **b** Cross section high-resolution High-Angle Annular Dark-Field Scanning Transmission Electron Microscopy (HAADF-STEM) image (scale bar, 5 nm). **c** Bright-field STEM image (scale bar, 5 nm) [59]

the light-sensitive area is in dark, the device is turned off. When the light-sensitive region is exposed to light, a small base current is generated that controls a much larger collector-to-emitter current.

In initial studies, MoS_2-based phototransistors were measured under illumination with photons of energies greater than the 1.9 eV bandgap [62]. Recently, few-layer WS_2 phototransistors, grown by CVD, have also been reported [63]. The wavelength-dependent studies show that the photocurrent roughly follows the absorption spectrum; this has led to the speculation of inter-band absorption and carrier separation as the dominant mechanism for photocurrent generation [64, 65].

A photodiode is a semiconductor device that converts light into current. The current is generated when photons are absorbed in the photodiode. A small amount of current is also produced when no light is present. It is reported that vertical Si/Monolayer MoS_2 heterostructures exhibit photodiode-like behavior and excellent photon response [66]. Similar results have been found on vertically stacked graphene-MoS_2-metal heterostructures [67].

6.6.3.2 Solar Cells and Light-Emitting Devices

The solar cell is one of the most widespread applications of p–n junctions. Semiconductors with high mobility and direct bandgap near 1.3 eV are desired for high-efficiency single junction photovoltaics. The monolayer TMDCs have excellent properties that make them excellent candidates for application in ultra-thin photovoltaics.

In 2013, calculations on a Schottky junction solar cell, consisting of a graphene-MoS_2 stack, suggested a maximum Power Conversion Efficiency (PCE) of ~1% while that of a type-II heterojunction between WS_2/MoS_2 was 1.5% [68]. Experimentally, an asymmetric M-S-M Schottky junction on a few-layer MoS_2 flake resulted in working photovoltaic devices with ~1% PCE in the same year [69]. Although the above results are encouraging, the thickness limited absorption is still a significant barrier to increase the efficiency.

LED, the light-emitting diode, is another promising application for MoS_2 and WS_2 p–n junctions. Their ultra-thin features and direct bandgap with desired energy makes them promising for ultra-thin, efficient and flexible LEDs. Excitonic electroluminescence has been observed from SL-MoS_2 heterojunctions [70]. However, the lack of controlled doping technologies makes it only possible to fabricate purely monolayers and limit their applications.

In recent years, extensive research has been performed on MoS_2; this is due to the abundance of molybdenite and the stable nature of bulk and monolayer MoS_2. MoS_2 exhibits physical properties that are most relevant in device applications such as Field-Effect Transistors (FETs), memory devices, photodetectors, solar cells, electrocatalysts for Hydrogen Evolution Reaction (HER) and lithium-ion batteries [71].

6.7 Physical Properties of $MoSe_2$ and WSe_2

6.7.1 Introduction

The transition metal dichalcogenides, MX_2, where “M” is a transition metal and “X,” a chalcogen, have been investigated for fullerene-like behavior due to their tendency to form layered structures. A single layer consists of a strongly bound X-M-X sandwich that is weakly stacked with other layers. The crystal structures vary in the number of layers per unit cell and the alignment of the layers with respect to one another [72].

6.7.2 Structure

$MoSe_2$ and WSe_2 are layered materials with strong in-plane bonding and weak out-of-plane interactions enabling exfoliation into two-dimensional layers of single unit cell thickness. Although TMDCs have been studied for decades, recent advances in nanoscale materials characterization and device fabrication have opened up new opportunities for two-dimensional layers of thin TMDCs in nanoelectronics and optoelectronics. $MoSe_2$ and WSe_2 have sizable bandgaps that change from indirect to direct in single layers, allowing applications such as transistors, photodetectors and electroluminescent devices. Adjacent layers are weakly held together to form the bulk crystal in a variety of polytypes, which vary in stacking orders and metal atom coordination. The overall symmetry of TMDCs is hexagonal or rhombohedral, and the metal atoms have octahedral or trigonal prismatic coordination [73].

Figure 6.26 depicts the structure of monolayer MoS_2 that has an interlayer spacing of 0.65 nm. $MoSe_2$ and WSe_2 have a physical structure similar to that of MoS_2, the only difference being the interlayer spacing. The difference in the atomic size of “Mo” and “W” results in a change in the interlayer spacing.

$MoSe_2$ and WSe_2 portray different properties at the monolayer and bulk levels. At the bulk level, these materials have an indirect bandgap, whereas at a monolayer level, the bandgap increases for both, $MoSe_2$ and WSe_2 [73]. Because of these layer dependent properties, the monolayer and bulk crystal structures of $MoSe_2$ and WSe_2 are examined. Monolayer WSe_2 has a thickness of ~0.7 nm and bulk WSe_2 has a thickness of ~20 nm. These are experimental values and have been obtained by various deposition techniques such as Chemical Vapor Deposition (CVD) and mechanical exfoliation of bulk WSe_2 crystals [74].

Figure 6.27 shows the crystal structure of monolayer WSe_2, the interlayer distance between Se-W-Se bonds is ~0.7 nm and is also the thickness of the monolayer. As seen in the figure, the Se-W-Se bonds form a hexagonal symmetry and is hence of 3H form. Bulk structure of WSe_2 is formed by stacking multiple monolayers of WSe_2; the thickness of these structures is generally ~20 nm. Usually, bulk WSe_2 crystals consist of weakly bonded Se-W-Se units (three atomic planes) [74].

The crystal structure of $MoSe_2$ is similar to that of WSe_2, the only difference being the transition metal atom, which leads to a minor change in the monolayer thickness. Just like monolayer WSe_2, in monolayer $MoSe_2$ "Mo" and "Se" atoms form a 2D hexagonal lattice with trigonal prismatic coordination as shown in Fig. 6.28 [76].

In $MoSe_2$, the "Mo" atoms occupy one type of sub-lattice of the hexagonal sheet and atoms of "Se" occupy the others. However, due to the chemical ratio of Mo: Se = 1:2, the sub-lattice layer of element "Mo" is sandwiched between two nearby "Se" sub-lattice layers. In a perfect $MoSe_2$ monolayer, "Mo" and "Se" atoms alternately connect each other, forming only Mo-Se bonds [77].

A few of the physical properties of $MoSe_2$ and WSe_2 such as melting point, lattice parameters, electron concentration, etc., are summarized in Table 6.5.

Fig. 6.26 Structure of monolayer MoS_2 [71]

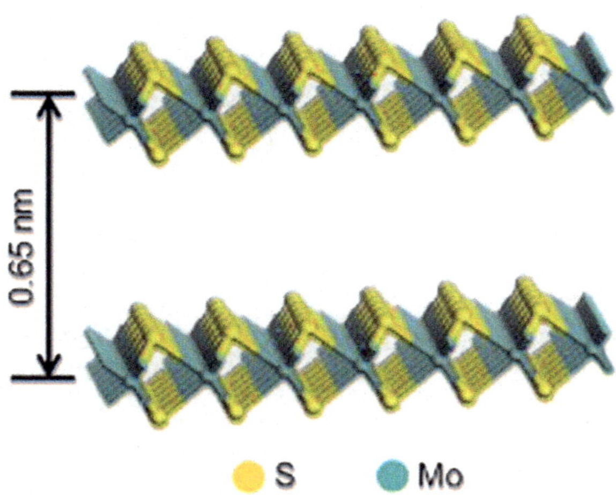

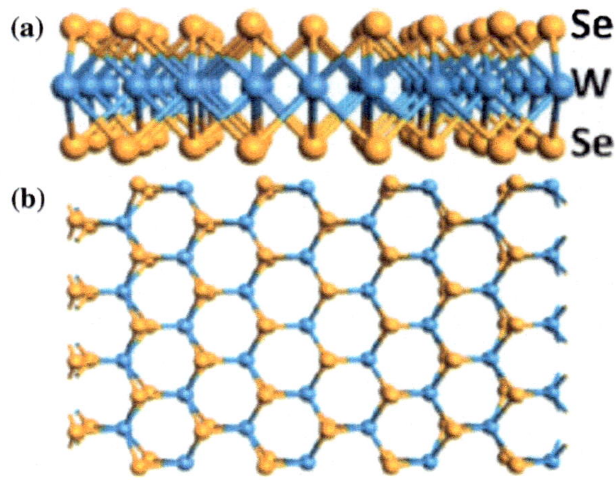

Fig. 6.27 Schematic of the crystal structure of monolayer WSe_2, **a** side view and **b** top view [75]

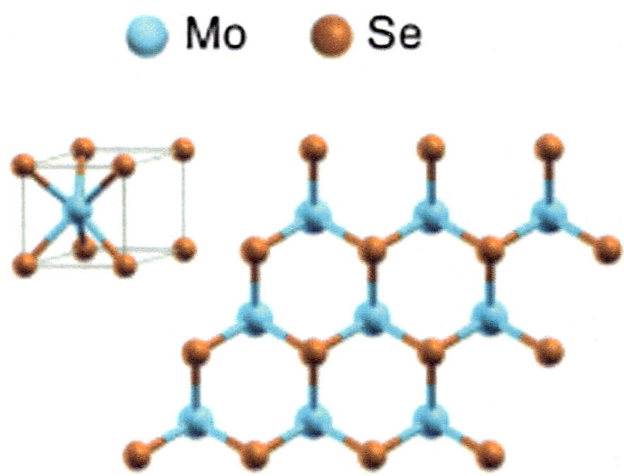

Fig. 6.28 Coordination structure and crystal structure of monolayer $MoSe_2$ [76]

Table 6.5 Physical properties of 2H-$MoSe_2$ and WSe_2

Material	Lattice parameter	Melting point	Energy gap	Effective hole mass	Electron concentration
2-H $MoSe_2$	$a = 3.288$ Å $c = 12.92$ Å	1473 K	1.60 eV (direct) 0.95 eV (indirect)	–	$0.35{-}1.6 \times 10^{17}$ cm^{-3} single crystal
2-H WSe_2	$a = 3.282$ Å $c = 12.937$ Å	1773 K	1.73 eV (direct) 1.33 eV (indirect)	0.01 m_0	1.25×10^{16} cm^{-3} single crystal
References	[78]	[79]	[80]	[81]	

6.8 Electronic properties of $MoSe_2$ and WSe_2

6.8.1 General Considerations

Understanding the electronic properties of a material opens a gateway to other properties that are exhibited by the material. It is important to know the bandgap, bandgap transitions, lattice structure and lattice parameters of the material. Based on the results obtained, we can further posit on its potential applications in various fields of science. In order to know if a material is a conductor, semiconductor or insulator, we must know the position/location of the bandgaps and the valleys present in it. The examination of quantum mechanical wave functions of a particular electron in a vast lattice of atoms or molecules forms the basis for determining the band structure.

In a band structure, the wave function is denoted with a quantum number "*n*" and the wave vector "*k*." Each value of "*k*" has a discrete spectrum of states, labeled by band index "*n*," the energy band numbering. The number of bands, in a band structure diagram, is equal to the number of atomic orbitals in the unit cell [82]. The overlap integral determines the width of the band. This is the difference in the energy between the lowest and highest points in a band. The greater the overlap between neighboring unit cells, the greater is the bandwidth and vice versa [82]. The wave vector "*k*" can take any value within the Brillouin zone. All the points in a Brillouin zone can be classified using the symmetry of the reciprocal lattice. Symmetric points or Lifschitz points [83], also called special/specific high symmetry points, are those points which remain fixed or transform into an equivalent one under a symmetry operation of the Brillouin zone. These points play a specific role in solid-state physics: (a) if two "*k*" vectors can be transformed into each other due to some set of symmetry elements, electronic energies at those *k*-vectors must be identical; (b) "wave functions can be expressed in a form such that they have definite transformation properties under symmetry operations of the crystal" [84]. Similarly, we can define symmetric lines and planes in the Brillouin zone. Customarily, Greek letters denote high symmetry points and lines inside the Brillouin zone while Roman letters denote those on the surface. The center of a Brillouin zone is always denoted by the Greek letter "Γ." The behavior of electrons in a solid can be studied microscopically from its electronic band structure [84].

There are various approaches to determine the band structure of the material. Some of them include the following: Density Functional Theory (DFT), Green—Kubo relations, Linear Density Approximations (LDAs), and so on. The Density Functional Theory is a computational quantum mechanical modeling method used in physics, chemistry and materials science to investigate the electronic structure (principally the ground state) of many-body systems, in particular atoms, molecules, and the condensed phases. Using this theory, the properties of a many-electron system can be determined by using functionals such as spatially dependent electron density. It is among the most popular and versatile methods available in condensed matter physics, computational physics, and computational chemistry.

In this section, the electronic properties of $MoSe_2$ and WSe_2 have been discussed. The electronic properties include electronic band structure, optical bandgap and the influence of temperature on the bandgap.

6.8.2 Electronic Band Structure of $MoSe_2$

The presence of a direct bandgap in monolayer $MoSe_2$ and WSe_2 makes them good candidates for applications in electronics and optoelectronics. Monolayer $MoSe_2$ has a direct bandgap at monolayer thickness, but the bandgap changes to an indirect bandgap with the increase in thickness of the monolayer. Hence, bulk $MoSe_2$ and WSe_2 have an indirect bandgap. This apparent transition in the bandgap causes a variation in the optical properties of monolayer and bulk TMDCs. The band structures presented in this study are a literature survey of existing research work on the electronic properties of TMDCs. Most of the simulations and modeling are done using the Linear Density Approximation (LDA) and Density Functional Theory (DFT) techniques.

The electronic structure of a material involves determining the lattice parameters, Brillouin zone identification and calculation of density of states. Lattice parameters of a monolayer and bulk orientations do not differ by much as monolayer TMDC is obtained from mechanical exfoliation of the bulk counterpart [21].

The electronic band structure, presented in Fig. 6.29, shows the results of the simulation by Kumar and Ahluwalia [21]. The Fermi level is set to 0 eV and all the partial density of states are multiplied by 1.5. As observed from the figure, the direct bandgap is found to be 1.58 eV and is close to the theoretical value of 1.60 eV stated in Table 6.5. The direct bandgap is observed at a "K" high symmetry point.

The electronic bands and density of states around -12 to -15 eV are mainly derived from the chalcogen "s" orbitals separated by a large gap from the group of electronic bands and density of states below the bandgap. The region up to 6–8 eV below the bandgap in the valence bands and the region above the bandgap are mainly contributed by "p" orbitals of chalcogens and "d" orbitals of transition metals [21]. The bands on each side of the bandgap originate primarily from the "d" states of transition metals. Strong hybridization between "d" states of metal atoms and s states of chalcogen atoms have been found below the Fermi energy for all the compounds considered in the study [21].

As reported by Kumar and Ahluwalia [21] and a few other papers, a bandgap transition is observed for bulk and monolayer $MoSe_2$. In its bulk phase, $MoSe_2$ is an indirect bandgap semiconductor. The authors also report a blue shift in its indirect bandgap while transitioning from the bulk phase to the monolayer limit. The bandgap of bulk $MoSe_2$ is in the infrared region. For the monolayer, it shifts toward the near-infrared and visible regions. Thus, a change is observed in the electronic properties of $MoSe_2$ from being an indirect bandgap semiconductor in the bulk phase to direct bandgap semiconductor in the monolayer limit [21].

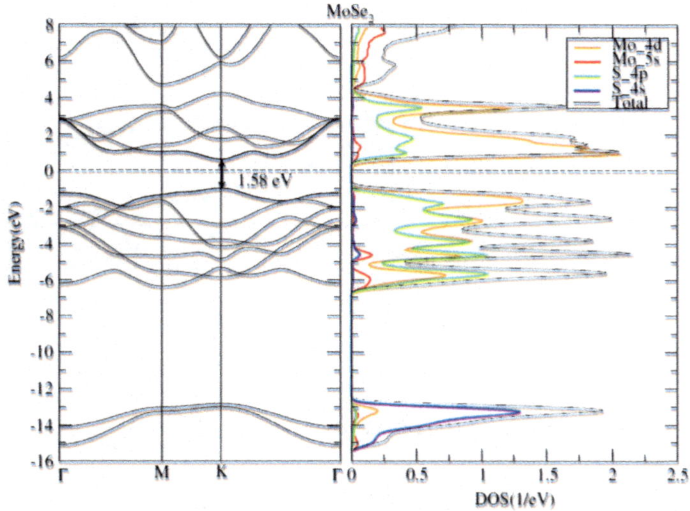

Fig. 6.29 Electronic band structure and corresponding total and partial density of states of 1H-$MoSe_2$ (Monolayer) [21]

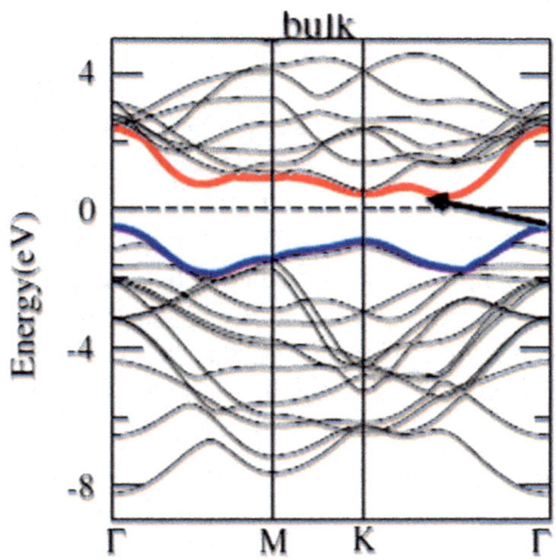

Fig. 6.30 Electronic band structure of 2H-$MoSe_2$ [21]

The electronic band structure of bulk $MoSe_2$ is shown in Fig. 6.30. The arrow points to the lowest value of the indirect or direct bandgap.

As observed in Fig. 6.31, the indirect bandgap of $MoSe_2$ is 0.80 eV [21]. The figure shows the variation in indirect bandgap energies (solid lines with triangles) and direct bandgap energies at the "K" point (solid line with circles). The horizontal dotted line and solid line indicates the direct gap at the "K" point and indirect gap, respectively, for bulk materials.

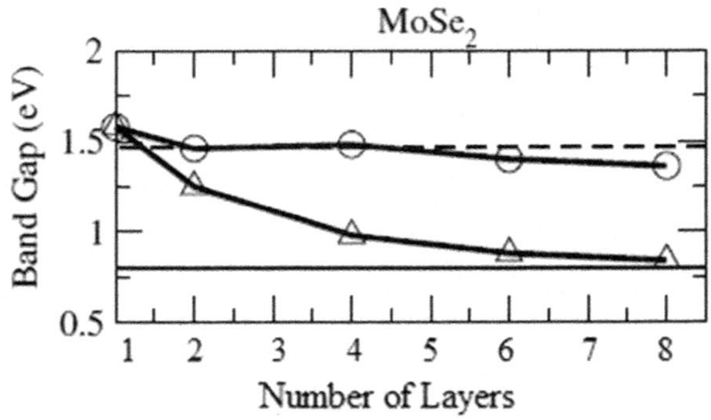

Fig. 6.31 Variation of indirect bandgap energies (solid lines with triangles) and direct bandgap energies at the "K" point (solid lines with circles) for $MoSe_2$ [21]

6.8.3 *Electronic Band Structure of WSe_2*

The evolution in the band structure of WSe_2 involves direct and indirect bandgap transitions. Monolayer WSe_2 has a direct bandgap while bulk WSe_2 as an indirect bandgap due to transition with increasing layer thickness. As mentioned in Table 6.5, its direct bandgap is 1.73 eV and indirect bandgap is 1.33 eV. However, these values differ from author to author and depend on the approximation techniques and theoretical approach used.

Figure 6.32 shows the electronic band structure and density of states of monolayer WSe_2. Linear density approximation was used to compute the electronic band structure. Direct bandgap was observed at 1.61 eV, which is close to the theoretical value of 1.73 eV mentioned in Table 6.5.

From Fig. 6.32, we observe that the valence band maximum and the conduction band minimum are located at the "K" point. The indirect–direct bandgap transitions have been attributed to the missing interlayer interactions in monolayer MX_2 [85]. It has been reported that monolayer WSe_2 has the largest splitting size amongst all the MX_2 semiconductors. Due to the strong spin–orbit coupling and inversion symmetry breaking, a spin-split structure arises, hence contributing to a key property of MX_2 semiconductors [86]. The spin–orbit coupling in tungsten is large compared to that of molybdenum, and hence, monolayer WSe_2 finds applications in the field of spintronics [87].

Figure 6.33 depicts the electronic band structure of bulk WSe_2; the arrow points to the lowest value of indirect or direct bandgap, while the bottom of the CB and top of VB are highlighted with red and blue, respectively [21].

Zhang et al. [87] reported the observation of direct–indirect transition for monolayer, bilayer and trilayer WSe_2. The use of Angle-Resolved Photoemission Spectroscopy (ARPES) enabled them to produce and focus on the band structure spectra of the valence band maximum. The top of the VB at the "Γ" point was slightly larger than the "K" point; this suggested a direct to indirect bandgap transition in WSe_2 between bilayer and trilayer levels [87]. The ARPES spectra also showed that the valence band at the "K" point split into two branches, which had an energy difference of ~475 meV, which was found to be higher than the values found for other monolayer TMDCs [87].

Figure 6.34 shows the variation of indirect bandgap energies and direct bandgap energies for WSe_2 at the "K" point. The horizontal dotted line and solid line indicate the direct gap at "K" point and indirect gap, respectively, for bulk materials. As reported by Kumar and Ahluwalia [21], the indirect bandgap energies in the bulk are blue-shifted relative to the direct bandgap in the monolayer limit. This leads to the tunability of the electronic bandgap [21].

6.8.4 Temperature Dependence of Energy Gap of Monolayer $MoSe_2$ and WSe_2

The energy gap of a semiconductor depends on temperature. In most semiconductors, the energy gap decreases with increase in temperature. There is an increase in atomic spacing due to increase in lattice vibrations because of an increase in thermal energy. Hence, an increased interatomic spacing decreases the potential seen by the electrons in the material, which in turn reduces the size of the energy gap [88].

The temperature dependence of the energy gap of $MoSe_2$ and WSe_2 were determined by using Eq. 6.1.

Fitting parameters, based on Eq. 6.1, are shown in Table 6.6. Monolayer $MoSe_2$ and WSe_2 were considered in this study.

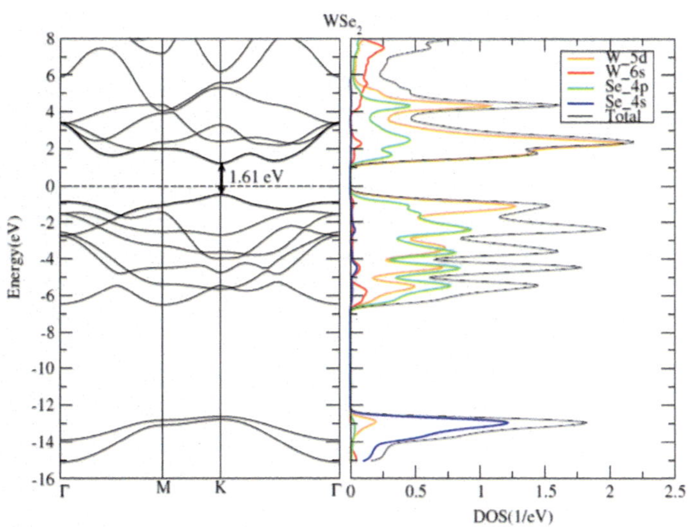

Fig. 6.32 Electronic band structure and corresponding total and partial density of states of 1H-WSe_2 [21]

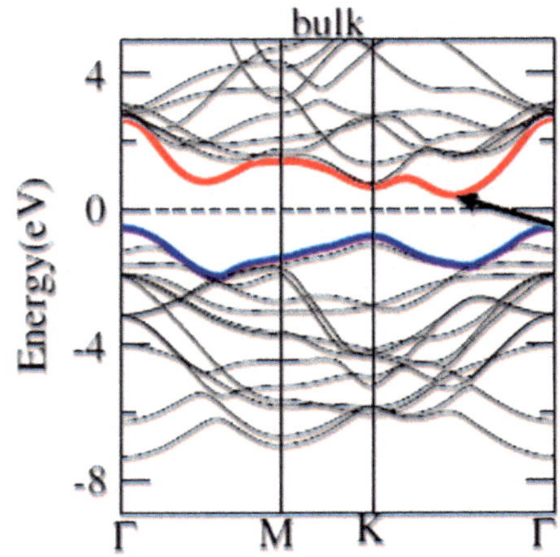

Fig. 6.33 Electronic band structure of bulk WSe_2 [21]

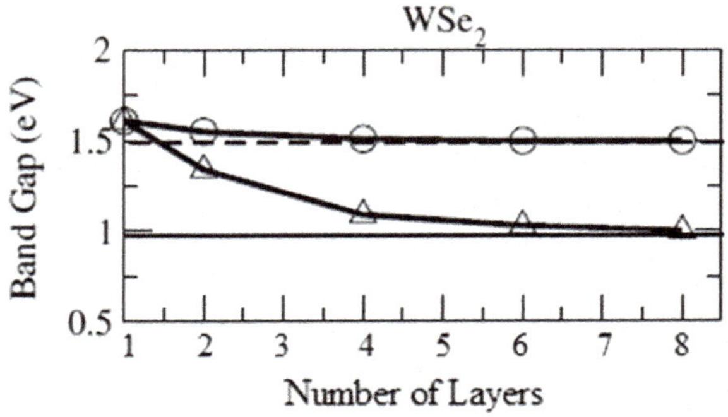

Fig. 6.34 Variation of indirect bandgap energies (solid lines with triangles) and direct bandgap energies at the "K" point (solid lines with circles) for WSe_2 [21]

Figure 6.35 shows the variation of bandgap with temperature while Fig. 6.36 shows the dE_g/dt plot of monolayer $MoSe_2$ and WSe_2. As observed from Fig. 6.35, the energy gap decreases with increase in temperature, which is generally the case for most semiconductors. Figure 6.36 shows that (dE_g/dT) of both materials is nonlinear with temperature; it is negative and decreases with increase in temperature.

Investigation of the monolayer and bulk electronic structures of $MoSe_2$ and WSe_2, in this section, provide an insight into the possible applications of these materials in the field of nanoelectronics. The large separation in valleys of k-space due to suppression of intervalley scattering could open the gates to a new field of electronics called valleytronics [90]. Bandgap engineering is an important field in electronics; the ability to control the bandgap of a semiconductor facilitates to create desirable

Table 6.6 Fitting parameters of E_g for monolayer $MoSe_2$ and WSe_2

Material	$E_g(0)$ (eV)	S	$\langle \hbar\omega \rangle$ (meV)	References
$MoSe_2$	1.640	1.93	1.16	[89]
WSe_2	1.742	2.06	1.50	[90]

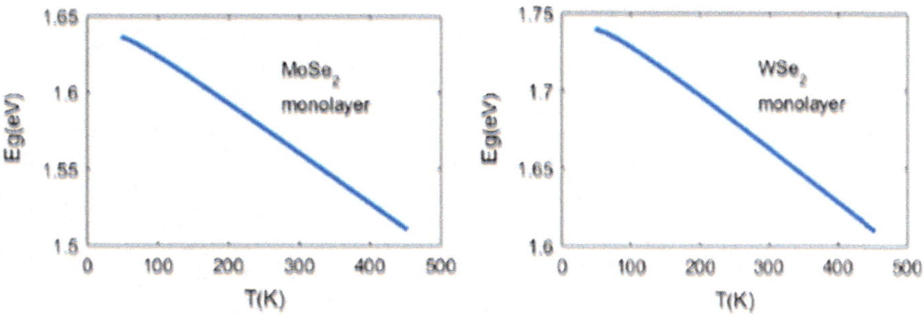

Fig. 6.35 Temperature dependence of monolayer $MoSe_2$ and WSe_2

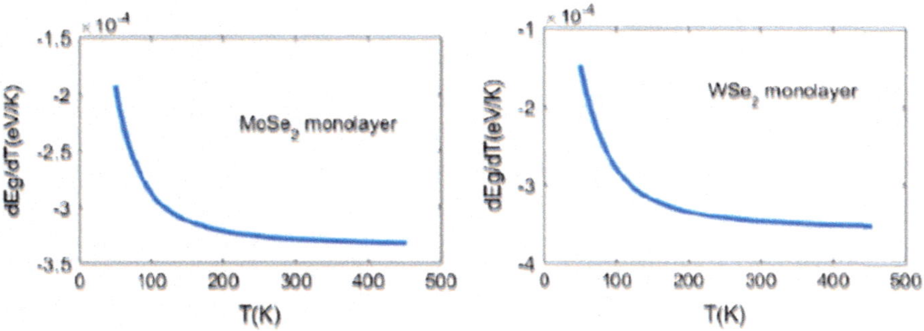

Fig. 6.36 $\left(\frac{\mathrm{d}E_g}{\mathrm{d}t}\right)$ of monolayer $MoSe_2$ and WSe_2

electrical and optical properties. Surface doping can control the band structure of WSe_2; this control permits to alter the size of the gap and the direct–indirect bandgap transition [87].

6.9 Optical Properties of $MoSe_2$ and WSe_2

6.9.1 Introduction

$MoSe_2$ and WSe_2 are direct bandgap semiconductors at monolayer thickness and indirect bandgap semiconductors at bulk. The bandgaps of these materials are in the visible region of the spectra (400–700 nm). Li et al. [91] have reported their studies of the dielectric constants for monolayer $MoSe_2$ and WSe_2 at room temperature (ε_1 and ε_2) from the experimental reflectance spectra by a constrained Kramers–Kronig analysis. These values have been used to calculate the photon energy-dependent refractive index and extinction coefficient. As reported by Mukherjee et al. [92], the values determined by Li et al. [91] were found to be better than the set of values determined by other authors and hence were chosen to analyze the optical properties.

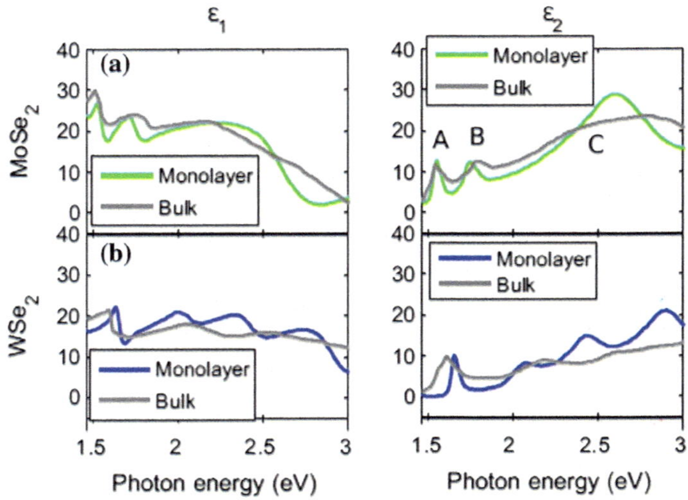

Fig. 6.37 Dielectric functions of monolayer and bulk $MoSe_2$ and WSe_2 [91]

6.9.2 Optical Constants of $MoSe_2$ and WSe_2

The optical constants of $MoSe_2$ and WSe_2 have been simulated using MATLAB. Figure 6.37 shows the "ε_1" and "ε_2" values reported by Li et al. [91]. These values were used to obtain the optical constants "n" and "k" in this work. Monolayer thickness of $MoSe_2$ and WSe_2 were taken as 0.70 nm [77], whereas bulk thickness of $MoSe_2$ and WSe_2 were taken as 20 nm, which seem to be the optimal value for the TMDCs [74].

Figures 6.38 and 6.39 show the simulated values of "n" and "k" for monolayer and bulk $MoSe_2$ and WSe_2.

The maximum value of refractive index (n) and the corresponding energy (E) for monolayer $MoSe_2$ is 5.20 (1.52 eV) and for bulk $MoSe_2$, it is 5.49 (1.51 eV). The maximum value of extinction coefficient (k) and the corresponding energy (E) for monolayer $MoSe_2$ is 3.29 (2.70 eV) and for bulk $MoSe_2$ is 3.04 (3.00 eV).

The maximum value of refractive index (n) and the corresponding energy (E) for monolayer WSe_2 is 4.72 (1.63 eV) and for bulk WSe_2 is 4.69 (1.62 eV). The maximum value of extinction coefficient (k) and the corresponding energy (E) for monolayer WSe_2 is 2.60 (2.94 eV) and for bulk WSe_2, it is 1.56 (3.00 eV).

The difference in the maximum values of n between monolayer and bulk is significant and hence we can conclude that a change in the thickness of the material does affect the refractive index. A similar trend is observed for the extinction coefficient and hence we conclude that the thickness of the material affects the extinction coefficient also. These values are in agreement with the values reported by Zhang et al. [93] and are very close to the values presented by Liu et al. [94]. In addition, the energies corresponding to the maximum refractive index differ significantly between the bulk

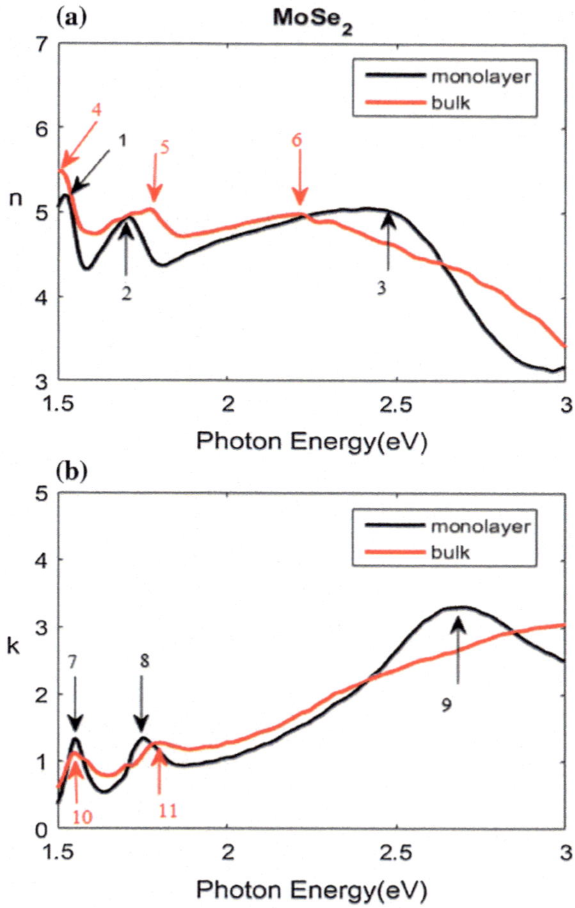

Fig. 6.38 Refractive index and extinction coefficient as a function of photon energy for suspended monolayer and bulk $MoSe_2$

and monolayer. It should be noted that the above analysis is based on the range of photon energies, 1.5–3.0 eV, considered in this study. The energies corresponding to the features in the variations of *n* and *k*, with energy, for monolayer and bulk $MoSe_2$ and WSe_2 are summarized in Table 6.7.

6.9.3 Optical Properties of Suspended Monolayer and Bulk $MoSe_2$ and WSe_2

This section of the chapter focuses on the simulation of optical properties such as reflectance, transmittance, and absorptance of suspended monolayer and bulk $MoSe_2$ and WSe_2.

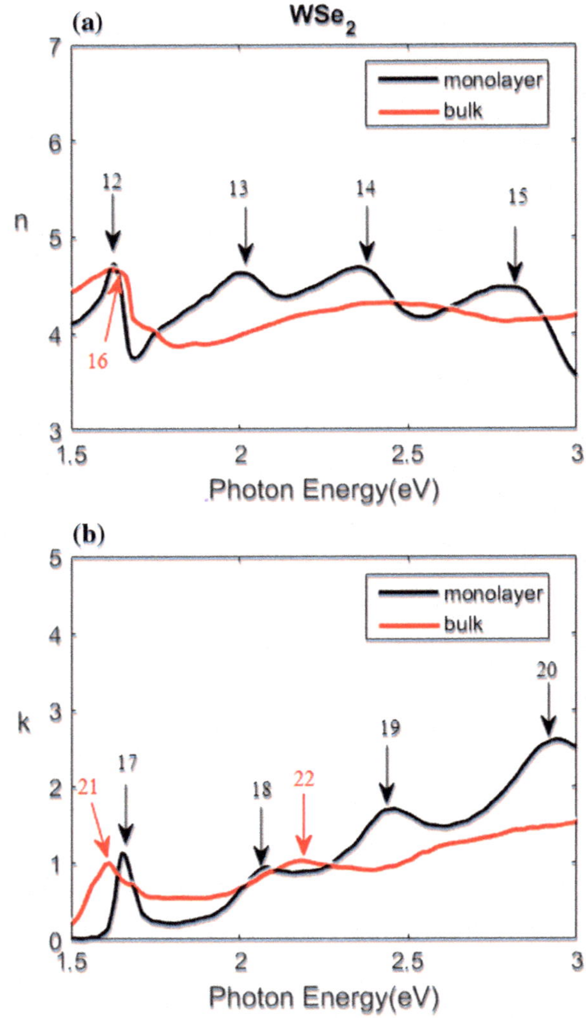

Fig. 6.39 Refractive index and extinction coefficient as a function of photon energy for suspended monolayer and bulk WSe_2

Figures 6.40 and 6.41 show the simulated *R*, *T*, and *A* of suspended monolayer and bulk $MoSe_2$ and WSe_2 at room temperature and normal incidence. The absorptance spectra of monolayer and bulk are separated due to the large change in their magnitude for monolayer and bulk. This change is reasonable due to the increase in the number of layers which leads to increase in absorptance by the material. For these materials, the two lowest energy peaks in the reflectance spectra correspond to the excitonic features that are associated with interband transitions in the K (K′) point in the Brillouin zone [28]. The two significant peaks in Fig. 6.40 can be attributed to the splitting of the valence bands by spin–orbit coupling [95]. At higher photon energies, a spectrally broad response is observed from higher lying interband transitions, including the transitions near the Γ point [28].

Table 6.7 Energies corresponding to features of $n(E)$ and $k(E)$ for suspended TMDCs

Material	No.	E (eV)	n	No.	E (eV)	k
$MoSe_2$ monolayer	1	1.52	5.20	7	1.55	1.33
	2	1.71	4.94	8	1.75	1.33
	3	2.41	5.05	9	2.70	3.30
$MoSe_2$ bulk	4	1.51	5.49	10	1.55	1.12
	5	1.77	5.04	11	1.80	1.27
	6	2.21	4.99			
WSe_2 monolayer	12	1.63	4.72	17	1.65	1.12
	13	2.10	4.47	18	2.08	0.94
	14	2.36	4.68	19	2.46	1.69
	15	2.77	4.69	20	2.94	2.60
WSe_2 bulk	16	1.63	4.73	21	1.61	1.03
				22	2.18	1.06

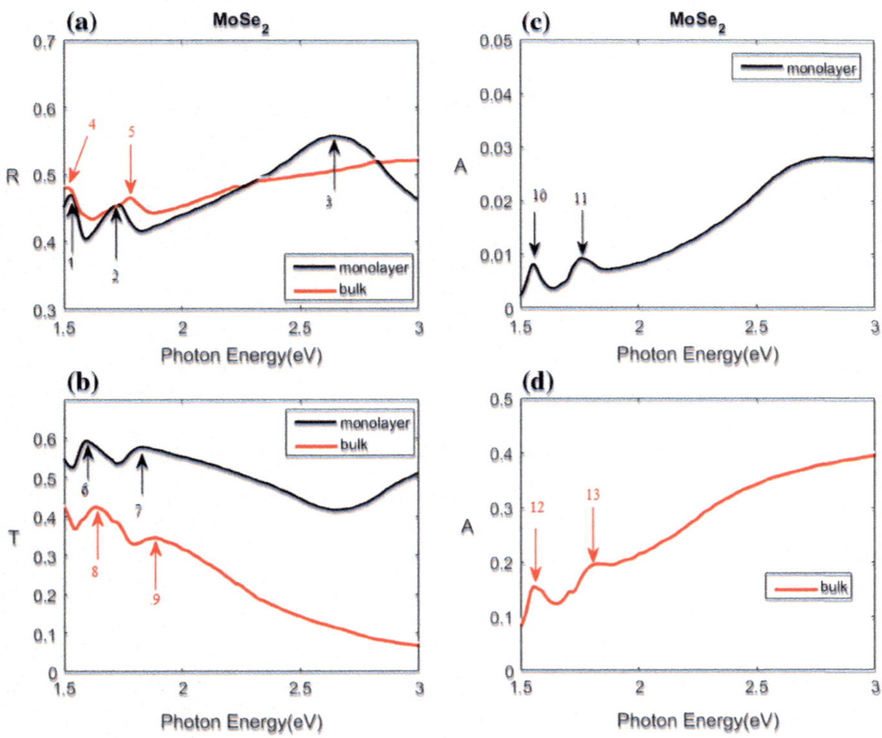

Fig. 6.40 Simulated R, T, and A of suspended monolayer and bulk $MoSe_2$

The maximum value of Reflectance (R) and the corresponding energy (E) for monolayer $MoSe_2$ is 55.7% (2.64 eV). Similarly, the maximum value of (R) for

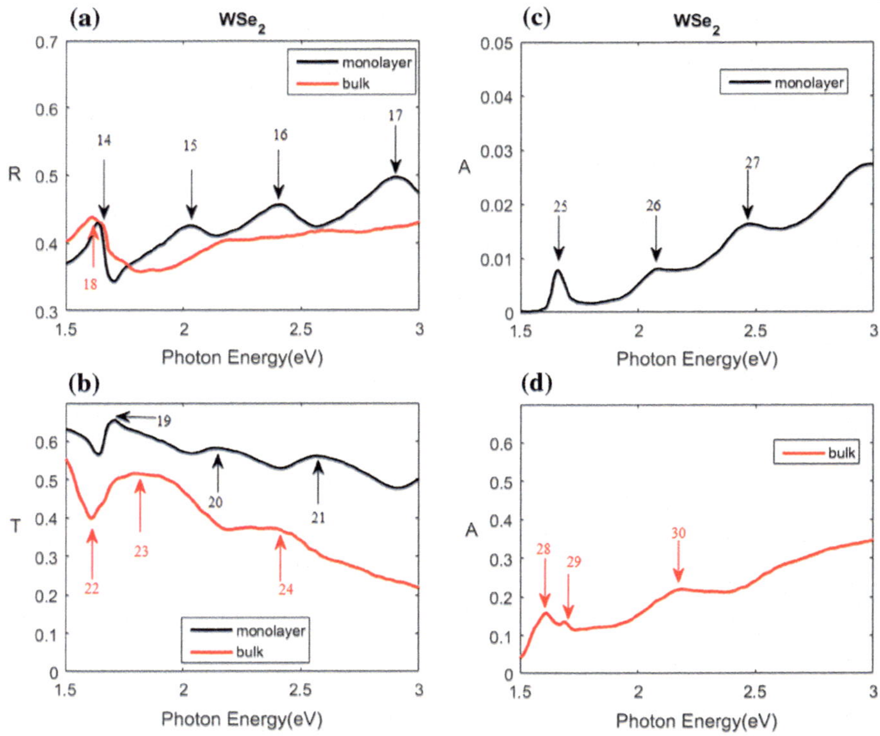

Fig. 6.41 Simulated *R*, *T*, and *A* for suspended monolayer and bulk WSe_2

bulk $MoSe_2$ is 52.6% (2.98 eV). The maximum value of Transmittance (*T*) and the corresponding energy (*E*) for monolayer $MoSe_2$ is 59.2% (1.60 eV). Similarly, the maximum value of (*T*) for bulk $MoSe_2$ is 43.1% (1.63 eV). The maximum value of Absorptance (*A*) and the corresponding energy (*E*) for monolayer $MoSe_2$ is 0.91% (1.76 eV). Similarly, the maximum value of (*A*) for bulk $MoSe_2$ is 20.2% (1.82 eV). From Fig. 6.40c, d, we observe that the location of peaks, in either case, remains relatively similar; these peaks are the A and B exciton absorption peaks, which originate from the spin-split direct gap transitions at the K point of the Brillouin zone.

The maximum value of Reflectance (*R*) and the corresponding energy (*E*) for monolayer WSe_2 is 49.7% (2.90 eV). Similarly, the maximum value of (*R*) for bulk WSe_2 is 43.9% (1.61 eV). The maximum value of Transmittance (*T*) and the corresponding energy (*E*) for monolayer WSe_2 is 65.4% (1.71 eV). Similarly, the maximum value of (*T*) for bulk WSe_2 is 51.9% (1.79 eV). The maximum value of Absorptance (*A*) and the corresponding energy (*E*) for monolayer WSe_2 is 1.62% (2.47 eV). Similarly, the maximum value of (*A*) for bulk WSe_2 is 22.2% (2.19 eV). All the energy values corresponding to the features, present in Figs. 6.40 and 6.41, are summarized in Table 6.8.

Table 6.8 Energies corresponding to features of *R*, *T*, and *A* for suspended TMDCs

Material	No.	*E* (eV)	*R*	No.	*E* (eV)	*T*	No.	*E* (eV)	*A*
$MoSe_2$ ML	1	1.53	0.469	6	1.60	0.592	10	1.56	0.00792
	2	1.72	0.457	7	1.83	0.577	11	1.76	0.00911
	3	2.64	0.557						
$MoSe_2$ bulk	4	1.51	0.484	8	1.63	0.432	12	1.55	0.159
	5	1.78	0.469	9	1.88	0.353	13	1.82	0.202
WSe_2 ML	14	1.64	0.430	19	1.71	0.654	25	1.66	0.00773
	15	2.03	0.425	20	2.14	0.582	26	2.09	0.00798
	16	2.41	0.456	21	2.56	0.561	27	2.47	0.01620
	17	2.90	0.497						
WSe_2 bulk	18	1.61	0.439	22	1.61	0.400	28	1.61	0.160
				23	1.79	0.519	29	1.69	0.137
				24	2.29	0.378	30	2.19	0.222

6.9.4 Optical Properties of Monolayer, Bulk $MoSe_2$, and WSe_2 on Various Substrates

In recent years, a large volume of research is being done on the applications of TMDCs as heterojunction and heterostructure devices. By exploring the optical properties of these materials on various substrates such as gold, silicon, and fused silica, potential applications of these materials in optoelectronics has been established. In this study, the thickness of silicon and fused silica substrate was assumed to be ~650 μm and the thickness of gold substrate was assumed to be ~10 μm.

6.9.4.1 Optical Properties of TMDCs on Gold Substrate

Figure 6.42 shows the simulated reflectance and absorptance spectra of monolayer and bulk $MoSe_2$ and WSe_2 on a gold substrate.

The reflectance tends to decrease with increase in photon energy, while the absorptance increases with the increase in photon energy. Due to the thickness of gold considered in the simulations, the transmittance of these multilayers is ~0 in this energy range; thus, reflectance and absorptance are complementary to each other.

As observed in the figures, the change in thickness of TMDC does not necessarily cause a significant change in the reflectance and absorptance of the TMDC/Au. No sharp peaks were observed in the reflectance and absorptance spectra. Maximum values of reflectance and absorptance for monolayer TMDC/Au and bulk TMDC/Au remained approximately the same, with maximum reflectance of ~90% and maximum absorptance of ~70%. It is to be noted within this context that, for the wavelength range considered in this study, the transmittance of gold is ~0 for film thickness of 10 μ.

6.9.4.2 Optical Properties of TMDCs on Silicon Substrate

Figure 6.43 shows the simulated reflectance and absorptance spectra of monolayer and bulk $MoSe_2$ and WSe_2 on a silicon substrate.

As observed in Fig. 6.43, the maximum values of Reflectance (*R*) for monolayer $MoSe_2$ and WSe_2 on silicon are as follows: 14.76% (2.84 eV) and 10.84%, respectively. Similarly, the maximum value of (*R*) for bulk $MoSe_2$ and WSe_2 on silicon are 14.94 and 3.80%, respectively. Unless otherwise specified, the maximum values of *R* occur at 3 eV. The maximum value of Absorptance (*A*) for monolayer $MoSe_2$ and WSe_2 on silicon are as follows: 98.2% (1.61 eV) and 99.84% (1.73 eV), respectively. The maximum value of *A* for bulk $MoSe_2$ and WSe_2 on silicon are 97.6% (1.62 eV) and 99.44% (1.90 eV), respectively. The average maximum value of reflectance has a very low value as compared to that of TMDC/Au. The absorptance of TMDC/Si is

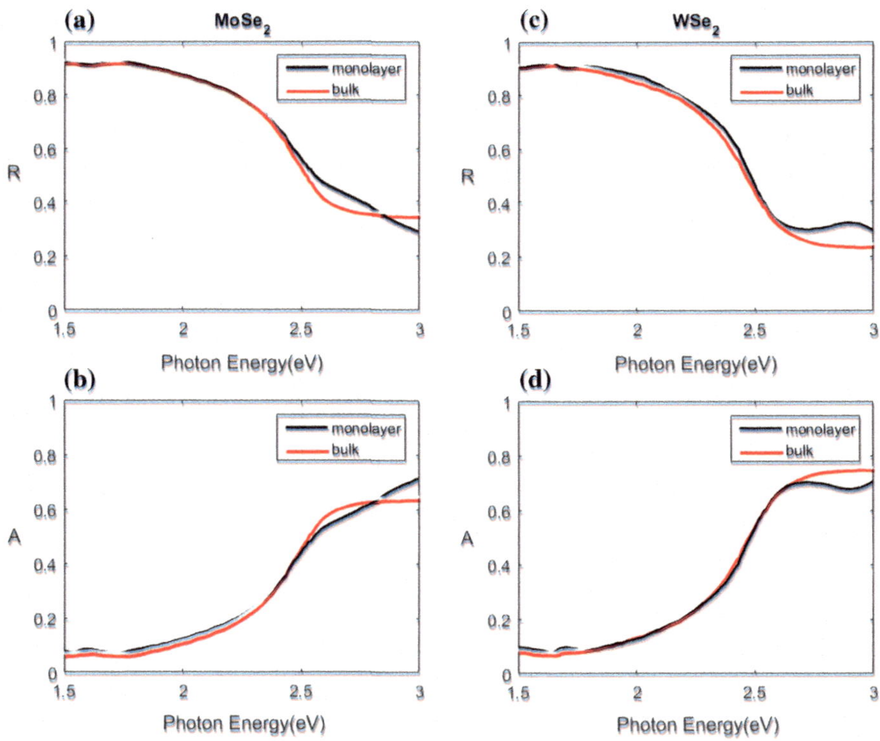

Fig. 6.42 Simulated reflectance and absorptance spectra of monolayer and bulk TMDCs/Au

very high compared to that of TMDC/Au. Similar to TMDC/Au, the transmittance of TMDC/Si is ~0 due to the large thickness of the silicon wafer.

6.9.4.3 Optical Properties of TMDCs on Fused Silica Substrate

Figures 6.44 and 6.45 show the simulated reflectance, transmittance, and absorptance spectra of monolayer and bulk $MoSe_2$ and WSe_2 on a fused silica substrate.

It is observed that the transmittance of TMDC is substantially high which could be attributed to the transparency of fused silica and TMDCs. As observed in Fig. 6.44, the maximum value of Reflectance (R) for monolayer $MoSe_2$ and WSe_2 on fused silica are as follows: 42.74% (2.65 eV) and 35.94% (2.90 eV), respectively. The maximum value of (R) for bulk $MoSe_2$ and WSe_2 on fused silica are 39.41% (3.00 eV) and 29.76% (1.61 eV), respectively. The maximum value of Transmittance (T) for monolayer $MoSe_2$ and WSe_2 on fused silica are, respectively, as follows: 73.37% (1.60 eV) and 80.26% (2.07 eV). The maximum value of "T" for bulk $MoSe_2$ and WSe_2 on fused silica at 1.50 eV, are 54.75 and 68.86%, respectively.

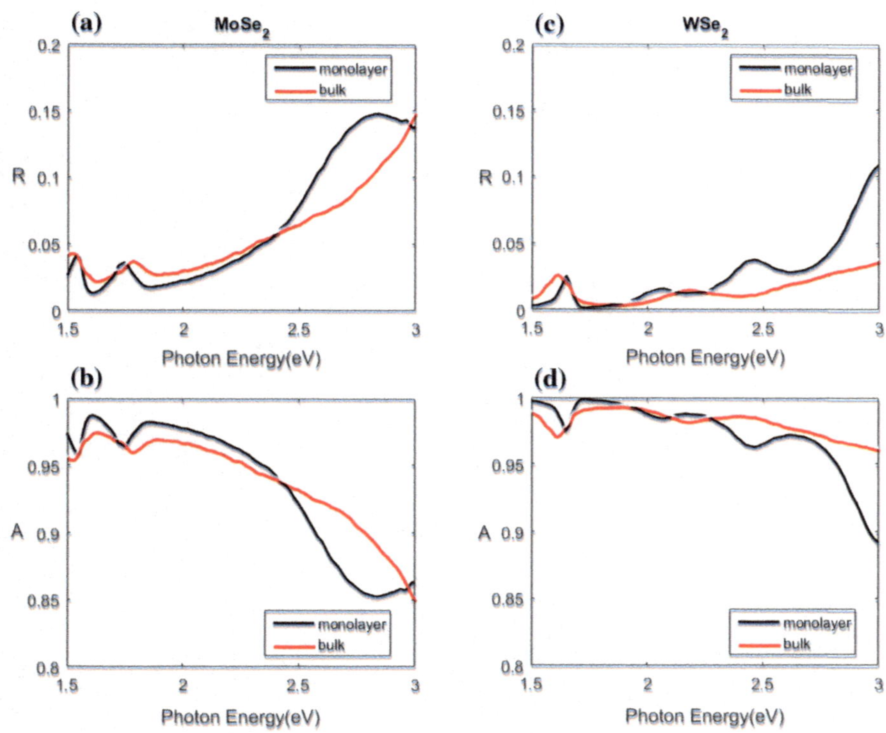

Fig. 6.43 Simulated reflectance and absorptance spectra of monolayer and bulk TMDCs/Si

As seen from Fig. 6.45, the maximum value of Absorptance (A) for monolayer $MoSe_2$ and WSe_2 on fused silica are 3.57 and 3.43%, respectively. The maximum value of A for bulk $MoSe_2$ and WSe_2 on fused silica are 51.06 and 43.7%, respectively. These values are almost in accordance with the optical properties obtained for suspended monolayer and bulk $MoSe_2$ and WSe_2. However, the presence of a substrate leads to a change in magnitude of R, T, and A for a multilayer case.

6.9.5 Optical Bandgap of Monolayer $MoSe_2$ and WSe_2

The nature of the bandgap, direct or indirect, is generally determined by absorption [96]. While a variety of approaches to the determination of energy gap of semiconductors are discussed in the literature [97], the various functional forms of the spectral dependence of the absorption coefficient have been utilized to determine the value of the bandgap as well as its nature—direct or indirect. If a plot of α^2 versus $\hbar\nu$ leads to a straight line, it is deemed a direct bandgap. The bandgap is evaluated

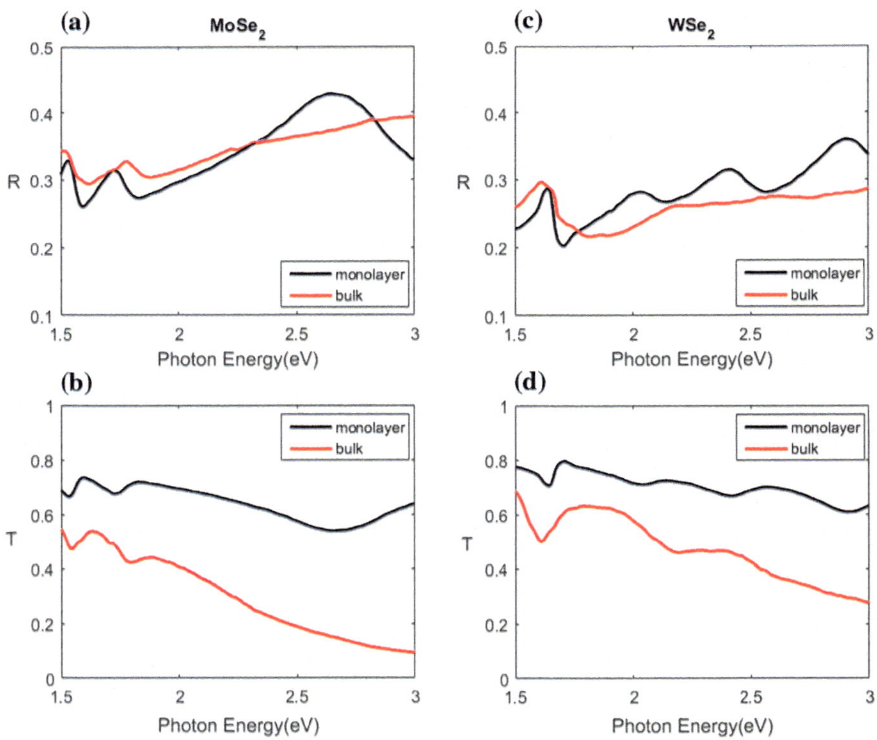

Fig. 6.44 Simulated reflectance and transmittance spectra of monolayer and bulk TMDCs on fused silica

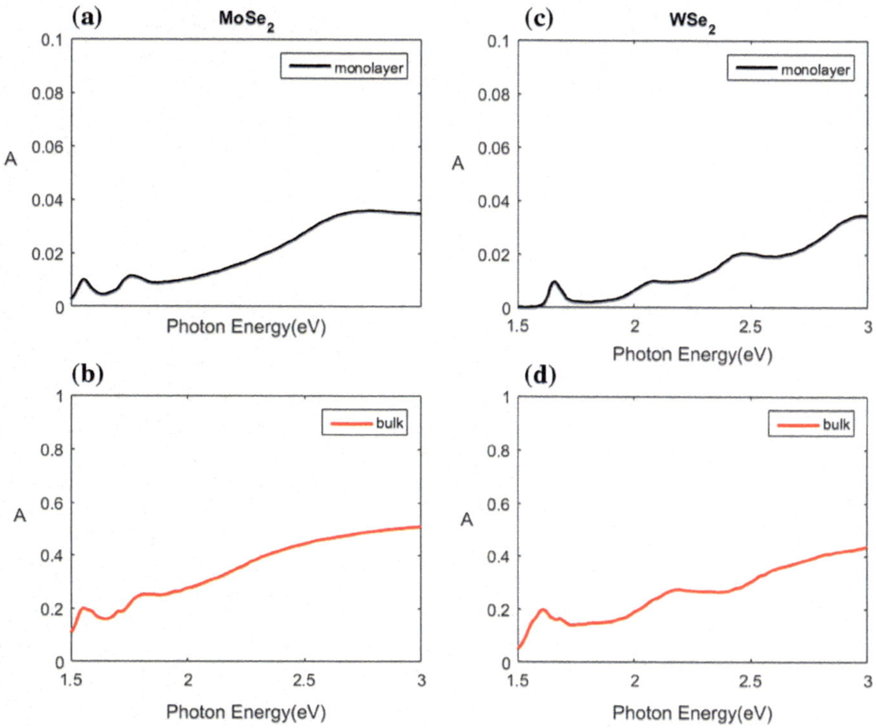

Fig. 6.45 Simulated absorptance spectra of monolayer and bulk TMDCs on fused silica

by extrapolating the α^2 versus $\hbar\nu$ straight line to $\alpha = 0$ axis. However, a plot of $\alpha^{1/2}$ versus $\hbar\nu$, leading to a straight line, is inferred as an indirect bandgap. This indirect bandgap is estimated by extrapolating the $\alpha^{1/2}$ versus $\hbar\nu$ straight line to $\alpha = 0$ axis.

In the earlier sections, the electronic bandgap of $MoSe_2$ and WSe_2 was discussed. The optical bandgap values, determined in this study, are smaller than the reported electronic bandgap values. This is due to the additional energy absorbed by an electron while making a transition from the valence band to the conduction band. This leads to a difference in the Coulomb energies of the two systems (excitation spectroscopy and tunneling spectroscopy) which therefore causes changes in the optical bandgaps and electronic bandgaps.

As seen from Fig. 6.46, a straight line is observed which corresponds to the first peaks of both the monolayer TMDCs; this shows that the monolayer TMDCs have direct bandgaps. The values of the optical bandgap of monolayer TMDCs were obtained by solving the equations to the straight lines. The optical bandgaps of monolayer $MoSe_2$ and WSe_2 were 1.51 and 1.62 eV, respectively.

Fig. 6.46 Optical bandgap of monolayer $MoSe_2$ and WSe_2

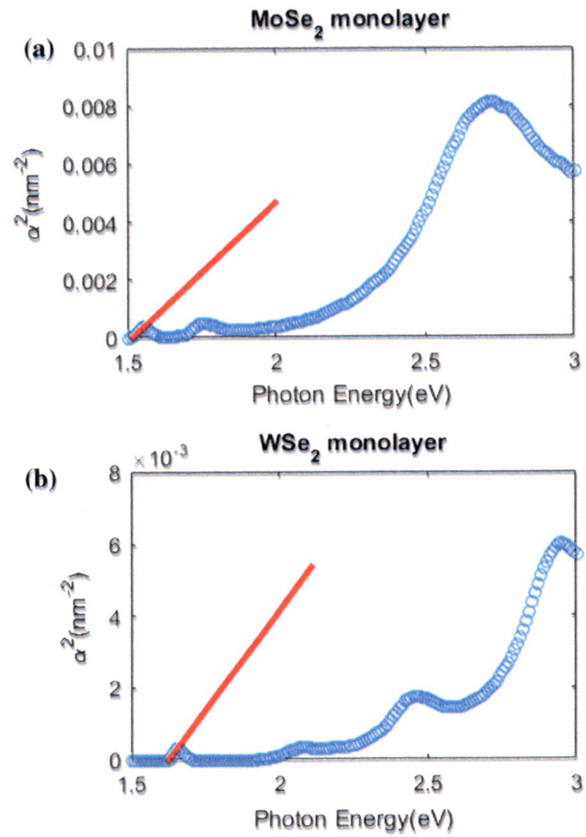

6.10 Electrical Properties of $MoSe_2$ and WSe_2

The bandgap analysis and temperature dependence study, discussed in the earlier Section, helps to understand the semiconductor behavior of $MoSe_2$ and WSe_2 at a fundamental level. The bandgap transitions in these monolayer and bulk TMDCs make them suitable candidates for applications in the semiconductor industry. This section primarily focuses on the electrical transport properties of $MoSe_2$ and WSe_2.

6.10.1 *Electrical Properties of $MoSe_2$ and WSe_2*

The electrical properties such as electron mobility, Hall effect, Hall coefficient, sheet resistance, etc., are very similar in $MoSe_2$ and WSe_2. This section is a compilation of experimental results and computational methods discussed in the literature. Brumme et al. [98], has presented a study of the structural, electronic, and transport properties of $MoSe_2$ and WSe_2 and examined these properties under field-effect doping for $MoSe_2$ and WSe_2 [98]. By using a theoretical approach to simulate and model the doping in field-effect devices, Brumma et al. applied this approach to the H polytypes of $MoSe_2$ and WSe_2 in addition to other TMDCs [98].

6.10.1.1 Effect of Doping

A FET setup is modeled to analyze the quantum capacitance and effect of doping on the band structure and electrical properties of TMDCs. Figure 6.47 depicts this FET structure and includes the equivalent circuit for overall capacitance seen at the gate electrode.

Doping was performed via the FET setup and it had a minor influence on the structure of $MoSe_2$ and WSe_2. Weak polarity of the bonds between the transition metal atom and chalcogen atom account for the minor influence on the structure [98]. The largest change in the polarity was determined for the thickness of the layer closest to the charged plane representing the gate electrode [98]. The layer thickness increased by ≈0.06 Å for a large electron doping of $n = -0.3$ e/unit cell ($n \approx -3.16 \times 10^{14}$ cm^{-2}) and decreased by ≈0.02 Å for a large hole doping of $n = +0.3$ e/unit cell. This change was mainly due to the increase/decrease of the chalcogen–transition metal bond length of those being closest to the gate. Bond lengths of ≈+0.04 Å for $n = -0.3$ e/unit cell and ≈−0.02 Å for $n = +0.3$ e/unit cell were reported. Accordingly, there was also a small change in the angle between the first chalcogen, the transition metal, and the second chalcogen of up to +0.9° (−0.4°) for large electron/hole doping [98].

6.10.1.2 Hall Effect Measurements

Figure 6.48 shows the ratio of inverse Hall coefficient to the doping charge concentration as a function of doping for the monolayer $MoSe_2$ and WSe_2. A comparison between curves at $T = 0$ K and $T = 300$ K is also shown. Temperature has a minor influence on the inverse Hall coefficient for $MoSe_2$ and WSe_2. It is interesting to

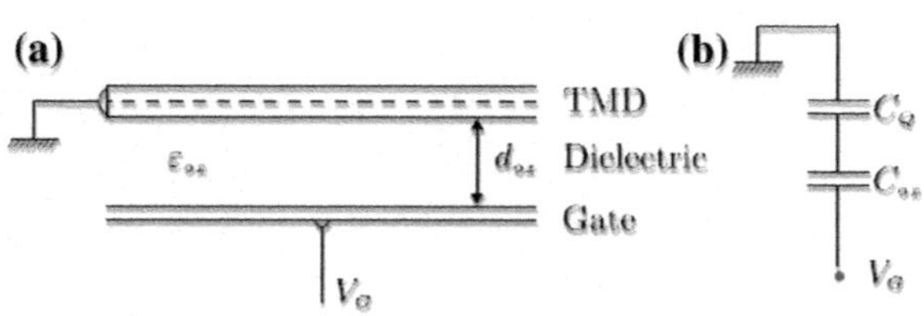

Fig. 6.47 **a** Schematic illustration of an FET setup in which the 2D metallic system is separated from the gate electrode by a dielectric with dielectric constant ε_{ox} of thickness d_{ox}. **b** Equivalent circuit for the overall capacitance seen at the gate electrode [98]

note that a peak is observed at $n \approx 0.1$ e/unit cell for both the materials. However, the inverse Hall coefficient of monolayer $MoSe_2$ was found to be slightly higher than that of monolayer WSe_2. The doping charge concentration calculated using the inverse Hall coefficient was reported to be 1.5 times larger than the real concentration. The deviation in the inverse Hall coefficient from the doping charge n increases with increasing doping of the valence band maximum at K [98].

Figure 6.49 shows the ratio of the inverse Hall coefficient to the doping charge as a function of doping for monolayer, bilayer and trilayers of $MoSe_2$ and WSe_2 at a temperature of $T = 300$ K. As observed from the figure, Monolayer $MoSe_2$ and WSe_2 have higher inverse Hall coefficient at the same levels of hole doping concentrations, i.e., $n \approx +1.4 \times 10^{14}$ cm^{-2} when compared to different layer thicknesses.

6.10.1.3 Resistance Measurements

Thakar et al. [99] have studied the variation of resistance with temperature for bulk $MoSe_2$ and WSe_2 crystals. Figure 6.50 shows the variation of resistance with temperature along the C axis for $MoSe_2$ and WSe_2 crystals.

As seen from Fig. 6.50, the resistance of $MoSe_2$ and WSe_2 crystals decrease exponentially with increase in temperature.

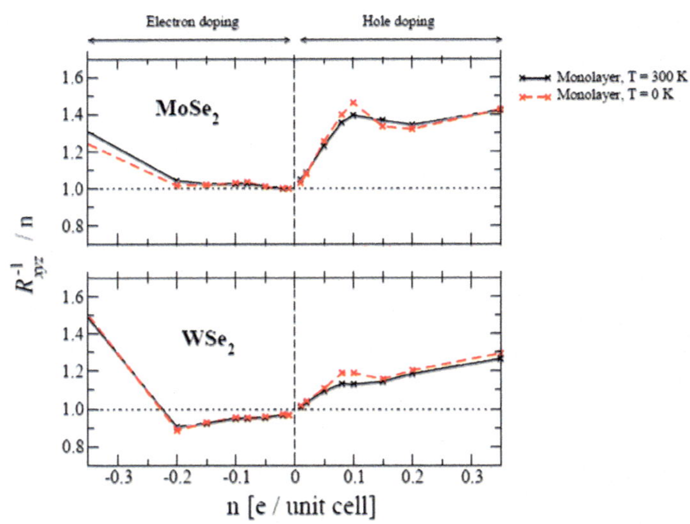

Fig. 6.48 Ratio of the inverse Hall coefficient to the doping charge as function of doping for monolayer $MoSe_2$ and WSe_2 at temperatures $T = 0$ K and $T = 300$ K [98]

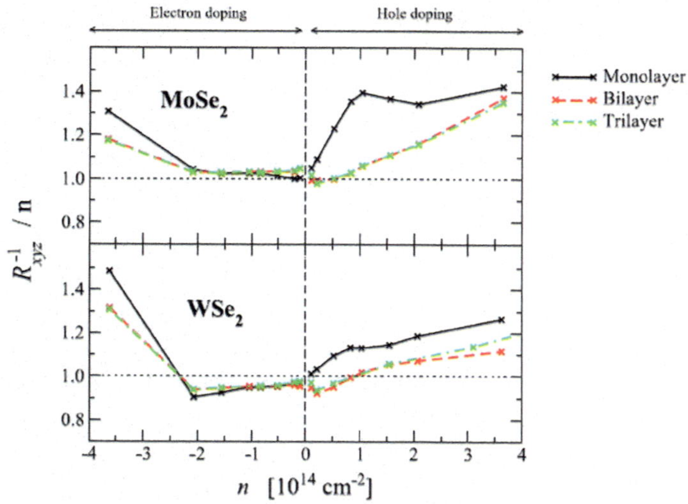

Fig. 6.49 Ratio of the inverse Hall coefficient to the doping charge as a function of doping for monolayer, bilayer, and trilayers of $MoSe_2$ and WSe_2 at a temperature of $T = 300$ K [98]

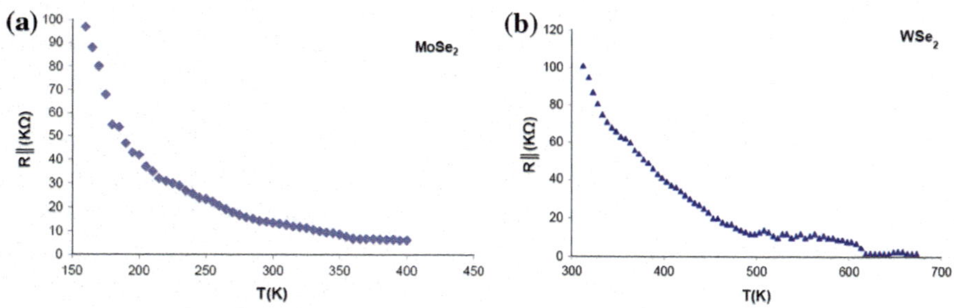

Fig. 6.50 Variation of resistance (along *c* axis) with temperature of **a** $MoSe_2$ crystal and **b** WSe_2 crystal [99]

6.11 Applications of $MoSe_2$ and WSe_2

The electronic, electrical, and optical properties help us to understand the potential applications of $MoSe_2$ and WSe_2 in various applications. The presence of a direct and indirect bandgaps make them ideal candidates for optoelectronic devices, field-effect transistors, heterostructure devices, etc. A few applications of these materials has been discussed in this section.

6.11.1 *Field-Effect Transistors*

Hang et al. [100] report their studies on transistors based on few-layer $MoSe_2$ nanoflakes that were fabricated to study their photoelectrical properties. Investigation of the photoelectrical properties of these structures facilitates to understand their applications as photodetectors. Their performance depends on properties such as high response time, On/Off ratio, and gate modulation [100].

$MoSe_2$ nanoflakes of thickness ~2 nm were mechanically exfoliated on a SiO_2/Si substrate. Earlier reported studies of monolayer and multilayer (bulk) $MoSe_2$ did not give the desired results required to produce an optimum output for photoelectrical applications; hence, few layers of $MoSe_2$ were used in this study. ~0.65 nm was considered to be the monolayer thickness; 2 nm corresponds to trilayers of $MoSe_2$.

Figure 6.51 shows the side elevation schematic of a fabricated back-gated few-layer $MoSe_2$ FET; highly p-doped silicon serves as back gate.

The wavelength of the incident light used in this study are 405, 450, 520, 638, and 785 nm, which correspond to the visible region of the spectra. The incident light intensity is 15 mW for each color [89].

According to this study, the device demonstrated a high On/Off ratio (~10^5) and reasonably high carrier mobility (1.79 $cm^2\ V^{-1}s^{-1}$). The Photoluminescence (PL) spectra (PL peak at 1.51 eV) obtained indicated a strong light–matter interaction in red and near-infrared light [100]. Photo response of the fabricated device was tested and calculated at different laser wavelengths. The device was most sensitive to red light, consistent with the conclusion from the PL spectra. For the 638 nm incident laser, the device shows high photo responsivity, quick response time, high EQE (External Quantum Efficiency), high detection rate and commendable linear working area [100].

Monolayer and few-layer WSe_2 also make good field-effect transistors; the mobility and On/Off ratio reported by Liu et al. [101] was exceptionally high compared to $MoSe_2$ results obtained by Hang et al. [100]. Figure 6.52 shows the schematic of a FET based on WSe_2 grown on BN.

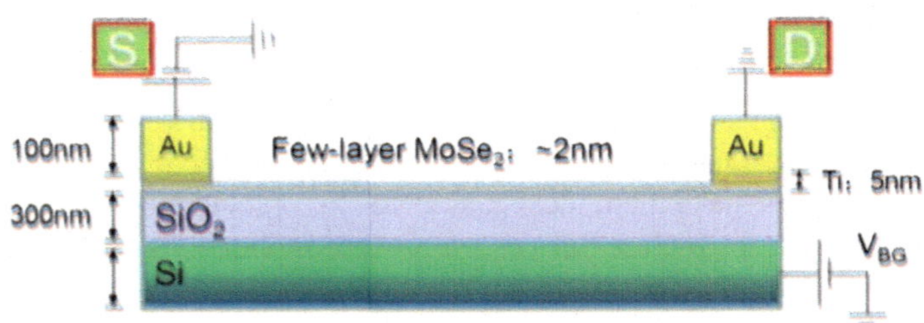

Fig. 6.51 Side elevation of fabricated back-gated few-layer $MoSe_2$ FET, highly p-doped silicon serves as back gate [100]

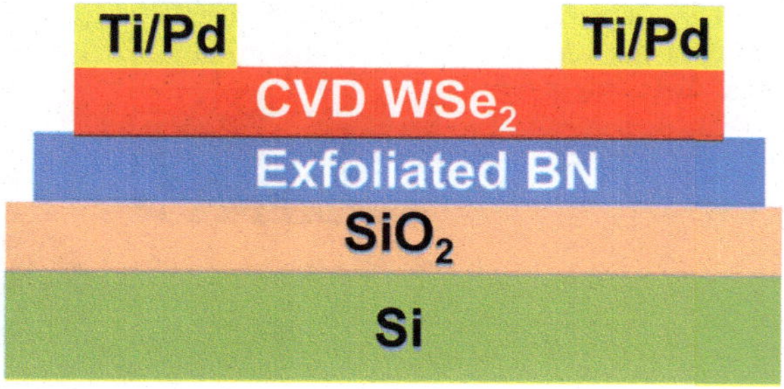

Fig. 6.52 Schematic of a FET based on WSe_2 grown on BN [101]

Boron Nitride (BN) was used as a substrate due to its good dielectric properties and its capability of providing dielectric interfaces with very little charge impurities in 2D materials [101].

6.11.2 *Optoelectronics*

Since it has been established that 2D TMDCs are optically active materials, a plethora of research is currently being carried out on their applications as optoelectronic devices. LEDs have emerged as a solution to most of the energy saving problems in the twenty-first century. Due to the ultra-thin nature and semiconducting properties of TMDCs, these materials are prospective candidates for LEDs.

Berraquero et al. [102] reported a study of the fabrication of LEDs by using monolayer and few-layer WSe_2 layered material. The fabricated LED structure was very similar to that reported by Liu et al. [101]. The active region of this device consisted of adjacent monolayer and bilayer active areas, both in contact with the ground electrode. Most of the injected current flow was through the bilayer region. Quantum LED (QLED) operation was observed in the form of highly localized light emission from both the monolayer and the bilayer WSe_2. These localized states were shown to lie within the bandgap of WSe_2 [102]. The observed emission wavelength range of WSe_2 was demonstrated to match rubidium transitions (~780 nm) which could help to open up quantum storage possibilities.

6.11.3 *Heterostructures*

Heterojunction is the interface that occurs between two layers or regions of dissimilar crystalline semiconductors. These semiconducting materials have unequal bandgaps as opposed to a homojunction. It is often advantageous to engineer the electronic energy bands in many solid-state device applications, including semiconductor lasers,

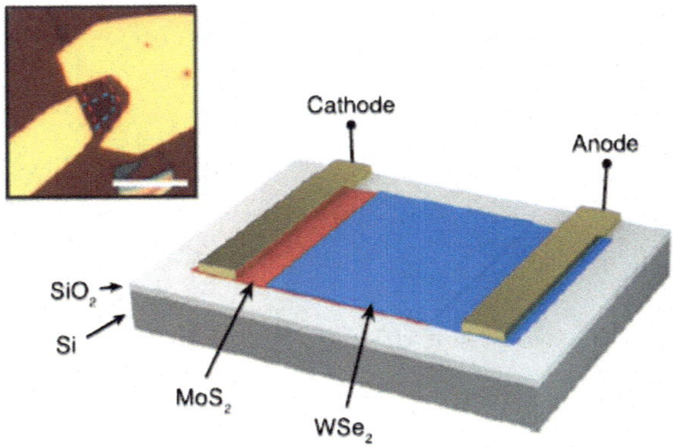

Fig. 6.53 Three-dimensional schematic view of the dual bandgap cells [104]

solar cells and transistors (heterotransistors) to name a few. The combination of multiple heterojunctions together in a device is called a heterostructure [103].

According to Furchi et al. [104], in the late 70s, TMDCs were used as electrodes in photoelectrovoltaic cells. Their bandgaps, which match well with the solar spectrum, as well as the covalent layer-type nature of these compounds made them promising candidates for efficient solar energy conversion. Solar to electrical power conversion efficiencies of up to 10.2% were achieved using WSe_2 and $MoSe_2$ crystals [105]. In this work, multiple van der Waals heterostructure-based solar cells were designed and fabricated. Figure 6.53 shows the schematic of the dual bandgap heterostructure.

The power conversion efficiency of the cell was found to be of the same order of magnitude (0.1–0.2%) which was in agreement with the predicted values [106]. The authors of this study expand the idea of bilayer TMDC heterostructures to a three-layer structure. This structure has higher efficiency compared to that of the double layer structure. The high absorption ability of $MoSe_2$ and WSe_2 make them good candidates for solar cell applications, as cited by Furchi et al. [104]. A monolayer of $MoSe_2$ (with a thickness of 0.65 nm) absorbs the same fraction of light as 15 nm of Gallium Arsenide (GaAs) or 50 nm of Silicon (Si) [104]. From this perspective, van der Waals heterostructures seem to be promising candidates. The amount of raw material needed for the active region is minimal; also, the power conversion densities are expected to be orders of magnitude higher than in conventional cells [104].

References

1. 2D Materials, http://iopscience.iop.org/journal/2053-1583
2. FlatChem, https://www.journals.elsevier.com/flatchem
3. Graphene, https://www.scirp.org/journal/graphene
4. Graphene Technology, https://link.springer.com/journal/41127
5. npj 2D Materials and Applications, https://www.nature.com/npj2dmaterials
6. Carbon, https://www.sciencedirect.com/journal/carbon
7. 2D Materials and Beyond, https://pubs.acs.org/page/vi/2Dmaterials.html

8. Focus issue: porous carbon and carbonaceous materials for energy conversion and storage. J Mater Res 33(9) (2018)
9. Cao T, Wang G, Han W, Ye H, Zhu C, Shi J et al (2012) Valley-selective circular dichroism of monolayer molybdenum disulphide. Nat Commun 3:887
10. Wang X, Xia F (2015) Van der Waals heterostructures: stacked 2D materials shed light. Nat Mater 14:264–265
11. Wang QH, Kalantar-Zadeh K, Kis A, Coleman JN, Strano MS (2012) Electronics and optoelectronics of two-dimensional transition metal dichalcogenides. Nat Nanotechnol 7:699–712
12. Wang Z, Su Q, Yin G, Shi J, Deng H, Guan J et al (2014) Structure and electronic properties of transition metal dichalcogenide MX_2 (M = Mo, W, Nb; X = S, Se) monolayers with grain boundaries. Mater Chem Phys 147:1068–1073
13. Mak KF, Lee C, Hone J, Shan J, Heinz TF (2010) Atomically thin MoS_2: a new direct-gap semiconductor. Phys Rev Lett 105:136805
14. Li H, Wu J, Yin Z, Zhang H (2014) Preparation and applications of mechanically exfoliated single-layer and multilayer MoS_2 and WSe_2 nanosheets. Acc Chem Res 47:1067–1075
15. Amin B, Kaloni TP, Schwingenschlögl U (2014) Strain engineering of WS_2, WSe_2, and WTe_2. RSC Adv 4:34561–34565
16. Schönfeld B, Huang J, Moss S (1983) Anisotropic mean-square displacements (MSD) in single-crystals of 2H-and 3R-MoS_2. Acta Crystallogr B 39:404–407
17. Haynes WM (2017) CRC handbook of chemistry and physics. CRC Press, Boca Raton
18. Wilson J, Yoffe A (1969) The transition metal dichalcogenides discussion and interpretation of the observed optical, electrical and structural properties. Adv Phys 18:193–335
19. Ye M, Winslow D, Zhang D, Pandey R, Yap YK (2015) Recent advancement on the optical properties of two-dimensional molybdenum disulfide (MoS_2) thin films. Photonics 2(1):288–307
20. Kumar A, Ahluwalia P (2012) Electronic structure of transition metal dichalcogenides monolayers 1H-MX_2 (M = Mo, W; X = S, Se, Te) from ab-initio theory: new direct band gap semiconductors. Eur Phys J B 85:1–7
21. Ravindra NM, Srivastava VK (1979) Temperature dependence of the energy gap in semiconductors. J Phys Chem Solids 40:791–793
22. Van Zeghbroeck B (2004) Principles of semiconductor devices. Colarado University, Denver
23. O'Donnell K, Chen X (1991) Temperature dependence of semiconductor band gaps. Appl Phys Lett 58:2924–2926
24. Varshni YP (1967) Temperature dependence of the energy gap in semiconductors. Physica 34(1):149–154
25. Tongay S, Zhou J, Ataca C, Lo K, Matthews TS, Li J et al (2012) Thermally driven crossover from indirect toward direct bandgap in 2D semiconductors: $MoSe_2$ versus MoS2. Nano Lett 12:5576–5580
26. He Z, Sheng Y, Rong Y, Lee G-D, Li J, Warner JH (2015) Layer-dependent modulation of tungsten disulfide photoluminescence by lateral electric fields. ACS Nano 9:2740–2748
27. Li Y, Chernikov A, Zhang X, Rigosi A, Hill HM, van der Zande AM et al (2014) Measurement of the optical dielectric function of monolayer transition-metal dichalcogenides: MoS_2, $MoSe_2$, WS_2, and WSe_2. Phys Rev B 90:205422
28. Rassay SS, Tang W, Ravindra NM (2016) Optical properties and temperature dependence of energy gap of transition-metal dichalcogenides. In: Proceedings MS&T, Salt Lake City, Utah, 23–27 Oct 2016
29. Tang W, Rassay SS, Ravindra NM (2017) Electronic and optical properties of transition-metal dichalcogenides. Madridge J Nano Tech 2(1):59–65
30. Mukherjee B, Tseng F, Gunlycke D, Amara KK, Eda G, Simsek E (2015) Complex electrical permittivity of the monolayer molybdenum disulfide (MoS_2) in near UV and visible. Opt Mater Express 5:447–455
31. Mak KF, Shan J, Heinz TF (2011) Seeing many-body effects in single-and few-layer graphene: observation of two-dimensional saddle-point excitons. Phys Rev Lett 106:046401

32. Arora A, Koperski M, Nogajewski K, Marcus J, Faugeras C, Potemski M (2015) Excitonic resonances in thin films of WSe_2: from monolayer to bulk material. Nanoscale 7:10421–10429
33. Mattheiss L (1973) Band structures of transition-metal-dichalcogenide layer compounds. Phys Rev B 8:3719
34. Carvalho A, Ribeiro R, Neto AC (2013) Band nesting and the optical response of two-dimensional semiconducting transition metal dichalcogenides. Phys Rev B 88:115205
35. Jin W et al (2013) Direct measurement of the thickness-dependent electronic band structure of MoS_2 using angle-resolved photoemission spectroscopy. Phys Rev Lett 111(10):106801
36. Fivaz R, Mooser E (1967) Mobility of charge carriers in semiconducting layer structures. Phys Rev 163:743
37. Radisavljevic B, Radenovic A, Brivio J, Giacometti IV, Kis A (2011) Single-layer MoS_2 transistors. Nat Nanotechnol 6(3):147–150
38. Novoselov K, Jiang D, Schedin F, Booth T, Khotkevich V, Morozov S et al (2005) Two-dimensional atomic crystals. Proc Natl Acad Sci USA 102:10451–10453
39. Ghatak S, Pal AN, Ghosh A (2011) Nature of electronic states in atomically thin MoS_2 field-effect transistors. ACS Nano 5:7707–7712
40. Radisavljevic B, Kis A (2013) Mobility engineering and a metal–insulator transition in monolayer MoS_2. Nat Mater 12:815–820
41. Kaasbjerg K, Thygesen KS, Jacobsen KW (2012) Phonon-limited mobility in n-type single-layer MoS_2 from first principles. Phys Rev B 85:115317
42. Liu L, Kumar SB, Ouyang Y, Guo J (2011) Performance limits of monolayer transition metal dichalcogenide transistors. IEEE Trans Electron Devices 58:3042–3047
43. Jo S, Ubrig N, Berger H, Kuzmenko AB, Morpurgo AF (2014) Mono-and bilayer WS_2 light-emitting transistors. Nano Lett 14:2019–2025
44. Ovchinnikov D, Allain A, Huang Y-S, Dumcenco D, Kis A (2014) Electrical transport properties of single-layer WS_2. ACS Nano 8:8174–8181
45. Podzorov V, Gershenson M, Kloc C, Zeis R, Bucher E (2004) High-mobility field-effect transistors based on transition metal dichalcogenides. Appl Phys Lett 84:3301–3303
46. Jariwala D, Sangwan VK, Late DJ, Johns JE, Dravid VP, Marks TJ et al (2013) Band-like transport in high mobility unencapsulated single-layer MoS_2 transistors. Appl Phys Lett 102:173107
47. Baugher BW, Churchill HO, Yang Y, Jarillo-Herrero P (2013) Intrinsic electronic transport properties of high-quality monolayer and bilayer MoS_2. Nano Lett 13:4212–4216
48. Zeng L, Xin Z, Chen S, Du G, Kang J, Liu X (2013) Remote phonon and impurity screening effect of substrate and gate dielectric on electron dynamics in single layer MoS_2. Appl Phys Lett 103:113505
49. Das S, Chen H-Y, Penumatcha AV, Appenzeller J (2012) High performance multilayer MoS_2 transistors with scandium contacts. Nano Lett 13:100–105
50. Popov I, Seifert G, Tománek D (2012) Designing electrical contacts to MoS_2 monolayers: a computational study. Phys Rev Lett 108:156802
51. Liu D, Guo Y, Fang L, Robertson J (2013) Sulfur vacancies in monolayer MoS_2 and its electrical contacts. Appl Phys Lett 103:183113
52. Walia S, Balendhran S, Wang Y, Ab Kadir R, Zoolfakar AS, Atkin P et al (2013) Characterization of metal contacts for two-dimensional MoS_2 nanoflakes. Appl Phys Lett 103:232105
53. Lee G-H, Yu Y-J, Cui X, Petrone N, Lee C-H, Choi MS et al (2013) Flexible and transparent MoS_2 field-effect transistors on hexagonal boron nitride-graphene heterostructures. ACS Nano 7:7931–7936
54. Shi H, Pan H, Zhang Y-W, Yakobson BI (2013) Quasiparticle band structures and optical properties of strained monolayer MoS_2 and WS_2. Phys Rev B 87:155304
55. Salvatore GA, Münzenrieder N, Barraud C, Petti L, Zysset C, Büthe L et al (2013) Fabrication and transfer of flexible few-layers MoS_2 thin film transistors to any arbitrary substrate. ACS Nano 7:8809–8815
56. Chang H-Y, Yang S, Lee J, Tao L, Hwang W-S, Jena D et al (2013) High-performance, highly bendable MoS_2 transistors with high-k dielectrics for flexible low-power systems. ACS Nano 7:5446–5452

57. Wang H, Yu L, Lee Y-H, Fang W, Hsu A, Herring P et al (2012) Large-scale 2D electronics based on single-layer MoS_2 grown by chemical vapor deposition, In: Presented at 2012 IEEE International Electron Devices Meeting (IEDM), San Francisco, CA, USA, 10–13 Dec 2012
58. Georgiou T, Jalil R, Belle BD, Britnell L, Gorbachev RV, Morozov SV et al (2013) Vertical field-effect transistor based on graphene-WS_2 heterostructures for flexible and transparent electronics. Nat Nanotechnol 8:100–103
59. Zhang W, Chuu C-P, Huang J-K, Chen C-H, Tsai M-L, Chang Y-H et al (2014) Ultrahigh-gain photodetectors based on atomically thin graphene-MoS_2 heterostructures. Sci Rep 4:3826
60. Geim AK, Grigorieva IV (2013) Van der Waals heterostructures. Nature 499:419–425
61. Choi W, Cho MY, Konar A, Lee JH, Cha GB, Hong SC et al (2012) High-detectivity multi-layer MoS_2 phototransistors with spectral response from ultraviolet to infrared. Adv Mater 24:5832–5836
62. Perea-López N, Elías AL, Berkdemir A, Castro-Beltran A, Gutiérrez HR, Feng S et al (2013) Photosensor device based on few-layered WS_2 films. Adv Func Mater 23:5511–5517
63. Yin Z, Li H, Li H, Jiang L, Shi Y, Sun Y et al (2011) Single-layer MoS_2 phototransistors. ACS Nano 6:74–80
64. Lee HS, Min S-W, Chang Y-G, Park MK, Nam T, Kim H et al (2012) MoS_2 nanosheet phototransistors with thickness-modulated optical energy gap. Nano Lett 12:3695–3700
65. Li Y, Xu C-Y, Wang J-Y, Zhen L (2014) Photodiode-like behavior and excellent photoresponse of vertical Si/monolayer MoS_2 heterostructures. Sci Rep 4:7186
66. Wi S, Chen M, Nam H, Liu AC, Meyhofer E, Liang X (2014) High blue-near ultraviolet photodiode response of vertically stacked graphene-MoS_2-metal heterostructures. Appl Phys Lett 104:232103
67. Bernardi M, Palummo M, Grossman JC (2013) Extraordinary sunlight absorption and one nanometer thick photovoltaics using two-dimensional monolayer materials. Nano Lett 13:3664–3670
68. Fontana M, Deppe T, Boyd AK, Rinzan M, Liu AY, Paranjape M et al (2013) Electron-hole transport and photovoltaic effect in gated MoS_2 Schottky junctions. Sci Rep 3:1634
69. Ye Y, Ye Z, Gharghi M, Yin X, Zhu H, Zhao M et al (2014) Exciton-related electroluminescence from monolayer MoS_2. In: Presented at the CLEO: science and innovations, San Jose, CA, USA, 8–13 Jun 2014
70. Li X, Zhu H (2015) Two-dimensional MoS_2: properties, preparation, and applications. J Materiomics 1(1):33–44
71. Wilson JA, Yoffe AD (1969) The transition metal dichalcogenides discussion and interpretation of the observed optical, electrical and structural properties. Adv Phys 18:193
72. Wang QH, Kalantar-Zadeh K, Kis A, Coleman JN, Strano MS (2012) Electronics and optoelectronics of two-dimensional transition metal dichalcogenides. Nat Nanotechnol 7(11):699–712
73. Arora A, Koperski M, Nogajewski K, Marcus J, Faugeras C, Potemski M (2015) Excitonic resonances in thin films of WSe_2: from monolayer to bulk material. Nanoscale 7(23):10421–10429
74. Liu W, Kang J, Sarkar D, Khatami Y, Jena D, Banerjee K (2013) Role of metal contacts in designing high-performance monolayer n-type WSe_2 field effect transistors. Nano Lett 13(5):1983–1990
75. Ross JS, Wu S, Yu H, Ghimire NJ, Jones AM, Aivazian G, Xu X (2013) Electrical control of neutral and charged excitons in a monolayer semiconductor. Nat Commun 4:1474
76. Wang Z, Su Q, Yin GQ, Shi J, Deng H, Guan J, Fu YQ (2014) Structure and electronic properties of transition metal dichalcogenide MX 2 (M = Mo, W, Nb; X = S, Se) monolayers with grain boundaries. Mater Chem Phys 147(3):1068–1073
77. Brixner LH (1962) Preparation and properties of the single crystalline AB_2-type selenides and tellurides of niobium, tantalum, molybdenum and tungsten. J Inorg Nucl Chem 24:257–263
78. Walker P, Tarn WH (1990) CRC handbook of metal etchants. CRC Press, Boca Raton, FL
79. Anedda A, Fortin E, Raga F (1979) Optical spectra in WSe_2. Can J Phys 57:368–374
80. El-Mahalawy SH, Evans BL (1977) Temperature dependence of the electrical conductivity and hall coefficient in 2H-MoS_2, $MoSe_2$, WSe_2, and $MoTe_2$. Phys Status Solidi (b) 79:713–722

81. Hoffmann R (1988) Solids and surfaces: a chemist's view of bonding in extended structures. VCH Publ, New York
82. Izyumov YA, Syromyatnikov VN (2012) Phase transitions and crystal symmetry, vol 38. Springer Science & Business Media, Berlin
83. Peter YU, Cardona M (2010) Fundamentals of semiconductors: physics and materials properties. Springer Science & Business Media, Berlin
84. Mak KF, Lee C, Hone J, Shan J, Heinz TF (2010) Atomically thin MoS_2: a new direct-gap semiconductor. Phys Rev Lett 105(13):136805
85. Zhu ZY, Cheng YC, Schwingenschlögl U (2011) Giant spin-orbit-induced spin splitting in two-dimensional transition-metal dichalcogenide semiconductors. Phys Rev B 84(15):153402
86. Zhang Y, Ugeda MM, Jin C, Shi SF, Bradley AJ, Martin-Recio A, Zhou B (2016) Electronic structure, surface doping, and optical response in epitaxial WSe_2 thin films. Nano Lett 16(4):2485–2491
87. Van Zeghbroeck B (2011) Principles of electronic devices. University of Colorado, Denver
88. Tongay S et al (2012) Thermally driven crossover from indirect toward direct bandgap in 2D semiconductors: $MoSe_2$ versus MoS_2. Nano Lett 12(11):5576–5580
89. Yuan H, Liu Z, Xu G, Zhou B, Wu S, Dumcenco D, Kandyba V (2016) Evolution of the valley position in bulk transition-metal chalcogenides and their monolayer limit. Nano Lett 16(8):4738–4745
90. Li Y, Chernikov A, Zhang X, Rigosi A, Hill HM, van der Zande AM, Heinz TF (2014) Measurement of the optical dielectric function of monolayer transition-metal dichalcogenides: MoS_2, $MoSe_2$, WS_2, and WSe_2. Phys Rev B 90(20):205422
91. Mukherjee B, Tseng F, Gunlycke D, Amara KK, Eda G, Simsek E (2015) Complex electrical permittivity of the monolayer molybdenum disulfide (MoS_2) in near UV and visible. Opt Mater Express 5(2):447–455
92. Zhang H, Ma Y, Wan Y, Rong X, Xie Z, Wang W, Dai L (2015) Measuring the refractive index of highly crystalline monolayer MoS_2 with high confidence. Sci Rep 5
93. Liu HL, Shen CC, Su SH, Hsu CL, Li MY, Li LJ (2014) Optical properties of monolayer transition metal dichalcogenides probed by spectroscopic ellipsometry. Appl Phys Lett 105(20):201905
94. Mattheiss LF (1973) Band structures of transition-metal-dichalcogenide layer compounds. Phys Rev B 8(8):3719
95. Ravindra NM, Narayan J, Ance C, Dechelle F, Ferraton JP (1986) Low-temperature optical properties of hydrogenated amorphous silicon. Mater Lett 4(8):343–349
96. Ravindra NM, Narayan J (1987) Optical properties of silicon related insulators. J Appl Phys 61(5):2017–2021
97. Brumme T, Calandra M, Mauri F (2015) First-principles theory of field-effect doping in transition-metal dichalcogenides: structural properties, electronic structure, Hall coefficient, and electrical conductivity. Phys Rev B 91(15):155436
98. Thakar BA (2011) Investigations of TMDCs and use in solar cell. Thesis. Hemchandracharya North Gujarat University, Patan, 2011. Shodhganga. Web. 23 Jun 2015. http://shodhganga.inflibnet.ac.in/handle/10603/43941
99. Hang Y, Li Q, Luo W, He Y, Zhang X, Peng G (2016) Photo-electrical properties of trilayer $MoSe_2$ nanoflakes. Nano 11(7):1650082
100. Liu B, Ma Y, Zhang A, Chen L, Abbas AN, Liu Y, Zhou C (2016) High-performance WSe_2 field-effect transistors via controlled formation of in-plane heterojunctions. ACS Nano 10(5):5153–5160
101. Berraquero CP, Barbone M, Kara DM, Chen X, Goykhman I, Yoon D, Ferrari AC (2016) Atomically thin quantum light emitting diodes. arXiv preprint arXiv:1603.08795
102. Colinge JP (2002) Physics of semiconductor devices. Springer, New York
103. Furchi MM, Zechmeister AA, Hoeller F, Wachter S, Pospischil A, Mueller T (2017) Photovoltaics in Van der Waals heterostructures. IEEE J Sel Top Quant Electron 23(1):1–11
104. Kline G, Kam K, Canfield D, Parkinson BA (1981) Efficient and stable photoelectrochemical cells constructed with WSe_2 and $MoSe_2$ photoanodes. Solar Energy Mater 4(3):301–308
105. Bernardi M, Palummo M, Grossman JC (2013) Extraordinary sunlight absorption and one nanometer thick photovoltaics using two-dimensional monolayer materials. Nano Lett 13(8):3664–3670

Chapter 7
Group II–VI Semiconductors

Bindu Krishnan, Sadasivan Shaji, M. C. Acosta-Enríquez, E. B. Acosta-Enríquez, R. Castillo-Ortega, MA. E. Zayas, S. J. Castillo, Ilaria Elena Palamà, Eliana D'Amone, Martin I. Pech-Canul, Stefania D'Amone and Barbara Cortese

B. Krishnan (✉) · S. Shaji
Universidad Autónoma de Nuevo León, San Nicolás de los Garza, Mexico
e-mail: kbindu_k@yahoo.com

S. Shaji
e-mail: sshajis@yahoo.com

M. C. Acosta-Enríquez (✉) · MA. E. Zayas · S. J. Castillo
Departamento de Investigación en Física, Universidad de Sonora,
Blvd. Luis Encinas y Rosales S/N, Col. Centro, C.P. 83000 Hermosillo, Sonora, Mexico
e-mail: milka.acosta@unison.mx

MA. E. Zayas
e-mail: mzayas@difus.uson.mx

S. J. Castillo
e-mail: semiconductores@difus.uson.mx

E. B. Acosta-Enríquez
Departamento de Física, Universidad de Sonora,
Blvd. Luis Encinas y Rosales S/N, Col. Centro, C.P. 83000 Hermosillo, Sonora, Mexico
e-mail: eulices_acosta@hotmail.com

R. Castillo-Ortega
Departamento de Ingeniería Industrial, Universidad de Sonora,
Blvd. Luis Encinas y Rosales S/N, Col. Centro, C.P. 83000 Hermosillo, Sonora, Mexico
e-mail: rcastillo@industrial.uson.mx

I. E. Palamà · E. D'Amone · S. D'Amone
Nanotechnology Institute, CNR-NANOTEC, Via per Monteroni, 73100 Lecce, Italy
e-mail: ilaria.palama@nanotec.cnr.it

E. D'Amone
e-mail: eliana.damone@nanotec.cnr.it

S. D'Amone
e-mail: stefania.damone@nanotec.cnr.it

M. I. Pech-Canul and N. M. Ravindra (eds.), *Semiconductors*,
https://doi.org/10.1007/978-3-030-02171-9_7

7.1 CdS and Related Binary, Ternary, and Quaternary Compounds

In recent years, II–VI semiconductors have been the focus of intense research and development due to their potential use in various applications. II–VI semiconductors are compound semiconductors formed by group IIB metallic elements (Cd, Zn, and Hg) with group VI nonmetallic elements (O, S, Se, and Te). Among such compounds, cadmium sulfide (CdS) has been of keen interest due to its extraordinary optoelectronic properties which has led to unique device applications in thin-film photovoltaics [1, 2], nanophotodetectors [3], lasers, etc. Zinc oxide (ZnO) is another important II–VI semiconductor with exceptionally good properties and wide applications in varistors, phosphors, sensors, and optoelectronic devices [4–6]. In this chapter, we describe the basic properties of CdS, ZnO, and related materials and their processing techniques, both in thin-film and nanostructured forms. A detailed description of the preparation methods, morphologies and optoelectronic properties is presented.

In this section, important growth techniques, properties, and applications of CdS and related ternary and quaternary compounds are described, beginning with those of CdS.

A. **Cadmium sulfide (CdS)**

Cadmium sulfide is a naturally existing yellow solid material (Fig. 7.1) with two different crystal structures, viz., hexagonal (greenockite) and cubic (hawleyite) forms (Fig. 7.2a–b); it transforms to NaCl-type (halite) structure at high pressure [7]. The cubic phase is less common and experiments showing the transformation of cubic to hexagonal phase have concluded that cubic CdS is metastable in the temperature range of 20–900 °C [8]. The values of lattice parameters of the hexagonal structure are $a = 4.1348$ Å, $c = 6.7490$ Å. In the bulk form, CdS exhibits high optical bandgap of 2.5 eV which can be increased to nearly 2.6 eV due to quantum confinement effect depending on the particle size [9]. Such properties and the development of technologies for the preparation of CdS thin films and nanostructures have triggered their wide range of device applications.

B. Cortese (✉)
Nanotechnology Institute, CNR-NANOTEC, University La Sapienza, P.zle Aldo Moro, Rome, Italy
e-mail: barbara.cortese@nanotec.cnr.it

M. I. Pech-Canul
Centro de Investigación y de Estudios Avanzados del Instituto Politécnico Nacional, Avenida Industria Metalúrgica No. 1062, Parque Industrial Saltillo-Ramos Arizpe, 25900 Ramos Arizpe, Coahuila, Mexico
e-mail: martin.pech@cinvestav.edu.mx

Fig. 7.1 Cadmium sulfide pellet

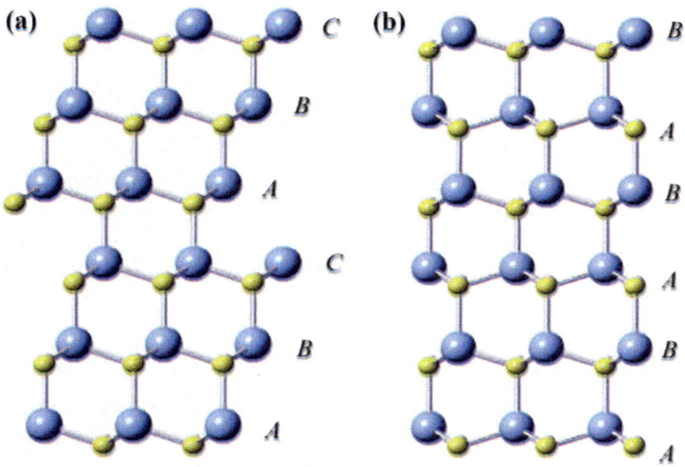

Fig. 7.2 Crystal structure of CdS [9]: **a** cubic and **b** hexagonal (with permission from IOP Science)

7.1.1 Processing Techniques

After the discovery of semiconducting properties of CdS in 1961, intense research has been going on worldwide and various methods have been developed including both physical and chemical methods. Nevertheless, the surge in interest in this material originated due to the invention of novel techniques to synthesize it in the form of thin films and nanoparticles. In general, CdS preparation techniques can be classified into two principal processes: (i) *physical* and (ii) *chemical* processes. Physical process is related to the techniques which depend on the evaporation of the material from a source. Chemical processes can be in gas phase and liquid phase. However, among the available technologies, chemical solution (liquid phase) based techniques are commonly used because of the extreme toxicity of cadmium especially in the vapor or gas form. In the following section, some of the important and established techniques of both physical and chemical methods to prepare good quality CdS thin films and nanostructures are discussed. Physical methods are discussed qualitatively, but a detailed description of chemical solution method is presented.

(i) ***Physical methods***:

Physical methods of deposition of thin films include vacuum-based approaches in which thin film deposition is by condensation of respective material vapor on substrates. Depending on the method of evaporation, they are named as thermal evaporation, electron beam evaporation, sputtering, laser ablation, etc. A detailed description of such methods to form CdS thin films/nanostructures is given below.

Thermal evaporation:

Usually, materials for evaporation are commercially available 99.999% pure CdS powder [10]. The powder can be evaporated in a molybdenum (Mo) or tungsten (W) boat loaded inside a high vacuum chamber maintained at a pressure in the range of 10^{-5}–10^{-6} Torr. In most cases, evaporation rate may be controlled using an evaporation control unit equipped with the thermal evaporation system. The final thin-film thickness is measured using quartz crystal thickness monitor coupled with the system. CdS thin films are deposited on various types of substrates that are kept at a certain distance from the evaporation source. Also, the substrate temperature can be varied. Pal et al. [10] used such a method to prepare CdS thin films and obtained hexagonal structured thin films.

RF Sputtering:

This is a technique in which, by creating a gaseous plasma and accelerating the ions from the plasma to a CdS target, CdS particles or clusters are eroded from the target to the substrates that are kept at a certain distance from the target. By this method, polycrystalline CdS thin films are formed using 50 W RF power at 1 mTorr Ar^+ plasma [11], at different substrate temperatures of 200–300 °C. In situ doping is also possible in this technique, using targets with suitable doping element. Also, co-sputtering of dopant material can be performed. Au-doped CdS has been prepared using CdS–Cd–Au targets.

There exist other evaporation methods in which modified techniques are used for CdS vapor production or transport to substrate to form CdS. They are flash evaporation [12, 13], electron beam [14], close space vapor transport [11], etc.

For preparing thin films of CdS, solution-based methods are preferred and are described:

(ii) ***Chemical solution methods***:

This method involves controlled reaction between dissolved precursor materials in aqueous solution at temperature below 100 °C [15–17] and precipitation of solid material in thin-film form or particles of varying size depending on the bath conditions

[9]. For the preparation of thin films of CdS, precursor materials containing cadmium and sulfur sources are used.

Chemical solution-based techniques, also known as chemical bath deposition (CBD), of CdS involve two types involving reaction mechanisms. One is the ion mechanism involving reaction between Cd^{2+} and S^{2-} ions in solution medium, a controlled reaction and precipitation of CdS on the substrates immersed in the solution. The other is the cluster (hydroxide, colloidal) mechanism in which homogeneous nucleation and formation of solid phase clusters in the solution is followed by subsequent growth, coagulation, and adsorption of the clusters on the substrate surface [9, 15].

Ion mechanism: In a supersaturated solution containing Cd^{2+} and S^{2-} ions, the precipitation of CdS occurs when the ionic product is greater than the solubility product. The controlled reaction between Cd^{2+} and S^{2-} ions is achieved by forming a cadmium complex to slow down the release of Cd^{2+} and thus the reaction with S^{2-} to form CdS in thin-film form. Such ion mechanism of the conventional CBD growth process of CdS thin films from an alkaline bath containing cadmium chloride ($CdCl_2$), a complexing agent (L), and thiourea ($SC(NH_2)_2$), can be described as follows [9, 15, 18, 19]:

(1) Formation and dissociation of a cadmium complex compound $[CdL_i]^{2+ik}$, where L denotes one or more ligands, i is the number of ligands and k is the charge of the ligand;

$$Cd^{2+} + iL^k \rightleftharpoons (CdL_i)^{2+k}$$

(2) Hydrolysis of the source of chalcogen; and

$$CS(NH_2)_2 + 2OH^- \rightleftharpoons CN_2H_2 + S^{2-} + H_2O$$

(3) Formation of solid metal chalcogenide.

$$Cd^{2+} + S^{2-} \rightleftharpoons CdS$$

Such an ion reaction mechanism of Cd^{2+} and S^{2-} ions to form solid CdS nucleus and its growth by further adsorption of the ions, followed by their coagulation by weak van der Waals forces to form uniform and homogeneous thin films of CdS is shown in Fig. 7.3 [9, 15]. Such a process leads to highly uniform and homogeneous CdS thin films.

In general, in any CBD process for CdS, various cadmium sources that are used are cadmium salts such as CdI_2, $CdSO_4$, $Cd(NO_3)_2$, $Cd(CH_3COO)_2$, and $CdCl_2$, and sulfur source is thiourea ($CS(NH_2)_2$) [15–21]. Many researchers have reported the synthesis of CdS thin films using these cadmium salts and studies using different salts are known in the literature [22]. In our laboratory, we have used the following solution composition to form hexagonal CdS: a bath containing $CdCl_2$, triethanolamine (TEA), ammonium hydroxide (NH_4OH) and Thiourea. The deposition process was

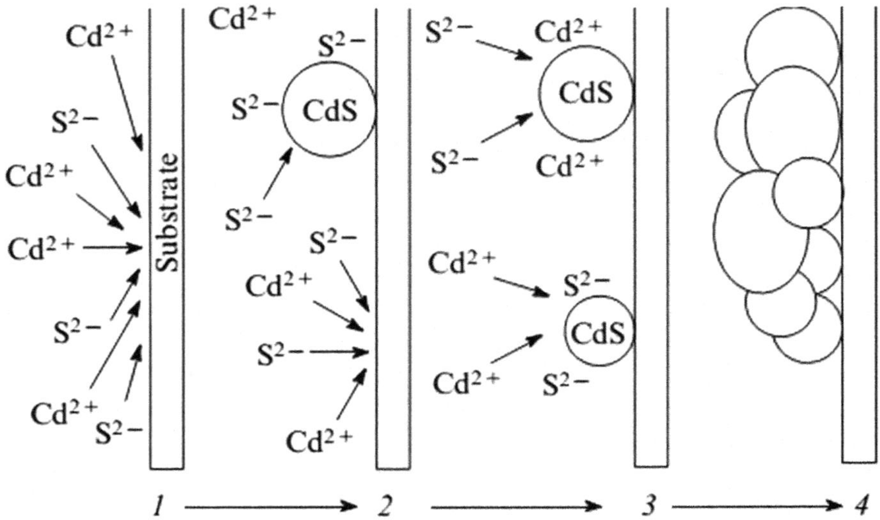

Fig. 7.3 Ion mechanism of CdS film formation on substrate [15]. (1, 2) Diffusion and reaction between Cd^{2+} and S^{2-} ions and formation of a nucleus of solid CdS on the substrate surface; (3) growth of CdS nucleus by adsorption of Cd^{2+} and S^{2-} ions; (4) coagulation of CdS particles due to van der Waals forces (with permission from Taylor and Francis Group LLC Books)

as follows: in a 100 ml beaker, 10 ml of 0.1 M $CdCl_2$, 5 ml of triethanolamine (50%), and 5 ml of ammonium hydroxide (NH_4OH) (15 M) were added and stirred well with 10 ml of 1 M thiourea followed by 65 ml of water at 70 °C [23]. Such solution growth process was accelerated using in situ irradiation of the solution by a continuous laser of wavelength 532 nm with regulated power (0–10 W, CNI Laser, Model MGL-W-532), with laser beam expanded to 5 cm diameter. Figure 7.4 shows the XRD patterns of CdS thin films deposited for 10 min by normal CBD process and laser-assisted (0.3 W/cm^2) CBD (LACBD) technique and 30 min LACBD (0.2 W/cm^2). Chemical bath deposition is a simple, low-cost, and industrially convenient method for the preparation of thin films on metallic, semiconductor, and insulating substrates [24].

The cluster or colloidal mechanism: It is also known as hydroxide mechanism which involves the formation of $Cd(OH)_2$ and its transformation to CdS:

$$Cd(OH)_2 + S_2^- \rightleftharpoons CdS + 2OH^-$$

This mechanism is shown in Fig. 7.5 [15].

The CBD process is considered as a simple and low-cost bottom-up process for the synthesis of CdS and related materials in their nanostructured forms [25]. These can be continuous or island-type films on surfaces of any size and configuration, core–shell structures, agglomerated polycrystals of regular shape, as well as isolated particles in colloidal solutions. Depending on the type of the source of sulfide ions, the chemical affinity and the possibility of competing in the formation of other solid phases, one can synthesize CdS particles of different sizes and different structures.

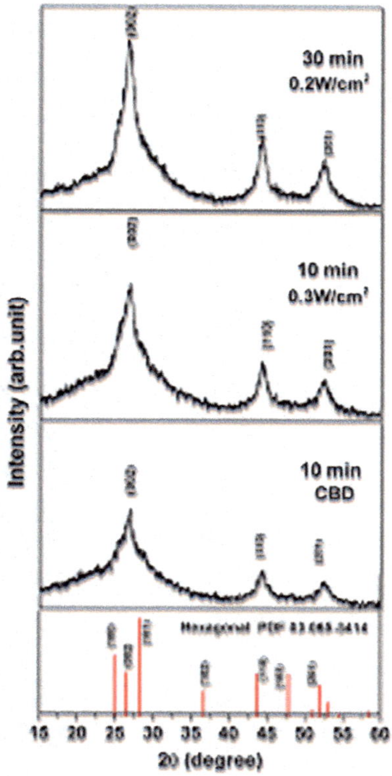

Fig. 7.4 X-ray diffraction patterns of CdS formed by laser-assisted CBD and **a** CdS deposited for 10 min by normal CBD. The standard pattern corresponding to hexagonal (PDF# 41-1049 Hex) form is included [24] (with permission from Elsevier)

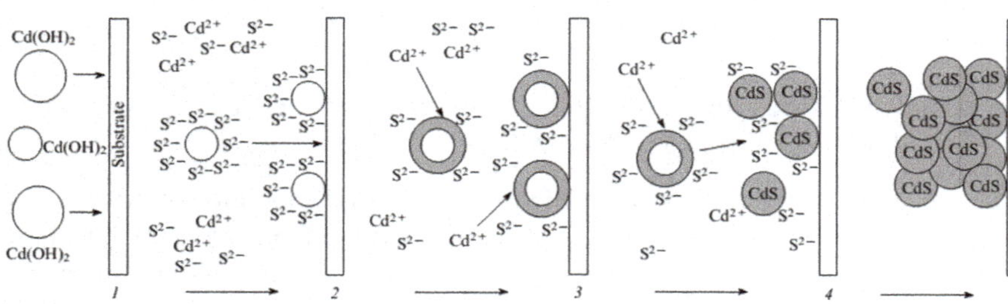

Fig. 7.5 Cluster (hydroxide, colloidal) mechanism of CdS film formation on substrate [15]. (1) Diffusion of $Cd(OH)_2$ colloidal particles; (2) their adsorption on the substrate; (3) interaction of particles with S^{2-} ions formed either upon homogeneous hydrolysis of the source of sulfide ions or in heterogeneous reaction; (4) and (5) exchange of hydroxide ions for sulfide ions proceeds on the substrate surface as well as in the solution to form CdS film (with permission from Taylor and Francis Group LLC Books)

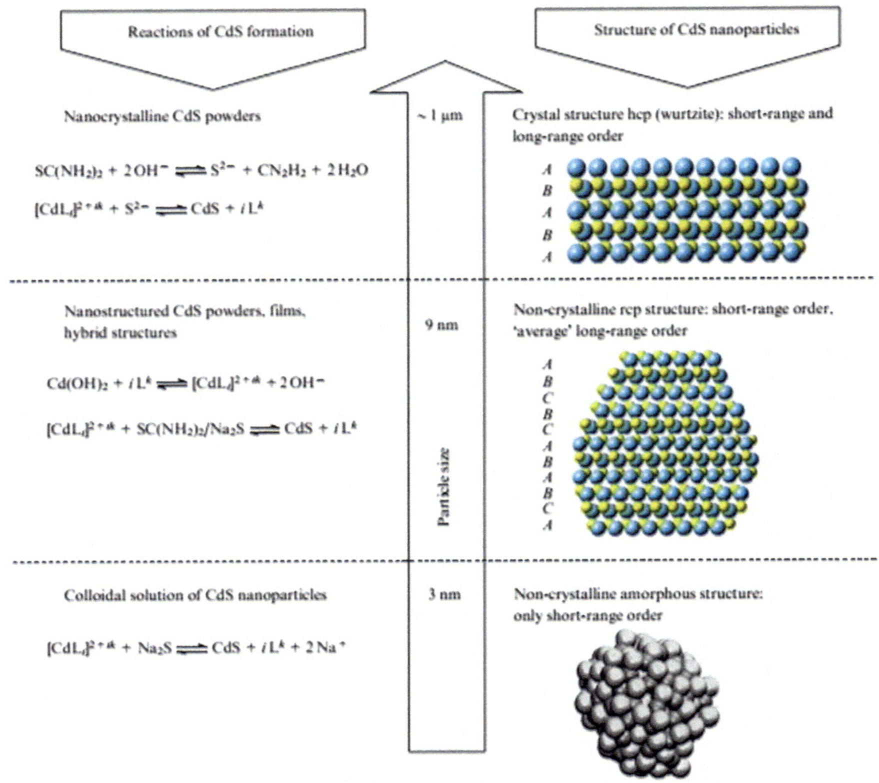

Fig. 7.6 Synthesis conditions for crystalline and noncrystalline CdS nanoparticles considering the type of sulfidizer and the possibility of competing formation of $Cd(OH)_2$ (L is complex-forming agent) [9] (with permission from IOP Science)

In a recently published review article, such conditions of chemical bath deposition were well illustrated as shown in Fig. 7.6 [9].

Experimental setup and procedure of chemical bath deposition technique
Chemical bath deposition (CBD) is a wet method used to fabricate materials in the form of powders or thin films. In this process, several aqueous solutions, including the precursors that react to form the required chemical compound, are mixed [26, 27].

The reactor consists of a glass beaker which contains the aqueous solution mix; it is immersed in a thermal reservoir in order to control the temperature, see Fig. 7.7.

In the CBD method, the required ions are in the aqueous mix, given that the chemical compound is produced by controlled precipitation. When equilibrium conditions are attained, a precipitate is generated and deposits thin films on the substrate immersed in the homogeneous mixing of the solutions.

This section focuses on the preparation of cadmium sulfide (CdS) thin films using CBD. The chemical procedure is as follows:

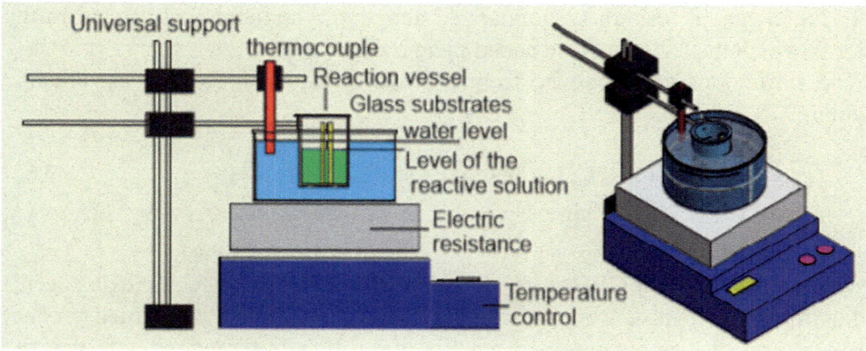

Fig. 7.7 Descriptive scheme of chemical bath deposition reactor

Cadmium chloride provides the cadmium ions to the solution. In order to control the reactivity of the Cd ions, it is necessary to use a complexing agent, i.e., a chemical compound to form a coordination of ionic structures from ions originating from the chemical precursors.

This can be achieved by adding $C_6H_5O_7Na_3$, KOH, NH_4OH or NH_4Cl to the solution. Otherwise, to favor the beginning of the formation of a CdS thin film, it is preferable that a small portion of $Cd(OH)_2$ acts on and around the substrate.

The semi-reactions related to Cd^{2+} ions are described by

$$CdCl_{2(ac)} \rightarrow Cd^{2+} + 2Cl^- \tag{7.1}$$

$$Cd^{2+} + C_6H_4O_7^{4-} \rightarrow [Cd(C_6H_4O_7)]^{2-} \tag{7.2}$$

$$KOH \rightarrow K^+ + OH^- \tag{7.3}$$

$$Cd^{2+} + 2(OH)^- \rightarrow Cd(OH)_{2(s)}$$

$$Cd^{2+} + 3\left(OH^-\right) \rightarrow \left[Cd(OH)_3\right]^-$$

$$NH_4Cl \rightarrow NH_4^+ + Cl^- \tag{7.4a}$$

$$NH_4OH \rightarrow NH_4^+ + OH^- \tag{7.4b}$$

$$NH_4^+ + OH^- \rightarrow NH_3 + H_2O$$

$$CdCl_{2(ac)} + 4NH_{3(ac)} \rightarrow \left[Cd(NH_3)_4\right]^{2+} + 2Cl^-$$

In the chemical reactions, summarized above, the numerals correspond to the order in which the chemicals are added in a process.

The sulfur ions can originate from the hydrolysis of thiourea in an alkaline medium:

$$(NH_2)_2CS_{(ac)} + OH^- \rightarrow CH_2N_2 + H_2O + HS^- \quad (7.5a)$$

$$HS^- + OH^- \rightarrow H_2O + S^{2-} \quad (7.5a)$$

These reactions in the CBD system contain both cadmium ion (complexing agent) and sulfur ion as byproducts; therefore, CdS can be obtained as the final product depending on the temperature and reaction time. It is important to note that the chemical reaction proceeds in an alkaline aqueous medium in which CdS is insoluble.

Preparation of CdS thin films by CBD

With the CBD technique, it is possible to obtain CdS thin films at reaction temperatures from 40 to 90 °C or even in an extended range. In this particular case, the CdS thin films were deposited at a temperature of 80 °C.

The main focus of this work is the design and implementation of a chemical formulation to produce CdS films, that consist in using a pH 10 buffer to preserve controlled pH of the solution. Buffer is a chemical compound with a clearly defined pH, used to control the pH in a solution without modifying the chemical reactions in the solution. For this reason, its use is proposed in the recipe for CdS, which according to the fundamentals of the CBD technique, is thought to have a positive influence on the formation of the films.

As a development of the study of the synthesized CdS films, the effect of heat treatments on its chemical composition was investigated. The CdS films were subjected to heat treatment in an oxidizing atmosphere of air at constant temperatures between 200 and 450 °C; a number of samples were prepared, analyzed and compared to untreated CdS film. The CdS films, protected by a layer of indium or copper, were also investigated. Due to the possibility of doping the films of CdS, in order to convert them into n-type or p-type semiconductors, a metal film of indium or copper was deposited on CdS, and subsequently, diffusion of metal into the CdS was performed by heat treatments in air. This process led to two sets of samples, whose properties were compared with thermally annealed undoped CdS films deposited at 80 °C.

In summary, a systematic study of the effect of thermal annealing, without controlled atmosphere at different temperatures, both on CdS films, as well as on CdS bilayers Indium (In/CdS) and CdS copper (Cu/CdS) was performed.

The compounds that are required to prepare the mixture of aqueous solutions in which thin films of Cds are deposited at 80 °C are listed below:

25 ml of cadmium chloride ($CdCl_2$) 0.1 M, metal source (Aldrich).
20 mL Sodium Citrate ($C_6H_5O_7Na_3$) 1 M, complexing agent (J. T. Baker).
5 ml of potassium hydroxide (KOH) 1 M, forming the metal hydroxide.
15 ml of Buffer NH_4Cl/NH_4OH, pH 10, controller alkaline pH (Aldrich).

10 ml of Thiourea °CS (NH_2) 2] 1 M, sulfur source (J. T. Baker).
c.b.p.100 ml of deionized H_2O.

The reaction container is placed in a water bath, the reaction temperature is at 80 °C, and then the reaction time is quantified; it may vary from 15 min to 3 h.

CBD process is well suited for the synthesis of CdS thin films and nanoparticles. In the next section, well-known techniques for the deposition of CdS derived ternaries and quaternaries are discussed.

B. **CdS-related ternary and quaternary compounds**:

(a) **Ternary compounds**

Typical CdS-related ternary compounds are solid solutions that are formed by substituting a fraction of S with Se [7, 28, 29, 30] or Te [31] and Cd substituted ternaries such as $Cd_{1-x}Zn_xS$. In this section, we describe important processing techniques for these ternaries, including both physical and chemical methods.

(i) $CdS_{1-x}Se_x$: These sulfoselenides are well known due to their wide range of applications and facile growth techniques, both in thin film and nanoparticles of different shapes and sizes.

Various chemical methods are known for the synthesis of such ternaries. A few of them are reviewed here. $CdS_{1-x}Se_x$ thin films were obtained from the chemical bath containing cadmium salt, thiourea, and sodium selenosulphate, covering total composition range from CdS ($x = 0$) to CdSe ($x = 1$) [18]. It was found that CdS films were cubic, while mixed $CdS_{1-x}Se_x$ and CdSe films were hexagonal. Also, this material was successfully formed using one-pot method with polymer capping for stability [29].

$CdS_{1-x}Se_x$ nanosheets were deposited by physical vapor transport method on Si substrates at various temperatures [32], in a quartz boat containing CdS and CdSe powders (99.999% pure). Mole fractions of CdS/CdSe powders were adjusted by varying the weight ratio between the two powder sources, and the total weight was kept at 100 mg. The evolution of the morphology of the nanosheets was also studied, by varying the growth temperature, as shown in Fig. 7.8.

By chemical vapor transport method, $CdS_{1-x}Se_x$ nanowires (NWs) and nanobelts (NB), with a controlled composition over the whole range ($x = 0.2$, 0.4, 0.6, 0.8 and 1), were prepared [33]. The SEM and TEM images of such NWs and NBs are shown in Fig. 7.9.

$Cd_{1-x}Zn_xS$ (CdZnS)

The well-known methods to synthesize the nanostructures and thin films of Cd substituted ternaries ($Cd_{1-x}Zn_xS$) are given here. A solution method using sulfur, cadmium stearate ($Cd(St)_2$), and zinc stearate ($Zn(St)_2$) as precursors and *N*-oleoylmorpholine as the reaction medium and solvent [34], is an example. In this process, the amount of S injected is equal to the total molar amount of metal precursors, and the resultant samples are CdZnS quantum dots, as shown in Fig. 7.10 [34].

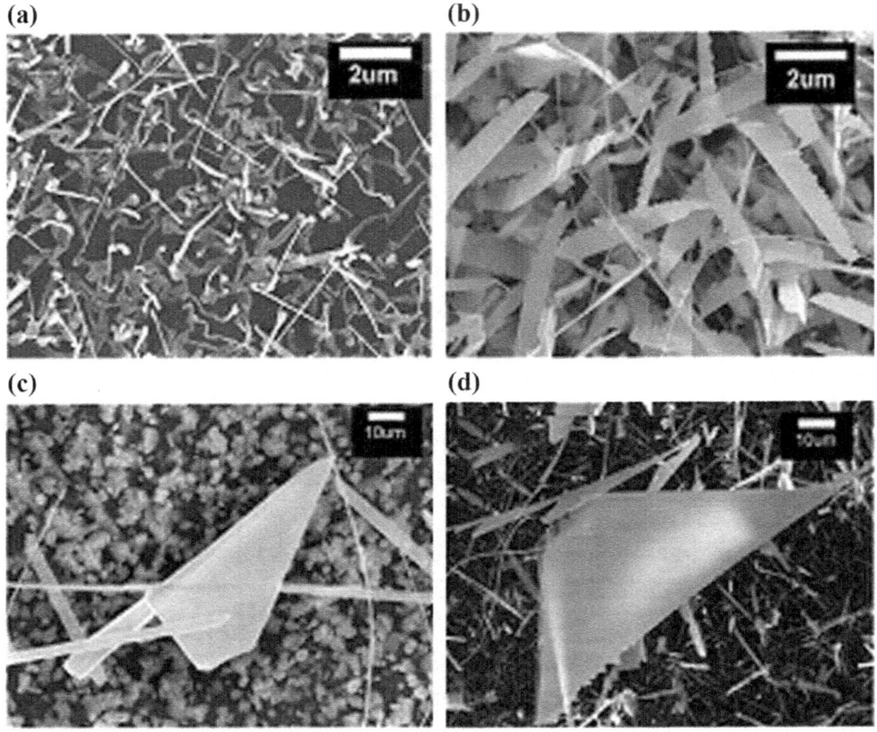

Fig. 7.8 SEM images of cadmium sulfoselenide nanosheets showing the morphological evolution depending on growth temperatures: 750 (**a**), 800 (**b**), 850 (**c**), and 900 (**d**) °C, growth time 1 h [32] (with permission from IOP Science)

Another method to form $Cd_{1-x}Zn_xS$ (CdZnS) in thin-film form is by the reactive diffusion of Zn in CdS [28]. Further, co-precipitation method from a solution of 0.02-M cadmium chloride and 0.15-M thiourea, 0.4 M sodium hydroxide under vigorous 264 stirring at 60 °C for 1 h followed by the addition of $ZnCl_2$ solution with stirring is known to produce CdZnS [35].

A recent report has shown the synthesis of CdZnS–CdS core–shell structures from an evacuated 100 ml solution mixture of 0.001 M cadmium oxide, 0.01 M zinc acetate, and OA (7 mmol) heated to 150 °C in nitrogen. Following this, 15 ml 1-octadecene (1-ODE) is added into the mixture and the temperature is further increased to 300 °C. Then, sulfur precursor, obtained by dissolving 0.002 M S powder into 3 ml 1-ODE, is quickly injected into the mixture. After 8 min of reaction at a constant temperature of 300 °C, 0.008 M S powder dissolved in TBP (Tributyl phosphate) is added into the mixture. The mixture is cooled to room temperature after the reaction ended. In this case, the formation of the ternary has been identified by measuring d-spacings of (002) planes with increase in molar fraction of Zn. The shift in (002) spacing, due to increase in the Zn molar fraction in $Cd_{1-x}Zn_xS$

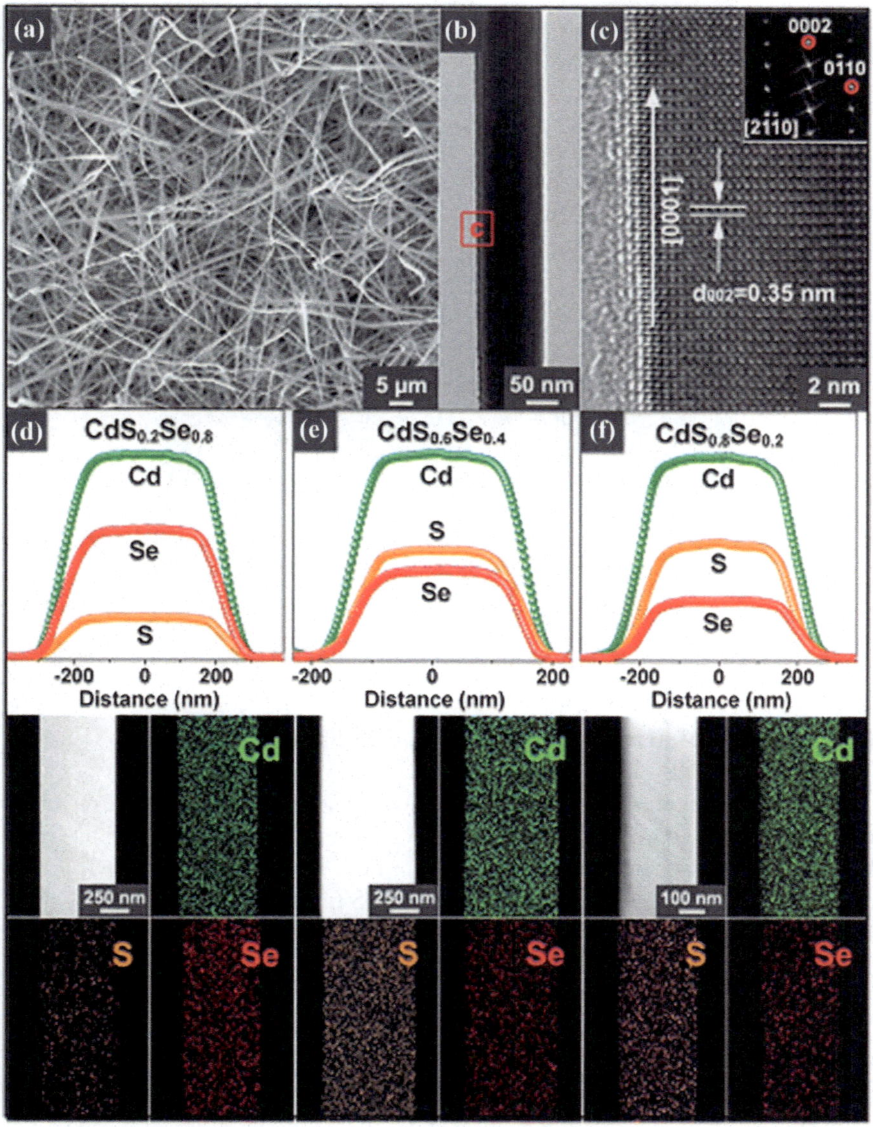

Fig. 7.9 **a** Scanning electron micrograph [33] and **b** transmission electron image of $CdS_{0.2}Se_{0.8}$NW/NB. **c** Lattice-resolved TEM image and its corresponding FFT EDX pattern. EDX line scan and elemental mapping, and STEM image of the selected. **d** $CdS_{0.2}Se_{0.8}$. **e** $CdS_{0.6}Se_{0.4}$, and **f** $CdS_{0.8}Se_{0.2}$NBs, revealing the homogeneous distribution of Cd, S, and Se elements over the whole NB (with permission from RSC)

thin films formed from zinc and cadmium complexes of bis (1,1,5,5-tetraalkyl-2-4-dithiobiurets), by varying the stoichiometries, is shown [7] in Fig. 7.11. The decrease in 2θ values, and hence the increase in the respective d values, shows the formation

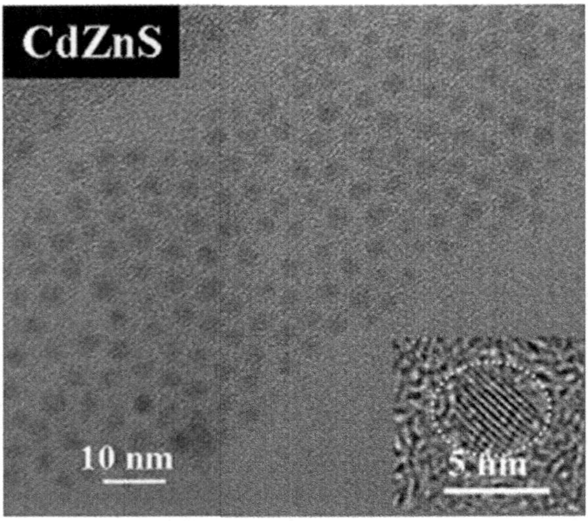

Fig. 7.10 TEM images of as-prepared CdZnS QDs. The inset shows high-resolution transmission electron image (HRTEM) of individual QDs [34] (with permission from Elsevier)

of solid solution of $Cd_{1-x}Zn_xS$ by substituting fraction of Cd atoms by smaller Zn atoms.

Effect of heat treatment on In/CdS and Cu/CdS bilayers
Heat treated In/CdS bilayer

The effect of heat treatment on In/CdS bilayer, in normal air atmosphere, is investigated. This is a continuation of the study of CdS films that were prepared by CBD. The objective of this study is to dope CdS with In.

50 nm thick In films, vacuum deposited on CdS films at a pressure of 10^{-6} Torr, were considered in this study. Despite the reflectance of indium, the In/CdS layer has some transparency, since the indium film is thin.

The X-ray diffraction patterns of these samples are illustrated in Fig. 7.12. The XRD pattern of the film without treatment shows peaks due to CdS and tetragonal peaks (101) and (002) due to In, at approximately $2\theta = 33$ and 36°, respectively. In this series of patterns, it is shown that the In layer is oxidized into cubic In_2O_3 from treatments at 200 °C. The peak appears at approximately 30.6° 2θ, corresponding to the crystalline planes (222) of cubic In_2O_3; this increases its intensity until it becomes maximum in the diffraction pattern of the sample with heat treatment of 350 °C as in the case of the CdS films at the temperature of 350 °C when it initiates the oxidation of Cd. This is based on the observed pattern of this film at about 38° signal, that is associated with the planes (200) of the CdO planes. Patterns of treated films at 400 and 450 °C exhibit reduced diffraction signal at 2θ ~33°, which is not associated with the planes (101) of tetragonal In but with the (111) planes of CdO since these temperatures represent the oxidation of Cd.

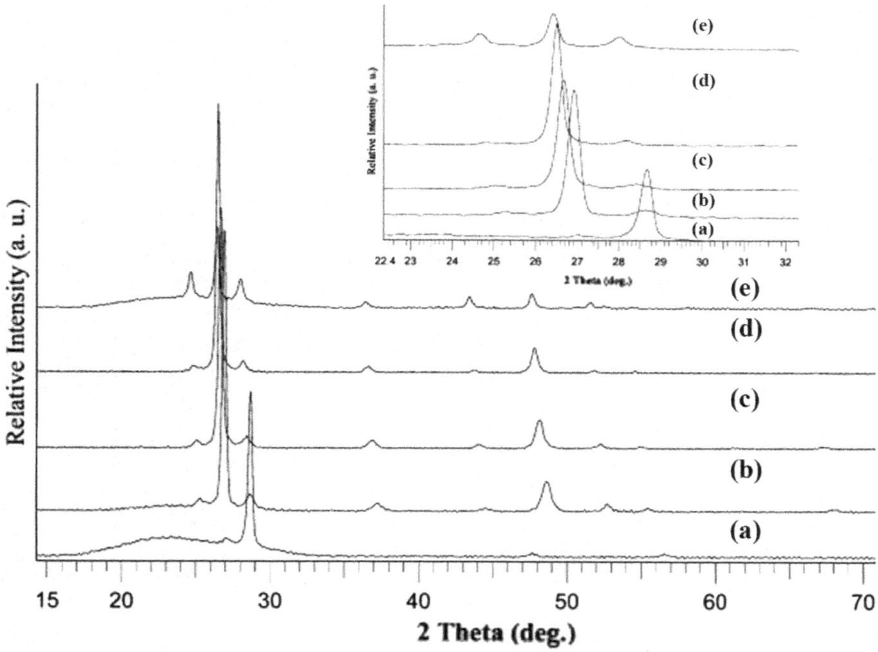

Fig. 7.11 XRD pattern of $Cd_{1-x}Zn_xS$ films deposited by varying Zn and Cd ratio in complex solutions. **a** Pure ZnS, **b** ZnS at (0.25), **c** ZnS at (0.50), **d** ZnS at (0.75) molar fraction of cadmium precursor, and **e** CdS films. Inset shows the variation of (002) plane with respect to Zn and Cd ratio [7] (with permission from American Chemical Society)

This figure also shows that, with treatment at 450 °C, the CdS film has not lost so much sulfur that is manifested in the peak intensity (002). This is because the In layer is oxidized in the surface forming a layer of In_2O_3, considerably reducing the release of sulfur. Moreover, this oxide layer also reduces the release of In out of the system because it forms a layer in the middle of the film and the oxide. This in turn results in the diffusion of In into the CdS film.

In order to remove the In_2O_3 layer formed in the heat-treated sample at 350 °C, it was pickled (it eroded) chemically by soaking in a solution of 5% HCl for 4–6 min. The results obtained were followed by X-ray diffraction measurements and are shown in Fig. 7.13. As can be observed, a decrease was obtained in the peak intensity (222) of the c-In_2O_3, showing an effectiveness in etching films for this system.

The measured electrical resistivity on etched samples, with heat treatments between 300 and 350 °C, was in the order of 10^{-1}–10^{-2} Ω cm. This result shows that the CdS film resistivity in which In is diffused is orders of magnitude between 9 and 10 compared to the film without treatment. This can be interpreted as an effect of doping In in the CdS film.

When the In/CdS bilayers are subjected to heat treatments, at temperatures as low as 100° C, the indium layer starts to lose brightness becoming opaque and even

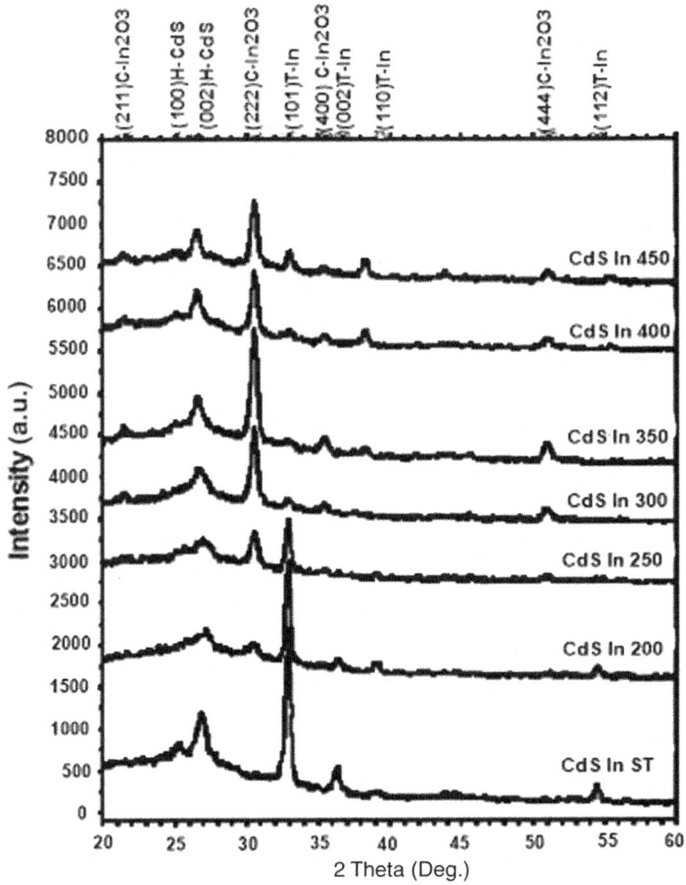

Fig. 7.12 Series of CdS films diffractograms with indium, treated at different temperatures

more around the edges; then this opacity is lost in a nonuniform manner becoming transparent but in a different color than the CdS. These changes are extended towards the center of the bilayer as the treatment temperature is increased up to 300 °C when the material reaches a uniform brown color and is completely transparent. This behavior is represented in Fig. 7.14 in which a decrease is observed in the optical density in the region of long wavelengths with increasing heat treatment temperature.

Data from the optical absorption spectra were used to estimate the Eg values of this series of materials, using the absorption edge region adjusted to the parabolic band model of a "direct bandgap" semiconductor for obtaining the width of forbidden energy bands (Eg) of the films. This was done by adjusting the absorption edge region of the spectrum [DO * E] 2 versus E to a straight line. Since in the spectra of the untreated samples and the ones treated at 200 and 250 °C are overlapped absorption (or reflection) of In, the Eg value, found for these CdS films by adjusting the "direct

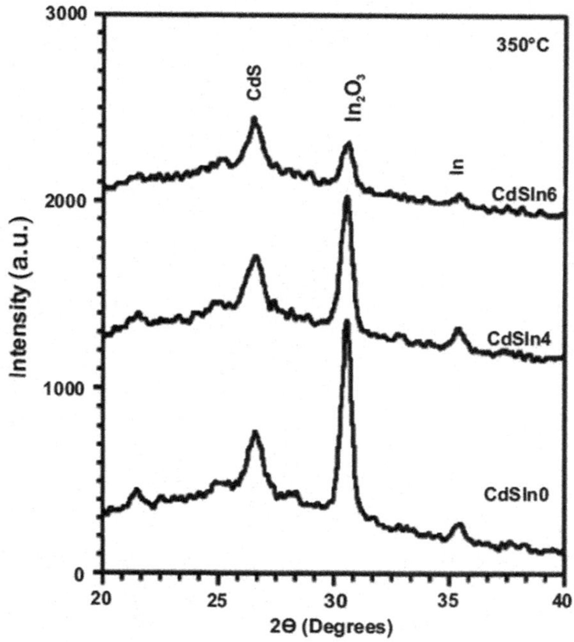

Fig. 7.13 CdSIn DRX-350 sample and the samples resulting from pickling for 4 and 6 min with a dilute HCl solution

Table 7.1 Variation of Eg of CdS substrate (ST) and In/CdS: with heat treatment

T (°C)	ST	200	250	300	350	400	450
Eg_{AL} (eV)	2.45	2.44	2.46	2.46	2.51	2.52	2.42
Eg_{D} (eV)	2.64	2.64	2.60	2.51	2.45	2.47	2.47

gap" model semiconductor with parabolic bands, is not accurate. For this reason, for comparison of Eg of CdS films that are heat treated, the criterion of the derivative values is used. Figure 7.15 shows the graphs of d (DO)/dλ versus wavelength (λ), where the Eg values for this series of samples were obtained (Table 7.1).

In Fig. 7.16, the Eg values of CdS films heat treated without and with the evaporated layer of In are compared. In both cases, the estimated value of Eg is plotted by the method of the derivative. Eg values of films with the In layer are greater throughout the temperature range. In both cases, E_g decreases to a minimum at a temperature of 350 °C. The variation of E_g in this range is about 0.2 eV. For higher temperatures, *Eg* virtually has no variation. If the decrease in E_g of CdS films with heat treatment is due to stress relaxation at the interfaces, then the difference between the E_g of the two kinds of films treated at the same temperature may be due to the additional stresses introduced in the CdS-In interface. Pure CdS film can relax in an easier way; the stresses with the substrate can be treated thermally, while the CdS film with the In layer relaxes to a lesser extent. Stresses maintain a compressed CdS network, which produces an increase in E_g.

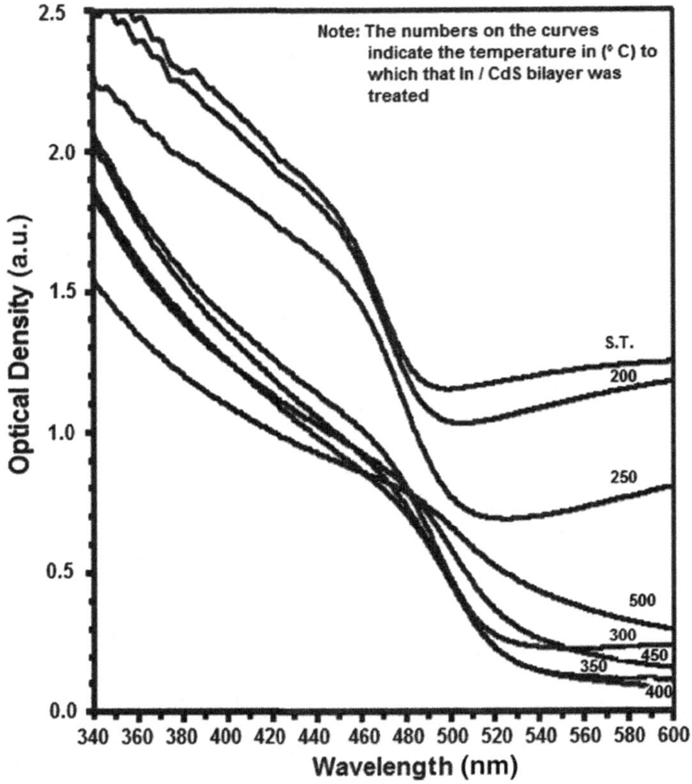

Fig. 7.14 Optical absorption spectra of heat-treated CdS series with indium

Figure 7.17 shows the two-dimensional images obtained by atomic force microscopy (AFM); the surface morphology for three samples of In/CdS, one without heat treatment and the other two with heat treatments of 300 and 400 °C are presented in this figure. The first image of this figure shows the semi-growth of In by way of clusters. This morphology is preserved on the surface of the samples treated at 300 and 350 °C, but is smaller. In these two images, the observed grains correspond to the In_2O_3 material, as will be demonstrated later.

Samples of this series were analyzed by spectroscopic ellipsometry to determine the diffused structure of the indium layer with thermal annealing. Analysis of the previous results and modeling preliminary tests with two layers led to a three-layer model, the upper layers and half (1 and 2) had to be set by considering as mixtures of two materials with dielectric functions ε_a and ε_b and with volume fractions f_a and $f_b = 1 - f_a$. One of these materials had to be different for the middle layer only in the sample treated at 250 °C. The dielectric functions that were used were taken from the literature, as indicated in [36, 37]. The distribution of layers proposed, adjusted by ellipsometric considerations, is presented in Fig. 7.18.

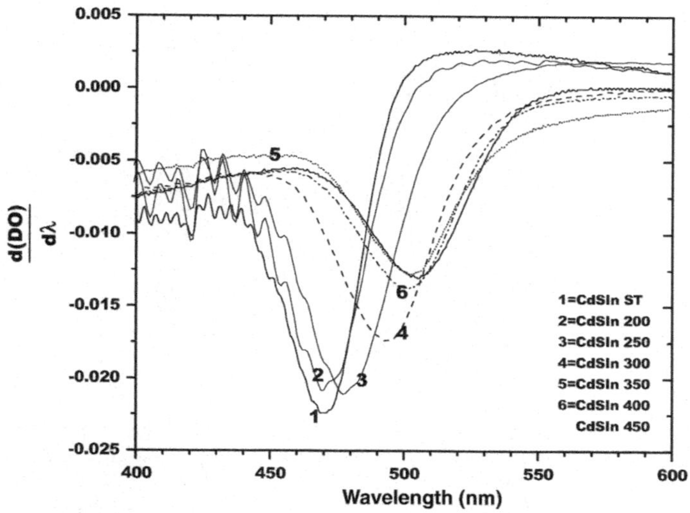

Fig. 7.15 Derivative of optical density versus wavelength

Fig. 7.16 Comparison of the variation of Eg between pure CdS films and CdS films with indium, heat treated at different temperatures

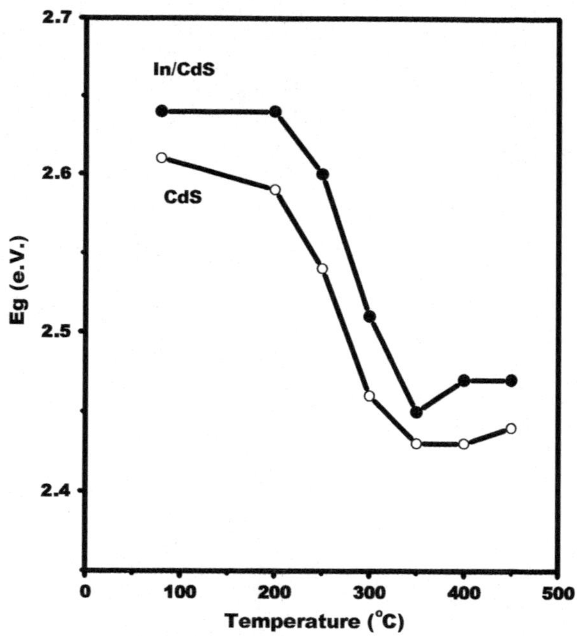

Figure 7.19 shows the two types of ellipsometry spectra that represent the In/CdS heat treated at different temperatures between 200 and 500 °C. Representative spectra correspond to samples treated at 250 and 300 °C; for the samples treated at temperature higher than 300 °C, interference oscillations are due to the effect of

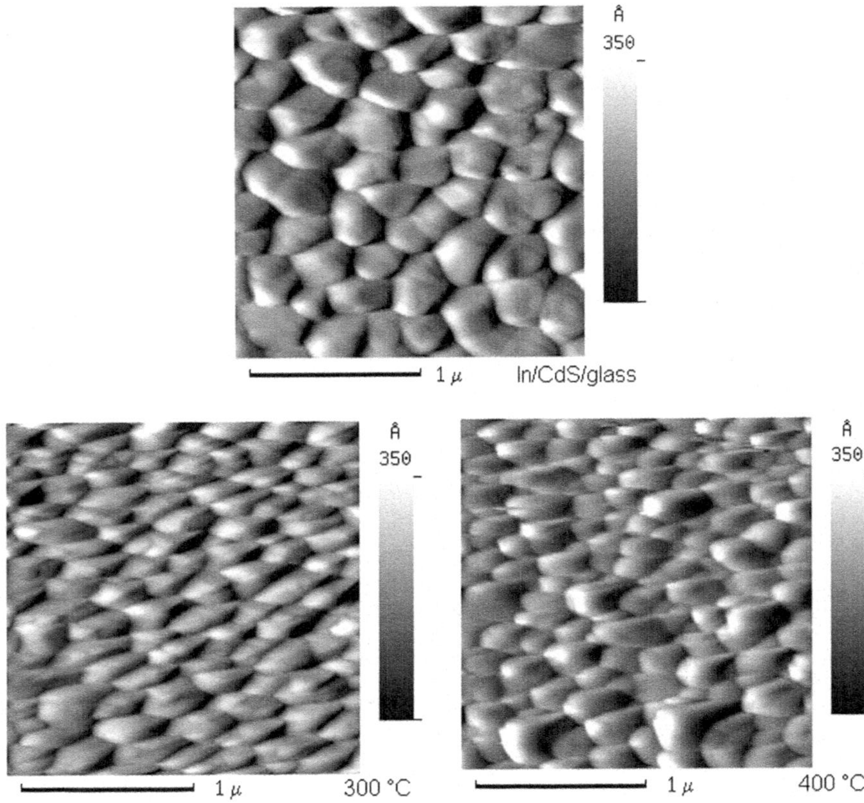

Fig. 7.17 CdS:In photographs by atomic force microscopy

Fig. 7.18 Multilayer structure for better ellipsometry fit for In/CdS system with heat treatment at 250, 300, 350, 400, and 450 °C

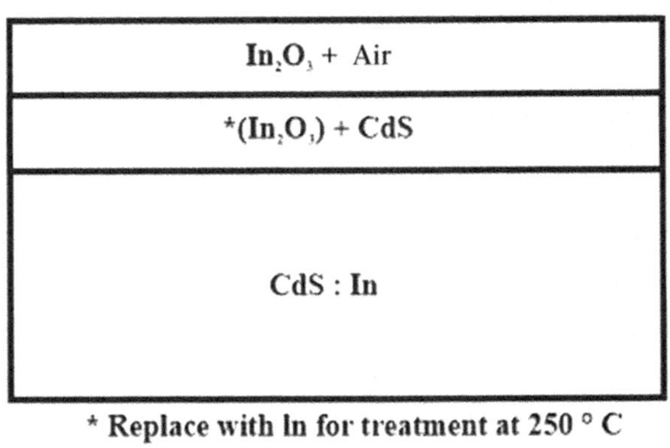

the film-substrate interface as well as the thickness and Eg of the films relative to the wavelengths. A semicontinuous surface morphology remains in the heat-treated

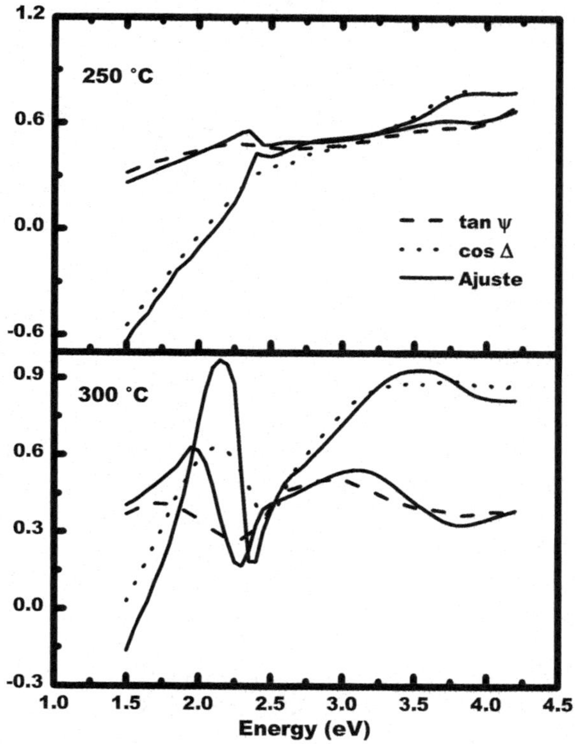

Fig. 7.19 Ellipsometric phase spectra of CdS:In materials: the spectra at the top correspond to the treatment at 250 °C and the ones at the bottom to the treatment at 300 °C

samples. For the samples heat treated at temperatures below 300 °C, the presence of a noticeable amount of indium in the middle layer (2) strongly inhibits retroreflection.

As a result, it follows that at a heat treatment temperature of 250 °C, most of the evaporated indium has been oxidized, leaving only a small amount in the transition zone (layer 2), which in the subsequent treatment is evolved into In_2O_3.

In Table 7.2, it can be seen that, with increasing heat treatment temperature, the upper layer (layer 1) decreases its thickness, while this is offset by the increased thickness of the middle layer (layer 2). The complete series of fitting parameters are shown in Table 7.2.

In conclusion, it has been possible to dope the CdS films formed by chemical bath with controlled pH, from the diffusion of In atoms. By utilizing Ellipsometry technique, the oxidation process and diffusion of metallic indium into the CdS film was modeled. A selective etching considerably reduced the indium oxide layer, allowing to obtain a significant approximation of the value of the electrical resistivity of the doped CdS, yielding a resistivity of the order of 10^{-1}–10^{-2} Ω cm for samples in a temperature range of 300 to 350 °C.

Heat treated Cu/CdS bilayer

Similar to the previous study of CdS doping with indium, this section focuses on a study of CdS doping with copper. In reference [38], a CdS solar cell homojunction

Table 7.2 Ellipsometric parameters obtained by modeling

T (°C)	Layer 1: In_2O_3 + Air		Layer 2: In/In_2O_3 + CdS		Layer 3: In + CdS	
	t_1 (Å)	f_1 (%) In_2O_3	t_2 (Å)	f_2 (%)	t_3 (Å)	f_3 (%) In
250	372	54.4	69	75.7 In	984	1.2
300	341	57.8	212	65.3 In_2O_3	977	4.3
350	280	49.3	239	90.0 In_2O_3	916	4.4
400	241	72.2	344	27.1 In_2O_3	916	4.4
450	233	55.4	463	28.8 In_2O_3	916	4.4

has been prepared, starting from a type-n CdS film which is deposited over a layer of 65 nm copper and is diffused by heat treatment. In this work, evidence is shown that copper diffuses 100–200 nm of CdS layer by heat treatment at 300 °C for 90 min in vacuum. In this case, in the intrinsic part of CdS films, a layer of 50 nm of Cu was evaporated and subjected to heat treatment in air for 1 h at temperatures between 150 and 300 °C. Before heat treatment, the Cu/CdS bilayers have reddish brilliance, characteristic of copper and are quite transparent as the thickness of the deposited copper film is quite small.

The diffractograms of this series of CdS with copper films are shown in Fig. 7.20. In the bilayer film without heat treatment, diffraction peaks are observed at $2\theta = 26.6°$ and $2\theta = 43.2°$ for hexagonal CdS and cubic Cu, respectively. For samples that are heat treated at 150 °C for one hour, it can be seen that copper reacts with oxygen, forming Cu_2O. Unlike the temperature at which CdO is formed in the films of pure CdS and in In/CdS exposed to heat treatment (at 350 °C, see Figs. 7.12 and 7.39), when CdS is shielded by copper, CdO starts to be detected from 250 °C. As the treatment temperature is increased, a decrease in the Cu_2O peak intensity is observed followed by the appearance of a peak for CuO, showing the natural evolution of copper oxidation: Cu → Cu_2O → CuO.

When the Cu/CdS bilayers are subjected to heat treatments, at temperatures as low as 100 °C, a slight opacity appears from the inside to the surface, then begins to get transparent, until toward 300 °C at which the material as a whole becomes completely translucent (same effect as in the case of the spread of indium); the color at this level is dark reddish yellow in contrast to the strong characteristic of CdS.

In Fig. 7.21, the transmission and reflection spectra of Cu/CdS system are presented. In all the cases, the transmittance of the samples is below 70%. This is due to the presence of copper and their oxides which are absorbers in the entire region of the spectrum [38]. The sample with the highest transmittance corresponds to the treatment temperature of 200 °C; this temperature is optimal for the formation of the Cu_2O phase which is an absorber at 550 nm; this temperature does not initiate the formation of the CuO phase which is an absorber at 800 nm [38]. In this spectra, the absorption edge of CdS is observed at about 500 nm. Reflectance spectra due to the effects of copper and their oxides in the samples are also manifested.

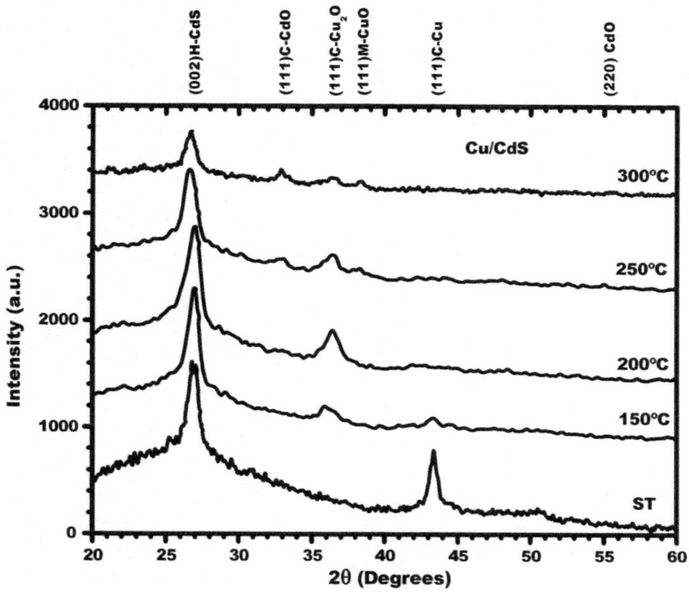

Fig. 7.20 X-ray diffraction patterns of CdS films with copper and those of specimens with heat treatment at different temperatures, as indicated in the diffractograms

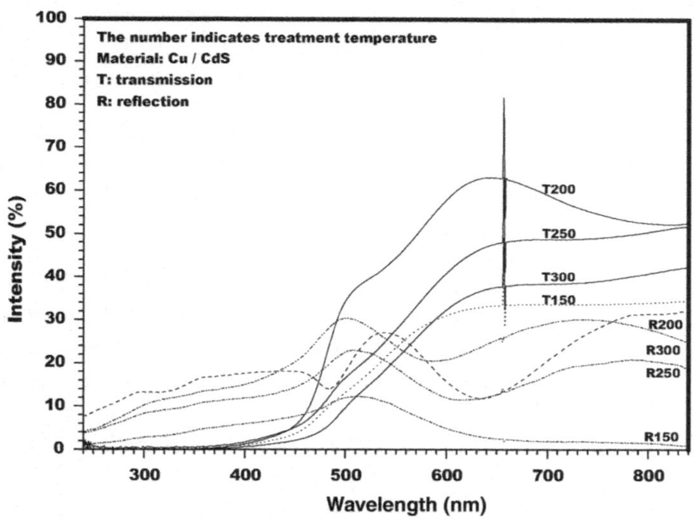

Fig. 7.21 Transmission and reflection spectra for a series of films of Cu/CdS, treated at different temperatures

Figure 7.22 shows the derivative of the transmission spectra of Fig. 7.21. In this figure, the shift to higher wavelengths representing the relative maximum absorption

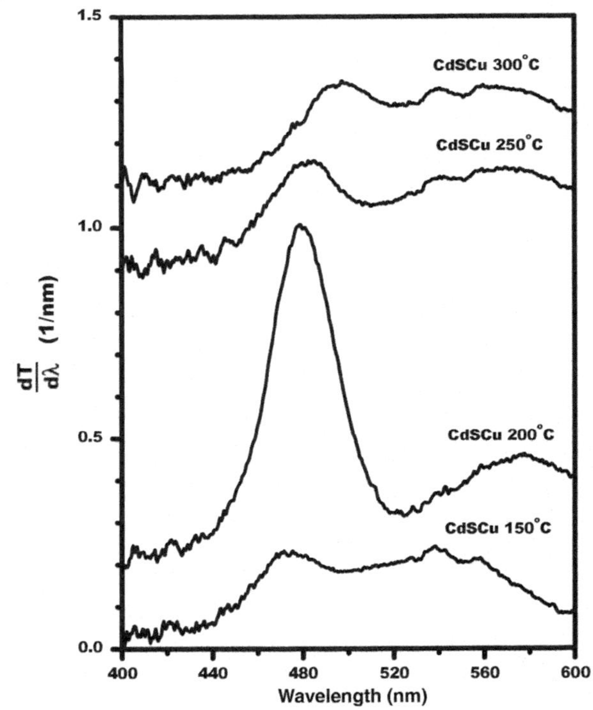

Fig. 7.22 Derivative of transmittance with wavelength for samples of Cu/CdS, treated at temperatures of 150, 200, 250, and 300 °C

Table 7.3 Variation of the values of forbidden energy gap of Cu/CdS system and those of the samples heat treated

T (°C)	Cu/CdS ST	150	200	250	300
Eg_D (eV)	2.64	2.62	2.59	2.56	2.49

edge of CdS is noticed. This indicates that the Eg of CdS increases with the treatment temperature and it occurs with films without and with the layer of evaporated In. At higher wavelengths, the absorption produced by the oxides of copper is observed. From these spectra, the Eg of each film in the series was determined. These values are summarized in Table 7.3.

Figure 7.23 shows the graph of variations of Eg of the series of bilayers Cu/CdS under different heat treatments. In order to compare their behavior, the variation of Eg for pure CdS films is also included. The same effect as in the case of CdS-In is observed; the Eg values of heat treated CdS-Cu films are greater than those for films of pure CdS, although in this case, the difference between Eg is less. In terms of stresses in the interfaces of CdS films, this result supports the conclusion that the stresses caused by copper oxides are lower than those produced by the In oxide allowing the intermediate film of CdS to be more easily stress relaxed.

Figure 7.24 shows the morphology of the films in the series of Cu/CdS with heat treatment, obtained by scanning electron microscopy.

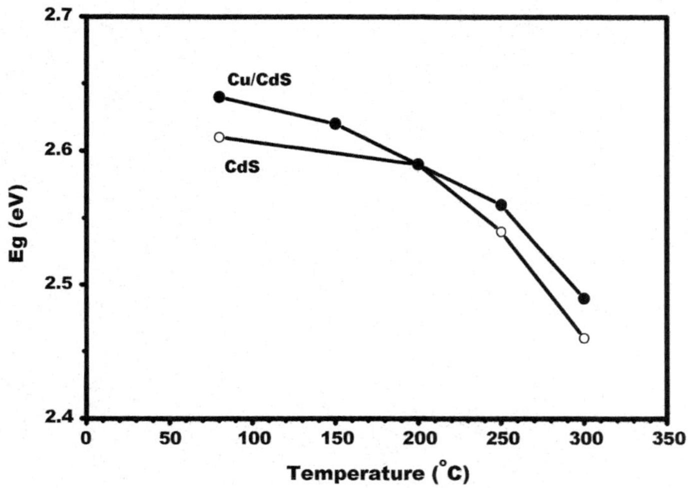

Fig. 7.23 Variation of E_g of films of CdS with copper, heat treated at different temperatures and compared with variations for pure CdS films

The morphology of this Cu layer on CdS is very similar to the one with the pure CdS, i.e., it also shows a flat bottom with some aggregates as islands of the order of 1 μm in diameter.

In the same way, as in the case of CdS-In films, the electrical resistivity of the CdS–Cu films was measured. It was found that the electrical resistivity of the CdS layer of the sample treated at 250 °C was approximately in the order of 10^{-1} Ω cm, indicating that CdS has been effectively doped with Cu.

In summary, doping of CdS films formed by a controlled pH chemical bath, from the diffusion of Cu atoms, has been demonstrated. A resistivity of the order of 10^{-1} Ω cm for these samples at a heat treatment temperature of 250 °C has been obtained.

(b) Quaternary compounds

The CdS-related quaternary compound is CdZnSSe which is known for its applications as photonic materials in lasing actions [39]. Due to the complexity in quaternary composition and structure, molecular beam epitaxy is the most appropriate technique for its synthesis [39]. Such a technique requires a crystalline substrate. Klude et al. prepared CdZnSSe on GaAS:Si coated substrates, in an MBE system consisting of growth chambers which are connected by a UHV transfer tube. The chamber was equipped with sources for Cd, Zn, S, and Se, and cells were Knudsen cells. The application of such material in lasing action is described in Sect. 7.1.3.

The properties of CdS and related ternaries and quaternaries are discussed in the following section.

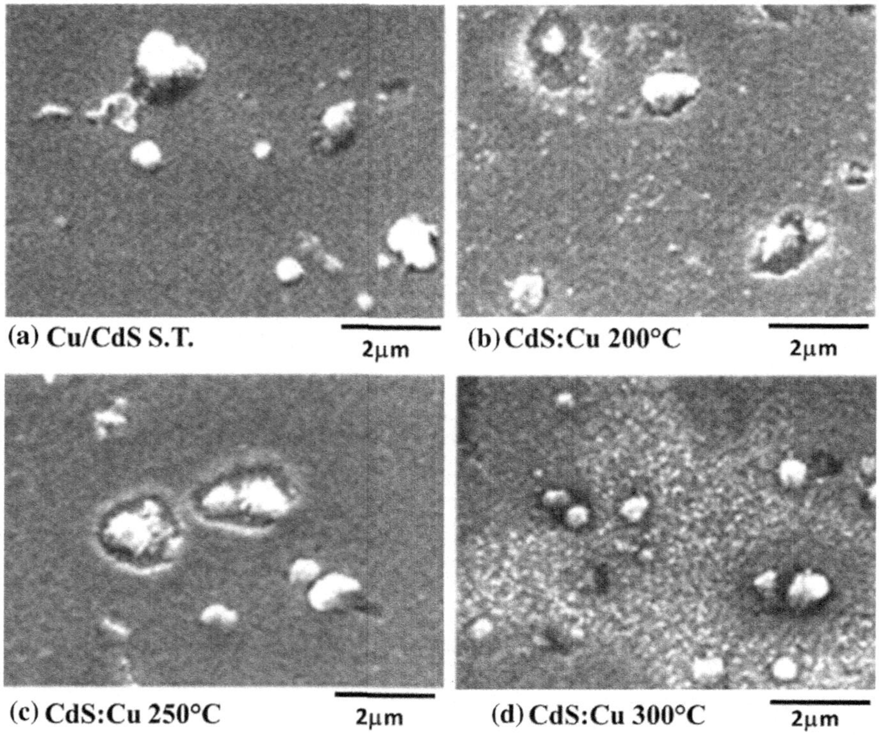

Fig. 7.24 SEM micrographs of Cu/CdS films heat treated at different temperatures

7.1.2 Properties

In this section, mainly, the optical and electrical properties of CdS and related materials formed at various conditions based on the published data are summarized, considering their applications.

After the discovery of bulk CdS, the basic research results have established that such properties depend on the microstructure and composition, atomic arrangement and size of the particle in one, two or three dimensions. Both optical and electrical properties of CdS and related ternaries and quaternaries are explained below. The variation in these properties depending on the sample conditions are discussed.

Optical properties: As mentioned at the beginning of this chapter, in the bulk form, (hexagonal) cadmium sulfide possesses a direct bandgap (E_g) of 2.5 eV corresponding to a fundamental optical absorption at the center of the Brillouin zone (Γ), based on band structure given in Fig. 7.25 [15].

A typical method to evaluate the bandgap of CdS thin films is by using their optical absorption spectra [11, 23]. In most cases, transmittance and reflectance spectra are measured using a spectrophotometer from which the absorption spectra

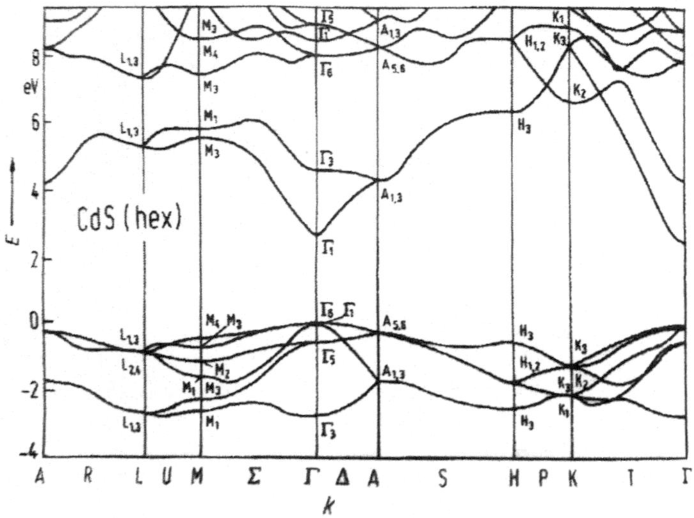

Fig. 7.25 Band structure of CdS (hexagonal) [15]. Symmetry points of the band structure are given and the center of the Brillouin zone Γ is marked (with permission from Springer)

can be evaluated. The absorption coefficient (α) for a specific absorbing frequency (ν) region is calculated from the equation:

$$\alpha = \frac{1}{d}\ln\left[\frac{(1-R)^2}{T}\right] \tag{7.6}$$

where d is the thin-film thickness. The $(1 - R)^2$ term accounts for the transmission through the front and back surfaces, while the ln factor gives the decrease in intensity due to the absorption according to Beer's law. Knowing the absorption spectra, for a direct bandgap material such as CdS, the E_g is related to α:

$$(\alpha h\nu)^2 = A(h\nu - E_g) \tag{7.7}$$

where A is a constant. According to Eq. (7.7), the plot $(\alpha h\nu)^2$ versus $h\nu$, also known as Tauc plot, is a straight line for all values of $h\nu > E_g$. Hence, the extrapolation of the straight line on to $h\nu$-axis gives the E_g value. Transmittance (T) and reflectance (R), the absorption spectra, and the evaluation of E_g for CdS thin films grown by CBD, as described in Sect. 7.1.1 (chemical solution deposition), are shown in Fig. 7.26. The calculated value of $E_g$is 2.58 eV.

It has been established that, as the particle size of CdS decreases to nanoscale level (less than 10 nm), the electronic structure and properties of the materials change due to the influence of two factors – the size confinement effect and an increase in the surface area-to-volume ratio [9, 15]. Theoretical studies show that the quantization

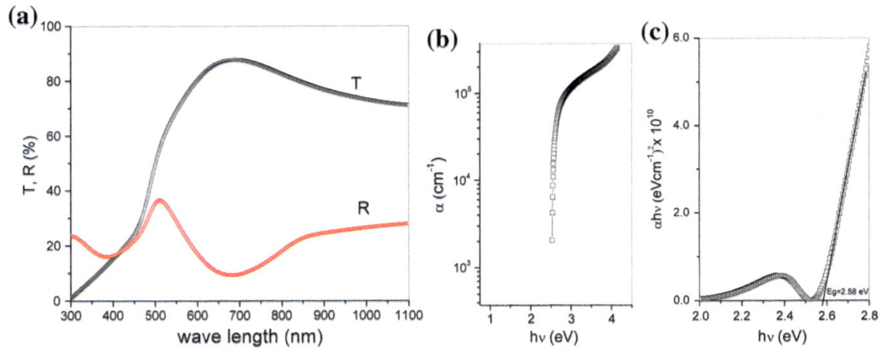

Fig. 7.26 **a** Transmittance (T) and reflectance (R) spectra. **b** Calculated absorption spectrum using Eq. 7.6. **c** Tauc plot and evaluation of Eg for CBD grown CdS thin films of thickness 100 nm (*based on our laboratory experiments*)

due to spatial confinement effect results in the absorption edge for CdS particles comparable or smaller than the exciton radius 5.8 nm, shifted by

$$E = E_g + \frac{0.71\pi^2\hbar^2}{2\mu r^2} \tag{7.8}$$

where E_g is the bandgap in the bulk crystal, and μ is the reduced mass of exciton given by

$$\frac{1}{\mu} = \frac{1}{m_e^*} + \frac{1}{m_h^*} \tag{7.9}$$

m_e^* and m_h^* are the electron and hole effective mass, respectively. For CdS $m_e^* = 0.2m_e$ and $m_h^* = 0.81m_e$, and m_e is the rest mass of the electron. Bandgap increase (blue shift) has been reported for thin films as well as nanoparticles depending on their size [9].

An experimental evidence [40] for the optical bandgap variation with particle size is well observed in CdS–CdSe core–shell nanoparticles with diameter of 1.7 nm (blue) up to 6 nm (red). This effect is illustrated in Fig. 7.27 in which the core–shell fluorescence along with schematic diagram of size effect and bandgap is presented. The difference in wavelength between absorption and fluorescence occurs due to the formation of exciton and subsequent relaxation to the new charge distribution. Thus, ternary compounds such as Cd(S,Se) and CdZnS show change in bandgap compared to CdS depending on Se or Zn doping concentration, respectively.

By varying the doping contents of Se, the change in optical transport properties of 1D Se-doped CdS nanostructures with different doping contents and/or crystallization degrees is measured [41]. The locally excited photoluminescence (PL) studies on the Se-doped CdS nanostructures such as nanobelts and nanowires show a significant redshift during transport along the long axis of the 1D structure and can

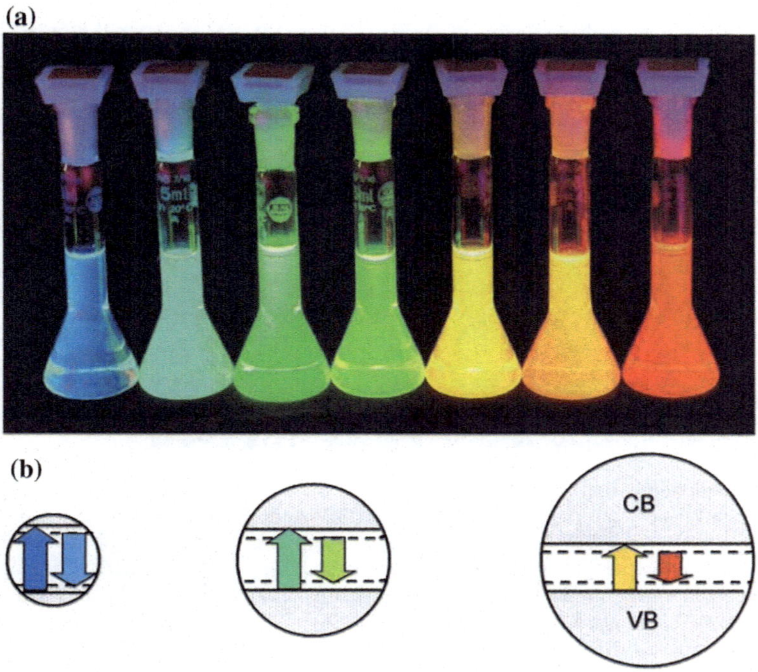

Fig. 7.27 **a** Fluorescence of CdSe–CdS core–shell nanoparticles with a diameter of 1.7 nm (blue) up to 6 nm (red), giving evidence of the scaling of the semiconductor bandgap with particle size (with permission from RSC [40]). **b** Schematic representation of the size effect on the gap between the valence band (VB) and the conduction band (CB) and the absorption (up arrow) and fluorescence (down arrow). Smaller particles have a wider bandgap

leave enough PL intensity for detection. The magnitude of the redshift is found to be highly dependent on the content of doping and the crystallization degree as shown in Fig. 7.28 [41].

Blueshift in bandgap for CdS and CdZnS nanoparticles, due to the variation in the particle size, is also known, as shown in Fig. 7.29 [34] and Fig. 7.30 [7], respectively.

Electrical properties: Electrical properties of CdS are governed by transport properties of electrons in the conduction band (Γ_7) and by holes in the valence band (Γ_9), from Fig. 7.25 [8]. Because of the compensating effects during the sample synthesis, CdS is normally n-type. Photoluminescence studies [42] have shown that the n-type defects can be sulfur vacancies and cadmium interstitials. Possible acceptor defects can be cadmium vacancies and sulfur interstitials. Such n-type defects are very shallow (0.7–0.9 eV) below conduction band and p-type defects are very deep (1.6–1.9 eV) above the valence band, hence unlikely in CdS samples. P-type conduction is observed in CdS only in a few special cases such as copper-doped samples, in vacuum co-evaporated CdS doped with copper (CdS:Cu) samples [43]. In such case, copper substituted cadmium (Cu_{Cd}) shallow acceptor levels of 0.69 eV were detected.

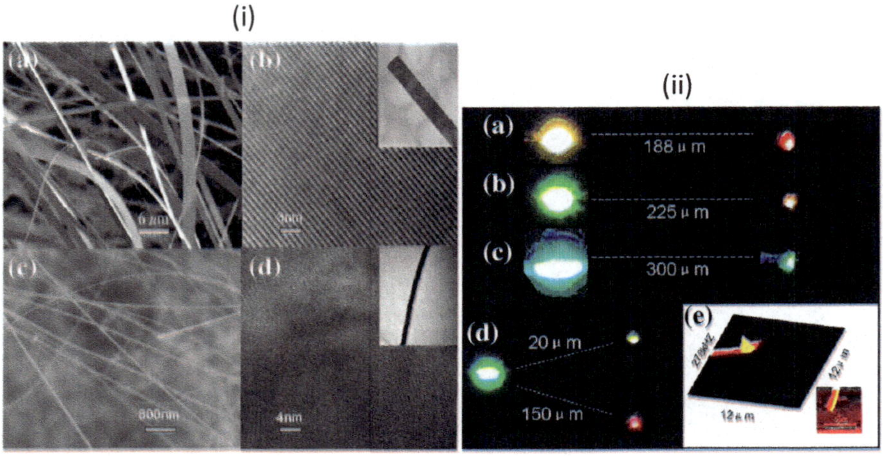

Fig. 7.28 (with permission from ACS): (i) **a** Typical SEM image of the as-prepared nanobelts by metal-catalyzed crystallization method, **b** the HRTEM lattice image of a representative belt as well as its corresponding TEM morphology image (inset), **c** the SEM image of the as-prepared Se-doped CdS nanowires grown at low deposition temperatures, and **d** the HRTEM image and the corresponding TEM morphology image (inset) of a representative wire [41]. (ii) **a**–**c** Far-field optical images of single excited $CdS_{0.65}Se_{0.35}$, $CdS_{0.82}Se_{0.18}$, and CdS nanobelts, respectively: left, the in situ PL images under laser illumination; right, the emission images of the examined ends. **d** The far-field PL image of an excited $CdS_{0.86}Se_{0.14}$ nanowire under laser illumination: left, the in situ PL image; right, the end PL images with an excitation–detection distance of 20 and 150 ím, respectively. **e** The three-dimensional near-field optical image of the end emission from the doped wire, with an excitation–detection distance of about 300 ím; inset, the corresponding morphology image of the end

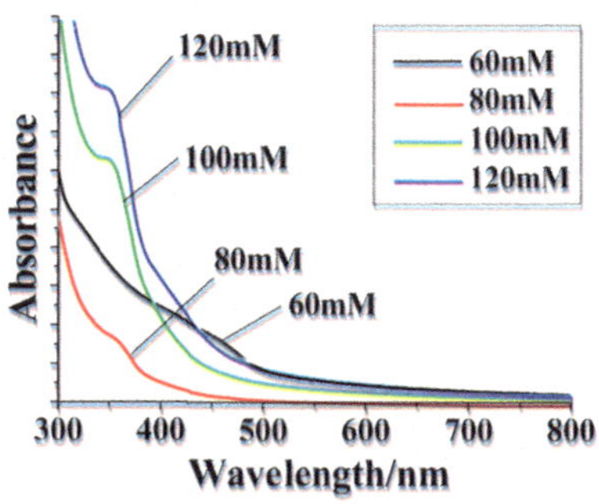

Fig. 7.29 **a** UV-vis spectra of CdS NPs with increasing concentration of the surfactant. The absorption peak of the nanoparticles shifts from 450 to 345 nm with increasing surfactant concentration from 60 to 120 mM resulting in decrease of the average diameter of the CdS NPs from 5 to 3 nm [34] (with permission from RSC)

Further, CdS thin films are well known for their high photoconductive nature [19]. Figure 7.31 shows the photocurrent measurements on CdS thin films prepared in our

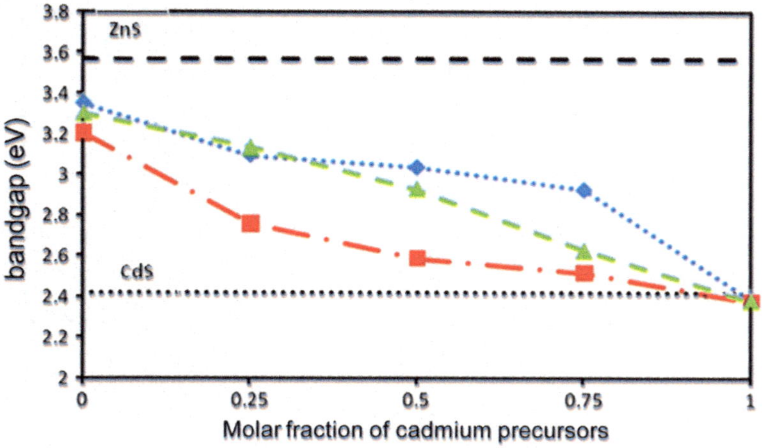

Fig. 7.30 Optical bandgap versus fraction of cadmium precursor **a** $Zn_xCd_{1-x}S$ films prepared using thio- and dithio-biuret zinc and cadmium complexes: 1,1,5,5-tetraalkyl-2-4-dithiobiurets [M(N-(SCNR2)2)2] and and 1,1,5,5-tetraalkyl-2-thiobiurets, M–Cd and Zn, R-methyl and iso-propyl. Dotted and dashed lines indicate the bandgap of hexagonal CdS and ZnS. Bandgap varies from 2.42 eV for pure CdS to 3.58 eV for pure ZnS [7] (with permission from American Chemical Society)

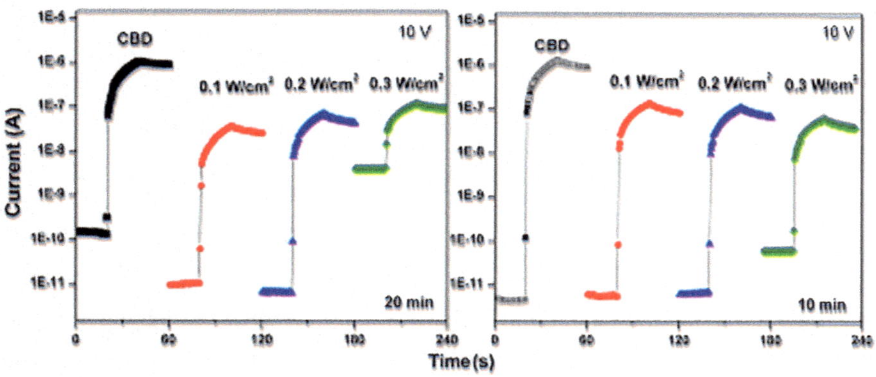

Fig. 7.31 Photocurrent curves for CdS thin films grown in 10 and 20 min by normal CBD and laser-assisted CBD (LACBD) techniques under laser power densities of 0.1, 0.2, and 0.3 W/cm^2 [24] (with permission from Elsevier)

laboratory by laser-assisted chemical bath deposition [24]. Under illumination, the current increases due to carrier generation in the thin films by absorption of light.

Electrical properties of ternaries and quaternaries have not shown much interesting results.

Case study: Optical and electrical properties of thermal annealed CdS thin films
Semiconductor films can be deposited on commercial glass, monocrystalline silicon, as well as electrically conductive glass (ITO); in this investigation, a first thin layer

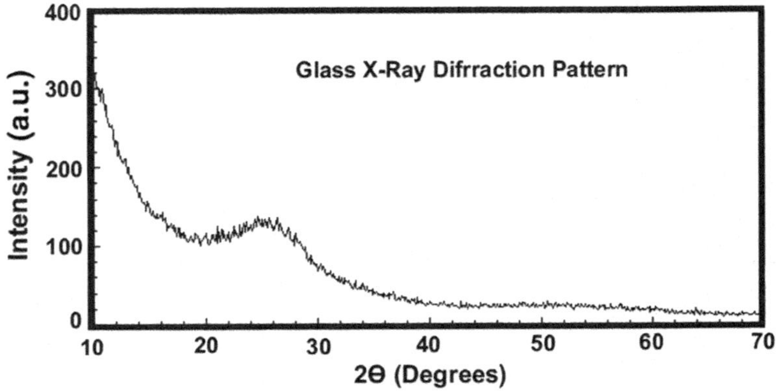

Fig. 7.32 Corning glass X-ray diffractogram

Fig. 7.33 Absorption spectrum of Corning glass

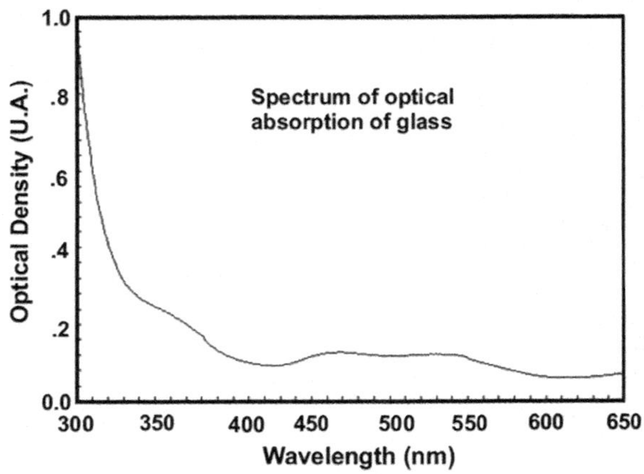

was deposited on a Corning glass substrate of dimensions 75 mm × 25 mm × 1 mm. These substrates are washed in a simple method with soap and water, rinsed with distilled water, and then dried at room temperature. This means that no treatment is required to activate the glass surface using reactants such as hydrofluoric acid, potassium permanganate or organosilanes.

Figures 7.32 and 7.33 show the responses to X-rays and the optical absorption of the Corning glass. This helps to justify that the amorphous X-ray signal may be due to the substrate, and to corroborate that this type of substrate absorbs in the UV region. A systematic study of the effects that may occur with the substrate has not been done, but if this is a single crystal, the orientation of the grown film could be influenced.

As ultraviolet radiation is considered to be between 200 and 400 nm and the visible between 400 and 800 nm, from Fig. 7.33, it can be observed that it is not advisable to use quartz in the studies in a range of 300–650 nm as the glass absorption in this

Table 7.4 Summary of some characteristics of CdS

Material	CdS
Semiconductor type	II–VI
Molecular weight (g/mol)	144.48
Density (g/cm^3)	4.49
Melting point (°C)	1750
Films and powder color	Yellow
Eg at 300 °K	2.50 eV
Eg at 0 °K	2.56 eV
Refraction index at wavelength of 400 nm	2.5
Structure	Hexagonal
Lattice parameters	a=3.82 Å c=6.26 Å
Solubility	Acid solutions
Solubility product K_{ps}	8×10^{-27}

range is very high; however, this signal due to glass is subtracted as a background signal or reference.

CdS film growth

In this section, CdS films grown at 80 °C and its modifications by heat treatments will be discussed. These results will be used later to compare with similar processes performed with a protective metal layer or copper or indium.

CdS thin films grown at 80 °C

CdS films obtained by the CBD process at 80 °C become excellently bonded to the substrate; they are homogeneous, transluscent and greenish yellow. Table 7.4 lists the characteristics of CdS.

Figure 7.34 shows the diffraction pattern of this film, where three peaks can be seen at 2θ =24.8, 26.4 and 43.8. These diffraction peaks correspond, respectively, to the (100), (002), and (110) crystallographic planes of the hexagonal structure of CdS (CdS-h). Due to the relative intensity corresponding to the (002) peak, it is established that there is a preferential polycrystalline growth in this direction. If we compare these CdS films with other films grown also by chemical bath in a very similar but without the buffer [44, 45] solution, it is noticed that the structure of the films grown unbuffered does not show a preferential orientation; therefore, the main effect of the buffer solution is precisely to produce such orientation. By using Debye–Scherrer formula [46, 47] for the determination of grain size of these films, an average value of 16 nm was obtained.

According to Fig. 7.35, these films have an order of 90% optical transmission for wavelengths of about 550 nm gradually decreasing to longer wavelengths up to ~75% at 850 nm. This high transmittance is a desirable property for applications

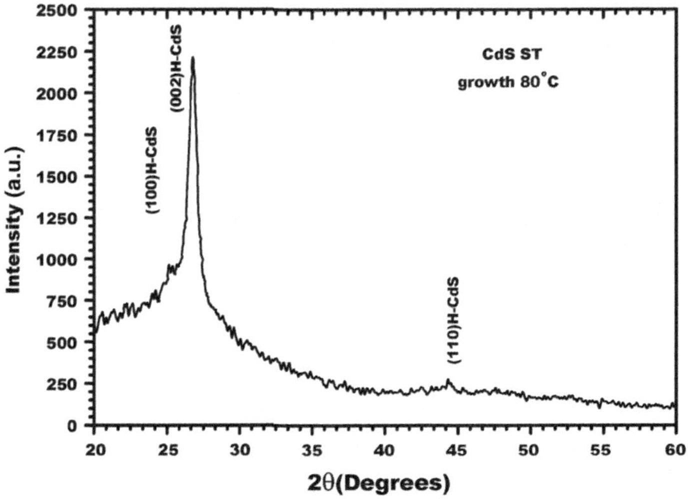

Fig. 7.34 Grown CdS film X-ray diffractogram on a Corning glass using CBD at a temperature of 80 °C with a reaction time of 1 h

such as transparent layer in a solar cell (optical window) [48]. It also shows that the absorption edge is around 480 nm which corresponds to energy of 2.58 eV. The absorption edge region was adjusted to the parabolic band model of a "direct bandgap" semiconductor [27] for obtaining the width of forbidden energy bands (Eg) of the film. This was done by adjusting the absorption edge region of the spectrum $[DO*E]^2$ versus E to a straight line. The result for this film was 2.58 eV. This value is greater than the Eg reported for CdS films deposited by CBD using other reactions as reported in references [49, 50]. This high value of Eg has been explained in terms of quantum confinement due to the small grain size of the films [51]. The reflectance spectra shows values lower than 23%, indicating a relative maximum in the transition region of the absorption edge.

In Fig. 7.36a–b, the morphology of CdS films without heat treatment, examined by atomic force microscopy (AFM), is presented. A flat bottom and over it, isolated aggregates of different sizes are observed. These aggregates are formed by groups of small CdS crystal grains, which is confirmed by scanning electronic microscopy (SEM) and energy dispersive X-ray spectroscopy (EDX). This type of morphology suggests that two types of growth are present, which determine the total thickness of the film: the ion by ion growth and the growth through aggregates [52]. At a higher resolution, the part b of Fig. 7.36 shows the morphology of the bottom where the aggregates are not found in the films. In the micrographs, it can be seen that some of the aggregates reach sizes of the order of 0.5 μm. In reference [53], an analysis on the influence of the reaction temperature on the size of this type of aggregates has been made. These aggregates have also been observed in CdS films deposited by CBD using different reagents, according to references [54, 55].

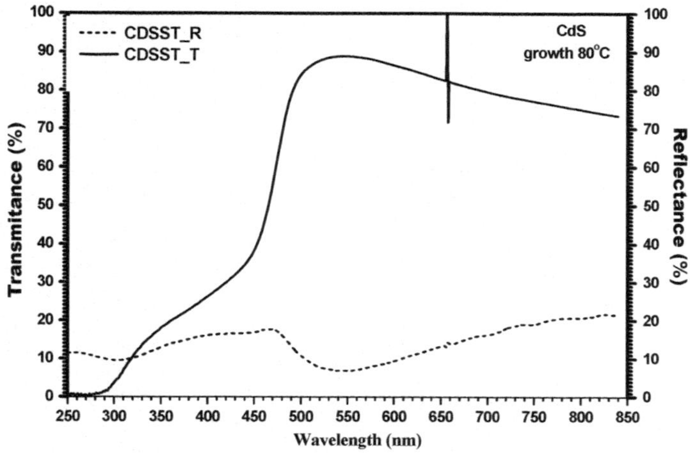

Fig. 7.35 Transmittance and reflectance spectra of CdS film on glass using CBD at a temperature of 80 °C with a reaction time of 1 h

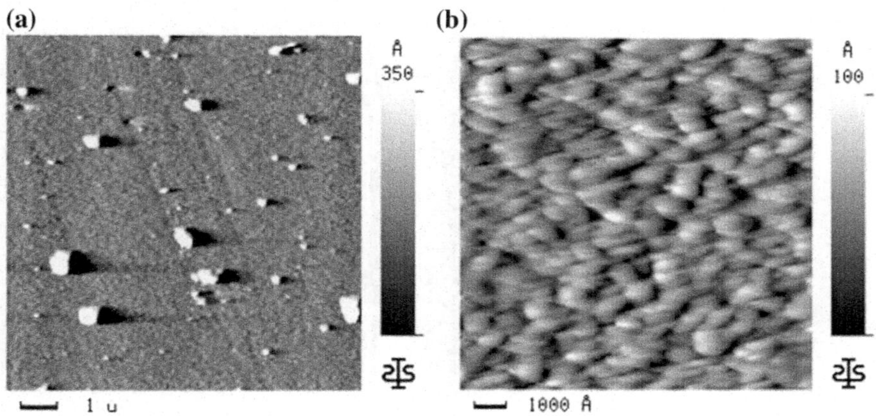

Fig. 7.36 AFM micrographs of a CdS film obtained at a temperature of 80 °C with a reaction time of 1 h: **a** scale of a micron and **b** scale of 1000 Å

The thickness of these CdS films for a time of one hour of reaction was of the order of 100 nm, see Fig. 7.37.

These films have electrical resistance in dark, in a square area of 0.25 cm^2 (R_{sh}, electrical sheet resistance), which is measured by two parallel electrodes of silver paint of length $l = 0.5$ cm and separated by a distance $d = 0.5$ cm; this resulted in $R_{sh} \sim 1 \times 10^{13}$ Ohms/square; its electrical resistivity is calculated (using the equation $\rho = leR_{sh}/d$, where (e) represents the thickness of the film, $\rho \sim 10^8$ Ω cm, a typical value in films deposited by CBD as those reported in references [51, 56].

The chemical composition of the surface of this film was analyzed qualitatively by Auger Electron Spectroscopy. The Auger spectra is presented in Fig. 7.38; and

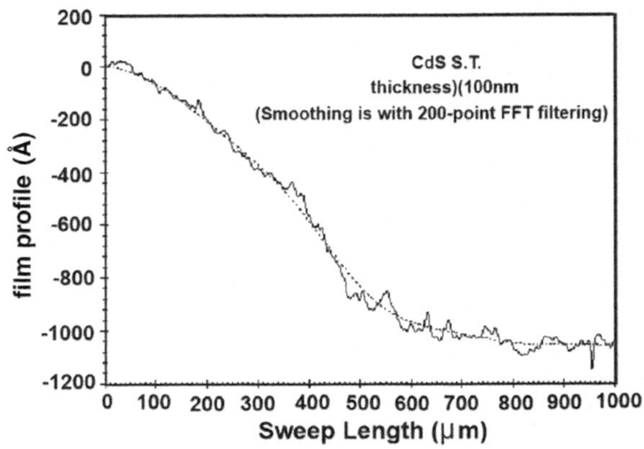

Fig. 7.37 A relief profile of CdS film deposited on glass using CBD at a temperature of 80 °C with a reaction time of 1 h

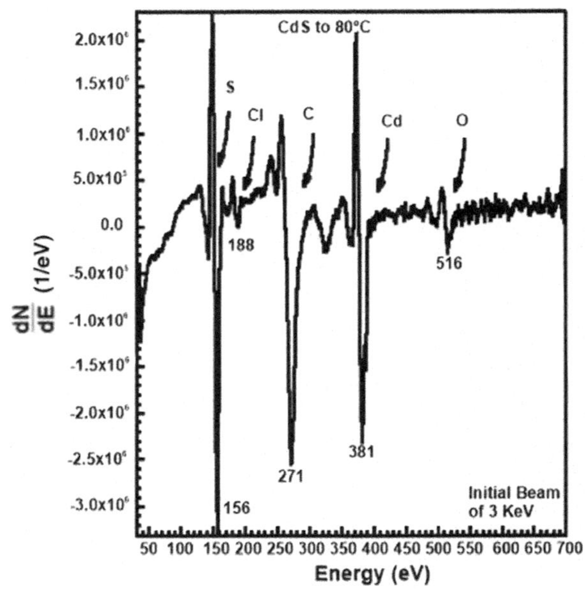

Fig. 7.38 Auger electron spectroscopy of CdS film prepared at 80 °C temperature with a reaction time of 1 h

the presence is observed of Cd and S in addition to the normal pollutants from the atmosphere (C, O, and Cl). In this spectra, it can be noticed that, excluding the atmospheric pollutants, CBD technique allows to obtain a residual contaminants free material in the chemical reaction.

In conclusion, these results demonstrate that the reaction formula implemented in this part of the work allows to obtain good quality CdS thin films. Among the properties that stand out, the process leads to a growth with high quality crystal preferential orientation and high transmittance values. For these films to be used in optoelectronic devices, it is necessary to decrease the electrical resistivity by effective doping. For this reason, the next section focuses on this direction.

Heat treated CdS thin film series

This section presents a study on the effect of heat treatment in air at different temperatures on the properties of CdS films described in the previous section. This allowed us to monitor the thermal oxidation process of these films. This analysis serves as a reference in the study of doping processes by thermal diffusion of impurities (Effect of heat treatment in In/CdS and Cu/CdS bilayers, Sect. 7.1.1). CdS films grown at 80 °C were subjected to heat treatments at temperatures in the range of 200–500 °C for one hour. A series of samples, to be analyzed considering the results of the previous section, was obtained.

Figure 7.39 shows the X-ray diffractograms of this set of sample. These diffraction patterns correspond to CdS films that were subjected to heat treatment of one hour at 200, 250, 300, 350, 400 and 450 °C in air. They exhibit the common peak corresponding to planes (002) of the hexagonal phase of cadmium sulfide (h-CdS); the heat treated films at 400 and 450 °C show peaks of the planes (111), (200), and (220) of cadmium oxide cubic phase (CdO-c) at $2\theta = 33$, 38.3, and 55.3°, positions, respectively. This sequence shows an increase in the peak intensity (200) of the h-CdS and a slight thinning at 400 °C, indicating an increase in grain size. It can also be seen in the sequence that the conversion from CdO to CdS process starts at 350 °C, since in the pattern of this sample, the signal appears very weak from the crystal planes of c-CdO (200). It is known that the sulfur evaporation process in CdS starts at temperatures greater than 300 °C in vacuum [57]. In air treatments, this process is expected to occur at higher temperatures. The oxidation of Cd occurs in decompensated Cd atoms when films release S atoms at temperatures of 350 °C onward. This effect is quite significant in the XRD pattern of the treated film at 450 °C showing a decrease in the peak intensity (002) of h-CdS.

In Fig. 7.40, the optical transmission and reflection spectra of the heat-treated films are shown. In the transmittance spectra, the characteristic wrapping corresponding to the absorption threshold of CdS, around 480 nm is observed. This absorption edge shifts to longer wavelengths with increasing treatment temperature. The transmittance remains over 80% in the absorption edge region, and above 70% in the 480–800 nm region of the treated films at 350 °C. Films treated at 400 and 450 °C show a decrease in transmittance in this region due to absorption of CdO [58]. The reflectance spectra show that the films reflect less than 20% of the incident light throughout the measured wavelength range. The spectra of the samples annealed at 350 and 400 °C exhibit a reflectance lower than 10%.

For all films, the values of Eg were estimated from the transmittance spectra. These calculations were performed by two methods: the adjustment to the satellite bands model, and the derivative criterion [36]. This last criterion is shown graphically in Fig. 7.41 where the changes in the temperature derivative of the transmittance of the CdS films, heat treated at different temperatures, is presented. According to this criterion, Eg is given by the maximum of the $\mathrm{d}T/\mathrm{d}\lambda$ curve as a function of wavelength. The obtained values of Eg are presented in Table 7.5. Both Eg estimates show a decrease with increasing treatment temperature as can be observed in Fig. 7.42. This behavior has been associated with two effects produced by heat treatment: the

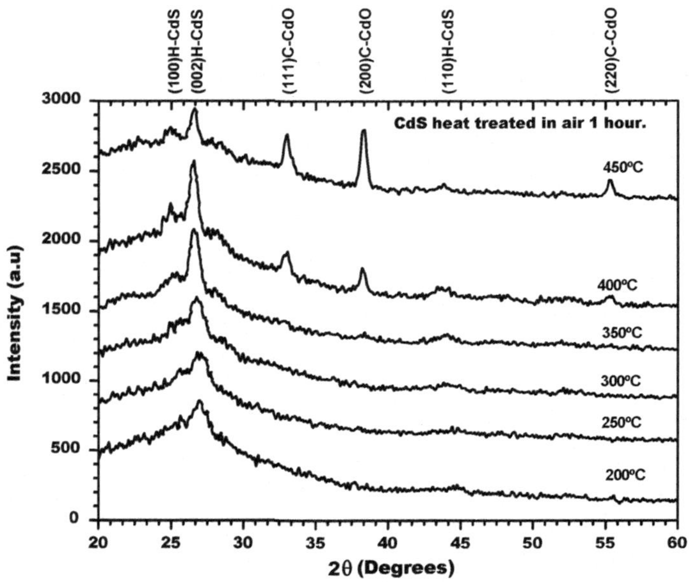

Fig. 7.39 X-ray diffraction patterns of a series of CdS films treated at various temperatures

Table 7.5 Variation of E_g of CdS by heat treatment

T (°C)	ST	200	250	300	350	400	450
Eg_{AL} (eV)	2.58	2.57	2.54	2.46	2.42	2.42	2.39
Eg_D (eV)	2.61	2.59	2.54	2.46	2.43	2.43	2.44

increased spacings between atoms [57] and the growth of CdS grain size of the films when thermally treated [49]. Both of the effects are noticeable in the diffraction patterns in Fig. 7.39.

Figure 7.43 shows the evolution of the morphology of this series of materials. It can be observed that, up to the temperature of 250 °C, the surface morphology is unaffected because it is very similar to the film without treatment as shown in Fig. 7.43a that exhibits a flat surface with isolated aggregates. Changes in both the roughness of the substance and the appearance of the aggregates can be perceived on the surface of the film with 350 °C treatment. Furthermore, it starts appearing as some shallow straight grooves. These changes are more noticeable on the surface of the sample heat treated at 400 °C. An interesting detail is that the direction of the grooves in this sample is approximately parallel. This may be an indication that these structures are fractures in the film along the direction of stress relaxation due to biaxial stresses from the film–substrate system. By increasing the treatment temperature to 450 °C, this effect disappears almost completely. Grooves are observed in some regions of the surface. The disappearance of the grooves coincides with the sharp

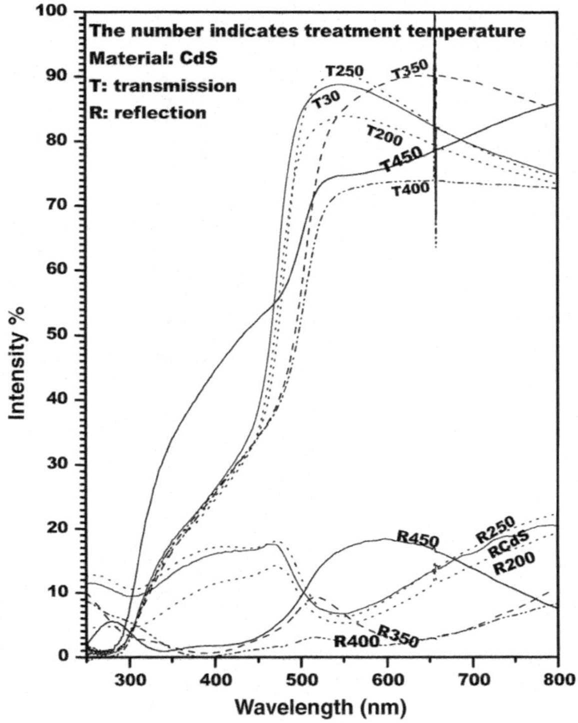

Fig. 7.40 Transmission and reflection spectra of CdS films heat treated at different temperatures

decrease in the crystalline phase in the CdS film and increased phase from the CdO film. This may be the cause of the disappearance of grooves at this temperature. Changes in surface morphology of the films produced by heat treatment can explain the optical reflectance behavior (Fig. 7.40). For example, the low reflectance of heat treated films at 350 and 400 °C is due to the high roughness induced by the scratching of the surface.

The electrical resistivity in dark, of the samples thermal annealed at 400 and 450 °C, varied from 10^1 to 10^{-2} Ω cm. This sharp increase in resistivity compared to the untreated CdS films, has been attributed to a conversion to a n-type doped semiconductor [59]. According to the previous analysis, the change in electrical resistivity is attributed to the presence of CdO in the films, which is a conductive material [58].

These results show that thermal annealing in CdS films for 1 h in air causes a decrease in the bandgap and a noticeable change in the surface morphology. The film oxidation starts from temperatures in the range of 350 to 450 °C; this leads to the presence of CdO in the composition of the film, producing a noticeable decrease in electrical resistivity.

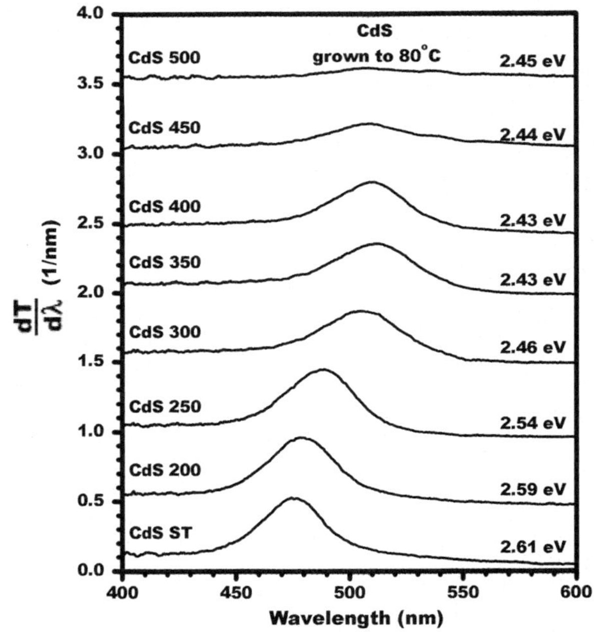

Fig. 7.41 Derivative of transmittance with wavelength versus wavelength of CdS films heat treated at various temperatures

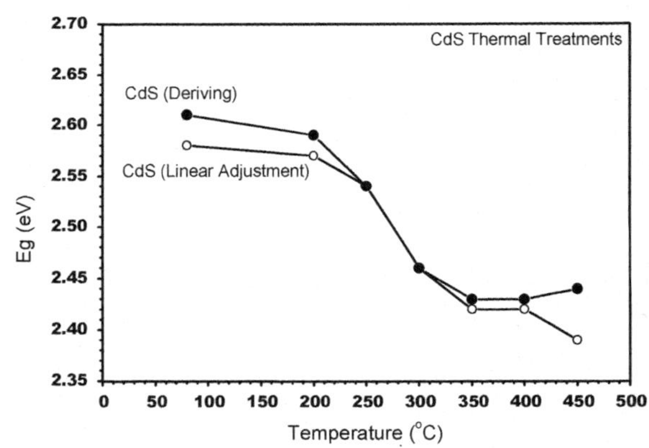

Fig. 7.42 Comparison of E_g of heat-treated CdS films obtained by linear fit of the data [DO * E] ^{2} versus E by the derivative method

7.1.3 Applications

CdS thin films and nanostructures have received much attention in recent years as building blocks due to their unique physical and chemical properties. The most important devices are CdS-based solar cells, lasers, and photodetectors. The device applications of CdS are described, followed by those of the ternaries and quaternaries.

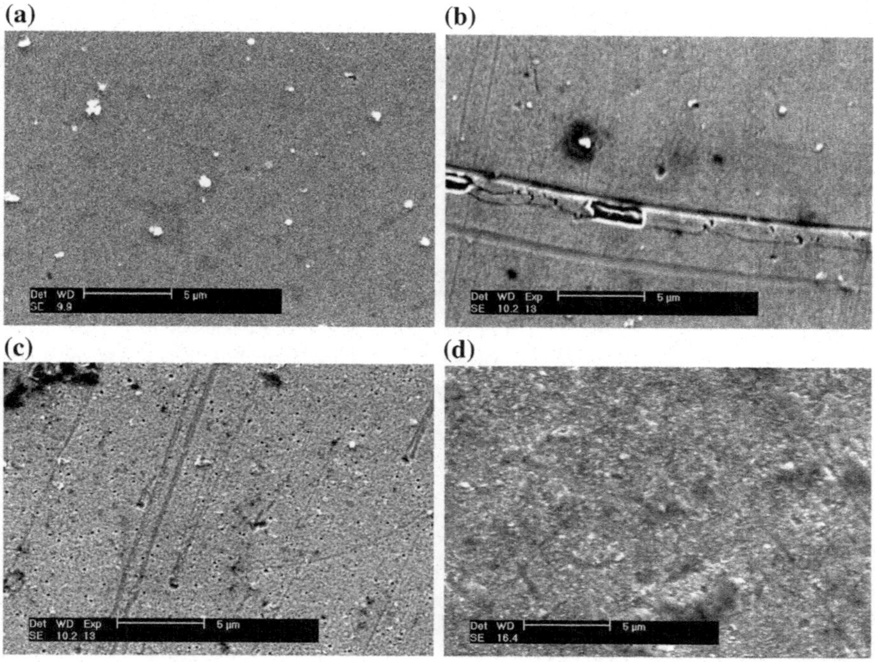

Fig. 7.43 SEM micrographs of CdS samples heat treated at temperatures: **a** 250, **b** 350, **c** 400, and **d** 450 °C

(a) **CdS in solar cells**

Since its discovery, CdS has been known for its application in solar cells due to its bandgap of 2.42 eV and the photocurrent properties. It has been suggested to use this material in different PV devices such as thin-film solar cells, dye-sensitized solar cells, and organic solar cells. However, after the development of thin-film technologies and thin-film solar cell technologies, CdS is established as the best window layer component in the most promising commercial solar cells namely: (i) CdS/CdTe and (ii) Cu(In, Ga)Se_2/CdS; both have achieved more than 20% efficiencies [60]. A recent analysis has helped to solve the conversion efficiency enhancement by utilizing a thin layer (200 Å) of CdS [1] in CdS/CdTe solar cells which is also applicable to Cu(In, Ga)Se_2/CdS PV devices. According to the report in the literature:

(a) For sufficient forward bias, there is no field quenching effect since the electric field on the CdS side of the junction varies with the voltage drop, and hence electrons can travel through the junction easily.
(b) The maximum field near the CdS/CdTe interface is too low to permit electron tunneling through the junction, which otherwise would shunt the junction and reduce the solar cell efficiency.
(c) The field quenching near the CdS/CdTe interface can limit the electron back-diffusion into the CdTe.

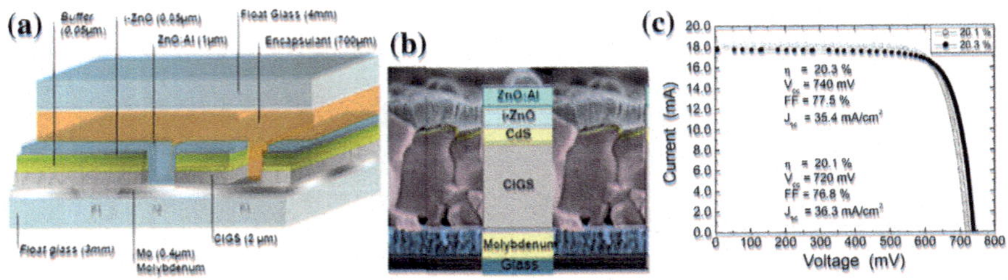

Fig. 7.44 Cu(In_{1-x},Ga_x) (CIGS) solar cell: **a** schematic structure of a typical PV device structure; **b** cross-sectional view of the device microstructure; **c** I/V curve of independently certified (by Fraunhofer ISE on 15th of April and 30th of June 2010) world record cells of 20.1% (cell area: 0.5 cm^2) and 20.3% (cell area: 0.5 cm^2) efficiency [62] (with permission from John Wiley & Sons, Ltd.)

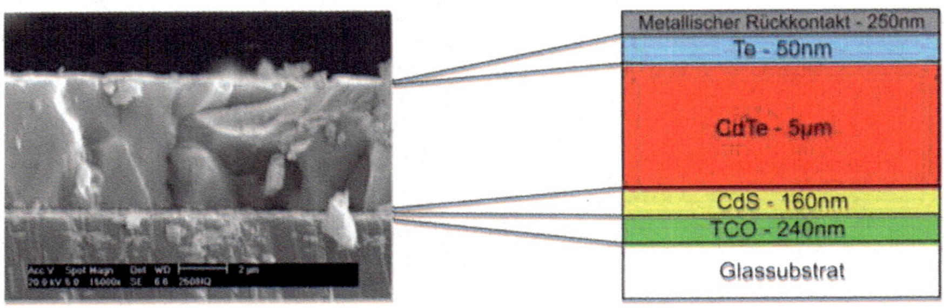

Fig. 7.45 CdTe solar cell **a** cross-sectional view of the device microstructure and schematic structure of a typical PV device structure [61] (with permission from Wolfram Jaegermann)

Figures 7.44 and 7.45 [61] present the schematic representation of solar cells in which CdS is used as a window layer and their performance, including record efficiencies, etc.

(ii) **CdS and related alloys in lasers**:

CdS nanowires are particularly interesting since optical excitation and electrical injection have been used to exceed the threshold for lasing applications. ZnSe-based laser diodes have the unique potential to deliver laser light in the entire short wavelength region of the visible spectrum. This is illustrated in Fig. 7.46 [39] which shows the photograph of three operating laser diodes. The emission color ranges from bluish green to yellow of operating ZnSe-based laser diodes. The different emission wavelengths of 505, 528 and 560 nm (from left to right) are achieved by varying the Cd content of the quantum wells.

(ii) **CdS and related alloys in photodetectors**:

Photodetectors exist in different configurations, namely, ohmic contact type (OC), Schottky (SC), and field-effect transistor contact (FEC) types. Nanostructured materials are preferred in general due to their high surface area-to-volume ratio and shrinking device. Nanoscale CdS photodetectors have gained attention recently due

Fig. 7.46 Photograph of operating ZnSe-based laser diodes. The different emission wavelengths of 505, 528, and 560 nm (from left to right) are achieved by varying the Cd content of the quantum wells [39] (with permission from Wiley)

to their improved response speed and spectral range, which is achieved by doping and bandgap engineering the basic CdS material. These aspects are widely discussed in a recently published review on CdS nanomaterials for photodetectors [3]. The summary of the important parameters that are required for the performance of nanoscale CdS-based photodetectors is given in the review work and is quoted in Table 7.6. Also, examples of photodetectors using CdS nanobelts and CdS_xSe_{1-x} nanowires, taken from the review work, are shown in Figs. 7.47 and 7.48.

Conclusions

The important processing techniques, properties, and applications of CdS and related compounds were reviewed in this section. Due to the unique and interesting properties of these compounds depending on their size, shape, and compositions, these materials remain a frontier area of intense research and development activities.

Table 7.6 Summary of important parameters of nanoscale photodetectors based on different semiconductors [3]

Materials	Contact type	Response range	Sensitivity	Decay time
Intrinsic CdS NB	QC	<500 nm	6	20 μs
Intrinsic CdS NW	SC	<500 nm	200	240 ms
Intrinsic CdS NB	MESFET	<500 nm	2.7 × 106	<0.3 s
CdS_xSe_{1-x} NW	SC	UV-visible	>100	240 ms
β-Ga $2Q_3$ NW	QC	UV	−1000	<0.09 s
ZnO NW	QC	UV	>100	33 s
ZnS NB	SC	UV	−2	<0.3 s

Abbreviations *NW* nanowire; *NB* nanobelt. (With permission from WILEY-VCH Verlag GmbH & Co. KGaA, Weinheim)

7.2 ZnO and Related Binary, Ternary, and Quaternary Compounds

Zinc oxide (ZnO) represents a key biosafe technological material which has attracted a great deal of attention due to its unique properties and potential applications in nanoscience and nanotechnology as sensors, transducers and catalysts, transparent electronics, ultraviolet (UV) light emitters, piezoelectric devices, chemical sensors, and spin electronics [6, 65]. In this section, the properties, processing techniques, and applications of ZnO are discussed.

7.2.1 *Processing Techniques*

Various techniques have been reported on the fabrication of ZnO either in the form of bulk nanocrystals or in thin-film form including bulk growth or substrate based growth. These methods can be classified as either physical vapor deposition (PVD), chemical vapor deposition (CVD), or solution-based chemistry.

7.2.1.1 Physical Vapor Deposition (PVD)

Physical vapor deposition (PVD) growth technique involves the solidification of a vapor directly onto a surface such that no chemical reaction takes place. In this way, the material is physically released from a source and transferred to a substrate. The three most common and important PVD technologies for ZnO deposition are thermal evaporation, pulsed laser deposition (PLD), and sputtering.

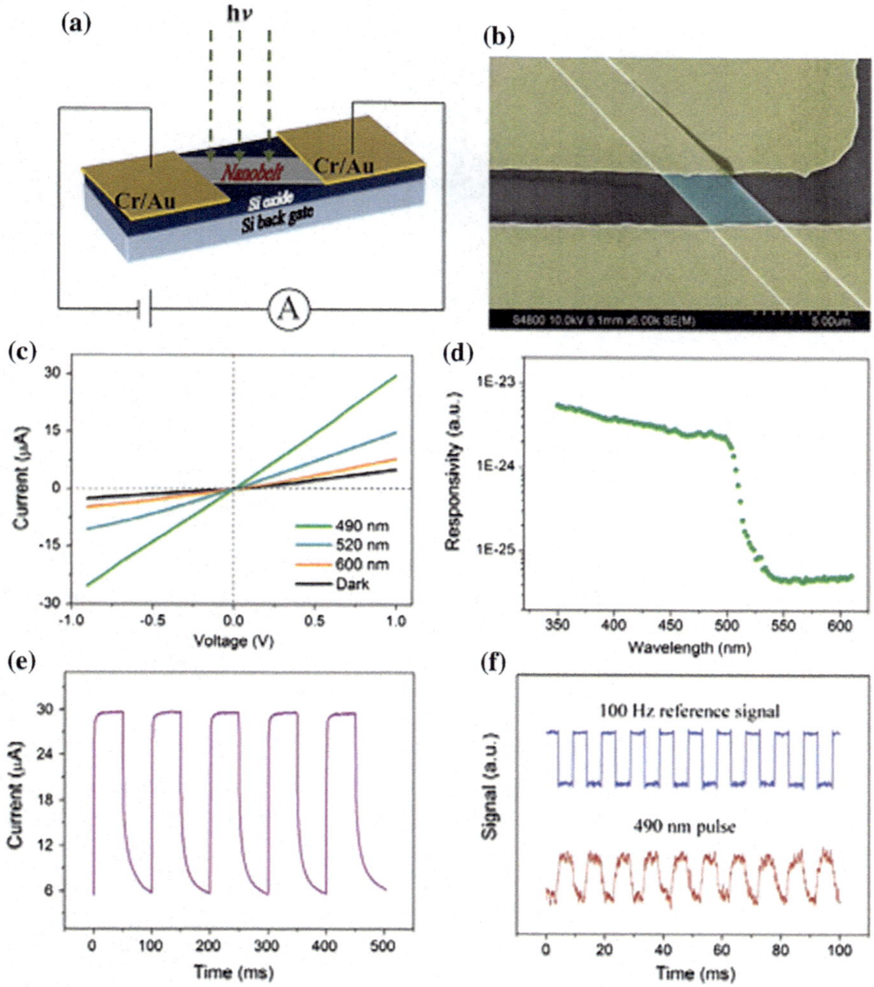

Fig. 7.47 **a** Schematics of a CdS-nanobelt photodetector [3, 63]. **b** A representative SEM image of an individual CdS-nanobelt device. **c** I–V curves of a CdS-nanobelt photodetector under illumination of different wavelengths or under dark conditions. **d** A typical spectral photoresponse of a device for different wavelengths. **e** The reproducible on/off switching upon 490 nm light illumination. **f** Time response by exposing nanobelts with 490 nm light pulse chopped at a frequency of 100 Hz (with permission from John Wiley and Sons)

Thermal Evaporation

The thermal evaporation basically utilizes ZnO powders inside a vacuum chamber, in which the vacuum reaches the range of 10^{-2} to 10^{-9} Torr. Once the appropriate pressure is reached, the ZnO powder is heated up to its vapor point and the vaporized ZnO condenses on the substrate as well as along the cooler surfaces in the vacuum chambers. By varying the experimental temperature and pressure in the chamber, various nanostructures and ZnO thin films can be produced. The heating sources

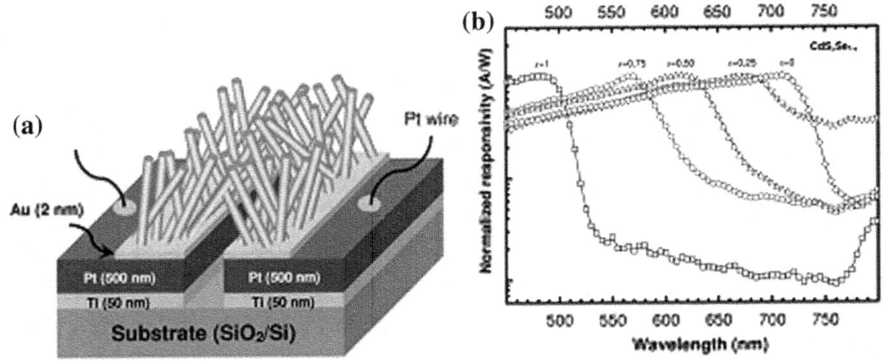

Fig. 7.48 CdS_xSe_{1-x} nanowire photodetector [3, 64]: **a** schematic illustrating a network structure of CdS_xSe_{1-x} nanowires floating above the SiO_2/Si substrate. **b** Normalized spectral responsivity curves of CdS_xSe_{1-x} nanowires with different x values. The applied bias to the device was 5 V (with permissions from IOP Publishing)

used are typically electron beam, radio frequency (RF) induction used to create localized heating of the source material so as to produce highly oriented ZnO thin films [66–68], and resistive heating commonly used in furnace applications which can create a specific temperature gradient along the tube, useful for nanostructure production [69–72].

Pulsed Laser Deposition (PLD)

Another common technique for the synthesis of thin films and nanostructure materials is PLD. This process consists of a laser beam focused onto the surface of a target material during the deposition in high vacuum. The laser ablation produces vapor plume that has the similar stoichiometry of the target material [73, 74]. Adequate high flux density and short pulse duration are needed to heat up the target rapidly to its evaporation temperature so as to form a vapor plume. Once formed, it accumulates onto a cooler substrate that promotes nucleation and growth of crystalline films. Nanostructure growth via the vapor–solid (VS) mechanism can be initiated by the gas phase formation of nanoparticles at high pressures within the ablation plume. Once deposited on the substrate, these nanoparticles act as nucleation centers and promote subsequent nanostructure growth. Epitaxial single-crystalline ZnO film with good quality can be produced by this method [75]. Doping can be performed and precisely controlled and monitored during the deposition by using targets with desired composition or a multi-target system [76, 77]. ZnO aligned ZnO nanorods are also synthesized by PLD [78]. Drawbacks of PLD are its relatively low duty cycle and incorporation of particulates in the deposited films. Also, it is not appropriate for deposition on large area substrates and cannot be scaled up for industrial production.

Sputtering

Sputtering is a relatively inexpensive and fast method of thin-film production consisting in the removal of atoms from the surface using high-energy ions [79,

80]. The current sputtering systems use a magnetron sputtering configuration, in which a strong magnetic field is applied in order to focus the plasma near the target to increase the deposition rate. The level of vacuum required is of 10^{-6} Torr, with working gas pressure at the level of 10^{-3} Torr. During growth, an inert gas, normally Ar as the working gas is introduced into the vacuum chamber where the source material target and substrates are present. Subsequently, a direct current (DC) or RF power supply is applied to ionize the inert gas so as to produce charged plasma. The ions are accelerated to the surface of the target and the atoms of the source material are bombarded out of the target. As a result, some of those atoms condense on the substrate to form films. A bias voltage can be applied to the substrate to assist the deposition and enhance the adhesion properties of the film to the substrate. The substrate can be heated to improve the crystal quality of the growing film. Sputtering is commonly used to produce polycrystalline thin film of ZnO at room temperature. RF sputtering is used to produce ZnO film with better n-type conductivity, while almost insulating ZnO film can be obtained by DC sputtering. Doping is possible to achieve with sputtering system by introducing gases containing dopant elements or mixing dopant materials into the target. Additional advantage is the scalability of the process with deposition on areas of up to several square meters in a single step which cannot be achieved by alternative methods such as PLD or MBE.

7.2.1.2 Chemical Vapor Deposition (CVD)

Chemical vapor deposition (CVD) makes use of chemical reactions between the vaporized source material and source gases, on a hot surface. The reaction product condenses within the reaction vessel where the pressure and gas flow are controlled, during the formation of a solid material proceeding, as schematically shown in Fig. 7.49.

Solidification occurs when the liquid seed layer is enriched until it is saturated. A large amount of experimental variables influence the growth mechanism and subsequently change the dimensions, crystal quality, and morphologies of the grown structures.

Conventional CVD techniques used to deposit ZnO are as follows: thermal CVD, low-pressure CVD (LPCVD), plasma-enhanced CVD (PECVD), laser CVD (LCVD), metal-organic CVD (MOCVD), MBE, and atomic layer deposition (ALD); these are classified by the modifications of the vacuum level, heating source and reactants gas types, and so on. All these methods require a vacuum chamber with gas flow control. Growth mechanisms are vapor–liquid–solid (V–L–S) and V–S. Within the V–L–S method, catalyst seeds in liquid form are used, so that the droplets serve as preferential sites for adsorption of the gas phase reactant. The catalyst is chosen on the basis that the component elements of nanostructures are soluble in it. When the droplets become supersaturated, crystallization and growth of nanostructures occur until the temperature drops below the eutectic temperature of the catalyst alloy or the reactant is no longer available [82]. Au and Sn are commonly used as a catalyst for ZnO nanowires and nanorods growth, respectively [69]. Parameters such as

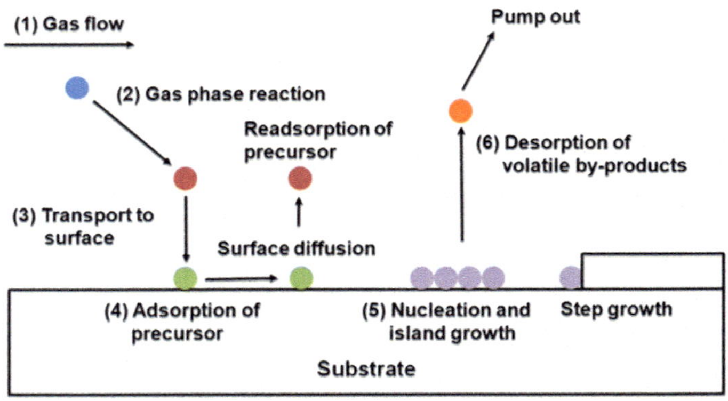

Fig. 7.49 Schematization of the CVD process steps. The precursor is transported by the gas inlet to reaction region (1), the chemical reaction in gas phase produces new reactive species (2), the reactive species are transported on the substrate surface (3), the reactive species are adsorbed on the substrate and surface diffusion occurs (4), nucleation and growth start film formation (5), and finally desorption of by-products and pumping out from the reaction region occurs (6) [81]

deposition temperature and pressure, oxygen partial pressure, carrier gas flux are key factors in the final structural characteristics of ZnO nanostructures [71, 83, 84].

Thermal CVD and LPCVD

With thermal CVD, high vacuum is not required as the pressure level is 10^{-3} Torr. The solid source is placed upstream and the solid is vaporized through heating, transported by a carrier gas which feeds the catalyst and source material. LPCVD deposition is a long-standing technique in industry for oxide film growth and sometimes confounded with the thermal CVD, but its working pressure range is slightly under 0.1 Torr. In the case of ZnO, both thermal CVD and LPCVD require lower operating temperature than thermal evaporation, but still over 900 °C. These techniques have been used for the production of various ZnO nanostructures [70, 84, 85].

PECVD and LCVD

PECVD offers several advantages that may improve the deposited film quality and morphology [86]. So far, however, PECVD of ZnO nanorods has been hardly performed [87, 88] and usually the materials growth involves the use of elevated temperatures (often 400 °C or higher), annealing, high input powers, and pyrophoric precursors. LCVD uses laser ablation as heating source with controlled doses, which leads to precise dimension and doping control for nanostructure formation.

Metal-organic chemical vapor deposition (*MOCVD*)

Metal-Organic Chemical Vapor Deposition (MOCVD) is traditionally used in thin-film growth. MOCVD uses metal-organic precursors to catalyze the chemical reaction and is carried out at a very high level of vacuum and elevated temperature (900 °C). The gas passes over a hot semiconductor substrate and breaks up, depositing atoms onto the surface, layer by layer. The metal can act either as an active site on the surface initiating the nanowire growth [89, 90] or as a particle guiding the growth at the nanostructure tip as in the VLS mechanism [91, 92]. Doping is relatively easy to be performed [93]. However, metal-organic compounds are usually more expensive than inorganic compounds.

MBE

MBE is renowned for preparing very high-quality single crystal ZnO films and nanostructures with excellent thickness control and composition uniformity. Growth is carried out in an ultrahigh vacuum environment, which minimizes the potential for contamination of films. The vapors are generated from element sources with high purity which are slowly heated in individual quasi-Knudsen cells, and released into the vacuum chamber by computer-controlled shutters. Upon condensation onto the single-crystal substrate, the elements originated from the vapors will start to interact with each other as well as with the substrates, and then atomic layers are epitaxially grown one layer at a time. The crystal quality and layer thickness can be monitored by reflective high-energy electron diffraction (RHEED) during the crystal growth. Drawbacks for using MBE are that the system is very expensive and the operating conditions must be accurately monitored and controlled. It is therefore more used for scientific research and not for scaling for use in industry [94].

7.2.1.3 Solution-Based Chemistry (SBC)

Among the various reports on ZnO fabrication, solution-based chemistry growth is by far the most promising technique for large-scale production, and more importantly, for its low cost as it does not require sophisticated equipment and is environment-friendly. Hydrothermal growth methods, also referred to as aqueous solution methods, occur at very low temperature and therefore hold great potentiality for synthesis on a variety of flexible/soft plastic and cellulose substrates [95, 96]. Vertically aligned nanorods can be successfully obtained (Fig. 7.50).

Different morphologies and length of the nanostructures can be readily obtained by tuning the hydrothermal conditions such as pH, concentration, temperature, and growth duration [97, 98]. One main advantage of the hydrothermal method is the use of a seed layer prior to the growth in the form of nanoparticles or thin films. This provides the nucleation sites for the growth of ZnO nanostructures. Not only as the seed layer contains the crystallites that can promote the textured growth, it is also critical in the growth of *c*-axis oriented ZnO hexagonal rods, controlling the degree of alignment and the packing density [99, 100]. Seed layers with a high concentration of zinc ions will lead to dense, well-packed, and highly aligned nanorods. Unseeded substrates will lead to anisotropic growth and orientation of ZnO nanorods. Therefore, the arrangement of nanorods can be organized by controlling the number of available nucleation sites. Moreover, the choice of the precursor (i.e., zinc acetate and zinc nitrate) can lead to differences in microstructure growth and kinetics. A zinc acetate precursor will lead to more aligned nanorods with a faster growth rate than a zinc

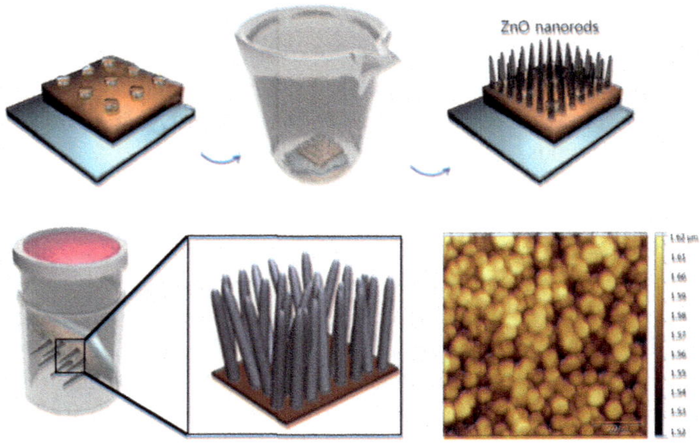

Fig. 7.50 Schematic description and AFM image of the preparation of ZnO nanorod arrays [96]

nitrate precursor. Bulk precipitates may occur on the films for long depositions. This can be avoided by draining and replacing with a freshly prepared solution [101]. Compared to other methods, hydrothermal methods are attractive. The hydrothermal method has been demonstrated as a powerful and versatile method for synthesizing metal oxide nanostructures and in the fabrication of electronic and sensing devices.

7.2.1.4 Melt Growth

Bulk ZnO can be grown using melt growth technique. This method of growth requires a high pressure and temperature (ZnO melting temperature ~1975 °C) crucible to contain the melt as it cools. However, crucibles may be problematic. An alternative for melt growth is skull melting [102, 103]. In skull melt, also known as cold crucible, the melt is contained in a thin solid shell of the same composition as the growth material. An RF heating energy source is coupled directly into the ZnO melt, surrounded by a cooled "skull" of solid ZnO powder (Fig. 7.51).

ZnO crystals grown by this method are chemically very pure since the material does not come into contact with any foreign substance. The atmosphere can be controlled in a very wide range, which is beneficial for the Zn/O stoichiometry of the grown crystals. However, this growth method results in a large temperature gradient from the shell (which must be kept cold and solid) to the melt which must be kept hot and as a liquid. This large thermal gradient results in a higher concentration of dislocation defects and limits the maximum size of the crystals.

There also exist many other methods for obtaining zinc oxide, including spray pyrolysis, sonochemical method, synthesis using microwaves, and many others. Each of these growth techniques has its own advantages and limitations. Generally, the gas phase approaches are preferred as a variety of high-quality, single-crystal ZnO

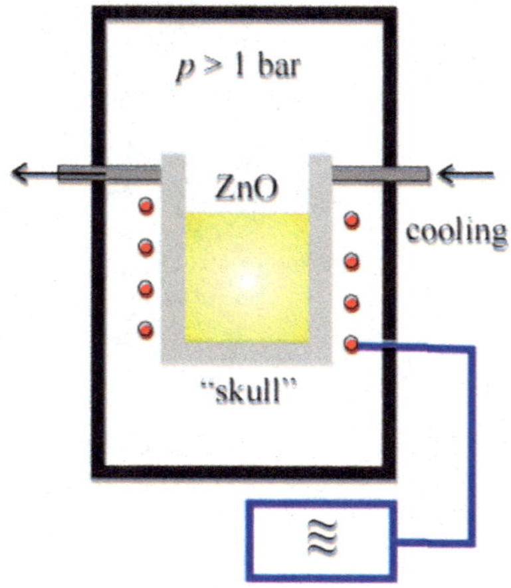

Fig. 7.51 Skull melting by RF energy coupling directly into ZnO melt [102]

nanostructures with doping can be relatively easily produced. However, the requirement of high vacuum and the elevated operation temperatures often restrict their use. The required instrument is also expensive, and it is hard to achieve mass production. Moreover, methods such as MOCVD use toxic chemicals and therefore cause environmental issues. In contrast, the solution-based methods are simpler and have minimum requirement for facilities. The operation temperatures are much lower than those of the gas phase methods, and thus are cost-efficient and economic energy-wise. These methods are also easy to scale up the production. However, the quality of the ZnO structures produced are lower and defects and impurities are almost always present. Furthermore, adhesion to the supporting substrates is not very good which is a problem for applications requiring long life cycle and good networks between nanostructures and their substrates.

7.2.2 *Properties*

ZnO is an II–VI n-type semiconductor, with a wide direct bandgap ($E_g = 3.4$ eV), with a high melting temperature ($T = 2300$ K) and a good capacity to carry large currents without material breakdown [104, 105]. Key advantages of using ZnO materials are their strong piezoelectric and pyroelectric properties ($e_{33} = 1.2$ C/m^2, among the highest of all semiconductors), high thermal conductivity of 0.54 (W cm^{-1} K^{-1}), largest exciton-binding energy of II–VI and III–V semiconductors (60 meV), excitonic stimulated light emission up to 550 K [6, 97, 98, 106]. ZnO also has a very large Shear Modulus (45.5 Gpa) (indicating the stability of the crystal) and its thin films are transparent in the visible region [107].

Knowing the surface structure of ZnO is the key to better understand its favorable properties and to further develop its applications. Structurally, ZnO crystallizes in

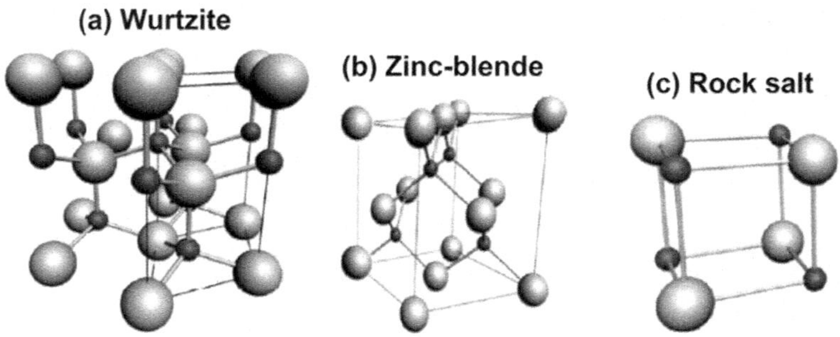

Fig. 7.52 Unit cells and the corresponding schematic diagrams with common crystal facets of **a** zinc blende structure, **b** wurtzite structure, and **c** rock salt structure

either cubic zinc blende or hexagonal wurtzite (Wz) or rock salt, structure [108], Fig. 7.52.

ZnO is most commonly crystallized in a hexagonal wurtzite structure, as it is thermodynamically stable due to its ionicity residing exactly at the borderline between the covalent and the ionic materials. The structure is a hexagonal unit with lattice parameters $a = 0.3296$ and $c = 0.52065$ nm and can be simplified by considering it as stack in which a number of alternating planes, composed of tetrahedrally coordinated O^{2-} and Zn^{2+} ions, are placed alternately along the c-axis (Fig. 7.53). Each oxygen anion is surrounded by four zinc cations at the corner of a tetrahedron and vice versa. The higher c/a ratio in the crystal is coupled with smaller ionicity and a lower stacking fault energy (i.e., the result of compressive and/or tensile strain, together with the imbalanced interface chemistry of the respective material system) [109]. Another important characteristic of ZnO is the polarity of its surfaces. While the whole unit cell of ZnO is neutral, some surfaces can be terminated entirely with anions or cations, resulting in positively or negatively charged surfaces, i.e., polar surfaces. The most common polar surface is the (0001), or basal plane. Additional common surfaces characteristic of the wurtzite structure are the positively charged Zn (0001) and negatively charged O $(000\bar{1})$ and the nonpolar $(11\bar{2}0)$ and $(10\bar{1}0)$ faces. The nonpolar faces enclose equivalent numbers of Zn and O atoms. The exact mechanisms for the stabilization of ZnO $\pm$ (0001) faces are still under debatable discussion and different mechanisms involved have been suggested (i.e., surface reconstruction and hydrogen adsorption) [110].The basal plane is the surface most commonly used for growth. The other two are important as they are employed as the primary directions in reflection high-energy electron diffraction (RHEED) in MBE growth, apart from being perpendicular to one another.

Additional to the wurtzite phase, ZnO has been identified to crystallize in the cubic zinc blende and rock salt (NaCl) structures. The zinc blende ZnO structure is thermodynamically a metastable structure which can be stabilized only by heteroepitaxial growth on cubic substrates, i.e., ZnS, GaAs/ZnS, and Pt/Ti/SiO_2/Si [111–114]. Compared to the wurtzite structure, the zinc blende phase has lower ionicity as a function of c/a ratio, indicating that these are absolutely covalent, and are therefore viable for

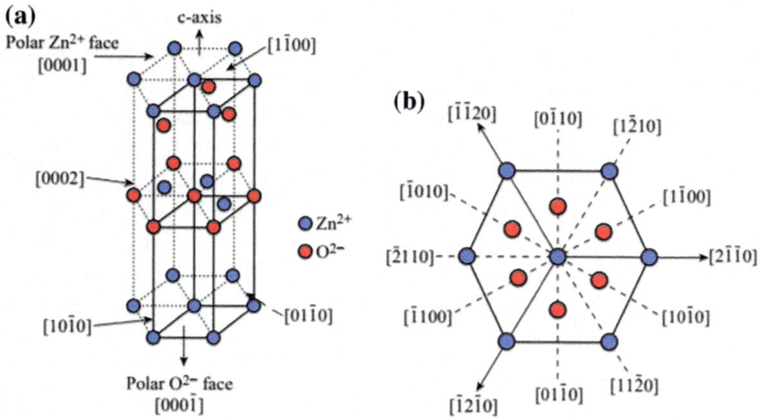

Fig. 7.53 **a** ZnO unit cell schematized in the wurtzite structure. **b** Various crystal planes of ZnO wurtzite structure

Table 7.7 Basic physical properties of ZnO at 300 K

Physical parameters	Values
Lattice parameters at 300 K	$a_0 = 0.32495$ nm
	$c_0 = 0.52069$ nm
	$a_0/c_0 = 1.602$ (ideally 1.633)
Density	5.606 g/cm^3
Molecular mass	81.389 g/mol
Stable phase at 300 K	Wurtzite
Melting point	1975 °C
Thermal conductivity	0.6, 1–1.2 W/Km
Specific heat	0.125 cal/g °C
Dielectric constant	8.66
Refractive index	2.008, 2.029
Energy gap	3.37 eV, direct
Excitation binding energy	60 meV
Electron mobility	100–200 cm^2/Vs
Hole mobility	5–50 cm^2/Vs
Electron/hole effective mass	0.28 m_0/0.59 m_0

application in semiconductor technology [115, 116]. ZnO can be also transformed to the rock salt (NaCl) structure at relatively modest external hydrostatic pressure (~8 to 10 GPa), and cannot be epitaxially stabilized [117]. The reason for this is that the reduction of the lattice dimensions causes the interionic Coulomb interaction to favor the ionicity more over the covalent nature. The basic physical properties of ZnO at 300 K are listed in Table 7.7.

Due to the tetrahedral coordination, ZnO results in a noncentral symmetric structure which consequently endows the piezoelectric and pyroelectric properties. Intrinsic optical properties of ZnO are the focal point of studies for implementation in photonic devices. Its wide bandgap shows promising prospective in the UV or blue spectral range, while the high exciton-binding energy allows efficient excitonic emission even at RT. Various techniques are widely reported for the optical transitions in ZnO such as reflection, photoreflection, transmission, optical absorption, photoluminescence (PL), cathodoluminescence, spectroscopic ellipsometry, and calorimetric spectroscopy. Red luminescence band for which doubly ionized oxygen vacancies are retained has also been reported.

7.2.3 Applications

ZnO encompasses a wide significant array of applications [118, 119]. The combination of unique properties such as UV absorption, antimicrobial properties, steady thermal, and optical properties makes it advantageous for various applications (i.e., ceramics, lubricants, ointments, adhesives, or the rubber industry). The wide bandgap and large excitation binding energy make ZnO viable to be applied as chemical sensors, solar cells, luminescence devices, etc. ZnO nanowires are also reported to have behaved like an n-type semiconductor due to the presence of interstitial defects and vacancies [120–122].

The high electron mobility endows ZnO as a candidate for use in dye-sensitized solar cells (DSCs). The key element to a highly efficient DSC is the interaction between the photosensitive dye and the charge carrier. Exposition of the dye to light generates an electron which is promoted to the conduction band and transported to the anode. Each feature such as surface area of the charge carrier (higher surface area means that more dye sensitive molecules can be adsorbed), transfer between the dye and the carrier, and the recombination rate (i.e., how fast the generated electron recombines with the positively charged dye molecule) will affect the efficiency of the DSC. Substrates comprised of oriented ZnO nanowires have shown potential applications in DSCs [123].

As gas sensors, ZnO films are highly sensitive and selective. The gas sensing performance of ZnO is affected by its crystal defects: the greater the extent of oxygen vacancies, the higher will be the gas response of ZnO [124]. In addition to this, the gas sensing performance of ZnO is also greatly found to be dependent on the size and morphology [125]. 1D ZnO nanostructures, for instance, nanorods, nanotubes, nanowires, and nanofibers, have been comprehensively investigated due to their high surface–volume ratio, crystallinity, and charge confinement ability. A two-dimensional nanostructure porous sensitive layer offers a comparatively larger surface–volume ratio and a larger number of reaction sites which enables higher and quicker sensor response. The porosity of the surface of thin films directly influences the gas diffusion rate and chemisorption of oxygen and analyte gas so as to enhance the gas sensing properties [126]. Other types of morphologies including hollow spheres [127], nanopyramids [128], and nanotetrapods [129] have also been utilized.

Properties such as antibacterial, disinfecting, and biocompatibility have enabled ZnO to be widely used for medical purposes (i.e., in dermatological substances against inflammation due to its properties which accelerate wound healing or peeling effects) [130, 131]. Among the various advantages is the strong antibacterial activity in small quantity and the effective biocidal action in the absence of light. It is also used in dentistry, mainly as a component of dental pastes, and for temporary fillings. Owing to its ultraviolet light absorbance, ZnO is also used in sunscreens and sunblocks to prevent sunburns.

Zinc oxide has been used as multifunctional coatings in textiles to impart self-cleaning, superhydrophobicity, and ultraviolet (UV) blocking properties [95]. Depending on the chemical and physical properties of ZnO/water interfacial interactions, ZnO thin films may have switchable wettability which holds great potential in many fields: i.e., sensors, intelligent membranes, microfluidic technology, industrial metal cleaning, marine antifouling coating, etc.

As Field-effect transistors (FETs), ZnO can function without even a p-n junction thus bypassing doping problems [132]. Some FETs use ZnO nanorods as conducting channels. A typical ZnO FET showed a gate threshold voltage of −15 V, a switching ratio of nearly 100 and a peak conductivity of $1.25 \times 10^{-3}\ \Omega^{-1}\ \mathrm{cm}^{-1}$. ZnO is also used in making Zn–C dry cells, Zn–Ag oxide batteries, and also some secondary batteries.

Another major area of ZnO application is in short wavelength light-emitting diodes (LEDs). However, the application of homojunction ZnO LEDs has been hindered due to the difficulties that are inherent in growing reproducible p-ZnO with high mobility and high hole concentration [133]. Recent progress has overcome these issues by using alternative approaches such as p-n heterojunction [134, 135] and organic-hybrid p-n junction [136, 137]. Although the stability of organic-hybrid p-n junctions decay at ambient conditions, p-n heterojunctions have received, all the same, a great deal of attention in recent years. Nonetheless, device performance is strongly dependent on the material properties and defects in the nanostructures.

7.3 HgTe and Related Binary, Ternary, and Quaternary Compounds

II–VI materials have drawn significant interest because of their capability to behave as a wide-bandgap (1.5–3.5 eV) material as well as a narrow-bandgap (0–1.5 eV) material. This has led to their uses for applications in devices, such as photoconduction, ultrasonic transducers, and also in Hall effect devices. In particular, mercury telluride (HgTe) has attracted interest due to its very high Hall mobility and mobility ratio. However, studies of mercury compounds are not frequent, due to the high toxicity of mercury. Moreover, the low interest in HgTe is attributed to difficulties in processing of HgTe-based devices and to obtain an exact control over the stoichiometry of samples. Wide-bandgap materials include zinc selenide (ZnSe), zinc telluride (ZnTe), cadmium zinc telluride (CdZnTe), zinc sulfide (ZnS), etc., whereas the narrow-bandgap materials include mercury cadmium telluride (HgCdTe) and mercury zinc telluride (HgZnTe). This section outlines the properties of HgTe and processing techniques.

7.3.1 Processing Techniques

Considerable attention has been given to the preparation of conventional and nanometer-scaled HgTe materials. Several methods have applied a mixture of the elements at elevated temperature [138], such as the molecular precursor method [139], and low-temperature organometallic vapor phase epitaxy growth [140, 141]. Samples grown by either the bulk or the liquid phase epitaxy (LPE) technique [142] present Hg vacancies, dislocations, and undesired impurities in several concentrations. These defects are normally formed during the crystal growth or during subsequent sample treatment.

Epitaxial growth is quite difficult, being very susceptible to thermally induced defects related to the growth temperature. Consequently, growth is usually carried out at reduced temperature. Molecular beam epitaxy (MBE) can give good control over the growth of thin layers. Moreover, growth of HgTe thin films can occur near thermal equilibrium (~200 °C); thus the Hg atoms can re-evaporate easily whereas the Te atoms remain very mobile on the surface [143]. For lower deposition rates, the crystals are allowed to grow self-organized during thin-film growth. HgTe has also been grown using laser or ultraviolet-assisted vapor phase epitaxy (VPE) [144] and plasma-enhanced VPE [145] using precracking techniques [146] and by using less stable Te precursors [147–149]. Precracking techniques are the most promising by using Te alkyls which are more readily cracked.

Of late, nanometer-scaled HgTe particles have attracted growing interest because of the quantum confinement effect which can dramatically increase the effective bandgap and give rise to near-infrared emission. This offers potential applications in opto-telecommunication devices [150–154]. Numerous studies on colloidal HgTe nanoparticles have been reported with hydrothermal methods at room temperature [112, 113, 150, 151, 155] as well as with sonochemical synthesis [156]. HgTe nanorods have also been synthesized in solution [156, 157], grown within single-wall carbon nanotubes [158] or by spray deposition of solvothermally synthesized iodine-doped HgTe nanoparticles [159].

Fewer reports related to colloidal quantum dots (QDs) have been investigated due to their instability and the difficulty involved in their synthesis [160]. HgTe quantum dots have been largely synthesized through solution chemistry although an additional phase transfer process is required. The particle size could be controlled by the aging temperature. Some research groups have reported the use of harmful precursors as H_2Te gas [161, 162]. However, a nonaqueous, non-Te-gas process using

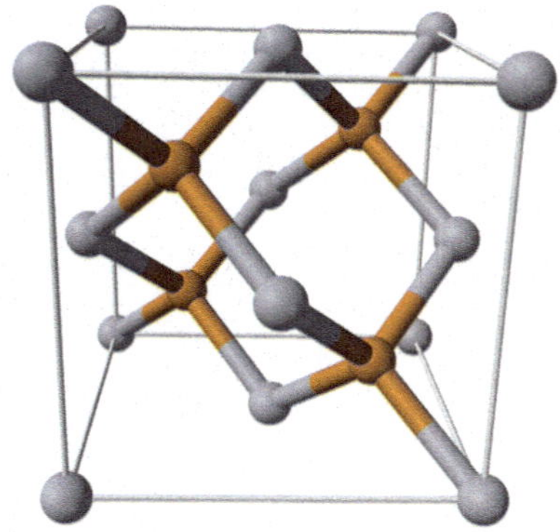

Fig. 7.54 Crystal structure of mercury telluride

an organic solvent based synthetic method which involved the use of Te powder has been reported, although the products showed poor stability [163, 164].

7.3.2 Properties

HgTe is an equiatomic binary compound of the elements of the II (Mercury)-VI (Tellurium) group which crystallizes into the zinc blende structure [165].

Furthermore, HgTe, as other II–VI compounds, is a tetrahedrally bonded semiconductor in which each atom is fourfold coordinated, i.e., each atom is surrounded by four nearest-neighbor atoms forming a tetrahedron (Fig. 7.54). Specifically, the zinc blende lattice involves two interpenetrating lattices: one sublattice of HgTe contains mercury atoms while the other contains tellurium atoms.

The ionicity of the covalent bonds in II–VI compounds increases the Coulomb interaction between the ions, leading to an increased bandgap in the electronic band structure. However, HgTe presents a negative narrow inverted bandgap. This means that the top of the valence band is higher than the bottom of the conduction band, as shown in Fig. 7.55. The conduction band of bulk HgTe is formed by p-states located at the group II atoms and the valence band by s-states at the group VI atoms, yielding an inverted band ordering around the Γ point of the Brillouin zone. For bulk HgTe, the Γ_8 band is p-type and lies energetically above the s-type Γ_6 bands. Therefore, the conduction and valence band consist of the light hole band LH1 and the heavy hole band H1 of the Γ_8 bands which yield a zero gap system.

In particular, the band inversion of the Γ_6 and Γ_8 bands, due to a combination of a high spin–orbit splitting in Te and large energy corrections (due to the mass–velocity term) to the energy levels in Hg, makes HgTe actually a semimetal.

The properties of HgTe are principally governed by the presence of defects (i.e., native point defects such as vacancies, interstitials, antisites and complexes, or extended defects as dislocations, grain boundaries, precipitates, melt spots) and undesired impurities. These defects are abundant in polycrystalline off-stoichiometric

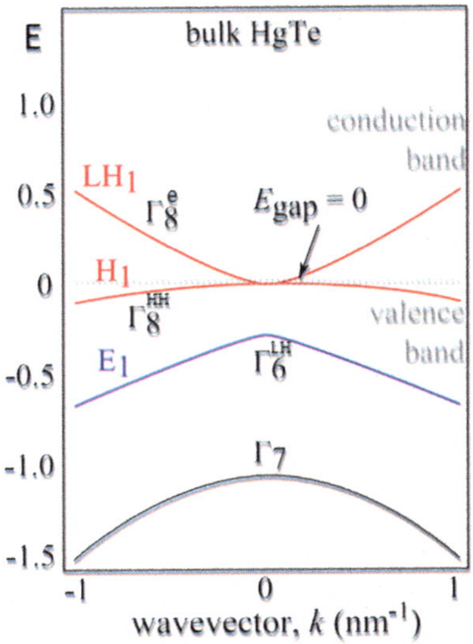

Fig. 7.55 Bandgap of HgTe (inverted bandgap) where the Γ_6 and Γ_8 points refer to the electron band and light hole (LH) and heavy hole (HH) band, respectively

HgTe. Defects may also be introduced intentionally during growth, in order to obtain the desired semiconductor properties. Most defects are mobile, even at relatively low temperatures.

HgTe can be obtained in n and p-conduction forms. N-type doping can be achieved using boron, aluminum, gallium, or indium, as well as iodine and iron. Mostly, HgTe grown at low temperatures is n-type. P-type doping can be achieved by introducing zinc, copper, silver, or gold; normally it is obtained with mercury vacancies. The donor and acceptor concentrations in HgTe are usually controlled by deviations from stoichiometry produced by heat treatments carried out at certain temperatures and mercury vapor pressures.

The understanding of the principal properties of HgTe has depended largely on the progress in the technology of preparation of this material. The importance of HgTe in semiconductor devices is due to the fact that, in its crystals, the carrier mobility is very high. However, due to higher volatile nature of Hg, imperfections are created in the crystals impeding the mobility. This can be improved by annealing the crystals in controlled Hg vapors [166–168]. A lower carrier mobility can be encountered in polycrystalline HgTe samples, which can be improved by annealing the samples in Hg atmosphere [138].

The first measurements of the electrical properties of HgTe were initially carried out in the fifties by the Russians who showed that the thermal width of the energy gap (the forbidden band) of this compound was remarkably small [169]. Subsequently, galvanomagnetic measurements established that the thermal gap of HgTe did not exceed several millielectron volts. As the effective electron mass is roughly proportional to the energy gap, this very narrow gap should correspond to a very small effective mass. For HgTe, the mass is found to be equal to the mass of electrons in indium antimonide whose gap is ~0.23 eV. Moreover, the mobility of electrons was found to be higher in the alloys of mercury and cadmium tellurides (with low cadmium concentrations) than in HgTe. This was quite a contradiction as disordered solid solutions have more defects and the mobility would be more reduced compared with the pure binary compound. However, this is explained considering the inverted band structure model proposed by Groves and Paul [170].

Another important property of the zinc blende structure is the absence of a center of inversion.

The electrical properties of polycrystalline HgTe are dominated by the defects present in the material. Abundant research has shown that most of these defects are electrically active, and their physical and electrical properties are still not well understood. The behavior of intrinsic defects, such as mercury vacancies, which act as shallow acceptors, is relatively less understood [171]. HgTe can also be considered as a thermoelectric material.

At high pressure (~1.25 GPa), HgTe undergoes a first-order phase transition from the zinc blende to the cinnabar structure with a volume change of ~8.4% [172]. In quite narrower pressure ranges, HgTe undergoes a sequence of transitions from a semimetal to a semiconductor and then again to a metal accompanied by analogous crystal–structural transitions from the zinc blende (semimetal) to the cinnabar (semiconductor), and to the rock salt (RS) structure (metal) [173].

Attractive material applications appear to be HgTe quantum dots (QDs) [174–176] due to their quantum confinement effect which can dramatically increase the effective bandgap and give rise to near-infrared emission, as unlike other materials, the bandgap of bulk HgTe is negative. However, few investigations on HgTe QDs are reported in the literature because of their instability and the difficulty involved in their synthesis.

HgTe-based ternaries of the II–VI group are made up of two elements in the II column of the periodic table, and one in the VI column. To complete the crystal structure, a II–VI ternary compound semiconductor must have equal parts of group II elements and group VI elements. The general composition of these compounds is denoted $Hg_{1-x}A_xTe$ where A is typically, though not always, a member of column II of the periodic table and $0 \leq x \leq 0.1$ The best known of these materials is Mercury Cadmium Telluride (HgCdTe, MCT); its energy gap varies from ~−0.3 to ~1.6 eV. Therefore, detectors with peak responses over a wide wavelength range can be fabricated using this alloy. Because HgTe is a semimetal and CdTe is a semiconductor, the carefully tuned mix of these two compounds can create a semiconductor with a very small bandgap corresponding to a relatively long cutoff wavelength. The practical upshot of this is that infrared detectors may be fabricated out of a material whose peak response can be adjusted over a wide range of wavelength by varying the parameter x. Another advantage in using MCT is that its lattice constant almost undergoes no change during the variation of the composition and the material presents direct bandgap transitions and large absorption coefficient which allow the quantum efficiency to be ~100%. Lastly, this material has long minority carrier lifetimes which results in low thermal noise allowing high-performance detectors to be made at the highest operating temperatures reported for infrared detectors of comparable wavelengths. Moreover, the energy band structure allows the material to be versatile in most of the applications, irrespective of the device architectures used.

Many potential alternatives, such as InAsSb, PhSnSe, and HgMnTe, have also been used [177, 178] for the fabrication of detectors.

Quaternary compounds consist of the replacement of one element with a couple of two elements of different groups of the periodic table keeping fixed the ratio between the number of atoms and valence electrons. The general composition is Cu_2–II–IV–VI4, where VI is a chalcogen element (S or Se) while II and IV represent divalent (Zn, Cd, Fe) and tetravalent (Sn, Ge, Si) elements, respectively. Among the possible Cu_2–II–IV–VI_4 compounds, kesterites, $Cu_2ZnSn(S,Se)_4$ (CZTS(Se)), have been the most studied in relation to photovoltaics. Kesterites are very promising materials, being low-cost, widely available, and nontoxic elements (even though selenium is sometimes added to the alloy), and show advantageous properties for thin-film photovoltaic applications. CZT(S,Se) films are spontaneously p-type semiconductors, with a direct bandgap between 1 eV (for the pure selenide $Cu_2ZnSnSe_4$ alloy) and 1.5 eV (for the pure sulfide Cu_2ZnSnS_4). The direct bandgap gives a high absorption coefficient, making these materials suitable as thin-film photoactive layers, but the device efficiency is low and must be improved to make it commercially viable [179–181].

7.3.3 *Applications*

II–VI semiconductors have attracted significant interest due to their increasingly wider applications. The narrow-bandgap materials have distinctive properties that can be utilized in novel device structures.

HgTe is, for example, an excellent candidate for spintronics applications. Nanocrystalline HgTe presents potential applications in the area of photoconductive/photovoltaic devices, IR detectors, and IR emitters. Among the field of narrow-bandgap II–VI semiconductors, MCT is the main infrared material.

Also, modern epitaxial techniques and the emergent interest in nanostructures have provided novel areas of applications that utilize some of the unique properties of these materials. Materials with narrow bandgap are able to detect the entire visible spectrum of light as well as infrared radiation. The small effective mass in these materials makes them ideal candidates to study quantum confinement. For example, quantum spin Hall Effect has been discovered in HgTe quantum wells [182]. Possible interesting HgTe and HgCdTe nanowire electro-optical applications include polarization-sensitive detectors, detectors with reduced cooling requirements, lasers, and LEDs in the infrared region of the spectrum.

References

1. Böer KW (2011) Cadmium sulfide enhances solar cell efficiency. Energy Convers Manag 52(1):426–430
2. Eskandari M, Ahmadi V et al (2015) Enhanced photovoltaic performance of a cadmium sulfide/cadmium selenide-sensitized solar cell using an aluminum-doped zinc oxide electrode. Ceram Int 41(2):2373–2380
3. Deng K, Li L (2014) CdS nanoscale photodetectors. Adv Mater 26(17):2619–2635
4. Janotti A, Van de Walle CG (2005) Oxygen vacancies in ZnO. Appl Phys Lett 87(12):122102
5. Lin TK, Chang SJ et al (2005) ZnO MSM photodetectors with Ru contact electrodes. J Cryst Growth 281(2–4):513–517
6. Özgür Ü, Alivov YI, Liu C et al (2005) A comprehensive review of ZnO materials and devices. J Appl Phys 98:041301–103
7. Ramasamy K, Malik MA et al (2011) Thio- and Dithio-Biuret precursors for zinc sulfide, cadmium sulfide, and zinc cadmium sulfide thin films. Chem Mater 23(6):1471–1481
8. Madelung O (2003) Semiconductors: data hand book. Springer, Germany
9. Kozhevnikova NS, Vorokh AS et al (2015) Cadmium sulfide nanoparticles prepared by chemical bath deposition. Russ Chem Rev 84(3):225
10. Pal U, Silva-González R et al (1997) Optical characterization of vacuum evaporated cadmium sulfide films. Thin Solid Films 305(1–2):345–350
11. Islam MA, Hossain MS et al (2013) Comparison of structural and optical properties of CdS thin films grown by CSVT, CBD and sputtering techniques. Energy Procedia 33:203–213
12. Canevari V, Romeo N et al (1984) Low resistivity CdS thin films grown by flash-evaporation at low substrate temperature (150–200 °C). J Vac Sci Technol, A 2(1):9–10
13. Murali KR, Kannan C et al (2010) Flash evaporated cadmium sulfide films. Mater Sci Semicond Process 13(5–6):356–359
14. Mohsin AK, Bidin N (2014) Effect of cadmium sulfide thickness on electron beam-deposited titania/cadmium sulfide nanocomposite films. Mater Sci Semicond Process 24:208–214
15. Hodes G (2003) Chemical solution deposition of semiconductor films. Marcel Dekker Inc, New York
16. Lo YS, Choubey RK et al (2011) Shallow bath chemical deposition of CdS thin film. Thin Solid Films 520(1):217–223
17. Mane RS, Lokhande CD (2000) Chemical deposition method for metal chalcogenide thin films. Mater Chem Phys 65(1):1–31
18. Mane RS, Lokhande CD (1997) Studies on chemically deposited cadmium sulphoselenide (CdSSe) films. Thin Solid Films 304(1–2):56–60
19. Nair PK, Nair MTS et al (1998) Semiconductor thin films by chemical bath deposition for solar energy related applications. Sol Energy Mater Sol Cells 52(3–4):313–344
20. Kozhevnikova NS, Rempel AA et al (2009) Structural study of the initial growth of nanocrystalline CdS thin films in a chemical bath. Thin Solid Films 517(8):2586–2589
21. Lisco F, Kaminski PM et al (2015) The structural properties of CdS deposited by chemical bath deposition and pulsed direct current magnetron sputtering. Thin Solid Films 582:323–327
22. Khallaf H, Oladeji IO et al (2008) Characterization of CdS thin films grown by chemical bath deposition using four different cadmium sources. Thin Solid Films 516(21):7306–7312
23. Gonzalez G, Krishnan B et al (2011) Modification of optical and electrical properties of chemical bath deposited CdS using plasma treatments. Thin Solid Films 519(21):7587–7591
24. Garcia LV, Mendivil MI et al (2015) CdS thin films prepared by laser assisted chemical bath deposition. Appl Surf Sci 336:329–334
25. Sharma AB, Sharma SK et al (2009) Structural and optical investigation of semiconductor CdSe/CdS core-shell quantum dot thin films. Spectrochim Acta A Mol Biomol Spectrosc 72(2):285–290
26. Emerson-Reynolds J (1884) On the synthesis of galena by means of thiocarbamide, and the deposition of lead sulphide as a specular film. J Chem Soc 45:162–165. https://doi.org/10.1039/CT8844500162

27. Hauser O, Biesalski E (1910) Chem-Ztg 34:1079
28. Dzhafarov TD, Ongul F et al (2006) Formation of CdZnS thin films by Zn diffusion. J Phys D Appl Phys 39(15):3221
29. Harrison MA, Ng A et al (2012) CdSSe nanocrystals with induced chemical composition gradients. Isr J Chem 52(11–12):1063–1072
30. Myung Y, Jang DM et al (2010) Composition-tuned ZnO – CdSSe core – shell nanowire arrays. ACS Nano 4(7):3789–3800
31. Ouendadji S, Ghemid S et al (2010) Density functional study of CdS1–xSex and CdS1–xTex alloys. Comput Mater Sci 48(1):206–211
32. Kim YL, Jung JH et al (2009) The growth and optical properties of CdSSe nanosheets. Nanotechnology 20(9):095605
33. Sung TK, Kang JH et al (2011) CdSSe layer-sensitized TiO_2 nanowire arrays as efficient photoelectrodes. J Mater Chem 21(12):4553–4561
34. Jiang J, He Y et al (2013) Synthesis of CdS nanoparticles in switchable surfactant reverse micelles. Chem Commun (Camb) 49(19):1912–1914
35. Peter AJ, Chang Woo L (2012) Electronic and optical properties of CdS/CdZnS nanocrystals. Chin Phys B 21(8):087302
36. Carballeda-Galicia DM, Castanedo-Perez R et al (2000) High transmittance CdO thin films obtained by the sol-gel method. Thin Solid Films 371:105–108. https://doi.org/10.1016/S0040-6090(00)00987-1
37. Ninomiya S, Adachi S (1995) Optical properties of wurtzite CdS Scilight relation icon. J App Phys 78:1183 https://doi.org/10.1063/1.360355
38. Varkey KP, Vijayakumar KP, Yoshida T, Kashiwaba Y (1999) Spray pyrolised thin film CdS homojunction solar cell with improved performance. Renew Energy 18:465–472
39. Klude M, Alexe G et al (2002) 500–600 nm laser emission from quaternary CdZnSSe quantum wells. Phys Status Solidi (b) 229(2):615–615
40. Roduner E (2006) Size matters: why nanomaterials are different. Chem Soc Rev 35(7):583–592
41. Pan A, Wang X et al (2007) Color-changeable optical transport through Se-doped CdS 1D nanostructures. Nano Lett 7(10):2970–2975
42. Mir FA, Chattarjee I et al (2015) Preparation and characterizations of cadmium sulfide nanoparticles. Optik - Int J Light Electron Opt 126(11–12):1240–1244
43. Xie H, Tian C et al (2010) Preparation of p-type CdS thin films and in situ dark conductivity in vacuum deposited CdS: Cu films. Appl Surf Sci 257(5):1623–1627
44. George PJ, Sánchez A, Nair PK, Nair MTS (1995) Doping of chemically deposited intrinsic CdS thin films to n type by thermal diffusion of indium Scilight relation icon. Appl Phys Lett 66:3624. https://doi.org/10.1063/1.113808
45. Pavlov PV, Jojlov AF (eds) (1987) Física del estado sólido. Editorial MIR, Moscú
46. Chopra KL, Das SR (eds) (1983) Thin film solar cells. Springer Science+Business Media New York
47. Potter BG, Simmons JH (1990) Quantum-confinement effects in CdTe-glass composite thin films produced using rf magnetron sputtering. J Appl Phys 68:1218. https://doi.org/10.1063/1.346720
48. Cullity BD (ed) (1978) Elements of X-RAY diffraction. Addison-Wesley Publishing Company, Inc., Massachusetts
49. Britt J, Ferekides C (1993) Thin-film CdS/CdTe solar cell with 15.8% efficiency Scilight relation icon. Appl Phys Lett 62:2851. https://doi.org/10.1063/1.109629
50. Ichimura M (1999) Structural and optical characterization of CdS films grown by photochemical deposition Scilight relation icon. J Appl Phys 85:7411. https://doi.org/10.1063/1.369371
51. Castillo SJ, Sotelo-Lerma M, Neyra IA et al (1998) Effects of reaction temperature on the physical properties of chemically deposited CdS films. Mat Sc Forum 287–288: 343–346. doi:MSF.287-288.343

52. Zelaya-Angel O, Esparza-Garcia AE, Falcony C et al (1995) Photoluminescence effects associated with thermally induced crystalline structure changes in CdS films. Sol Stat Comm 94:81–85
53. Pandya DK, Chopra KL (1980) Growth kinetics and polymorphism of chemically deposited CdS films. J Electrochem Soc 127:943–948. https://doi.org/10.1149/1.2129792
54. Orozco-Terán RA, Sotelo-Lerma M et al (1999) PbS-CdS bilayers prepared by the chemical bath deposition technique at different reaction temperatures. Thin Sol Films 343–344:587–590. https://doi.org/10.1016/S0040-6090(98)01719-2
55. Sebastian PJ, Campos J, Nair PK (1993) The effect of post-deposition treatments on morphology, structure and opto-electronic properties of chemically deposited CdS thin films. Thin Sol Films 227:111–228
56. Doña JM, Herrero J (1995) Chemical bath codeposited CdS-ZnS film characterization. Thin Solid Films 268:5–12
57. Hernández L, De Melo O, Zelaya-Angel O et al (1994) Electro-optical characterization of sulfur-annealed chemical-bath deposited CdS films. J Electrochem Soc 141(11):3238–3241
58. Lozada-Morales R, Rubín-Falfán M, Zelaya-Angel O, Ramírez-Bond R (1998) Characterization of cubic CdS thin films annealed in vacuum. J Phys Chem Sol 59(9):1393–1398. https://doi.org/10.1016/s0022-3697(98)00242-x
59. Nair MTS, Nair PK (1994) Conversion of chemically deposited photosensitive CdS thin films to n-type by air annealing and ion exchange reaction Scilight relation icon. J App Phys 75:1557. https://doi.org/10.1063/1.356391
60. Green MA, Emery K et al (2013) Solar cell efficiency tables (version 42). Prog Photovoltaics Res Appl 21(5):827–837
61. Wolfram J CdTe solar cells. Retrieved 8 Oct 2015, from http://www.mawi.tu-darmstadt.de/of/of/forschung_2/solarzellen/cdte/index.en.jsp
62. Jackson P, Hariskos D et al (2011) New world record efficiency for Cu(In, Ga)Se_2 thin-film solar cells beyond 20%. Prog Photovoltaics Res Appl 19(7):894–897
63. Li L, Wu P, Fang X, Zhai T, Dai L, Liao M, Koide Y, Wang H, Bando Y, Golberg D (2010) Single-Crystalline CdS nanobelts for excellent field-emitters and ultrahigh quantum-efficiency photodetectors, Adv Mater 22(29):3161–3165. https://doi.org/10.1002/adma.201000144
64. Young-Jin C, Kyung-Soo P, Jae-Gwan P (2010) Network-bridge structure of CdSxSe1−x nanowire-based optical sensors, Nanotechnol 21(50):505605. https://doi.org/10.1088/0957-4484/21/50/505605
65. Segets D, Gradl J, Taylor RK et al (2009) Analysis of optical absorbance spectra for the determination of ZnO nanoparticle size distribution, solubility, and surface energy. ACS Nano 3:1703–1710
66. Choi WS, Kim EJ, Seong SG et al (2009) Optical and structural properties of ZnO/TiO_2/ZnO multi-layers prepared via electron beam evaporation. Vacuum 83:878–882
67. Sahu DR, Lin S-Y, Huang J-L (2007) Study on the electrical and optical properties of Ag/Al-doped ZnO coatings deposited by electron beam evaporation. Appl Surf Sci 253:4886–4890
68. Subrahmanyam A, Barik UK (2005) Synthesis of P-type transparent conducting silver:indium oxide (AIO) thin films by reactive electron beam evaporation technique. J Phys Chem Solids 66:817–822
69. Gao P, Wang ZL (2002) Self-assembled nanowire-nanoribbon junction arrays of ZnO. J Phys Chem B 106:12653–12658
70. Greyson EC, Babayan Y, Odom TW (2004) Directed growth of ordered arrays of small-diameter ZnO nanowires. Adv Mater 16:1348–1352
71. Song JH, Wang XD, Riedo E et al (2005) Systematic study on experimental conditions for large-scale growth of aligned ZnO nanwires on nitrides. J Phys Chem B 109:9869–9872
72. Wu JM, Shih HC, Wu WT et al (2005) Thermal evaporation growth and the luminescence property of TiO_2 nanowires. J Cryst Growth 281:384–390
73. Hu WS, Liu ZG, Sun J et al (1997) Optical properties of pulsed laser deposited ZnO thin films. J Phys Chem Solids 58:853–857

74. Lorenz M (2008) Pulsed Laser Deposition of ZnO-Based Thin Films. Transparent Conductive Zinc Oxide 104:303–357 Springer Series in Materials Science
75. Nakamura T, Minoura H, Muto H (2002) Fabrication of ZnO(0001) epitaxial films on the cubic(111) substrate with C6 symmetry by pulsed laser ablation. Thin Solid Films 405:109–116
76. Franklin JB, Zou B, Petrov P et al (2011) Optimised pulsed laser deposition of ZnO thin films on transparent conducting substrates. J Mater Chem 21:8178–8182
77. Zhao J-L, Li X-M, Bian J-M et al (2005) Structural, optical and electrical properties of ZnO films grown by pulsed laser deposition (PLD). J Cryst Growth 276:507–512
78. Sun Y, Fuge GM, Ashfold MNR (2004) Growth of aligned ZnO nanorod arrays by catalyst-free pulsed laser deposition methods. Chem Phys Lett 396:21–26
79. Dang WL, Fu YQ, Luo JK, Flewitt AJ, Milne WI (2007) Deposition and characterization of sputtered ZnO films. Superlattices Microstruct 42:89–93
80. Ellmer K (2000) Magnetron sputtering of transparent conductive zinc oxide: relation between the sputtering parameters and the electronic properties. J Phys D Appl Phys 33:R17–R32
81. Ohring M (2002) Materials science of thin films. Academic Press, USA, p 279
82. Wagner RS, Ellis WC (1964) Vapout-Liquid-Solid mechanism of dingle crystal growth. Appl Phys Lett 4:89–90
83. Nicolay S, Benkhaira M, Ding L et al (2012) Control of CVD-deposited ZnO films properties through water/DEZ ratio: decoupling of electrode morphology and electrical characteristics. Sol Energy Mater Sol Cells 105:46–52
84. Ramgir NS, Late DJ, Bhise AB et al (2006) ZnO multipods, submicron wires, and spherical structures and their unique field emission behavior. J Phys Chem B 110:18236–18242
85. Wu JJ, Liu SC (2002) Low-temperature growth of well-aligned ZnO nanorods by chemical vapor deposition. Adv Mater 14:215–218
86. Haga K, Kamidaira M, Kashiwaba Y et al (2000) ZnO thin films prepared by remote plasma-enhanced CVD method. J Cryst Growth 214:77–80
87. Han N, Hu P, Zuo et al (2010) Photoluminescence investigation on the gas sensing property of ZnO nanorods prepared by plasma-enhanced CVD method. Sens. Actuators B 145:114–119
88. Pedersen JD, Esposito HJ, Teh KS (2011) Direct synthesis and characterization of optically transparent conformal zinc oxide nanocrystalline thin films by rapid thermal plasma CVD. Nanoscale Res Lett 6:568
89. Park WI, Lee C-H, Chae JH et al (2009) Ultrafine ZnO nanowire electronic device arrays fabricated by selective metal-organic chemical vapor deposition. Small 5:181–184
90. Zhang G, Nakamura A, Aoki T et al (2006) Au-assisted growth approach for vertically aligned ZnO nanowires on Si substrate. Appl Phys Lett 89:113112
91. Falyouni F, Benmamas L, Thiandoume C et al (2009) Metal organic chemical vapor deposition growth and luminescence of ZnO micro- and nanowires. J Vac Sci Technol, B 27:1662–1666
92. Montenegro DN, Souissi A, Martínez-Tomá C et al (2012) http://www.sciencedirect.com/science/article/pii/S002202481200601X - aff1 Morphology transitions in ZnO nanorods grown by MOCVD. J Crystal Growth 359:122–128
93. Park WI, Kim DH, Jung SW et al (2002) Metalorganic vapor-phase epitaxial growth of vertically well-aligned ZnO nanorods. Appl Phys Lett 80:4232–4234
94. Elshaer AAM (2008) Molecular beam epitaxy growth and characterization of ZnO-based layers and heterostructures. Cuvillier Verlag, 138 pages
95. Palamà IE, D'Amone S, Arcadio V et al (2015) Underwater Wenzel and Cassie Oleophobic behaviour. J Mater Chem A 3:3854–3861
96. Palamà IE, D'Amone S, Biasiucci M et al (2014) Bioinspired design of a photoresponsive superhydrophobic/oleophilic surface with underwater superoleophobic efficacy. J Mater Chem A 2:17666–17675
97. Wang ZL (2004) Zinc oxide nanostructures: growth, properties and applications. J Phys-Condens Matter 16:R829–R858
98. Wang ZL, Xu S, Lao C et al (2008) Density-controlled growth of aligned ZnO nanowire arrays by seedless chemical approach on smooth surfaces. J Mater Res 23:2072–2077

99. Greene LE, Law M, Tan DH et al (2005) General route to vertical ZnO nanowire arrays using textured ZnO seeds. Nano Lett 5:1231–1236
100. Yamabi S, Imai H (2002) Growth conditions for wurtzite zinc oxide films in aqueous solutions. J Mater Chem 12:3773–3778
101. Lee S, Roy BK, Cho J (2013) Vertically aligned ZnO nanorods grown by low-temperature solution processing. Jpn J Appl Phys 52:1–9
102. Nause J, Nemeth B (2005) Pressurized melt growth of ZnO boules. Semicond Sci Technol 20:S45–S48
103. Nause JE (1999) ZnO broadens the spectrum. III-Vs Rev 12:28–31
104. Bekermann D, Gasparotto A, Barreca D et al (2010) Highly oriented ZnO nanorod arrays by a novel plasma chemical vapor deposition process. Cryst Growth Des 10:2011–2018
105. Morkoç H, Özgür Ü (2009) Index, in Zinc Oxide: fundamentals, materials and device technology. Wiley-VCH Verlag GmbH & Co. KGaA, Weinheim
106. Wei A, Pan L, Huang W (2011) Recent progress in the ZnO nanostructure-based sensors. Mater Sci Eng, B 176:1409–1421
107. Verghese PM, Clarke DR (2000) Piezoelectric contributions to the electrical behavior of ZnO varistors. J Appl Phys 87:4430–4438
108. Chang J, Waclawik ER (2014) Colloidal semiconductor nanocrystals: controlled synthesis and surface chemistry in organic media. RSC Adv 4:23505–23527
109. Ashrafia A, Jagadish C (2007) Review of zincblende ZnO: stability of metastable ZnO phases. J Appl Phys 102:071101–071112
110. Goniakowski J, Finocchi F, Noguera C (2008) Polarity of oxide surfaces and nanostructures. Rep Prog Phys 71:016501
111. Ashrafi ABMA, Ueta A, Avramescu et al (2000) Growth and characterization of hypothetical zinc-blende ZnO films on GaAs (001) substrates with ZnS buffer layers. Appl Phys Lett 76:550–552
112. Kim H, Cho K, Song H et al (2003) Photocurrent mechanism in a hybrid system of 1-thioglycerol-capped HgTe nanoparticles. Appl Phys Lett 83:4619–4621
113. Kim SK, Jeong SY, Cho CR (2003) Structural reconstruction of hexagonal to cubic ZnO films on Pt/Ti/SiO_2/Si substrate by annealing. Appl Phys Lett 82:562–564
114. Kogure T, Bando Y (1998) Formation of ZnO nanocrystallites on ZnS surface by electron beam radiation. J Electron Microsc 47:135–141
115. Murayama M, Nakayama T (1994) Chemical trend of band offsets at wurtzite/zinc-blende heterocrystalline semiconductor interfaces. Phys Rev B 49:4710–4724
116. Zhang L, Huang H (2007) Structural transformation of ZnO nanostructures. Appl Phys Lett 90:023115
117. Bates CH, White WB, Roy R (1962) New high-pressure polymorph of zinc oxide. Science 137:993
118. Kołodziejczak-Radzimska A, Jesionowski T (2014) Zinc oxide—from synthesis to application: a review. Materials 7:2833–2881
119. Moezzi A, McDonagh AM, Cortie MB (2012) Zinc oxide particles: synthesis, properties and applications. Chem Eng J 185–186:1–22
120. Baruah S, Dutta J (2009) Hydrothermal growth of ZnO nanostructures. Sci Technol Adv Mater 10:013001
121. Fan Z, Lu JG (2005) Zinc oxide nanostructures: synthesis and properties. J Nanosci Nanotechnol 5:1561–1573
122. Subramanyam TK, Srinivasulu Naidu B, Uthanna S (2000) Physical properties of zinc oxide films prepared by DC reactive magnetron sputtering at different sputtering pressures. Cryst Res Technol 35:1194–1202
123. Law M, Greene LE, Johnson JC et al (2005) Nanowire dyesensitized solar cells. Nat Mater 4:455–459
124. Zhang Y, Xu J, Xiang Q (2009) Brush-like hierarchical ZnO nanostructures: synthesis, photoluminescence and gas sensor properties. J Phys Chem C 113:3430–3435

125. Li C, Li L, Du Z et al (2008) Rapid and ultrahigh ethanol sensing based on Au-coated ZnO nanorods. Nanotechnology 19:035501
126. Bai S, Sun C, Guo T et al (2013) Low temperature electrochemical deposition of nanoporous ZnO thin films as novel NO_2 sensors. Electrochim Acta 90:530–534
127. Li X, Feng W, Xiao Y et al (2014) Hollow zinc oxide microspheres functionalized by Au nanoparticles for gas sensors. RSC Adv 4:28005–28010
128. Ahmad MZ, Chang J, Ahmad MS et al (2013) Non-aqueous synthesis of hexagonal ZnO nanopyramids: gas sensing properties. Sens Actuators B: Chem 177:286–294
129. Calestani D, Zhaa M, Mosca R (2010) Growth of ZnO tetrapods for nanostructure-based gas sensors. Sens Actuators B: Chem 144:472–478
130. Xie Y, He Y, Irwin PL et al (2011) Antibacterial activity and mechanism of action of zinc oxide nanoparticles against Campylobacter jejuni. Appl Environ Microbiol 77:2325–2331
131. Zhang Y, Nayak TR, Hong H et al (2013) Biomedical applications of zinc oxide nanomaterials. Curr Mol Med 13:1633–1645
132. Fleischhaker F, Wloka V, Hennig I (2010) ZnO based field-effect transistors (FETs): solution-processable at low temperatures on flexible substrates. J Mater Chem 20:6622–6625
133. Xiang B, Wang PW, Zhang XZ, Dayeh SA, Aplin DPR, Soci C et al (2007) Rational synthesis of p-type zinc oxide nanowire arrays using simple chemical vapor deposition. Nano Lett 7:323–328
134. Chen M-T, Lu M-P, Wu Y-J et al (2010) Near UV LEDs made with in situ doped p-n homo-junction ZnO nanowire arrays. Nano Lett 10:4387–4393
135. Zhao JZ, Liang HW, Sun JC et al (2008) Electroluminescence from n-ZnO/p-ZnO: Sb homo-junction light emitting diode on sapphire substrate with metal–organic precursors doped p-type ZnO layer grown by MOCVD technology. J Phys D Appl Phys 41:195110
136. Baltakesmez A, Tekmen S, Tuzemen S (2011) ZnO homojunction white light-emitting diodes. J Appl Phys 110:054502
137. Hsu YF, Xi YY, Tam KH et al (2008) Undoped p-type ZnO nanorods synthesized by a hydrothermal method. Adv Funct Mater 18:1020–1030
138. Nath T, Roy S, Saxena P et al (1990) Effect of annealing on the electrical properties of polycrystalline intermetallic compound HgTe. J Appl Phys 67:826–831
139. Steigerwald ML, Sprinkle CR (1987) Organometallic synthesis of II-VI semiconductors. 1. Formation and decomposition of bis(organotelluro)mercury and bis(organotelluro)cadmium compounds. J Am Chem Soc 109:7200–7201
140. Ghandhi SK, Bhat IB, Ehsani H et al (1989) Low-temperature growth of HgTe and HgCdTe using methylallyltelluride Appl. Phys Lett 55:137–139
141. Korenstein R, Hoke WE, Lemonias PJ (1987) Metalorganic growth of HgTe and CdTe at low temperatures using diallyltelluride. J Appl Phys 62:4929–4931
142. Suh SH, Moon SW, Kim JS et al (1992) Slider liquid phase epitaxial growth of Hg0.8Cd0.2Te, Hg0.7Cd0.3Te and Hg0.3Cd0.7 Te with precise control of alloy compositions. J Crysal Growth 121:417–422
143. Selvig E, Tonheim CR, Kongshaug KO et al (2008) Defects in CdHgTe grown by molecular beam epitaxy on (211) B-oriented CdZnTe substrates. J Vac Sci Technol, B 26:525–533
144. Irvine SJC (1987) UV photo-assisted crystal growth of II-VI compounds. Crit Rev Solid State Mater Sci 13:279–309
145. Williams LM, Lu PY, Chu SNG et al (1987) Multilayers of HgTe-CdTe grown by low temperature metalorganic chemical vapor deposition. J Appl Phys 62:295–297
146. Williams LM, Lu PY, Wang CH et al (1987) Plasma enhanced chemical vapor deposition of epitaxial mercury telluride. Appl Phys Lett 51:1731
147. Hoke WE, Lemonias PJ (1985) Metalorganic growth of CdTe and HgCdTe epitaxial films at a reduced substrate temperature using diisopropyltelluride. Appl Phys Lett 46:398–400
148. Kisker DW, Steigerwald ML, Kometani T-Y et al (1987) Low-temperature organometallic vapor phase epitaxial growth of CdTe using a new organotellurium source. Appl Phys Lett 50:1681–1683

149. Lichtmann LS, Parsons JD, Cirlin EH (1988) Temperature-independent unassisted pyrolytic MOCVD growth of cadmium telluride at 250 °C using 2,5-dihydrotellurophene. J Cryst Growth 86:217–221
150. Green M, Wakefield G, Dobson PJ (2003) A simple metalorganic route to organically passivated mercury telluride nanocrystals. J Mater Chem 13:1076–1078
151. Harrison MT, Kershaw SV, Rogach AL et al (2000) Wet chemical synthesis of highly luminescent HgTe/CdS core/shell nanocrystals. Adv Mater 12:123–125
152. Kazes M, Lewis DY, Ebenstein Y et al (2002) Lasing from semiconductor quantum rods in a cylindrical microcavity. Adv Mater 14:317–321
153. Kovalenko MV, Kaufmann E, Pachinger D et al (2006) Colloidal HgTe nanocrystals with widely tunable narrow band gap energies: from telecommunications to molecular vibrations. J Am Chem Soc 128:3516–3517
154. Olk P, Buchler BC, Sandoghdar V (2004) Subwavelength emitters in the near-infrared based on mercury telluride nanocrystalsAppl. Phys Lett 84:4732–4734
155. Sangsefidi FS, Salavati-Niasari M, Esmaeili-Zare M (2014) Synthesis and characterization of mercury telluride nanoparticles using a new precursor. J Ind Eng Chem 20:3415–3420
156. Song H, Cho K, Kim H et al (2004) Synthesis and characterization of nanocrystalline mercury telluride by sonochemical method. J Cryst Growth 269:317–323
157. Qin A-M, Fang Y-P, Su C-Y (2007) Hydrothermal synthesis of HgTe rod-shaped nanocrystals. Mater Lett 61:126–129
158. Carter R, Sloan J, Kirkland AI et al (2006) Correlation of structural and electronic properties in a new low-dimensional form of mercury telluride. Phys Rev Lett 96:215501
159. Ranga R, Dutta V (2006) Nanotubes in spray deposited nanocrystalline HgTe:I thin films. Mater Res Symp Proc 901E, 0901-Ra11-19 - Rb11-19.1- 0901-Ra11-19 - Rb11- 19.6, 0272-9172
160. Kim S, Kim T, Im SH et al (2011) Bandgap engineered monodisperse and stable mercury telluride quantum dots and their application for near-infrared photodetection. J Mater Chem 21:15232–15236
161. Harrison MT, Kershaw SV, Burt MG et al (1999) Investigation and factors affecting the photoluminescence of colloidally-prepared HgTe nanocrystals. J Mater Chem 9:2721–2723
162. Rogach AL, Kershaw SV, Burt MG et al (1999) Colloidally prepared HgTe nanocrystals with strong room-temperature infrared luminescence. Adv Mater 11:552–555
163. Li LS, Wang H, Liu Y et al (2007) Room temperature synthesis of HgTe nanocrystals. J Colloid Interface Sci 308:254–257
164. Piepenbrock MOM, Stirner T, Kelly SM et al (2006) A low-temperature synthesis for organically soluble HgTe nanocrystals exhibiting near-infrared photoluminescence and quantum confinement. J Am Chem Soc 128:7078–7090
165. Berchenko NN, Pashkovski MV (1976) Mercury telluride—a zero-gap semiconductor. Sov Phys Usp 19:462–480
166. Giriat W (1962) Electrical properties of mercury telluride. Br J Appl Phys 15:151–156
167. Harman TC, Logan MJ, Goering HL (1958) Preparation and electrical properties of mercury telluride. J Phys Chem Solids 7:228–235
168. Rodot H, Triboulet R (1962) C R Acad Sci 248:938
169. Mokrovski˘ι NP, Regel' AR (1952) O Svyazi mezhdu izmeneniyami plotnosti i elektronnoi provodimisti pri plavenii veshchestv so strukturoi tipa almaza ili tsinkovoi obmanki. Zh Tekh Fiz 22:1281–1289
170. Groves SH, Paul W (1963) Band structure of gray tin. Phys Rev Lett 11:194–196
171. Nath T, Tyagu S, Saxena P et al (1993) Electrical behaviour of off-stoichiometric polycrystalline mercury telluride. J Mater Sci Lett 12:1193–1196
172. Bridgman PW (1940) The compression of 46 substances to 50,000 kg/cm^2. Proc Am Acad Art Sci 74:21–51
173. Ohtani A, Seike T, Motobayashi M, Onodera A (1982) The electrical properties of HgTe and HgSe under very high pressure. J Phys Chem Solids 43:627–632

174. Chen M, Shao L, Kershaw SV et al (2014) Photocurrent enhancement of HgTe quantum dot photodiodes by plasmonic gold nanorod structures. ACS Nano 8:8208–8216
175. Kershaw SV, Susha AS, Rogach AL (2013) Narrow bandgap colloidal metal chalcogenide quantum dots: synthetic methods, heterostructures, assemblies, electronic and infrared optical properties. Chem Soc Rev 42:3033–3087
176. Keuleyan S, Lhuillier E, Brajuskovic V et al (2011) Mid-infrared HgTe colloidal quantum dot photodetectors. Nat Photon 5:489–493
177. Capper P (2007) Narrow-bandgap II–VI semiconductors: growth, in Springer handbook of electronic and photonic materials Springer-Verlag, Berlin, Germany, pp 303–324
178. Capper P, Garland J, Baker IM (2010) HgCdTe photovoltaic infrared detectors. Mercury Cadmium Telluride. Wiley, Hoboken, pp 447–467
179. Katagiiri H, Jimbo K, Maw WS et al (2009) Development of CZTS-based thin film solar cells. Thin Solid Films 517:2455–2460
180. Kato T, Hiroi H, Sakai N et al (2012) Characterization of front and back interfaces on Cu2ZnSnS4 thin-film solar cells. In Proceedings of the 27th EUPVSEC
181. Wang W, Winkler MT, Gunawan O et al. (2013) Device characteristics of CZTSe thin-film solar cells with 12.6 efficiency. Adv Energy Mater 4:1301465
182. Konig M, Wiedmann S, Brune C et al (2007) Quantum spin hall insulator state in HgTe quantum wells. Science 318:766–770

Chapter 8
Other Miscellaneous Semiconductors and Related Binary, Ternary, and Quaternary Compounds

Dongguo Chen and Nuggehalli M. Ravindra

8.1 Introduction

8.1.1 Semiconductors and Their Alloys

A semiconductor is defined as a material with electrical conductivity in the range of 10^{-9}–10^2 S/m. Alternatively, it can be defined as a material with an energy gap between 0 and 6 electron volts (eV). For example, Si and Ge have small energy gaps of 1.12 and 0.67 eV, respectively, while ZnS and diamond have large band gaps of 3.60 and 5.50 eV, respectively [1]. Materials with zero band gaps are metals or semimetals and those with band gap larger than 6 eV are considered as insulators. According to the types of band structures, materials can also be characterized as direct or indirect band gap semiconductors.

In the development of semiconductor technologies, semiconductor alloys have attracted much attention because of their ability to tailor the optoelectronic properties, such as the band gap with the alloy compositions. For example, the band gap of $Al_xGa_{1-x}N$ varies from 3.42 to 6.20 eV when the Al composition x is increased from 0 to 1. The most basic alloys are formed by two group IV elements, such as Si_xGe_{1-x} and Ge_xSn_{1-x}. Ternary alloys (AB_xC_{1-x}) are formed by substituting some atoms B in binary compound AB by C atoms, which are usually in the same column with B in the periodic table, such as the common-cation alloy CdS_xTe_{1-x} and the common-anion alloy $Ga_xIn_{1-x}P$. Similarly, quaternary alloys, such as $Al_xGa_{1-x}N_yAs_{1-y}$ have been synthesized in experiments [2]. For some quaternary alloys, the atom can be replaced by other atoms from the neighboring columns in the periodic table. For example, Cu_2ZnGeS_4 can be formed by replacing Ga atom in $CuGaS_2$ compound by Zn and Ge atoms. These multi-ternary alloys will, in general, inherit the structures of

D. Chen · N. M. Ravindra (✉)
New Jersey Institute of Technology, Newark, NJ 07102, USA
e-mail: nmravindra@gmail.com

M. I. Pech-Canul and N. M. Ravindra (eds.), *Semiconductors*,
https://doi.org/10.1007/978-3-030-02171-9_8

their parent alloys and short or long range of subordering can be formed depending on the growth conditions.

8.1.2 Growth Techniques

The recent developments in crystal growth techniques have greatly catalyzed the experimental and theoretical studies on semiconductor alloys. Techniques such as Atomic Layer Deposition (ALD), Metal-Organic Chemical Vapor Deposition (MOCVD) and Molecular Beam Epitaxy (MBE) allow crystals to be deposited on substrates at fixed compositions within very small deviations. These techniques have made it possible to synthesize semiconductor devices with complicated structures, such as MOS transistors and CIGS solar cells. Some growth techniques are briefly introduced here.

8.1.2.1 Growth of Bulk Crystals

Czochralski (CZ) method [3] is the most important and common method used in the growth of semiconductor bulk crystals because it is capable of producing large diameter crystals. In this technique, the raw material is melted in a crucible. A small single-crystal seed is then lowered into the melt. The melt will be gradually pulled out and solidified onto the seed. The crystal orientation of the seed will determine the orientation of the resulting pulled crystals.

During the CZ crystal growth, the seed and crucible are normally rotated in the opposite directions to promote more uniform growth. However, the opposite rotations increase the corrosion of the crucible by the melt. The main unexpected impurities of the CZ technique are from the crucible material and the gas surrounding the melt. For example, in the growth of bulk crystals of Si, carbon evaporation from the graphite susceptor of quartz crucible and melt source results in the carbon incorporation in the crystal, typically at levels of 10^{15}–10^{16} cm^{-3} [4].

Another bulk crystal growth technique is the Bridgman method, which is similar to the CZ method. The differences are the following: (I) A temperature gradient is implemented along the crucible for a better solidification; (II) The crucible can be positioned either vertically or horizontally to control convection flow. The Bridgman method is a popular way of producing certain crystals, such as GaAs, for which the CZ technique is more difficult.

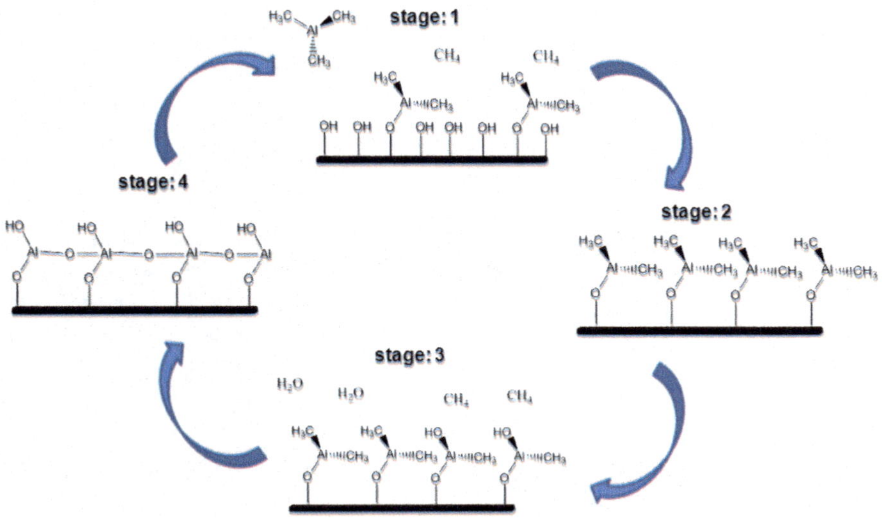

Fig. 8.1 ALD growth cycle includes four stages: (1) Exposure of the first precursor, (2) purge of the reaction chamber, (3) exposure of the second precursor, and (4) a further purge of the reaction chamber [5]

8.1.2.2 Growth of Thin Films

Atomic Layer Deposition

Atomic Layer Deposition (ALD) is fundamentally a Chemical Vapor Deposition (CVD) process with self-limiting growth [5]. It is controlled by a chemical reaction that consists of two separate half reactions; the film is formed in a growth cycle. Throughout the reaction, the precursors need to be separate. A growth cycle occurs via the following four stages: (1) Exposure of the first precursor, (2) Purge of the reaction chamber, (3) Exposure of the second precursor, and (4) A further purge of the reaction chamber (Fig. 8.1) [5 and references therein].

Chemical Vapor Deposition

Epitaxial films are usually grown by Chemical Vapor Deposition (CVD) method. In CVD, gases containing the required chemical elements are introduced into the deposition chamber to react and form the desired film on the surface of the substrate. The semiconductors formed during the gas reaction are deposited as a thin film on a substrate inside the reactor.

Figure 8.2 shows the growth of ternary semiconductor HgCdTe alloys (MCT) using Metal-Organic Chemical Vapor Deposition (MOCVD) [6]. In this reactor, two separated gas inlet channels are provided to prevent premature gas reactions and dust

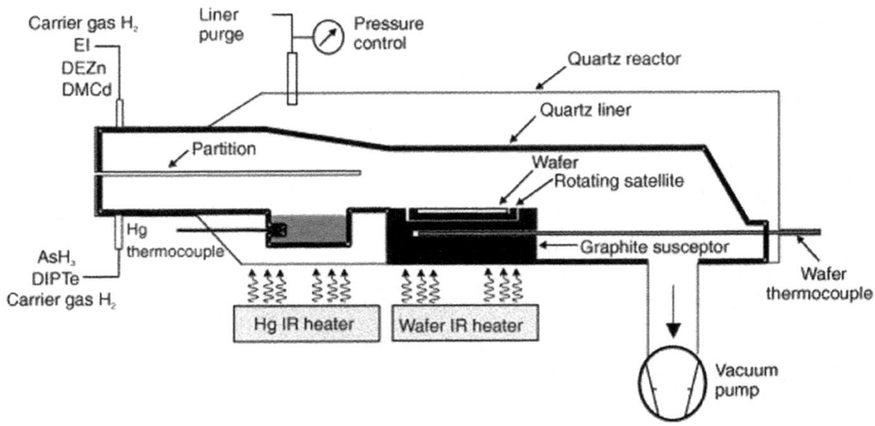

Fig. 8.2 Schematic drawing of Aix-200 II–VI MOCVD reactor for HgCdTe epitaxial growth. The system is equipped with additional mercury bath [6]

formation. The precursors, dimethylcadmium (DMCd) and diethylzinc (DEZn), and the donor doping source ethyl iodine (EI), together with the carrier gas H_2, are delivered into the reactor through the upper channel. The precursor, diisopropyltelluride (DIPTe), and the acceptor doping source, arsine AsH_3, together with the carrier gas H_2, are delivered into the reactor through the lower channel. The mercury source is held in the quartz container inside the reactor with the temperature maintained at 200–220 °C. The MCT epilayer is formed on the substrate at around 360–410 °C.

Molecular Beam Epitaxy (MBE)

A typical disadvantage with the CVD technique is that the reactor may contain a high concentration of contaminants in the form of residual gases. This problem can be avoided in the Molecular Beam Epitaxy (MBE) method (Fig. 8.3) [7]. The principle underlying MBE growth is relatively simple: a molecular beam is created in an effusion cell with a very small orifice by heating a source material until vaporization. The molecular beams from effusion cells at different angles then migrate to a UHV environment and impinge on a hot substrate surface, where the molecules can diffuse and eventually incorporate into the growing film. Despite the conceptual simplicity, it is difficult to control the crystal stoichiometry in MBE growth. However, it is possible to monitor the surface quality and epilayer growth conditions using some electron or ion-based techniques, such as Auger Electron Spectroscopy (AES) and Reflection High-Energy Electron Diffraction (RHEED) [4].

Some other techniques are also employed to grow semiconductor alloys. For example, the Liquid Phase Epitaxy (LPE) growth technique is extensively used in the growth of GaAs samples and devices [8]. The Ga metal is utilized as a solvent for As. When the solvent is cooled in contact with a GaAs substrate, it becomes super-

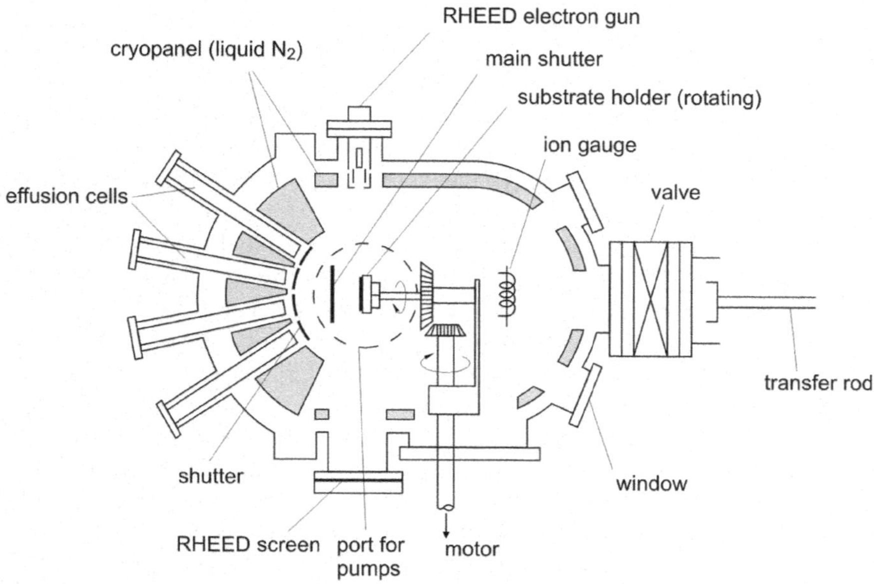

Fig. 8.3 Top cross-view of a MBE system [7]

saturated with As and the nucleation of GaAs starts on the substrate. The advantage of this method is that the equipment is inexpensive and easy to setup. However, it is difficult to control the growth conditions.

8.1.3 *Wide Band Gap Alloys and Their Applications*

Wide band gap (larger than 2.5 eV) semiconductors are experiencing extensive developments because of their thermal conductivities, breakdown electric fields, thermal and chemical stabilities at high temperatures and high powers [9]. A wide range of materials have been investigated as wide band gap semiconductors, most notably 6H-SiC, III–V nitrides and ZnS-based II–VI alloys.

The wide band gap energy and low intrinsic carrier concentration of SiC [10] will allow SiC semiconductor device to function at much higher temperatures. For example, SiC device operation, at 600 °C, has been experimentally demonstrated on a variety of SiC devices [10]. Moreover, the high breakdown field and high thermal conductivity of SiC coupled with high operational junction temperature theoretically permits extremely high-power densities and efficiencies to be realized in SiC devices. The high breakdown field of SiC relative to silicon enables the blocking voltage region of a power device to be roughly 10× thinner and 10× heavier doped, permitting a roughly 100-fold beneficial decrease in the blocking region resistance at the same voltage rating [11]. Finally, while SiC's smaller on-resistance and faster switching help to minimize energy loss and heat generation, SiC's higher thermal conductivity enables more efficient removal of waste heat energy from the active device.

The wide band gap III–V nitrides have been the subject of significant interest because of their applications in the blue light emitting diodes and lasers as well as solar-blind UV photodetectors. Recently, much effort has been directed toward a p-i-n photodiode involving solely high Al content $Al_xGa_{1-x}N$ layers [12]. The choice of the p-i-n photodiode designed for solar-blind UV detectors is driven by its intrinsic advantages: (I) a very low dark current due to large potential barrier; (II) a high speed of operation; (III) a direct control of the quantum efficiency and speed through controlling the thickness of the intrinsic layer, and (IV) the device can operate under low to no bias. One of the greatest obstacles [13, 14] in these photodetectors is that the high activation energy of the impurities (such as Mg and Zn) results in a low concentration of free holes at room temperature.

The II–VI ZnS wide band gap semiconductor alloys have wide applications in fabricating technologically important solid-state devices. The ZnS_xSe_{1-x} and ZnS_xTe_{1-x} semiconductor alloys are used to fabricate optoelectronic devices in blue-green spectral region since they have direct band gaps in the range of 2.75–3.66 eV and 2.10–3.66 eV [15], respectively. These II–VI semiconductors can be grown lattice-matched on substrates such as Si, GaAs and GaP, making them potential candidates for high efficiency multi-junction solar cells and other applications in silicon technology. Due to the fact that 80% of the band offset between the corresponding binary constituents of the ternary alloy is in the conduction band when quantum wells are formed, $Zn_xCd_yMg_{1-x-y}S$-based quantum structures have recently found important applications in inter-subband devices, such as quantum cascade lasers and quantum well photodetectors [16].

8.1.4 Narrow Band Gap Alloys and Their Applications

The narrow band gap alloy refers to the material with a band gap that is comparatively smaller than that of silicon. These band gaps are typically in the infrared region, i.e., near infrared with wavelength in the range of 0.78–3 μm, mid-infrared with wavelength in the range of 3–50 μm, and far infrared with wavelength in the range of 50–1000 μm. The narrow band gap alloys, such as GaAsSb and HgCdTe can be used for infrared detectors and thermoelectrics.

One particular important narrow band gap semiconductor is the GaAsSb system, which allows Heterojunction Bipolar Transistors (HBTs) to operate in the terahertz (THz) range of frequencies. For example, the staggered band lineup at the GaAsSb/InP base–collector junction not only eliminates the electron blocking problem but also allows electrons to be launched from the base to the collector at very high initial energies [17]. Moreover, GaAsSb grown at low temperature can also be used as a substitute for those traditional materials in THz applications.

The field of narrow band gap II–VI semiconductors is almost dominated by HgCdTe (MCT), although many potential alternatives, such as InAsSb, PbSnSe, and

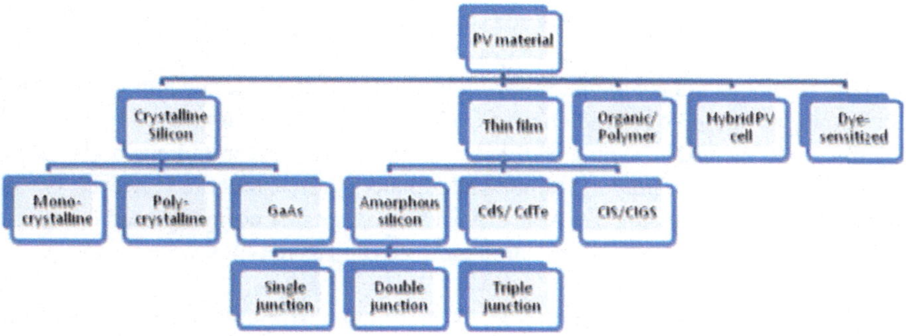

Fig. 8.4 Chart of PV solar cell materials [20]

HgMnTe have been suggested and discussed [18, 19]. The reasons that MCT is still the main infrared material are the following: (I) The band gap of MCT can be made to cover all infrared regions by varying the composition; (II) The lattice constant of MCT almost undergoes no change during the variation of the composition; (III) The material has direct band gap transitions and large absorption coefficient which allow the quantum efficiency to be 100%; (IV) Long minority carrier lifetimes result in low thermal noise allowing high-performance detectors to be made at the highest operating temperatures reported for infrared detectors of comparable wavelengths. All these advantages originate from the energy band structures of the material and they apply to all the situations irrespective of the device architectures that are used.

8.1.5 Solar Cell Materials

Solar cell is one very important application of semiconductors and their alloys. A brief overview of the materials for solar cell production is given in Fig. 8.4 [20]. Silicon is the leading material in solar cell production due to its abundance, maturity of technology and high conversion efficiency. However, due to its limitations as an indirect band gap semiconductor, many more materials, such as CdTe and $Cu(In_{1-x}Ga_x)(S_{1-y}Se_y)_2$ (CIGS), are proposed and discussed.

Silicon is the second easiest raw material that can be found on earth and has been widely used in developing solar cells. Monocrystalline silicon-based solar cells have been reported to have the highest efficiency of more than 20%. In order to reduce the cost of monocrystalline silicon, polycrystalline silicon cell is developed which has minor flaws in the metal contamination and crystal structure and therefore the efficiency of polycrystalline silicon cell is relatively lower. In thin film technology, amorphous silicon is more popular than other materials, such as CdTe and CIGS, due to its higher efficiency. Amorphous silicon is a noncrystalline allotropic form of silicon and has a 40 times higher rate of light absorption than monocrystalline silicon. The advantage of its disordered structure is the high band gap of about 1.7 eV [20].

GaAs is another solar cell material which has high efficiency, lower thickness, and ideal band gap of 1.43 eV. Efficiency of GaAs solar cell can be improved by alloying it with certain materials, such as Al, P, and Sb. GaAs is lighter and has higher heat resistance compared to polycrystalline and monocrystalline silicon. However, GaAs material and manufacturing can be costly and therefore it is normally used for space applications due to its superior radiation hardness and excellent device performance at high temperatures [21].

CdTe has an ideal band gap of 1.45 eV and high absorption coefficient and has thus produced a high efficiency of 15% [22]. It is also known for its stability for a long duration. However, this technology faces some problems that are related to the environment as well as the lack of Tellurium (Te). As an *n*-type semiconductor [23], it has been shown by experiments that doping of copper (Cu) into CdS layer will improve the photoconductivity. Moreover, the CdSTe interlayer formed between *n*-CdS and *p*-CdTe is believed to benefit the solar cell performance including its efficiency.

CIGS solar cell is still in its developing phase and is set to compete with silicon solar cells in performance. An efficiency of 20% for CIGS solar cell has been recorded [24]. Its band gap can be tuned to be close to ideal value by adjusting the composition of In or Ga atoms. For example, the band gap of ternary compounds $CuInSe_2$, $CuGaSe_2$, and $CuInS_2$ are 1.05, 1.68, and 1.65 eV, respectively. The absorption coefficient of $CuInSe_2$ is greater than 10^5 cm^{-1} [20]. One disadvantage with CIGS solar cell is that it contains expensive element, and its window layer, CdS, contains toxic element Cd. Recently, new quaternary semiconductor alloys, such as $Cu_2ZnSn(S,Se)_4$, and $Cu_2ZnGe(S,Se)_4$ have been proposed for the substitution of CIGS since they have similar ideal band gaps and high absorption coefficient, but contain only inexpensive elements [25, 26]. However, many fundamental properties of these quaternary semiconductor alloys are not yet known.

8.2 Theory of First-Principles Calculations

8.2.1 Theoretical Background

The basic ideas of the First-principles calculations of the electronic properties of materials are based on the Born–Oppenheimer Approximation and the Hartree Fock Theory.

8.2.1.1 Born–Oppenheimer Approximation

The Hamiltonian of a many-body system can be written as [3],

$$H = \sum_{I=1}^{N} \frac{\vec{P}_I^2}{2M_I} + \sum_{i=1}^{N_e} \frac{\vec{P}_i^2}{2m} + \sum_{i>j} \frac{e^2}{\left|\vec{r}_i - \vec{r}_j\right|} + \sum_{I>J} \frac{Z_I Z_J e^2}{\left|\vec{R}_I - \vec{R}_J\right|} - \sum_{i,j} \frac{Z_I e^2}{\left|\vec{R}_I - \vec{r}_i\right|}, \quad (8.1)$$

where $\vec{r}_i, \vec{r}_j$ are the positions of ith and jth electrons. $\vec{R}_I, \vec{R}_J$ are the positions of Ith and Jth ions. Z_I, Z_J are the atomic numbers of Ith and Jth ions. $\vec{p}_i$ and $\vec{P}_I$ are the momentum of ith electron and Ith ion. $m,\ M_I$ are the masses of electron and Ith ion. e is the electron charge. In Eq. (8.1), the first and second terms are the kinetic energy of ions and electrons. Third, fourth, and last terms are the Coulomb interactions between electron and electron, ions and ion, electron, and ion. In order to solve this many-body Hamiltonian, the Born–Oppenheimer Approximation [3] states the following: (I) The electronic wavefunction depends on the nuclear positions $\vec{R}_I$ but not on their velocities, that is, the nuclear motion is so much slower than the electron motion that the nuclear component can be treated as fixed; (II) The nuclear motion sees a smeared out potential from the speedy electrons. From this approximation, the electron Hamiltonian can be decoupled out as follows:

$$H = \sum_{i=1}^{N_e} \frac{\vec{p}_i^2}{2m} + \sum_{i>j} \frac{e^2}{\left|\vec{r}_i - \vec{r}_j\right|} - \sum_{i,j} \frac{Z_I e^2}{\left|\vec{R}_I - \vec{r}_i\right|} \quad (8.2)$$

8.2.1.2 Hartree Fock Theory

To solve the simplified many-body Hamiltonian, Hatree Fock theory [27] assumes that the wavefunction is given by a single Slater determinant of N spin-orbits as

$$\Psi = \begin{vmatrix} \varphi_1(x_1) & \varphi_1(x_2) & \cdots & \varphi_1(x_N) \\ \varphi_2(x_1) & \varphi_2(x_2) & \cdots & \varphi_2(x_N) \\ \vdots & \vdots & \ddots & \vdots \\ \varphi_N(x_1) & \varphi_N(x_2) & \cdots & \varphi_N(x_N) \end{vmatrix} \quad (8.3)$$

where the variable x is the coordinates of the electrons. φ is the normalized electron wavefunction. Note that Ψ is antisymmetric with respect to an interchange of locations of any two electrons. Inserting this wavelength into Eq. (8.2) and minimizing the Hamiltonian using Lagrange multiplier ε with respect to φ will yield

$$\frac{\delta}{\delta\varphi}\left[<\hat{H}> - \sum_j \varepsilon_j \int \left|\varphi_j\right|^2 \mathrm{d}\vec{r} \right] = 0 \quad (8.4)$$

Further simplification of the equation yields a set of one-electron equations as follows

$$-\frac{1}{2}\Delta\varphi_i(\vec{r}) + V_{\text{ion}}(\vec{r})\varphi_i(\vec{r}) + U(\vec{r})\varphi_i(\vec{r}) = \varepsilon_i\varphi_i(\vec{r}) \tag{8.5}$$

where V_{ion} is the local ionic potential and $U(\vec{r})$ is a nonlocal potential. The full Hartree Fock Equations are given by

$$e^2\sum_{j\neq i}\int\frac{|\varphi_j(\vec{r}')|}{|\vec{r}-\vec{r}'|}d^3\vec{r}'\varphi_i(\vec{r}) - e^2\sum_{j,j\neq i}\Delta_{S_{zi}S_{zj}}\int\frac{\varphi_j^*(\vec{r}')\varphi_i(\vec{r}')}{|\vec{r}-\vec{r}'|}d^3\vec{r}'\varphi_j(\vec{r})$$
$$+\left(-\frac{\hbar^2}{2m}\Delta - \frac{Ze^2}{\vec{r}}\right)\varphi_i(\vec{r}) = \varepsilon_i\varphi_i(\vec{r}) \tag{8.6}$$

The left-hand side of the equations consists of four terms. The third and fourth terms denote the kinetic energy and electron–ion potential. The first term, or Hartree term, is simply the electrostatic potential originating from the charge distribution of N electrons. The second term is the exchange term. The equation set (8.6) converts the many-body Hamiltonian into a number of one-electron Hamiltonians and neglects the correlation between electrons. The Hartree Fock Equations, in general, yield too wide band gaps and too small band widths in semiconductor band structure calculations.

8.2.2 *Density Functional Theory (DFT)*

The fundamental principle underlying the Density Functional Theory (DFT) is that the ground state energy of a many-electron system can be represented by a functional of its electron density and is obtained by minimizing the energy with respect to the density.

8.2.2.1 Hohenberg–Kohn Theorem

The core of Hohenberg–Kohn theorem [28] is that the many-electron system can be determined by the particle density $n(\vec{r})$. The original framework of Density Functional Theory was proposed by Thomas and Fermi in 1927, known as the Thomas–Fermi model [28]. Hohenberg and Kohn greatly improved the DFT theory by the following two theorems: (I) The ground state energy of a many-body system is a unique functional of the particle density $E_0 = E\left[n(\vec{r})\right]$; (II) The functional $E\left[n(\vec{r})\right]$ has its minimum relative to variations of the particle density $\delta n(\vec{r})$ at the equilibrium condition $n_0(\vec{r})$.

Compared to Hartree Fock Equations, the Hohenberg–Kohn theorem considers the correlation between electrons. Its mathematical form can be written as follows

$$E\left[n(\vec{r})\right] = F\left[n(\vec{r})\right] + \int n(\vec{r})V_{\text{ext}}(\vec{r})\text{d}\vec{r} \tag{8.7}$$

8.2.2.2 Kohn–Sham Equation

In order to find the solution of Eq. (8.7), Kohn and Sham [28] separated $F\left[n(\vec{r})\right]$ into three distinct parts as follows

$$E\left[n(\vec{r})\right] = T_s\left[n(\vec{r})\right] + \frac{1}{2}\iint \frac{n(\vec{r})n(\vec{r}')}{|\vec{r}-\vec{r}'|}\mathrm{d}\vec{r}\mathrm{d}\vec{r}' + E_{\mathrm{XC}}\left[n(\vec{r})\right] + \int n(\vec{r})V_{\mathrm{ext}}(\vec{r})\mathrm{d}\vec{r} \quad (8.8)$$

where $T_s\left[n(\vec{r})\right]$ is the kinetic energy of a noninteracting electron gas with density $n(\vec{r})$. $E_{\mathrm{XC}}\left[n(\vec{r})\right]$ is the exchange-correlation energy function. Using Lagrange multiplier and the normalization constraint on the electron density, $\int n(\vec{r})\mathrm{d}\vec{r} = N$, the energy function $E\left[n(\vec{r})\right]$ can be rewritten as

$$\frac{\delta}{\delta n(\vec{r})}\left[E\left[n(\vec{r})\right] - \mu\int n(\vec{r})\mathrm{d}\vec{r}\right] = 0 \quad (8.9)$$

Combining Eq. (8.8) with Eq. (8.9), one can obtain

$$\frac{\delta T_s\left[n(\vec{r})\right]}{\delta n(\vec{r})} + V_{\mathrm{eff}}(\vec{r}) = \mu \quad (8.10)$$

$$V_{\mathrm{eff}}(\vec{r}) = V_{\mathrm{ext}}(\vec{r}) + \int \frac{n(\vec{r}')}{|\vec{r}-\vec{r}'|}\mathrm{d}\vec{r}' + \frac{\delta E_{\mathrm{XC}}\left[n(\vec{r})\right]}{\delta n(\vec{r})} \quad (8.11)$$

With the above Eqs. (8.8)–(8.11), in order to find the ground state energy, E_0, and the ground state particle density, n_0, of a many-body system, it is only needed to solve the one-electron Schrodinger Equation

$$\left(-\frac{1}{2}\Delta + V_{\mathrm{eff}}(\vec{r})\right)\varphi_i(\vec{r}) = \varepsilon_i\varphi_i(\vec{r}) \quad (8.12)$$

Kohn–Sham Equations provide a theoretically exact solution for finding the ground state energy of an interacting many-body system. The only remaining question here is that the form of the exchange-correlation functional $E_{\mathrm{XC}}\left[n(\vec{r})\right]$ is unknown.

8.2.3 *Exchange-Correlation Functional*

Currently, the exchange-correlation functional $E_{\mathrm{XC}}\left[n(\vec{r})\right]$ is most commonly approximated with the Local Density Approximation (LDA) [29] and Generalized-Gradient Approximation (GGA) [30].

In LDA, the exchange-correlation functional is given by

$$E_{XC}^{LDA}\left[n(\vec{r})\right] = \int n(\vec{r})\varepsilon_{XC}^{homo}(n(\vec{r}))d\vec{r} \tag{8.13}$$

where $\varepsilon_{XC}^{homo}(n(\vec{r}))$ is the exchange-correlation energy of a homogeneous electron gas with density $n(\vec{r})$ and can be calculated from Quantum Monte Carlo methods. The LDA often works surprisingly well in more homogeneous systems and over-binds molecules and solids. However, it is very inaccurate in strongly correlated systems where an independent particle picture breaks down.

In order to better simulate the charge density distribution, GGA improves the exchange-correlation functional of LDA by including the charge density gradient,

$$E_{XC}^{GGA}\left[n(\vec{r})\right] = \int n(\vec{r})\varepsilon_{XC}^{homo}(n(\vec{r}))d\vec{r} + \int f\left[n(\vec{r})\right], \nabla n(\vec{r})d\vec{r} \tag{8.14}$$

where f is the correction chosen to satisfy one or several known limits for E_{XC}. There are two strategies for determining f: (I) Adjust f such that it satisfies most known properties of the exchange-correlation hole and energy; (II) Fit f to a large data-set with exactly known binding energies of atoms and molecules. Strategy (I) is to be preferred. Since GGA involves the charge density gradient, it predicts better results than the LDA for strongly correlated systems.

8.2.4 Pseudopotentials

The electron wavefunctions have rapid oscillations at the region close to the nuclei and consequently, very large cutoff energy and basis set which will make the computation prohibitive. However, studies show that only the valence electrons participate in the correlations between atoms and core electrons can be assumed to be fixed. Based on this assumption, a pseudopotential can be constructed with two parts as shown in Fig. 8.5. In the region near the nuclei, a pseudowavefunction, with no rapid oscillation, is used to replace the real electron wavefunction and in the valence electron region, it maintains the same form as the real electron wavefunction.

There are three commonly used pseudopotentials: norm-conserving pseudopotentials [32], ultrasoft pseudopotentials [33], and Projected Augmented Wave (PAW) method [34]. The norm-conversing pseudopotentials are constructed such that they match the true potentials in the valence electron region and yield the same charge densities. In addition, the integral of the squared amplitudes of the real and pseudopotential in the core region must be identical. This type of pseudopotentials usually requires high cutoff energies. The ultrasoft pseudopotentials are constructed such that they match the true pseudopotentials at the valence electron region. However, in the core region, instead of norm conservation, a generalized orthonormality condition is introduced to make the pseudowavefunctions as soft as possible. Therefore, the cutoff energies using ultrasoft pseudopotentials are dramatically reduced. The PAW method is a combination of pseudopotential and linear augmented plane wave

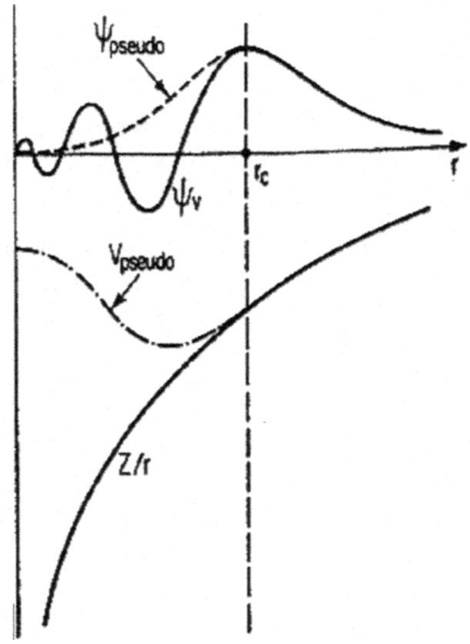

Fig. 8.5 A schematic illustration of all-electron potentials (solid lines) and pseudopotentials (dashed lines) and their corresponding wavefunctions. The all-electron potentials and pseudopotentials match at the region $r > r_c$ [31]

method. It treats the shallow core electrons, such as *d* and *f* electrons, as valence electrons. In this chapter, the ultrasoft pseudopotential and PAW method are used to predict the electronic and optical properties of ternary and quaternary semiconductor alloys.

8.3 Chapter Framework

As has been outlined, semiconductor multi-ternary (binary, ternary and quaternary) alloys have been extensively used in electronic and optoelectronic devices, such as solar cells, Light Emitting Diodes (LEDs), detectors, solid-state lasers, and Optoelectronic Integrated Circuits (OEICs). The quality and efficiency of devices are determined by the properties of these alloys. This chapter presents a study of the crystal structures, elastic/mechanical, electronic and optical properties of some multi-ternary alloys. This chapter is organized as follows:

In Sect. 8.4, the existing model, in the literature, that relates the average shear modulus with the bond length and Phillips' ionicity has been extended to derive new models for the average Young's modulus, shear modulus, and Young's modulus on (111) plane for diamond and zincblende crystals. The models are also used to predict the elastic constants of ternary semiconductors.

Section 8.5 investigates the properties of GaP_xSb_{1-x} and InP_xSb_{1-x} for various structures and compositions using first-principles method. The structure relaxation and band structure parameters, such as crystal field splitting and band gap are calculated and compared with the experimental values.

In Sect. 8.6, studies of the properties of the CdS_xTe_{1-x} interlayer in CdS/CdTe solar cells, using first-principles calculations, are presented. The properties, including crystal field splitting, spin-orbit splitting, density of states and bowing parameters are calculated and compared between different ordered and disordered structures.

Section 8.7 presents the effects of spontaneous Y2 ordering on the optical fingerprints of five III–V ternary semiconductor alloys. The trends of these optical fingerprints are qualitatively explored. The results presented in this section can be used to predict the degree of ordering for ternary semiconductor alloys.

In Sect. 8.8, a general expression for the pressure-dependent energy gap of a number of III–V and II–VI ternary semiconductor alloys are derived based on the previous work of Phillips and Van Vectchen. The trends of the pressure coefficient are analyzed with respect to the bond length and ionicity.

In Sect. 8.9, the electronic and optical properties of quaternary semiconductor alloys Cu_2ZnGeS_4, $Cu_2ZnGeSe_4$ and $Cu_2ZnGeTe_4$ are calculated. These materials are believed to be new substitutes for CIGS solar cells. This is the first work to systematically explore the basic properties, such as crystal structures, electronic band structures, and optical spectra of these compounds.

In Sect. 8.10, the electronic and optical properties of Cu_2ZnSiS_4 and $Cu_2ZnSiSe_4$ in wurtzite-kesterite and wurtzite-stannite structures are studied using first-principles calculations. Optical transitions at high symmetry points between valence bands and conduction bands are discussed. Band structures are presented to analyze other properties. Optical constants and dielectric function spectra are investigated.

8.4 Elastic Properties of Binary and Ternary Semiconductors

8.4.1 Overview

As one of the most fundamental properties of materials, the research on elastic modulus, i.e., bulk modulus, shear modulus, and Young's modulus is critical in order to understand the physical structures and mechanical behavior of materials. In fact, the ongoing research on superhard materials makes the study of elastic modulus more attractive because it is generally believed that harder materials should also have larger elastic modulus as in the case of diamond. It is also suggested that the ratio of bulk modulus to shear modulus is a critical parameter to address the fragility and brittle–ductile transition properties of materials [35]. Similarly, Young's modulus plays a significant role in the assessment of the fracture mechanics of materials. As semiconductor research advances from three dimensions (bulk) to two dimensions (thin films) and to one dimension (nanowire), shear modulus and Young's modulus become more important.

Interdependencies of these elastic moduli are through elastic constants which are the fundamentals of almost all properties of materials. In experiments, these elastic constants are usually extrapolated from the experimental results of elastic modulus. In theoretical calculations, only first-principles calculations are generally used to evaluate these elastic constants. Due to the computational complexities and the associated cost of first principle calculations, researchers are always interested in developing simple analytical models for the elastic modulus.

In this section, the existing models on the bulk and average shear modulus are introduced. New expressions are proposed for shear modulus and Young's modulus on the crystal (111) plane as well as the average Young's modulus for diamond and zincblende crystals. The proposed models are used to predict the elastic constants of III–V and II–VI ternary semiconductor alloys. This study on the elastic properties is one of the important prerequisites for the discussion of electronic properties and the derivation of the pressure coefficient of energy gaps in later sections.

8.4.2 *Existing Models on Bulk and Shear Modulus*

One semiempirical approach to determine the bulk modulus, B, of diamond and zincblende semiconductors was proposed by Cohen [36, 37] as the following:

$$B = (1972 - 220I)d^{-3.5} \tag{8.15}$$

where d (in Å) is the bond length between two nearest neighbor atoms. I is an empirical ionicity factor defined by Cohen to account for the reduction in bulk modulus arising from increased charge transfer. The values of I are 0, 1, and 2, respectively, for group IV, III–V, and II–VI semiconductors. Results of the calculation, based on Eq. (8.15), yield a surprisingly good agreement with the experimental data [36, 37].

Recently, Kamran et al. [38] replaced the empirical ionicity factor I with a more physics-based Phillips' ionicity [39] and obtained a similar expression as in Eq. (8.15). They also extended the ideas and developed similar expressions for the average shear modulus:

$$B = \frac{\kappa_2 - \lambda_2 f_i}{d^{3.5}} \tag{8.16}$$

$$G = \frac{\kappa_1 - \lambda_1 f_i}{d^{5.5}} \tag{8.17}$$

where κ_1, κ_2 and λ_1, λ_2 are constants. In the calculations of average shear modulus, Kamran et al. [38] proposed that diamond and zincblende group IV, III–V, and II–VI covalent crystals can be split into two groups (group a and b) with different fitting coefficients. In their work, diamond and grey tin are placed in group a, while Si and Ge belong to group b. AlN and ZnS are put into group a, while GaN, InN, ZnSe, and ZnTe are put into group b.

8.4.3 Development of New Expressions

Following the ideas of the existing models on bulk and shear modulus, new expressions can be derived for shear modulus and Young's modulus on (111) plane as well as the average Young's modulus for diamond and zincblende crystals.

Young's modulus (E) can be experimentally determined from the slope of a stress–strain curve obtained from tensile tests performed on a material. The stress is the derivative of the elastic potential energy of the internal forces with respect to the strain. In elastic deformation, the potential energy is the same as the internal energy. Therefore, Young's modulus at the equilibrium state can be expressed in terms of the second derivative of the internal energy with respect to strain as follows:

$$E = \frac{\partial^2 U}{\partial \gamma^2} \tag{8.18}$$

where U is the internal energy and γ is the strain. Similar expression for shear modulus G has also been developed by Roundy et al. [40] in their study on Al and Cu.

In general, the elastic deformation generated during the tensile tests will change the bond strengths of both core and valence electrons and, consequently, the internal energy levels. However, the bond strength in the core regime is so tightly bonded that the effect of the small elastic deformation on the inner core bond strength is negligible compared to that on the bond strength of valence electrons. Phillips [39] proposed a bonding–antibonding average energy gap E_g to describe the bond strength of valence electrons. This energy gap is separated into two parts: the homopolar energy gap E_h which characterizes the strength of pure covalent bond and the heteroploar energy gap C which characterizes the strength of pure ionic bond. Hence, the average energy gap is given by

$$E_g^2 = E_h^2 + C^2 \tag{8.19}$$

Thus, $E_g = E_h$ for group IV covalent crystals/elements such as diamond and Si. Generally speaking, the ionic bonding arises from the long-range electrostatic force which undergoes no variation under small elastic deformation [39]. Cohen [36, 37] showed the independence of ionic energy gap C from the lattice constant as well as the elastic modulus. Therefore, the internal energy in Eq. (8.18), as stated above, characterized by valence bond strengths, can be simply replaced by E_h which, in Phillips' argument [39], is given as follows:

$$E_h = \frac{39.74}{d^{2.5}} \tag{8.20}$$

where d is the bond length in Å. E_h is in units of eV. The strain, γ, in Eq. (8.18), is usually expressed in a Taylor expansion formalism which, under small elastic deformation, depends linearly on lattice constant and bond length. Based on the above argument, the following relation can be obtained for the Young's modulus:

$$E \propto d^{-4.5} \tag{8.21}$$

Noting that the Cohen's ionicity factor is empirical, it is convenient to adopt Phillips' ionicity [39] which is shown to have a decreasing linear trend with elastic modulus [41]. Based on this argument, the expression for Young's modulus is developed as follows:

$$E = \frac{\alpha - \beta f_{\mathrm{i}}}{d^{4.5}} \tag{8.22}$$

where α and β are constants, and f_{i} is Phillips' ionicity, defined by the terms in Eq. (8.19) as $f_{\mathrm{i}} = C^2/E_{\mathrm{g}}^2$. Similar derivation was applied to bulk modulus and average shear modulus by Cohen [36] and Kamran et al. [38], respectively. This theory on homopolar energy gap and ionicity for covalent crystals has also been adopted to develop relations for other mechanical properties such as hardness [42] and significant success has been demonstrated. However, it is found that Kamran et al. [38] used the experimental data of many crystals with hexagonal structure or the data along typical crystal directions to derive the coefficients in Eq. (8.17). Therefore, the work in this chapter revised their results and extracted coefficients that are more reliable for materials.

The procedure to obtain the general expressions for elastic modulus is the following: first, the constants in Eqs. (8.17) and (8.22) are interpolated using some experimental data of covalent crystals with diamond and zincblende structures. The obtained expressions are then tested by comparing their predicted results with other available experimental or theoretical data. Good accord with experiment indicates that the interpolated equations do reflect the correlations between bond length, ionicity, and elastic modulus. The resulting expression for average Young's modulus and shear modulus are given as follows:

$$E_V = \frac{\alpha_V - \beta_V f_{\mathrm{i}}}{d^{4.5}} = \frac{8539 - 6105.56 f_{\mathrm{i}}}{d^{4.5}} \tag{8.23a}$$

$$G_V = \frac{\kappa_V - \lambda_V f_{\mathrm{i}}}{d^{5.5}} = \frac{8591.48 - 6035.47 f_{\mathrm{i}}}{d^{5.5}} \tag{8.23b}$$

It is worthwhile to mention that these two average elastic moduli in Eqs. (8.23a) and (8.23b) are typically for polycrystals, since, in a polycrystal material, the anisotropy of single crystals is averaged out to yield an isotropic material. Significant attention has been given to the isotropy of materials because it is believed that superhard materials should be homogenous and have an isotropic lattice with small bond length. In reality, all diamond and zincblende covalent crystals are anisotropic. However, the three-fold symmetry on (111) plane in these crystals leads to the assumption of fairly isotropic properties on this plane and, therefore, provides a path to study the

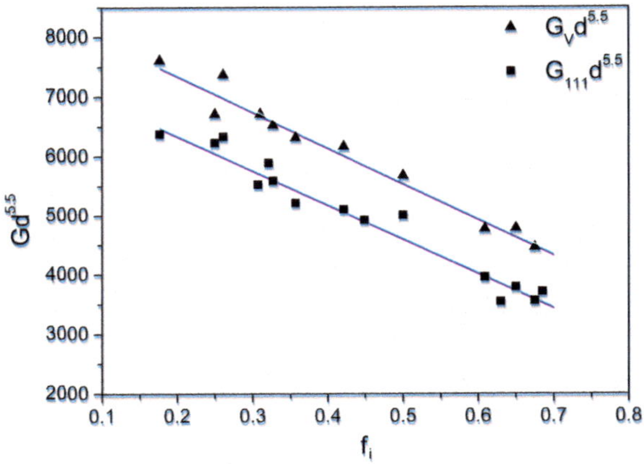

Fig. 8.6 Correlations between $Gd^{5.5}$ and ionicity $f_{\rm i}$. Triangle and square points are experimental data. Solid lines are plotted using Eqs. (8.23b) and (8.24b)

isotropic properties of these crystals. It has been verified that the Young's modulus and shear modulus are isotropic on silicon (111) plane [43]. Recent work also shows that the elastic properties on the (111) plane are planar isotropic [44]. These factors enable the application of Eqs. (8.17) and (8.22) to determine the Young's modulus and shear modulus of diamond and zincblende covalent crystals on the (111) plane. The resulting expressions are as follows:

$$E_{111} = \frac{\alpha_{111} - \beta_{111} f_{\rm i}}{d^{4.5}} = \frac{9166.18 - 7845.87 f_{\rm i}}{d^{4.5}} \tag{8.24a}$$

$$G_{111} = \frac{\kappa_{111} - \lambda_{111} f_{\rm i}}{d^{5.5}} = \frac{7539.44 - 5807.16 f_{\rm i}}{d^{5.5}} \tag{8.24b}$$

The units for elastic modulus are in GPa and bond lengths are in Å. The correlations between elastic modulus, bond length, and ionicity are plotted in Figs. 8.6 and 8.7. The comparisons between the values from the modeled equations and other experimental and calculated data, based on first-principles, are listed in Table 8.1.

In this table, G_{111} values are calculated from elastic constants. Also listed is the heteropolar energy gap C calculated from Eq. (8.19). Elastic modulus is in units of GPa, bond length is expressed in Å.

8.4.4 Comparison with Experimental Data

In general, the calculated values of the elastic properties from Eqs. (8.23a), (8.23b), (8.24a), and (8.24b) show good accord with the experimental data and other calculated results for zincblende group III–V and II–VI crystals. However, for group IV materials, a large discrepancy is observed. This is because of the fact that the Phillips' expression for $E_{\rm h}$ in Eqs. (8.19) and (8.20) is the optimized result for all the available cubic, hexagonal covalent materials and it is generalized to characterize the

Table 8.1 Calculated elastic moduli and comparison with experimental results

Crystal	d [1]	f_i [39]	C [39]	G_V (exp) [45]	G_V (cal)	G_{111} (exp) [45]	G_{111} (cal)	E_V (exp) [45]	E_V (cal)	E_{111} (exp) [1]	E_{111} (cal)
C	1.55	0	0	535.7	771.34 (725.61)	508.93	676.88 (634.79)	1144.6	1188.27 (1131.41)	1165	1275.55 (1134.92)
Si	2.35	0	0	70.8	78.20 (70.99)	62	68.62 (62.10)	171.8	182.64 (168.91)	169	196.06 (169.43)
Ge	2.45	0	0	56.6	62.18 (56.45)	49.47	54.57 (49.38)	136	151.41 (140.03)	137.1	162.53 (140.46)
SiC	1.89	0.177	3.85	146 [46] 231 [47]	228.37	193.67 [47]	197.66	363 [46] 424 516 [47]	427.40	423 [46] 581 [24] 603 [47]	445.69
AlN	1.90	0.449	7.30		172.32	144.33 [48]	144.50		322.74	302 [49] 345	314.16
AlP	2.37	0.307	3.14	56.18 [48]	59.11	48.5 [48]	50.50	138.7 [48]	138.32	138	140.24
AlAs	2.45	0.274	2.67		49.94		42.81		121.20	117.9	123.86
AlSb	2.66	0.25	2.07	31.1	32.83	28.89	28.22	79.4	86.37	84.7	88.73
GaN	1.96	0.5	7.64	140.2 [48]	137.64	123.67 [48]	121.12	267 [50]	265.53	267	265.76
GaP	2.36	0.327	3.30	58	58.80	49.71	50.12	142.9	137.21	144	138.43
GaAs	2.45	0.31	2.90	48.8	48.86	41.57	41.73	120.4	118.30	121.3	119.86

(continued)

Table 8.1 (continued)

Crystal	d [1]	f_i [39]	C [39]	G_V (exp) [45]	G_V (cal)	G_{111} (exp) [45]	G_{111} (cal)	E_V (exp) [45]	E_V (cal)	E_{111} (exp) [1]	E_{111} (cal)
GaSb	2.64	0.261	2.10	35.4	33.70	30.43	28.93	88	88.06	89.1	90.25
InN	2.16	0.578	6.78		73.85		60.53		156.60		144.76
InP	2.54	0.421	3.34	36.5	35.80	30.2	30.14	93.8	89.75	91.7	88.17
InAs	2.62	0.357	2.74	31.4	31.99	25.92	27.17	80.4	82.91	79.3	82.98
InSb	2.81	0.321	2.10	24.2	22.85	20.23	19.49	61.9	63.39	62.1	64.05
MgS	2.43	0.639	7.10		35.85		28.99		85.32	76.3	76.40
ZnS	2.34	0.623	6.20	35.5	44.73	28.47	36.31	92.6	102.71	86.4	92.80
ZnSe	2.45	0.63	5.60	32.9	34.29	25.4	27.79	83.3	82.48	78.6	74.23
ZnTe	2.60	0.609	4.48	24.8	25.55	20.57	20.81	64	65.20	63.2	59.35
CdS	2.52	0.685	5.90	25.98 [51]	27.51	22.9 [51]	21.98	67.99 [51]	67.81	50.9	59.02
CdSe	2.63	0.699	5.50	23.34 [51]	21.38	20.63 [51]	17.02	61.2 [51]	54.96	46.5	47.37
CdTe	2.81	0.675	4.90	15.3	15.50	12.21	12.42	41	42.54	40	37.27
HgS	2.53	0.79	7.30		22.99		17.75		56.61	50.4	45.22
HgSe	2.63	0.68	5.00		21.81		17.45		56.16	42.2	49.04
HgTe	2.78	0.65	4.00	16.7 [52]	16.29	13.23 [52]	13.14	44.82 [52]	44.62	40.3	39.70

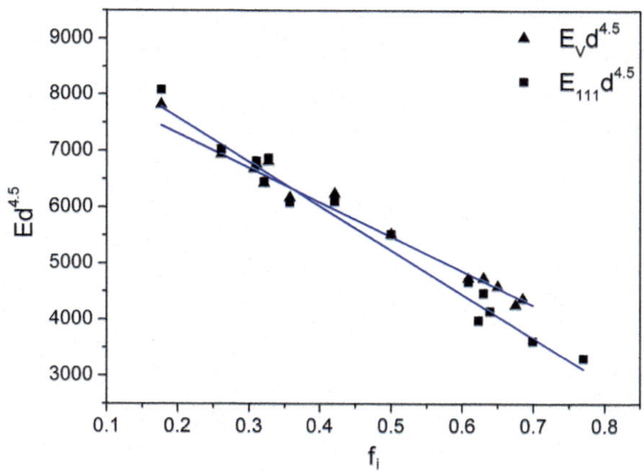

Fig. 8.7 Correlations between $Ed^{4.5}$ and ionicity f_i. Triangle and square points are experimental data. Solid lines are plotted using Eqs. (8.23a) and (8.24a)

average optical gap of the material [39]. This optimization is not required for originally pure covalent group IV materials, such as diamond, Si, and Ge, since $C = 0$ in Eq. (8.19). Here, based on the power relations between elastic moduli and bond length as derived in Eqs. (8.23a), (8.23b), (8.24a), and (8.24b), new expressions can be proposed for all the four elastic moduli for group IV crystals. For average shear modulus: $G_V = 7799.55d^{-5.5}$, for shear modulus on (111) plane: $G_{111} = 6823.33d^{-5.5}$, while for average Young's modulus: $E_V = 7897d^{-4.5}$, and for Young's modulus on (111) plane: $E_{111} = 7921.51d^{-4.5}$. The coefficients of these four expressions for elastic modulus can be retrieved from the homopolar energy gap values determined from dielectric functions and lattice constants [36]. The calculated results are listed in parentheses in Table 8.1. For Si and Ge, the differences between experimental and calculated values are all within 3%. For diamond, Young's modulus differs from an experiment by 2.6% while the shear modulus differs by up to 35%. One possible reason is that the small bond length of diamond makes it incomparable to other heavier crystals in the periodic table [38, 41] or the s-p^3 hybridization and the high isotropy of diamond introduces certain different angular properties in diamond than in other materials. The anomaly in diamond suggests that the ionicity may not be the best parameter to characterize the elastic modulus of various covalent crystals. One solution [58, 59] is to correlate the elastic modulus with the internal ionization energy because, in general, elastic deformation requires the bound electrons to cross the energy gap to contribute to excited states resulting in a reduction in the local bonding strength of valence electrons and, consequently, increase the polarizability and decrease the elastic modulus.

One interesting material is 3C–SiC which, in accordance with the composition of elements, belongs to group IV in the periodic table. According to the theory of ionicity and covalency, 3C–SiC is closer to group III–V partial covalent crystals. Experimental determination of the elastic constants of 3C–SiC is not even complete in the literature. This is due to the unavailability of single crystals of 3C–SiC material of the required size [47]. First-principles calculations of the elastic modulus of 3C–SiC are extremely scattered as listed in Table 8.1. In order to verify the properties of 3C–SiC, the elastic moduli of 3C–SiC have been calculated using Eqs. (8.23a), (8.23b), (8.24a), and (8.24b). The results are listed in Table 8.1. The equations for group IV crystals as proposed above are also used to calculate the elastic moduli of 3C–SiC. The results are as follows: $G_V = 236.76$ GPa, $G_{111} = 207.13$ GPa, $E_V = 452.54$ GPa, and $E_{111} = 453.95$ GPa. Comparison of these two sets of calculated results with other available data is prone to show that 3C–SiC is much more similar to group III–V and II–VI partial covalent crystals than pure covalent group IV materials.

Since Phillips' homopolar energy gap E_{h} in Eq. (8.19) is generalized to include broader range of crystals, the results of the calculations of group III–V and II–VI crystals, as shown in Table 8.1, are in excellent agreement with the available experimental and theoretical data based on first-principles methods in the literature. Some trends can be observed from the calculated results. For common-cation system, ionicity decreases with increasing bond length. On the contrary, for common-anion system, ionicity increases with increasing bond length. These have also been investigated in the earlier work on group III–V and II–VI ternary semiconductors [53]. Despite different trends of the ionicity with bond length in various systems, elastic modulus always decreases with increasing bond length due to the large exponent in Eqs. (8.23a), (8.23b), (8.24a), and (8.24b). Therefore, elastic modulus is predominated by bond length and ionicity only plays an auxiliary influence. It is noteworthy that ZnS is an exception to the trend of the variation of ionicity with bond length in common-cation system. According to the trend, ZnS should have larger ionicity than ZnSe and ZnTe while its ionicity is much smaller. This is the reason for the calculated results of the elastic modulus for ZnS to be always larger than the experimental values even within a tolerable range.

8.4.5 *Elastic Constants of Ternary Semiconductors*

The proposed models for elastic modulus can be utilized to study the elastic properties of more complex materials with tetrahedral bonding. One good example is the ternary semiconductor alloys. They have attracted significant interest in recent years because of their applications in optoelectronics, including, photovoltaics. For example, CdZnTe has been studied extensively for use as an X-ray and Gamma-ray detector due to its high photon attenuation coefficient and good charge transport properties. It is generally accepted that the isotropic elastic modulus are usually expressed by elastic constants through the following expressions:

$$B = 1/3(C_{11} + 2C_{12}) \tag{8.25a}$$

$$G_V = 1/5(C_{11} - C_{12} + 3C_{44}) \tag{8.25b}$$

$$G_{111} = 1/3(C_{11} - C_{12} + C_{44}) \tag{8.25c}$$

where C_{11}, C_{12} and C_{44} are the independent second-order elastic constants of cubic crystals. Based on Eqs. (8.25a), (8.25b), and (8.25c), these elastic constants can be resolved in terms of bulk modulus B, average shear modulus G_V, and shear modulus on (111) plane G_{111} as follows:

$$C_{11} = 1/3(3B + 9G_{111} - 5G_V) \tag{8.26a}$$

$$C_{12} = 1/6(6B - 9G_{111} + 5G_V) \tag{8.26b}$$

$$C_{44} = 1/2(5G_V - 3G_{111}) \tag{8.26c}$$

Equations (8.23b), (8.24b), (8.26a), (8.26b), and (8.26c) are used to extrapolate the elastic constants for group III–V and II–VI ternary semiconductor alloys. Bulk modulus can be calculated either from Eq. (8.16) or taken from the available experimental data. Results of the calculations of elastic properties and comparison with the experimental data are listed in Table 8.2. Good accord with the experimental data indicates that this approach can be applied to provide a theoretical prediction of the elastic properties of other semiconductor alloys or more complex diamond or zincblende structure materials whose experimental data are unknown.

8.4.6 Summary

In this section, the existing work on average shear modulus has been corrected. New expressions are developed for average Young's modulus as well as shear modulus and Young's modulus on (111) plane for diamond and zincblende structure group IV, III–V and II–VI semiconductors. Analyses of the results of the new expressions show that the bond length dominates the elastic modulus while ionicity only plays an auxiliary role. The material 3C–SiC is shown to have elastic properties similar to that of partial covalent crystals other than pure covalent crystals. The corrected and newly proposed models on elastic modulus can be applied to derive elastic constants of some ternary alloys.

8.5 Properties of III–V Ternary Alloys

8.5.1 Overview

When two zincblende binary compounds AB and AC are mixed to form an alloy AB_xC_{1-x}, the parent zincblende structure is generally inherited. However, the arrangement of atoms B and C can form periodic orderings. The types of orderings depend on the variables in the experiments such as growth temperature, growth

Table 8.2 Calculations of elastic constants of some ternary semiconductors for composition 0.5 and comparison with available experimental data

	d [53]	f_i [53]	C [53]	B	C_{11} (exp)	C_{11} (cal)	C_{12} (exp)	C_{12} (cal)	C_{44} (exp)	C_{44} (cal)
GaInP [54]	2.456	0.378	3.333	81.95 [55]	124	121.34	62	62.26	59	55.40
GaInAs [54]	2.534	0.395	3.198	67.42 [55]	95	99.83	48	51.21	45	45.99
GaInSb [56]	2.725	0.235	1.827	46.4	74.6	72.88	32.2	33.16	34.5	34.96
CdZnTe [57]	2.725	0.604	4.086	45	60	60.53	38	37.23	23.8	25.47

rates, substrate orientation, and so forth. One might expect important variations in the properties due to the different orderings.

In the previous section, models of the mechanical properties of ternary alloys, irrespective of the crystal structures, were presented. In this section, the crystal ordered III–V ternary alloys, GaP_xSb_{1-x} and InP_xSb_{1-x}, are studied at compositions of 0.25, 0.5, and 0.75. Numerous properties are addressed, including the following: (I) Qualitative relationship between structural relaxation and lattice mismatch; (II) Formation enthalpies; (III) Ordering-induced crystal field splitting; (IV) Alloy band gap and bowing coefficient.

8.5.2 Theoretical Background

Group III–V ternary semiconductor alloys have been studied for decades due to their technological applications such as in optoelectronic devices and high-speed, low-power logic circuits. For example, the adjustable band gap of GaP_xSb_{1-x} with composition makes it a potentially useful material in fiber optic communication systems [53]. The expected resonance enhancement of the hole impact ionization rate makes GaP_xSb_{1-x} a useful material for low-noise avalanche photodiodes utilizing hole injection [60]. InP_xSb_{1-x} is an interesting material for optical devices in the mid-infrared. The first mid-infrared lasers, using InPSb layers, have been reported by some groups [61, 62].

The III–V and II–VI ternary alloys can be divided into two categories: conventional alloys and unconventional alloys. The conventional alloys have small lattice mismatch between the binary constituents. For example, the lattice mismatch between AlAs and GaAs in compound $Al_xGa_{1-x}As$ is 0.14%. On the contrary, the unconventional alloys have large lattice mismatch between the binary constituents. For example, the lattice mismatch for GaP and GaSb in compound GaP_xSb_{1-x} is 11.2%. The unconventional alloys are expected to have some anomalous properties [63–65], such as follows: (I) Structural anomaly at the percolation composition threshold; (II) Large and composition-dependent bowing parameter; and (III) Composition-dependent interband transition intensities.

The research in the literature has focused on the properties of conventional alloys for various structures and properties of unconventional alloys for random structures. As unconventional ternary alloys, the crystal ordering structures have been observed in both GaP_xSb_{1-x} and InP_xSb_{1-x}. Therefore, this section investigates the properties of these two alloys for all possible ordered structures.

8.5.3 Ordered Structures of Ternary Alloys

The Landau–Lifshitz theory [66] on phase transitions provides rules to select a number of ordered structures which can interconvert, under well-defined constraints, into

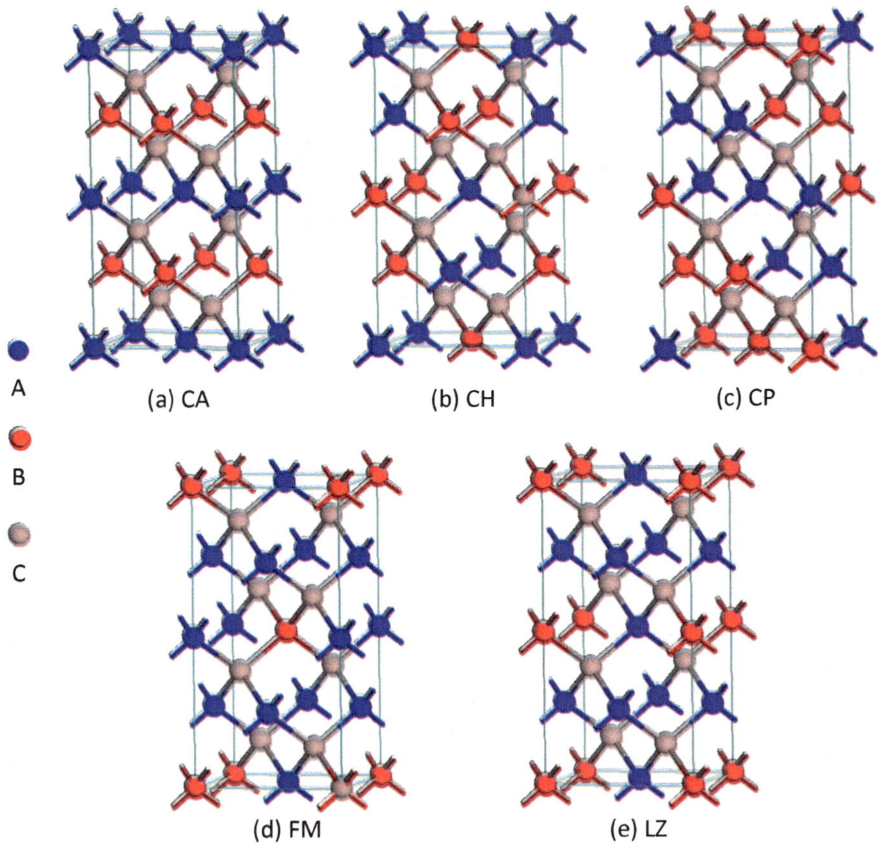

Fig. 8.8 Crystal structures of ordered ternary semiconductor alloys $A_xB_{1-x}C$ at compositions 0.25, 0.5 and 0.75

disordered phases of the same composition. The selection rules are as follows: (I) The space group of the ordered structure must be a subgroup of that of the disordered alloy; (II) The ordered structure must be associated with an ordering vector located at a special k point of the parent space group. For a ternary semiconductor alloy, AB_xC_{1-x}, there are five generally observed ordered structures derived from zincblende structure of the binary constituents (AB and AC): For composition 0.5–0.5, layered tetragonal CuAu-I like (CA, $P\bar{4}m2$, No. 115) structure, layered trigonal CuPt (CP, $R3m$, No. 160) structure and chalcopyrite (CH, $I\bar{4}2d$, No. 122) structure exist while for composition 0.25–0.75 and 0.75–0.25, Famatinite (FM, $I\bar{4}2m$, No. 121) structure and Luzonite like (LZ, $P\bar{4}3m$, No. 215) structure are formed. Details of these structures can be found in Refs. [67, 68]. Figure 8.8 shows the crystal structures of these five orderings.

The sets of structures in Fig. 8.8 can be organized into three groups according to the ordering vectors: (I) For the (0, 0, 1) ordering vector, there is CA structure at composition 0.5 and LZ structure at composition 0.25 and 0.75; (II) For the (2, 0, 1) ordering vector, there is CH structure at composition 0.5 and FM structure at composition 0.25 and 0.75; (III) For the (1, 1, 1) ordering vector, there is CP structure at composition 0.5.

8.5.4 Structural Properties

The first-principles total energy minimization approach is applied to obtain the structural parameters of each ordered structure. The calculated lattice constants of the alloys, in accord with experimental data [69, 70], follow a linear function with the alloy compositions. Results are given in Tables 8.3 and 8.4. Some interesting features can be found in these following calculations:

8.5.4.1 Lattice Relaxation

The standard zincblende structure has ideal tetragonal distortion parameter $\eta = 1$ and cell internal relaxation parameter $\mu = 0.25$, corresponding to fully relaxed structure. The deviations of these two parameters from their ideal values reflect how well the alloy is structurally relaxed. The calculated results, in Tables 8.3 and 8.4, show that the deviations of these two parameters from their ideal values in GaPSb compounds are larger than those in InPSb compounds. This is due to the fact that the lattice mismatch between the binary constituents in GaPSb (11.2%) is larger than that in InPSb (9.8%). This indicates that alloys with larger lattice mismatch between their constituents will be less relaxed.

8.5.4.2 Bimodal Behavior

It is observed that the bond length in ordered conventional alloys such as GaAsSb and AlGaAs is not uniformly distributed. But, instead, it exhibits a bimodal behavior [71]. In this work, the bond lengths in GaPSb and InPSb compounds have been calculated. The results of bond lengths in In alloys are listed here (units: Å):

Instead of average, the bond lengths exhibit a bimodal distribution. The short bonds in CA and CH structures are, in general, smaller while the long bonds are greater than those in the corresponding zincblende binary constituents. This local structure property indicates the importance of distortion and internal relaxation parameters in releasing the cell internal strain energy. Four different bonds (singlet and triplet) are observed in CP structure because there are more degrees of freedom to vary in that crystal relaxation. Similarly, singlet and doublet bonds are found in FM structure for composition 0.25 and 0.75.

Table 8.3 LDA calculated properties of GaP_xSb_{1-x}

Alloys	Structure	a (Å)	η	μ	ΔH (eV/atom)	Δ_{CF} (eV)	E_g (eV)	E_g (eV) (LDA + C)	b (eV)
$GaP_{0.75}Sb_{0.25}$	FM	5.554	0.998	0.268	0.039	0.042	0.832	1.56	3.306
$GaP_{0.75}Sb_{0.25}$	LZ	5.561	1.000	0.263	0.060	0	0.327	1.06	5.996
$GaP_{0.5}Sb_{0.5}$	CA	5.725	1.004	0.224	0.071	0.265	0.156	0.80	3.377
$GaP_{0.5}Sb_{0.5}$	CH	5.711	0.994	0.213	0.029	−0.082	0.891	1.52	0.438
$GaP_{0.5}Sb_{0.5}$	CP	5.735	1.002	0.234	0.084	0.523	−0.731	−0.10	6.923
$GaP_{0.25}Sb_{0.75}$	FM	5.882	0.998	0.231	0.027	0.012	0.365	0.90	0.98
$GaP_{0.25}Sb_{0.75}$	LZ	5.889	1.000	0.236	0.048	0	0.147	0.68	2.143

LDA + C refers to the LDA corrected band gap

Table 8.4 LDA Calculated Properties of InP_xSb_{1-x}

Alloys	Structure	a (Å)	η	μ	ΔH (eV/atom)	Δ_{CF} (eV)	E_g (eV)	E_g (eV) (LDA + C)	b (eV)
$InP_{0.75}Sb_{0.25}$	FM	5.981	0.998	0.268	0.026	0.017	0.203	0.81	0.884
$InP_{0.75}Sb_{0.25}$	LZ	5.987	1.000	0.262	0.041	0	0.048	0.65	1.711
$InP_{0.5}Sb_{0.5}$	CA	6.142	1.003	0.225	0.049	0.195	−0.290	0.25	1.738
$InP_{0.5}Sb_{0.5}$	CH	6.132	0.995	0.217	0.020	−0.082	0.067	0.60	0.313
$InP_{0.5}Sb_{0.5}$	CP	6.147	1.002	0.234	0.059	0.517	−0.841	−0.31	3.946
$InP_{0.25}Sb_{0.75}$	FM	6.292	0.998	0.233	0.019	−0.001	−0.161	0.29	0.438
$InP_{0.25}Sb_{0.75}$	LZ	6.297	1.000	0.237	0.033	0	−0.291	0.16	1.129

LDA + C refers to the LDA corrected band gap

ZB	R(In−P) = 2.525	R(In−Sb) = 2.795
CA	R(In−P) = 2.576	R(In−Sb) = 2.754
CH	R(In−P) = 2.539	R(In−Sb) = 2.773
CP	R(In−P) = 2.500	R(In−Sb) = 2.728 (×3)
	R(In−P) = 2.611 (×3)	R(In−Sb) = 2.820

8.5.5 Formation Enthalpies

The formation enthalpy ΔH of alloy AB_xC_{1-x} is defined in terms of the fully relaxed total energies E of the alloy and binary components as follows:

$$\Delta H(x) = E(AB_xC_{1-x}) - xE(AB) - (1-x)E(AC) \quad (8.27)$$

The calculated results of ΔH are presented in Tables 8.3 and 8.4. The following features can be found:

8.5.5.1 Structure Dependence

The calculated formation enthalpies show the trend $\Delta H(\text{CP}) > \Delta H(\text{CA}) > \Delta H(\text{CH})$ for composition $x = 0.5$, suggesting that the alloys favor CH structure as their stable ground state structure rather than CA and CP structures. In the experiment, the Transmission Electron Diffraction (TED) pattern [72] shows a CA ordering mixed with disordered structure in GaPSb alloy. To verify this partial ordering structure, the formation enthalpy of disordered $GaP_{0.5}Sb_{0.5}$ alloy has been calculated using Special Quasirandom Structures (SQS8) model [73]. It is found that $\Delta H(\text{CA}) > \Delta H(\text{SQS8}) > \Delta H(\text{CH})$, indicating, as in experiment, a high possibility of finding a CA and random mixed structure. Similarly, $\Delta H(\text{LZ}) > \Delta H$ the (FM) suggests that FM structure is more stable than the LZ structure. For a given ordering, such as CA and LZ along (100) direction, the dependence of formation enthalpy on composition is also observed.

8.5.5.2 Trend from Ga to In Cation

For a given structure, the calculated values show that the formation enthalpy decreases from GaPSb to InPSb compounds. This is due to the smaller lattice mismatch and smaller bulk modulus [43] in InPSb compounds. This smaller lattice mismatch leads to smaller strain energy, and thus smaller formation enthalpy. This trend is consistent with the conclusion of lattice relaxation in Sect. 8.5.4.1 that alloys with larger lattice mismatch will be less relaxed.

8.5.5.3 Alloy Mixture

The calculated results in Tables 8.3 and 8.4 show that the formation enthalpy does not monotonically decrease from GaP to GaSb. Instead, ΔH is smaller in $GaP_{0.25}Sb_{0.75}$ and $GaP_{0.75}Sb_{0.25}$ and larger in $GaP_{0.5}Sb_{0.5}$. The same phenomenon is also found in In compounds. This is because the equal amount mixture of binary constituents induces more strain energy in ternary systems and therefore larger formation enthalpy. This also explains the observation of large-range miscibility gap in experiments.

8.5.6 Electronic Properties and Bowing Parameter

Tables 8.3 and 8.4 also show the LDA calculated electronic properties (i.e., crystal field splitting, band gap) and bowing parameters. Together listed are the LDA corrected [74] band gaps in order to compare with experimental values since LDA underestimates the band gap. The bowing parameters are calculated using the uncorrected band gaps.

8.5.6.1 Crystal Field Splitting

According to the perturbation theory, for a given structure, an alloy with larger valence band offset between its binary components will have larger crystal field splitting. For a given alloy, the crystal field splitting induced by structure effect can be described by $(\Delta V)^2/[\varepsilon_m(k_1) - \varepsilon_n(k_2)]$ [71]. Here, ΔV is the coupling matrix determined by the potential difference and bond length mismatch in the binary constituents. The energy denominator refers to the energy difference between the binary components' unperturbed states before folding. The details of the folding relation can be found in Ref. [71]. With these two facts, the calculated results can be explained as follows: (I) Amongst the different structures, the following relation holds: Δ_{CF}(CP) > Δ_{CF}(CA) > Δ_{CF}(CH). This is because the energy difference $\varepsilon_m(k_1) - \varepsilon_n(k_2)$ of these structures follows an opposite sequence. For example, the energy difference ($\Gamma_{15v} - L_{3v} =$ 0.925 eV) of CP structure of alloy $GaP_{0.5}Sb_{0.5}$ is much smaller than the energy difference ($\Gamma_{15v} - X_{5v} = 2.56$ eV) of its CA structure. The available experimental energy levels [1] have been listed in Table 8.5; (II) For a given structure, there always is Δ_{CF} (GaPSb) > Δ_{CF} (InPSb) due to the same trend in band offset: GaPSb (1.04 eV) > InPSb (0.92 eV); (III) The negative crystal field splitting in CH structure is because the stronger ($\Gamma_{15v} - W_{3v}$) coupling than ($\Gamma_{15v} - X_{5v}$) coupling makes the Γ_{4v} above Γ_{5v} state [75].

Table 8.5 Experimental valence band energy levels (in eV) of binary constituents, GaP, GaSb, InP, and InSb

	GaP	GaSb	InP	InSb
Γ_{15v}	−0.08	−0.75	−0.11	−0.80
L_{3v}	−1.13	−1.55	−1.00	−1.40
X_{5v}	−2.85	−3.10	−2.00	−2.40
Γ_{1c}	2.90	0.81	1.43	0.24
L_{1c}	2.64	1.10	2.04	1.00
X_{1c}	2.35	1.72	2.30	1.70

8.5.6.2 Band Gap

The calculated values of direct Γ point band gaps for all the calculated alloy compounds have the following trend: E_g(CH) > E_g(CA) > E_g(CP), and E_g(FM) > E_g(LZ). This structure-induced difference in band gap can also be explained in terms of the perturbation theory used for crystal field splitting. However, the denominator energy difference will be mainly contributed by the conduction energy levels (Table 8.5) [1]. The energy difference in CP structure ($\Gamma_{1c} - L_{1c} = 0.015$ eV) in $GaP_{0.5}Sb_{0.5}$ is smaller than the energy difference in CA structure ($\Gamma_{1c} - X_{1c} = 0.18$ eV), resulting in a greater band gap narrowing and hence a smaller band gap in CP phase. Similarly, the largest band gap in CH structure is due to the greatest ($\Gamma_{1c} - W_{1c}$) difference compared to CA and CP structure. The same mechanism can also explain the band gap comparison between FM and LZ structures. Comparison between Tables 8.3 and 8.4 also shows that the band gaps of GaPSb compound are always larger than those of InPSb compounds.

8.5.6.3 Bowing Parameter

The band gap $E_g(x)$ of a random AB_xC_{1-x} alloy is described by a bowing function as follows:

$$E_g(x) = xE_g(AB) + (1-x)E_g(AC) - bx(1-x) \quad (8.28)$$

where, b is the bowing parameter. $E_g(AB)$ and $E_g(AC)$ are the band gaps of binary constituents AB and AC, respectively. Results in Tables 8.3 and 8.4 show that the bowing parameters for these two systems depend strongly on the structures, that is, b(CP) > b(CA) > b(CH) and b(LZ) > b(FM). This is due to the different identities of the repelling states and the different symmetry properties. The calculations of GaPSb compound ordering along (001) plane shows that the bowing parameter increases by 58% from $GaP_{0.25}Sb_{0.75}$ to $GaP_{0.5}Sb_{0.5}$, and 78% from $GaP_{0.5}Sb_{0.5}$ to $GaP_{0.75}Sb_{0.25}$. For InPSb compound, the bowing parameter increases by 54% from $InP_{0.25}Sb_{0.75}$ to $InP_{0.5}Sb_{0.5}$. The reason for this large and composition-dependent bowing parameter is attributed to the large differences between the two constituents in their lattice constants and energy levels.

8.5.7 Comparison with Experiments and Other Calculations

For the formation enthalpy, Fedders et al. [76] derived a model in terms of bond distortions and macroscopic elastic properties which predicts results that are in good accord with experiments. The model finds the formation enthalpies of $GaP_{0.5}Sb_{0.5}$ to be 0.053 eV and $InP_{0.5}Sb_{0.5}$ as 0.056 eV, in good agreement with the calculated values of CA structure. This comparison of formation enthalpy may suggest that the CA structure can serve as the representative structure other than CH and CP structures.

For GaPSb compounds, Stringfellow group reported [70] a low temperature photoluminescence peak at 1.394 eV for sample $GaP_{0.73}Sb_{0.27}$, and a fitted bowing parameter of 3.8 eV. Absorption spectra measurements [77] from the same group observed single-line peaks at 1.14 eV for $GaP_{0.53}Sb_{0.47}$ and 1.625 eV for $GaP_{0.76}Sb_{0.24}$ with bowing parameter of 3.11 eV. In these calculations, the CA and LZ structures of GaPSb compounds suggest that $GaP_{0.5}Sb_{0.5}$ and $GaP_{0.75}Sb_{0.25}$ should have band gaps of 0.8 and 1.06 eV and bowing parameter of 3.377 eV. The reason that their reported band gaps are larger than the predicted values is due to the fact that the structures of their samples are CA and disorder mixture and the band gap of random structure is larger than that of CA structure [75]. Room temperature photoluminescence and optical transmission measurements [78] exhibited a peak at energy of 0.845 eV which is identified as band gap transition $E_0(\Gamma_v - \Gamma_{1c})$, in agreement with the calculated data.

For InPSb compound, Jou et al. [79] found, via low temperature photoluminescence measurements, a direct band gap transition at an energy of 0.62 eV for $GaP_{0.69}Sb_{0.31}$ compared with the predicted value of 0.65 eV for composition 0.75–0.25. Their fitting to all the experimental data yields a band gap of 0.35 eV for compound $InP_{0.5}Sb_{0.5}$ and bowing parameter of 1.83 eV, in accord with the predicted value of 0.25 eV for band gap and 1.738 eV for bowing parameter. Similarly, photoluminescence and absorption spectra measurements, by Reihlen et al. [80], suggest a band gap of 0.445 eV for $InP_{0.577}Sb_{0.423}$ with bowing parameter of 1.52 eV. It is to be noted here that neglecting the spin-orbital interaction makes the calculated band gap to deviate from the experimental values.

8.5.8 Summary

In this section, the crystal relaxations, formation enthalpies and electronic properties of GaP_xSb_{1-x} and InP_xSb_{1-x} are discussed for various structures and compositions. GaP_xSb_{1-x} is found to be less relaxed than InP_xSb_{1-x} compounds. The formation enthalpy is found to be maximum at composition $x = 0.5$. The crystal field splitting and band gap are larger in GaP_xSb_{1-x} than in InP_xSb_{1-x}. This has been explained in terms of the energy repulsion rules. All the properties are found to be strongly structure and composition dependent. Good accord between the calculated results and experimental values are obtained.

8.6 Properties of II–VI Ternary Alloys

8.6.1 Overview

Two ordered III–V ternary alloys (GaP_xSb_{1-x} and InP_xSb_{1-x}) have been studied in Sect. 8.5. Properties that are investigated include: structure, formation enthalpy, crystal field splitting, energy gap and optical bowing parameters. In this section, the study is extended to the II–VI ternary alloys. The alloy CdS_xTe_{1-x} is used as an example. Different from the two alloys studied in the last Section, CdS_xTe_{1-x} contains heavy atoms Cd and Te. Therefore, the Spin-Orbit (SO) splitting and its bowing parameter will be considered in this section. Moreover, a new Y2 ordered structure is included in the calculation. All the properties of the ordered structures are compared with those of the disordered structure.

8.6.2 Theoretical Background

Thin film CdS/CdTe solar cells have ideal band gap and high absorption coefficient. High energy conversion efficiencies up to 16% [81] have been achieved. The inter-diffusion between the CdS window layer and CdTe absorber layer leads to the formation of a mixed CdS_xTe_{1-x} interfacial layer. This layer is generally believed to be beneficial to the solar cell performance. Through interfacial layer formation, (I) the large strain energy due to the lattice mismatch (10.7%) between CdS and CdTe can be largely relieved; (II) the degree of inter-diffusion will certainly shift the electrical junction away from the metallurgical interface and reduce the defect density at the interface [82]; (III) the adjustable band gap (E_g) with respect to the alloy composition will result in changes in the open-circuit voltage $V_{oc,}$ and thus the efficiency of the solar cells. Despite the significant benefits from CdS_xTe_{1-x} alloy, many fundamental properties of this system are not yet understood. These fundamental properties include the following:

(I) Possible sublattice crystal orderings and their effects on the properties of alloys: A short-range ordering in CdS_xTe_{1-x} has been observed recently by a room temperature Raman spectroscopy measurement [83]. However, theoretical studies have only discussed the properties of this alloy for random structure. Therefore, it is worthwhile to address the possible crystal orderings and their effects on the properties of this system.

(II) The sign of the SO splitting (Δ_{SO}) bowing parameter $b(\Delta_{SO})$: The composition variation of the spin-orbit splitting can be fitted to the form:

$$\Delta_{SO}(x) = \overline{\Delta}_{SO}(x) - x(1-x)b(\Delta_{SO}) \tag{8.29}$$

where $\overline{\Delta}_{SO}$ is the concentration-weighted average SO splitting. Controversy has existed for a long time on the sign of $b(\Delta_{SO})$. Most of the early experimental studies [71] reported positive values, e.g., GaPAs (0.175 eV), InPAs (0.357 eV), GaInP (0.101 eV), and GaInAs (0.144 eV). Later on [84, 85], some negative values have been reported, e.g., ZnSeTe (−0.59 eV) and GaInP (−0.05 eV). Therefore, it is interesting to find out the SO splitting and its bowing parameter of CdS_xTe_{1-x} alloys.

(III) Dependence of band gap and its bowing parameter b on the alloy composition: For most semiconductor alloys, the bowing parameter is nearly independent of composition x. However, for alloys with large size and chemical disparity between its constituents, the bowing parameter could be strongly composition dependent [86]. Therefore, it is interesting to see whether the bowing parameter varies with composition in CdS_xTe_{1-x} compounds.

In order to answer these questions, this section presents a systematic study of the electronic structures of CdS_xTe_{1-x} compounds. Calculation involves the properties including the following: (I) The crystal orderings and alloy formation energy; (II) Crystal field (CF) splitting Δ_{CF} and SO splitting Δ_{SO}; (III) The alloy band gaps and bowing parameters; (IV) Density of states.

8.6.3 Special Quasirandom Structures

Structural models such as Monte Carlo approach and large-scale pseudopotential method, used in the calculations of properties of random alloys, have to involve either rather large number of configurations or large cell sizes to mimic the randomness. These models attempt to approach the random correlation functions between atoms by statistical means. However, these techniques are impractical for first-principles calculations.

In the Special Quasirandom Structures (SQS) theory [73], the first step is to specify a set of correlation functions that mimics the random alloy in a hierarchical manner, and then find the structures corresponding to that set of correlation functions. This theory, therefore, can mimic the random alloys using periodic structures with small number of atoms. Based on the level of randomness, the SQS method generated cells can be characterized as SQSN, where N is the number of atoms in the primitive cells. Specifically, SQS2 refers to zincblende structure; SQS4 refers to Y2 ordered structure (Fig. 8.9a). Amongst the generated SQSN structures, SQS8 (Fig. 8.9b) is accurate enough to mimic the behavior of random alloys. In this section, the properties of ordered CdS_xTe_{1-x} for composition 0.5, i.e., CA, CH, CP, and Y2 are compared

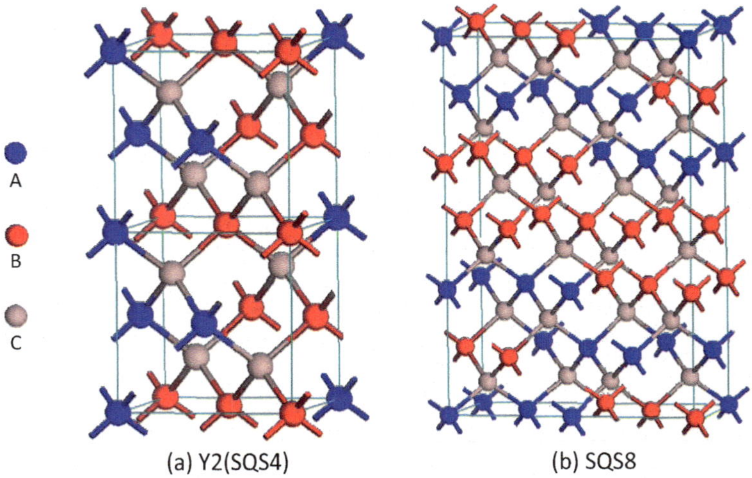

Fig. 8.9 The crystal structures of alloy $A_xB_{1-x}C$ at composition 0.5: **a** Y2 ordered structure and **b** SQS8 random structure

with the properties of random SQS8 structure. The properties of LZ structure are compared with those of SQS8 structure for composition 0.25 and 0.75.

8.6.4 Ground State Structure

The calculated properties of ordered and disordered $CdS_{0.5}Te_{0.5}$ are listed in Table 8.6. It is found that the formation energy, defined as, $\Delta E = E(x) - \overline{E}(x)$, follows the trend: ΔE(CP) > ΔE(CA) > ΔE(SQS8) > ΔE(Y2) > ΔE(CH), indicating that CH is the ground state structure and contains lower strain energy and Madelung energy. It is noteworthy that the difference in the formation energies between Y2 ordering and disordered SQS8 structures can be as small as 2 meV/atom. This suggests a possible ordered and disordered mixed structure. Indeed, Raman Spectroscopy measurements [83] have reported a short-range ordering in random CdS_xTe_{1-x} system.

8.6.5 Crystal Field Splitting

It is known that the CF splitting Δ_{CF} and SO splitting Δ_{SO} separate the triply degenerate valence band maximum states into singly and doubly degenerate states. Δ_{CF} and Δ_{SO} of all the ordered and disordered structures have been extracted, as given in Table 8.6, using the Hopfield quasicubic model [87]. Δ_{CF} is defined to be positive if the doubly degenerate states are above the nondegenerate state. According

Table 8.6 Calculated properties of $CdS_{0.5}Te_{0.5}$ at various structures

Structure	a (Å)	η	ΔE (eV/atom)	Δ_{CF} (eV)	Δ_{SO} (eV)	$b\,(\Delta_{SO})$ (eV)	E_g (eV)	b (eV)
CA	6.0961	0.996	0.038	0.230	0.566	−0.437	0.189 (1.37)	1.843
CH	6.0893	0.997	0.019	−0.124	0.497	−0.161	0.524 (1.70)	0.506
CP	6.1002	1.002	0.051	0.720	0.678	−0.885	−0.471 (0.71)	4.485
Y2	6.0889	0.995	0.027	−0.422	0.551	−0.377	0.202 (1.38)	1.792
SQS8	6.0914	0.995	0.029	−0.219	0.417	0.161	0.270 (1.45)	1.518

Listed properties in this table include the following: lattice constant, tetragonal distortion parameter, formation energy, CF splitting, SO splitting and its bowing parameter, band gap and optical bowing parameter. The HSE06 corrected band gaps are given in parenthesis

Table 8.7 LDA calculated valence band energy levels of CdS and CdTe, relative to valence band maximum

	Γ_{15v}	W_{3v}	X_{5v}	L_{3v}
CdS	0.00	−1.95	−1.66	−0.65
CdTe	0.00	−2.05	−1.70	−0.68

to the perturbation theory and the folding relations, Δ_{CF} is inversely proportional to the difference in the unperturbed energy levels of the end-point binary constituents before folding into new states of the ordered ternary compounds [71]. Based on this theory, the crystal ordering effects on the CF splitting can be explained in terms of the energy levels of the binary constituents, as listed in Table 8.7 [88]. From Table 8.6, one can find the following: (I) For the ordered structures, the following relation holds: Δ_{CF}(CP) > Δ_{CF}(CA) > Δ_{CF}(CH) > Δ_{CF}(Y2). This is because the energy difference follows the opposite trend. For example, the energy difference in CH ($\Gamma_{15v} - W_{3v} = 2$ eV) is larger than $\Gamma_{15v} - X_{5v} = 1.68$ eV in CA and $\Gamma_{15v} - L_{3v} = 0.67$ eV in CP ordering; (II) The CF splitting in CH ordering is small and negative due to the fact that the stronger ($\Gamma_{15v} - W_{3v}$) coupling than ($\Gamma_{15v} - X_{5v}$ r) coupling makes the Γ_{4v} above Γ_{5v} state; (III) Different from CA, CP, and CH, the doubly degenerate state in Y2 ordering is further split by a small amount into two nondegenerate states due to the yet lower symmetry.

8.6.6 Spin-Orbit Splitting and Band Gap

Earlier perturbation theory [89] treats the SO splitting Δ_{SO} as a disorder-induced effect and relates its bowing $b(\Delta_{SO})$ to the difference in *s*-*p* interaction of the alloy constituents. This theory predicts positive $b(\Delta_{SO})$. In order to explain some experimentally observed negative bowing values [84, 85], Wei et al. [90] propose that the interband *p*-*p* coupling enhances Δ_{SO} and dominates the value of $b(\Delta_{SO})$. From the calculated values in Table 8.6 for CdS_xTe_{1-x}, one can find the following: (I) The SO splitting shows strong ordering dependence. This indicates that the ordering geometry strongly affects the *s*-*p* and *p*-*p* coupling; (II) Ordered structures yield negative $b(\Delta_{SO})$ and disordered structures yield a positive value. The ordering-induced negative $b(\Delta_{SO})$ are consistent with the results of Wei et al. [90] and can be attributed to the intraband *p*-*p* coupling. The positive $b(\Delta_{SO})$ for disordered structure is due to the fact that the disorder effect mixes *d* state at the top of valence band with *p* states and this *p*-*d* hybridization reduces Δ_{SO}; (III) Largest $b(\Delta_{SO})$ is found in CP ordering and smallest is found in SQS8 structure. This suggests that the *p*-*d* coupling is strongest in SQS8 and weakest in CP ordering.

The calculated results also show that CP ordering has the smallest band gap (0.71 eV). This is due to the fact that the smallest $(\Gamma_{15v} - L_{3v})$ energy difference causes strongest repulsion in its energy levels. This repulsion lowers Γ_{1c} and raises Γ_{15v} states, and thus results in smallest band gap. Similarly, CH ordering has the largest band gap relative to other structures. In principle, an ideal solar cell material should have a direct band gap around 1.3–1.5 eV. Therefore, experimental conditions should be controlled to avoid the formation of CP ordering.

8.6.7 Dependence on Alloy Composition

The effects of composition of the alloy can be studied by calculating the properties of CdS_xTe_{1-x} for composition 0.25, 0.5, and 0.75. The calculated lattice constants follow a simple linear function of composition x. The interaction parameter, $\Omega = \Delta E/x(1-x)$, increases with increasing Te concentration. The SO splitting increases monotonically when the anion atomic number increases from S to Te. This is because the valence band has large anion *p* character, and the atomic SO splitting of the anion valence *p* state increases with the atomic number [91]. The change in Δ_{SO}, however, is not a linear function of composition. Its bowing parameter $b(\Delta_{SO})$, as listed in Table 8.8, shows significant composition dependence. The calculations of band gap E_g show that initially, adding S into CdTe will actually reduce the band gap. Further increase in S concentration will eventually increase the band gap. This is due to the fact that at low S concentration, the bowing parameter b is larger than the band gap difference between CdS and CdTe. Results show that b and $b(\Delta_{SO})$ are both strongly composition dependent. However, b increases as $b(\Delta_{SO})$ decreases, unlike the scaling assumption used in the *s*-*p* model [89, 92]. The effects of crystal ordering

Table 8.8 Calculated properties of CdS_xTe_{1-x} at various compositions along (100) ordering

Composition	a	Ω	$b(\Delta_{SO})$	b
$x = 0.25$	6.2589	0.140	−0.265	1.291
$x = 0.5$	6.0961	0.152	−0.437	1.843
$x = 0.75$	5.9352	0.163	−0.466	1.819

The calculated properties include: lattice constant, interaction parameter, bowing parameter of SO splitting and band gap

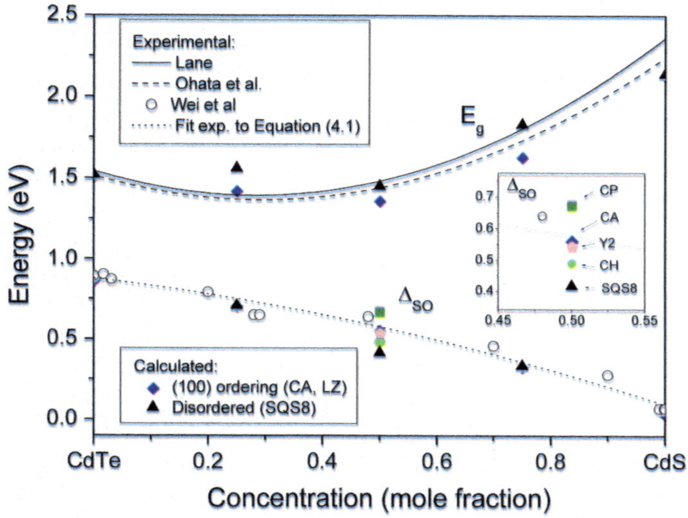

Fig. 8.10 The effects of crystal ordering and composition on the spin-orbit splitting and band gap of CdS_xTe_{1-x}. Experimental values are plotted to compare with the calculated results

and composition on the spin-orbit splitting and band gap of CdS_xTe_{1-x} are shown in Fig. 8.10.

The compositions and crystal orderings also have effects on the variations of the partial and total density of states (DOS) of CdS_xTe_{1-x}. As shown in Fig. 8.11, for pristine CdS and CdTe, the top of the valence band is dominated by S 3*p* and Te 5*p* states, respectively, and the bottom of the conduction band is mainly derived from Cd 5*s* state. The DOS of ternary CdS_xTe_{1-x} can be seen as the combination of CdS and CdTe. With increasing Te concentration, the magnitude of Te states increases while the magnitude of S states decreases. It can also be found that: (I) the main Cd 4*d* peak red shifts from CdS to CdTe, implying that *p*-*d* coupling becomes weaker which explains the increase in SO splitting and the reduction in band gap; (II) the valence bandwidth increases with the mixing of CdS and CdTe and it reaches maximum at $CdS_{0.5}Te_{0.5}$, indicating the formation of CdS_xTe_{1-x}. It increases the mobility of holes generated by light irradiation and hence improves the solar cell performance. Calculation of DOS for various orderings (not presented) for composition 0.5 shows

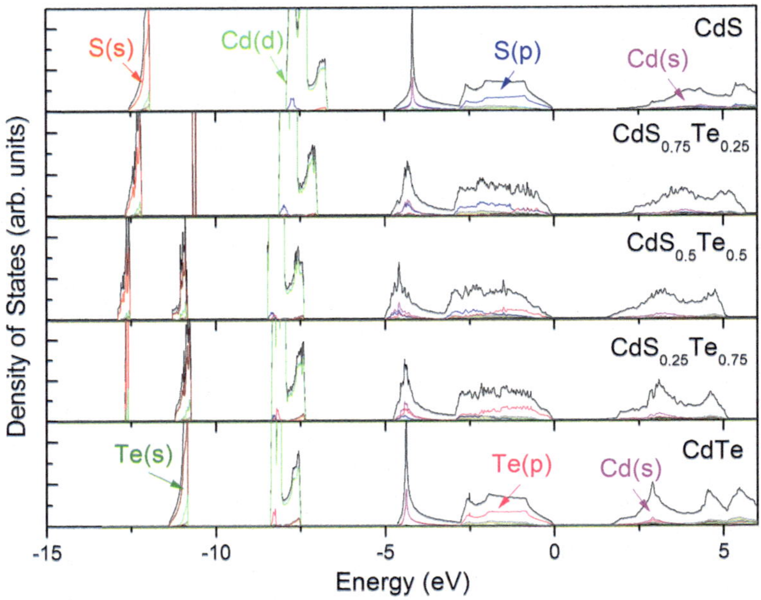

Fig. 8.11 The calculated partial and total density of states of CdS_xTe_{1-x} at various compositions

that the CP ordering has widest valence bandwidth, followed by Y2 and disordered phases. Shifts in Cd 4*d* state shows that the *p-d* coupling is strongest in CH and disordered and weakest in CP ordering.

8.6.8 Comparison with Experiments

The calculated SO splitting and band gaps are compared with the experimental data in Fig. 8.10. Good agreement is found throughout the entire composition range. The only available data [93] on SO splitting was measured by performing ellipsometry at room temperature. Fitting the experimental values to Eq. (8.29) will yield a negative $b(\Delta_{SO})$ of -0.408 eV. Therefore, it is expected that the experimental samples are at least partially ordered. In order to have a direct comparison of the calculated band gaps with experiments [53, 94, 95], the LDA calculated results are corrected according to the Heyd-Scuseria-Ernzerhof hybrid functional calculations (HSE06) [96, 97] since LDA underestimates the absolute band gaps. Comparison shows the following: (I) Amongst the ordered structures, (100) ordering (CA and LZ) is the best representation of the random alloy; (II) A small amount of Te in CdS can drastically reduce its band gap. This is because the impurity limit of Te substitution on S site leads to a localized isovalent impurity level [98]. The experimentally reported optical bowing parameters are around 1.7–1.88 eV [53, 95], well consistent with the calculated disordered and representative (100) ordering results.

8.6.9 Summary

In this section, the properties of CdS_xTe_{1-x} alloys have been investigated for various crystal orderings and compositions. Good agreement is seen between the calculated results and experimental data. The following results are found: (I) CH is the ground state structure. Y2 ordering may occur in disordered structure due to small energy difference; (II) Ordering can significantly affect the SO splitting and energy gap; (III) Negative bowing parameter of spin-orbit splitting is found in ordered structure while positive value is found in disordered structure; (IV) The bowing parameters of energy gap and SO splitting are both strongly ordering and composition dependent. However, the bowing parameter of energy gap increases with decreasing bowing parameter of SO splitting.

8.7 Fingerprints of Y2 Ordering in III–V Ternary Alloys

8.7.1 Overview

The properties of fully ordered and fully disordered III–V and II–VI ternary semiconductor alloys have been studied in the previous two sections. However, due to the experimental conditions, alloys are usually synthesized at a partially ordered structure. In this section, a method is provided to determine the properties of an alloy at any degree of ordering. Five ternary alloys with partial Y2 ordering are discussed, i.e., $Al_xGa_{1-x}As$, $Ga_xIn_{1-x}As$, $Ga_xIn_{1-x}P$, $GaAs_xSb_{1-x}$, and InP_xSb_{1-x}. Reported here are the fingerprints, including valence band splittings and band gap narrowing (the ordering-induced band gap reduction relative to the random alloy), $\Delta E_g(\eta) = E_g(\eta) - E_g(0)$, of these five Y2 ordered compounds at partial and full degree of ordering. The physical factors that affect the fingerprints will be identified and the trends will be graphically represented. In order to generalize the research in Sect. 8.6, the properties of materials in Y2 ordering will also be compared with properties of these materials in other observed orderings in a brief manner. The calculated data in this section can be useful in analyzing the experimental observations and deriving the ordering parameters of partially ordered samples.

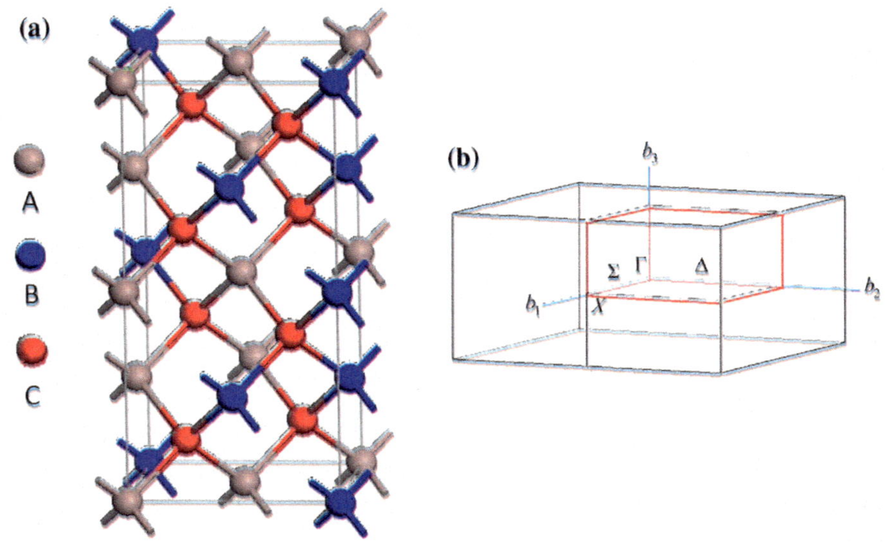

Fig. 8.12 **a** The crystal structure of ternary alloy $A_xB_{1-x}C$ in Y2 ordering; **b** the Brillouin zone of the Y2 ordered superlattice

8.7.2 Spontaneous Y2 Ordering

Spontaneous Y2 ordering of isovalent $A_xB_{1-x}C$ semiconductor alloys has been observed in vapor phase growth of several III–V systems [99–104]. However, the fundamental properties of this ordering have not been systematically studied. The ordered phase consists of alternate cation monolayer planes $A_{x+\eta/2}B_{1-x-\eta/2}$ and $A_{x-\eta/2}B_{1-x+\eta/2}$ stacked along the [110] direction, where $0 \leq \eta \leq 1$ is the long-range order parameter. $\eta = 1$ corresponds to the fully ordered phase (Fig. 8.12a) and $\eta = 0$ corresponds to the fully disordered phase. The degree of ordering depends on the experimental conditions such as the growth temperature and pressure, growth rates, and substrate orientation, etc.

When a zincblende disordered alloy forms the long-range ordered Y2 phase, the unit cell is increased and the Brillouin zone is reduced. The point group symmetry is changed from T_d to C_{2v}. These lead to a series of experimentally observable changes in materials properties, including new photoluminescence and electroreflectance peaks [101, 102], new x-ray diffraction spots at (1/2, 1/2, 0) [99, 100], new pressure deformation potential [105], and the shift in absorption edge [99]. In this work, the study focuses on the changes in optical properties near the absorption edge. These changes are due to the fact that, in the ordered phase, the Γ, X and Σ points in the zincblende binary constituents all fold into the Γ point at the Y2 Brillouin zone (Fig. 8.12b). These folding relations couple the states that have the same symmetry and this coupling splits the degenerate states in the random alloys.

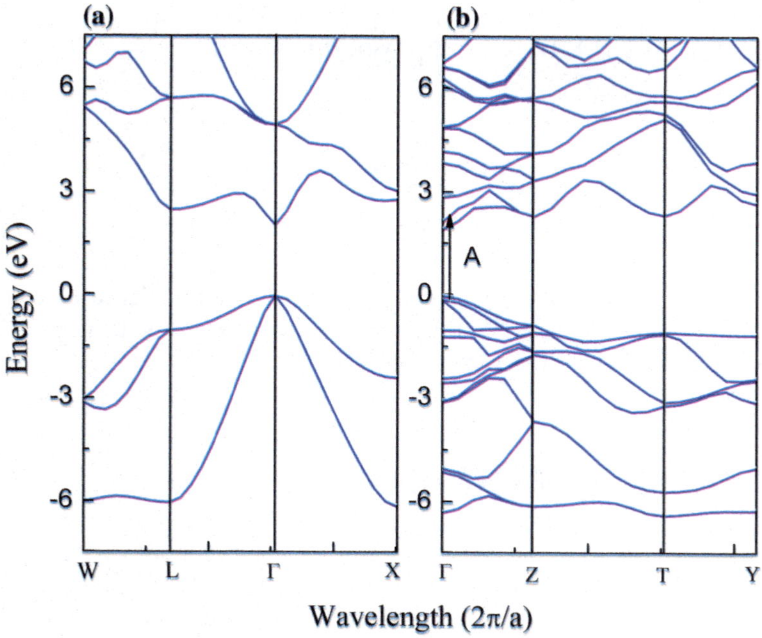

Fig. 8.13 The band structures for the **a** random and **b** Y2 ordered $Ga_{0.5}In_{0.5}P$ alloy are plotted along the high symmetry lines in the Brillouin zone. The arrow A denotes the interband transition responsible for the anomalous peak at 2.2 eV in the electroreflectance spectra

8.7.3 Hopefield Quasicubic Model

In the absence of spin-orbit (SO) splitting, the valence band maximum (VBM) of the random alloy is a triply degenerate state with Γ_{15v} symmetry (Fig. 8.13a). In Y2 ordering, this state splits into a single state Γ_{1v} and a doubly degenerate state. The doubly degenerate state splits further into two single states Γ_{2v} and Γ_{3v} due to the yet lower symmetry (Fig. 8.13b). In the presence of SO splitting, the amount of splitting becomes more significant.

The valence band splittings can be expressed, in terms of the energies of the top three valence band states, $E_1(\Gamma_{1v})$, $E_2(\Gamma_{2v})$, and $E_3(\Gamma_{3v})$, as follows:

$$\Delta E_{12}(\eta) = E_1(\Gamma_{1v}) - E_2(\Gamma_{2v})$$
$$\Delta E_{13}(\eta) = E_1(\Gamma_{1v}) - E_3(\Gamma_{3v}) \quad (8.30)$$

The Hopfield quasicubic model [87] states that $\Delta E_{12}(\eta)$ and $\Delta E_{13}(\eta)$ can be expressed by

$$\Delta E_{12}(\eta) = \frac{1}{2}[\Delta_{SO}(\eta) + \Delta_{CF}(\eta)] - \frac{1}{2}\left\{[\Delta_{SO}(\eta) + \Delta_{CF}(\eta)]^2 - \frac{8}{3}\Delta_{SO}(\eta) + \Delta_{CF}(\eta)\right\}^{\frac{1}{2}} \quad (8.31a)$$

$$\Delta E_{13}(\eta) = \frac{1}{2}[\Delta_{SO}(\eta) + \Delta_{CF}(\eta)] + \frac{1}{2}\left\{[\Delta_{SO}(\eta) + \Delta_{CF}(\eta)]^2 - \frac{8}{3}\Delta_{SO}(\eta) + \Delta_{CF}(\eta)\right\}^{\frac{1}{2}} \quad (8.31b)$$

where $\Delta_{SO}(\eta)$ is the SO splitting and $\Delta_{CF}(\eta)$ is the ordering-induced Crystal Field (CF) splitting in the absence of SO splitting. $\Delta_{CF}(\eta)$ is defined to be negative if the doubly degenerate state is below the single state.

8.7.4 Fingerprints of Y2 Ordering

For a spontaneously formed partially ordered semiconductor alloy with $n \leq 1$, the physical properties $P(x, \eta)$ such as the CF splitting $\Delta_{CF}(\eta)$, the SO splitting $\Delta_{SO}(\eta)$, and the band gap $E_g(\eta)$, at composition x can be described by [106, 107]:

$$P(x, \eta) = P(x, 0) + \eta^2[P(X_\sigma, 1) - P(X_\sigma, 0)] \quad (8.32)$$

This equation shows that the property $P(x, \eta)$ of a semiconductor alloy $A_xB_{1-x}C$ can be calculated by (I) the corresponding properties $P(x, 0)$ of the random structure at the same composition x, (II) the degree of ordering η, and (III) the difference in the property, $P(X_\sigma, 1) - P(X_\sigma, 0)$, between the fully ordered structure and random structure at composition $X_\sigma = 0.5$.

According to Eqs. (8.30)–(8.32), if valence band splitting, $\Delta E_{12}(\eta)$, $\Delta E_{13}(\eta)$, and band gap, $E_g(\eta)$, are known independently, for example, from electroreflectance or photoluminescence spectra, the SO splitting $\Delta_{SO}(x, \eta)$ and CF splitting $\Delta_{CF}(x, \eta)$ can be derived according to Eq. (8.31a, 8.31b). Then the theoretically calculated differences in the SO splitting, $[\Delta_{SO}(1) - \Delta_{SO}(0)]$, CF splitting, $[\Delta_{CF}(1) - \Delta_{CF}(0)]$, and the band gap narrowing, $\Delta E_g(\eta)$, can be used to derive the ordering parameter η using Eq. (8.32). On the other hand, if η is available independently from experiment, such as x-ray diffraction, one can assess the valence band splitting, $\Delta E_{12}(\eta)$, $\Delta E_{13}(\eta)$, and band gap narrowing $\Delta E_g(\eta)$.

The results of the GGA calculations are shown in Table 8.9. The following results can be found:

(I) Ordering induces a decrease in band gap and CF splitting, but an increase in SO splitting in all the five alloy systems.
(II) $[\Delta_{SO}(1) - \Delta_{SO}(0)]$ is always positive. This is due to the fact that the VBM wave function of the ordered compounds, relative to the random alloy, is more localized on the cation atom with larger atomic number [91]. For common-anion systems, the two binary constituents have similar Δ_{SO}. Therefore, the

Table 8.9 GGA calculated optical fingerprints of five III–V alloys

Alloys	$Al_{0.5}Ga_{0.5}As$	$Ga_{0.5}In_{0.5}As$	$Ga_{0.5}In_{0.5}P$	$GaAs_{0.5}Sb_{0.5}$	$InP_{0.5}Sb_{0.5}$
$\Delta E_{12}(1)$	0.009	0.129	0.091	0.188	0.357
$\Delta E_{13}(1)$	0.318	0.385	0.169	0.643	0.701
$\Delta_{SO}(1) - \Delta_{SO}(0)$	0.001	0.005	0.013	0.021	0.052
$\Delta_{CF}(1) - \Delta_{CF}(0)$	−0.004	−0.094	−0.102	−0.213	−0.385
$\Delta E_g(1)$	−0.034	−0.093	−0.202	−0.274	−0.423

The calculated properties include the following: valence band splitting, $\Delta E_{12}(1)$ and $\Delta E_{13}(1)$, changes in spin-orbit splitting $[\Delta_{SO}(1) - \Delta_{SO}(0)]$, crystal field splitting $[\Delta_{CF}(1) - \Delta_{CF}(0)]$ and band gap $\Delta E_g(1)$. Values are given in units of eV

ordering-induced difference $[\Delta_{SO}(1) - \Delta_{SO}(0)]$ is rather small. However, the common-cation systems (e.g., $GaAs_{0.5}Sb_{0.5,}$ and $InP_{0.5}Sb_{0.5}$) are relatively larger $[\Delta_{SO}(1) - \Delta_{SO}(0)]$ because they have larger anion atom Sb. The SO splitting increases monotonically when anion atomic number increases [91].

(III) As shown in Fig. 8.14a, the CF splitting $[\Delta_{CF}(1) - \Delta_{CF}(0)]$ increases in the following sequence: $Ga_{0.5}In_{0.5}As \rightarrow Ga_{0.5}In_{0.5}P \rightarrow GaAs_{0.5}Sb_{0.5} \rightarrow InP_{0.5}Sb_{0.5}$. According to the perturbation theory, Δ_{CF} is proportional to the valence band offset and inversely proportional to the difference between the symmetric energy levels of binary constituents [71]. The band offset of a semiconductor alloy AB_xC_{1-x} refers to the relative alignment of the valence band maxima of the corresponding constituents AB and AC. This can explain the trend in CF splitting. For example, the band offset [108] between GaAs and GaSb for $GaAs_{0.5}Sb_{0.5}$ (0.57 eV) is much larger than that between GaAs and InAs for $Ga_{0.50.5}As$ (0.06 eV). Therefore, the perturbation and CF splitting in the valence bands are larger in $GaAs_{0.5}Sb_{0.5}$ than in $Ga_{0.50.5}As$. Note that the band offset between AlAs and GaAs for $Al_{0.5}Ga_{0.5}As$ is rather large (0.51 eV). However, it is the smallest $[\Delta_{CF}(1) - \Delta_{CF}(0)]$ at Γ point of its Brillouin zone, as shown in Table 8.9 and Fig. 8.14a. This is because $Al_{0.5}Ga_{0.5}As$ compound has an indirect band in this ordering.

(IV) The band gap narrowing $\Delta E_g(\eta)$ increases with the increasing of the alloy lattice mismatch between the binary constituents, as shown in Fig. 8.14b. For example, the lattice mismatch between the binary constituents for $Al_{0.5}Ga_{0.5}As$ and $Ga_{0.5}In_{0.5}As$ are 0.14 and 6.92%, respectively, smaller than that of 9.88% between InP and InSb for $InP_{0.5}Sb_{0.5}$ compound. During the formation of the lattice-mismatched alloys, the structure relaxation tends to shift the charge from the long bond to the short bond, and thus reduce the repulsion between the symmetric energy levels. This repulsion lowers the Γ_{1c} state and raises the Γ_{VBM} state, resulting in band gap narrowing. In alloys with larger lattice mismatch between the constituents, more charge is transferred and therefore the band gap narrowing is larger.

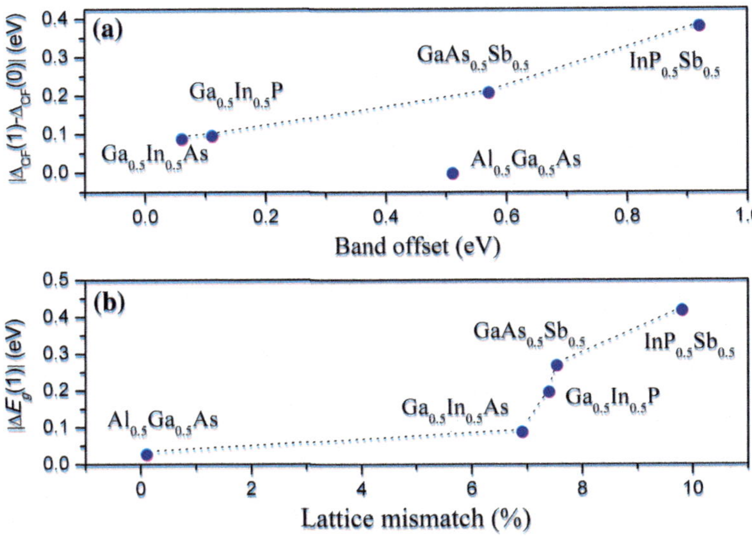

Fig. 8.14 **a** Variation of the crystal field splitting $[\Delta_{CF}(1) - \Delta_{CF}(0)]$ with the band offset between alloys' binary constituents. **b** Variation of the band gap narrowing $\Delta E_g(1)$ with the alloys lattice mismatch

8.7.5 Comparison with Other Orderings

Numerous studies [109–112] on the ordering of the alloy $Ga_{0.50.5}P$ have reported the CuPt structure. However, the Y2 ordering has generally been ignored in the literature. This is due to the similarity between CuPt and Y2 orderings. They are built from the same (001) plane and differ only in the stacking of the subsequent planes. In fact, some reports [113, 114] have mistakenly attributed the extra features in the spectra originating from Y2 ordering into CuPt ordering. Moreover, the small difference between the formation enthalpies of CA and Y2 ordering in some alloys may also cause the coexistence of CA and Y2 orderings [100]. In view of these facts, the optical fingerprints of Y2 ordering are compared here with those of CA, CP, and CH structures. Results are summarized in Table 8.10. The following results are found:

(I) Relative to other orderings, Y2 ordering has large and negative CF splitting Δ_{CF}. As has been highlighted earlier, the ordering separates the triply degenerate states in random alloy into a single state and a doubly degenerate state. In CA and CP orderings, the doubly degenerate state is above the single state, resulting in a positive CF splitting. However, in CH and Y2 ordering, the doubly degenerate state is below the single state, resulting in a negative CF splitting. In Y2 ordering, the doubly degenerate state splits further into two single states. Due to the smaller difference in the symmetric energy levels of the binary constituents in Y2 ordering than in CH ordering, the CF splitting is larger in Y2 ordering.

Table 8.10 Calculated properties of five fully Y2 ordered compounds

Alloys	Structure	a	Δ_{CF}	Δ_{SO}	$b(\Delta_{SO})$
$Ga_{0.5}In_{0.5}P$	Y2	5.6599	−0.102	0.104	−0.053
$Ga_{0.5}In_{0.5}P$	CA	5.6599	0.199	0.097	−0.023
$Ga_{0.5}In_{0.5}P$	CH	5.6599	−0.015	0.093	−0.008
$Ga_{0.5}In_{0.5}P$	CP	5.6599	0.232	0.103	−0.047
$GaAs_{0.5}Sb_{0.5}$	Y2	5.8746	−0.213	0.539	−0.084
$GaAs_{0.5}Sb_{0.5}$	CA	5.8927	0.085	0.549	−0.10
$GaAs_{0.5}Sb_{0.5}$	CH	5.8922	−0.013	0.521	−0.01
$GaAs_{0.5}Sb_{0.5}$	CP	2.8974	0.230	0.605	−0.33
$Al_{0.5}Ga_{0.5}As$	Y2	5.5659	−0.004	0.316	−0.005
$Ga_{0.5}In_{0.5}As$	Y2	5.8558	−0.094	0.343	−0.019
$InP_{0.5}Sb_{0.5}$	Y2	6.1740	−0.385	0.467	−0.208

Results of CA, CH, and CP orderings of GaInP and GaAsSb are also listed for comparison. The units are Å for lattice constant a and eV for crystal field splitting Δ_{CF}, spin-orbit splitting Δ_{SO} and its bowing parameter $b(\Delta_{SO})$

(II) The bowing parameter of SO splitting is negative in Y2 ordering. SO splitting reflects the way that bonding in solids redistributes the charge around the atomic cores of the constituents [39]. The sign of the SO splitting bowing parameter reflects the alloy environment and acts to enhance or diminish the magnitude of SO splitting beyond the linear average of the constituents. The calculation shows that the formation of Y2 ordering enhances the magnitude of Δ_{SO} and yields a negative SO splitting bowing parameter $b(\Delta_{SO})$. This is consistent with the result of CdS_xTe_{1-x} in the previous section and the upward concave bowing is attributed to the intraband *p-p* coupling.

8.7.6 Comparison with Experiments

Using electroreflectance spectroscopy method, Kurtz [101] found an anomalous peak at about 2.2 eV in spontaneously ordered $Ga_{0.50.5}P$ alloy. This peak is attributed to the *X* point folding to Γ point in the first Brillouin zone due to the Y2 ordering. According to the calculated band structures, this peak corresponds to the transitions (denoted as A in Fig. 8.13b) from the second and third VBM states to the second conduction band minimum (CBM) state. The calculated value for transition A is 2.27 eV, in good agreement with the experimental result of 2.2 eV. The band diagram, produced by Kurtz [101], using virtual crystal approximation method, suggests that Y2 ordering in $Ga_{0.5}In_{0.5}P$ shall result in a band gap narrowing of 0.17 eV. This value is close to the GGA calculated data of 0.202 eV. Using transmission electron microscopy and photoluminescence methods, Gomyo et al. [99] reported a band gap narrowing of 0.05 eV for $Ga_{0.5}In_{0.5}P$ due to the partial Y2 ordering. According to the calculations, the sample should have ordering η of around 0.5.

The calculated values of valence band splittings, $\Delta E_{12}(1) = 0.091$ eV and $\Delta E_{13}(1) = 0.169$ eV, for fully Y2 ordered $Ga_{0.5}In_{0.5}P$ are consistent with the results, 0.10 and 0.15 eV, reported by Lee et al. [115]. The calculated values of CF splitting and SO splitting are also in good accord with the available experimental results. For example, the reported [116] SO splitting for $Ga_{0.5}In_{0.5}As$ are 0.345 and 0.33 eV while the calculations obtained 0.343 eV. The calculated bowing parameter of SO splitting for $Ga_{0.5}In_{0.5}P$ is −0.053 eV. This value is very close to the measured result of −0.05 eV using the electroreflectance and wavelength modulation methods [85].

Recently, Wu et al. [103] reported the observation of Y2 ordering in $InP_{0.52}Sb_{0.48}$. Using reciprocal space mapping and extended x-ray absorption fine structure method, they found the structure parameter $c/a = 1.009$. This is considerably larger than the predicted value of 0.997 (not listed) for $InP_{0.5}Sb_{0.5}$. However, they find that the strong distorted In-P and In-Sb bonds prevent the crystal lattice from complete relaxation. This may explain the difference between the calculated results and their measured values.

8.7.7 Summary

This section presents the calculations of the Y2 ordering-induced changes in the optical fingerprints, including crystal field splitting, spin-orbit splitting, band gap, and valence band splittings, for $Al_xGa_{1-x}As$, $Ga_xIn_{1-x}As$, $Ga_xIn_{1-x}P$, $GaAs_xSb_{1-x}$ and InP_xSb_{1-x} using first-principles calculations. These values for the five materials are provided as a function of the degree of long-range order η. For the partially ordered samples, the trends of the changes in the crystal field splitting and band gap narrowing are explained. The change in spin-orbit splitting is found to be positive and small. For the fully ordered samples, Y2 ordering is compared with other orderings. It is found that Y2 has a large and negative crystal field splitting and negative spin-orbit bowing parameter. The calculated data in this section can be useful in analyzing the experimental results and deriving the ordering parameters of partially ordered samples.

8.8 Pressure Dependence of Energy Gap of III–V and II–VI Ternary Semiconductors

8.8.1 Theoretical Background

Previous sections, in this chapter, have been dedicated to the studies of all fundamental (i.e., elastic, electronic, and optical) properties of III–V and II–VI binary and ternary semiconductors. During the fabrication of semiconductor devices and the industrial applications of sensors, the pressure dependence of the energy gaps of ternary alloys is always of great interest. In general, there has been very little data on the pressure dependence of the energy gap of ternary semiconductors and even within the limited available experimental data, there is a significant variation. For example, the pressure coefficient of the band gap of $Ga_{0.5}IN_{0.5}P$, reported by Hakki and coworkers [117], is 13 meV/kbar, in contrast with the 8.4 meV/kbar obtained by Chen et al. [118]. Thus a theoretical approach is required to analyze the problem.

Except the first-principles calculations, empirical approaches have been developed to address some of these problems. The transition from binary compound semiconductors to ternary compound semiconductors requires the understanding of the bowing parameter [119]. As given in Sect. 8.5, the expression for the bowing parameter, c_{ABC}, of ternary compound AB_xC_{1-x} can be rewritten as

$$E_g(x) = xE_g^{AB} + (1-x)E_g^{AC} - c_{ABC}x(1-x) \tag{8.33}$$

where E_g^{AB} and E_g^{AC} are the energy gaps of binary compounds AB and AC, respectively. Hill [119] has ascribed the physical meaning of the bowing parameter to the nonlinear dependence of the crystal potential on the properties of the component ions and derived the following expression:

$$c_{ABC} = \frac{Zer_{BC}}{4\pi\epsilon_0}\left[\frac{1}{r_B} - \frac{1}{r_C}\right]^2 \exp\left(-\frac{1}{2}sr_{BC}\right) \tag{8.34}$$

in which $Z = Z_B = Z_C$ is the valence number of ions B and C, r_B and r_C are the covalent radii of B and C, $r_{BC} = r_B + r_C$ and $s = 0.25$ is a screening constant. Differentiating Eq. (8.34) with respect to pressure, Hill and Pitt [120] obtained the following expression:

$$\frac{\mathrm{d}c_{ABC}}{\mathrm{d}P} = \frac{1}{2\sqrt{3}}\frac{c_{ABC}}{r_B - r_C}\left[\frac{r_C}{r_B}a_{AB}\chi_{AB} - \frac{r_B}{r_C}a_{AC}\chi_{AC}\right] \tag{8.35}$$

where a_{AB}, a_{AC} and χ_{AB}, χ_{AC} are lattice constants and compressibilities of compounds AB and AC, respectively. Based on this model, Hill and Pitt [120] calculated the pressure coefficients of the bowing parameters for a number of ternary semiconductors. However, in order to calculate the pressure-dependent energy gap of ternary semiconductor AB_xC_{1-x} using this model, one has to use the experimental data for band gap pressure coefficients of binary compounds AB and AC, because, according to Eq. (8.33):

$$\frac{\mathrm{d}E_g(x)}{\mathrm{d}P} = x\frac{\mathrm{d}E_g^{AB}}{\mathrm{d}P} + (1-x)\frac{\mathrm{d}E_g^{AC}}{\mathrm{d}P} - \frac{\mathrm{d}c_{ABC}}{\mathrm{d}P}x(1-x) \tag{8.36}$$

Van Vechten [121, 122] proposed a dielectric theory for tetrahedral compounds based on Phillips' spectroscopic theory of electronegativity difference [39]. The theory was successfully applied and generalized to a variety of areas in materials science, including, band structures, alloy bowing parameters, elastic constants, and so forth. Camphausen et al. [123] used this model to calculate the pressure coefficients of band gaps of nineteen binary semiconductors and appeared to yield good agreement between theoretical expectations and experimental results.

In this section, the Van Vechten's theory is modified to calculate the pressure dependence of energy gap of a number of group III–V and II–VI zincblende ternary semiconductors. The calculated results are compared with the available data in the literature. The trends in the variations of the band gap and its pressure dependence will be discussed.

8.8.2 Modeling Pressure Dependent Band Gap of Ternary Alloys

In Van Vechten's dielectric theory [121, 122], the energy gap between the minimum conduction band and the maximum valence band, if the effect of d-state core is involved, is expressed as

$$E'_{\mathrm{g}} = \left[E_{\mathrm{g,h}} - (D_{av} - 1)\Delta E_{\mathrm{g}}\right]\left[1 + (C/E_{\mathrm{g,h}})^2\right]^{\frac{1}{2}} \tag{8.37}$$

In this expression, E'_{g} is used to differentiate from the energy gap in Eq. (8.33). $E_{\mathrm{g,h}}$ is the homopolar gap for transition corresponding to particular energy gap and is assumed to be a power function of the nearest neighbor distance r given by $E_{\mathrm{g,h}} \propto r^{s_1}$ where $s_1 = -2.75$. D_{av} is the factor that describes the lowering of s-like conduction band states caused by the effect of d states and the value is the skewed average of D_{av} values of the crystals containing the constituent atoms and the atom from the same row in the periodic table. For the first three rows in the periodic table, D_{av} is unity and Eq. (8.37) reduces to Phillips' pseudopotential theory [39]. ΔE_g is the correction related factor given by $\Delta E_g \propto r^{s_2}$, where $s_2 = -5.07$. C is the heteropolar gap produced by the antisymmetric potential in the corresponding binary compounds. For binary compound AB, C is given by [121]:

$$C = b_p e^2\left[\frac{Z_A}{r_A} - \frac{Z_B}{r_B}\right]\exp(-k_s r) \tag{8.38}$$

For a ternary compound AB_xC_{1-x}, it can be generalized as follows:

$$C = b_p e^2\left[\frac{Z_A}{r_A} - x\frac{Z_B}{r_B} - (1-x)\frac{Z_C}{r_C}\right]\exp(-k_s r) \tag{8.39}$$

where the pre-factor b_p is constant around 1.5, k_s is the radius-dependent Thomas–Fermi screening wave number, $r = r_A + xr_B + (1-x)r_C$ is the nearest neighbor distance.

From Eqs. (8.37)–(8.39), the pressure coefficients of the energy gap can be given by

$$\begin{aligned}\frac{\mathrm{d}E'_{\mathrm{g}}}{\mathrm{d}P} &= \left[1 + \left(C/E_{\mathrm{g,h}}\right)^2\right]^{\frac{1}{2}}\left[\frac{\mathrm{d}E_{\mathrm{g,h}}}{\mathrm{d}P} - \Delta E_{\mathrm{g}}\frac{\mathrm{d}(D_{av}-1)}{\mathrm{d}P} - (D_{av}-1)\frac{\mathrm{d}\Delta E_{\mathrm{g}}}{\mathrm{d}P}\right] \\ &\quad + \left[\frac{E_{\mathrm{g}}}{1 + \left(E_{\mathrm{g,h}}/C\right)^2}\right]\left[\frac{1}{C}\frac{\mathrm{d}C}{\mathrm{d}P} - \frac{1}{E_{\mathrm{g,h}}}\frac{\mathrm{d}E_{\mathrm{g,h}}}{\mathrm{d}P}\right]\end{aligned} \tag{8.40}$$

This Eq. (8.40) is the same as the one derived by Camphausen et al. [123] and is applicable to both binary and ternary compounds. Camphausen et al. have shown that even in nonionic materials, $\frac{dC}{dP}$ is sufficiently small and can be considered to be negligible while compared with the other terms in the expression. They further pointed out the expression for the correction term to be $D_{av} - 1 \propto r^y(1 - f_i)^z$, in which, f_i is the Phillips' ionicity [39] as introduced in Sect. 8.4. E_h is the average homopolar energy gap given as $E_h \propto r^{s_3}$, where $s_3 = -2.48$. s_1, s_2, and s_3 are constants determined from experimental values of two group IV elements [122]. Camphausen et al. [123] found that $y = 13$ and $z = 2.4$ by fitting the pressure coefficient of the energy gap of Ge.

In their later study, Van Vechten and Bergstresser [124] pointed out that the bowing parameter c_{ABC} of ternary compound AB_xC_{1-x} comprises of two parts. The first part, intrinsic bowing parameter c_i, originates from the variation of the average crystal potential under virtual crystal approximation which assumes the periodic potential in the crystal. If one calculates the energy difference using Eq. (8.37), the result, $E_g'^{ABC}$, will be different from the compositionally weighted average energy of the corresponding two binary compounds. This difference is the intrinsic bowing parameter. Another part, c_e, the extrinsic bowing parameter, arises from the real short-range aperiodicity and is the small deviation of the real potential from virtual periodic potential. In terms of this theory, the real energy gap in Eq. (8.37) becomes the following:

$$E_g(x) = E_g'^{ABC} - c_e x(1 - x) \tag{8.41}$$

where the intrinsic bowing parameter is included in the first term; the extrinsic bowing parameter is proposed by Van Vechten and Bergstresser as [124] as:

$$c_e = \frac{C_{BC}^2}{A} \tag{8.42}$$

$$C_{BC} = b_p e^2 \left| \frac{Z_B}{r_B} - \frac{Z_C}{r_C} \right| \exp(-k_s r) \tag{8.43}$$

In Eq. (8.42), the bandwidth parameter A is a constant for all the compounds and is found to be 0.98 eV by fitting the extrinsic bowing parameter with C_{BC} for the ZnSTe system. C_{BC} is the fluctuation of the actual potential in the virtual crystal approximation which is different from C in Eq. (8.39).

By summarizing the aforementioned equations, the following expression can be obtained for the pressure-dependent band gap of ternary semiconductor AB_xC_{1-x}:

$$\frac{dE_g(x)}{dP} = \frac{1}{3B}\left\{\left[1 + \left(C/E_{g,h}\right)^2\right]^{\frac{1}{2}}\left[s_1 E_{g,h} - (D_{av} - 1)\Delta E_g\left(y + \frac{2zs_3}{1 + (E_h/C)^2} + s_2\right)\right]\right.$$
$$\left. - \frac{s_1 E_g(x)}{1 + (E_h/C)^2} - 2x(1 - x)c_e r\left(\frac{1}{b_p}\frac{db_p}{dr} - \frac{k_s}{4} - \frac{1}{r}\right)\right\} \tag{8.44}$$

In this expression, the first two terms on the right-hand side stem from the pressure dependence of the band gap in virtual crystal approximation. The last term is the pressure dependence of the extrinsic bowing parameter. B is the bulk modulus of the ternary compound semiconductor.

8.8.3 Results and Discussion

The bowing parameter, energy gap and their pressure coefficients of III–V and II–VI ternary semiconductors, calculated from the above theory, are summarized in Table 8.11 together with other parameters that are relevant to the present calculations.

8.8.3.1 Dependence of Pre-factor b_p on Pressure

In Van Vechten's dielectric theory [121], the pre-factor, b_p, is introduced to balance the overestimate of the Thomas–Fermi effect on dielectric screening at short distances. The dependence of this pre-factor on pressure, i.e., $\frac{r}{b_p}\frac{db_p}{dr}$, was shown to be approximately 2.0–2.5 for materials with ionicity larger than 0.93, while for partial covalent materials ($f_i < 0.93$), this dependence is much weaker. Since for all the ternary compounds considered, in this section, ionicity is less than 0.72, $\frac{1}{b_p}\frac{db_p}{dr} = 0$ is assumed throughout the entire calculations.

8.8.3.2 Bulk Modulus of Ternary Alloys

As has been studied in Sect. 8.4, the bulk modulus of binary and ternary semiconductors can be expressed as

$$B = kr^{-3.48} \tag{8.45}$$

In order to obtain the bulk modulus of ternary semiconductors, Eq. (8.45) is fitted to all the available experimental data of binary compounds in the same group as those of the ternaries. For group III–V ternary semiconductors, it is found that the constant coefficient $k = 1726$ while for group II–VI, $k = 1491$. From the above equation, it is seen that the bulk modulus is inversely proportional to the nearest neighbor distance.

8.8.3.3 Trends in Pressure Coefficients of Energy Gap

For the common-cation system, for example, GaInP, GaInAs, GaInSb (Ga: In::0.5:0.5), the pressure coefficient of the band gap increases with increasing nearest neighbor distance (Columns 13 and 2, respectively, in Table 8.11). In Eq. (8.44), the pressure coefficient of the band gap is inversely proportional to the bulk modulus

Table 8.11 Calculated properties of III–V and II–VI ternary semiconductors at composition 0.5

Alloy	r [121] (A)	D_{av} [122]	C (eV)	c_e (eV)	c_{ABC} (eV)	c_{ABC} (exp) [125] (eV)	f_i	E_g (eV)	E_g (exp) [126] (eV)	$\frac{dc_{ABC}}{dP}$ (meV/kbar)	$\frac{dc_{ABC}}{dP}$ [120] (meV/kbar)	$\frac{dE_g}{dP}$ (meV/kbar)	$\frac{dE_g}{dP}$ (exp) (mev/kbar)
GaPAs	2.404	1.183	3.090	0.186	0.399	0.175–0.21 0.54	0.319	2.217	2.048 2.15 [127]	0.523	0.4	9.005	
GaPSb	2.508	1.223	2.327	1.647	2.768	2.7	0.247	1.244	0.845 [78] 1.06	3.486	2.5	9.736	
GaAsSb	2.552	1.266	2.159	0.802	1.093	1.0–1.2 1.42–1.44	0.235	1.031	0.81 [128] 0.763	2.726	2.5	11.400	12.25 [129]
InPAs	2.586	1.308	3.445	0.119	0.174	0.09–0.38	0.455	1.039	0.819	−0.620	0.3	10.623	
InPSb	2.678	1.345	2.725	1.294	1.814	1.2–2.0	0.384	0.478	0.48 [130] 0.36	3.423	1.2	11.277	
InAsSb	2.712	1.388	2.758	0.774	0.89	0.58–0.7	0.405	0.319	0.12	4.505	−1.0	12.271	
GaInP	2.456	1.203	3.333	0.315	0.737	0.39–0.76	0.378	2.026	1.98 [127] 1.9 2.19 [131]	−2.026	0.3	8.939	8.4 [118] 8.8 [132] 13 [117]
GaInAs	2.534	1.288	3.198	0.307	0.527	0.32–0.6	0.395	1.056	0.813 [133] 0.75	1.165	0.3	10.880	10.95 [134]
GaInSb	2.730	1.366	1.827	0.177	0.283	0.36–0.43	0.235	0.587	0.34 0.43 [135]	1.649	−0.6	13.894	16 [136]
AlGaN	1.964	1.070	10.624	0.097	4.051	0.25-1.78	0.670	8.358	4.48 [137] 3.12 [138]	−1.131		3.607	3.24 [138] 4 [137]
AlGaP	2.361	1.070	3.215	0.009	−0.031	0, 0.49	0.317	4.051	3.34 2.38 [139]	−0.339		5.608	

(continued)

Table 8.11 (continued)

Alloy	r [121] (A)	D_{av} [122]	C (eV)	c_e (eV)	c_{ABC} (eV)	c_{ABC} (exp) [125] (eV)	f_i	E_g (eV)	E_g (exp) [126] (eV)	$\frac{dc_{ABC}}{dP}$ (meV/kbar)	$\frac{dc_{ABC}}{dP}$ [120] (meV/kbar)	$\frac{dE_g}{dP}$ (meV/kbar)	$\frac{dE_g}{dP}$ (exp) (mev/kbar)
AlGaAs	2.441	1.173	2.795	0.014	0.0126	−0.127–1.183	0.293	2.337	2.94 2.158 [140]	0.654		8.887	9.15 [141] 10.85 [142]
ZnSSe	2.398	1.153	5.885	0.349	0.506	0.456–0.68	0.627	3.743	3.08	3.118	0.8	6.180	
ZnSTe	2.493	1.169	4.905	2.476	3.144	2.4–3 [143] 3.75 [126]	0.586	2.626	2.36 [144] 2.061	7.387	6.6	5.633	
ZnSeTe	2.549	1.191	4.623	1.041	1.189	1.23–1.7 [143]	0.584	2.606	2.3 [145] 2.12	4.658	1.6	6.920	7.6 [146]
CdSSe	2.590	1.247	6.416	0.291	0.399	0.53 [143]	0.745	2.670	2.25 [147] 1.95	−0.997	1.6	5.576	4–6 [148]
CdSTe	2.674	1.265	5.511	2.345	2.696	1.73–1.84 [143]	0.716	1.810	1.58	6.212	2.3	4.751	6.2 [149]
CdSeTe	2.722	1.287	5.159	1.001	1.04	0.755 [143] 0.87 [143]	0.708	1.864	1.047 1.48 [150]	2.453	0.3	6.221	
CdZnS	2.442	1.179	6.524	0.170	0.491	0.3, 0.6 [143] 0.83 [126]	0.693	3.611	2.89	−0.809		5.792	5.3 [151]
CdZnSe	2.546	1.223	5.801	0.136	0.36	0.3,0.35 [143] 0.387 [126]	0.687	2.775	2.1	0.584		6.579	
CdZnTe	2.725	1.255	4.086	0.090	0.197	0.153–0.463 [143]	0.604	2.113	1.85 [152] 1.7	−0.646		8.356	

r is the nearest neighbor distance. D_{av} is the d-state effect parameter. C is the heteroplar energy gap. c_e is the external bowing parameter. f_i is the ionicity. c_{ABC} and $\frac{dc_{ABC}}{dP}$ is the bowing parameter and its pressure coefficient. E_g and $\frac{dE_g}{dP}$ is the energy gap and its pressure coefficient which, according to Eq. (8.45), is inversely proportional to the nearest neighbor distance. Therefore, the pressure coefficient of the band gap will increase with increasing nearest neighbor distance due to the decrease in bulk modulus. In general, pressure will cause a dilation of the lattice and will lead to changes in its potential energy resulting in overlap of the energy levels which will subsequently lead to change in the energy gap. However, for the common-anion system, this trend is not so significant. For instance, the pressure coefficient increases in group III–V in the following order of common group V elements: (GaPAs, InPAs), (GaPSb, InPSb), (GaAsSb, InAsSb) while it decreases in group II–VI in the following order of common group VI elements: (ZnSSe, CdSSe), (ZnSTe, CdSTe), (ZnSeTe,CdSeTe). As discussed earlier, the decrease in bulk modulus will result in an increase in the pressure coefficient of the energy gap. In the studies on the predicted pressure coefficient of the energy gap, Wei and Zunger [141] have found that the *s-s* and *p-p* coupling will enhance while *p-d* coupling will reduce the pressure coefficients of the energy gap. Thus, it may be concluded that the trend in group III–V common-anion system is because the effect of *s-s*, *p-p* coupling, and bulk modulus is stronger than the effect of *p-d* coupling, and vice versa for group II–VI common-anion system.

Another trend is that the pressure coefficient of the band gap decreases with increasing ionicity (Columns 13 and 8, respectively, in Table 8.11). In order to verify this correlation, comparison is made amongst compounds with similar bulk modulus due to similar nearest neighbor distance (Column 2 in Table 8.11), for example, GaPAs (2.404 Å) with ZnSSe (2.398 Å). For GaPAs and ZnSSe, the corresponding ionicities are 0.319 and 0.586 and pressure coefficients are 9.005 and 6.18 meV/kbar, respectively. Similar trends are also found in other comparisons. Combining these results with the above analysis, this trend also indicates that coupling effects could be reflected from ionicity.

Exceptions to the trend in the variation of the pressure coefficient of the energy gap with ionicity and nearest neighbor distance appear in zinc and cadmium chalcogenide common-anion systems. This is due to the large bowing parameter (Column 6 in Table 8.11) and its pressure coefficient (Column 11 in Table 8.11) of ZnSTe and CdSTe. The bowing parameter pressure coefficient is 7.387 meV/kbar in ZnSTe compared with 3.118 and 4.658 meV/kbar in ZnSSe and ZnSeTe systems, respectively. Similarly, the bowing parameter pressure coefficient for CdSTe is 6.212 meV/kbar compared with -0.997 meV/kbar for CdSSe and 2.453 meV/kbar for CdSeTe. These exceptions reflect the importance of bowing parameters in determining the electronic properties of ternary compounds and the invalidity of the well-accepted linear interpolation rule which can be used to obtain the physical properties of ternary compounds from the linear interpolation of two binary compounds.

8.8.4 Comparison with Experiments

The agreement between the calculated results and the experimental data are generally good. All the calculations, presented in this section, have been performed for composition $x = 0.5$. However, some experimental values are only available for other compositions. These experimental values are listed and compared with the calculated results for the corresponding compositions in Table 8.12. For example, Zhao et al. [155] found that the pressure coefficient of band gap for $Cd_{0.73}Zn_{0.27}Se$ is 3.54 meV/kbar and the calculated value at this composition is about 6.586 meV/kbar. This difference may arise from the wurtzite structure of their experimental sample while all the calculations assume zincblende structures. The mechanism of the real difference of structure in determining the pressure coefficients is not yet theoretically well understood. However, the available data show that the pressure coefficient of wurtzite structure is generally less than that of zincblende structure. For example, in this case, experimental pressure coefficient of wurtzite ZnSe is around 4.5 meV/kbar [161] while for zincblende ZnSe, the available data is 7.0–7.5 meV/kbar [153, 154] and the calculated result is 7.564 meV/kbar. The pressure coefficient of wurtzite CdSe is around 4.3 meV/kbar [162] and for zincblende structure is around 5.8 meV/kbar [141, 153] and the calculation shows 5.886 meV/kbar. Based on this analysis, the results of the experiment by Zhao et al. may not be very accurate since their band gap pressure coefficient of wurtzite CdSe is 2.84 ± 0.6 meV/kbar, much smaller than the generally accepted results. From the perspective of the above-analyzed trends with respect to nearest neighbor distance and ionicity in the order: CdZnS, CdZnSe, CdZnTe, the result is also more reasonable. For ternary compound $ZnS_{0.3}Te_{0.7}$, Fang et al. [158] found the band gap pressure coefficient to be about 6.2 meV/kbar. From their results, the pressure coefficient is almost invariable with respect to composition. This study calculates the system at $x = 0.3$ and finds that the value is 5.732 meV/kbar which is very close to their data within experimental uncertainty. Moreover, the band gap pressure coefficient for $GaAs_{0.88}Sb_{0.12}$ is reported by Prins et al. [159] as 9.5 meV/kbar and the calculated result shows value of 11.516 meV/kbar, close to GaAs.

In Table 8.12, the calculated AlGaN pressure coefficient 3.607 meV/kbar is very close to the reported experimental value of 3.24 meV/kbar and 4 meV/kbar. However, the calculated binary AlN coefficient of 1.423 meV/kbar is much smaller than that of Wei's [141] results of 4.7 meV/kbar based on first-principles calculations. Similarly, the value for AlP is 3.385 meV/kbar compared with their 11.1 meV/kbar and for AlAs, this calculation yields 7.49 meV/kbar while the available experimentally obtained [153] value is 10.2 meV/kbar. The reason for this discrepancy has not yet been found.

A close investigation of the calculated energy gaps (Column 10 in Table 8.11) shows that they are, in general, larger than the experimental values and this discrepancy is even larger for Group II–VI than Group III–V compounds. One possible origin for this result is the expressions of $E_{g,h}$ and ΔE_g in Eq. (8.37) are obtained by fitting to the experimental data of nonionic group IV materials. Similar as in the trend in the pressure coefficients, the data shows that the energy gap, in general, increases with decreasing nearest neighbor distance and increasing ionicity. Since the ionicity increases from group IV to III–V to II–VI, the calculated band gap values of these III–V and II–VI ternary semiconductors will be enhanced and are larger than the experimental values. The ionicity of group III–V ternary compounds is in the range of 0.23–0.45, and the difference between the calculated results and experimental

Table 8.12 Comparison between calculated pressure coefficients of band gap for some ternary compounds for compositions $\neq 0.5$ and Al-compounds

	x	$\frac{dE_g}{dP}$ (meV/kbar)	$\frac{dE_g}{dP}$ (exp) [153] (meV/kbar)
$Cd_xZn_{1-x}Se$	0	7.564	7.2–7.5, 7.0 [154]
	1	5.886	5.8, 5.5 [141]
	0.73	6.586	3.54 [155]
ZnS_xTe_{1-x}	0	8.606	10.5 [156], 11.5
	1	6.355	5.8, 6.4 [154], 6.7 [157]
	0.3	5.732	6.2 [158]
$GaAs_xSb_{1-x}$	0	13.552	14.0
	1	10.61	8.5–12.6
	0.88	11.516	9.5 [159]
$Al_xGa_{1-x}N$	0	5.224	3.6 [141], 4.0 [160]
	1	1.423	4.7 [141]
	0.5	3.607	3.24 [138], 4 [137]
$Al_xGa_{1-x}P$	0	7.662	9.7
	1	3.385	11.1 [141]
	0.5	5.608	
$Al_xGa_{1-x}As$	0	10.61	8.5–12.6
	1	7.49	10.2
	0.5	8.887	9.15 [141], 10.85 [142]

data is approximately in the range of 0–0.25 eV. The ionicity of group II–VI ternary compounds is around 0.58–0.72, and correspondingly, the energy gap discrepancy is about 0.3–0.8 eV.

The model proposed by Hill and coworkers has been discussed in the first section. Their calculated results for bowing parameter pressure coefficients are listed in Table 8.11 (Column 12) and comparisons show that their results are generally much smaller than those in the present work. This may be because they take the screening wave number in Eq. (8.34) as a constant 0.25 which may not affect the accuracy of calculating the bowing parameter but will certainly affect the accuracy of the pressure coefficient of the bowing parameter. It is also noted that they take a set of approximations in their calculations which may also result in the difference. The reason that they could fit their results to GaInP system is because the pressure coefficient of the bowing parameter in this system, as described in Eq. (8.36), is much smaller than its binary band gap coefficients which the authors took from experiments.

8.8.5 Temperature Coefficients of Ternary Alloys

Methods that are similar to this theory cannot be applied to temperature coefficients of ternary compounds by relating them to thermal expansion coefficients. This is because the temperature coefficient could be expressed as two terms: the effect of volume expansion which could be similarly derived from this theory and explicit temperature coefficient at constant volume which has to be calculated by other means. Yu and Cardona [163] have shown that the first term only contributes less than 20% of the total temperature coefficients.

8.8.6 Summary

The pressure dependence of the energy gap of a series of group III–V and II–VI ternary semiconductor compounds have been calculated in terms of a generalized expression of Van Vechten's dielectric theory. Good agreement is obtained between the calculated values and the available experimental data and other calculated results. The calculation shows the following: (I) The pressure coefficient of the energy gap increases with increasing nearest neighbor distance in common-cation systems; (II) The pressure coefficient of the energy gap decreases with increasing ionicity; (III) The energy gap increases with decreasing nearest neighbor distance and increasing ionicity; (IV) The theory shows certain discrepancy in calculating the energy gap due to its built-in assumptions.

8.9 Electronic and Optical Properties of I_2–II–IV–VI_4 Quaternary Semiconductors

8.9.1 Theoretical Background

One of the most important applications of multi-ternary semiconductor alloys is in the fabrication of solar cells. The history of solar cell development is the history of searching for materials that can be utilized to produce better solar cells that can absorb most of the solar radiation. These materials include Si, GaAs, CdTe, and $CuIn_{1-x}Ga\ Se_2$ (CIGS). CIGS is thought to be the only substitute for Si solar cell due to its high efficiency. However, CIGS contains expensive element Indium (In) and also its window layer CdS contains the toxic element Cadmium (Cd).

Recently, the I_2–II–IV–VI_4 series of quaternary chalcogenide semiconductors have gained broad interest in their potential applications as photovoltaic absorbers [164–168], optoelectronic and thermoelectric materials [169–171]. For instance, Cu_2ZnSnS_4-based thin film solar cells have attained conversion efficiencies in excess of 6.7% [165]. A non-vacuum, slurry-based coating method and particle-based deposition have enabled the fabrication of $Cu_2ZnSn(S, Se)_4$ devices with over 9.6% efficiency [171]. Compared to the conventional CIGS absorbers, Cu_2ZnSnS_4 and $Cu_2ZnSnSe_4$ compounds only contain abundant, inexpensive and nontoxic elements and their band gaps are close to 1.5 eV, which is ideal for solar cell applications.

These wide applications have led to increased interest in studying many other members in the I_2–II–IV–VI_4 family, such as the Ge-compounds: Cu_2ZnGeS_4, $Cu_2ZnGeSe_4$ and $Cu_2ZnGeTe_4$ In experiments, the physical properties [25, 172–175] such as crystal orderings and lattice constants of these compounds have been studied using x-ray diffraction method. Transmission [176] and absorption measurements [177] have been performed to study the band gaps. Recently, the optical constants of Cu_2ZnGeS_4 have been reported by ellipsometry measurements [178]. In theoretical work, the structural and electronic properties of some compounds have been studied using first-principles calculations [179, 180]. However, the optical properties of these compounds have not yet been systematically addressed.

In this section, the electronic and optical properties of Cu_2ZnGeX_4 (X=S, Se and Te) quaternary compounds are investigated through first-principles calculations. The electronic structures and density of states will be first calculated because it is known that the structures in the optical spectra are directly related to the band structure of the material itself. Then, the optical properties will be presented, including the dielectric function, refractive index, optical reflectivity, and absorption spectra. The trends in the variation of the electronic and optical properties with the crystal structure and the group VI anion atomic number are explored qualitatively.

8.9.2 Crystal Structures

In Fig. 8.15, the kesterite (KS) and stannite (ST) structures [181] are presented for Cu_2ZnGeX_4. KS structure ($I\bar{4}$, No. 82, Fig. 8.15a) has its conventional unit cell with four Cu atoms on the Wyckoff position $2a$ and $2c$, two Zn atoms on position $2d$, two Ge atoms on position $2b$, and eight X atoms on the $8g$ position. The cation positions have all S_4 point group symmetry, and X has C_1 symmetry. ST structure ($I\bar{4}2m$, No. 121, Fig. 8.15b) has the equivalent Cu atoms on Wyckoff $4d$ position, two Zn atoms on $2a$, two Ge atoms on $2b$ and eight X atoms on the $8i$ position. The Cu atoms have S_4 point group symmetry. The Zn and Ge atoms have D_{2d} point group symmetry. Eight X atoms have C_s point group symmetry.

It is to be noted that many of the experimental reports [25, 172–175] on these compounds claim that the synthesized samples have ST structure. For example, Parasyuk and coworkers [174, 175] synthesized Cu_2ZnGeS_4, $Cu_2ZnGeSe_4$, and $Cu_2ZnGeTe_4$ samples from high-purity elements and determined their crystal constants by X-ray diffraction. They report that all the three samples crystallize in the ST structure. It is found that the calculated c parameters (Table 8.13) are in general larger than their reported values. This discrepancy between the experiments and the calculations is caused by the fact that the similarity of Cu and Zn atoms lead to the cation disorder in their experimental structures. Because the atomic numbers of Cu, Zn and Ge are

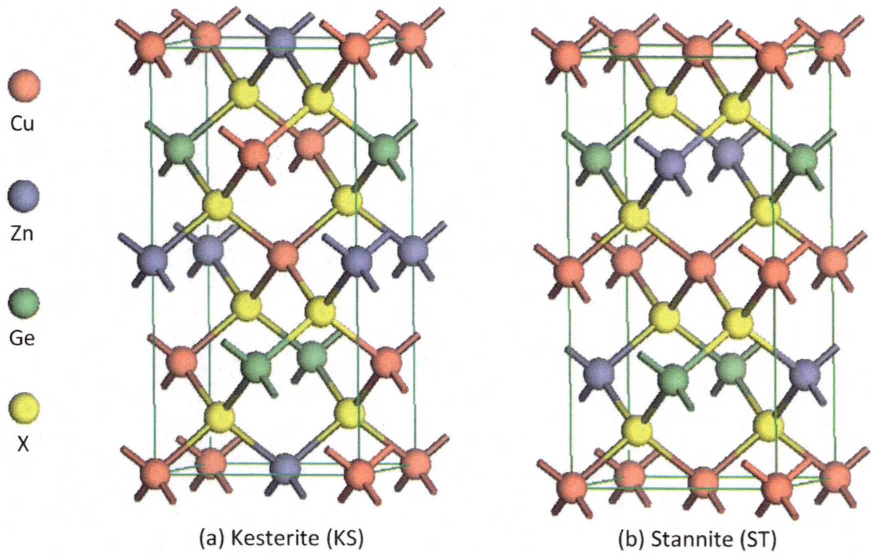

Fig. 8.15 The crystal Structures of **a** kesterite (KS) and **b** stannite (ST) Cu_2ZnGeX_4

close in the periodic table, experimentally, it is very difficult to detect the cation disorder by X-ray diffraction. In fact, the partial cation disorder has been observed in Cu_2ZnSnS_4 sample instead by a recent neutron-diffraction measurement [167].

8.9.3 Electronic Properties

The calculated properties of Cu_2ZnGeS_4, $Cu_2ZnGeSe_4$, and $Cu_2ZnGeTe_4$ of KS and ST structures are listed in Table 8.13. Together listed are the available experimental values for comparison.

8.9.3.1 Band Structures

The band structures for the three compounds in KS structure are shown along the T(Z) $\rightarrow$ Γ $\rightarrow$ N(A) lines in Fig. 8.16. All the band gap values are given in Table 8.13. It is found that the band structures of all the compounds are rather comparable. The lowest conduction band (CB) is a sole band at about 1–3 eV. This is characteristic of the I_2–II–IV–VI_4 family. It is also different from that of chalcopyrite CIGS and CIGSe compounds which have overlapping conduction bands [181]. The calculations indicate that the band gap decreases with the increasing anion atomic numbers [53]. For example, the band gap is 2.27 eV for KS-Cu_2ZnGeS_4 compared to 1.50 eV for KS–$Cu_2ZnGeSe_4$ and 0.81 eV for KS–$Cu_2ZnGeTe_4$. This is because the valence

Table 8.13 Calculated properties of Cu_2ZnGeX_4 (X=S, Se and Te)

	Cu_2ZnGeS_4			$Cu_2ZnGeSe_4$			$Cu_2ZnGeTe_4$		
	KS	ST	EXP.	KS	ST	EXP.	KS	ST	EXP.
a (Å)	5.264	5.328	5.270, 5.342	5.602	5.583	5.606, 5.610	6.102	6.094	5.954, 5.999
c (Å)	10.843	10.741	10.509,10.516, 10.540	11.259	11.325	11.04	12.126	12.220	11.848, 11.918
E_g (eV)	2.27	2.06	2.15, 2.28, 2.04	1.50	1.32	1.52, 1.29	0.81	0.55	–
E_{1A} (eV)	2.832	2.588	2.85, 2.87, 2.88	2.258	1.838	–	1.139	0.659	–
E_{1B} (eV)	3.946	3.751	4.03, 4.28, 4.34	3.483	3.171	–	2.214	1.793	–
n_0	2.61	2.63	–	2.90	3.00	–	3.73	4.23	–
ε_0	6.84	6.89	–	8.41	9.01	–	13.89	17.93	–
ε_∞	0.49	0.50	0.47, 0.49, 0.76	0.52	0.53	–	0.50	0.54	–

Calculated lattice constant a and c, band gap E_g and critical point threshold energy E_{1A} and E_{1B}, static optical constants. Experimental data (EXP) are listed for comparison, whereas "–" means no experimental data are currently available

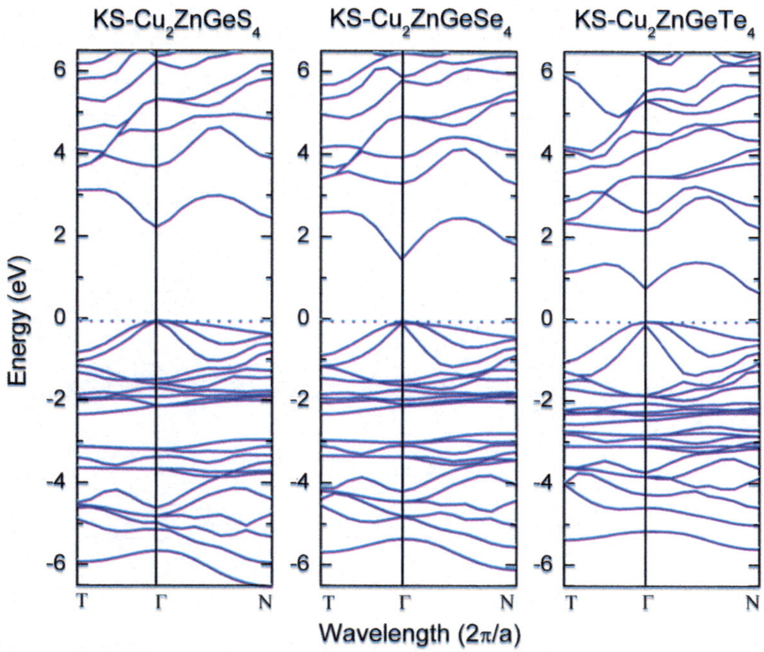

Fig. 8.16 Calculated band structures along the high symmetry lines in the first Brillouin zone: T(Z): $2\pi/a$ (0, 0, 0.5) $\rightarrow$ Γ : $2\pi/a$ (0, 0, 0) $\rightarrow$ N(A): $2\pi/a$ (0.5, 0.5, 0.5), for Cu_2ZnGeS_4, $Cu_2ZnGeSe_4$ and $Cu_2ZnGeTe_4$ in KS structure

band maximum (VBM) is composed of hybridized Cu 3*d* and group VI *p* states. The shallower atomic level of heavy anion atom results in higher VBM states, and therefore smaller band gaps. Comparison of the band structures between KS and ST structures shows that band gaps of the KS structure are in general larger than those of the ST structure. This is due to the fact that the KS structure has larger anion displacements. For example, the anion displacement in $Cu_2ZnGeSe_4$ system is 0.2542 for KS structure compared to 0.2479 for ST structure [180]. The band gaps of these compounds range from 0.55 to 2.27 eV, covering a wide range of the solar spectra. In particular, $Cu_2ZnSnSe_4$, with band gap values of 1.5 eV for KS and 1.32 eV for ST structure, is a potential candidate for photovoltaic applications.

8.9.3.2 Density of States

The density of states (DOS) of Cu_2ZnGeS_4 in KS and ST structures, $Cu_2ZnGeSe_4$ and $Cu_2ZnGeTe_4$ in KS structure are shown in Fig. 8.17. From the calculated DOS, it can be seen that the DOS of the KS and ST structures are quite similar. The upper VB DOS contains mainly the hybridization between *p* states from the anion atoms and 3*d* state from Cu atoms while the lower CB DOS consists mainly of the hybridization

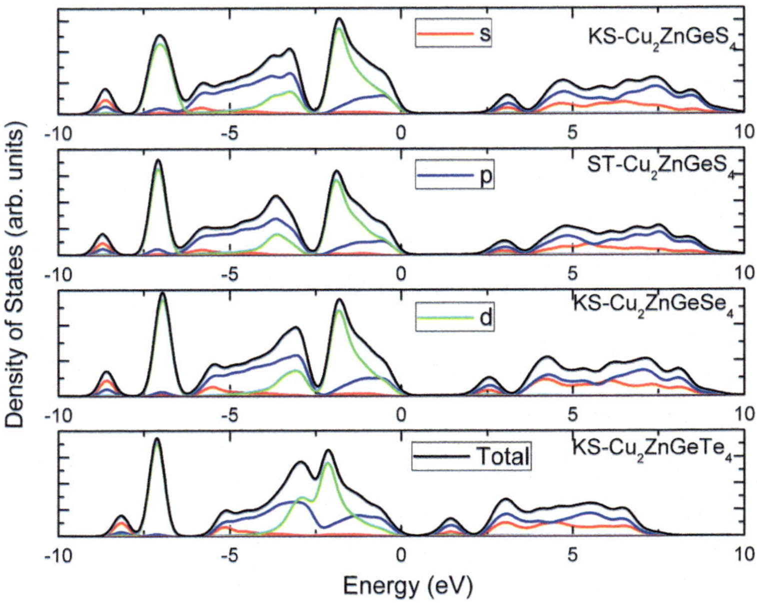

Fig. 8.17 The partial and total DOS of Cu_2ZnGeS_4 in KS and ST structures, $Cu_2ZnGeSe_4$ and $Cu_2ZnGeTe_4$ in KS structure

between cation *s* states and anion *p* states. Comparing the DOS between different structures and compounds, it is found that: (I) the valence band width of KS structure is slightly narrower than that of the ST counterpart. This is because the KS structure has longer Cu-VI (VI=S, Se and Te) bonds and hence larger anion displacement than the corresponding ST structure. Therefore, the hybridization between Cu 3*d* state and anion *p* state is weaker in the KS structure, leading to a narrower band width; (II) Analyzing the band structure in Fig. 8.16 together with the DOS in Fig. 8.17 shows that the lowest solo conduction band is derived from the Ge 4 *s* and anion *p* states. The conduction band shifts to the lower energy when the anion atomic number increases from S to Te; (III) In the valence band region, from −6 to −2.5 eV, there are bonding states consisting of anion *p* states hybridized with Cu 3*d* state. From −2.5 to 0 eV, there are antibonding states consisting of anion *p* states hybridized with Cu 3*d* state. The overlapping (at around −2.5 eV) between the *p-d* bonding and antibonding states increases when the group VI anion atomic number increases from S to Te.

8.9.4 Optical Properties

8.9.4.1 Dielectric Functions and Interband Transitions

The dielectric function $\varepsilon(\omega) = \varepsilon_1(\omega) + i\varepsilon_2(\omega)$ of all the three compounds in KS and ST structures is presented in Fig. 8.18. Overall, the three compounds show similar dielectric functions over a broad range of energy. The main difference is that the spectrum shifts to lower energy region when the anion atomic number increases. In the lower energy region, the spectrum of $Cu_2ZnGeTe_4$ compound is above the other two materials while the spectrum of Cu_2ZnGeS_4 compound is above the others in the higher energy region. The spectra exhibit some critical point (CP) structures E_{1A}, E_{1B} labeled in Fig. 8.18 and listed in Table 8.13. The E_{1A} and E_{1B} energy thresholds can be attributed to transitions at the high CPs N(A) and T(Z) of the first Brillouin zone. According to the band structures in Fig. 8.16, it is found that E_{1A} and E_{1B} are 2.83 and 3.95 eV, respectively, for KS-Cu_2ZnGeS_4 while an ellipsometry measurement [178] shows 2.85–2.88 eV for E_{1A} and 4.03–4.34 eV for E_{1B}, as listed in Table 8.13. The calculated results are in good agreement with the experimental values. It is also interesting to analyze the shift in the spectra as a function of the anion atomic number in the three compounds. As has been stated, the conduction band is derived from the hybridization of the Ge 4 s and anion p states. When the anion atomic number increases (e.g., S → Se → Te), the Ge-VI hybridization becomes higher which shifts downward the CBM, and hence the spectrum moves toward the lower energy regime.

8.9.4.2 Refractive Index

The optical complex refractive index, $\tilde{n} = n + ik$ that is of interest for the design of optoelectronic devices, can be computed from dielectric functions [182]. Figure 8.19 presents the energy-dependent n and k values of all the three compounds in KS and ST phases. In experiment, n and k of Cu_2ZnGeS_4 have been reported in the energy range from 1.4 to 4.7 eV [178]. The calculated results are in good accord with the experimental values. For example, the calculated n at energy 1.4 eV is 2.73 compared to the corresponding 2.65 from experiment. The peak values of n from experiment in the energy range of 2.4–2.9 eV is 3.02 compared to the calculated 3.05–3.10 in Fig. 8.19. It is found that the static refractive index (n_0 in Table 8.13) increases from sulfide to telluride compound and increases from KS to ST structure.

8.9.4.3 Absorption and Reflectivity

In Fig. 8.20, the calculated results of the absorption coefficient α and the normal incident reflectivity R for all the cases are presented. They represent the linear optical response from the VBs to the lowest CBs. Due to the fact that the absorption

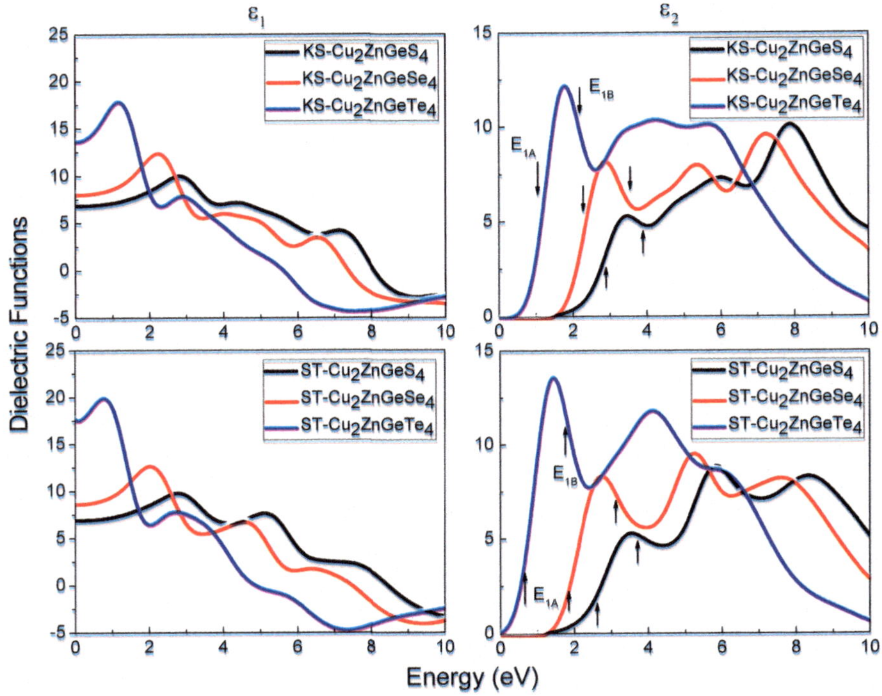

Fig. 8.18 The dielectric function $\varepsilon(\omega) = \varepsilon_1(\omega) + i\varepsilon_2(\omega)$ of Cu_2ZnGeS_4, $Cu_2ZnGeSe_4$ and $Cu_2ZnGeTe_4$ in KS and ST structures. The left panels represent the real part $\varepsilon_1(\omega)$ and the right panels represent the imaginary part $\varepsilon_2(\omega)$. The optical transitions E_{1A} and E_{1B} are labeled in the $\varepsilon_2(\omega)$ spectra

coefficient and reflectivity are obtained from the dielectric function [182], all the compounds in this study have similar absorption spectra, although with different energy for the onset to absorption (i.e., the band gap energy). It is found that the Cu_2ZnGeS_4 compound has large band-edge absorption coefficient (about 5×10^4 cm^{-1}). At a given photon energy, the $Cu_2ZnGeTe_4$ compound has the largest absorption coefficient while the Cu_2ZnGeS_4 compound has the smallest value. Comparison with the calculated spectra of other materials [181, 183] shows that the absorption coefficient of Cu_2ZnGeX_4 is smaller than that of Cu_2ZnSnX_4 and Cu_2ZnTiX_4 has the largest absorption coefficient. This might indicate that the compounds with heavier group IV elements should have higher light transformation efficiency. It is noticed that, in the energy range of 1.5–4.0 eV, the reflectivity and absorption coefficient decreases for all the compounds. This energy region corresponds to the gap between the lowest solo CB and the upper CBs of the band structures in Fig. 8.16, since the upper CBs do not contribute to the optical absorption in the low energy regime. This conduction band gap is a disadvantage for the band-edge absorption efficiency in KS and ST structures.

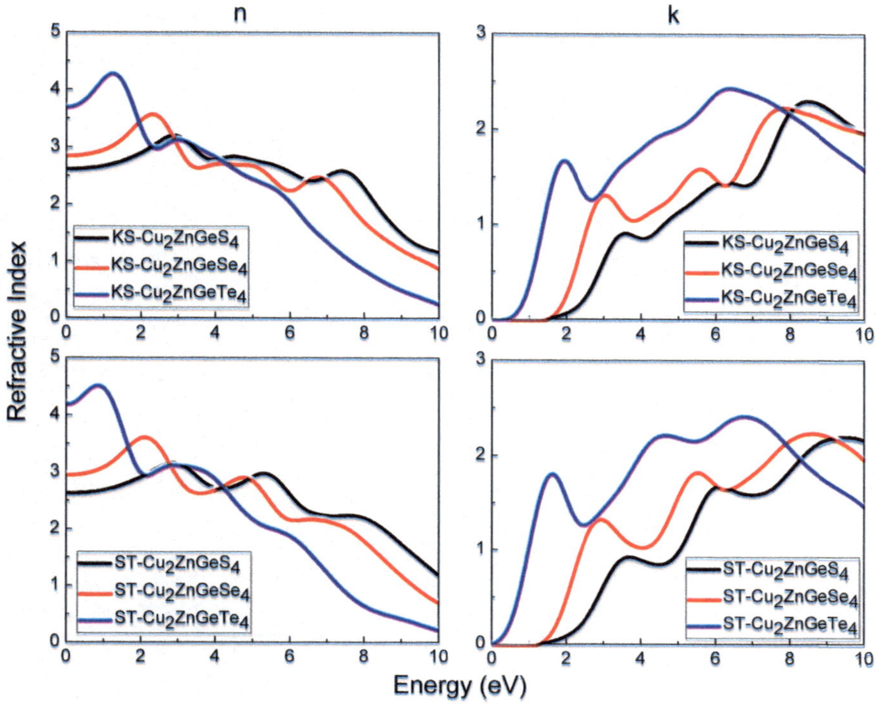

Fig. 8.19 The complex refractive index of Cu_2ZnGeS_4, $Cu_2ZnGeSe_4$ and $Cu_2ZnGeTe_4$ in KS and ST structures. The left panels represent the refractive index *n* and the right panels represent the extinction coefficient *k*

8.9.5 Summary

In this section, the electronic and optical properties of Cu_2ZnGeS_4, $Cu_2ZnGeSe_4$, and $Cu_2ZnGeTe_4$ in KS and ST structures are studied. Band structures and optical spectra such as the dielectric function, refractive index, absorption coefficient and reflectivity have been determined. It is found that the conduction band shifts downward and the overlapping between *p-d* bonding and antibonding states in the valence band increases when the system changes from Cu_2ZnGeS_4 to $Cu_2ZnGeSe_4$ and then $Cu_2ZnGeTe_4$. Some critical points in the optical spectra are assigned to the interband transitions in accordance with the calculated band structures. The electronic structures and optical spectra are rather similar in shape for all the compounds. When anion atomic number increases from S to Te, the optical spectra shift to the low energy regime. A good agreement between the calculated results and the experimental data has been obtained.

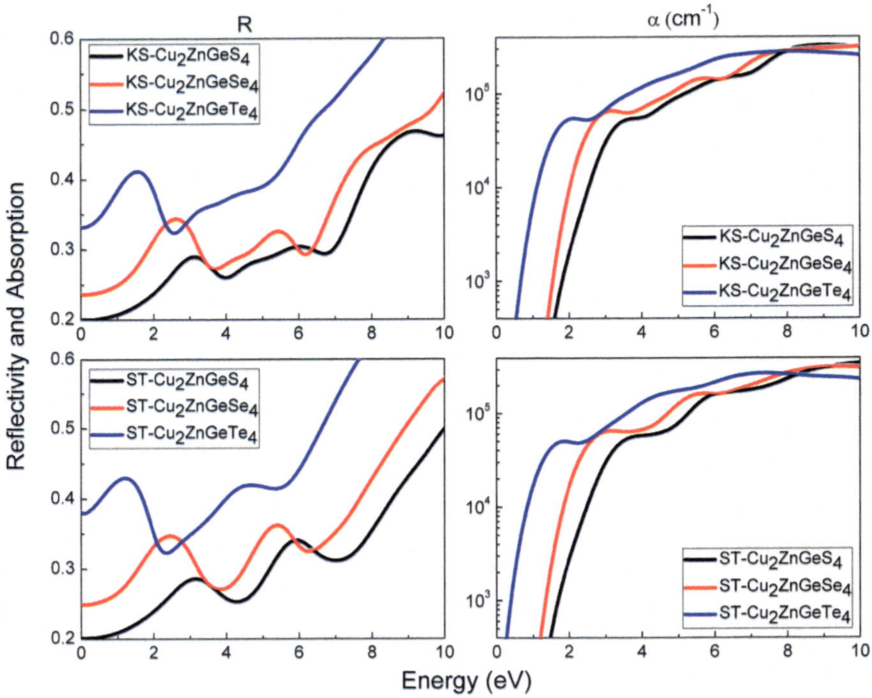

Fig. 8.20 The normal incident reflectivity and absorption coefficient of Cu_2ZnGeS_4, $Cu_2ZnGeSe_4$ and $Cu_2ZnGeTe_4$ in KS and ST structures. The left panels represent the reflectivity R and the right panels represent the absorption coefficient $\alpha(\mathrm{cm}^{-1})$. The absorption coefficient is plotted in logarithm scale

8.10 Electronic and Optical Properties of Wurtzite-Derived Semiconductors—Cu_2ZnSiS_4 and $Cu_2ZnSiSe_4$

8.10.1 Abstract

The electronic and optical properties of Cu_2ZnSiS_4 and $Cu_2ZnSiSe_4$ in wurtzite-kesterite and wurtzite-stannite structures are studied using first-principles calculations. Optical transitions at high symmetry points between valence bands and conduction bands are discussed in this section. Band structures are presented to analyze the related properties. Optical constants and dielectric function spectra are investigated.

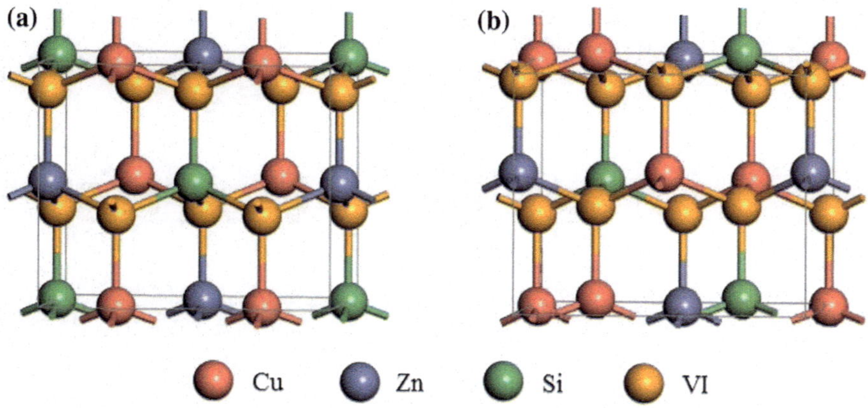

Fig. 8.21 Crystal structure of $Cu_2ZnSiVI_4$ (VI=S, Se) in **a** WKS, and **b** WST structures

8.10.2 Background

In recent years, the Cu_2–II–IV–VI_4 series of quaternary chalcogenide semiconductors have drawn significant attention for their potential applications in several areas [184 and references therein]. For example, Cu_2ZnSnS_4 and $Cu_2ZnSnSe_4$ are excellent candidates as solar cell absorbers since the elements are inexpensive and are environmental-friendly. $Cu_2ZnSn(S, Se)_4$ based solar cells, with over 9.6% efficiency, have been fabricated using a non-vacuum, slurry-based coating method and particle-based deposition. Cu_2CdSnS_4 and $Cu_2ZnSiSe_4$ have been demonstrated to be promising nonlinear optical materials for use in the infrared region. Lithium-based chalcogenides, $Li_2CdGeSe_4$ and Li_2CdSnS_4, have been shown to exhibit strong second harmonic generation (SHG) responses of 70X and 100X α-quartz, respectively.

8.10.3 Crystal Structure

Experimentally, Cu_2ZnSiS_4 and $Cu_2ZnSiSe_4$ are found to exhibit orthorhombic(wurtzite-derived) structure. In Fig. 8.21, two wurtzite-derived crystal structures are presented for Cu_2ZnSiS_4 and $Cu_2ZnSiSe_4$. Wurtzite-kesterite (WKS) structure, as shown in Fig. 8.21a has space group *Pc* (Monoclinic, no. 7) and Wurtzite-stannite(WST) structure, as shown in Fig. 8.21b has space group $Pmn2_1$ (Orthorhombic, no. 31) [184 and references therein]. The lattice constants of these compounds in both structures are summarized in Table 8.14 along with the experimental values. Comparison shows that WKS and WST structures have similar crystal volumes and calculated results are in accord with experimental values.

In order to compare the stability between different structures, the total energies for both Cu_2ZnSiS_4 and $Cu_2ZnSiSe_4$ have been calculated. As can be noted from

Table 8.14 Calculated properties of Cu_2ZnSiS_4 and $Cu_2ZnSiSe_4$ in WKS and WST structures—compared with the available experimental values

	Cu_2ZnSiS_4			$Cu_2ZnSiSe_4$		
	WKS	WST	EXP.	WKS	WST	EXP.
a (Å)	7.430	7.376	7.436[a], 7.44[b], 7.435[c]	7.760	7.763	7.826[a], 7.83[b], 7.823[c]
b (Å)	6.421	6.458	6.398[a], 6.39[b], 6.396[c]	6.794	6.773	6.727[a], 6.73[b], 6.72[c]
c (Å)	6.157	6.161	6.137[a], 6.13[b], 6.135[c]	6.455	6.462	6.455[a], 6.44[b]
ΔE (meV/atom)	0	1.54	–	0	4.21	–
E_g (eV)	1.68 (3.09)	1.37 (2.71)	3.04[a], 3.25[b], 2.94[d]	0.84 (1.99)	0.56 (1.69)	2.20[a], 2.33[b], 2.42[e], 2.08[d]
n	2.36	2.40	–	2.61	2.64	–
ε_0	5.56	5.58	–	6.83	6.94	7.04[e]
ε_∞	0.51	0.51	–	0.54	0.54	0.43[e]

a, b, c are lattice constants. ΔE is the total energy per atom relative to WKS structure. E_g is the band gap with corrected values listed in the parenthesis. n and ε_0, ε_∞ are refractive index and dielectric constants*
*[184 and references therein]

Table 8.14, WST always has higher energy than WKS structure which indicates that WKS is more stable than WST structure. The energy difference between these two structures is small, i.e., only 1.54 meV/atom for Cu_2ZnSiS_4. This suggests a high possibility of mixing structures in this material. The energy difference between these two structures is larger in Se compound than in S compound. This can be attributed to the fact that Se compound has large lattice mismatch and therefore large strain energy. Experiments also report that Cu_2ZnSiS_4 is much more stable than $Cu_2ZnSiSe_4$.

8.10.4 Band Structures

The band structures of Cu_2ZnSiS_4 in WKS and WST structures are plotted in Fig. 8.22. It may be noted that both WKS and WST structures have direct band gaps. Figure 8.22 also shows that the band structures have a clear gap in the valence band region. The bands above the gap, i.e., from −2 to 0 eV, are occupied by antibonding states consisting of anion *p* states hybridized with Cu *d* states while the bands below the gap are made by bonding states consisting of anion *p* states hybridized with Cu *d*

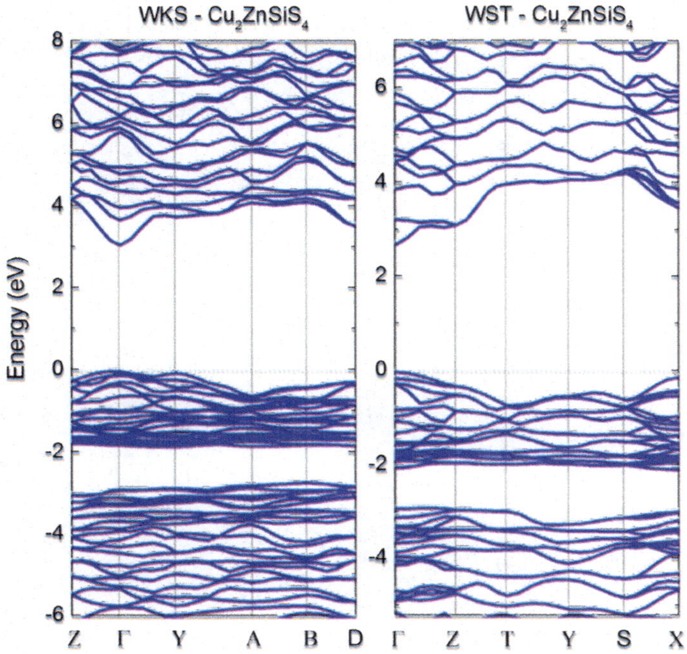

Fig. 8.22 Band structures of Cu_2ZnSiS_4 in WKS and WST structures

states. The gap indicates the overlapping between the *p-d* bonding and antibonding states. It is also observable that the valence band width above the gap is broader in WST structure than in WKS structure. This is due to the fact that WST has shorter Cu-anion bonds and hence smaller anion displacement than WKS structure. Thus, the hybridization between the Cu *d* state and anion *p* state is stronger in the WST structure. The band structures are found to be similar for S and Se compound since they are similar in chemical compositions and crystal structures.

The calculated band gaps are summarized in Table 8.14. The band gap values show some trends for different compounds and structures. For example, for a given structure, S compound has larger band gap than Se compound. This is because the top valence band is composed of hybridized Cu *d* states and anion *p* states. The shallower atomic level of heavy anion atom results in higher top valence band states and therefore smaller band gaps. It is also observed that, for a given material, WKS structure has larger band gap than that of WST structure. This is because WKS structure has longer Cu-anion bonds and hence larger anion displacement.

In order to understand the band engineering of these two compounds, the exciton transitions, along two base *X*, *Y* vectors in reciprocal space as well as the center Γ point of the Brillouin zone, are presented in Table 8.15. The available experimental data is listed for comparison. The experimental data were obtained by comparing the optical spectrum with simplified band structure diagram.

Table 8.15 Optical transitions between valence bands and conduction bands at high symmetry points for Cu_2ZnSiS_4 and $Cu_2ZnSiSe_4$ in WKS and WST structures along with the available experimental data*

Transitions	Cu_2ZnSiS_4			$Cu_2ZnSiSe_4$		
	WKS	WST	EXP.[a]	WKS	WST	EXP.[d]
$\Gamma: v_1 - c_1$	3.090	2.707	3.345, 3.323, 2.97–3.08[b]	1.99	1.687	2.326, 2.348[a]
$\Gamma: v_2 - c_1$	3.188	2.844	3.432, 3.413, 3.32–3.40[b]	2.073	1.883	2.410, 2.406[a]
$\Gamma: v_3 - c_1$	3.350	2.95		2.335	1.92	2.601, 2.605[a]
$X: v_1 - c_1$	3.821	3.629	3.44[c]	3.0326	2.91	3.602, 3.578
$X: v_2 - c_1$	3.988	3.909		3.184	3.176	3.852
$Y: v_1 - c_1$	4.541	4.604		3.755	3.78	4.137, 4.240
$Y: v_2 - c_1$	4.818	4.417	4.59[c]	4.022	3.855	4.384, 4.480

*[184 and references therein]

8.10.5 Optical Properties

As potential optical materials, the optical constants of Cu_2ZnSiS_4 and $Cu_2ZnSiSe_4$ are critical in applications. In Table 8.14, the reflective index n and dielectric constants ε_0 at infinite wavelength and ε_∞ at zero wavelength, along with the available experimental results, are presented. As is generally known, we find that refractive index n has an opposite trend with band gap [185]. For instance, n increases from 2.235 to 2.613 from S to Se compound in WKS structure while the band gap decreases from 1.677 to 0.835 eV. n increases from WKS structure to WST structure while the band gap decreases correspondingly. The dielectric constant ε_0 also presents similar trends while the values of ε_∞ are almost structure-independent. For example, ε_∞ is 0.51 for S compound and 0.54 for Se compound regardless of the structures.

The spectra of dielectric constant in the range of 2–10 eV for these compounds are plotted in Fig. 8.23. As can be seen in this figure, the shape of the dielectric function is quite similar for all compounds. However, the spectrum shifts to lower energy region when the anion atom changes from S to Se. This red shift can be attributed to the following: The conduction band is derived from the hybridization of the Si s and anion p states. When the VI atom changes from S to Se, the Si-VI hybridization becomes stronger which lowers the conduction band minimum. Therefore, the spectra move toward the lower energy region. The effect of the structure on the dielectric spectra is relatively small. The peaks for WKS and WST structures have nearly the same locations.

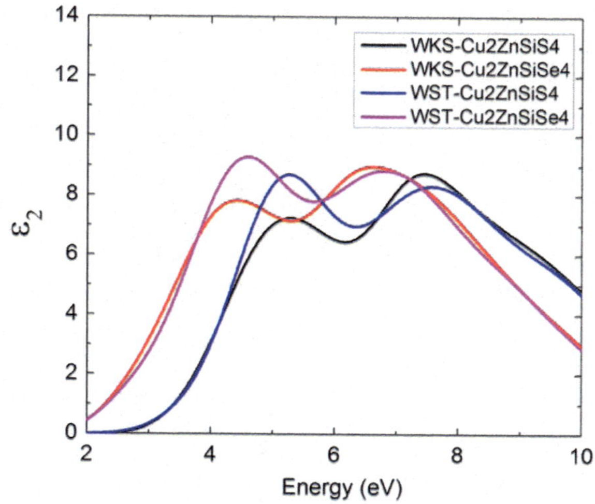

Fig. 8.23 Comparison of the imaginary component of the dielectric functions in the energy range of 2–10 eV for Cu_2ZnSiS_4 and $Cu_2ZnSiSe_4$ in WKS and WST structures

8.10.6 Summary

Using first-principles density functional methods, the electronic and optical properties of these Cu_2ZnSiS_4 and $Cu_2ZnSiSe_4$ in wurtzite-derived structures have been investigated. The wurztie-kesterite structure is found to have lowest energy and the S compounds are more stable than Se compounds. These materials are found to have direct band gaps. The optical transitions between valence bands and conduction bands are reported along the high symmetry points in the Brillouin zone.

References

1. Adachi S (2004) Handbook on physical properties of semiconductors, vol 2. Kluwer Academic, New York
2. Elyukhin VA, Sanchez-R VM, Elyukhina OV (2004) Self-assembling in $Al_xGa_{1-x}N_yAs_{1-y}$ alloys. Appl Phys Lett 85(10):1704–1706
3. Peter YY, Cardona, M (2005) Fundamentals of semiconductors: physics and materials properties, 3rd edn. Springer, Berlin, Germany
4. Plummer JD, Deal MD, Griffin PB (2000) Silicon VLSI technology: fundamentals, practice and modeling. Prentice Hall, Upper Saddle River, New Jersey
5. Moshe H, Mastai Y (2013) Atomic layer deposition on self-assembled monolayers, chapter 3, materials science—Advanced topics, InTech, pp 63–184. http://dx.doi.org/10.5772/54814
6. Piotrowski A, Madejczyk P, Gawron W, Klos K, Pawluczyk J, Rutkowski J, Piotrowski J, Rogalski A (2007) Progress in MOCVD growth of HgCdTe heterostructures for uncooled infrared photodetectors. Infrared Phys Technol 49(3):173–182
7. Ibach H, Lüth H (2003) Solid state physics, 3rd edn. Springer, Berlin, Germany
8. Hicks HGB, Manley DF (1969) High purity GaAs by liquid phase epitaxy. Solid State Commun 7(20):1463–1465
9. Moustakas TD, Pankove JI, Hamakawa Y (1992) Wide band gap semiconductors. Materials Research Society, Pittsburgh, Pennsylvania

10. Neudeck PG (1995) Progress in silicon carbide semiconductor electronics technology. J Electron Mater 24(4):283–288
11. Bhatnagar M, Baliga BJ (1993) Comparison of 6H–SiC, 3C–SiC, and Si for power devices. IEEE Trans Electron Devices 40(3):645–655
12. Kung P, Yasan A, McClintock R, Darvish S, Mi K, Razeghi M (2002) Future of $Al_xGa_{1-x}N$ materials and device technology for ultraviolet photodetectors. In: SPIE proceedings, vol 4650. pp 199–206
13. Saxler A, Mitchel WC, Kung P, Razeghi M (1999) Aluminum gallium nitride short-period superlattices doped with magnesium. Appl Phys Lett 74(14):2023–2025
14. Suzuki M, Nishio J, Onomura M, Hongo C (1998) Doping characteristics and electrical properties of Mg-doped AlGaN grown by atmospheric-pressure MOCVD. J Cryst Growth 189–190:511–515
15. Haase M, Qiu J, DePuydt JM, Cheng H (1991) Blue-green laser diodes. Appl Phys Lett 59(11):1272–1274
16. Zeng L, Cavus A, Yang BX, Tamargo MC, Bambha N, Gray A, Semendy F (1997) Molecular beam epitaxial growth of lattice-matched $Zn_xCd_yMg_{1-x-y}Se$ quaternaries on InP substrates. J Cryst Growth 175–176(1):541–545
17. Pavlidis D (2006) *Wide-and narrow-bandgap semiconductor materials*. Thema Forschung 2:38–41
18. Capper P, Garland J, Baker IM (2010) HgCdTe photovoltaic infrared detectors. In: Mercury cadmium telluride. Wiley Hoboken, New Jersey, pp 447–467
19. Capper P (2007) Narrow-bandgap II–VI semiconductors: growth. In: Springer handbook of electronic and photonic materials. Springer, Berlin, Germany, pp 303–324
20. Tyagi VV, Rahim NAA, Rahim NA, Selvaraj JAL (2013) Progress in solar PV technology: Research and achievement. Renew Sustain Energy Rev 20:443–461
21. Iles PA (2001) Evolution of space solar cells. Sol Energy Mater Sol Cells 68(1):1–13
22. Britt J, Ferekides C (1993) Thin-film CdS/CdTe solar cell with 15.8% efficiency. Appl Phys Lett 62(22):2582–2851
23. Hegedus SS, McCandless BE (2005) CdTe contacts for CdTe/CdS solar cells: effect of Cu thickness, surface preparation and recontacting on device performance and stability. Sol Energy Mater Sol Cells 88(1):75–95
24. Repins I, Contreras MA, Egaas B, DeHart C, Scharf J, Perkins CL, To B, Noufi R (2008) 19.9%-efficient $ZnO/CdS/CuInGaSe_2$ solar cell with 81.2% fill factor. Prog Photovoltaics Res Appl 16(3):235–239
25. Doverspike K, Dwight K, Wold A (1990) Preparation and characterization of copper zinc germanium sulfide selenide ($Cu_2ZnGeS_{4-y}Se_y$). Chem Mater 2(2):194–197
26. Guo Q, Hillhouse HW, Agrawal R (2009) Synthesis of Cu_2ZnSnS_4 nanocrystal ink and its use for solar cells. J Am Chem Soc 131(33):11672–11673
27. Szabo A, Ostlund NS (1989) Modern quantum chemistry: introduction to advanced electronic structure theory. Dover, New York
28. Parr RG, Yang W (1989) Density-functional theory of atoms and molecules. Oxford University Press, Oxford, UK
29. Kohn W, Sham LJ (1965) Self-consistent equations including exchange and correlation effects. Phys Rev 140(4A):A1133–A1138
30. Perdew JP, Wang Y (1992) Accurate and simple analytic representation of the electron-gas correlation energy. Phys Rev B 45(23):13244–13249
31. Payne MC, Teter MP, Allan DC, Arias TA, Joannopoulos JD (1992) Iterative minimization techniques for ab initio total-energy calculations: molecular dynamics and conjugate gradients. Rev Mod Phys 64(4):1045–1097
32. Hamann DR, Schlüter M, Chiang C (1979) Norm-conserving pseudopotentials. Phys Rev Lett 43(20):1494–1497
33. Laasonen K, Pasquarello A, Car R, Lee C, Vanderbilt D (1993) Car-Parrinello molecular dynamics with Vanderbilt ultrasoft pseudopotentials. Phys Rev B 47(16):10142–10153
34. Blöchl PE (1994) Projector augmented-wave method. Phys Rev B 50(24):17953–17979

35. Greaves GN, Greer AL, Lakes RS, Rouxel T (2011) Poisson's ratio and modern materials. Nat Mater 10(11):823–837
36. Cohen ML (1985) Calculation of bulk moduli of diamond and zinc-blende solids. Physical Review B 32(12):7988–7991
37. Cohen ML (1993) Predicting useful materials. Science 261(5119):307–308
38. Kamran S, Chen K, Chen L (2008) Semiempirical formulae for elastic moduli and brittleness of diamondlike and zinc-blende covalent crystals. Phys Rev B 77(9):094109–094113
39. Phillips JC (1973) Bonds and bands in semiconductors. Academic Press, New York
40. Roundy D, Krenn CR, Cohen ML, Morris JW (1999) Ideal shear strengths of fcc aluminum and copper. Phys Rev Lett 82(13):2713–2716
41. Martin RM (1970) Elastic properties of ZnS structure semiconductors. Phys Rev B 1(10):4005–4011
42. Gao F, He J, Wu E, Liu S, Yu D, Li D, Zhang S, Tian Y (2003) Hardness of covalent crystals. Phys Rev Lett 91(1):015502–015505
43. Chen D, Ravindra N (2013) Elastic properties of diamond and zincblende covalent crystals. Emerg Mater Res 2(1):58–63
44. Lee DN (2003) Elastic properties of thin films of cubic system. Thin Solid Films 434(1):183–189
45. Simmons G, Wang H (1971) Single crystal elastic constants and calculated aggregate properties: a handbook, 2nd edn. MIT Press, Cambridge, Massachusetts
46. Lee DH, Joannopoulos JD (1982) Simple scheme for deriving atomic force constants: application to SiC. Phys Rev Lett 48(26):1846–1849
47. Lambrecht WRL, Segall B, Methfessel M, Schilfgaarde M (1991) Calculated elastic constants and deformation potentials of cubic SiC. Phys Rev B 44(8):3685–3694
48. Azuhata T, Sota T, Suzuki K (1996) Elastic constants of III–V compound semiconductors: modification of keyes' relation. J Phys Condens Matter 8(18):3111–3119
49. Sherwin M, Drummond T (1991) Predicted elastic constants and critical layer thicknesses for cubic phase AlN, GaN, and InN on β-SiC. J Appl Phys 69(12):8423–8425
50. Xiong Q, Duarte N, Tadigadapa S, Eklund PC (2006) Force-deflection spectroscopy: a new method to determine the Young's modulus of nanofilaments. Nano Lett 6(9):1904–1909
51. Deligoz EK, Colakoglu K, Ciftci Y (2006) Elastic, electronic, and lattice dynamical properties of CdS, CdSe, and CdTe. Physica B 373(1):124–130
52. Alper T, Saunders G (1967) The elastic constants of mercury telluride. J Phys Chem Solids 28(9):1637–1642
53. Chen D, Ravindra NM (2012) Pressure dependence of energy gap of III–V and II–VI ternary semiconductors. J Mater Sci 47(15):5735–5742
54. De Bernabe A, Prieto C, Gonzalez L, Every AG (1999) Elastic constants of $In_xGa_{1-x}As$ and $In_xGa_{1-x}P$ determined using surface acoustic waves. J Phys: Condens Matter 11(28):L323–L327
55. Yeh CY, Chen AB, Sher A (1991) Formation energies, bond lengths, and bulk moduli of ordered semiconductor alloys from tight-binding calculations. Phys Rev B 43(11):9138–9151
56. Bouarissa N (2003) Compositional dependence of the elastic constants and the Poisson ratio of $Ga_xIn_{1-x}Sb$. Mater Sci Eng B 100(3):280–285
57. Maheswaranathan P, Sladek RJ, Debska U (1985) Elastic constants and their pressure dependences in $Cd_{1-x}Mn_xTe$ with $0 < x < 0.52$ and in $Cd_{0.52}Zn_{0.48}Te$. Phys Rev B 31(8):5212–5216
58. Brinck T, Murray JS, Politzer P (1993) Polarizability and volume. J Chem Phys 98(5):4305–4306
59. Gilman JJ (2003) Electronic basis of the strength of materials. Cambridge University Press, Cambridge, UK
60. Hildebrand O, Kuebart W, Pilkuhn M (1980) Resonant enhancement of impact in $Ga_{1-x}Al_xSb$. Appl Phys Lett 37(9):801–803
61. Kurtz SR, Biefeld RM, Dawson LR, Baucom KC, Howard AJ (1994) Midwave (4 μm) infrared lasers and light-emitting diodes with biaxially compressed InAsSb active regions. Appl Phys Lett 64(7):812–814

62. Menna RJ, Capewell DR, Martinelli RU, York PK, Enstrom RE (1991) 3.06 μm InGaAsSb/InPSb diode lasers grown by organometallic vapor-phase epitaxy. Appl Phys Lett 59(17):2127–2129
63. Bellaiche L, Wei SH, Zunger A (1996) Localization and percolation in semiconductor alloys: GaAsN versus GaAsP. Phys Rev B 54(24):17568–17576
64. Bellaiche L, Wei SH, Zunger A (1997) Composition dependence of interband transition intensities in GaPN, GaAsN, and GaPAs alloys. Phys Rev B 56(16):10233–10240
65. Wei SH, Zunger A (1996) Giant and composition-dependent optical bowing coefficient in GaAsN alloys. Phys Rev Lett 76(4):664–667
66. Landau LD, Lifshits EM (1969) Statistical physics. Pergamon Press, Oxford, UK
67. Mbaye AA, Wood DM, Zunger A (1988) Stability of bulk and pseudomorphic epitaxial semiconductors and their alloys. Phys Rev B 37(6):3008–3024
68. Yeo YC, Li MF, Chong TC, Yu PY (1997) Theoretical study of the energy-band structure of partially CuPt-ordered $Ga_{0.5}In_{0.5}P$. Phys Rev B 55(24):16414–16419
69. Behet M, Stoll B, Heime K (1992) Lattice-matched growth of InPSb on InAs by low-pressure plasma MOVPE. J Cryst Growth 124(1):389–394
70. Jou MJ, Cheng YT, Jen HR, Stringfellow GB (1988) Organometallic vapor phase epitaxial growth of a new semiconductor alloy: $GaP_{1-x}Sb_x$. Appl Phys Lett 52(7):549–551
71. Wei SH, Zunger A (1989) Band gaps and spin-orbit splitting of ordered and disordered $Al_xGa_{1-x}As$ and $GaAs_xSb_{1-x}$ alloys. Phys Rev B 39(5):3279–3304
72. Stringfellow GB (1989) Ordered structures and metastable alloys grown by OMVPE. J Cryst Growth 98(1):108–117
73. Wei SH, Ferreira LG, Bernard JE, Zunger A (1990) Electronic properties of random alloys: Special quasirandom structures. Phys Rev B 42(15):9622–9649
74. Bechstedt F, Del Sole R (1988) Analytical treatment of band-gap underestimates in the local-density approximation. Phys Rev B 38(11):7710–7716
75. Wei SH, Zunger A (1990) Band-gap narrowing in ordered and disordered semiconductor alloys. Appl Phys Lett 56(7):662–664
76. Fedders PA, Muller MW (1984) Mixing enthalpy and composition fluctuations in ternary III–V semiconductor alloys. J Phys Chem Solids 45(6):685–688
77. Reihlen EH, Jou MJ, Jaw DH, Stringfellow GB (1990) Optical absorption and emission of $GaP_{1-x}Sbx$ alloys. J Appl Phys 68(2):760–767
78. Shimomura H, Anan T, Sugou S (1996) Growth of AlPSb and GaPSb on InP by gas-sthece molecular beam epitaxy. J Cryst Growth 162(3):121–125
79. Jou MJ, Cheng YT, Jen HR, Stringfellow GB (1988) OMVPE growth of the new semiconductor alloys $GaP_{1-x}Sb_x$ and $InP_{1-x}Sb_x$. J Cryst Growth 93(1):62–69
80. Reihlen EH, Jou MJ, Fang ZM, Stringfellow GB (1990) Optical absorption and emission of $InP_{1-x}Sb_x$ alloys. J Appl Phys 68(9):4604–4609
81. Aramoto T, Kumazawa S, Higuchi H, Arita T, Shibutani S, Nishio T, Nakajima J, Tsuji M, Hanafusa A, Hibino T, Omira K, Ohyama H, Murozono M (1997) 16.0% efficient thin-film CdS/CdTe solar cells. Jpn J Appl Phys 36(10):6304–6305
82. Wang D, Hou Z, Bai Z (2011) Study of interdiffusion reaction at the CdS/CdTe interface. J Mater Res 26(05):697–705
83. Fischer A, Anthony L, Compaan AD (1998) Raman analysis of short-range clustering in laser-deposited CdS_xTe_{1-x} films. Appl Phys Lett 72(20):2559–2561
84. Ebina A, Yamamoto M, Takahashi T (1972) Reflectivity of $ZnSe_xTe_{1-x}$ single crystals. Phys Rev B 6(10):3786–3791
85. Lange H, Donecker J, Friedrich H (1976) Electroreflectance and wavelength modulation study of the direct and indirect fundamental transition region of $In_{1-x}Ga_xP$. Physica Status Solidi (B) 73(2):633–639
86. Chen D, Ravindra NM (2013) Structural, thermodynamic and electronic properties of GaP_xSb_{1-x} and InP_xSb_{1-x} alloys. Emerg Mater Res 2(2):109–113
87. Hopfield JJ (1960) Fine structure in the optical absorption edge of anisotropic crystals. J Phys Chem Solids 15(1):97–107

88. Zakharov O, Rubio A, Blasé X, Cohen ML, Louie SG (1994) Quasiparticle band structures of six II–VI compounds: ZnS, ZnSe, ZnTe, CdS, CdSe, and CdTe. Phys Rev B 50(15):10780–10787
89. Chadi DJ (1977) Spin-orbit splitting in crystalline and compositionally disordered semiconductors. Phys Rev B 16(2):790–796
90. Wei SH, Zunger A (1989) Negative spin-orbit bowing in semiconductor alloys. Phys Rev B 39(9):6279–6282
91. Carrier P, Wei SH (2004) Calculated spin-orbit splitting of all diamondlike and zinc-blende semiconductors: Effects of $p_{1/2}$ local orbitals and chemical trends. Phys Rev B **70**(3):035212–035212-9
92. Van Vechten JA, Berolo O, Woolley JC (1972) Spin-orbit splitting in compositionally disordered semiconductors. Phys Rev Lett 29(20):1400–1403
93. Wei K, Pollak FH, Freeouf JL, Shvydka D, Compaan AD (1999) Optical properties of $CdTe_{1-x}S_x(0 \leq x \leq 1)$: experiment and modeling. J Appl Phys 85(10):7418–7425
94. Lane DW (2006) A review of the optical band gap of thin film CdS_xTe_{1-x}. Sol Energy Mater Sol Cells 90(9):1169–1175
95. Ohata K, Saraie J, Tanaka T (1973) Optical energy gap of the mixed crystal CdS_xTe_{1-x}. Jpn J Appl Phys 12(10):1641–1642
96. Heyd J, Scuseria GE, Ernzerhof M (2003) Hybrid functionals based on a screened Coulomb potential. J Chem Phys 118(18):8207–8215
97. Heyd J, Scuseria GE, Ernzerhof M (2003) Erratum: Hybrid functionals based on a screened Coulomb potential. J Chem Phys 118:8207. J Chem Phys 124(21):219906–219906-1
98. Wei SH, Zhang SB, Zunger A (2000) First-principles calculation of band offsets, optical bowings, and defects in CdS, CdSe, CdTe, and their alloys. J Appl Phys 87(3):1304–1311
99. Gomyo A, Suzuki T, Kobayashi K, Kawata S, Hino I, Yuasa T (1987) Evidence for the existence of an ordered state in $Ga_{0.5}In_{0.5}P$ grown by metalorganic vapor phase epitaxy and its relation to band gap energy. Appl Phys Lett 50(11):673–675
100. Kuan TS, Kuech TF, Wang WI, Wilkie EL (1985) Long-range order in $Al_xGa_{1-x}As$. Phys Rev Lett 54(3):201–204
101. Kurtz SR (1993) Anomalous electroreflectance spectrum of spontaneously ordered $Ga_{0.5}In_{0.5}P$. J Appl Phys 74(6):4130–4135
102. Ruvimov S, Werner P, Scheerschmidt K, Gosele U, Heydenreich J, Richter U, Ledentsov NN, Grundmann M, Bimberg D, Ustinov VM, Yu Egorov A, Kopev PS, Alferov ZI (1995) Structural characterization of (In, Ga)As quantum dots in a GaAs matrix. Phys Rev B 51(20):14766–14769
103. Wu C, Feng Z, Chang W, Yang C, Lin H (2012) Bond lengths and lattice structure of $InP_{0.52}Sb_{0.48}$ grown on GaAs. Appl Phy Lett 101(9):091902–091902-4
104. Zhong Z, Li JH, Kulik J, Chow PC, Norman AG, Mascarenhas A, Bai J, Golding TD, Moss SC (2001) Quadruple-period ordering along [110] in a $GaAs_{0.87}Sb_{0.13}$ alloy. Phys Rev B 63(3):033314
105. Franceschetti A, Zunger A (1994) Pressure dependence of optical transitions in ordered GaP/InP superlattices. Appl Phys Lett 65(23):2990–2992
106. Wei SH, Laks DB, Zunger A (1993) Dependence of the optical properties of semiconductor alloys on the degree of long-range order. Appl Phys Lett 62(16):1937–1939
107. Wei SH, Zunger A (1993) Erratum: Dependence of the optical properties of semiconductor alloys on the degree of long-range order. Appl Phys Lett 62:1937. Appl Phys Lett **63**(9):1292
108. Wei SH, Zunger A (1998) Calculated natural band offsets of all II–VI and III–V semiconductors: chemical trends and the role of cation d orbitals. Appl Phys Lett 72(16):2011–2013
109. Baxter CS, Broom RF, Stobbs WM (1990) The characterization of the ordering of MOVPE grown III–V alloys using transmission electron microscopy. Surf Sci 228(1):102–107
110. Baxter CS, Stobbs WM, Wilkie JH (1991) The morphology of ordered structures in III–V alloys: inferences from a TEM study. J Cryst Growth 112(2):373–385
111. Morita E, Ikeda M, Kumagai O, Kanedo K (1988) Transmission electron microscopic study of the ordered structure in GaInP/GaAs epitaxially grown by metalorganic chemical vapor deposition. Appl Phys Lett 53(22):2164–2166

112. Suzuki T, Gomyo A, Iijima S, Kobayashi K, Kawata S, Hino I, Yuasa T (1988) Band-gap energy anomaly and sublattice ordering in GaInP and AlGaInP grown by metalorganic vapor phase epitaxy. Jpn J Appl Phys 27(11):2098–2106
113. Kondow M, Kakibayashi H, Minagawa S, Inoue Y, Nishino T, Hamakawa Y (1988) Influence of growth temperature on crystalline structure in $Ga_{0.5}In_{0.5}P$ grown by organometallic vapor phase epitaxy. Appl Phys Lett 53(21):2053–2055
114. Nishino T, Inoue Y, Hamakawa Y, Kondow M, Minagawa S (1988) Electroreflectance study of ordered $Ga_{0.5}In_{0.5}P$ alloys grown on GaAs by organometallic vapor phase epitaxy. Appl Phys Lett 53(7):583–585
115. Lee K, Lee S, Chang KJ (1995) Optical properties of ordered $In_{0.5}Ga_{0.5}P$ alloys. Phys Rev B 52(22):15862–15866
116. Nee TW, Green AK (1990) Optical properties of InGaAs lattice-matched to InP. J Appl Phys 68(10):5314–5317
117. Hakki BW, Jayaraman A, Kim CK (1970) Band structure of InGaP from pressure experiments. J Appl Phys 41(13):5291–5296
118. Chen J, Sites JR, Spain IL, Hafich MJ, Robinson GY (1991) Band offset of GaAs/$In_{0.48}Ga_{0.52}P$ measured under hydrostatic pressure. Appl Phys Lett 58(7):744–746
119. Hill R (1974) Energy-gap variations in semiconductor alloys. J Phys C: Solid State Phys 7(3):521–526
120. Hill R, Pitt GD (1975) The pressure and temperature dependence of electron energy-gaps in semiconductor alloys. Solid State Commun 17(6):739–742
121. Van Vechten JA (1969) Quantum dielectric theory of electronegativity in covalent systems. I. Electronic dielectric constant. Phys Rev 182(3):891–905
122. Van Vechten JA (1969) Quantum dielectric theory of electronegativity in covalent systems. II. Ionization potentials and interband transition energies. Phys Rev 187(3):1007–1020
123. Camphausen DL, Connell GAN, Paul W (1971) Calculation of energy-band pressure coefficients from the dielectric theory of the chemical bond. Phys Rev Lett 26(4):184–188
124. Van Vechten JA, Bergstresser TK (1970) Electronic structures of semiconductor alloys. Phys Rev B 1(8):3351–3358
125. Vurgaftman I, Meyer JR, Mohan LR (2001) Band parameters for III–V compound semiconductors and their alloys. J Appl Phys 89(11):5815–5875
126. Adachi S (2009) Properties of semiconductor alloys: group-IV, III–V and II–VI semiconductors, vol 28. Wiley, Hoboken, New Jersey
127. Nicklas JW, Wilkins JW (2010), Accurate ab initio predictions of III–V direct-indirect band gap crossovers. Appl Phys Lett 97(9):091902–091902-3
128. Tsang WT, Chiu TH, Chu SNG, Ditzenberger JA (1985) GaSbAs/AlGaSbAs superlattice lattice matched to InP prepared by molecular beam epitaxy. Appl Phys Lett 46(7):659–661
129. Teissier R, Sicault D, Harmand JC, Ungaro G, Le Roux G, Largeau L (2001) Temperature-dependent valence band offset and band-gap energies of pseudomorphic GaAsSb on GaAs. J Appl Phys 89(10):5473–5477
130. Drews D, Schneider A, Werninghaus T, Behres A, Heuken M, Heime K, Zahn DRT (1998) Characterization of MOVPE grown InPSbInAs heterostructures. Appl Surf Sci 123–124:746–750
131. Alibert C, Bordure G, Laugier A, Chevallier J (1972) Electroreflectance and band structure of $Ga_xIn_{1-x}P$ alloys. Phys Rev B 6(4):1301–1310
132. Uchida K, Yu PY, Noto N, Weber ER (1994) Pressure-induced Γ-X crossover in the conduction band of ordered and disordered GaInP alloys. Appl Phys Lett 64(21):2858–2860
133. Alavi K, Aggarwal RL, Groves SH (1980) Interband magnetoabsorption of $In_{0.53}Ga_{0.47}As$. Phys Rev B 21(3):1311–1315
134. Lambkin JD, Dunstan DJ (1988) The hydrostatic pressure dependence of the band-edge photoluminescence of GaInAs. Solid State Commun 67(8):827–830
135. Desplanque L, Vignaud D, Godey S, Cadio E, Plissard S, Wallart X, Liu P, Sellier H (2010) Electronic properties of the high electron mobility $Al_{0.56}In_{0.44}Sb/Ga_{0.5}In_{0.5}Sb$ heterostructure. J Appl Phys 108(4):043704–043704-6

136. Bouarissa N, Atheag H (1995) Band structure calculations of $In_xGa_{1-x}Sb$ under pressure. Infrared Phys Technol 36(6):973–980
137. Shan W, Ager JW, Yu KM, Walukiewicz W, Haller EE, Martin MC, Mckinney WR, Yang W (1999) Dependence of the fundamental band gap of $Al_xGa_{1-x}N$ on alloy composition and pressure. J Appl Phys 85(12):8505–8507
138. Dridi Z, Bouhafs B, Ruterana P (2002) Pressure dependence of energy band gaps for $Al_xGa_{1-x}N$, $In_xGa_{1-x}N$ and $In_xAl_{1-x}N$. New J Phys 4(1):94.1–94.15
139. Chen A, Woodall JM (2009) Photodiode characteristics and band alignment parameters of epitaxial $Al_{0.5}Ga_{0.5}P$. Appl Phy Lett 94(2):021102–021102-3
140. Bosio C, Stachli JL, Guzzi M, Burri G, Logan RA (1988) Direct energy gap dependence on Al concentration in $Al_xGa_{1-x}As$. Phys Rev B 38(5):3263–3268
141. Wei SH, Zunger A (1999) Predicted band gap pressure coefficients of all diamond and zinc-blende semiconductors: the chemical trends. Phys Rev B 60(8):5404–5411
142. Adachi S (1994) GaAs and related materials: bulk semiconducting and superlattice properties. World Scientific, Singapore
143. Tamargo MC (2002) II–VI Semiconductor materials and their applications. Taylor & Francis, New York
144. Yang XD, Xu ZY, Sun Z, Sun BQ, Li GH, Sou IK, Ge WK (2005) Recombination kinetics of Te isoelectronic centers in ZnSTe. Appl Phys Lett 86(5):052107–052107-3
145. Seong MJ, Alawadhi H, Miotkowski I, Ramdas AK, Miotkowska S (1999) The anomalous variation of band gap with alloy composition: cation vs anion substitution in ZnTe. Solid State Commun 112(6):329–334
146. Wu J, Walukiewicz W, Yu KM, Shan W, Ager, JW III, Haller, WK, Miotkowski I, Ramdas AK, Su CH (2003) Composition dependence of the hydrostatic pressure coefficients of the bandgap of $ZnSe_{1-x}Te_x$ alloys. Phys Rev B 68(3):033206–033206-4
147. Murali KR, Thilagavathy K, Vasantha S, Gopalakrishnan P, Oommen PR (2010) Photoelectrochemical properties of CdS_xSe_{1-x} films. Sol Energy 84(4):722–729
148. Azhniuk YM, Lopushansky VV, Hutych YI, Prymak MV, Gomonnai AV, Zahn DRT (2011) Precipitates of selenium and tellurium in II–VI nanocrystal-doped glass probed by Raman scattering. Physica Status Solidi (b) 248(3):674–679
149. Zerroug S, Ali Sahraoui F, Bouarissa N (2007) Structural parameters and pressure coefficients for CdS_xTe_{1-x}: FP-LAPW calculations. Eur Phys J B, 57(1):9–14
150. Muthukumarasamy N, Balasundaraprabhu R, Jayakumar S, Kannan MD (2007) Photoconductive properties of hot wall deposited $CdSe_{0.6}Te_{0.4}$ thin films. Mater Sci Eng B 137(1–3):1–4
151. Beliveau A, Carlone C (1989) Pressure study of the direct band gap of $Zn_xCd_{1-x}S$ mixed crystals. Semicond Sci Technol 4(4):277–279
152. Olego DJ, Faurie JP, Sivananthan S, Raccah PM (1985) Optoelectronic properties of $Cd_{1-x}Zn_xTe$ films grown by molecular beam epitaxy on GaAs substrates. Appl Phys Lett 47(11):1172–1174
153. Madelung O, Von Der Osten W, Rossler U (1986) Landolt-Bornstein: numerical data and functional relationships in science and technology. Springer, Berlin, Germany
154. Reimann K, Haselhoff M, St. Rubenacke S, Steube M (1996) Determination of the pressure dependence of band-structure parameters by two-photon spectroscopy. Physica Status Solidi (b) 198(1):71–80
155. Zhao Z, Zeng J, Ding Z, Wang X, Hou J (2007) High pressure photoluminescence of CdZnSe quantum dots: Alloying effect. J Appl Phys 102(5):053509–053509-3
156. Gil B, Dunstan DJ (1991) Tellurium-based II–VI compound semiconductors and heterostructures under strain. Semicond Sci Technol 6(6):428–438
157. Gonzalez J, Perez FV, Moya E, Chervin JC (1995) Hydrostatic pressure dependence of the energy gaps of CdTe in the zinc-blende and rocksalt phases. J Phys Chem Solids 56(3–4):335–340
158. Fang ZL, Li GH, Liu NZ, Zhu ZM, Han HX, Ding K, Ge WK, Sou IK (2002) Photoluminescence from $ZnS_{1-x}Te_x$ alloys under hydrostatic pressure. Phys Rev B 66(8): 085203–085203-6

159. Prins AD, Dunstan DJ, Lambkin JD, O'Reilly EP, Adams AR, Pritchard R, Truscott WS, Singer KE (1993) Evidence of type-I band offsets in strained $GaAs_{1-x}Sb_x$/GaAs quantum wells from high-pressure photoluminescence. Phys Rev B 47(4):2191–2196
160. Teisseyre H, Kozankiewicz B, Leszczynski M, Grzegory I, Suski T, Bockowski M, Porowski S, Pakula K, Mensz PM, Bhat IB (1996) Pressure and time-resolved photoluminescence studies of Mg-doped and undoped GaN. Physica Status Solidi (b) 198(1):235–241
161. Cardona M (1963) Band parameters of semiconductors with zincblende, wurtzite, and germanium structure. J Phys Chem Solids 24(12):1543–1555
162. Shan W, Walukiewicz W, Ager JW, Yu KM, Wu J, Haller EE (2004) Pressure dependence of the fundamental band-gap energy of CdSe. Appl Phys Lett 84(1):67–69
163. Yu PY, Cardona M (1970) Temperature coefficient of the refractive index of diamond- and zincblende-type semiconductors. Phys Rev B 2(8):3193–3197
164. Fernandes PA, Salomé PMP, Da Cunha AF (2011) Study of polycrystalline Cu_2ZnSnS_4 films by Raman scattering. J Alloy Compd 509(28):7600–7606
165. Katagiri H, Jimbo K, Yamada S, Kamimura T, Maw WS, Fukano T, Ito T, Motohiro T (2008) Enhanced conversion efficiencies of Cu_2ZnSnS_4-based thin film solar cells by using preferential etching technique. Appl Phys Express 1(4):041201–041202
166. Moholkar AV, Shinde SS, Babar AR, Sim K, Lee HK, Rajpure KY, Patil PS, Bhosale CH, Kim JH (2011) Synthesis and characterization of Cu_2ZnSnS_4 thin films grown by PLD: solar cells. J Alloy Compd 509(27):7439–7446
167. Schorr S, Hoebler HJ, Tovar M (2007) A neutron diffraction study of the stannite-kesterite solid solution series. Eur J Mineral 19(1):65–73
168. Shavel A, Arbiol J, Cabot A (2010) Synthesis of quaternary chalcogenide nanocrystals: Stannite $Cu_2Zn_xSn_ySe_{1+x+2y}$. J Am Chem Soc 132(13):4514–4515
169. Liu ML, Chen IW, Huang FQ, Chen LD (2009) Improved thermoelectric properties of Cu-doped quaternary chalcogenides of $Cu_2CdSnSe_4$. Adv Mater 21(37):3808–3812
170. Sevik C, Çağın T (2010) Ab initio study of thermoelectric transport properties of pure and doped quaternary compounds. Phys Rev B 82(4):045202–045202-6
171. Todorov TK, Reuter KB, Mitzi DB (2010) High-efficiency solar cell with earth-abundant liquid-processed absorber. Adv Mater 22(20):E156–E159
172. Matsushita H, Ichikawa T, Katsui A (2005) Structural, thermodynamical and optical properties of Cu_2–II–IV–VI_4 quaternary compounds. J Mater Sci 40(8):2003–2005
173. Matsushita H, Maeda T, Katsui A, Takizawa T (2000) Thermal analysis and synthesis from the melts of Cu-based quaternary compounds Cu–III–IV–VI4 and Cu_2–II–IV–VI4 (II=Zn, Cd; III=Ga, In; IV=Ge, Sn; VI=Se). J Cryst Growth 208(1–4):416–422
174. Parasyuk OV, Olekseyuk ID, Piskach LV (2005) X-ray powder diffraction refinement of $Cu_2ZnGeTe_4$ structure and phase diagram of the Cu_2GeTe_3–ZnTe system. J Alloy Compd 397(1):169–172
175. Parasyuk OV, Piskach LV, Romanyuk YE, Olekseyuk ID, Zaremba VI, Pekhnyo VI (2005) Phase relations in the quasi-binary Cu_2GeS_3-ZnS and quasi-ternary Cu_2S-Zn(Cd)S-GeS_2 systems and crystal structure of Cu_2ZnGeS_4. J Alloy Compd 397(1–2):85–94
176. Yao GQ, Shen HS, Honig ED, Kershaw R, Dwight K, Wold A (1987) Preparation and characterization of the quaternary chalcogenides $Cu_2B(II)C(IV)X_4$ [B(II)=Zn, Cd; C(IV)=Si, Ge; X=S, Se]. Solid State Ionics 24(3):249–252
177. Schleich DM, Wold A (1977) Optical and electrical properties of quarternary chalcogenides. Mater Res Bull 12(2):111–114
178. León M, Levcenko S, Serna R, Gurieva G, Nateprov A, Merino JM, Friedrich EJ, Fillat U, Schorr S, Arushanov E (2010) Optical constants of Cu_2ZnGeS_4 bulk crystals. J Appl Phys 108(9):093502–093502-5
179. Chen S, Gong XG, Walsh A, Wei SH (2009) Electronic structure and stability of quaternary chalcogenide semiconductors derived from cation cross-substitution of II–VI and I–III–VI_2 compounds. Phys Re B 79(16):165211–1652211-10
180. Zhang Y, Sun X, Zhang P, Yuan X, Huang F, Zhang W (2012) Structural properties and quasiparticle band structures of Cu-based quaternary semiconductors for photovoltaic applications. J Appl Phys 111(6):063709–063709-6

181. Persson C (2010) Electronic and optical properties of Cu_2ZnSnS_4 and $Cu_2ZnSnSe_4$. J Appl Phy 107(5):053710–053710-8
182. Lamsal C, Chen D, Ravindra NM (2012) Optical and electronic properties of AlN, GaN and InN: An Analysis. In: Supplemental proceedings: materials processing and interfaces, vol 1. pp 701–713
183. Wang X, Li J, Zhao Z, Huang S, Xie W (2012) Crystal structure and electronic structure of quaternary semiconductors $Cu_2ZnTiSe_4$ and Cu_2ZnTiS_4 for solar cell absorber. J Appl Phys 112(2): 023701–023701-4
184. Zhang X, Rao D, Lu R, Deng K, Chen D (2015) First-principles study on electronic and optical properties of $Cu_2ZnSiV\ I_4$ (VI=S, Se, and Te) quaternary semiconductors. AIP Adv 5:057111. https://doi.org/10.1063/1.4920936
185. Ravindra NM, Ganapathy P, Choi J (2007) Energy gap—Refractive index relations in semiconductors—An overview. Infrared Phys Technol 50:21–29

Chapter 9
Organic Semiconductors

Josefina Alvarado Rivera, Amanda Carrillo Castillo and María de la Luz Mota González

9.1 Processing Techniques

One of the most exciting opportunities in electronics, optoelectronics or flexible electronics is to be able to make devices based on organic semiconductors. Organic active materials can exhibit many advantages such as lower demands on processing technology with less sensitivity to the processing environment, flexibility, and the opportunity to apply the simplicity of organic synthesis to tailoring the properties of the materials for specific applications [1].

Depending on their vapor pressure and solubility, organic semiconductors are deposited either from a vapor or solution phase. In this section, some of the organic semiconductor deposition methods are discussed.

J. Alvarado Rivera (✉)
Conacyt—Departamento de Física, Universidad de Sonora, Hermosillo, Mexico
e-mail: jalvaradori@conacyt.mx

A. Carrillo Castillo
Instituto de Ingeniería y Tecnologia, Universidad Autónoma de Ciudad Juárez, Chihuahua, Mexico
e-mail: amanda.carrillo@uacj.mx

M. L. Mota González
Conacyt—Universidad Autónoma de Ciudad Juárez, Chihuahua, Mexico
e-mail: mdllmotago@conacyt.mx

M. I. Pech-Canul and N. M. Ravindra (eds.), *Semiconductors*,
https://doi.org/10.1007/978-3-030-02171-9_9

9.1.1 Vacuum Deposition

9.1.1.1 Vacuum Thermal Evaporation

The most common way to deposit small organic molecules in thin film form is by vacuum thermal evaporation (VTE), which involves heating a solid material inside a high vacuum chamber, and taking it to a temperature that achieves the desired vapor pressure (see Fig. 9.1).

In vacuum, even a relatively low vapor pressure is sufficient to raise a vapor cloud inside the chamber. This evaporated material now constitutes a vapor stream, which traverses the chamber and strikes the substrate, sticking to it as a coating or film. For small organic molecules and oligomers that are solution insoluble, vacuum thermal evaporation is an ideal deposition method. Some organic semiconductors have been deposited using this method. Currently, the best mobility for organic semiconductors has been reported for vacuum-deposited pentacene films [2].

Multilayer deposition and co-deposition of several organic semiconductors are possible without delamination or dissolution of the previous layers during subsequent deposition steps. Measures may have to be taken to assure film adhesion, as well as control various desired film properties. Several process parameters can be adjusted in the design of evaporation systems; this allows process engineers to achieve the desired film properties such as thickness, uniformity, adhesion strength, stress, grain structure, optical, or electrical properties [3, 4].

Other vacuum-based thin-film deposition methods employed for processing organic semiconductors include organic vapor phase deposition (OVPD), organic molecular beam deposition (OMBD), and laser evaporation.

(a) **Organic vapor phase deposition (OVPD)**

Some of the problems of VTE are addressed by OVPD. In this deposition technique, the organic compound is thermally evaporated into a diluting, not as reactive gas

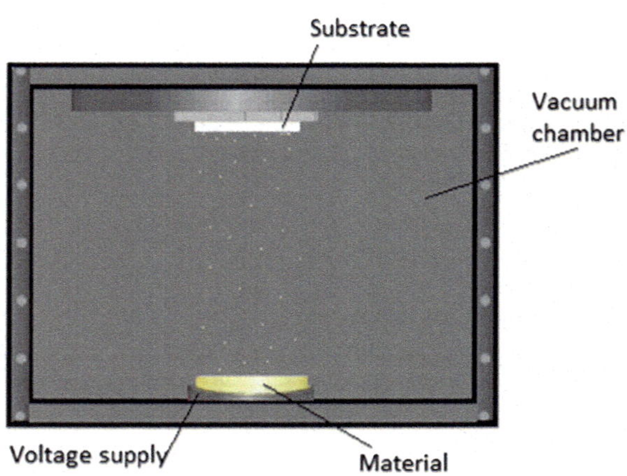

Fig. 9.1 Schematic diagram of a thermal evaporation system (modified from [106])

stream which is convectively transported through a hot-walled reactor, and finally, it condenses onto a substrate [5]. The quality of the deposited material is primarily dependent on the purity of the feeding gas rather than the chamber base pressure.

OVPD, in general, has the advantages listed below [4, 6, 7]:

- High material utilization efficiency and thickness uniformity.
- Low vacuum operation.
- High deposition rates.
- Tool design flexibility.

(b) **Organic vapor jet printing (OVJP)**

OVJP is a natural extension of OVPD and VTE, where it is not necessary to use shadow masks to deposit a patterned film [5, 8]. Similar to inkjet printing, the deposition of individual pixels on extended plastic substrates, continuously deployed near the localized jet of gas, can be achieved. The difference from solution-based inkjet printing of polymers is that the solvent in OVJP is a gas. Hence, it is easily volatilized during growth to leave a uniform film of the desired organic material. This process, invented at Princeton, has the possibility of growth of small molecule organic thin films by rapidly and simply depositing ultra-small patterns of organic thin-film materials or precursors.

As an application of this technique, the OVJP tool has been used to produce continuous films of pentacene, which are used to make organic thin-film transistors (TFTs) with hole mobilities of 0.2 cm^2/Vs, comparable to that achieved by VTE-grown films. More implementations of OVJP require the deposition to be carried out in a chamber evacuated to ≤ 1 Torr of absolute pressure. OVJP can, however, be used in an atmosphere with an inert N_2 guard flow around the jet of depositing vapor [9, 10].

(c) **Organic molecular beam deposition (OMBD)**

Organic molecular beam deposition (OMBD) is a vacuum deposition technique based on the direction of a molecular beam from a crucible in an evaporator in ultra-high vacuum (UHV) onto a substrate. Whereas, the deposition rate of conventional vacuum deposition using a boat-type source container in a low vacuum is usually too high and far from optimum, the deposition rate can be very small in UHV because the residual gas pressure is very low and no impurities are incorporated in the film. If the temperature of the source and substrate are adequately controlled, it is possible to deposit various materials by OMBD onto various substrates under optimum conditions [11]. One of the merits of OMBD is that it is not necessary to modify source materials for the film formation.

"To control the flux of each organic molecular beam, the effusion cells are provided with mechanical shutters. Reproducible organic thin film growth is controlled with the main shutter in front of the sample. Quasi-epitaxial thin films are deposited at growth rates of about 0.1 nm/s measured by a quartz oscillator thickness controller. To obtain layers with high optical and electrical quality, the crystal growth of organic semiconductors requires substrate temperatures usually between 77 and 200 K during the deposition process. Therefore, the substrate holder can be cooled with liquid nitrogen" [12].

"Finally, the only limitation on materials used in OMBD is that they must have an appreciable vapor pressure at a temperature low enough to not decompose. Fortunately, under UHV, the temperature necessary to keep the deposition rate constant can be lower than in low vacuum. Therefore, decomposition of source material in the crucible is minimized in UHV. OMBD has been applied to molecules as large as metallonaphthalocyanines" [13].

9.1.1.2 Laser Deposition

The study and implementation of new methods for electronic component manufacture, particularly on flexible substrates, represents a critical stage in developing plastic microelectronics.

As discovered, the laser evaporation of some low molecular organic compounds led to the formation of their nanoclusters. Moreover, a high-temperature crystal phase was formed at room temperature [14, 15]. Laser deposition of clustered multicomponent films, based on polymer matrix, presents one promising research topic for optoelectronics and photonics applications, particularly for sensors.

Laser-based processes [16] offer versatile alternatives for deposition of thin films, in developing organic devices that operate on flexible supports [13]. The most common laser-based techniques are the pulsed laser deposition (PLD), the matrix-assisted pulsed laser evaporation (MAPLE), and the laser-induced forward transfer (LIFT) [17–20].

(a) **Pulsed laser deposition (PLD)**

"PLD is a growth technique in which the photon energy of a laser, characterized by pulse duration and laser frequency, interacts with a bulk material [21–23]. As a result, the material is removed from the bulk depending on the absorption properties of the target materials" [24].

PLD allows thin epitaxial or large grain-oriented films, heterostructures, and films with "step-like" morphology to grow. It also allows synthesizing metastable, small grain and even nanocrystalline films to be deposited, and composite materials consisting of different constituents to be fabricated (Fig. 9.2).

(b) **Matrix-assisted pulsed laser evaporation (MAPLE)**

"The MAPLE technique is emerging as an alternative route to the conventional methods for depositing organic materials, although the MAPLE-deposited films very often present high surface roughness and characteristic morphological features" [25].

MAPLE is derived from the PLD technique; however, in MAPLE, the target is not a solid material as such, it is a frozen solution instead. The molecules to be deposited are dissolved in a highly volatile and light absorbing solvent. When the pulsed laser beam impacts the target, the laser energy absorbed by the solvent is converted into

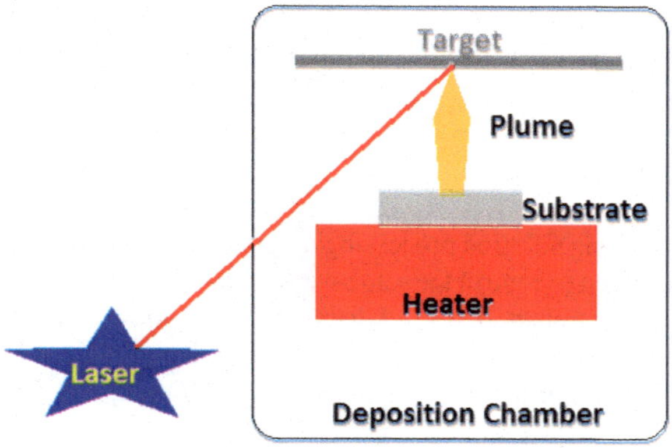

Fig. 9.2 PLD scheme (substantially modified from [107])

thermal energy by photochemical processes. Thus, the solvent vaporizes while it mixes with the polymer molecules. A plume is formed, which is composed of the volatile solvent molecules and the polymer solute molecules. The volatile solvent molecules are pumped away by a vacuum pump while the organic/polymer molecules are deposited onto the substrate [26].

(c) **Laser-induced forward transfer (LIFT)**

LIFT is a direct-write technique that allows the selective transfer of many materials on a micrometer scale [27–29]. In the LIFT process, a thin film serves as a donor material that is to be transferred, which is referred to as the donor layer. A pulsed laser is focused on the carrier–donor interface to induce the necessary energy to transfer the donor material onto a receiving substrate, referred to as the receiver. The latter is separated from the donor with a typical gap size in the order of 0–100 μm [30].

The LIFT technique offers several advantages compared to other conventional transfer and/or printing techniques. With printing velocities up to 4 m/s [31], high precision and accuracy in energy delivery, LIFT has become a promising answer to this challenge [32]. Also, LIFT processes offer versatile alternatives for the deposition of thin films, to create organic electronic devices operating on flexible supports. The LIFT technique is an alternative way for the fabrication of organic electronic components when conventional techniques cannot be considered [19].

9.1.2 *Solution Deposition*

Vacuum, vapor thermal, or laser deposition is one of the most popular methods for the fabrication of the active functional layer of organic semiconductors. However, these techniques are limited by critical drawbacks as high manufacturing costs. In this way, organic semiconductor materials are considered to be promising for the development of electronic and flexible electronic devices having the advantages of being solution processable, lightweight, low cost, and flexible. This section describes an array of processing methods that use organic semiconductor solutions.

9.1.2.1 Dip Coating

"Among solution processing approaches, dip coating technique has been extensively utilized to control the self-assembly of organic semiconductors which is important for carrier transport" [33]. Dip coating is a simple technique that guarantees the fast and high-throughput process for the selective growth of highly ordered molecular films in the hydrophobic regions with solution preparation, used for OFETs applications [32, 34, 35].

In this process, the surface to be coated is initially immersed in the coating fluid and then withdrawn. One of the advantages of this process is the ability to coat irregularly shaped substrates [36, 37].

The main parameter monitors are withdrawal velocity and substrate/solution temperature, which influence the development of concentration gradients and fluid flow within the meniscus. Depending on the solvent evaporation rate and the substrate speed, wet films of varying thicknesses are achievable and can produce aligned crystalline domains in the dried films.

The free (liquid–air) and fixed (liquid–solid) interfaces are also relevant boundary conditions when considering the fluid mechanics of these systems. Solvent choice is especially important because of its effect on the rate of solvent evaporation [38]. Binary azeotropic solvent mixtures, for example, have been exploited, yielding improvements to both the film morphology and the performance of devices incorporating deposited organic semiconductor thin films. For instance, the performance properties of optical fibers, including abrasion, resistance, and strength are strongly influenced by adding a coated layer [39] (Fig. 9.3).

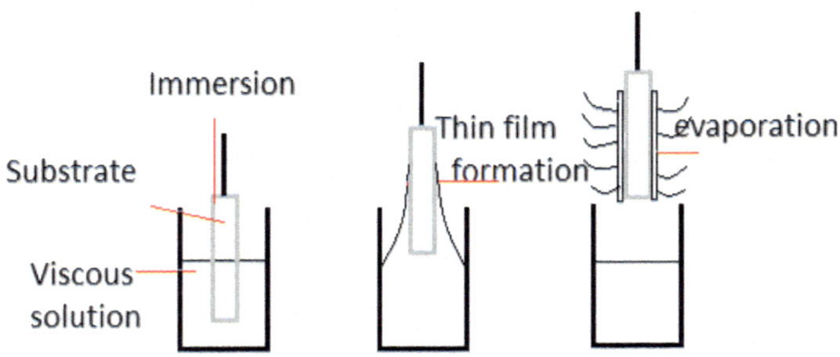

Fig. 9.3 Stages of the dip coating process: dipping the substrate into the coating solution, wet layer formation by withdrawing the substrate, and gelation of the layer by solvent evaporation (substantially modified from [108])

9.1.2.2 Spin Coating

There are two approaches to fabricate organic devices based on the materials employed: thermal vapor deposition and solution-based process. Spin coating or inkjet printing of polymers has been well studied since 1990 [40], and screen printing can efficiently be used to fabricate large area, high-resolution full-color flat panel displays [41, 42].

Spin coating is commonly used for solution deposition. This technique can be thought of as having three important stages: (1) dispense the coating solution, (2) fluid flow dominated thinning, and (3) solvent evaporation and coating "set." These three aspects will be described in sequence, though the stages are not always very distinctly separated in time [38, 43]. In the spin coating method, the solvent dries relatively fast allowing less time for molecular ordering compared to that in solution casting (Fig. 9.4).

The solution is dropped onto a substrate, and the substrate is accelerated to a high angular velocity to spread the liquid and evaporate the solvent simultaneously. The thickness of the wet film is inversely related to the spin speed and also depends on the solution concentration and viscosity. The films deposited by spin coating technique can be annealed afterward to improve molecular ordering. For materials with tendencies to form a highly ordered molecular packing, it was found that even spin-coated films can achieve very high mobilities [38, 44]. For solution deposition, semiconductor concentration, solvent evaporation rate, the solubility of the semiconductor, and the nature of the substrate surface play essential roles in the quality of the resulting semiconductor films.

Unfortunately, this method is not scalable to large volume production. Each substrate has to be handled individually and patterning the layers would require costly

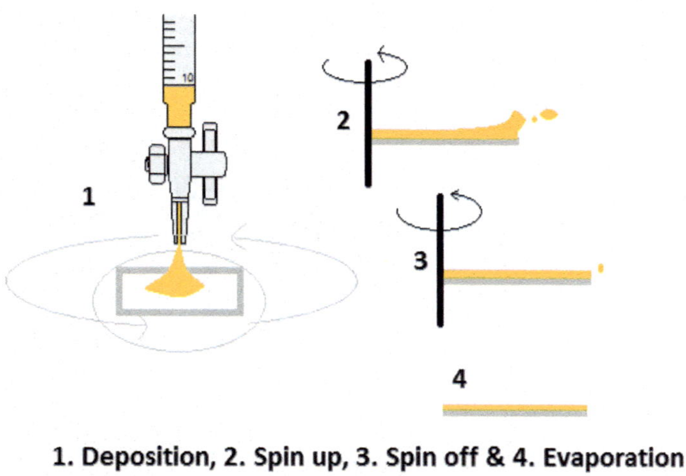

Fig. 9.4 Schematic illustrating the spin coating process

subtractive patterning steps such as lithography. Additionally, the most applied solution is wasted, and only a small amount is eventually used to coat the substrate.

9.1.2.3 Printing

Driven by the needs of device qualities such as low cost, large area, and flexibility, significant progress has been made recently in developing suitable printing methods for the fabrication of organic devices. Methods such as screen printing and inkjet printing have already been widely used to produce other devices [38, 45].

(a) **Inkjet printing**

Among several solution processes, inkjet printing, in which desired patterns can be fabricated by printing functional inks on a substrate [46–49], has been considered as a promising technique for the implementation of solution processed OTFTs in the field of printed electronics. The significant advantage of inkjet printing is the development of a fabrication method for semiconductor films with uniform and well oriented crystalline structures (Fig. 9.5).

In inkjet printing, "a droplet is ejected at the nozzle orifice. As it hits the surface of the substrate, the droplet spreads and wets the substrate. Spreading and wetting depends on the surface tension of both the fluid as well as the substrate. Film formation occurs by the coalescence of adjacent droplets. A liquid bulk is built that shows particular drying kinetics controlled by the solvent properties. Inkjet printing utilizes low viscosity inks, of viscosity 1^{-10} mPa-s, or 1^{-10} cP (centipoise). Precursor drying, surface tension, and ink stability must be engineered carefully to enable uniformly printed electronics with predictable material properties" [50].

(b) **Screen printing**

Screen printing is a relatively simple printing method that has been used for printing circuit boards [51]. It has been used to print all the active layers in OTFTs

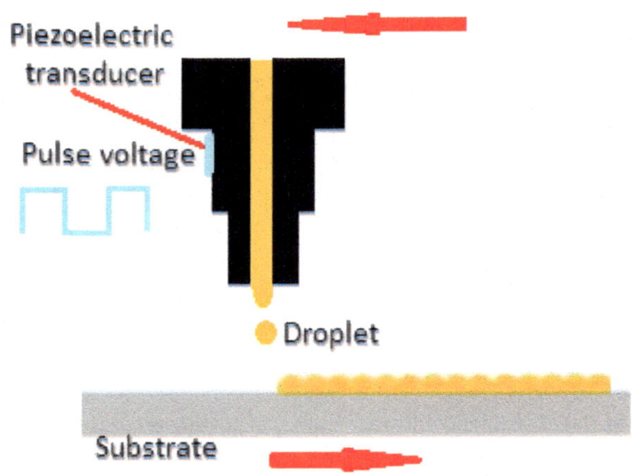

Fig. 9.5 Piezo inkjet printing (substantially modified from [109])

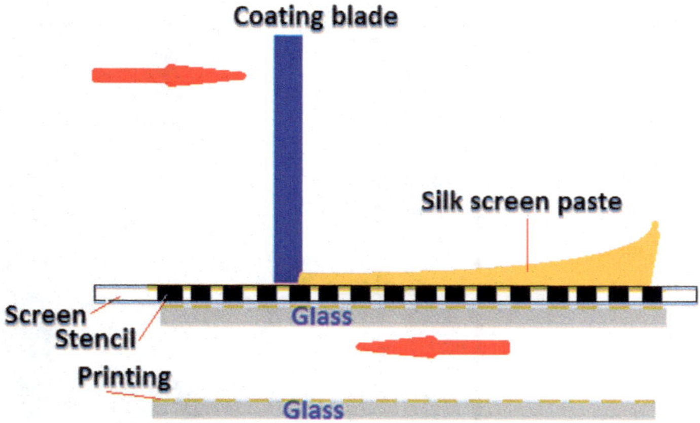

Fig. 9.6 Schematic illustrating the screen printing process (modified from [110])

[52]. The popularity of the technique lies mostly in its simplicity and compatibility with various materials and surfaces of substrates, which do not need to be planar. A conventional screen printer mainly consists of screen, stencil, squeegee, and ink to achieve uniform patterns [53]. This process can create films up to 100 μm thick with relative ease. In its simplest configuration, screen printing consists of creating an ink containing the active powder, an organic carrier fluid (such as pine oil), dispersing agent and ceramic binder. The ink is printed onto the substrate and dried to remove the carrier fluid. The active powder, along with any nonvolatile materials, is then left behind in a loosely bound state. A high-temperature sintering stage is used to densify the powder and bond it to the surface of the substrate (Fig. 9.6).

Some research groups have presented screen printing as a deposition method for conjugated polymer-based solar cells and obtained a maximum power conversion efficiency of 1.25% under AMI.5G spectra at 100 mW/cm^2 intensity [54, 55]. All these results prove the potential of screen printing for manufacturing OPVs. Moreover, screen printing is an elegant and fast technology that is compatible with roll-to-roll processing [50].

9.1.2.4 Spray Coating

The deposition of organic functional materials based on small molecules of polymers [56] on solid surfaces is of paramount importance for applications in various scientific and technological fields ranging from organic electronics to paints and protective coatings [57, 58]. Several techniques such as inkjet and screen printing have endeavored to deliver fast deposition over large areas. Within the framework of these efforts, spray coatings capture increasing attention with organic semiconductor deposition [59–61].

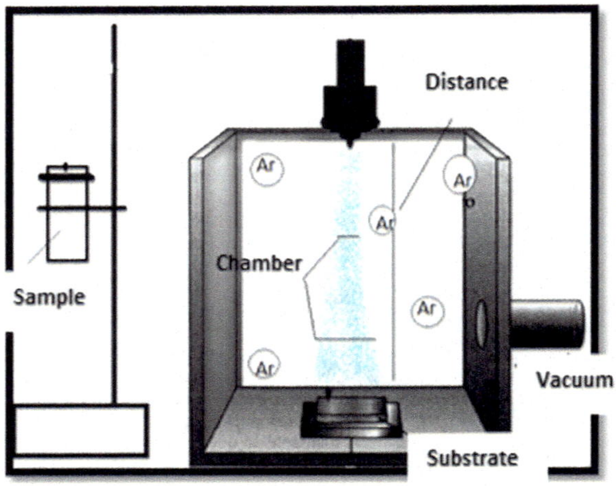

Fig. 9.7 Spray coating scheme

Spray coating was recently reported as a convenient technique for the fabrication of bulk heterojunction (BHJ) devices. This technique has also dominated aerosol deposition. The spray deposition process can be broken down into several subprocesses that consist of generation of the aerosol, transport of the aerosol droplets to the substrate, nucleation, and aggregation of the aerosol droplets in the thin film, and thermal processing to yield the final composition and structure (Fig. 9.7).

Single droplets are deposited by the transfer gas pressure with a high velocity onto the substrate. The droplets become immediately dry upon reaching the substrate surface. This results in a different morphology compared to more conventional techniques such as inkjet printing.

There are several techniques by which aerosols are generated for thin-film deposition processes. These techniques are based on the transfer of pneumatic, ultrasonic, or electrostatic energy to bulk liquid in sufficient magnitude to break liquid tension forces.

9.2 Properties

Organic semiconductors can be divided into two main groups, oligomers, and polymers and both have been of great scientific and technological interest over the past decades [62]. Oligomers are organic materials of low molecular weight and like polymers, have the alternated π-conjugated bonds along the carbon backbone. The conduction phenomena in organic materials occur at molecular orbitals of much weaker interactions (van der Waals bonds) than those of their counterpart inorganic semiconductors. In this respect, conduction phenomena of inorganic crystals have been widely and thoroughly studied, resulting in the band theory of solids, which has served as a base to explain the electrical conduction in organic semiconductors leading to several theories and models. Nonetheless, charge transport in organic molecules is strongly dependent on its conjugation degree and structure. In this section, the general aspects of the chemical and structural properties of organic semiconductors, as well as the accepted theories of conduction mechanisms, will be revisited.

9.2.1 Conjugated Polymers

It has been almost four decades since the serendipitous discovery of Alan J. Heeger, Alan MacDiarmid, and Hideki Shirakawa about the high conductivity of oxidized iodine-doped polyacetylene, a new realm in organic materials research was born [63, 64]. Nowadays, there are a vast number of electronic devices that are constituted with organic materials like field-emission transistors, solid-state lasers, organic light-emitting diodes (OLEDs), and solar cells [65–68].

Semiconducting polymers are macromolecular materials constituted by a carbon backbone with single, double or triple bonds alternated along the chain; the polymers with this particular characteristic are called conjugated polymers. The alternate bond arrangement leads to an unbalanced bond distance along the polymer chain and, as a consequence, the electrons are delocalized showing mobility along the shared bonds. This union is well known as **π-conjugated**, which is formed by the delocalization of unpaired electrons from p orbitals as shown in Fig. 9.8. When a photon is absorbed, a π-electron is promoted from the π to the π^* band. The electrons move along π-electron clouds thus allowing electronic conduction and σ electrons are not affected and remain in their molecular orbital.

Semiconducting polymers can be classified into two types, **intrinsic** and **extrinsic**. Intrinsic polymers are those with the before mentioned carbon backbone structure, and their semiconducting properties are a function of the doping, structure, and crystallinity. These types of polymers have been widely studied, and some of the most used are polyacetylene, polypyrrole, polythiophene, and polyaniline. Extrinsic polymers are composite materials composed of a polymer matrix and a conductor material like metallic powders or wires, and the semiconducting properties are determined by the inorganic material concentration.

Intrinsic polymers can be modified to improve their conductivity by the addition of a new chemical species into the polymer chain or “doping.” However, the process of charge transfer occurs differently from inorganic semiconductors. Instead, it can be carried out in four forms, chemical, electrochemical, photochemical, and interfacial [69]. Moreover, depending on the dopant charge, two types of semiconductor polymers can be achieved which are n-type and p-type. Partial oxidation produces p-type polymers and can be obtained chemically or electrochemically, while n-type polymers are produced by partial reduction by adding cations to maintain electroneutrality.

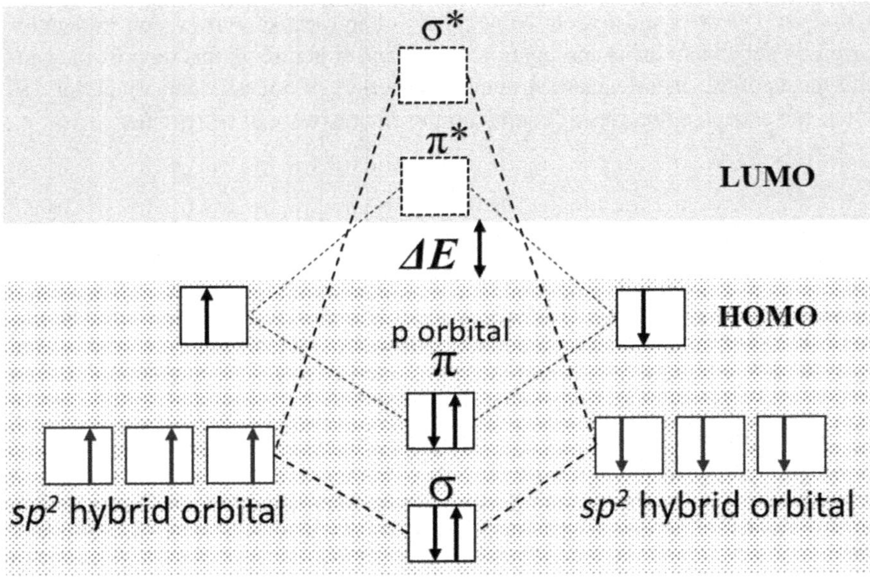

Fig. 9.8 Molecular orbital diagram of carbon sp^2 hybridization

9.2.2 Conduction Phenomena

It is worth mentioning that polymers are versatile materials that change their conduction behavior from that of an insulator to a conductor or semiconductor by the addition of dopants; therefore, a wide range of possible combinations to produce electronic devices that are mostly composed of organic materials has been opened.

Previously, it was mentioned that the electron delocalization of the π-electrons is the main cause of electric conduction and it is also accompanied by electron–phonon coupling and energetic and positional disorder. Moreover, the mobility of π-electrons along the chain is not sufficient for macroscopic conduction; chain-to-chain mobility is a must considering that the charge carriers are spinless [69]. In polymers, there are no valence and conduction bands as in inorganic semiconductors. We are dealing with molecules that are formed from covalently bonded atoms sharing electrons and forming molecular orbitals. The molecular orbitals afar from the nucleus are called frontier orbitals; there, the highest occupied molecular orbital (HOMO) and the lowest unoccupied molecular orbital (LUMO) are found (see Fig. 9.8). The difference in energy between these orbitals is called HOMO–LUMO gap and can be vaguely related to the energy bandgap of inorganic semiconductors. Also, there is the single occupied molecular orbital (SOMO) which refers to a molecular orbital occupied by a single electron. In the ground state, HOMO is fully occupied while LUMO is empty, and under excitation by a photon with the needed energy, will promote an electron from the HOMO to the LUMO, similarly as in inorganic semiconductors.

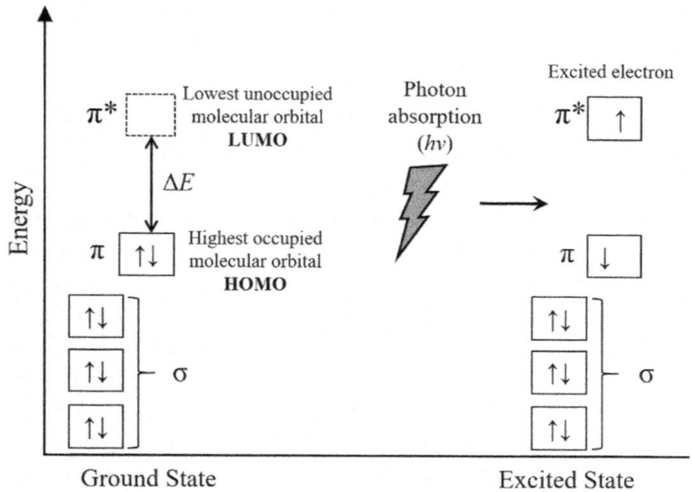

Fig. 9.9 Schematic representation of the HOMO and LUMO in organic molecules

However, this process is not that simple, the position of these levels defines the conjugated polymer bandgap as well as its spectroscopic and redox properties [70]. The energy gap in organic semiconductors has values that are in the interval of 1.5–3 eV, which falls in the visible spectral range [71]. The added charge is localized on the chain, and it generates a local molecular distortion. This promotes localized electronic states in the gap and an energy shift in both HOMO and LUMO levels. When an electron is removed from the chain, it forms what we call a cation; the energy associated with the removed electron lowers the ionization energy, and the cation has a higher energy than that of the valence band. Thus, the radical energy level is localized between bands and at the same time, it is surrounded by the molecular distortion over a segment of the chain polymer and is called a polaron [69, 72]. The positions of HOMO and LUMO levels are very important for semiconducting polymer applications, i.e., for polymer and semiconductor nanocrystal hybrid materials, there must be a proper alignment of the donor (electron donor) and acceptor (nanocrystal) LUMO levels to ensure an effective charge separation for photovoltaic devices [70] (Fig. 9.9).

From solid-state physics, a polaron is a molecule (part of an oligomer or polymer) with different ionization energy than that of the molecule at ground state; thus, anions or cations can be formed depending on the added dopant. If another electron is removed from the chain and moves with the polaron, then a bipolaron is formed. In this case, two cations or a dication are formed [69]. Otherwise, if an extra electron is removed from a different segment of the polymer chain, then, two polarons are formed. However, bipolarons are thermodynamically more stable because of the higher ionization energy than that of two polarons despite Coulomb repulsions. For example, polyacetylene has a degenerate ground state; this is two resonant forms of equal total energy. The difference among them arises from the exchange in single and double bonds length, and the resulting bipolaron splits into two independent cationic units which are solitons.

9.2.2.1 p-Type Organic Semiconductors

Cation formation leads to an electron-rich HOMO level leading to a hole transport material, which are the charge carriers responsible for conductivity in p-type semiconductors. There is a wide variety of hole transport in organic materials available that has been applied in different types of organic devices such as organic light-emitting diodes (OLEDs), organic thin-film transistors (OTFTs) and organic solid-state photovoltaic devices (OPVs) [73, 74]. Polythiophene and its derivatives, especially Poly(3-hexylthiophene-2,5-diyl) or P3HT, have been the subject of intense research around the world for their good charge carrier mobility and a high degree of crystallization. From the family of acenes, one of the first studied conjugated oligomers as a p-type semiconductor was pentacene; it was demonstrated to have efficient charge transport when used in OTFTs, derived from its high molecular order and large grain size formation [75]. There are several types of oligomers and their derivatives like oligothiophenes, fused, unsubstituted and substituted; co-oligomers such as thiophene-thiazole, thiophene-phenylene, thiophene-fluorene, thiophene-acene, selenophene; N-heterocyclic oligomers and tetrathiafulvalene derivatives [74].

9.2.2.2 n-Type Organic Semiconductors

Since most of the development in organic semiconductors has been directed to hole transport materials, n-type or electron transport polymers have been set aside for a while and did not receive much interest. However, there are some polymers commercially available that are n-type semiconductors; among them, we can count Poly(benzimidazobenzophenanthroline), Poly(2,5-di(3,7- dimethyloctyloxy)cyanoterephthalylidene), Poly(2,5-di(hexyloxy)cyanoterephthalylidene), and Poly(2,5-di(octyloxy)cyanoterephthalylidene). These polymers are electron deficient, that is, anions are formed by electron hopping from molecule to molecule. Also, development of this type of organic semiconductors is intimately related with its final electronic application, and there is no such thing as an "all-purpose" n-type semiconductor regarding the electronic device in which it will be used (Organic light-emitting diodes, OLEDs, Organic field-effect transistors, OFETs, organic photovoltaics, OPVs). Moreover, the interaction of the n-type organic semiconductor has to match the properties of the other materials used to form the heterojunction. Thus, HOMO and LUMO levels of both donor and acceptor have to meet determined energies for transport to occur. If the n-type organic material will be subjected to electrical charge injection, the LUMO energy is kept lower (<4.0 eV) to avoid the reduction of the molecules due to its electron deficient nature [76, 77]. In this respect, conjugated polymers are being replaced by oligomers, which have shown properties like electron field-effect conduction and high electron affinity that makes them interesting as n-type semiconductors [76, 78]. Nevertheless, there are other requirements like the ease of synthesis, high transport mobility, and stability during processing and affinity to inject charge at the interface that the material has to meet before it is applied in an electronic device. For a more detailed review of the particular aspects of the design of small molecule organic n-type semiconductors for some applications like OTFTs and OLEDs, the reader can consult the references cited here [67, 77–79].

9.2.3 *Charge Carrier Mobility*

It was mentioned previously that for organic semiconductors, molecular orbitals and radical molecules were involved, and either holes or electrons are transported by means of a redox reaction from molecule (M) to molecule generating M^+ or M^-. This process is regarded as hopping since the charge is strongly localized in one molecule and then transfers to the next neutral molecule and so on. Electron transport occurs as follows: in hole transport materials, electrons are transferred from M to a cation M^+ through the HOMO; and for electron transport materials, electrons transfer from an anion M^- to M through the LUMO. Some molecular crystals with high purity can present band-to-band transitions at low temperature as they occur in inorganic crystals [71]. Moreover, mobility is affected by structural features like degree of crystallization and morphology as well as if the material is a molecular crystal or an oligomer; it is also important to take into account if the charge is photoinduced or injected.

One of the most extended methods to determine charge carrier mobility is time of flight technique, which consists of measuring the carrier transport time (τ) of a photogenerated sheet of charge carrier that drifts across the sample under an applied electric field. The transient time is then expressed as:

$$\mu = \frac{d^2}{V\tau}$$

where μ is the charge carrier drift mobility, d is the sample thickness, and V the applied voltage. The transient photocurrent is plotted as a function of time, and depending on the degree of order, dispersion occurs, decreasing the velocity of the charge carrier as it travels through the layer. In a typical nondispersive curve, the value of τ is determined from the cusp of the transient photocurrent; while in a dispersive curve, there is no cusp, and it is necessary to convert both photocurrent and time axis to logarithmic scales to reveal the onset in the graph. Some disadvantages of this method are the need for a transparent electrode and a sample thickness restriction in the range of 5–20 μm. Other techniques used to estimate τ are space charge limited current (SCLC), carrier extraction by linearly increasing voltage (CELIV), double injection (DoI), and impedance spectroscopy [80].

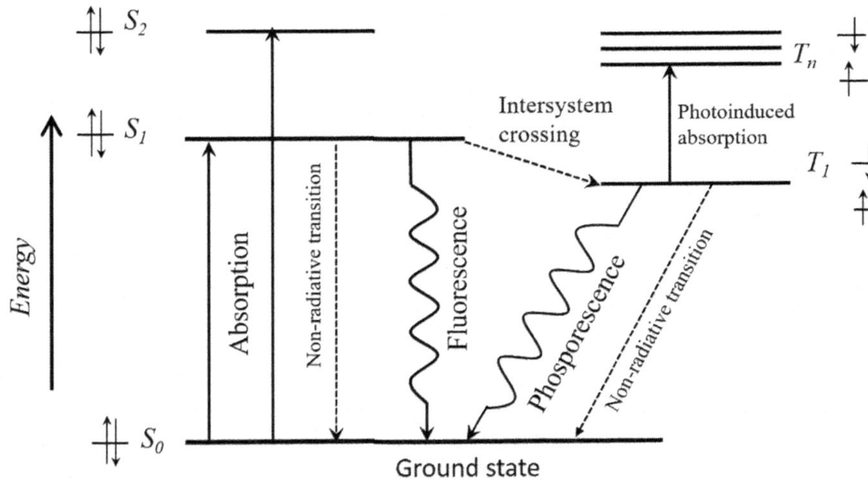

Fig. 9.10 Jablonski diagram of possible transitions in an organic molecule

9.2.4 Optical Properties

The bandgap of an organic material is often described by the energy difference between HOMO and LUMO levels having values lower than 3.0 eV [81]; it determines its application as well. The HOMO level is usually referred to as the ground state, and its net spin is zero if photoexcitation of the molecule occurs, then an electron is promoted to the LUMO level. The excited states are characterized by either net spin zero or one, which are labeled as singlet states, S_n; or triplet states, S_T, respectively. The transition of an electron from HOMO to LUMO in semiconducting polymers is accompanied by other vibrionic phenomena due to the high delocalization of the charge.

In Fig. 9.10, the Jablonski diagram for an organic molecule shows different possible transitions. When a photoluminescent material is subjected to UV light radiation, an electron of the outer shell of the molecule at S_0 is promoted to a singlet state (S_1 for example); then it drops off to ground state releasing energy by emitting a photon through a phenomenon called fluorescence. Furthermore, a promoted electron can relax in other ways by releasing the extra energy dissipating heat to the media (solvent), this is a "non-radiative relaxation." If the electron goes from a singlet to triplet state, which has a different energy, eventually the electron will loose its energy causing phosphorescence or another non-radiative relaxation. It can occur that another molecule is in the proximity interacting with the excited molecule, causing a quenching of the fluorescence [82].

The proximity of other molecules tends to separate HOMO and LUMO levels inducing the formation of vibrational sublevels by reorganization of the intermolecular distances. The promoted electron still has a strong attraction with the formed hole left in the HOMO level, leading to the formation of an exciton, which is basically described as the electron-hole pair formed by the coulombic attraction between them. In organic semiconductors, the exciton is not associated with a single molecule but to a collection of molecules or monomers, and it is strongly bonded and has a high binding energy characteristic of a Frenkel exciton [81, 83]. The exciton binding energy is the work required to separate the hole-electron pair and determines the color (wavelength) of the polymer or oligomer emission when it undergoes photoexcitation; and if it is large enough, the material can be used for organic photovoltaic devices (OPV) [72, 84]. From Fig. 9.10, excitons can be singlet and triplet excitons, S_n and T_n, respectively; usually, the binding energies of singlet excitons are in the range of 0.3–0.5 eV [85, 86].

We have mentioned that the absorption of light (photoexcitation) promotes an electron from the HOMO to LUMO forming an excited molecule. There is another path to achieve an excited state, and this is through electrical excitation. In this case, it is necessary to inject a charge (hole or electron) to the HOMO for hole transport and the LUMO for electron transport. This process is carried out by having the polymer film in a diode configuration sandwiched by two electrodes which inject the necessary charge to excite the organic molecules [81].

9.3 Applications

Similar to its inorganic counterparts, organic semiconductors have been the subject of extensive research to produce organic electronic devices such as organic photovoltaic cells (OPV), organic field-effect transistors (OFET), and organic light-emitting diodes (OLED) [2, 3, 73–77, 82]. However, organic semiconductors have certain limitations such as a short lifetime, degradation by UV light, temperature sensitivity, low efficiency compared to inorganic semiconductors, and not well understood charge transfer mechanisms. Despite these limitations, advantages like their lightweight, transparency, flexibility, and lower production cost make them candidates for the development of novel electronic devices fomenting research in this area. It is worthwhile to note that organic semiconductors have been combined with other carbon nanomaterials like carbon nanotubes, fullerenes, and graphene, to improve their charge carrier mobility, which is one of the limitations of polymers and oligomers [87, 88].

9.3.1 Organic Field-Effect Transistors

In general, a transistor works like an on/off switch and is an essential component of circuits in several electronic devices like displays; in the case of OFETs, they offer the possibility to fabricate flexible and transparent displays, sensors and memories [68, 76, 77, 79]. Field-effect transistors (FET) are constituted of several layers of conductor and insulator materials operating as a capacitor, where the active material is sandwiched between two ohmic contacts, the source and drain electrodes and the gate. [73, 89]. Indium tin oxide (ITO) is the most used transparent semiconductor as the anode, and gold, silver or aluminum as the cathode. A significant proportion of OFETs are of thin-film transistor type, and with two geometries, top contact or bottom contact as depicted in Fig. 9.11. Each of them has its advantages and disadvantages regarding the possibility to incorporate the process of organic semiconductor deposition to large scale manufacturing and performance. In the case of top contact, good charge mobility can be achieved but it presents shadowing effect that limits the reduction of channel dimensions; while bottom contact presents the challenge to deposit a homogeneous organic semiconductor layer over a non-smooth surface because the electrodes are deposited over the insulator layer [74]. If a negative voltage is applied to the electrodes, then hole transport in the organic layer will take place, and if a positive voltage is applied, then electron transport occurs. In the first case, we will have a p-channel device (hole transport) and in the second, an n-channel device (electron transport) (Fig. 9.11).

As was mentioned in the previous section, one of the key properties of organic semiconductors to be applied in OFETs is the field-effect mobilities. In the past, research on organic semiconductors was directed to achieve mobilities comparable to that of amorphous silicon (~1 $cm^2V^{-1}s^{-1}$). Nowadays, organic semiconductor materials are reaching and even surpassing that value [73, 77, 79, 90]. The most common polymer or oligomer deposition techniques are vacuum and solution deposition; the insulator layer can be an organic dielectric or a thin film of SiO_2 [90].

There is an extensive range of organic molecules that can be used in OFETs; among them are acenes, which consist of a series of aligned benzene rings conforming the backbone and are p-type semiconductors. Anthracene, tetracene, and pentacene are examples of crystalline organic materials; pentacene is one of the acenes that has good charge mobilities in the range of 5–40 $cm^2V^{-1}s^{-1}$, and it is a p-type semiconductor [73, 74, 87, 90]. Hexacene (a chain of six benzene rings) is a single crystal synthesized by physical vapor transport (PVT) that shows higher mobilities than that of pentacene when measured in a FET [91]. However, high number acenes are

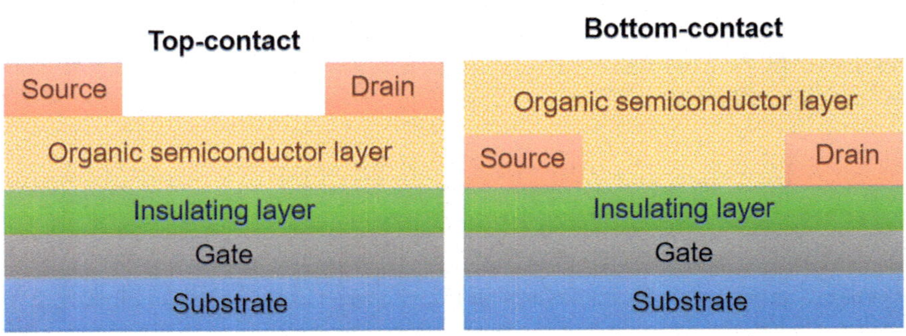

Fig. 9.11 Schematic representation of the top and bottom contact distribution of conductor, semiconductor and insulating layers in thin-film OFETs

unstable and suffer from oxidation and are insoluble in common organic solvents; one option is to substitute one of its rings with other functional groups like isopropylsilylethynyl pentacene (TIPS-pentacene), where isopropyl groups are substituted in the central ring making it more soluble and reducing its photo-oxidation [74, 92]. In general, acenes are functionalized, substituted (like in the case of pentathioacene) or both at the same time, and combined with other molecules like fluorene and thiophenes to improve their mobilities [93]. Polythiophenes are the oligomers that have attracted most of the attention for their ability to be processed by solution methods. Poly(3-hexylthiophene) or P3HT is the most used and studied semiconducting polymer not only for OFETs but also for OLEDs and OPVs in bulk heterojunction, and is commonly combined with other organic semiconductors, inorganic semiconductor nanocrystals, fullerenes, and so on.

Another application of OFETs is as point-of-care testing (POCT) systems, where a biochemical reaction is monitored through the detection of a biological response that is converted into a signal by a transducer making it measurable and be able to monitor at any time. In this type of device, a bio-recognition layer that can be directly deposited on the organic semiconductor or the dielectric layer, or in a configuration where there is a biological receptor over an electrolyte layer that is deposited on the organic semiconductor, is used. The configuration will depend on the analyte that is measured and how it will be measured regarding bio-recognition [94].

9.3.2 Organic Light-Emitting Diodes

OLEDs have been an intense research topic for the last two decades with the goal to develop flat panel displays and lightweight flexible electronic devices with real-life technological applications [82, 95]. In the past few years, OLEDs have become a marketable technology used to produce thin TV displays with ultra-high resolution; several companies have now implemented this technology [82, 87]. OLEDs are charge injection devices, where recombination occurs in the organic semiconductor layer producing either singlet or triplet excitons, which hop from molecule to molecule until its radiative decay and spontaneous light emission; the color of the light will correspond to the energy gap between the LUMO and HUMO of the organic material [72, 96–98].

These types of devices can have different structures regarding the number of electroluminescent semiconductor layers; for example, a white light-emitting OLED will have three electroluminescent layers, red, green, and blue light that is combined and seen as white light. Like OFETs, OLEDs are constructed in layers of several components including the electrodes (ITO as the anode and Al/Ca as the cathode) deposited over a substrate, which could be a flexible polymer. The first fabricated organic LED consisted of a Polyphenylene vinylene (PPV) layer sandwiched between ITO and aluminum layers [96]. Afterwards, small organic molecules were used as an electroluminescent layer and are what we call OLEDs; if a polymer is used as an emissive layer, then the device is denominated as polymer light-emitting diode

(PLED). In general, an OLED can be single-, double-, three-, or multilayer in which the following types of layers can be found: a hole injection layer, a hole transport layer, and an electron transport layer without counting the anode and the cathode [82].

In Fig. 9.12, the different types of OLEDs are schematically represented. The ITO anode forms a heterojunction with the emitting organic semiconductor layer; the cathode must have a low work function to allow the charge to be injected, calcium fulfills this requirement, and it usually substitutes aluminum [96]. The addition of another layer like the hole transport layer (HTL), electron transport layer (ETL), and the hole injection layer (HIL) aid to improve the charge injection and confine the charge carriers into the emitting organic layer [87], as presented in Fig. 9.12. A double-layer OLED has an HTL deposited over the ITO anode; this layer consists of a polymer with a higher work function than that of the anode which injects holes onto the HOMO of the HTL. The most used material is Poly(3,4-ethylendioxithiophen) polystyrene sulfonate (PEDOT:PSS) which injects hole charge carriers into the polymer incrementing the efficiency and the lifetime of the OLED [99]. On the other hand, ETLs serve the purpose of aiding the cathode for injecting electrons to the emitting layer; the electrons are injected from the cathode into the LUMO of the HTL material [98–100]. Both the HTL and ETL inject charge to the electroluminescent organic molecule or polymer under an applied electric field, electrons and holes transport toward the side with opposite charge by hopping through the layers until recombination in the emissive layer as explained above [100]. Materials like 2-(4-biphenylyl)-5-(4-tert-butylphenyl)-1,2,4-oxadiazole (PBD) are used as ETL, oxadiazole molecules and its dendrimers and polymers are the most employed organic molecules for this purpose, other materials are metal chelates, azole-based materials, triazines, polybenzobizasoles, and phenanthrolines to name a few [82, 101].

In recent days, OLEDs are commercially used in TV and cell phone displays and are classified as follows: passive-matrix OLED (PMOLED), active-matrix OLED (AMOLED), transparent OLED, top-emitting OLED, bottom-emitting OLED, foldable OLED, and white OLED (WOLED) [82]. Except for the last two types of OLEDs, organic molecules are the emissive layer. In foldable and white OLEDs, polymers and conjugated and metal complexes and organic dyes are used.

9.3.3 Organic Photovoltaic Devices

Crystalline silicon is the preferred material for solar cells followed by CdTe and Cu(In,Ga)Se_2. One of the challenges in solar energy conversion is to lower the production cost, and this is where organic semiconductors can play a role because they are cost-effective and can be processed at room temperature [102]. Moreover, the use of organic materials in photovoltaic cells (PV) opens the door to the integration of solar energy harvesting in large area structures like windows in a building. However, there are challenges that involve their stability, environmental degradation and

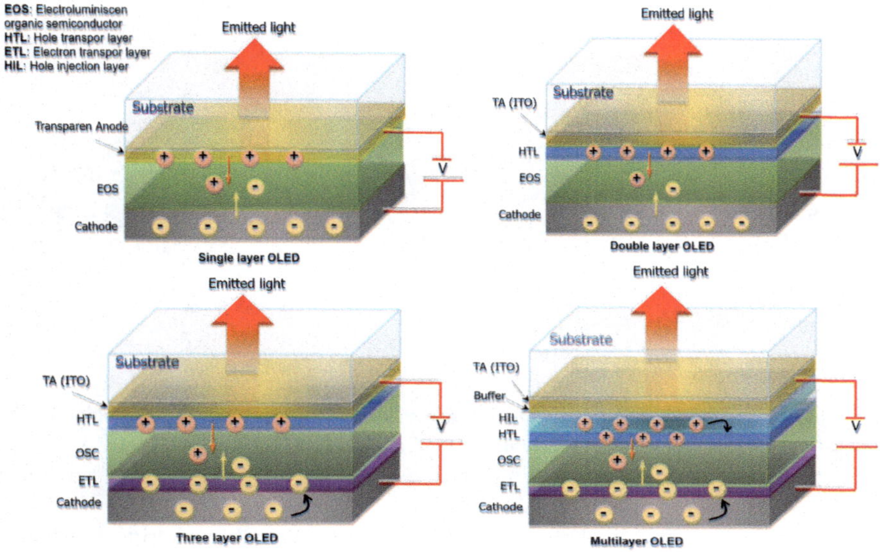

Fig. 9.12 Schematic representation of the different layer configurations of OLEDs

operational lifetime, even when the cost is lower than that of the silicon-based solar cells; the high durability and stability of silicon is preferred.

In an OPV device, the incident photons lead to the formation of excitons (electron-hole pairs), which migrate to zones where dissociation occurs; then the separated charges travel toward the electrodes and thus charge transport within the photoactive material is produced [103]. A basic structure of OPVs consist of a bulk heterojunction conformed by a mixture of nanoscale domains of the donor and acceptor semiconductors sandwiched between the electrodes, a hole extraction layer or HTL is deposited between the organic material and the anode as schematically represented in Fig. 9.13. The light conversion mechanism is as follows:

(i) Upon light absorption, electrons in the HOMO of the donor material undergo photo excitation and are promoted to the LUMO forming excitons.
(ii) The exciton usually has a binding energy in the range of 0.5–1 eV [104].
(iii) Excitons migrate toward the interface between the donor and acceptor before recombination occurs (fullerenes are the most used acceptor materials) [102, 104].
(iv) At the donor–acceptor interface, electron transfer process takes place forming charge transfer (CT) excitons, the hole-electron pair remains at the interface by coulombic attraction.
(v) CT exciton separates into electrons and holes and then moves toward the electrodes driven by the built-in electric field and a gradient occurs in the charge carrier concentration [102, 104].

In planar heterojunctions, the donor and acceptor materials are stacked in layers of a very small thickness, and this is because the formation and dissociation of excitons occurs around 10–20 nm at the interface [103]. This type of arrangement is

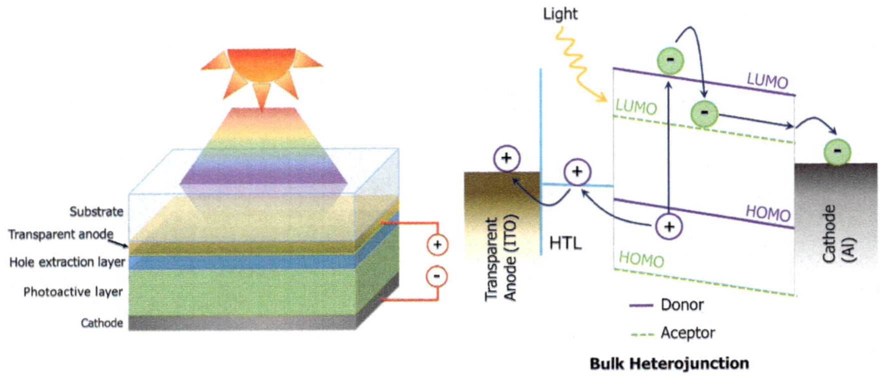

Fig. 9.13 Typical configuration of an OPV device as well as the schematic representation of the charge carrier transport by photoexcitation

less efficient due to limited charge generation around the mentioned critical distance from the donor–acceptor interface, and the photons absorbed to form excitons further away from the interface are lost within the materials. It is worth mentioning that in bulk heterojunctions, the donor and acceptor regions need to have a proximity in the range of 20 nm for an efficient energy conversion (Fig. 9.13).

OPVs have several challenges to overcome before commercialization among which are the tailoring of low energy bandgap polymers that can absorb a wider range of the spectrum and reach higher efficiencies. The most used organic materials are the polymer P3HT, and copper and zinc phthalocyanines as donors, fullerenes, and their derivatives like [6,6]-phenyl-C61-butyric acid methyl ester ($PC_{61}BM$) and [6,6]-phenyl-C71-butyric acid methyl ester ($PC_{71}BM$) as acceptors [102, 105]. Further research is ongoing to develop new materials as well as to address the chemical, physical, interface chemistry, synthesis and production challenges for practical application of OPVs. It is clear that in the future when this technology translates into commercial use, a great step into clean energy generation will be reached.

References

1. Angus R (2008) The materials science of semiconductors. Springer, London
2. Ling MM, Bao Z (2004) Thin film deposition, patterning, and printing in organic thin film transistors. Chem Mater 16(23):4824–4840. https://doi.org/10.1021/cm0496117
3. Perepichka DF, Perepichka IF, Meng H, Wudl F (2007) Light-emitting polymers. In: Li Z, Meng H (eds) Organic light-emitting materials and devices. CRC Press, Boca Raton, Fla, pp 45–293
4. Rolin C, Vasseur K, Genoe J, Heremans P (2010) Growth of pentacene thin films by in-line organic vapor phase deposition. Org Electron 11(1):100–108. https://doi.org/10.1016/j.orgel.2009.10.005
5. Baldo M, Deutsch M, Burrows P, Gossenberger H, Gerstenberg M, Ban V, Forrest S (1998) Organic vapor phase deposition. Adv Mater 10(18):1505–1514. https://doi.org/10.1002/(SICI)1521-4095(199812)10:18%3c1505:AID-ADMA1505%3e3.0.CO;2-G

6. Forrest SR (2004) The path to ubiquitous and low-cost organic electronic appliances on plastic. Nature 428:911–918. https://doi.org/10.1038/nature02498
7. Mahon K, Zhou TXSR, Forrest M, Schwambera N Meyer (2002) Organic VPD shows promise: for OLED volume production. Solid State Technol 45(7):131–139
8. McGraw GJ, Forrest S (2012) Fluid dynamics and mass transport in organic vapor jet printing. J Appl Phys 111(4):043501. https://doi.org/10.1063/1.3680881
9. Shtein M, Peumans P, Benziger JB, Forrest SR (2004) Direct, mask-asks solvent-solvent printing of molecular organic semiconductors. Adv Mater 16(18):1615–1620. https://doi.org/10.1002/adma.200400260
10. Biswas S, Pipe KP, Shtein M (2010) Solvent-free, direct printing of organic semiconductors in atmosphere. Appl Phys Lett 96(26):125
11. Yamashita A, Hayashi T (1996) Organic molecular beam deposition of metallophthalocyanines for optolelectronics applications. Adv Mater 8(10):791–799. https://doi.org/10.1002/adma.19960081004
12. Rompf C, Kowalsky W (1995) Organic molecular beam deposition (OMBD) for photonic and electronic devices. Annual report 1995, Institut fur Hochfrequenztechnik, TU Braunschweig. https://www.tu-braunschweig.de/Medien-DB/ihf/p045-047.pdf 30 July 2016
13. Constantinescu C, Rapp L, Rotaru P, Delaporte P, Alloncle AP (2015) Pulsed laser processing of poly (3, 3‴-didodecyl quarter thiophene) semiconductor for organic thin film transistors. Chem Phys 450:32–38. https://doi.org/10.1016/j.chemphys.2015.02.004
14. Gritsenko KP, Tolstopyatov EM, Krasovsky AM (2001) Thin film deposition by laser ablation of polymer and dye in a vacuum. Optical Mem Neural Netw 10(3):173–194
15. Gritsenko KP, Krasovsky AM (2003) Thin-film deposition of polymers by vacuum degradation. Chem Rev 103(9):3607–3650. https://doi.org/10.1021/cr010449q
16. Eason R (ed) (2007) Pulsed laser deposition of thin films: applications-led growth of functional materials. Wiley, Hoboken
17. Piqué A (2011) The matrix-assisted pulsed laser evaporation (MAPLE) process: origins and future directions. Appl Phys A Matter Sci Process 105(3):517–528. https://doi.org/10.1007/s00339-011-6594-7
18. Delaporte Ph, Ainsebaa A, Alloncle AP, Benetti M, Boutopoulos C, Cannata D, Di Pietrantonio F, Dinca V, Dinescu M, Dutroncy J, Eason R, Feinaugle M, Fernández-Pradas JM, Grisel A, Kaur K, Lehmann U, Lippert T, Loussert C, Makrygianni M, Manfredonia I, Mattle T, Morenza JL, Nagel M, Nüesch F, Palla-Papavlu A, Rapp L, Rizvi N, Rodio G, Sanaur S, Serra P, Shaw-Stewart J, Sones CL, Verona E, Zergioti I (2013) Applications of laser printing for organic electronics. In: Proceeding of SPIE 8607, Laser applications in microelectronic and optoelectronic manufacturing (LAMOM). https://doi.org/10.1117/12.2004062
19. Rapp L, Constantinescu C, Delaporte P, Alloncle AP (2014) Laser-induced forward transfer of polythiophene-based derivatives for fully polymeric thin film transistors. Org Electron 15(8):1868–1875. https://doi.org/10.1016/j.orgel.2014.04.029
20. Shaw-Stewart J, Lippert T, Nagel M, Nüesch F, Wokaun A (2012) A simple model for flyer velocity from laser-induced forward transfer with a dynamic release layer. Appl Surf Sci 258(23):9309–9931. https://doi.org/10.1016/j.apsusc.2011.08.111
21. Chrisey D, Huber GK (eds) (1994) Pulsed laser deposition of thin films. Wiley, New York
22. Willmott PR, Huber JR (2000) Pulsed laser vaporization and deposition. Rev Mod Phys 72(1):315. https://doi.org/10.1103/RevModPhys.72.315
23. Yang X, Tang Y, Yu M, Qin Q (2000) Pulsed laser deposition of aluminum tris-8-hydroxyquinline thin films. Thin Solid Films 358(1):187–190. https://doi.org/10.1016/S0040-6090(99)00687-2
24. Chneider CW, Lippert T (2010) Laser ablation and thin film deposition. Laser processing of materials, vol 139. Springer, Heidelberg, pp 89–112
25. Caricato AP, Anni M, Cesaria M, Lattante S, Leggieri G, Leo C, Martino M, Perulli A, Resta V (2015) MAPLE-deposited PFO films: influence of the laser fluence and repetition rate on the film emission and morphology. Appl Phys B 119(3):453–461. https://doi.org/10.1007/s00340-015-6078-0

26. Bloisi F, Pezzella A, Barra M, Alfè M, Chiarella F, Cassinese A, Vicari L (2011) Effect of substrate temperature on MAPLE deposition of synthetic eumelanin films. Appl Phys A 105(3):619–627. https://doi.org/10.1007/s00339-011-6603-x
27. Zergioti I, Mailis S, Vainos NA, Papakonstantinou P, Kalpouzos C, Grigoropoulos CP, Fotakis C (1998) Microdeposition of metal and oxide structures using ultrashort laser pulses. Appl Phys A-Matter 66(5):579–582. https://doi.org/10.1007/s003390050717
28. Yang L, Wang CY, Ni XC, Wang ZJ, Jia W, Chai L (2006) Microdroplet deposition of copper film by femtosecond laser-induced forward transfer. Appl Phys Lett 89(16):161110. https://doi.org/10.1063/1.2364457
29. Dinca V, Kasotakis E, Catherine J, Mourka A, Mitraki A, Popescu A, Dinescu M, Farsari M, Fotakis C (2007) Development of peptide-based patterns by laser transfer. Appl Surf Sci 254(4):1160–1163. https://doi.org/10.1016/j.apsusc.2007.08.042
30. Pohl R, Jansink M, Römer GRBE, Huis AJ (2015) Solid-phase laser-induced forward transfer of variable shapes using a liquid-crystal spatial light modulator. Appl Phys A 120(2):427–434. https://doi.org/10.1007/s00339-015-9212-2
31. Rapp L, Biver E, Alloncle AP, Delaporte P (2014) High-Speed laser printing of silver nanoparticles ink. J Laser Micro/Nanoeng 9(1):5–9. https://doi.org/10.2961/jlmn.2014.01.0002
32. Hennig G, Baldermann T, Nussbaum C, Rossier M, Brockelt A, Schuler L, Hochstein G (2012) Lasersonic® LIFT process for large area digital printing. J Laser Micro/Nanoeng 7(3):299–305. https://doi.org/10.2961/jlmn.2012.03.0012
33. Li M, An C, Pisula W, Mullen K (2014) Alignment of organic semiconductor microstripes by two-phase dip-coating. Small 10(10):1926–1931. https://doi.org/10.1002/smll.201303182
34. Wang B, Zhu T, Huang L, Tam TLD, Cui Z, Ding J, Chi L (2015) Addressable growth of oriented organic semiconductor ultra-thin films on hydrophobic surface by direct dip-coating. Org Electron 24:170–175. https://doi.org/10.1016/j.orgel.2015.05.038
35. Shan L, Liu D, Li H, Xu X, Shan B, Xu JB, Miao Q (2015) Monolayer field-effect transistors of nonplanar organic semiconductors with brickwork arrangement. Adv Mater 27(22):3418–3423. https://doi.org/10.1002/adma.201500149
36. Kistler SF, Schweizer PM (eds) (1997) Liquid film coating. Chapman & Hall, London
37. Peralta JM, Meza BE, Zorrilla SE (2014) Mathematical modeling of a dip-coating process using a generalized Newtonian fluid. 1. Model development. Ind Eng Chem Res 53(15):6521–6532. https://doi.org/10.1021/ie500407t
38. Diao Y, Shaw L, Bao Z, Mannsfeld SCB (2014) Morphology control strategies for solution-processed organic semiconductor thin films. Energy Environ Sci 7:2145–2159. https://doi.org/10.1039/C4EE00688G
39. Rogowski RZ, Dzwilewski A, Kemerink M, Darhuber AA (2011) Solution processing of semiconducting organic molecules for tailored charge transport properties. J Phys Chem C 115(23):11758–11762. https://doi.org/10.1021/jp201219h
40. Burroughes JH, Bradley DDC, Brown AR, Marks RN, Mackay K, Friend RH, Holmes AB (1990) Light-emitting diodes based on conjugated polymers. Nature 347(6293):539–541
41. Sonoyama T, Ito M, Seki S, Miyashita S, Xia S, Brooks J, Brown JJ (2008) Inkjet printable phosphorescent organic light emitting diode devices. J Soc Inf Display 16(12):1229–1236. https://doi.org/10.1889/JSID16.12.1229
42. Xia S, Cheon KO, Brooks JJ, Rothman M, Ngo T, Hett P, Kwong MC, Inbasekaran M, Brown JJ, Sonoyama T, Ito M, Seki S, Miyashita S (2009) Printable phosphorescent organic light emitting devices. J Soc Inf Display 17(2):167–172. https://doi.org/10.1889/JSID17.2.167
43. Schneller T, Waser R, Kosec M, Payne D (eds) (2013) Chemical solution deposition of functional oxide thin films. Springer, New York, pp 233–261
44. Sirringhaus H, Brown PJ, Friend RH, Nielsen MM, Bechgaard K, Langeveld-Voss BMW, Spiering AJH, Janssen RAJ, Meijer EW, Herwing P, de Leeuw DM (1999) Two-dimensional charge transport in self-organized, high-mobility conjugated polymers. Nature 401(6754):685–688. https://doi.org/10.1038/44359
45. Rogers J, Katz H (1999) Printable organic and polymeric semiconducting materials and devices. J Mater Chem 9(9):1895–1904. https://doi.org/10.1039/A902652E

46. Sirringhaus H, Kawase T, Friend RH, Shimoda T, Inbasekaran M, Wu W, Woo EP (2000) High-resolution inkjet printing of all-polymer transistor circuits. Science 290(5499):2123–2126. https://doi.org/10.1126/science.290.5499.2123
47. Chen J, Tee CK, Shtein M, Martin DC, Anthony J (2009) Controlled solution deposition and systematic study of charge-transport anisotropy in single crystal and single-crystal textured TIPS pentacene thin films. Org Electron 10(4):696–703. https://doi.org/10.1016/j.orgel.2009.03.007
48. Li H, Tee BC, Cha JJ, Cui Y, Chung JW, Lee SY, Bao Z (2012) High-mobility field-effect transistors from large-area solution-grown aligned C60 single crystals. J Am Chem Soc 134(5):2760–2765. https://doi.org/10.1021/ja210430b
49. Pierre A, Sadeghi M, Payne MM, Facchetti A, Anthony JE, Arias AC (2014) All-Printed flexible organic transistors enabled by surface tension-guided blade coating. Adv Mater 26(32):5722–5727. https://doi.org/10.1002/adma.201401520
50. Hoth CN, Schilinsky P, Choulis SA, Balasubramanian S, Brabe CJ (2013) Solution processed organic photovoltaics. In: Cantatore E (ed) Applications of organic and printed electronics. Springer, Boston, pp 27–56
51. Gilleo K (1996) Polymer thick films. Van Nostrand Reinhold, New York
52. Pola J, Kupcik J, Durani SM, Khavaja EE, Masoudi HM, Bastl Z, Šubrt J (2003) Laser ablative structural modification of poly (ethylene-a lt-maleic anhydride). Chem Mater 15(20):3887–3893. https://doi.org/10.1021/cm034262m
53. Krebs FC (2009) Processing and preparation of polymer and organic solar cells. Sol Energ Mat Sol C 93(4):394–412. https://doi.org/10.1016/j.solmat.2008.10.004
54. Aernouts T, Vanlaeke P, Geens W, Poortmans J, Heremans P, Borghs S, Mertens R, Andriessen R, Leenders L (2004) Printable anodes for flexible organic solar cell modules. Thin Solid Films 451:22–25. https://doi.org/10.1016/j.tsf.2003.11.038
55. Krebs FC (2008) Air stable polymer photovoltaics based on a process free from vacuum steps and fullerenes. Sol Energ Mat Sol C 92(7):715–726
56. Grimsdale AC, Mullen K (2005) The chemistry or organic nanomaterials. Angew Gchem Int Ed 44(35):5592–5629. https://doi.org/10.1002/anie.200500805
57. Rabe JP, Buchholz S (1991) Commensurability and mobility in two-dimensional molecular patterns on graphite. Science 253(5018):424–427. https://doi.org/10.1126/science.253.5018.424
58. Leclère P, Surin M, Brocorens P, Cavallini M, Biscarini F, Lazzaroni R (2006) Supramolecular assembly of conjugated polymers: from molecular engineering to solid-state properties. Mater Sci and Eng R 55(1):1–56. https://doi.org/10.1016/j.mser.2006.12.001
59. Büchele P, Morana M, Bagnis D, Tedde SF, Hartmann D, Fischer R, Schmidt O (2015) Space charge region effects in bidirectional illuminated P3HT: PCBM bulk heterojunction photodetectors. Organic Electron 22:29–34. https://doi.org/10.1016/j.orgel.2015.03.027
60. Azarova NA, Owen JW, McLellan CA, Grimminger MA, Chapman EK, Anthony JE, Jurchescu OD (2010) Fabrication of organic thin-film transistors by spray-deposition for low-cost, large-area electronics. Org Electron 11(12):1960–1965. https://doi.org/10.1016/j.orgel.2010.09.008
61. Treossi E, Liscio A, Feng X, Palermo V, Müllen K, Samorì P (2009) Large-area bi-component processing of organic semiconductors by spray deposition and spin coating with orthogonal solvents. Appl Phys A 95(1):15–20. https://doi.org/10.1007/s00339-009-5123-4
62. Coropceanu V, Cornil J, da Silva Filho DA, Olivier Y, Silbey R, Brédas JL (2007) Charge transport in organic semiconductors. Chem Rev 107:926–952. https://doi.org/10.1021/cr050140x
63. Shirakawa H, Louis EJ, MacDiarmid AG, Chiang CK, Heeger AJ (1997) Synthesis of electrically conducting organic polymers: halogen derivatives of polyacetylene, $(CH)_x$. Chem Commun https://doi.org/10.1039/c39770000578
64. Chinag CK, Fincher CR, Park YW Jr, Heeger AJ, Shirakawa H, Louis EJ (1978) Phys Rev Lett. https://doi.org/10.1103/PhysRevLett.40.1472
65. McGehe MD, Heeger AJ (2000) Semiconducting (conjugated) polymers as materials for solid-state Lasers. Adv Mater 12:1655–1668. https://doi.org/10.1002/1521-4095(200011)12:22%3c1655:AID-ADMA1655%3e3.0.CO;2-2

66. McCulloch I, Heeney M, Chabinyc ML, DeLongchamp D, Kline RJ, Cölle M, Duffy W, Fischer D, Gundlach D, Hamadami B, Hamilton R, Richter L, Salleo A, Shkunov M, Sparrowe D, Tierney S, Zang W (2009) Semiconducting Thienothiophene copolymers: design, synthesis, morphology, and performance in thin-film organic transistors. Adv Mater 21:1091–1109. https://doi.org/10.1002/adma.200801650
67. Zhao X, Zhan X (2011) Electron transporting semiconducting polymers in organic electronics. Chem Soc Rev 40:3728–3743. https://doi.org/10.1039/C0CS00194E
68. Chi HY, Hsu HW, Tung SH, Liu CL (2015) Nonvolatile organic field-effect transistors memory devices using supramolecular block copolymer/functional small molecule nanocomposite electret. ACS Appl Mater Interfaces 7:5663–5673. https://doi.org/10.1021/acsami.5b00338
69. Heeger AJ, Semiconducting and metallic polymers: the fourth generation of polymeric materials. Chin J Polym Sci 9(6):545–572
70. Reiss P, Couderc E, De Girolamo J, Pron A (2011) Conjugated polymers/semiconductor nanocrystals hybrid materials – preparation, electrical transport properties and applications. Nanoescale 3:446–489. https://doi.org/10.1039/C0NR00403K
71. Brütting W (2005) Physics of organic semiconductors. WILEY-VCH Verlag GmbH & Co, KGaA, Weinheim, Germany
72. Brédas JL, Beljonne D, Coropceaunu V, Cornil J (2004) Chem Rev 2004(104):4971
73. Shirota Y, Kageyama H (2007) Charge carrier transporting molecular materials and their applications in devices. Chem Rev 107:953–1010. https://doi.org/10.1021/cr050143+
74. Murphy AR, Fréchet JMJ (2007) Organic semiconducting oligomers for use in thin film transistors. Chem Rev 107(4):1066–1096. https://doi.org/10.1021/cr0501386
75. Meyer Zu Heringdorf FJ, Reuter MC, Tromp RM (2001) Growth dynamics of pentacene thin films. Nature 412:517–520. https://doi.org/10.1038/35087532
76. Dimitrakopoulos CD, Malenfant PRL (2002) Organic thin film transistors for large area electronics. Adv Mater 14(2):99–117. https://doi.org/10.1002/1521-4095(20020116)14:2%3c99:AID-ADMA99%3e3.0.CO;2-9
77. Gao X, Hu Y (2014) Development of n-type organic semiconductors for thin film transistors: a viewpoint of molecular design. J Mater Chem C. https://doi.org/10.1039/C3TC32046D
78. Anthony JE, Facchetti A, Heeney M, Marder SR, Zhan X (2010) n-Type organic semiconductors in organic electronics. Adv Mater 22:3876–3892. https://doi.org/10.1002/adma.200903628
79. Sirringhaus H (2005) Device physics of solution-processed organic field-effect transistors. Adv Mater 17:2411–2425. https://doi.org/10.1002/adma.200501152
80. Kokil A, Yang K, Kumar J (2012) Techniques for characterization of charge carrier mobility in organic semiconductors. J Polym Sci Polym Phys 50:1130–1144. https://doi.org/10.1002/polb.23103
81. Costa JCS, Taveira RJS, Lima CFRAC, Mendes A, Santos LMNBF (2016) Optical band gaps of organic semiconductor materials. Opt Mater 58:51–60. https://doi.org/10.1016/j.optmat.2016.03.041
82. Thejo Kalyani N, Dhobleb SJ (2012) Organic light emitting diodes: Energy saving lighting technology—a review. Renew Sust Energ Rev 16:2696–2723. https://doi.org/10.1016/j.rser.2012.02.021
83. Scheblykin IG, Yartsev A, Pullerits T, Gulbinas V, Sundstro V (2007) Excited state and charge photogeneration dynamics in conjugated polymers. J Phys Chem B 111:6303–6321. https://doi.org/10.1021/jp068864f
84. Knupfer M (2003) Exciton binding energies in organic semiconductors. Appl Phys A 77(5):623–626. https://doi.org/10.1007/s00339-003-2182-9
85. Ruini A, Caldas MJ, Bussi G, Molinari E (2003) Solid state effects on exciton states and optical properties of PPV. Phys Rev Lett. https://doi.org/10.1103/PhysRevLett.88.206403
86. Conwell EM (1996) Definition of exciton binding energy for conducting polymers. Synth Met 83:101–102
87. Ahmad S (2014) Organic semiconductors for device applications: current trends and future prospects. J Polym Eng 34(4):279–338

88. Hatton RA, Miller AJ, Silva SRP (2008) Carbon nanotubes: a multi-functional material for organic electronics. J Mater Chem 18:1183–1192
89. Horowitz G (1998) Organic field-effect transistors. Adv Mater 10:365–377
90. Mei J, Diao Y, Appleton AL, Fang L, Bao Z (2013) Integrated materials design of organic semiconductors for field-effect transistors. J Am Chem Soc 135(18):6724–6746
91. Watanabe M, Chang YJ, Liu SW, Chao TH, Goto K, Islam MM, Yuan CH, Tao YT, Shinmyozu T, Chow TJ (2012) The synthesis, crystal structure and charge transport properties of hexacene. Nat Chem 4:574–578
92. Stewart Z (2013) Organic thin-film transistors and TIPS-pentacene. Honors Program Senior Capstone Collection. Paper 11. http://uknowledge.uky.edu/honprog/11 Accessed on May 2017
93. Yamashita Y (2009) Organic semiconductors for organic field-effect transistors. Sci Technol Adv Mater. https://doi.org/10.1088/1468-6996/10/2/024313
94. Torsi L, Magliulo M, Manoli K, Palazzo G (2013) Organic field-effect transistors sensors: a tutorial review. Chem Soc Rev 42:8612–8628
95. Mahon JK, Zhou T, Forrest SR, Shwambera M, Meyer N (2012) Organic VPD shows promise for OLED volume production. Solid State Technol 45(7):131–139
96. Köhler A, Bässler H (2009) Triplet states in organic semiconductors. Mater Sci Eng, R 66:71–109. https://doi.org/10.1016/j.mser.2009.09.001
97. Geffroy B, le Roy P, Prat C (2006) Review Organic light-emitting diode (OLED) technology: materials, devices and display technologies. Polym Int 55:572–582. https://doi.org/10.1002/pi.1974
98. Braun D (2002) Semiconducting polymer LEDs. Mater Today 5:32–39. https://doi.org/10.1016/S1369-7021(02),00637-5
99. So F, Kondakov D (2010) Degradation mechanisms in small-molecule and polymer organic light-emitting diodes. Adv Mater 22:3762–3777
100. Karzazi Y (2014) Organic light emitting diodes: devices and applications. J Mater Environ Sci 5:1–12
101. Kulkarni AP, Tonzola CJ, Babel A, Jenekhe SA (2004) Electron transport materials for organic light-emitting diodes. Chem Mater 16:4556–4573
102. Cao W, Xue J (2014) Recent progress in organic photovoltaics: device architecture and optical design. Energy Environ Sci 7:2123–2144
103. Günes S, Neugebauer H, Sariciftci NS (2007) Conjugated polymer-based organic solar cells. Chem Rev 107(4):1324–1338
104. Su YW, Lan SC, Wei KH (2012) Organic photovoltaics. Mater Today 15:554–562
105. Liang Y, Wu Y, Feng D, Tsai ST, Son HJ, Li G, Yu L (2009) Development of new semiconducting polymers for high performance solar cells. J Am Chem Soc 131(1):56–57
106. The University of Texas at El Paso (s.f.) Fabrication Techniques recovered on July 30, 2016. http://www.ece.utep.edu/research/cdte/Fabrication/index.htm
107. Institut für Materialphysik, Complex thin films/Pulsed laser deposition on Oct 2018. http://mp8.material.physik.unigoettingen.de/index.php?site=krebs_forschung&lang=en
108. Schmidt H, Mennig M (2018) The Sol-Gel Getaway, Wet Coating Technologies for Glass. Institut für Neue Materialien, Germany on Oct 2018. http://www.solgel.com/articles/nov00/coating.htm
109. Lin X, Kavalakkatt J, Lux-Steiner MCh, Ennaoui A (2015) Inkjet-printed $Cu_2ZnSn(S, Se)_4$ solar cells. Adv Sci 2:1500028. https://doi.org/10.1002/advs.201500028
110. Perrin Manufacturing (s.f.) Silk Screen and Graphics Systems recovered on Oct 2018. https://www.perrinmfg.com/manufacturing#d3

Chapter 10
Emerging Opportunities and Future Directions

Martin I. Pech-Canul, Socorro Valdez Rodríguez, Luis A. González and Nuggehalli M. Ravindra

10.1 Introduction

The semiconductor industry is perhaps the most influential driver of growth in today's society with its impact on all walks of life. From mobile phones, home appliances, personal computers, to automotive, navigation, aviation, and military equipment, as wells as in medical diagnostic instruments, semiconductors play almost a ubiquitous role. Attempting to address all the possible related topics on semiconductors in a book is a challenging task. However, the constant evolution and the rapidly growing amount of information on the wide variety of topics in semiconductors makes it necessary to document them. This is in addition to the well-known textbooks in the field [1] as well as specialized journals on semiconductors, and more general journals in the fields of electrical engineering, physics, materials science and engineering. In textbooks on "materials science and engineering", the subject of semiconductors occupies a preeminent position [2, 3]. The topics that are typically covered include the

M. I. Pech-Canul (✉) · L. A. González
Centro de Investigación y de Estudios Avanzados del Instituto Politécnico Nacional, Avenida Industria Metalúrgica No. 1062, Parque Industrial Saltillo-Ramos Arizpe, 25900 Ramos Arizpe, Coahuila, Mexico
e-mail: martin.pech@cinvestav.edu.mx

L. A. González
e-mail: luis.gonzalez@cinvestav.edu.mx

S. Valdez Rodríguez
Instituto de Ciencias Físicas-UNAM, Av. Universidad S/N, Col. Chamilpa, 062210 Cuernavaca, Morelos, Mexico
e-mail: svaldez@fis.unam.mx

N. M. Ravindra
New Jersey Institute of Technology, University Heights, Newark NJ 07102-1982, USA
e-mail: n.m.ravindra@njit.edu; nmravindra@gmail.com

M. I. Pech-Canul and N. M. Ravindra (eds.), *Semiconductors*,
https://doi.org/10.1007/978-3-030-02171-9_10

energy band model for electrical conduction, intrinsic and extrinsic semiconductors, compound semiconductors, semiconductor devices, and microelectronics. Specific annual or periodic reports prepared by associations in the semiconductor industry are, by all means, excellent information sources.

In the vision report of March, 2017, presented by the Semiconductor Industry Association (SIA) and the Semiconductor Research Corporation (SRC), it was recognized that, to continue to achieve significant advancements in the semiconductor industry, a comprehensive approach that considers all aspects of semiconductor technology, including novel materials, new manufacturing techniques, new structures, systems architecture, and applications, should be utilized [4]. Fourteen research areas that are essential to progress were identified and outlined by a group of industry experts and leaders in the semiconductor industry during the period 2016–2017 [4]. The identified areas are:

1. Advanced Devices, Materials, and Packaging
2. Interconnect Technology and Architecture
3. Intelligent Memory and Storage
4. Power Management
5. Sensor and Communication Systems
6. Distributed Computing and Networking
7. Cognitive Computing
8. Bio-influenced Computing and Storage
9. Advanced Architectures and Algorithms
10. Security and Privacy
11. Design Tools, Methodologies, and Test
12. Next-Generation Manufacturing Paradigm
13. Environmental Health and Safety Environmental health and safety: Materials and Processes
14. Innovative Metrology and Characterization.

According to the report, until recently, a significant proportion of the semiconductor research has focused on materials, processes, and a variety of technologies that are needed to continue scaling, or reducing the size of transistors, in accordance with "Moore's Law." However, concerning scaling, the experts have recommended a shift to new research directions, exploring a variety of strategies. It was emphasized that it is essential to conduct fundamental research in a variety of areas beyond the existing technology [4].

The topics covered in this book, except Chap. 1—Semiconductor Fundamentals—address the processing, microstructure, properties, and applications. Along with highlighting some of the current research opportunities, this chapter suggests that, regardless of the crucial or emerging research topic, all should come together into the need of addressing the environmental-, health-, and safety-related aspects. Section 10.3 is explicitly devoted to this key aspect.

Vanadium dioxide (VO_2) thin films have been demonstrated to be promising for electronic and photonic applications such as uncooled microbolometers, nano-oscillators, optical switches, or coatings with a modulated transmission. These potential applications come from the fact that they can undergo abrupt and extreme resistivity changes—of up to five orders of magnitude—over the metal–insulator transition (MIT), along with the onset of unusual nonlinear electrical and optical properties (e.g., negative differential resistance or negative differential emissivity in the far-infrared domain) [5]. Several processing routes for VO_2 have been used in recent investigations, including Sol-gel [6] Reactive Pulsed Laser Deposition [7], and Atmospheric Pressure Chemical Vapor Deposition [8]. Coatings for smart glazing application [6] and thermochromic VO_2 thin films for intelligent window applications have been produced by the aforementioned routes [8].

For about 15 years, a sheer number of scientific and technological investigations worldwide have revolved around graphene. Its integration into microelectronic devices has the potential for numerous applications, and due to its extraordinary electronic, optical, mechanical and thermoelectric properties, graphene is found to be an excellent material for applications in optoelectronics. Nonetheless, a current challenge being addressed is the modification of its properties from semimetal to semiconductor behavior. Accordingly, intense efforts have been made to prepare large area doped graphene films to tune the structure-related optical and electrical properties [9]. In a recent investigation by Tang et al. [9], monolayer and bilayer graphene were prepared by CVD techniques while graphene oxide (GO) was prepared by the "Tang-Lau" method. From the work of [9], it is apparent that processing routes for the variety of graphene materials, including Cl, S, and K doped graphene quantum dots, still constitute key research windows of opportunity. In accord with a recent review by Ruhl et al. [10], it is suggested that while for the most part, synthesis challenges have already been resolved, the main problem is the accessibility to suitable techniques for the production and deposition of graphene on dielectric substrates. Furthermore, maintaining graphene's properties by encapsulating it for further processing requires more research. Nonetheless, there is significant progress concerning graphene/metal contact, finding it suitable for electronic devices.

In recent years, 2D and 3D graphene-based materials have found applications in the production of fuel cells, supercapacitors, and photovoltaic devices. Recent investigations have reported the preparation of a range of graphene-based materials by a variety of routes, from raw graphene or graphene oxide, including the metal and metal oxide additions, to graphene doped with foreign elements [11]. The synthesis routes include the chemical methods and the physical deposition of carbon vapor under extreme conditions. According to Ruhl et al. [10], nowadays, the critical issues that have been delaying graphene integration are known, and the corresponding research tasks to address them are clearly defined.

As organic semiconductors find applications in photovoltaic devices, thin film transistors, and organic-light-emitting diodes, there is plenty of opportunities in research and development. Highlighting the current research interests, a recent review aims at providing a comprehensive overview of the molecular packing, morphology and charge transport features of organic semiconductor crystals, the control of crystallization for achieving high-quality crystals and the device physics in Field-Effect Transistors (OFETs), Light-Emitting Diodes (OLEDs), and photovoltaic cells (OPVs) [12].

Another window of opportunity that has been gaining research interest in recent years is the integration of experimental and computational approaches in materials and manufacturing. In the case of semiconductors, it is not uncommon to find reports in the literature, but they are not as abundant either. In the work of Mei et al. [13], the authors provide a historical view of the development of Organic Field-Effect Transistors (OFETs) and introduce the concepts of integrated molecular design and process engineering approaches, identifying the breakthroughs and the future challenges to using OFETs in real engineering applications. It has been remarked that organic circuits are generally featured as low-cost electronics, even though there is still a lack of a comprehensive and accessible analysis of cost structures of OFET devices for various applications. Moreover, since Organic Semiconductors (OSCs) are aromatic compounds and likely to be carcinogens, they require detailed environmental assessment, particularly if OFETs are intended to be used in disposable devices [13].

10.2 Chemical Bath Deposition

Chemical Bath Deposition (CBD) is recognized as a low-temperature, economical, and large area compatible processing technique for the growth of high-quality semiconductor layers. Investigations on the chemical deposition of CdS and PbS-CdS thin films, studying the effect of reaction temperature and post-deposition treatments on the structure, morphology, and optoelectronic properties were reported about 25 years ago [14, 15, 16]. Several advantages to its use have been reported recently, including the production of CdS thin films in reaction solution without using the harmful ammonia [17, 18], and its versatility of doping the thin films with one single atom type [19] or with different agents simultaneously, for instance, during the preparation of ZnS films [20]. It also allows the preparation of nanoparticles, such as pure cadmium sulfide with diamagnetic properties [21].

One of the most recent advances using CBD reports the in situ doping of ZnS films with Cu and Mn ions [20]. X-ray photoelectron spectroscopy has shown simultaneous incorporation of Cu and Mn ions and defect formation in the crystalline structure of ZnS films. Moreover, room temperature photoluminescence spectroscopy has confirmed the effective activation of the ZnS films with Cu and Mn ions by observing a shift to longer wavelengths (green-orange emissions centered at 530 nm) in the luminescence emission centered at 470 nm. The original process has proved promising for the preparation of ZnS films with a mixture of other doping agents.

Zinc-doped CdS thin films prepared by the chemical bath deposition technique have proven to be attractive for the fabrication of Thin Film Transistors (TFTs) exhibiting Negative Differential Resistance (NDR). The incorporation of Zn in CdS active layers produces lattice distortions that induce charge carrier trapping sites. NDR devices are used conveniently to simplify the structure and circuit complexity of oscillators and logic memories [19].

Despite the attributes of CBD as mentioned above, the use of ammonia as a complexing and pH fine-tuning agent and the amount of Cd-containing toxic wastes that result during CdS thin film production are identified as the two main drawbacks of using the CBD method. Favorably, these issues are being addressed currently from two different perspectives. While the first lies in proposing processing routes that avoid using ammonia [17, 18, 22], the second focuses on recycling ammonia and recovering cadmium from chemical bath solutions [23, 24].

In harmony with an environmentally friendly philosophy at the laboratory and industrial levels, much interest has aroused in substituting other complexing agents such as sodium citrate and nitrilotriacetic acid for the harmful ammonia or ammonia compounds. For instance, glycine has been used as a complexing agent for the preparation of CdS films by the CBD process [22]. Moreover, although ammonia compounds such as NH_4OH and NH_4Cl are still used as buffer solutions, efforts are being made currently to produce CdS films in an ammonia-free process. As an alternate approach, alanine has been used as a complexing agent for the preparation of high-quality nanocrystalline CdS films [18]. The influence of the amino acids (AAs) on the growth, microstructure, and morphology, elemental content and optical properties of the CdS films has been discussed in the work of [18].

In a rationale based on an environmental and economic standpoint, Malinowska et al., argue that the ammonia and cadmium waste chemicals from the CBD process must be recovered, and if possible recycled [23, 24]. Accordingly, they have implemented a process for recovering ammonia and confining cadmium, using a closed reactor to entrap ammonia as it is vaporized during the deposition step. A designed ex situ process allows recovering most of the ammonia, obtaining concentrated solutions of ammonia (>10 mol/ml). Cadmium, on the other hand, is precipitated mainly as cadmium sulfide. According to the authors, the global process recovers at least 99.999% of cadmium and generates only solid sulfur and a liquid effluent containing traces of cadmium (< 10 μg/l) [24]. The recovery and recycling of hazardous wastes produced in the research laboratory or in industrial facilities is in harmony with the modified central paradigm of materials science and engineering [25], in which the recover/recycle aspect is introduced as a key component in between the already existing ones—processing, structure, properties, and performance—in the paradigm.

The chemical waste generation issue during the preparation of new materials is perhaps one of the inevitable challenges that the new developments will have to factor into from the outset. In the case of the CdS thin films by the CBD process, the ammonia and cadmium issues have been addressed successfully from both perspectives, (i) eliminating or reducing the use of ammonia, and (ii) the recovery/recycling of ammonia and cadmium. However, there is still scope for research.

10.3 Environmental, Health and Safety-Related Aspects

An evident emerging opportunity is in the field of development of processes and semiconductor materials that are friendly with the environment and safe to the manufacturer, employees and to the product consumer. On that score, the possible toxic effect of the use of nanomaterials and the waste management of chemicals are two aspects that are already being addressed from scientific and technological approaches. It has been considered that the exceptional performance of nanomaterials (in reactivity, conductivity, and optical sensitivity, amongst others) may have unwanted results consisting of dangerous interactions with biological systems and the environment, with the potential to produce toxicity [26]. The urgent necessity to develop and implement a rational, science-based approach to nanotoxicology, was suggested to be feasible, thus allowing to ensure safe manufacturing and utilization of engineered nanoproducts [26].

Efforts on identifying the long-term effects on Environment, Health and Safety implications in the semiconductor process development have been made more than two decades ago [27]. According to the report, back then, Motorola's Advanced Process Research and Development Laboratory (APRDL), in the Semiconductor Product Sector (SPS), established its environmental group to implement a Design for EHS (DFEHS) strategy.

There is an urgent need for identifying from the outset, the advances that can be made by an increase in device efficiency and improving cost-effectiveness, without compromising the environment and health. Undoubtedly, it is believed that, these days, there is no need for making the associated safety and health issues, but of acknowledging it in the research proposal outline, and complying with it throughout the project development.

Within the "Fifteen Core Strategies" stated in the Environmental, Safety, and Health Charter and Mission of the International Roadmap for Devices and Systems (IRDS) [28] Edition manifest, the efforts of 20+ years of Semiconductor Roadmap, internationally, has been documented [28]. These strategies are in accord with the full range of technology applications and devices envisioned to 2030. All the 15 strategies target addressing the EHS criteria, but strategy numbers 1 and 10 emphasize by identifying critical gaps and opportunities in ESH (and sustainability), and by encouraging the industrial use of materials and their by-products that are less hazardous, respectively [28].

A straightforward way of considering the EHS criteria during R&D of semiconductors is by incorporating the recycling/reusing philosophy in processing. This is in accord with the modified central paradigm of materials science and engineering, which seeks after sustainable processing [25]. In the modified central paradigm of materials science and engineering, the recycling/reusing aspects are introduced into the cycle involving steps from processing to material performance.

10.4 Summary and Concluding Remarks

On account of the sheer number of research prospects in the field of semiconductors and the accelerated pace at which developments occur, discerning the current opportunities and bringing them into focus at once is a challenging task. Research opportunities are focused on two-dimensional (2D) and three-dimensional (3D) structured graphene-based materials for applications in fuel cells, supercapacitors, and photovoltaic devices. Along the same lines, investigations on raw graphene or graphene doped with foreign elements, as well as graphene oxide, or graphene loaded with metals or metal oxides, are highly encouraged for the same purposes. Investigations on vanadium dioxide (VO_2) thin films are promising for electronic and photonic applications such as uncooled microbolometers, nano-oscillators, optical switches, or coatings with a modulated transmission. Suitable and cost-effective processing routes constitute salient windows of opportunity. With respect to Chemical Bath Deposition (CBD) route, the in situ doping of thin films, such as ZnS films doped with Cu and Mn, and Zn-doped CdS thin films for potential use in Thin Film Transistors (TFTs), is an emerging theme with a new scope for research. Although with some significant advances, the handling and confinement of waste chemicals and the recycling/reuse of individual chemical elements, like cadmium, this CBD related issue can still be considered an area of opportunity for research, until all associated procedures become a common practice in the lab and manufacturing facilities.

In harmony with the ever-growing environmental concerns, regarding advancements in the field of semiconductors, in all likelihood, there are emerging opportunities in the design and development of fabrication routes and materials/products that meet the Environment, Health and Safety (EHS) criteria. While there is cognizance that significant efforts have been made at the corporate level for more than 20 years, the current literature manifests that research groups are still tackling the topic of the use of non-harmful reactants and processes, from the research stage in the laboratory, as in the case of the use of ammonia-free processes. There is still scope for research in this area. The need for identifying at the earliest possible stages the implications of a given process/material development on EHS can be addressed if, for instance, the research and development proposals include them. The nanomaterial-related issues and the recycling/reusing/confinement of waste chemicals could be described in detail. In the case of thesis related research projects, description of ES&H effects could be an essential component of the proposal, even perhaps as a prerequisite. This would allow addressing the pin-pointed issues well in advance before a process may be transferred from lab to pilot plant and finally to the manufacturing facilities. All in all, the semiconductor field constitutes an open-ended area with enormous scope for research. The strength and certainty of the future contributions lie in the confluence of interests from cross-disciplinary research.

Acknowledgements The authors gratefully acknowledge Cinvestav IPN (Center for Research and Advanced Studies of the National Polytechnic Institute, in México) for the support and encouragement in the R&D in the semiconductor field, emphasizing the Environment, Health and Safety aspects.

References

1. Yu PY, Cardona M (2010) Fundamentals of semiconductors, physical and materials properties, 4th edn. Springer. ISBN: 978-3-642-00709-5
2. Callister WD (ed) (1997) Materials science and engineering: an introduction, 4th edn. Wiley, New York, ISBN: 0-471-13459-7
3. Smith WF (ed) (1990) Principles of materials science and engineering, 2nd edn. McGraw-Hill Publishing Company, New York. ISBN: 0-071-059169-5
4. Isaacs D et al (2017) SIA SRC Vision Report 2017. https://www.semiconductors.org/…/SIA%20SRC%20Vision%20Report%203.30.17.pd
5. Théry V, Boulle A, Crunteanu A, Orlianges JC, Beaumont A et al (2017) Structural and electrical properties of large area epitaxial Vo" films grown by electron beam evaporation. J Appl Phys Am Inst Phys 121(5):055303
6. Seyfouri MM, Binions R (2017) Sol-gel approaches to thermochromic vanadium dioxide coating for smart glazing application. Sol Energy Mater Sol Cells 159:53–65
7. Maldiba IG et al (2017) Effects of gamma irradiations on reactive pulsed laser deposited vanadium dioxide thin films. Appl Surf Sci 411:271–278
8. Malarde D et al (2017) Optimized atmospheric-pressure chemical vapor deposition thermochromic VO_2 thin films for intelligent window applications. ACS Omega 2(3):1040–1046
9. Tang L, Ji R, Tian P, Kong J, Xiang J (2017) Functionalization of graphene by size and doping control and its optoelectronic applications. In: Proceeding of SPIE 10177, Infrared Technology and Applications, XLIII, 101770B
10. Ruhl G, Wittmann S, Koeing M, Neumaier D (2017) The integration of graphene into microelectronic devices. Beilstein J Nanotechnol 8:1056. https://doi.org/10.3762/bjnano.8.107
11. Tsang ACH, Kwok HYH, Leung DYC (2017) The use of graphene based materials for fuel cell, photovoltaics and supercapacitor electrode materials. Solid State Sci 67:A1–A14
12. Wang C, Dong H, Jian L, Hu W (2018) Organic semiconductor crystals. Chem Soc Rev 47:222. https://doi.org/10.1039/c7cs00490g
13. Mei J, Diao Y, Appleton AL, Fang L, Bao Z (2013) Integrated materials design of organic semiconductors for field effect transistors. J Am Chem Soc 135:6724–6746, https://doi.org/10.1021/ja400881n
14. Nair MTS, Nair PK (1994) Conversion of chemically deposited photosensitive CdS thin films to n-type by air annealing and ion exchange reaction. J Appl Phys 75:1557. https://doi.org/10.1063/1.356391
15. Orozco-Terán RA, Sotelo-Lerma M et al (1999) PbS-CdS bilayers prepared by the chemical bath deposition technique at different reaction temperatures. Thin Solid Films 343–344:587–590. https://doi.org/10.1016/S0040-6090(98)01719-2
16. Sebastian PJ, Campos J, Nair PK (1993) The effect of post-deposition treatments on morphology, structure and opto-electronic properties of chemically deposited CdS thin films. Thin Solid Films 227:111–228
17. Carreón-Moncada I, González LA, Pech-Canul MI, Ramírez-Bon R (2013) $Cd_{1-x}1Zn_xS$ thin films with low Zn content obtained by an ammonia-free chemical bath deposition process. Thin Solid Films 548:270–274
18. Carreón-Moncada I, González LA, Rodríguez-Galicia JL, Rendón-Angeles JC (2016) Chemical deposition of CdS films by an ammonia-free process with amino acids as complexing agents. Thin Solid Films 599:166–173
19. González LA, Carreón-Moncada I, Quevedo-López MA (2017) Negative differential resistance as effect of Zn doping of chemically processed CdS thin film transistors. Mater Lett 192:161–164
20. Alvarez-Coronado AG, González LA, Rendón-Angeles JC, Ramírez-Bon R (2018) Study of the structure and optical properties of Cu and Mn in situ doped ZnS films by chemical bath deposition. Mater Sci Semicond Process 81:68–74
21. Sathiya PN, Shalini Packiam Kumala S, Anbarasu V et al (2018) Characterization of CdS thin films and nanoparticles by a simple chemical bath technique. Mater Lett 220:161–164

22. Mendivil-Reynoso T, Berman-Mendoza D, González LA, Castillo SJ, Apolinar-Iribe A, Gnade B, Quevedo-López MA, Ramírez-Bon R (2011) Fabrication and electrical characteristics of TFYs based on chemically deposited CdS films using glycine as a complexing agent. Semicond Sci Tecnol 26(115010):1–6
23. Malinowska B, Rakib M, Durand G (2001) Ammonia recycling and cadmium confinement in chemical bath deposition of CdS thin layers. Prog Photovolt: Res Appl 9:389–404
24. Malinowska B, Rakib M, Durand G (2002) Ammonia recycling and cadmium confinement in chemical bath deposition of CdS thin layers. Prog Photovolt: Res Appl 10:215–228
25. Pech-Canul MI, Kongoli F (2016) The modified central paradigm of materials science and engineering in the development of new and recycled materials. J Mineral Process Extr Metall 125(4):238–241
26. Nel A, Xia T, Mädler L, Lil N (2006) Toxic potential materials at nanolevel. Science 311:622–627
27. Mendicino L, Beu L (1997) Addressing environment, health and safety in semiconductors process development. In: IEEVCPMT International electronics manufacturing technology symposium, pp 129–133
28. International Roadmap for devices and systems, 2016, https://irds.ieee.org/images/files/pdf/2016_MM.pdf

Index

M. I. Pech-Canul and N. M. Ravindra (eds.), *Semiconductors*,
https://doi.org/10.1007/978-3-030-02171-9

Printed by Printforce, the Netherlands